SYSTEM-BASED VISION FOR STRATEGIC AND CREATIVE DESIGN

PROCEEDINGS OF THE SECOND INTERNATIONAL CONFERENCE ON STRUCTURAL
AND CONSTRUCTION ENGINEERING, 23–26 SEPTEMBER 2003, ROME, ITALY

System-based Vision for Strategic and Creative Design

Edited by

Franco Bontempi
University of Rome "La Sapienza", Rome, Italy

Volume 3

A.A. BALKEMA PUBLISHERS   LISSE / ABINGDON / EXTON (PA) / TOKYO

Published by: A.A. Balkema, a member of Swets & Zeitlinger Publishers
www.balkema.nl and www.szp.swets.nl

For the complete set of three volumes: ISBN 90 5809 599 1
Volume 1: 90 5809 600 9
Volume 2: 90 5809 601 7
Volume 3: 90 5809 630 0

Printed in the Netherlands

System-based Vision for Strategic and Creative Design, Bontempi (ed.)
© 2003 Swets & Zeitlinger, Lisse, ISBN 90 5809 599 1

Table of Contents

VI

4. Cost control and productivity management

5. Human factor, social–economic constraints on the design process

6. Structural optimization and evolutionary procedures

7. Shaping structures and form finding architectures

8. Issues in computational mechanics of structures

9. Advanced modelling and non-linear analysis of concrete structures

10. Concrete and masonry structures

11. Steel structures

Volume 2

12. Bridges and special structures

13. Precast structures

14. Earthquake and seismic engineering

15. Geotechnical engineering and tunnelling

19. Innovative methods for repair and strengthening structures

20. Artificial intelligence in civil engineering

21. Knowledge management

22. Quality and excellence in constructions

23. Sustainable engineering

24. Life cycles assessment

25. New didactical strategies and methods in higher technical education

26. Durability analysis and lifetime assessment

Volume 3

27. High performance concrete

28. Concrete properties and technology

29. Innovative tools for structural design

30. Issues in the theory of elasticity

31. Special session on "Structural monitoring and control"

32. Special session on "New didactical strategies"

33. Special session on "Information and decision systems in construction project management"

34. Special session on "Innovative materials and innovative use of materials in structures"

35. Invited session on "Recent advances in wind engineering"

36. Special session on "Geomechanics aspects of slopes excavations and constructions"

37. Special session on "Fastening technologies for structural connections"

38. Special session on "Advanced conceptual tools for analysis long suspended bridges"

Preface

The present Proceedings collect over 360 papers that will be presented during ISEC-02, the 2nd International Conference on Structural and Construction Engineering, hold in Rome, Italy, on September 23–26, 2003.

The Theme of the Conference, *System-based vision for strategic and creative design*, includes the following concepts:

- *systemic framework*: it includes the capacity to see all the aspects and the connections of any problem and its solution;
- *creative work*: the tension to develop something new;
- *design*: one of the methods to improve the world and make it better.

The current practice and research in structural and construction engineering is characterized by increasing levels of complexity and interaction.

This is due to several reasons. First, the transition to a global, high technology environment demands sound safety requirements in all aspects of human life and activities. With respect to what concerns buildings, structures and infrastructures, a century of continuous progress in knowledge, materials and technology should have reduced occurrences of damage, failure, and misconception by large magnitudes. In spite of this expectation, small and large structural and construction deficiencies are still common episodes. This is, probably, due to the more ambitious objectives our society intends to pursue. But, without doubt, there are still problems and mismatches along the engineering path from concept development to practical realization. Secondly, there is special attention being paid in the world today to the interaction, full of uncertainties and constraints, of the structure with the environment. In this sense, competitiveness and sustainability require a systems approach in which research activities support the development of coherent, interconnected and ecologically-efficient civil engineering structural systems, responding to both market and social needs. Finally, there is the necessity to answer to socioeconomic needs, by stimulating holistic approaches and heuristic techniques, by strengthening the innovative capacity, and by fostering the creation of businesses and services built on emerging technologies and market opportunities. Research will turn into environmentally and consumer friendly processes, products, and services and will contribute to improve the quality of life and working conditions.

With this point of view, the main objective of the Conference will be to define knowledge and technologies needed to design and develop project processes and produce high-quality, competitive, environment- and consumer-friendly structures and constructed facilities. This goal is clearly connected with the development and reuse of quality materials, excellence in construction management, and reliable measurement and testing methods.

Franco Bontempi

Professor of Structural Analysis and Design,
Faculty of Engineering, University of Rome "La Sapienza", Rome (ITALY)
Postgraduate School of Reinforced Concrete Structures "F.lli Pesenti",
Polytechnic of Milan, Milan (ITALY)

Acknowledgements

The Editor gratefully acknowledges the promotion, the sponsorship, and the endorsement of the following organizations:

Promoting Association:
– CTE, Italian Association for Building Industrialization

Co-sponsors:
– AICAP, Italian Association for Reinforced and Prestressed Concrete
– CIB, International Council for Research and Innovation in Building and Construction
– ACI, American Concrete Institute
– ASCE, American Society of Civil Engineers
– IABMAS, International Association for Bridge Maintenance and Safety

Endorsers:
– Faculty of Engineering of the University of Rome "La Sapienza"

From an institutional point of view, the following persons will be gratefully remembered: Giandomenico Toniolo, President of CTE, Emanuele Filiberto Radogna, President of AICAP, Giselda Barina, Executive Secretary – CTE, Vivetta Bianconi, Executive Secretary of AICAP, Tullio Bucciarelli, Chairman of the Faculty of Engineering of the University of Rome "La Sapienza", Fabrizio Vestroni, Head of the Department of Structural Engineering, Amarjit Singh, Chair of ISEC-01.

The specific financial support of University of Rome "La Sapienza", Ministry of Instruction, University and Research (MIUR), Societa' Stretto di Messina S.p.A. and HILTI Italia S.p.A. are gratefully recognized.

The scientific framework of the Conference finds its origin within interesting discussions with Remo Calzona, Fabio Casciati and Pier Giorgio Malerba, whose experience and advices are, for the Editor, of the utmost importance.

The conceptual organization and the operative development of such a complex and large international conference couldn't have been possible without the outstanding commitment of several people. Specifically, the Editor wants to recognize the contribute, always made with efficacy and positive tense, of the following persons, at the same time friends and colleagues: Fabio Biondini, Luciano Catallo, Pier Luigi Colombi, Ezio Dolara, Elsa Garavaglia, Cristina Jommi, Simone Loreti, Giuseppe Parlante, Flavio Petrilli, Paola Provenzano, Luca Sgambi, Maria Silvestri, Angelo Simone, Maria Laura Vergelli, Paola Tamburrini.

The Proceedings are dedicated to Professor Francesco Martinez y Cabrera.

Rome, September 2003

Organisation

International Scientific and Technical Committee:

Franco Bontempi, Chair, University of Rome "La Sapienza", Rome, Italy

Amarjit Singh, Chair of past ISEC Conference, University of Hawaii, USA
Hojjat Adeli, the Ohio State University, USA
Chimay J. Anumba, Loughborough University, UK
Ghassan Aouad, the University of Salford, UK
Larry Bergman, University of Illinois at Urbana-Champaign, USA
Fabio Casciati, University of Pavia, Italy
John Christian, University of New Brunswick, Canada
Richard Fellows, the University of Hong Kong, Hong Kong
Dan Frangopol, University of Colorado at Boulder, USA
Ian Gilbert, New South Wales University, Australia
Paul Grundy, Monash University, Australia
Takashi Hara, Tokuyama University of Technology, Japan
Makarand Hastak, Purdue University, USA
Osama Ahmed Jannadi, King Fahd University of Petroleum & Minerals, Saudi Arabia
Vladimir Kristek, Czech Techinical University in Prague, Czeck Republic
In-Won Lee, Korea Advanced Institute of Science and Technology, Korea
Anita Liu, The University of Hong Kong, Hong Kong
Pier Giorgio Malerba, Technical University of Milan, Italy
Giuseppe Mancini, Technical University of Turin, Italy
Indu Patnaikuni, RMIT University, Australia
Takahiro Tamura, Tokuyama University of Technology, Japan
Francis Tin-Loi, New South Wales University, Australia
Ali Touran, Northeastern University, USA
Richard N. White, Cornell University, USA
Frank Yazdani, North Dakota State University, USA

Local Scientific and Technical Committee

Nicola Pio Belfiore, University of Rome "La Sapienza"
Remo Calzona, University of Rome "La Sapienza"
Claudio Ceccoli, University of Bologna
Carlo Cecere, University of Rome "La Sapienza"
Mario Como, University of Rome, "Tor Vergata"
Antonio D'Andrea, University of Rome "La Sapienza"
Bruno Della Bella, Precompressi Centro Nord S.p.A, Novara
Alessandro De Stefano, Technical University of Turin
Valter Esposti, ICITE Director, CIB Treasurer, S.Giuliano Milanese
Roberto Guercio, University of Rome "La Sapienza"
Luigi Annibale Materazzi, University of Perugia
Antonino Musso, University of Rome "Tor Vergata"
Emanuele Filiberto Radogna, University of Rome "La Sapienza", AICAP President
Alessandro Ranzo, University of Rome, "La Sapienza"

Luca Romano, Consultant Engineer, Albenga
Franco Rovelli, Gecofin Prefabbricati S.p.A, Verona
Marco Savoia, University of Bologna
Renato Sparacio, University of Napoli, "Federico II"
Paolo Spinelli, University of Florence
Franco Storelli, University of Rome "La Sapienza"
Giandomenico Toniolo, Technical University of Milan, CTE President

Local Organizing Committee

Giselda Barina, (Executive Secretary), CTE
Fabio Biondini, Technical University of Milan
Fabio Bongiorno, CTE
Luciano Catallo, University of Rome "La Sapienza"
Francesco Chillè, ENEL Hydro, Milan
Pier Luigi Colombi, Technical University of Milan
Ezio Dolara, University of Rome "La Sapienza"
Elsa Garavaglia, Technical University of Milan
Claudia Gomez, Consulting Engineer, Milan
Simone Loreti, University of Rome "La Sapienza"
Corrado Pecora, Consulting Engineer, Milan
Flavio Petrilli, Consulting Engineer, Rome
Paola Provenzano, University of Rome "Tor Vergata"
Luca Sgambi, University of Rome "La Sapienza"
Maria Silvestri, University of Rome "La Sapienza"
Angelo Simone, Delft University of Technology
Maria Laura Vergelli, University of Rome "La Sapienza"
Paola Tamburrini, University of Rome "La Sapienza"
Giuseppe Parlante, Content Manager and Web Master

27. High performance concrete

Biaxially loaded high strength concrete columns for strength design

I. Patnaikuni
RMIT University, Melbourne, Australia

A.S. Bajaj & P. Mendis
The University of Melbourne, Melbourne, Australia

ABSTRACT: This paper presents the details of a project to investigate the behaviour of high strength concrete (HSC) columns under biaxial bending. The numerical results of a case study are presented to show the differences in failure surfaces obtained with the conventional rectangular stress block, the revised stress blocks suitable for high strength concrete and with Modified Scott Model (MSMC). Finally the proposed approach (PA), together with other rectangular stress block (RSB) models is compared against a range of experimental data. All revised RSB models suitable for HSC show a reasonable correlation between them. In this paper, the observations in the studies discussed above have been used to rationalise the suitability of revised RSB models for these types of columns.

1 INTRODUCTION

Applications of HSC in structures are increasing day by day. The use of HSC is very common in heavily loaded lower storey columns in buildings. Biaxial bending occurs in corner columns of these buildings and is also present in the exterior and interior columns due to load imbalance between the adjacent spans. The research at the University of Melbourne and elsewhere has shown that simply extrapolating the constitutive equations for normal strength concrete (NSC) to HSC could either be unsafe or uneconomical. However, to a large extent practice has preceded theory with constitutive equations simply being extrapolated to higher strengths. A comprehensive literature survey and other work done at the University of Melbourne have shown the inadequacy of the design rules in biaxial bending. The important findings up to now are presented in this paper.

2 RECTANGULAR STRESS BLOCK MODELS USED IN CASE STUDIES

Presently most codes for concrete structures define the RSB for concrete (ACI 318-1989, AS 3600 – 1994) as given below and is mainly used for NSC with concrete compressive strength f'_c up to 50 MPa. The normal conventional rectangular stress block (NRSB) has been defined by two parameters $\beta = 0.85$ as shown in Figure 1, a horizontal constant (which is to determine the intensity of the stress block) and γ (the ratio of the depth of the stress block to the depth of the neutral axis d_n from the top) is given below:

$$\gamma = 0.85 - 0.007\left(f'_c - 28\right), \text{ where } f'_c \text{ is in MPa} \tag{1}$$

The research projects conducted around the world have shown that the shape of the stress–strain curve for high-strength concrete is different. Therefore the currently used parameters for an equivalent rectangular stress block are not applicable to HSC (Pendyala & Mendis 1997).

The revised stress block (RSB-1) suggested by Pendyala & Mendis (1997) is given below:

$$\gamma = \left[0.85 - 0.007\left(f'_c - 28\right)\right]$$
$$0.85 \text{ for } f'_c \leq 60 \text{ MPa} \tag{2a}$$

$$\gamma = \left[0.65 - 0.00125\left(f'_c - 60\right)\right], 60 < f'_c < 100 \text{ MPa} \tag{2b}$$

$$\beta = 0.85 \text{ for } f'_c \leq 60 \text{ MPa} \tag{3a}$$

$$\beta = 0.85 - 0.0025\left(f'_c - 60\right) \text{ for } 60 < f'_c \leq 100 \text{ MPa} \tag{3b}$$

The above-mentioned stress-block is a combination of rectangular and triangular stress blocks. More details are given by (Pendyala & Mendis 1997).

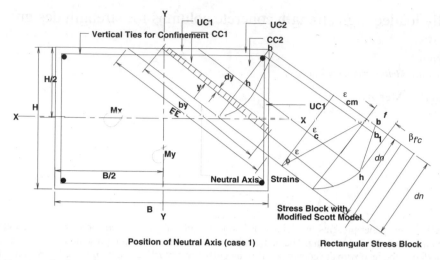

Figure 1. Stress–strain distribution for concrete with MSMC and rectangular stress block.

For comparison, the proposed revised stress block (RSB-2) for normal and high strength concrete up to 100 MPa based on experimental testing of uniaxial eccentrically loaded columns by (Ibrahim & Macgregor 1997) is also considered (Equation 4(a) & 4(b)).

$$\gamma = 0.95 - \frac{f'_c}{400} \geq 0.70, \ f'_c \text{ in MPa} \qquad (4a)$$

$$\beta = 0.85 - \frac{f'_c}{800} \geq 0.725, \ f'_c \text{ in MPa} \qquad (4b)$$

And for further comparison, another proposed revised stress block (RSB-3) for normal and high strength concrete up to 120 MPa based on experimental studies of HSC in consideration of spalling effects of concrete cover on axial strength of concentric and uniaxial eccentrically loaded columns by (Foster 1998) is also considered (Equation 5(a) & 5(b)).

$$\gamma = 1.046 - 0.007\,f'_c, \ 0.67 \leq \gamma \leq 0.85, \ f'_c \text{ in MPa}$$
$$(5a)$$

$$\beta = 1.02 - 0.0032\,f'_c, \ 0.64 \leq \beta \leq 0.85, \ f'_c \text{ in MPa}$$
$$(5b)$$

3 THEORETICAL STRESS–STRAIN MODEL

For establishing strength and design relationships, the MSMC (Pendyala & Mendis 1997) has been recommended as the most suitable model to predict the full

range stress–strain behaviour of concrete strength (f'_c) between 25 MPa and 10 MPa. A computer program has been developed to incorporate the full-integrated equations for all the possible cases of neutral axis for biaxial bending. The concrete compressive force, P_C, can be calculated as explained by (Bajaj & Mendis 1999) and is reproduced in Equation 6. Figure 1 shows the first case of neutral axis with MSMC used to denote the stress–strain relationship for confined concrete. These cases are described in detail elsewhere.

Total concrete compressive force, P_C

Unconfined compressive force in stages 1 & 2 = $P_{UC} = UC1 + UC2$

Confined compressive force in stages 1 & 2 = $P_{CC} = CC1 + CC2$

Therefore, $P_C = P_{UC} + P_{CC}$ \qquad (6)

4 EXPERIMENTAL WORK

A total of 8, 150 × 150 columns with concrete compressive strengths (f'_c) of 50 and 100 MPa representing normal-strength and high-strength concrete respectively were cast and tested in this study. The geometry of the specimens and the test set-up is shown in Figure 2. First six specimens were tested with biaxial eccentricities (e_x, e_y) of (27,100), (57,100) and (100,100) corresponding to α values of 15 or 75, 30 or 60 and 45 degrees ($\alpha = \tan^{-1} e_y/e_x$). Results for these tests are shown in Table 1. More details are available in (Bajaj & Mendis 1999).

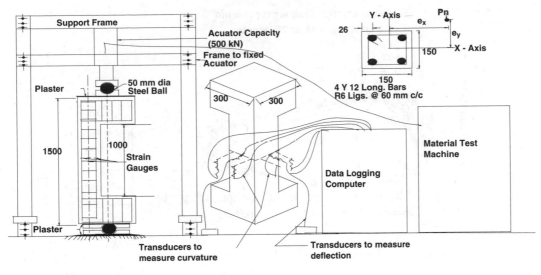

Figure 2.   Test set-up (Not to scale).

Table 1.   Comparison of nominal failure load capacity for tested columns. (Particulars of columns tested (150 × 150 mm) with clear cover = 20 mm.)

	(Average cylinder strength) f'_c (MPa)	Tested eccentricity (mm)		Nominal failure loads capacity at maximum concrete strain ($\varepsilon_{cm} = 0.003$)(kN)						
		e_x	e_y	NRSB(kN)	RSB-1	RSB-2	RSB-3	MSMC	PA	Expt.
1	56.8	27	100	201.3	201.3	203.0	201.5	210.9	185.6	184.1
2	98	27	100	255.9	243.5	243.0	240.6	250.9	222.4	221.2
3	64.9	57	100	171.7	170.0	171.5	169.6	175.3	157.1	172.0
4	96	57	100	207.9	191.9	195.8	192.4	209.5	178.7	190.0
5	50.6	100	100	113.3	113.7	116.4	113.7	118.7	106.2	111.0
6	90	100	100	142.0	133.0	139.4	134.7	148.3	130.1	134.0

5   CASE STUDIES

In biaxial bending, the inclination and depth of neutral axis will vary depending upon the magnitude of applied axial load or moment or both. A computer program (Ehsani 1986) has been modified by the authors to incorporate the revised rectangular stress blocks. The computer program performs exact analysis of short reinforced concrete columns subjected to axial load and biaxial moment. As an example, the column cross-section used in the experimental program (Fig. 2) with same amount of reinforcement has been analysed. In Figure 2, the moments about axes x and y are M_x and M_y and nominal axial load is P_n acting at a point such that eccentricity, $e_x = M_y/P_n$ and $e_y = M_x/P_n$. Therefore angle of rotation is also defined as angle of eccentricity, $\alpha = \tan^{-1} e_y/e_x$. When the angle of eccentricity is either 0° or 90°, the corresponding moments will be uniaxial bending moments. The design of a column

under an axial load associated with bending moments about both axes is tedious.

The ultimate load and moments for certain values of eccentricities is influenced by such factors as dimensions of cross section, percentage of reinforcement, number and arrangement of bars, yield stress of steel, compressive strength of concrete and cover to the steel reinforcement. The square columns with $f'_c = 56.8$, 98, 64.9, 96, 50.6 and 90 MPa have been analysed by ORSB, RSB-1, RSB-2, RSB-3, MSMC and with proposed approach (PA) based on Modified Scott Model. PA is to limit the maximum cylinder stress (f'_c) in concrete for the tested column specimen stress to $C_1 f'_c$, where C_1 is defined in Equation 7. Figure 3 shows a typical interaction diagram for the square column with PA, MSMC, conventional and revised rectangular stress blocks respectively. As seen from the diagram, the rectangular stress blocks overestimate the moment capacity of these biaxially loaded columns. There is a

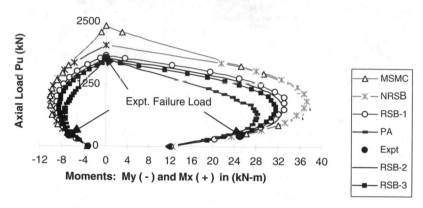

**Biaxial Interaction Diagram for column
at Alpha = 15 or 75 Degree**

Figure 3. Interaction diagrams for square column with $f'_c = 98$ MPa.

need to redefine a separate position factor in the "Revised Stress Block" for high strength concrete biaxial loaded columns. In the PA, maximum concrete strength, f, (as shown in stress block of Figure 1), has been taken as $C_1 f'_c$ instead of $\beta C_1 f'_c$.

$$C_1 = 1.1 - 0.004 \, f'c \le 0.75, \text{ where } f'c \text{ is in MPa} \quad (7)$$

6 ANALYSIS AND DISCUSSION OF RESULTS

The numerical and experimental analyses are useful to derive the ultimate strength parameters of these short reinforced concrete columns subjected to biaxial bending. The computer program is based on proposed approach and uses theoretical stress–strain model as shown in Figure 1. The computer program provides failure load capacity with respect to both principal axes at given eccentricities. The computer analysis ignores effects of creep, shrinkage, and reduction in strength due to spalling of concrete cover and deflections along both axes.

The concrete codes of various countries including ACI do not provide empirical equations for the design of columns subjected to biaxial bending. However, for checking the design adequacy based on Bresler Reciprocal Load Method (Bresler 1960) of these types of columns as used by CRSI handbook (Courtesy of Concrete Reinforcing Steel Institute) is considered satisfactory. The equation is given below.

$$\frac{P_n}{P_{nox}} + \frac{P_n}{P_{noy}} - \frac{P_n}{P_{no}} \le 1.0 \quad (8)$$

In the above Equation 8, P_n is the nominal failure load capacity, P_{no} is the pure axial load or squash load of the column, which ignores the confining capacity of the lateral reinforcement and is normally taken as $0.85 f'_c A_g + A_s f_y$, where A_g is the gross concrete area, f_y is the yield stress of longitudinal steel, and A_s is the total area of longitudinal steel. P_{nox} is the axial load capacity at an eccentricity $e_x = 0$ with $e_y = 100$ mm. P_{noy} is the axial load capacity at an eccentricity $e_x = 27$ mm with $e_y = 0$. Equation 8 has been found to be reasonably accurate for design purposes provided $P_n \ge 0.10 \, P_{no}$.

Considering the outcome of Bresler load formula mentioned in Equation 8 as equal to X, which is given below:

$$X = \frac{P_n}{P_{nox}} + \frac{P_n}{P_{noy}} - \frac{P_n}{P_{no}}$$

For satisfactory results, the outcome of Equation 8 (X), must be less than or equal to one. Test results presented in Tables 1 and 2 have shown that biaxial columns fabricated with HSC fail at premature loads than their predicted theoretical loads calculated with different rectangular stress block models.

But, outcome of the Bresler load equation (X), shown in Table 2 gives satisfactory results though corresponding loads calculated from various models, which are higher than experimental values, have been considered in this equation.

It does not seem safe to apply Bresler load equation for checking the design strength for columns having concrete strength higher than 50 MPa.

When Bresler introduced the equation, the results were based on low strength concrete columns. Table 2

Table 2. Load capacity parameters with different methods for test column 2 ($f'_c = 98\,\text{MPa}$).

Load parameters	Different methods of analysis with load in kN for (P_n, P_{no}, P_{nox} and P_{noy}). And values of X and ratios are constant						
	NRSB	RSB-1	RSB-2	RSB-3	MSMC	PA	Experiment
P_n	255.9	243.0	243.5	240.6	250.9	222.4	221.2
Test/Predicted	0.86	0.91	0.908	0.919	0.88	0.995	1.0
P_{nox}	278.2	261.2	257.3	249.7	302.2	265.1	–
P_{noy}	1363.1	1190.3	1164.7	1104.0	1706.0	1246.3	–
P_{no}	2031.1	1825.8	1766.4	1750.2	2355.2	1724.6	1759.9
P_n/P_{no}	0.126	0.133	0.139	0.137	0.107	0.129	0.126
Outcome of Bresler reciprocal load equation 8							
X	0.98	1.0	1.02	1.04	0.87	0.89	N/A

shows load capacity parameters for a typical specimen (No. 2 of Table 1), tested at eccentricities $e_x = 27\,\text{mm}$ and $e_y = 100\,\text{mm}$. As seen from Tables 1 and 2, good agreement is obtained between the experimental strengths and the analytical values calculated using the proposed approach at maximum concrete strain of 0.003.

7 CONCLUSIONS

(1) Different failure surfaces are obtained with the proposed approach and rectangular stress blocks suitable for normal and high strength concretes.

(2) It is evident from the columns tested in the experimental program that the rectangular stress blocks suitable for high strength concrete overestimate the load and moment capacity of these types of columns. A position factor relating to the angle of eccentricity in the RSB must be introduced for these columns.

(3) Bresler reciprocal load equation does not provide reliable results in checking the design adequacy of these types of columns having concrete strengths higher than 50 MPa.

(4) Experimental data was limited to 4 bars system. More validation of the PA is recommended in biaxial columns with multiple bars system and with concrete strength over 50 MPa.

REFERENCES

Australian Standard Concrete Structures Code. 1994. *Standards Australia,* NSW, Australia, AS 3600–1994, pp. 93–102.

Bajaj, A.S. and Mendis, P.A. 1999. Biaxial Bending of High Strength Concrete Columns. *Proc. Of the 16th Australasian Conference on the Mechanics of Structures and Materials,* Sydney, Australia, 8–10 December, pp. 121–126.

Bresler, B. 1960. Design Criteria for Reinforced Columns under Axial Load and Biaxial Bending. *Journal Of The American Concrete Institute,* pp. 481–490.

Building Code Requirements for Reinforced Concrete. 1989. *ACI Committee,* American Concrete Institute, Detroit, ACI 318–89, pp. 318–353.

Ehsani, R.M. 1986. CAD for Columns, *Concrete International,* pp. 43–47.

Foster, S.J., "Design of HSC columns for strength", *International Conference on HPHSC,* Perth, Australia, 1998, pp. 409–423.

Hisham, H.H., Ibrahim and Macgregor, J.G. 1997. Modification of the ACI Rectangular Stress Block for High-Strength Concrete. *ACI Structural Journal,* pp. 40–48.

Pendyala, R.S. and Mendis, P.A. 1997. Rectangular Stress Block For High Strength Concrete. *Civil Engineering Transactions, IE Australia,* pp. 135–142.

System-based Vision for Strategic and Creative Design, Bontempi (ed.)
© *2003 Swets & Zeitlinger, Lisse, ISBN 90 5809 599 1*

Testing eccentrically loaded externally reinforced high strength concrete columns

M.N.S. Hadi
University of Wollongong, Wollongong, Australia

ABSTRACT: This paper investigates the performance of externally confined high strength concrete columns subjected to eccentric loading and evaluates the effectiveness of two confinement materials, Carbon fibre and E-glass. Seven HSC columns were cast, two of which had internal reinforcement and the other five were plain concrete but externally reinforced. The columns had circular cross section and were haunched in one direction at either end to allow the application of eccentric loading on the middle section of the columns. Different number of wrapping layers with different layouts were used to wrap the columns. Results from the experimental program indicate that the enhancement of the strength of the plain column specimens under eccentric loading is not so pronounced as for the reinforced concrete specimens under concentric loading however, the wrapping materials provided considerable increase in strength and the different wrapping materials used in this study offer more favourable long term properties as compared to reinforcing steel.

1 INTRODUCTION

With the development of technology, the use of high-strength concrete members has proved most popular in terms of economy; superior strength; stiffness and durability. With the increase of concrete strength, the ultimate strength of the columns increases, but a relatively more brittle failure occurs. The lack of ductility of high strength concrete results in sudden failure without warning, which is a serious drawback of high strength concrete. Previous studies have shown that addition of compressive reinforcement and confinement will increase the ductility as well as the strength of materials effectively. Concrete, confined by transverse ties, develops higher strength and to a lesser degree of ductility. Studies conducted by some investigators on the improvement of the ductility of high strength concrete members have proven that the use of the spiral confinement is more effective and beneficial in the improvement of performance of concrete members (Razvi & Saatcioglu 1994).

In recent years, FRP wrapping in lieu of steel jacket has become an increasingly popular method for external reinforcement in which FRP offers improved corrosion and fatigue resistance compared to the steel reinforcement. The high tensile strength and low weight make FRP ideal for use in the construction industry. Another attractive advantage of FRP over steel straps as

external reinforcement is its easy handling, thus minimal time and labour are required to implement them (Demer & Neale 1999). However, research studies conducted so far on external confinement of concrete columns have mainly concentrated on concentric loading. In practice, structural concrete columns axially compressed (i.e., concentrically) rarely occur. Even in a column nominally carrying only axial compression, bending action is almost always present due to unintentional load eccentricities and possible construction error. Also, there are many columns where an eccentric load is deliberately applied. Therefore, the studies for concrete columns under eccentric loading are essential for the practical use.

This study experimentally investigates the benefits of external confinement using FRP on high strength concrete columns under eccentric loading and compares the effectiveness of two types of external reinforcement.

2 EXTERNAL CONFINEMENT WITH FRP

The application of FRP in the construction industry can eliminate some unwanted properties of high strength concrete, such as the brittle behaviour of high strength concrete. FRP is particularly useful for strengthening columns and other unusual shapes.

Parameters that affect the strength and ductility of FRP confined concrete include concrete strength, type of fibres and resin and thickness of FRP (different layers). The experimental program conducted by Houssam & Balauru (1999) showed that the compressive strength improved by approximately 200 % due to confinement with Carbon fibre and by approximately 100 % due to Glass fibre. Also, it is well known that the shape of cross section and the spacing of FRP straps can directly impact the confinement effectiveness of FRP wrapping in the confinement. The orientation of fibres is another contributing factor to the mechanical performance of a composite. Fibres oriented in one direction give very high stiffness and strength in that direction. If the fibres are oriented in more than one direction, such as in a mat, there will be high stiffness and strength in the direction of the fibre orientation (Autar 1997).

3 PRELIMINARY TESTING

In order to have a better insight about the contribution of FRP on confinement and to be able to design the column specimens in the main testing program of this study, it was essential to carry out preliminary testing on the materials used in this study. Therefore, the properties of fibre-reinforced polymers could be investigated more closely. Furthermore, the layout of the FRP could be optimized too.

3.1 Tensile testing of FRP

FRP specimens were tested in tension in order to determine their tensile strength and ductility. Two types of reinforced-reinforced polymers materials used in this study included unidirectional Carbon and plain weave E-glass. Unidirectional layout means that approximately 90% of fibers are aligned parallel to each other, the remaining fibers are woven at right angles to hold the main fibers together. In this case a few strands of E-glass fibers are used to hold the unidirectional Carbon fibers together at certain spacing. The E-glass fibers are manufactured with equal number of tows of fibers in the warp and weft directions.

3.2 Specimen preparation and testing

One layer, three layers and five layers of FRP were tested. The tensile testing on the reinforced-reinforced specimens was conducted using the 500 kN Instron 8033 tensile testing machine.

Once the specimens were ready to be tested, the dimensions of each specimen were measured. Then, the specimen was clamped in the grips of the testing machine, with a testing length between the two grips of 160 mm. Then, the tensile load was applied by the load cell until the failure of each specimen occurred.

Table 1. Tensile testing results on FRP coupons.

Specimen		Thick. (mm)	Max. load (kN)	Max. elongation (mm)	Max. tensile stress (MPa)
One layer	1	0.61	20.06	10.07	822.13
Carbon	2	0.61	18.66	6.62	829.00
Three layer	1	1.53	51.20	12.65	906.88
Carbon	2	1.77	49.80	6.30	810.82
Five layer	1	2.36	71.60	18.35	781.93
Carbon	2	2.70	74.06	13.80	704.37
One layer	1	0.43	1.50	4.01	90.03
E-glass	2	0.37	1.45	3.96	101.00
Three layer	1	1.15	3.45	3.20	77.32
E-glass	2	0.74	3.55	3.60	123.64
Five layer	1	1.16	6.03	3.62	135.77
E-glass	2	1.15	5.90	3.70	133.61

3.3 Testing results

Table 1 summarizes the results of the tensile tests conducted on the FRP coupons. From the testing results, it is clear that the ultimate load-carrying capacity of three-layer and five-layer Carbon specimens are 150% and 250% higher than that of one-layer specimens, which shows the tensile strength is significantly increased with the increase of the number of layers. The E-glass specimens presented the same manner in terms of the load carrying capacity. However, the tensile strength of E-glass fibers is much lower than that of Carbon fibers. Therefore, Carbon fibers provide the most effective confinement if the fibers are arranged under tension. This arrangement was used to confine the columns in the main testing program.

4 EXPERIMENTAL PROGRAM

The objective of the experimental program in this study is to investigate the behaviour of external reinforced high strength concrete columns (no internal reinforcement) subjected to eccentric loading and to evaluate the effectiveness of external confinement with FRP composites. Proven by previous studies, the major parameters affecting the behaviour of concrete columns confined with external FRP are the type of fibers; the number of layers and the shape of cross-section. As the influence of cross-section is already well known, this study is limited to circular columns under eccentric loading. The testing variables selected for this study are: (1) the type of reinforcement: internal and external, (2) the number of layers of FRP, (3) the type of wrapping materials: unidirectional Carbon and plain weave E-glass.

4.1 Column's details

Seven high strength concrete columns were designed for testing. Each column was designed to have a diameter, D, of 235 mm for both the haunched ends and 150 mm in the test region, and an overall length, H, of 1400 mm. The clear distance between the haunched ends was 620 mm. The dimensions were selected to be compatible with the capacity of the testing machine. There are two major amounts of reinforcement designed for the two internally reinforced specimens. Six RW10 bars were equally spaced around the inside circumference of ϕ110 helix with a pitch of 60 mm through the entire length of specimens and three RW8 bars confined by circular ties are spaced in equal distances at both ends. The geometry and dimension and internal reinforcement details of column are shown Figure 1.

Five specimens wrapped continually with FRP had the following configurations: one-layered and three-layered Carbon fibres and one-layered, three-layered and five-layered E-glass. The other two specimens were internally reinforced. The only difference between these two columns was that one specimen was continually wrapped with three-layer E-glass fibres. The testing matrix is summarised in Table 2.

4.2 Eccentric loading

In this study, all the seven columns were tested under eccentric loading, which was achieved by the introduction of haunched ends to each column. This can be seen clearly in Figure 1. When the concentric loading was applied to the top haunched ends of the column specimen, an eccentricity, e, of 42.5 mm, was achieved in the test region of each column. The large haunched ends were introduced in the configuration of the column specimens in order to prevent premature failure and to allow for eccentric loading.

A steel plate and a knife edge were used on the top surface of the column in order to provide an accurate concentric loading to the haunched end and to facilitate the adjustment on the direction of loading.

4.3 Specimen preparation

All the seven column specimens were cast in the Engineering Laboratory of the University of Wollongong. The target strength for both batches of concrete was 100 MPa. 103.1 MPa and 95.9 MPa determined by compressive tests were achieved for the two batches of concrete respectively.

After removal of the moulds, two internally reinforced columns were found with significant defects as shown in Figure 2, which were probably caused due to insufficient vibration. Then, column C1-1 was decided to be wrapped with three-layer Carbon fibre at both

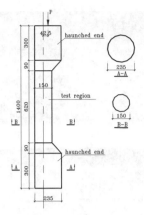

(a) Column geometry

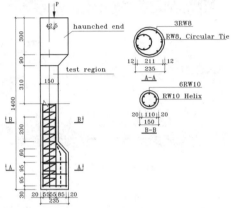

(b) Internal reinforcement details

Figure 1.   Column details.

Table 2.   Testing matrix on column specimens.

| Column | Diameter, mm | | Internal reinf. | Configurations |
	End	Middle		
C1-1	235	150	Yes	3-layered Carbon (ends)
C1-2	235	150	Yes	3-layered E-glass
C1-3	235	150	No	3-layered E-glass
C1-4	235	150	No	5-layered E-glass
C1-5	235	150	No	3-layered Carbon
C2-6	235	150	No	1-layered E-glass
C2-7	235	150	No	1-layered Carbon

haunched ends to prevent premature failure outside the test region.

The resin was prepared by mixing with slow hardener according to 5:1 ratio and firstly applied to the concrete surface. Then, the first layer of FRP was

Figure 2.   Columns with significant defects.

Table 3.   Column test results

Col.	Ultimate load (kN)	Axial stress (MPa)	Axial deflection (mm)	Max. compressive stress (MPa)	Max. tensile stress (MPa)
C1-1	836.4	47.33	0.753	−154.67	60.00
C1-2	525.5	29.74	1.983	−97.14	37.66
C1-3	601	34.01	2.4	−111.10	43.08
C1-4	736.8	41.69	1.45	−136.20	52.82
C1-5	791.5	44.79	1.405	−146.31	56.73
C2-6	669	37.86	1.771	−123.67	47.95
C2-7	644.6	36.48	–	−119.16	46.20

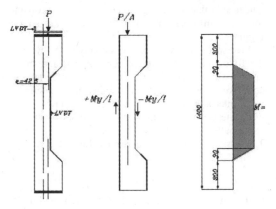

Figure 3.   Bending moment produced by the eccentric loading.

applied to the column with an overlap of 25 mm in each revolution. After wrapping the first layer, the second coating of epoxy was applied on the surface of the first layer to allow the second layer of FRP to be applied. This process was repeated until the desired layers of FRP were wrapped. Finally, the final layer of epoxy resin was applied on the surface of the wrapped specimens. The wrapped column specimens were left at room temperature for about 2 weeks for the epoxy system to harden adequately before the testing.

4.4   Test specimens

Seven columns were tested to failure using the 900 kN strong floor testing machine of the Engineering Laboratory at the University of Wollongong. The load eccentricity is 42.5 mm, which resulted in big e/r (eccentricity/column radius) ratio of 0.57.

5   OBSERVED BEHAVIOUR AND TEST RESULTS

All the columns showed similar behaviour under the eccentric loading. Although sounds of snapping of the fibres could be heard near the failure load, the failure of the column specimens in all cases was characterised by a very loud and explosive failure. Results from the experiments conducted on the seven column specimens are shown in Table 3.

Figure 3 shows how the eccentric loading was achieved and produced an axial load combined with the bending moment. From the bending moment diagram shown in Figure 3, it can be seen the maximum moment occurred right at the joint between the haunched ends and the test region, which was exactly the same as occurred in the experiment as shown in Figure 4: all columns failed at the upper part of the test region except C1-2 due to the significant defects in it.

5.1   Internally reinforced columns

The loading on the internally reinforced specimen wrapped with Carbon at both ends resulted in the spalling of the concrete cover. The final sudden failure of this column was due to the yielding of steel reinforcement. Although defects existed in the haunched ends of this column, the failure of this column did occur in the test region as designed, which proved

Figure 4. Columns after failure.

the effectiveness of wrapping using Carbon fibres at the ends.

For the internally reinforced column with E-glass wrapping, ultimate failure occurred in the patched location. This confirmed that the final failure was marked by the fracture of the E-glass fibres as a result of lateral expansion under axial eccentric loading, preceded by the crushing of concrete of the patching part. The results shown in Table 2 confirm that the influence of defects on the load carrying capacity, which is much lower than that of the internally reinforced column.

5.2 E-glass wrapped columns

The failure of all E-glass specimens was marked by the rupture of E-glass fibres. However the externally wrapped E-glass was ruptured in hoop direction only for 3-layered E-glass column. While for 1-layered and 5-layered specimens, the fibres were torn in multi-direction and longitudinal direction besides hoop direction, respectively. Approaching failure load, the appearance of white patches can be discerned, which indicated the yielding of E-glass and resin. The snapping sounds were heard before the ultimate failure, revealing the yielding of FRP composites and debonding between the layers of wrapping.

Regarding the one-layered E-glass column, the inner side of wrapping bound together with concrete even after failure, indicated that this column achieved the best bond effect between the concrete and FRP. And this is a possible explanation for this column having higher ultimate load carrying capacity than the single layered Carbon column.

5.3 Carbon wrapped columns

With a slight delamination of fibres between layers and accompanied by a simultaneous fracture of Carbon fibres and the concrete core, the final failure of both two Carbon wrapped columns was more explosive and sudden when compared to the E-glass wrapped specimens.

It is important to note that the ultimate load of single layered carbon is slightly lower than that of the single layered E-glass specimen. This can be attributed to the best bond effect achieved by the single layered E-glass specimen. Another reason is contributed to the layout of fibres: the tensile strengthening by the unidirectional fibres is not as effective as that provided by plain weave fibres.

6 COMPARISONS AND ANALYSIS

The experimental results from the six wrapped columns shows that the Carbon wrapped columns out-performed the E-glass wrapped columns. The three-layered Carbon specimen exhibited 7% and 23% increase over the five-layered and three-layered E-glass specimens, respectively. The three-layered and single layered Carbon columns exhibited 7.4% and 7.2% increase in ultimate load over the five-layered and three-layered E-glass columns, respectively. This proves that Carbon fibres are more effective than E-glass for external confinement. However, the single layered E-glass column achieved higher ultimate load than the single layered Carbon column due to the better bond effect and possible less eccentricity.

A comparison in terms of the maximum compressive stress and maximum tensile stress made between columns C1-3, C1-4 and C2-6 show that increasing the number of layers leads to higher load carrying capacity of wrapped column generally. The five-layered E-glass column achieved 22% increase compared to the three-layered E-glass column. However, this is not the case for the single layered E-glass specimen due to the possible better bond effect and less eccentricity. As the steel plate and knife edge on the top surface of column could not easily be centred accurately, which could introduce less eccentricity, then higher ultimate load was reached.

The comparison made between the two Carbon wrapped columns shows that increasing the number of layers from 1 to 3 increased the ultimate load by 23%. This again indicates that higher ultimate load could be achieved by increasing the number of layers.

In order to evaluate the effectiveness of external confinement under eccentric loading as opposed to the internal reinforcement, the comparison between C1-1, C1-3, C1-4, C1-5 and C2-7 was made as well. Although C1-2 is one of the internally reinforced columns, it was not used here for comparison due to the significant defects that existed in this column. The three-layer Carbon wrapped column achieved the ultimate load of 791.5 kN, which is just 5% lower than the high strength concrete column internally reinforced with high strength steel. This confirms that the external confinement with three-layer Carbon is as effective as the internal reinforcement with high strength steel. While for the E-glass wrapped columns under eccentric loading, the compressive stress and tensile stress of three-layered and five-layered columns were decreased by approximately 28% and 12% respectively. The ultimate load achieved by the single layered Carbon was decreased by 23% compared to the internally reinforced column.

7 CONCLUSIONS

The experimental program involved in this study is mainly to evaluate the effectiveness of external and internal reinforcement as well as the contribution of two types of external reinforcement – Carbon and E-glass to high strength concrete columns under eccentric loading. Based on the results of testing the seven column specimens, it can be concluded that:

(a) The experimental results clearly demonstrate that composite wrapping can enhance the structural performance of concrete columns under eccentric loading to some extent. However, the enhancement is not as significant as that of columns under concentric loading as suggested by previous studies. This is attributed to the fact that an eccentric loading was engaged, which induced the columns not only under axial compression, but under bending action too;

(b) For the circular specimens under concentric or eccentric loading, the number of layers of FRP materials is one of the major parameters having significant influence on the behaviour of specimens. However, the influence of the number of layers of FRP on the specimens under eccentric loading is not so pronounced as that of the specimens under concentric loading;

(c) The fibre layout is one of major factors that affect the effectiveness of confinement especially when eccentricity is introduced. Plain weave fibres are effective both for flexural and compressive reinforcements. The behaviour of structural members can be markedly improved by using unidirectional fibres applied in right direction, which means the fibres are orientated in the direction where the higher tensile strength of FRP can be utilised;

(d) Taking the expensive costs involved into consideration, external confinement with Carbon fibres is not suggested for strengthening of columns under eccentric loading at a larger eccentricity ratio.

REFERENCES

Autar, K.K. 1997. *Mechanics of Composite Materials*. CRC Press, Boca Raton, New York.

Demer, M. & Neale, K.W. 1999. Confinement of Reinforced Concrete Columns with Fibre-reinforced Composite Sheets – an Experimental Study. *Canadian Journal of Civil Engineering* 26(2):226–241.

Razvi, S.R. & Saatcioglu, M. 1994. Strength and Deformability of Confined High-Strength Concrete Columns. *ACI Structural Journal*, 91(6):678–687.

Toutanji, H.A. & Balaguru, P. 1999. Effects of Freeze-Thaw Exposure on Performance of Concrete Columns Strengthened with Advanced Composites. *ACI Materials Journal* 96(6):605–610.

System-based Vision for Strategic and Creative Design, Bontempi (ed.)
© 2003 Swets & Zeitlinger, Lisse, ISBN 90 5809 599 1

Helically reinforced high strength concrete beams

M.N.S. Hadi
Faculty of Engineering, University of Wollongong, Australia

ABSTRACT: Current developments of the construction industry have led to the continuous improvement of construction materials. However, continuous improvements in a material's strength capacity are often burdened by a decrease in ductility. To increase the ductility will allow new materials to be used, as well as harness the full potential flexural strength of reinforced concrete beams. Improved ductility through the incorporation of helical reinforcement located in the compression region is investigated herein. The helix must successfully confine the inner core in order for substantial ductility improvements to occur. In this study, different helical configurations are investigated, in order to determine the full implications of their use. This paper considers the strength gain and ductility of high strength reinforced concrete beams incorporating high strength tensile and helical reinforcement. The effect on ductility through the application of helical reinforcement located in the compressive region of the beam is highlighted. Beams of span four meters were subjected to flexural loading, with an emphasis placed on the mid span deflection. The results indicate that helical reinforcement is an effective way of increasing ductility in high strength reinforced concrete beams.

1 INTRODUCTION

Current developments of the construction industry have led to the continual improvement of construction materials. However, continual improvements in a material's strength capacity are often burdened by a decrease in ductility. To increase the ductility will allow new materials to be used, as well as harness the full potential flexural strength of reinforced concrete beams. Improved ductility through the incorporation of helical reinforcement located in the compression region is investigated herein. Helical reinforcement has been successfully used to improve the ductility of high strength concrete columns, but is yet to be implemented into beams. The helix must successfully confine the inner core in order for substantial ductility improvements. In this work, the effect of increasing the tensile reinforcement in the beam is investigated, while keeping the amount of helical reinforcement constant. Four four-metre span high strength concrete beams were cast and tested. One beam (reference beam) had no helical reinforcement and was reinforced at the maximum allowable tensile reinforcement to prevent brittle failure of the beam. The second beam exactly the same reinforcement as the first beam but had helical reinforcement added. The third and fourth beams had 1.5 and 2, respectively, the maximum allowed reinforcement. All beams were provided with the same adequate amount of shear reinforcement. The helical reinforcement was kept constant in the three helically reinforced beams. All beams were tested under the three point loading regime.

2 HIGH STRENGTH CONCRETE CHARACTERISTICS

High strength concrete has become increasingly popular in recent years. The Australian Standard (AS 3600) recognises the use of concrete up to 55 MPa. Recently, concretes of 70, 80 and 90 MPa have been used on a small number of projects throughout Australia, whilst in other countries, compressive strengths up to 130 MPa have been used (Webb 1993). As a result of its highly brittle nature, its immediate use throughout the industry is restricted by limitations placed by the Australian Standard. A full understanding into its behaviour is required before it can be successfully incorporated into present and future structures.

The high compressive strength of high strength concrete can be obtained from a low water to cement ratio and the use of potential high strength cements, such as silica fume (Cement and Concrete Association of Australia and National Ready Mixed Concrete Association of Australasia 1992).

Silica fume is a by-product from the production of silicon and ferrosilicon alloys in an arc furnace.

The resulting powder is extremely fine (surface area of approximately 20,000 m²/kg) and highly reactive (Cement and Concrete Association of Australia and National Ready Mixed Concrete Association of Australasia, 1992). The fine cementitious material will completely react, filling any small voids and providing a highly dense material.

Incorporation of a low water-cement ratio (≈ 0.25) and an increase in water demand from the silica fume, results in a highly unworkable mix. To improve the workability a superplasticiser may be incorporated. The superplasticiser increases the slump of the mix, but does not affect the composition or characteristics of the concrete. However, the period of high workability is limited to around an hour.

3 DUCTILITY INVESTIGATIONS – HELICAL REINFORCEMENT

Lateral reinforcement will provide confinement to the concrete, and hence increase the compressive strength of the member. It achieves this confinement by resisting the lateral expansion due to the Poisson effect upon loading, and thus acts on the concrete with a lateral compressive force. This allows the core of the member to be subjected to a beneficial multi-axial compression. In this state, both the deformation capacity and strength of the concrete are enhanced (Pessiki & Pieroni 1997). A helical configuration will provide a reasonably even distribution of lateral forces to the enclosed core of concrete.

Experiments preformed by Pessiki & Pieroni (1997) looked at the axial loading of large-scale spirally reinforced concrete columns. During the experimentation, it was observed that large cracks occurred along the length of the concrete, and nearing the ultimate loading capacity the concrete cover split and buckled, and fell away. The closely spaced reinforcement physically separated the concrete cover from the core, causing the early failure of the cover. It was then suggested that a decrease in spiral pitch was required for the high strength column to achieve the same ductility as that of normal strength columns.

Helix reinforcement is used mainly in columns, however it may be incorporated into other applications to improve ductility or compressive strength. To be effective, the helical reinforcement must be placed in the compression zone and the pitch must be small enough to confine the concrete core. Herein, the exploration of helical reinforcement in the compressive region of the reinforced concrete beams is considered, so as to determine whether the helix will increase the strength and ductility of the beams.

The helical pitch size is a major influence on the effectiveness of the confinement. If the concrete is sufficiently confined, only the cover will be shed near

the ultimate loading capacity. However, if the pitch is increased the concrete core may not be effectively confined, causing some of the core to shed with the concrete cover. The reduction in cross sectional area will then affect the ultimate load capacity.

4 EXPERIMENTAL PROGRAM

4.1 Beam composition

Four beams were subjected to three point flexural loading. All beams had a cross sectional area of 200 mm by 300 mm. Table 1 shows a summary of the beams' composition. As shown in Table 2, Beam 1 was reinforced with four deformed bars of 28 mm diameter and 400 MPa tensile strength. This amount being maximum allowed tensile reinforcement to prevent brittle failure of the beam. Likewise Beam 2 was reinforced with exactly the same amount of tensile reinforcement but with helical reinforcement added. Beam 3 had 1.5 times the maximum allowed amount of reinforcement. This amount of reinforcement was provided by 5 deformed bars of 28 mm diameter. The last beam was reinforced with five deformed bars of 32 mm diameter. This amount was the practically maximum amount of reinforcement to fit in the beam. Shear reinforcement was provided in all beams at a spacing of 100 mm and they were composed of 10 mm plain 250 MPa bars. Helical reinforcement of 12 mm deformed bars with 400 MPa tensile strength

Table 1. Summary of beam composition.

No.	Concrete strength (MPa)	Tensile steel	Helical reinforcement	
			Bar size	Pitch P (mm)
1	51.1	4Y28	–	–
2	67	4Y28	Y12	50
3	67	5Y28	Y12	50
4	51.1	5Y32	Y12	50

Table 2. Summary of beam composition.

No.	Concrete strength (MPa)	Tensile steel	$\dfrac{A_{st}}{bd}$ %
1	51.1	4Y28	5.17
2	67	4Y28	5.17
3	67	5Y28	7.76
4	51.1	5Y32	8.45*

* This is the maximum amount of steel that is practical for this beam.

was provided in beams 2, 3 and 4. The outside diameter of the helices was 150 mm with a 50 mm pitch. Photographs of the beams cross section are shown in Figure 1.

The beams had a cross sectional dimensions of 200 mm by 300 mm and had a span of four metres. The loading was applied midspan of the beam. Cover was kept constant at 20 mm.

4.2 Beam fabrication

The reinforcing cages were constructed through the use of wire ties. Deformed steel bars located in the compression region of the beam were not considered in resisting flexural loads, but were used to support the shear stirrups and helical reinforcement. The helical reinforcement were purchased in their coiled form.

(a) Beam 1 (b) Beam 2

(c) Beam 3 (d) Beam 4

Figure 1.   Beams' cross section.

Figure 2.   Loading arrangement.

The concrete was purchased from a local concrete supplier. The beams were quickly poured and compacted through the use of an internal vibrator. All beams were left in their moulds until adequate strength was obtained, and were cured by covering with wet hessian bags. To obtain a comparative concrete strength, the test cylinders were cured in the same manner.

4.3 Test details

As shown in Figure 2, all beams were tested in the strong floor with deflection controlled loading system in the University of Wollongong Laboratory. Strain gauges were used to monitor strains in the midspan longitudinal steel and helical reinforcement. A deflection gauge was situated at midspan to measure vertical deflection. Demec gauges were approximately placed at the location of the longitudinal steel and 25 mm from the top of the cross section at midspan.

A plotter was used to graph the load versus the deflection of the beam during loading. Loading of the beam was ceased upon failure, or due to the inability to continue effective loading. The total load versus the midspan deflection of all tested beams are shown in Figure 3.

5   EXPERIMENTAL RESULTS

5.1   Flexural behaviour

Table 3 presents the yield strength and corresponding deflection for the four tested beams. Ultimate strength and corresponding deflection for the four tested beams are presented in Table 4. The maximum deflection that the beams underwent during the tests and the corresponding load are presented in Table 5.

Ductility of a beam is taken here as the ratio of the maximum deflection divided by the deflection at yield load. Ductilities of all tested beams are shown in Table 6.

6   EXPERIMENTAL ANALYSIS

6.1   Ductility of increased tensile reinforced beams

As shown in Table 4 an increase in tensile reinforcement increases by about 197% the load resistance of the beam. The large increase in strength was accomplished by a simple increase in tensile steel to the next available deformed bar. Also. As shown in Table 6 this increase in strength produced a 235% increase in structural ductility, which is defined in this paper as the measured deflection between the yield point and final failure regardless of load capacity.

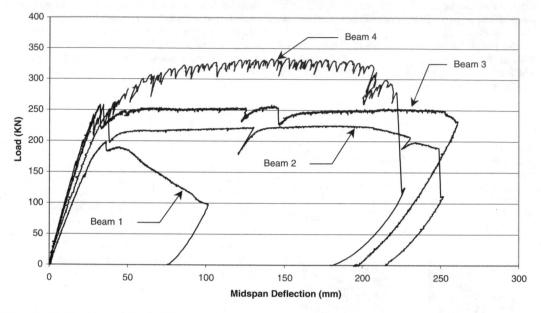

Figure 3. Load–midspan deflection of tested beams.

Table 3. Yield strength and corresponding deflection of tested beams.

No.	Helical reinf.	Yield load (kN)	Yield ratio	Yield defl. (mm)	Rel. defl.
1	No	198.5	1	36.3	1
2	Yes	245.9	1.24	37.8	1.04
3	Yes	258.4	1.30	32.6	0.90
4	Yes	253.1	1.28	34.6	0.95

Table 5. Maximum deflection and corresponding load of tested beams.

No.	Helical reinf.	Max defl. (mm)	Rel.	Corres. load (kN)	Rel.
1	No	101.5	1	96.7	1
2	Yes	251.9	1.27	111.2	1.15
3	Yes	261	1.31	231.9	2.32
4	Yes	227.3	1.15	125.8	1.30

Table 4. Ultimate load and corresponding deflection of tested beams.

No.	Helical reinf.	Ultimate load (kN)	Ultimate ratio	Ultimate defl. (mm)	Rel. ult. defl.
1	No	198.5	1	36.3	1
2	Yes	225.1	1.13	182.4	5.02
3	Yes	252.1	1.27	213.6	5.88
4	Yes	332.5	1.97	176.4	4.86

Table 6. Ductility of tested beams.

No.	Helical reinf.	Yield defl. (mm)	Max defl. (mm)	Ductility	Rel.
1	No	36.3	101.5	2.80	1
2	Yes	37.8	251.9	6.66	2.38
3	Yes	32.6	261	8.01	2.86
4	Yes	34.6	227.3	6.57	2.35

If the amount of tensile steel is further increased, the member will surpass the suggested k_u (the ratio of the depth to the neutral axis from the extreme compressive fibre to the distance to the resultant tensile force in the reinforcing steel) limit of 0.4 (AS3600, 2001).

The strength of the member will become enhanced, however the ductility will be significantly reduced. To achieve a suitable value of ductility, the current method of limiting the amount or strength of tensile steel restricts the optimum potential to be gained from the use of reinforced high strength concrete.

6.2 Ductility of helical beams

The application of helical reinforcement made a substantial difference in the flexural behaviour, by significantly increasing the beams' ductility. The flexural strengths of the helical beams were significantly increased as observed in Table 6.

Both the plain and helical beams, behaved in an identical manner prior to the ultimate load. Where the plain beams deflected and then ultimately failed, the helical beams shed their cover and continued to deflect at a substantial load. However, the maximum loading plateau achieved was more for the helical beams. This was due to the concrete cover being separated by the presence of the helical reinforcement.

Once the concrete cover was shed, the flexural capacity of the beam was decreased. As the beams were small, compared with that used in current structures, the concrete cover of 20 mm was a significant portion of the total area as stated earlier. This left a small amount of concrete to resist all compressive forces. The reduction of flexural loading would not be as predominant in larger sized beams.

6.3 Overall behaviour of beams

The elastic region of all the beams provides an interesting trend. The slope of the ascending line in each beam progressively increases indicating an increase in stiffness from beam to beam.

The overall total load carried by the Beam 2 increased from 198.5 kN in Beam 1 to 225.1 kN. This is a 13% increase in load carrying capacity. The plastic region of Beam 1 allows for much less deflection as compared to Beam 2 and in fact the other two beams. Beam 2, which can be directly compared to Beam 1 as it contains identical reinforcing steel, has out performed beam 1 in strength and ductility. The higher strength concrete in Beam 2 may have contributed to the increase in load. This higher strength should have resulted in Beam 2 being less ductile. This was not the case and the increase in ductility may be due to the use of the helix as this was the only physical difference between the two beams.

Beam 2 has a yield strength of 245.9 kN. An increase of 5.1% to 258.4 kN is noted when comparing Beam 2 to Beam 3. The overall shape of the load versus deflection curve for Beam 2 and 3 is very similar. The main difference is in the elastic region. Beam 3 has a larger load carrying capacity compared to Beam 2. Both Beams 2 and 3 exhibit ductile properties with an increase in the deflection of the beam in the plastic region of 214.1 mm and 228.4 mm, respectively. The maximum deflection varies less than 5% between Beams 2 and 3. This slight increase may be due to the fact that the testing of Beam 3 was stopped due to

unsafe conditions and so the final deflection of the beam could not be achieved.

Beam 4 showed an increase in ultimate load of 31% to 332.5 kN as compared to Beam 3. This increase in load has reduced the deflection of the Beam 4 by approximately 17% as compared to Beam 3. Although this is quite a significant increase the overall performance of the beam is still excellent. Beam 4 still behaved in a ductile manner and exhibited a 123% increase in deflection as compared to Beam 1, which was designed with a more ductile failure in mind.

The incorporation of helical reinforcement of an appropriate pitch significantly increases the ductility of a reinforced concrete beam. If an effective pitch is obtained for complete concrete confinement, the beams will continue to deflect and resist loading at constant load. The improvement in ductility allows an increase in strength through increasing the tensile reinforcement. This new approach differs from the conservative methods of AS3600, and offers the opportunity to utilise the full potential of high strength reinforced concrete beams.

7 CONCLUSIONS

Based on this research, the following observations can be made on the basis of the specimens considered:

An increase in tensile steel of otherwise plain high strength concrete beams increased the strength and caused significant increase in ductility.

The use of a 3 point loading system has not affected the performance of the helix in confining the concrete. This was indicated by an increase in the load carrying capacity of each beam with a helix as compared to the control beam. In terms of load carrying capacity (LCC) there was an increase in LCC with an increase in the tensile steel. This increase in LCC resulted, in the case of Beam 4, in a decrease in ductility compared with Beams 2 and 3. This decrease in ductility still outperformed the ductility of Beam 1. For this reason it can be said that the inclusion of a helix in the beam design will increase the ductility. Indeed, a beam that has been designed to fail in a very brittle manner displays deflections consistent with ductile design.

There is great deal of stress in the steel helix. The design of the helix may limit the performance of the beam. Two out of the three beams containing a helix failed due to the helix breaking. The third beam was not tested to failure.

The results from this research are encouraging. To improve further the effects of helical reinforcement on beam ductility, additional investigations into helical configuration and structural behaviour is required.

ACKNOWLEDGMENT

The author wishes to acknowledge the University of Wollongong small grant, which helped in conducting this research.

REFERENCES

AS3600. 2001. *Concrete Structures*. Standards Australia.
CCAA. 1992. *High-Strength Concrete*. Cement and Concrete Association of Australia and National Ready Mixed Concrete Association of Australasia.

Pessiki, S. & Pieroni, A. 1997. Axial Behaviour of Large-Scale Spirally-Reinforced High-Strength Concrete Columns, *ACI Structural Journal*, May–June. 94(3):304–314.
Webb, J. 1993. High-Strength Concrete: Economics, Design and Ductility. *Concrete International: Design and Construction*, ACI, Jan. 15(1):27–32.

System-based Vision for Strategic and Creative Design, Bontempi (ed.)
© 2003 Swets & Zeitlinger, Lisse, ISBN 90 5809 599 1

Time series modelling of high performance concrete column shortening due to creep and shrinkage

C.Y. Wong
Post Graduate Student, RMIT University, Australia

K.S.P. de Silva
Senior Lecturer, RMIT University, Australia

ABSTRACT: The work presented in this paper is the first stage of a research project undertaken at RMIT University, Australia. The project investigates the time series modelling of high performance concrete column shortening in tall buildings, due to elastic, creep and shrinkage, and their impact on the residual stresses induced within the floor systems. The first stage of the program, on which this paper is based on, is a parametric evaluation of the current time series modelling techniques and their net impact on the relative shortening of columns. This paper presents the results of a parametric study which analyses a typical column of a 30, 60 and 90 story building with concrete strengths of 60 and 80 MPa. Second stage involves the development of an analytical and decision support tool to be used in the industry. The proposed software tool is capable of estimating relative shortening of the columns, moment re-distribution and the resulting residual stresses within the floor system using Australian, ACI and CEB-FIP standards.

1 INTRODUCTION

1.1 *Column shortening problem*

Column shortening becomes a prime design consideration for buildings over 30 stories. Ghali (1978) concluded that buildings taller than thirty stories would experience column shortening of 3 mm per floor. High Performance Concrete (HPC) is the preferred construction material used in buildings taller than 30 stories. Combined effect of elastic, shrinkage and creep strains contributes to the overall shortening of the columns, which varies with age of concrete, environmental factors, loading history and the material qualities of concrete. Differential shortening of columns and other vertical load bearing elements such as load bearing walls, if not carefully engineered, is the most significant contributing factor for many early age distresses experienced in tall buildings especially at upper stories. Relative shortening can cause non-structural damage to mechanical systems, claddings, partitions, vertical pipes and lifts. It also can cause structural distresses resulting from moment redistribution and residual stresses within floor plates, especially in pre-stressed floor system.

2 TIME SERIES ANALYSIS

2.1 *Time series models*

Accurate time series modelling of built columns and the prediction of column shortening at a given time is an extremely difficult task. This is because:

1. The creep and shrinkage characteristics of concrete are time dependent variables
2. Loading sequence is a highly variable function which depends on the contractors method of working and capabilities
3. Built columns are exposed to uncontrolled environments.

A significant research effort has improved the understanding of this complex phenomenon (Gilbert 1988, Smith 1988). The theoretical basis of the shrinkage and creep models adopted around the world is based on the following methods, which are modified over the time to give better estimations.

1. Effective Modulus Method (Faber 1927)
2. Age-adjusted Effective Modulus Method (AEMM)
   – (Bazant 1972)

3. Rate of Creep Method – (Whitney 1932)
4. Improved Dischinger Method – (Nielsen 1978)
5. Superposition – (Ghali et al. 1967).

2.2 Design standards

The current design practices are adopting these modified models especially the Effective Modulus Method. Mathematical models such as power, hyperbolic and exponential functions are adopted to extrapolate long term creep coefficient and shrinkage strain from short-term test results (Gilbert 1988). A parametric study has been carried out using the shrinkage and creep time series models adopted by the following three standards.

i.   AS3600 (2001)
ii. ACI 318R-02 (2002)
iii. CEB-FIP (1990)

As the research project is primarily targeting the Australian building industry the current AS3600 (2001) results are benchmarked against two widely accepted models of ACI 318 and CEB-FIP.

2.2.1 Australian standard – AS3600 (2001)
2.2.1.1 Creep
Equation 1 gives the time series function of creep, $\phi(t,t_o)$ adopted by the current AS3600, which is also referred to as the creep factor coefficient. The creep factor coefficient is a function of time since the application of load (sustained loading)

$$\phi_{cc} = k_2 k_3 \phi_{cc.b} \tag{1}$$

k_2 and k_3 factors specified in the current AS3600 are based on the Gilbert (1988) as given in Equation 2 and Equation 3 respectively

$$k_2 = \frac{k_7 k_8 t^{0.7}}{t^{0.7} + k_9} \tag{2}$$

with

$$k_7 = 0.76 + 0.9e^{-0.008t_h}$$
$$k_8 = 1.37 - 0.011RH$$
$$k_9 = 0.15t_h \tag{2a}$$
$$t_h = 2(A_g / U)$$

where t_h = hypothetical thickness; RH = relative humidity; A_g = cross sectional area (mm^2) of member; U = perimeter of the exposed cross section (mm)

$$k_3 = 0.9 \ \ for \ \frac{f_{cm}}{f'_c} \geq 1.4$$

$$= -0.5\left(\frac{f_{cm}}{f'_c}\right) + 1.6 \ \ for \ 1.0 \leq \frac{f_{cm}}{f'_c} < 1.4$$

$$= -0.8\left(\frac{f_{cm}}{f'_c}\right) + 1.9 \ \ for \ 0.5 \leq \frac{f_{cm}}{f'_c} < 1.0 \tag{3}$$

where f_{cm} = mean value of the compressive strength of concrete at the relevant age; f'_c = characteristic compressive strength of concrete.

The factor k_2 takes into account the duration of sustained loading, hypothetical thickness and the environment. Factor k_3 indirectly considers the age of the concrete at the time of load application t_o. The basic creep factor, $\phi_{cc.b}$ is the ratio between ultimate creep strain and the initial elastic strain of a specimen loaded at 28 days to a stress of $0.4f'_c$. Values for basic creep factor can be obtained from Table 6.1.8.1 in AS 3600.

2.2.1.2 Shrinkage
Equation 4 gives the time series function of the design shrinkage strain, ε_{cs}

$$\varepsilon_{cs} = k_1 \varepsilon_{cs.b} \tag{4}$$

where $\varepsilon_{cs.b}$ = basic shrinkage strain.

The k_1 factor given in the current AS3600 is based on Gilbert (1988) and given in Equation 5. The k_1 factor is a function of the concrete age, environmental conditions and the size and shape of the concrete member

$$k_1 = \frac{k_4 k_5 t^{0.7}}{t^{0.7} + k_6} \tag{5}$$

with

$$k_4 = 0.62 + 1.5e^{-0.005t_h}$$
$$k_5 = \frac{4 - 0.04RH}{3} \tag{5a}$$
$$k_6 = \frac{t_h}{7}$$

where t_h = hypothetical thickness; RH = relative humidity; A_g = cross sectional area (mm^2) of member; U = perimeter of the exposed cross section (mm).

Basic shrinkage strain, $\varepsilon_{sh.b}$ must be based on test data. In the absence of any experimental data a median value of 850×10^{-6} is to be assumed.

2.2.2 ACI building code requirements for structural concrete – ACI 318 (2002)
2.2.2.1 Creep
Equation 6 gives the generalized creep time series function proposed in the ACI 209R (1982), which is

the basis of ACI 318 approach. This time series function is a combination of hyperbolic and power functions

$$\phi(t,\tau) = \frac{(t-\tau)^{0.6}}{10+(t-\tau)^{0.6}} \quad (6)$$

where τ = age of concrete at first loading (days); $(t - \tau)$ = duration of loading (days).

The time series is then modified by a correction factor, $\phi^*(\tau)$ to estimate the creep coefficient as shown in Equation 7. The correction factor is used to account for various specific attributes such as age, environmental factors, curing method and etc.

$$\phi(t,\tau) = \frac{(t-\tau)^{0.6}}{10+(t-\tau)^{0.6}}\phi^*(\tau) \quad (7)$$

$$\phi^*(\tau) = 2.35\gamma_1\gamma_2\gamma_3\gamma_4\gamma_5\gamma_6 \quad (8)$$

= final creep coefficient

γ_1 = correction factor for age of loading
= $1.25(\tau)^{-0.118}$, for moist cured concrete, $\tau > 7$ days
= $1.13(\tau)^{-0.094}$, for steam cured concrete, $\tau > 3$ days
γ_2 = correction factor for relative humidity
= $1.27 - 0.0067\,RH$, for $RH > 40$
γ_3 = correction factor for size and shape of member based on average thickness of members
= $2/3[1 + 1.13\,e^{-0.0213v/s}]$, for $h_o \geqslant 380\,mm$
= $1.14 - 0.00092\,h_o$, for $150\,mm < h_o < 380\,mm$, $(t - \tau) \leqslant 1$ year
= $1.10 - 0.00067\,h_o$, for $150\,mm < h_o < 380\,mm$, $(t - \tau) > 1$ year
= Table 1, for $h_o \leqslant 150\,mm$
γ_4 = correction factor for slump
= $0.82 + 0.00264\,s$
γ_5 = correction factor for cement content
= $0.88 + 0.0024\,\psi$
γ_6 = correction factor for air content
= $0.46 + 0.09a < 1.0$

where τ = age of concrete at first loading (day); RH = relative humidity in percentage; v/s = volume to surface ratio; h_o = 4(v/s); s = slump of the fresh concrete in mm; ψ = ratio of the fine aggregate to total aggregate by weight in percent; a = air content in percent.

Then, under a constant stress σ_0 first applied at age τ, the load-dependent strain (elastic strain and creep strain) at time t would be

$$\varepsilon(t) = \frac{\sigma_0}{E_c(\tau)}[1+\phi(t,\tau)] \quad (9)$$

Table 1. Creep correction factors for average thickness of members, h_o.

h_o (mm)	50	75	100	125	150
γ_3	1.30	1.17	1.11	1.04	1.00

Table 2. Values of the constants α and β.

	Curing condition	
Cement type	Moist cured	Steam cured
Type I	$\alpha = 4.00, \beta = 0.85$	$\alpha = 1.00, \beta = 0.95$
Type III	$\alpha = 2.30, \beta = 0.92$	$\alpha = 0.70, \beta = 0.98$

with $E_c(\tau) = 0.043\rho^{1.5}\sqrt{f_{cm}}$

where ρ = density of concrete

The concrete strength at age τ may be obtained from the 28 days strength as below

$$f'_c(\tau) = \frac{\tau}{\alpha + \beta\tau}f'_c(28) \quad (10)$$

with α and β depending on the cement type and the curing condition respectively as shown in Table 2.

2.2.2.2 Shrinkage
ACI 209R (1982) recommends different hyperbolic time series functions to estimate the shrinkage coefficient depending on the curing method. Equations (11a) and (11b) give the generalized shrinkage time series function to estimate the shrinkage coefficient for moist-cured and steam-cured conditions respectively

$$\varepsilon_{shk}(t_d) = \frac{t_d}{35+t_d} \quad (11a)$$

$$\varepsilon_{shk}(t_d) = \frac{t_d}{55+t_d} \quad (11b)$$

with t_d = age of concrete after drying (days).

The time series is then modified by a correction factor, ε^*_{shk} to estimate the creep coefficient as shown in Equation (12a) and (12b). The correction factor is used to account for various specific attributes such as slump, environmental factors, size and shape of the member and etc.

$$\varepsilon_{shk}(t_d) = \frac{t_d}{35+t_d}\varepsilon^*_{shk} \quad (12a)$$

$$\varepsilon_{shk}(t_d) = \frac{t_d}{55+t_d}\varepsilon^*_{shk} \quad (12b)$$

with $\varepsilon_{shk}^{*} = 780\kappa_1\kappa_2\kappa_3\kappa_4\kappa_5\kappa_6\kappa_7$ (13)

= final shrinkage coefficient at time infinity

κ_1 = correction factor for relative humidity
   = $1.40 - 0.01 RH$ for $40 \leqslant RH \leqslant 80$
   = $3.00 - 0.03 RH$ for $80 < RH \leqslant 100$
κ_2 = correction factor for size and shape of member based on average thickness of members, h_o
   = $1.2\,e^{-0.00472\,V/S}$ for $h_o > 380$ mm
   = $1.23 - 0.0015\,h_o$ for 150 mm $< h_o \leqslant 380$ mm and $t \leqslant 1$ year
   = $1.17 - 0.0011\,h_o$, for 150 mm $< h_o \leqslant 380$ mm and $t > 1$ year
   = Table 3 for 50 mm $\leqslant h_o \leqslant 150$ mm
κ_3 = correction factor for slump
   = $0.89 + 0.00161\,s$
κ_4 = correction factor for ratio of fine aggregate to total aggregate content
   = $0.30 + 0.014\,\psi$ for $\psi \leqslant 50\%$
   = $0.90 + 0.002\,\psi$ for $\psi > 50\%$
κ_5 = correction factor for air content
   = $0.95 + 0.008\,a$
κ_6 = correction factor for cement content
   = $0.75 + 0.00061\,c$
κ_7 = correction factor for initial moist curing time, $T_c = 1.0$, for steam cured between 1 to 3 days
   = Table 4, for moist cured

where RH = relative humidity in percentage; v/s = volume to surface ratio; $h_o = 4(v/s)$; s = slump of the fresh concrete in mm; ψ = ratio of the fine aggregate to total aggregate by weight in percent; a = air content in percent; c = cement content in kg/m^3.

2.2.3 CEB-FIP Model Code (1990)

This design code is a comprehensive revision to the original 1978 model code, which was produced jointly by the Comité Euro-International du Béton (CEB) and the Fédération International de la Précontrainte (FIP). The original 1978 CEB-FIP Model Code has a great impact on the national design codes in many countries. Eurocode 2 for example is one of the national codes, which used Model Code 1978 as its basic

Table 3. Shrinkage correction factors for average thickness of members, h_o.

h_o (mm)	50	75	100	125	150
γ_3	1.35	1.25	1.17	1.08	1.00

Table 4. Shrinkage correction factors for initial moist curing.

T_c	1	3	7	14	28	90
κ_7	1.20	1.10	1.00	0.93	0.86	0.75

reference document. Model Code (1990) is more comprehensive in providing guidelines and explanations.

2.2.3.1 Creep

This model is valid for ordinary structural concrete with compressive strength less than 80 MPa and subjected to compressive stress less than $0.4\,f_c'$ at the age of loading. The stress dependent strain, ε_σ may be expressed as

$$\varepsilon_\sigma(t,\tau) = \sigma_c(\tau)\left[\frac{1}{E_c(\tau)} + \frac{\phi(t,\tau)}{E_{ci}}\right] = \sigma_c(\tau)J(t,\tau)$$
(14)

where $J(t,\tau)$ = creep function or creep compliance; $E_c(\tau)$ = modulus of elasticity at time of loading, τ.

The creep coefficient is calculated as

$$\phi(t,\tau) = \phi_o\,\beta_c(t-\tau)$$
(15)

where ϕ_0 = notional creep coefficient; β_c = coefficient describing the development of creep with time after loading; t = age of concrete; τ = age of concrete when loaded.

The notional creep coefficient is estimated as

$$\phi_0 = \phi_{RH}\,\beta(f_{cm})\beta(\tau)$$
(16)

with

$$\phi_{RH} = 1 + \frac{1 - RH/RH_0}{0.46(n/n_0)^{1/3}}$$

$$\beta(f_{cm}) = \frac{5.3}{(f_{cm}/f_{cmo})}$$

$$\beta(\tau) = \frac{1}{0.1 + (\tau/t_1)^{0.2}}$$

$$h = \frac{2A_c}{U}$$

where f_{cm} = 28 days mean compressive strength of concrete; $f_{cmo} = 10$ MPa; RH = relative humidity; $RH_0 = 100\%$; n = notational size of member (mm); $n_o = 100$ mm; A_c = column's cross section; U = perimeter of the member in contact with the atmosphere; $t_1 = 1$ day.

The development of creep with time is expressed

$$\beta_c(t-\tau) = \left[\frac{(t-\tau)/t_1}{\beta_H + (t-\tau)/t_1}\right]^{0.3}$$
(17)

with

$$\beta_H = 150\left\{1+\left(1.2\frac{RH}{RH_0}\right)^{1.8}\right\}\frac{n}{n_0}+250 \le 1500$$

where $t_1 = 1$ day; $RH_0 = 100\%$; $n_0 = 100$ mm.

In some cases, type of cement would influence the creep coefficient of the concrete. Equation 18 would take this effect into account by modifying the age at loading

$$\tau = \tau_t\left[\frac{9}{2+(\tau_t/\tau_{1,t})^{1.2}}+1\right]^\alpha \ge 0.5 \ days \qquad (18)$$

where τ_t = age of concrete at loading; $\tau_{1,t} = 1$ day; $\alpha = -1$ for slow hardening cements, SL, 0 for normal or rapid hardening cements N and R, and 1 for rapid hardening high strength cements RS.

2.2.3.2 Shrinkage
The shrinkage or swelling strains is estimated as

$$\varepsilon_{cs}(t,\tau_d) = \varepsilon_{cso}\beta_s(t-\tau_d) \qquad (19)$$

where ε_{sso} = notional shrinkage coefficient; β_s = coefficient describing the development of shrinkage with time; t = age of concrete; τ_d = age of concrete at the beginning of shrinkage or swelling.

The notional shrinkage coefficient is expressed as

$$\varepsilon_{cso} = \varepsilon_s(f_{cm})\beta_{RH} \qquad (20)$$

with $\varepsilon_s(f_{cm}) = [160 + 10\beta_{sc}(9 - (f_{cm}/f_{cmo}))]x\ 10^{-6}$

$\beta_{RH} = -1.55\beta_{sRH}$ for $40\% \le RH < 99\%$
$\quad\quad = +0.25$ for $RH \ge 99\%$

$$\beta_{sRH} = 1-\left(\frac{RH}{RH_0}\right)^3$$

where f_{cm} = 28 days mean compressive strength of concrete; $f_{cmo} = 10$ MPa; β_{sc} = coefficient influenced by type of cement which is equal to 4 for slow hardening cements, SL, 5 for normal or rapid hardening cements, N and R and 8 for rapid hardening high strength cements, RS; RH = relative humidity; $RH_0 = 100\%$.

The development of shrinkage with time is as below

$$\beta_s(t-\tau_d) = \left[\frac{(t-\tau_d)/\tau_1}{350(n/n_o)^2+(t-\tau_d)/\tau_1}\right]^{0.5} \qquad (21)$$

where $t_1 = 1$ day; $n_0 = 100$ mm.

3 PARAMETRIC STUDY

A parametric evaluation of shortening in tall building columns using the creep and shrinkage time series

models described in Section 2 is presented here. A computer program, Column Shortening Analytical Tool (CSAT) was developed with built in capabilities to simulate different scenarios using loading sequences, material properties and environmental factors. Program has the option of selecting AS3600 (2001), ACI 318 (2002) or CEB-FIP (1990) recommended procedure. The current AS3600 have been modified to cater for the use of HPC.

The time series modelling of typical internal columns in a 30, 60 and 90 story office tower is attempted using the program CSAT. All these columns support a 250 mm × 6 m × 6 m flat slab at each floor level. Concrete strengths of 60 MPa and 80 MPa were used in the comparison. All the columns are assumed to have been loaded by a 7-day floor cycle with a 49-day lag in imposing service dead loads. Live load is assumed as being imposed after the completion of the building. Column sizes are selected and designed to have a maximum compression stress of $0.4f'_c$. The creep and shrinkage time series results output from program CSAT are tabulated in Table 5.

Figure 1 compares the results of total column shortening due to elastic, shrinkage and creep strains. The estimates were based on ACI, AS and CEB time series models for all three buildings using concrete strengths of 60 MPa.

Figure 2 compares the 60 MPa and 80 MPa columns of the 90-storey building. As the number of stories increase, the estimated total column shortening values for the three models appears to diverge. As the concrete strength increases from 60 MPa to 80 MPa, the total shortening of the column reduces as expected. Concrete strength plays a smaller influence in the ACI estimates compared to CEB and AS models.

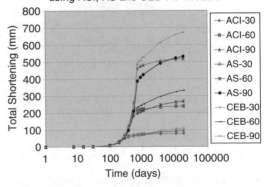

Total Shortening of 30, 60 and 90 storey building using ACI, AS and CEB-FIP models

Figure 1. Total estimated shortening of 30, 60 and 90 storey columns using ACI, AS and CEB-FIP models. Concrete Strength 60 MPa.

Table 5. Creep and shrinkage time series results of 30, 60 and 90 story columns (CSAT output).

Time (day)	Code	Number of storey (mm) 30		60		90	
		Creep	Shrink	Creep	Shrink	Creep	Shrink
7 days	ACI	0	0	0	0	0	0
	AS	0	0	0	0	0	0
	CEB	0	0	0	0	0	0
28 days	ACI	0	1	0	1	0	1
	AS	0	1	0	1	0	1
	CEB	0	0	0	0	0	0
90 days	ACI	2	5	2	5	1	5
	AS	1	4	1	4	0	4
	CEB	2	2	2	2	2	1
1 year	ACI	22	23	43	34	38	34
	AS	14	22	23	31	12	31
	CEB	29	12	56	15	44	10
3 years	ACI	25	27	98	53	195	79
	AS	20	33	67	64	75	92
	CEB	36	21	139	41	242	35
10 years	ACI	27	28	106	55	213	83
	AS	25	40	83	80	96	119
	CEB	40	34	157	68	275	61
30 years	ACI	28	28	110	56	221	84
	AS	27	44	92	88	105	131
	CEB	41	45	163	90	287	81
50 years	ACI	28	28	111	56	223	84
	AS	28	45	94	90	108	135
	CEB	42	48	164	97	289	87

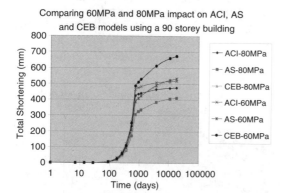

Figure 2. Comparison of total column shortening of the 90 storey building for 60 MPa and 80 MPa strength.

4 CONCLUSIONS

- The combined elastic, creep and shrinkage shortening in columns which support over 30-stories, exceed 100 mm and therefore cannot be ignored in the design.

- There is a noticeable difference in the estimated values of column shortening based on ACI, AS and CEB models. This becomes more apparent with the increase in the number of stories.
- The apparent scatter is a direct result of the long-term predictions based on short-term observed data.
- The current state of knowledge warrants further research in this area. Instrumentation of real buildings to obtain long-term results should be encouraged where possible.
- Effective analytical tools can be developed using the current knowledge, which should then calibrate as the data become available.

REFERENCES

ACI Committee 209, Subcommittee II. 1982. *Prediction of creep, shrinkage, and temperature effects in concrete structures*. Detroit, Michigan: American Concrete Institute.

Bazant, Z.P. 1972. Prediction of concrete creep effects using age-adjusted effective modulus method. *ACI Journal* 69: 212–217.

Comite Euro-International du Beton. 1990. *CEB-FIP model code 1990*. London: Thomas Telford.

Faber, O. 1927. Plastic yield, shrinkage and other problems of concrete and their effects on design. *Minutes of Proc. Of the Inst. Of Civil Engineers Part 1, London*: 27–73.

Ghali, A., Neville, A.M. & Jha, P.C. 1967. Effects of elastic and creep recoveries of concrete on loss of prestress. *ACI Journal* 64: 802–810.

Ghali, A. 1978. Creep, Shrinkage and Temperature Effects. *Structural design of tall concrete and masonry buildings, Council on Tall Buildings and Urban Habitat*. New York: American Socicty of Civil Engineers.

Gilbert, R.I. 1988. *Time Effects in Concrete Structures*. Amsterdam: Elsevier Science Publishers B. V.

Nielsen, L.F. 1978. The improved Dischinger method, *ACI Symposium on design for creep and shrinkage in concrete structures, Houston, Texas, Oct–Nov.*

AS3600. 2001. *Australian standard for concrete structures*. Sydney: Standards Association of Australia.

Smith, B.S. & Coull, A. 1988. Tall building structures: analysis and design. John Wiley & Sons Inc.: New York.

Whitney, C.S. 1932. Plain and reinforced concrete arches. *ACI Journal* 28: 479–519.

System-based Vision for Strategic and Creative Design, Bontempi (ed.)
© 2003 Swets & Zeitlinger, Lisse, ISBN 90 5809 599 1

An experimental study of basic creep of High Strength Concrete

J. Anaton, S. Setunge & I. Patnaikuni
School of Civil and Chemical Engineering, RMIT University, Melbourne, Australia

ABSTRACT: Most of the reported work on creep of High Strength Concrete (HSC) covers measurements of total creep on standard specimens kept in a standard environment of 50% RH and 23°C. The results of these studies are used to establish basic creep factor of HSC assuming environmental effect on creep of HSC is similar to that for normal strength concrete (NSC). However, some recent work has shown that the effect of environment on creep of HSC is different to that observed for NSC, which indicates that a realistic creep model for HSC needs to be developed based on a more fundamental material behaviour.

This paper reports the details and the results of an experimental study conducted at RMIT University to investigate the basic creep of HSC and equivalent mortar. The results indicate that existing models for basic creep of NSC cannot be used to predict the behaviour of HSC. A more fundamental analysis and a mathematical model are proposed.

1 INTRODUCTION

Basic creep behavior of High Strength Concrete (HSC) is important for estimating the total long-term deformation of concrete structures due to creep effect. Persson (1999) has investigated basic creep of high strength concrete under compressive load for long-term creep. Ward et al (1969) have carried out experimental research on basic creep of normal strength concrete (NSC) and mortar. The majority of reported work on creep of concrete has been concerned with the total creep behavior of HSC in a standard environment conditions such as 50% RH and 23°C.

According to some recent work by Setunge & Toyne (1999), the effect of environment on creep of HSC is different to that observed for NSC. Calculated basic creep factor for a 103 MPa concrete was 0.5 when the concrete was drying at 30% RH whereas the same concrete at 50% RH yielded basic creep factors close to 1. Results of the above study clearly indicated the need for the development of a fundamental material model for basic creep.

This paper reports the experimental results of basic creep behavior of HSC. The results of the experiments have been used to validate the basic creep model proposed by Anaton and Setunge (2002), based on the model B_3^* of Bazant and Baweja (1995). The model uses the age adjusted effective modulus method to account for the stress-history, a composite material model to represent the composite behaviour of concrete and the solidification theory of cement paste to allow for creeping of mortar.

2 EXPERIMENTAL PROGRAM

2.1 Materials

The materials used in the experimental program consisted of the following:

Ordinary Portland cement (General Purpose) complying with the requirements of AS 3972 (1997), which was mostly used to produce HSC successfully in laboratory or at sites, was selected for all the mixes. Condensed silica fume complying with the requirements of AS 3582.3 (1994) was selected to produce HSC. High strength concrete mix was designed by replacing 8% to 10% by weight of cement with silica fume. Natural sand with a saturated surface dry density of 2.65 tonnes/m^3 and water absorption of 0.22 was selected as fine aggregate. The ratio of fine to total aggregates was kept at 37% for all concrete mixes. Maximum size of 12–14 mm Basalt type crushed rocks was selected as coarse aggregate. Coarse aggregate was tested according to Australian Standards (AS 1141.6.2-1996) and it was found that it has a saturated surface dry density of 2.89 tonnes/m^3 and water absorption of 0.66. Three core samples from the quarry of crushed coarse aggregate were tested to determine the Modulus of Elasticity of Coarse Aggregate, which was established as 80 GPa.

Table 1. Composition of concrete and mortar mixes.

Mix	w/b	Cement	SF	FA	CA	W	SP
C1	0.25	12.96	1.17	17	31.6	3.39	0.29
C2	0.35	10.73	.93	17.3	32.13	4.00	0.17
C3	0.45	9.85	0	17.7	32.9	4.41	0
M1	0.25	22.93	2.06	30.06	0	5.81	0.50
M2	0.35	19.24	1.67	31.02	0	6.90	0.31
M3	0.45	18.01	0	32.43	0	7.96	0

C–Concrete; M–Mortar.

To achieve the workability of the mixes with w/b ratio below 0.45, high range water reducers (Rheobuild T1000 superplasticisers) were used. In designing mortar mix, an amount of water equivalent to $10 \, L/m^3$ of coarse aggregate was reduced from the concrete mix design to account for the moistening effect on coarse aggregate according to the recommendation of de Larrard (1988).

2.2 Design mix proportions

The mix designs are similar to those used by Setunge (1993). Based on the experiments performed by Setunge (1993) to determine the modulus of concrete in terms of modulus of mortar and aggregate modulus, three different concrete mixes were selected with water/binder (w/b) ratios of 0.25 to 0.45 to produce concrete and mortar with 28-day strength of 100 MPa to 50 MPa. Table 1 shows the compositions of the concrete and mortar mixes used in the experiment. w/b ratio is taken by weight of water to binder. Cement, silica fume(SF), fine aggregates (FA) and coarse aggregate (CA) are given in kg. Water (W) and superplasticisers (SP) are given in litres.

2.3 Testing standards

Australian Standards 1012.9-1999 and 1012.16-1996 were used for all testing procedures. 100 mmφ × 200 mm high specimens were used to determine the compressive strength, creep and shrinkage. All specimens were cured for 28 days. A high-strength Sulphur capping compound was used for capping of specimens. An epoxy resin used for water proofing (Bote-Cote) was used for sealing of specimens. All tests were carried out after 28 days and stress/strength ratio was kept constant as 40%. A loading frame operated by a hydraulic arrangement was used for creep test. MTS machine was used for compressive strength test. Demec gauges were used for measuring creep and shrinkage strains.

2.4 Experimental equipment

A forced mixer with a flat bottom pan rotating clockwise, and mixing blades mounted on a vertical shaft

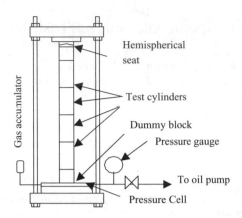

Figure 1. Setup of creep test.

rotating anticlockwise was used to prepare the concrete mix in the laboratory. 100 mmφ × 200 mm long steel cylinder moulds were prepared for casting specimens and a vibrating table was used for compaction. Humidifier was set at 95% RH and at 23°C for the storage of specimens for 28 days. A 1000 kN capacity MTS machine, which was connected with computer interface, was used to test the cylinders for compressive strength.

Creep tests were performed in a modified experimental setup that was developed by Setunge based on the basic guidelines of AS1012.16-1996 (Figure 1). A uniaxial compressive load is applied by means of a hydraulic system, which consists of a pump, an accumulator, a pressure cell and a loading frame. The calculated load is applied to the pressure cell at the bottom of the frame using a hydraulic pump. Once the desired load is obtained, the oil pump is disconnected by means of a valve. The pressure is read using the pressure gauge, and is adjusted whenever necessary over the course of time. In order to limit the number of adjustments, a high pressure accumulator was connected to the pressure cell. This accumulator was filled with oil to about half of its volume, while other half was filled with nitrogen gas. With constant pressure maintained in the cell, a constant load is applied through the steel rods and steel plates to the specimen.

A demountable mechanical gauge, which was first calibrated for the gauge factor for the calculations of strains from deformation readings, was used to measure the strain. Two sets of demec points were fixed in the mid height of each specimen in diametrically opposite sides and the gauge length was fixed as 150 mm.

2.5 Sampling in laboratory

Since the moisture content of aggregates varied significantly, aggregates were oven dried for 12 hours and then were allowed to cool for 12 hours in the natural environment before the quantities were measured. Mixing water was adjusted to allow for absorption.

First the measured coarse and fine aggregates were mixed in a dry state for 2 minutes in a mixer. After that, the measured cement and silica fume, if any, were mixed and added to the mixer and mixed by shovel. Then the mixer was operated for another 2 minutes. The measured water was gradually added during the next 3 minutes. First half of the measured superplasticisers, if any, were also added within the next 2 minutes and the mixing was continued for another 2 minutes or until workability of concrete is achieved.

Altogether 12 specimens ($100\,\text{mm}\phi \times 200\,\text{mm}$) were cast immediately after mixing by filling the cylindrical moulds in two layers and compacted by the use of a vibrating table for not more than 2 minutes and before the final setting time of 20 minutes judged from the time when the water was added. They were then covered with polythene bags to avoid any moisture loss.

2.6 Curing and storage of specimens

After 24 hours, specimens were demoulded and were stored in the humidifier with 95% relative humidity and 23°C setting and left for curing for 28 days.

2.7 Preparation of specimens for testing

After 27 days from the day of casting, specimens were taken out from the humidifier for sample preparation for creep and compression tests. In order to minimize the influence of water exposure during Sulphur capping and sealing process, they were sealed in the polythene bags until they were used. On the sides of the cylinders, minimum of two sets of gauge points were attached to measure the strain by using Demec gauge. The gauge length of those gauge points was fixed at 150 mm. After that, the cylinders were capped with a sulphur capping compound and then two layers of epoxy coating was applied for sealing the surface to prevent moisture transfer. After the sample preparation, specimens were again placed into the humidifier for 24 hours for curing.

2.8 Compressive strength test

A minimum of three specimens from each set were tested in the MTS machine for compressive strength at 28 days. The loading rate was set to 20 MPa per minute and the loading was applied using stroke mode. Then the strength was calculated as the average of 3 specimens. The failure pattern of the crushed cylinders was also observed to see whether there are any honeycombs or end capping failure.

2.9 Creep and shrinkage test

Creep tests were performed at the age of 28 days with a compressive load with all concrete and mortar

Table 2. Compressive strength (f_m), applied load ($\sigma(t_0)$) and initial elastic modulus ($E(t_0)$).

Mix	f_m MPa	$\sigma(t_0)$ MPa	$E(t_0)$ GPa
C1	98.4	39.4	39.9
C2	91.8	36.7	38.7
C3	53.1	21.2	30.5
M1	101.9	40.8	36.3
M2	78.9	31.5	34.4
M3	60.2	24.1	26.3

specimens. There were many precautions taken to ensure the safety of the testing apparatus and the personnel. Loading rate of creep specimens in all cases was maintained at 40% of the strength at 28 days and the loads are shown in Table 2. For the creep test, 4 cylinder specimens were used in the creep rig. After all specimens were set up in the creep rig, they were sealed together with epoxy again.

The experiments also included three control specimens unloaded for measuring shrinkage, but completely sealed with 2 layers of epoxy coats. This experimental setup was kept in a natural environment (23°). However, the temperature measurements were monitored daily. The specimens were sealed and therefore, effect of humidity on the specimens was negligible.

The axial deformations on both sides of the loaded and unloaded specimens were measured by using demec gauges on two gauge lines fixed on the sides of the cylinders. Measurements were recorded just before the loading and 15 minutes (0.01 day) after the loading. The initial elastic modulus was calculated by using these readings. In addition, the shrinkage of unloaded specimens was determined on similar companion specimens simultaneously. Basic creep was determined as time dependent deformation of loaded specimens corrected for the deformation of unloaded specimens.

The measurements on loaded and unloaded specimens were taken at regular intervals after 1 day and then every day for a week and thereafter every week for a month and then monthly for the maximum time period of 150 days.

3 EXPERIMENTAL RESULTS

None of the tests caused failure of the concrete or mortar. The test of M2 was repeated as inconsistent results were observed. A significant fluctuation of load has lead to fluctuation of results in M2. The problem was identified as a malfunction of the hydraulic system. Results of the repeat test were found to be satisfactory.

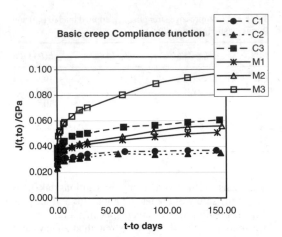

Basic creep Compliance function

Figure 2. Basic creep as compliance function for mortar and concrete.

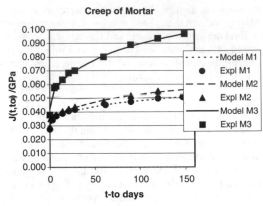

Creep of Mortar

Figure 3. Compliance function of mortar.

Figure 2 shows the basic creep as a compliance function with age at loading for tests where the applied load was 40% of 28-day compressive strength. The test was continued for 150 days to establish the probable trends.

Similar to the results from Persson(1999), the experimental results revealed a high rate of creep of concrete within the first 7 to 28 days after loading. As the concrete aged to 150 days the creep rate diminished. However, the creep of mortar was observed to be still increasing after 150 days although this rate was less compared to the rate, which was observed from 7 to 28 days. It was observed from the curves that the rate of creep of mortar and concrete is significantly different.

It can be seen from the experimental results that the basic creep also increases as the w/b ratio increases. The equivalent mortar of the concrete, which has the same w/b ratio and same mix proportions but without coarse aggregates shows higher basic creep than the concrete. This is an expected result, since presence of coarse aggregate reduces the creep.

Compliance function $J(t,t_0)$ is defined as strain at time t by unit uniaxial constant stress applied at time t_0. $t - t_0$ is given in days and $J(t,t_0)$ is given per GPa.

4 VALIDATION OF THE PROPOSED MODEL

4.1 Creep of mortar

Model proposed by Bazant & Prasannan (1989) based on solidification theory is used to predict the basic creep of mortar.

$$J(t,t_0) = q_1 + q_2 Q(t,t_0) + q_3 \ln(1 + (t-t_0)^n) + q_4 \ln(t/t_o) \quad (1)$$

$$Q(t,t_0) = Q_f(t_0)\left[1 + \left(\frac{Q_f(t_0)}{Z(t,t_0)}\right)^{r(t_0)}\right]^{\frac{-1}{r(t_0)}} \quad (2)$$

$$Q_f(t_0) = \left(0.086 t_0^{2/9} + 1.21 t_0^{4/9}\right)^{-1} \quad (3)$$

$$Z(t,t_0) = t_0^{-m}\left[1 + (t-t_0)^n\right] \quad (4)$$

$$r(t_0) = 1.7(t_0)^{0.12} + 8 \quad (5)$$

Bazant and Baweja (1995) have proposed a Model B_3^* similar to Equation 1 and the material parameters are given as follows:

$$q1 = 0.6 \times 10^6/E_{28}, \ E_{28} = 57000(f_c')^{1/2} \quad (6)$$

$$q_2 = 451.1 \ c^{0.5} \ (f_c')^{-0.9} \quad (7)$$

$$q3 = 0.29 \ (w/c)^4 \ q_2 \quad (8)$$

$$q4 = 0.14 \ (a/c)^{-0.7} \quad (9)$$

where, f_c' is the characteristic compressive strength of concrete, w/c is the water to cement ratio, c is the cement content and a/c is the aggregate to cement ratio.

Figure 3 shows the model (Equation 1) for creep of mortar. Material parameters presented in Model B_3^*, which was based on solidification theory does not fit with the experimental results of the HSC, which is understandable since the model was developed for NSC.

Table 3. Model and fitted parameters for mortar.

	Bazant's model			Fitted model		
	M1	M2	M3	M1	M2	M3
$q1$	0.0184	0.0194	0.0253	0.0184	0.0194	0.0253
$q2$	0.0678	0.0761	0.0904	0.0678	0.0761	0.0904
$q3$	0.0001	0.0003	0.0011	0.008	0.006	0.018
$q4$	0.0168	0.0145	0.0135	0.007	0.010	0.021

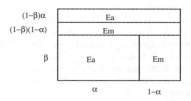

Figure 4. Composite model used by Setunge (1993).

Table 3 shows the difference of model parameters ($q1$ to $q4$) to the fitted parameters. Fitted parameters for the experimental results of creep of mortar show the difference of behaviour of HSC. Since parameter $q2$ is associated with the aging viscoelastic effect, which is a function of the cement content used and $q1$ is determined from the initial static modulus, these parameters can be used from Bazant's equation. But $q3$ and $q4$ are determined from the experimental results. However, more experiments must be performed to get a quantitative idea about parameters $q3$ and $q4$. Observed behavior shows that the parameters have to be redefined for HSC.

4.2 Creep of concrete

Experimental results were used to validate the composite model proposed by Anaton and Setunge (2002).

Figure 5 shows the basic creep curves for concrete using the proposed model (Anaton and Setunge, 2002). Creep of concrete follows the same trend as the creep of mortar although much smaller as expected. However, due to the initial modulus of elasticity of mortar(E_m), aggregate modulus(E_a) and age effective modulus of concrete($E_m^"$), the relationship between concrete creep and mortar creep is non-linear. Thus, the creep behavior cannot be related only to the initial elastic modulus of concrete.

$$J(t,t_0) = \frac{\beta}{\alpha E_a + (1-\alpha)E_m^0}\left[1+(1-\alpha)\frac{E_m^"}{E_{am}^"}\phi_m(t,t_0)\right]$$
$$+ (1-\beta)(1-\alpha)J_m(t,t_0) + \frac{(1-\beta)\alpha}{E_a}$$
$$(10)$$

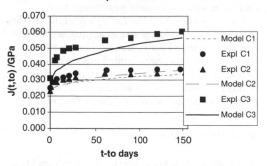

Figure 5. Compliance function of concrete as a composite model.

$$\phi_m(t,t_0) = E_m(t_0)J_m(t,t_0) - 1 \qquad (11)$$

$$E_m^" = \frac{E_m(t_0) - R_m(t,t_0)}{\phi_m(t,t_0)} \qquad (12)$$

$$E_{am}^" = \alpha E_a + (1-\alpha)E_m^" \qquad (13)$$

$$R(t,t_0) = \frac{0.992}{J(t,t_0)} - \frac{0.115}{J_0}\left(\frac{J(t_0+\Delta,t_0)}{J(t,t-\Delta)}-1\right) \qquad (14)$$

where $J_0 = J(t_0+\Delta,t_0+\Delta-1)$ & $\Delta = \dfrac{t-t_0}{2}$ (15)

Figure 5 shows that the proposed composite model gives a reasonable prediction of creep of the three concrete mixes studied in this research program. This was achieved by setting the parameter β as 0.1. It was extremely interesting to observe that β for elastic behaviour of concrete was 0.45 and for creep behaviour was 0.1 consistently. Authors believe that this is due to the inadequate representation of the bond between the aggregate and mortar in the composite model. Under sustained load, deterioration of the bond may have contributed towards the reduction in β. Further work is being planned to explore this phenomenon.

5 CONCLUSIONS

The experimental study shows that the proposed composite model can be used to predict basic creep behaviour of both high and normal strength concrete when the parameter β of the composite model is taken as 0.1. The study also indicates that there is no simple relationship between the basic creep of concrete and the basic creep of mortar. Also, these experiments show that the basic creep is dependent on aggregate content regardless of same water/cement ratio.

It was also observed that for high modulus and low volume concentration of coarse aggregates, effect of aggregate is less. Finally, it should be noted that this research only represents limited amounts of experimental results, and therefore conclusions should be regarded as preliminary investigations and more experiments are needed for confirmation of the results.

ACKNOWLEDGEMENTS

We would like to acknowledge the support from Master Builders Technologies and Blue Circle Cement Ltd., in providing the materials, towards this research project. RMIT University provided scholarship for one year to the first author.

REFERENCES

Anaton, J & Setunge, S. 2002. Basic creep of high strength concrete, *Proceedings of the 17th conference on the mechanics of structures and materials*, ACMSM-17, Gold Coast, Queensland, 12–14, June 2002, pp. 265–269.

AS1012.9, 1999. Methods of testing concrete: Compressive strength, *Standards Association of Australia*, Australia.

AS1012.16, 1996. Methods of testing concrete: Creep, *Standards Association of Australia*, Australia.

AS1141.6.2, 1996. Methods for sampling and testing aggregates: Coarse aggregate, *Standards Association of Australia*, Australia.

AS3600, 1994. Concrete Structures, Australian Standard, *Standards Association of Australia*, Sydney.

AS 3582.3, 1994. Supplementary cementitious materials for use with Portland cement – Silica fume, Australian Standard, *Standards Association of Australia*, Sydney.

AS 3972, 1997. Portland and blended cement, *Standards Association of Australia*, Australia.

Bazant, Z.P. & Baweja, S. 1995. Creep and shrinkage prediction model for analysis and design of concrete structures model B_3^*, *Materials and Structures*, No. 28: pp. 357–365.

Bazant, Z.P. & Prasannan, S. 1989. Solidification theory for concrete creep. *Journal of Engineering Mechanics* 115(8): pp. 1961–1725.

Granger, L.P. & Bazant, Z.P. 1995. Effect of composition on basic creep of concrete and cement paste. *Journal of Engineering Mechanics* 121(11): pp. 1261–1270.

de Larrard, F. 1988. A mix design method for high strength concrete, Paper presented at the first international symposium on utilization of high strength concrete, Scavenger – Norway.

Persson, B. 1999. Influence of maturity on creep of high performance concrete with sealed curing, *Materials and Structures*, Vol.32: pp. 506–519.

Setunge, S. 1993. *Structural properties of very high strength concrete*, PhD Thesis, Dept. of Civil Eng., Monash University, Australia.

Setunge S. & Toyne B. 1999. Effect of relative humidity of the environment and the loads level on the creep and shrinkage of high performance concrete, *Innovation in Concrete Structures*, edited by R.K. Dhir and M.R. Jones, Thomas Telford, pp. 267–276.

Ward, M.A., Neville, A.M. & Singh, S.P. 1969. Creep of air entrained concrete, *Magazine of Concrete Research* 21(690): pp. 205–210.

Double – tee prestressed unit made from high strength concrete

E. Dolara
University of Rome "La Sapienza", Rome, Italy

G. Di Niro
University of Strathclyde, Glasgow, UK

ABSTRACT: The objectives and results of a large experimentation on HSC (High Strength Concrete) are reported in this paper. The aim of the investigation is to check the possibility of introducing the use of HSC in a pre-cast site plant for production of prestressed structures. During this research projects the physico-chemical characteristics of the aggregate used, the properties of both the wet and hardened HSC have been determined. A mix design procedure enable HSC to attain the design target strength has been checked. The mechanical behaviour of HSC has been investigated. Test on specimens made from HSC have been performed. Full scale tests on prestressed double – tee unit made from HSC have been carried out. Pre-cast units were manufactured up to 2400 mm with a shallow flange depth of 50 mm and a total length of 16 800 mm. The results show that it is practicable to make prestressed concrete elements using HSC in a pre-cast plant during normal production cycles and that these elements can have satisfactory and predictable mechanical performance.

1 INTRODUCTION

Today economic and technological reasons made concrete perhaps the widely diffused construction material in the world because is able to meet almost all the project requirements.

In a structural analysis the physical and mechanical characteristics of concrete are normally represented by the Poisson and Young modules, compressive strength, tensile strength, etc.

However the macroscopic properties are strictly related to the micro-structure of concrete. Modifying the structure of concrete at microscopic level is possible to improve the mechanical properties, related to the strength of concrete, and the physico-chemical characteristics, related to the durability.

For this reason in the last few years researcher interests have been concentrated on HSC (High Strength Concrete) and HPC (High Performance Concrete). These are concretes with high compressive strength, normally $60 \div 150 \, \text{N/mm}^2$, and good performances to chemical attack coming from environmental conditions.

2 PROPERTIES OF RECYCLED AGGREGATE

Double tee slab made from HSC and from Normal Concrete (NC) have been casted and tested. Therefore different behaviour of fresh and hardened NC and HSC and mechanical behaviour of prestressed elements made from NC and HSC have been deeply analysed. The casting and testing procedures for HSC specimens have been the same adopted for NC specimens.

2.1 Grading, particle shape and texture

During this research projects the physico-chemical characteristics of the normal aggregate used in the pre-cast plant, the properties of both the wet and hardened concrete have been determined. The mechanical behaviour of concrete used in a pre-cast plant for normal production has been investigated.

Test on specimens made from six different mixes of NC have been performed. In Table 1 the exact mix proportions for the mix MX 6S, used to cast the double tee slab, are reported. $5.4 \, \text{l/m}^3$ of a superplasticizer has been added to this mix. Slump tests have been carried out before casting each double tee slab. Specimens for evaluation of the mechanical properties were casted together with the slabs. In Table 2 are reported the results of these tests for the slabs.

2.2 "Equivalent in sand" test

High quality natural aggregate is required for HSC production. Natural crushed basalt aggregate from Colleferro quarry has been adopted.

Table 1. Mix proportions (1 m³).

Mix code	W/C ratio	Cement kg	Water l	Fine aggregate 0–4 mm (kg)		Coarse aggregate 4–13 mm (kg)	
				Uncrushed	Crushed	Uncrushed	Crushed
MX6S	0.45	450	203	448	269	716	358

Table 2. Mechanical properties of the mix used to produce the double tee slab (air cured).

Mix code	W/C ratio	Slump mm	28-Day strength, Mpa			Modulus of elasticity MPa
			Compressive	Tensile	Flexural	
MX65	0,45	210	66,80	3,40	4,10	42 750

Figure 1. Natural basalt aggregate: grading a) 0–3 mm, b) 3–8 mm, c) 8–13 mm.

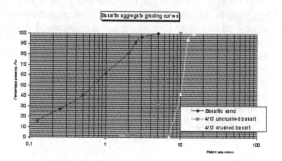

Figure 2. Grading curve.

The physico-chemical characteristics of the basalt aggregate used, the properties of both the wet and hardened concrete have been determined.

The mechanical behaviour of concrete used has been investigated. Test on specimens made from eight different mixes of HSC have been performed.

In Figure 1 the grading of basalt aggregate adopted are reported. In Figure 2 grading curve from sieve analysis is plotted.

In Table 3 the exact mix proportions for the eight trial mixes are reported. A superplasticizer has been added to this mix.

Slump tests have been carried out on each mixes. Test results are reported in Table 4. In Figure 3 slump tests for HSC 1 and MX 6S mixes are reported.

It can be noticed from Figure 3 different color for two mixes. Dark color for HSC 1 depends on basalt aggregate and silica fume. Different spreading is due to the higher stickiness for HSC 1 mix. This depends on the presence of silica fume, the high dosage of superplasticizer and high quantity of cement. In fact this higher stickiness has been noticed for mixes with 700 kg/m³ of cement.

It has been noticed that increasing cement content decrease superplasticizer demand (Fig. 4) for HSC 1, HSC 5 and HSC 8 mixes with the same a/c ratio, the same percentage of silica fume on cement content and the same slump figure. It could be concluded that workability depends on the presence of very fine aggregate in the batch as cement and silica fume.

Silica fume presence in the batch avoid bleeding and efflorescence effects on specimens surface due to the high dosage of superplasticizer.

Mechanical properties of HSC specimens are reported in Table 4 for each trial mixes.

3 TEST SET-UP AND TESTING PROCEDURE

Full scale tests on two pre-cast double tee slab made from NC and HSC have been carried out. MX 6S and HSC 2 have been the two trial mixes chosen to cast the double tee slab. Mix design characteristics and mechanical properties are reported in Table 3 and 4.

Table 3. Mix proportions (1 m³).

Mix code	w/c ratio	Cement kg	Silice fume kg	Water l	Super plasticizer l	Slump	Fine aggregate 0–4 mm (kg)		Coarse aggregate 4–16 mm (kg)	
							Basalt	Limestone	Basalt	Limestone
HSC 1	0.29	700	70	203	18 R	215	777	–	689	–
HSC 2	0.29	500	50	145	30 R	250	954	–	846	–
HSC 3	0.25	600	60	150	35 R	250	954	–	846	–
HSC 4	0.29	600	60	174	18 R	190	–	1037	–	558
HSC 5	0.29	600	60	174	20 R	215	872	–	773	–
HSC 6	0.29	600	30	174	19 R	220	893	–	792	–
HSC 7	0.29	600	90	174	2 1G	210	851	–	755	–
HSC 8	0.29	500	50	145	27 G	210	959	–	850	–

Table 4. Mechanical properties of HSC specimens.

Mix code	w/c ratio	Compressive strength MPa			Tensile strength 28 days (MPa)	Flexural strength 28 days (MPa)	Modulus of elasticity 28 days (MPa)
		24 h	7 days	28 days			
HSC 1	0.29	75.3	104.5	128.7	5.7	4.7	43 914
HSC 2	0.29	55.0	106.3	135.1	6.5	5.6	53 035
HSC 3	0.25	42.5	110.0	140.0	6.4	4.7	40 441
HSC 4	0.29	68.8	91.5	108.4	–	–	–
HSC 5	0.29	68.5	91.7	135.4	6.4	5.8	–
HSC 6	0.29	82.5	112.7	129.4	–	–	–
HSC 7	0.29	81.3	107.0	128.1	6.0	5.4	–
HSC 8	0.29	52.5	114.8	128.9	6.4	5.7	–

(a)   (b)

Figure 3. Slump test for mixes: a) HSC1 and b) MX 6S.

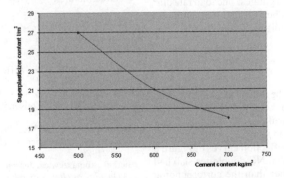

Figure 4. Cement and superplasticizer content relationship.

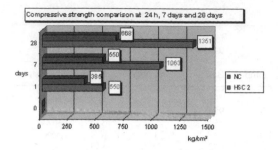

Figure 5.   Compressive strength comparison.

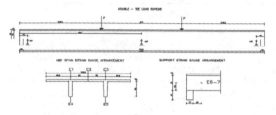

Figure 6.   Double tee cross section and load scheme.

Figure 7.   Test equipment.

parts of the slabs were also measured at discrete points with the use of strain gauges. The load rate during the test was chosen to be 500 kg/3 min. Different load cycles have been performed for each slab.

4   TEST RESULTS

Load deflection curves for the double tee slab have been plotted in Figure 8. The maximum deflection reached was lower than the theoretical one. All slabs crack patterns have been followed and recorded during the test.

Comparing the load – deflection curves of Figure 8 and 9 it could be noticed that NC double-tee slab had a midspan deflection higher than the corresponding HSC double-tee slab.

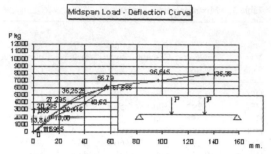

Figure 8.   Load deflection curve for NC double tee slab.

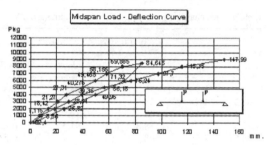

Figure 9.   Load deflection curve for HSC double tee slab.

5   CONCLUSIONS

From the first results of tests it could be concluded that double – tee prestressed unit made from NC failures with a load and deflection lower than the corresponding double – tee unit made from HSC.

The full scale test results show that the double tee slab made from HSC, during normal cycle production, can have a satisfactory and predictable mechanical performance.

In fact a comparison with theoretical modelling using Finite Element Analysis showed that double – tee slab made from HSC behaved in a numerical predictable manner.

REFERENCES

A.S. Ngab, F.O. Slate, A.H. Nilson, *"Behavior of High Strength Concrete Under Sustained Compressive Stress"*, Research Report No. 80-2, Department of Structural Engineering, Cornell University, Ithaca, pp.201, February (1980).

Aictin P.C., A. Neville, *"High Performance Concrete Demystified"*, Concrete International, January 1993, pp.21–26.

Bjerkeli, Tomaszewicz, Jensen, *"Deformation Properties and Ductility of High Strength Concrete"*, 2nd Int. Symposium on Application of High Strength Concretes, Berkley (1990).

Carrasquillo et al., *"Microcraking and Behaviour of HSC subject to Short-term Loading"*, ACI-Material Journal, May–June (1981).

Carrasquillo, Slate, Nilson, *"Properties of High Strength Concrete Subjected to Short-Term Loading"*, ACI Material Journal, May–June (1981).

CEB Bullettin d'Information n° 228, *"High Performance concrete: Recommended extensions to the Model Code 90. Research Needs"*, Luglio (1995).

D. Richard e M.H. Cherezy, *"Reactive Powder Concrete whit High Ductility and 200–800 Mpa Compressive Strength"*, "Concrete Technology: Past, Present and Future", Ed. P.K. Metha, ACI SP-144, S. Francisco (1994).

D.M. Roy, *"Advanced Cement Systems Including CBC, DSP, MDF"*, 9th International Congress on the Chemistry of Cement Vol. 1, pp. 357–380, New Delhi (1992).

H.H. Bache, *"Densified Cement Ultra-Fine Particle Based Materials"*, Second International Conference on Superplasticizers in Concrete, 35 pagine, Ottawa (1981).

S. Ahmad, S.P. Shah, *"High Strength Concrete – A Review. First Symposium on Utilization of High Strength Concrete"*, Stavanger (1988).

State of the Art report on High Strength Concrete. Reported by ACI Committee 363, ACI Material Journal.

Tognon, Ursella, Coppetti, *"Design and Properties of Concretes with Strength over 150 kg/cm^2"*, ACI Material Journal, May–June (1981).

Efficiency of repair of HSC beams subjected to combined forces by GFRP sheets

H.A. Anwar & W. El-Kafrawy
Housing & Building Research Center (HBRC), EGYPT

ABSTRACT: The introduction of fiber reinforced composite materials in civil engineering has progressed at a very rapid rate in the recent years. The high-performance materials, which consist of high strength fibers, embedded in the plastic matrix, possess unique properties, which make them extremely attractive for a wide range of structural applications. An area where the use of fiber reinforced composites has attracted considerable interest is the repair and retrofitting of R.C. structural elements, but what about elements cast using high strength concrete. The global aim for this study was finding the structural efficiency restored after increasing lateral confinement of the high strength concrete box – section beams. In this study, glass fiber plastic sheets were used in retrofitting of structurally deficient R.C. beams due to combined shear and torsion, e.g. spandrel beams – curved beams -spiral staircases. Such beams failed in a brittle manner. The experimental program contained seven damaged beams wrapped by two layers of unidirectional GFRP sheets. Parameters investigated were reinforcement percentage (whether under or over reinforcement according to the (ACI 318-95) and the accompanied applied shearing force level. Beams dimension were 25×40 cm with a box-section wall thickness of 6.5 cm and tested zone of 100 cm. The efficiency of repairing such beams by this wrapping system using GFRP sheets was investigated and the percentages of restored characteristics were calculated after being tested. Behavioral characteristics were recorded under monotonically increasing torque until failure. Load – strain relationships were illustrated and conclusions are presented. Comparison between beams cast using normal and high strength concrete are presented also, the experimental work was held in the reinforced concrete lab through a scheme for "Application of Modern Composites in Repair and strengthening of Structural Elements" sponsored by the Housing and Building Research Center (HBRC).

1 INTRODUCTION

There is a need to establish suitable techniques that can provide sufficient rehabilitation efficiency for high strength R.C. structural members. It is well known that making use of the confinement effect induced by lateral reinforcement can effectively restore section strength and endurance of such members. Applying high-strength, corrosion – resistant fiber polymers to lateral reinforcing of concrete is a research topic with great activity and promise. Recently, particular attention has been paid to the retrofitting techniques with fiber reinforced plastic epoxy bonded (GFRP) sheet for addressing this problem. Very little information is available on the behavior and performance of repaired high strength R.C. structural members under combined forces of shear and torsion. The mechanism of fiber sheets confinement and its applications has not been fully studied yet. Retrofitting of R.C. beams with box – section started by repairing its original uncracked box dimensions.

2 EXPERIMENTAL PROGRAM

2.1 Research significance

The aim of this experimental study was to investigate the efficiency of wrapping sheets in confinement of reinforced concrete beams subjected to constant shear and torsion forces. The goal here was to restore the torsion capacities and sections' structural performance for R.C. beam damaged by an increasing out of plane torque. The study was mainly concerned with the influence of both the main reinforcement ratio and the level of previously applied constant loads in the first test on the ultimate torsion capacity of the repaired beams.

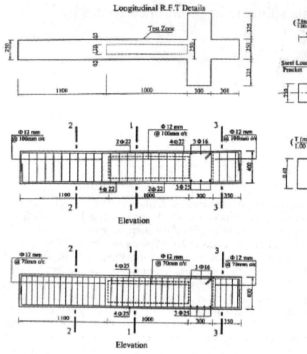

Figure 1. Reinforcement detailing for tested beams.

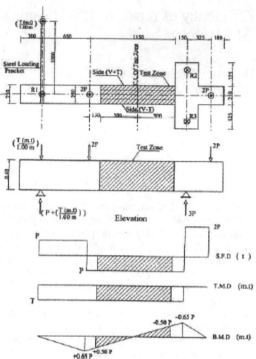

Figure 2. Experiment applied forces & straining actions.

2.2 Reference groups and materials used

Seven structurally deficient HSC beams with rectangular box – section having cross – section of (25 × 40) cm and wall thickness of 6.5 cm were taken as reference beams. The tested zone was of length of 100 cm. These beam were previously damaged under constant shear force (P) and monotonically increasing torsion moment (torque) to failure. The beams were cast and cured for 28 days in laboratory temperature. Target cube compressive strength was 750 kg/cm² after 28 days. Specimens were divided into two groups namely (I & II) to cover the investigated parameters. Group (I) is the under-reinforced group that contained three specimens referred to by (HU1, HU2&HU3). Group (II) is the over reinforced group that contained four specimens referred to by (HO1, HO2, HO3&HU4). The applied constant shear force (P) was taken as a ratio of the nominal shear strength provided by concrete as calculated (Vc = 6.46), which is based on ACI 318-95 section 11.3.1.1. Reinforcement detailing for the two tested groups are shown in Fig (1). Also plan of tested specimens showing straining actions are shown in Fig (2).

Locally produced available materials were used. Ordinary Portland cement confirming to the Egyptian Standard Specifications was used (OPC type 1). Natural siliceous sand and crushed dolomite (with maximum nominal size of 10 mm) were used as fine and coarse aggregate, respectively. Silica fume and super-plasticizer (BVF) were used. Quality control tests were carried out on aggregate specimens in order to assure their compliance with the Egyptian Standard Specifications. Clean tap water was used in all test specimens. The concrete mix proportions by weight were 1: 1.1: 2.8: 0.22: 0.038 (Cement: Sand: dolomite: Silica: BVF) and the water to cement ratio was 0.3. The concrete was mixed in revolving drum-type electrical mixer of capacity 0.1 m³ and vibrated carefully after being placed in the molds. All specimens were removed from the molds after 24 hr. They were cured under wet burlap for two weeks, after which they were left to dry in the laboratory conditions for another two weeks. Table (1). Present the reinforcement detailing and applied forces for the tested specimens. Two LVDT spaced 50 cm were used to measure vertical displacements placed at the end of the tested box zone, for calculating the twist angle [θ (rad/m × 10⁻³)]. The two-shear constant loads [P] are manually controlled. The torsion moment [T] was applied using double acting hydraulic jack at the end of one-meter lever arm. The loading was computerized with the use of a data acquisition system and the lab view software program.

Table 1. Reinforcement detailing and applied forces.

G	Sp. No.	Reinforcement		Shear force	
		Main	Stirrups/m	P (ton)	P/Vc
G(I)	HU1	2 F 22	11 F 12	0	0
	HU2	2 F 22	11 F 12	12.6	1.95
	HU3	2 F 22	11 F 12	24.0	3.72
G(II)	HO1	4 F 25	16 F 12	0	0
	HO2	4 F 25	16 F 12	12.6	1.95
	HO3	4 F 25	16 F 12	16.8	2.6
	HO4	4 F 25	16 F 12	24.0	3.72

Table 2. Experimental test results.

G	Sp. No.	Ultimate stage		Failure mode
		T (m.t)	T	
G(I)	HU1	6.2	45.7	Under reinforced
	HU2	5.7	34.4	
	HU3	4.9	32.9	
G(II)	HO1	7.5	35.69	Over reinforced
	HO2	6.8	34.59	
	HO3	6.3	39.4	
	HO4	5.6	34.6	

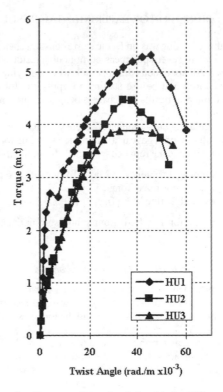

Figure 3. Torque–twist angle relationship for G(I).

Applying the constant shear force caused shear cracks before applying the torsion force. Specimens test results are summarized in Table (2).

The torque versus twist angle relationships for the two groups (I, II) are presented in Figures (3,4).

2.3 Repaired groups and materials used

The same seven beams were internally and externally repaired according to the following steps. Cracked high strength concrete was replaced by non-shrinkage grout (cement concrete) with compressive cube strength of 750 kg/cm². Structural corners were rounded to a radius of 15 mm. Specimens were cured in water for 7 days. The concrete surface was glued by the epoxy resin, then two strips of width 40 cm were used to cover the sides of the box section zone with an extension of 20 cm. Then the second layer of the glass fiber strips wrapped the outer perimeter of the beam with overlap of 15 cm. After placing the fiber sheet, it was well saturated with the epoxy resin to have two layer of glass fiber matrix confining the section perimeter. Glass laminates used were unidirectional with unit weight 365 gm/m². The adhesive epoxy resin used was commercially known as Sikadure-330.

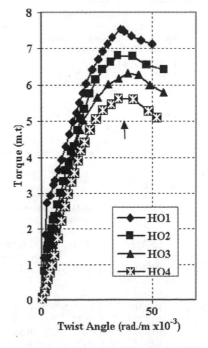

Figure 4. Torque–twist angle relationship for G(II).

3 DISCUSSION OF EXPERIMENTAL RESULTS

As the re-testing was under the same constant shearing force for all seven specimens, comparative results were drawn. All specimens did not crack when applying the shear force and before starting to apply the torque. Experimental results for repaired beams are presented in Table (3).

Calculating the wrapping efficiency for both groups and stiffness index gave us a better understanding for the performance of the repaired specimens for both groups.

1) Wrapping efficiency = Max. Torque after repair [Tr]/previous max. Torque [T]
2) Torsional Stiffness = [Tr/θr]/[T/θ]

Results for repaired groups are summarized in Table (4).

3.1 *Group RG(I) under reinforced beams*

For specimen previously tested under zero constant shearing force (pure torsion), it was noticed that wrapping efficiency was 144% and the torsion stiffness was 110%.

For specimen previously tested under P = 1.95Vc constant shearing force, it was noticed that wrapping efficiency was 110% and the torsion stiffness was 80%.

Table 3.   Experimental results for repaired beams.

| G | Sp. No. | Ultimate stage | | |
		Pr (ton)	Tr (ton.m)	?r
RG(I)	RHU1	12.6	8.9	59.77
	RHU2	12.6	6.3	47.57
	RHU3	12.6	1.9	25.1
RG(II)	RHO1	12.6	5.5	35.45
	RHO2	12.6	4.9	34.11
	RHO3	12.6	4.5	41.34
	RHO4	12.6	3.9	34.87

Table 4.   Wrapping efficiency.

| G | Sp. No. | Ratio of restored characteristics | |
		Tr/T	[Tr/θr]/[T/θ]
RG(I)	RHU1	1.44	1.1
	RHU2	1.1	0.8
	RHU3	Fail	—
RG(II)	RHO1	0.73	0.74
	RHO2	0.72	0.73
	RHO3	0.7	0.68
	RHO4	0.7	0.69

For specimen previously tested under P = 3.72Vc constant shearing force, it was noticed that the wrapping system used in this study failed to restore any behavioral characteristics for this specimen.

In Fig (5), it could be noticed that, all specimens exhibited ductile flexure failure and horizontal rupture in the vertical fibers of the sheet in the tension side at 95% of the maximum torque reached.

3.2 *Group RG(II) over reinforced beams*

For specimen previously tested under zero constant shearing force (pure torsion), it was noticed that wrapping efficiency was 73% and the torsion stiffness was 74%.

For specimen previously tested under P = 1.95Vc constant shearing force, it was noticed that wrapping efficiency was 72% and the torsion stiffness was 73%.

For specimen previously tested under P = 2.6Vc constant shearing force, it was noticed that wrapping efficiency was 70% and the torsion stiffness was 68%.

For specimen previously tested under P = 3.72Vc constant shearing force, it was noticed that wrapping efficiency was 70% and the torsion stiffness was 69%.

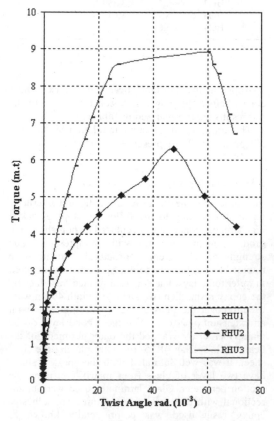

Figure 5.   Torque–twist Angle for RG(I).

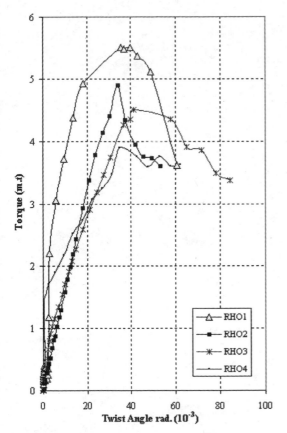

Figure 6.   Torque–twist Angle for RG(II).

In Fig (6), it could be noticed that, all specimens exhibited the same shear flexure performance irrespective of the previous level of constant shearing force; flexure performance till rupture of the vertical fiber then successive concrete shear cracks, then failure occurred.

4   CONCLUSIONS

Torsional behavior of seven repaired high strength reinforced concrete beams with box section and different reinforcement ratio was investigated. From the experimental program, the following conclusions can be summarized as follows:

– Tests performed in this study indicate satisfactory results when using externally bonded GFRP sheets as a repair and retrofitting material in regaining torsion capacity of under reinforced high strength reinforced concrete beams carrying constant shear force. This was due to the compatibility of confined repaired concrete section and the already stressed steel in withstanding the increasing torsion moment till failure.

– The efficiency of wrapping in restoring torsion capacity was more for under-reinforced beams than for over-reinforced ones. Its efficiency in restoring torsion capacity ranged from 140% till 110% for under reinforced beams while it did not exceed 75% for over reinforced beams under combined shear and torsion.

– Torque stiffness restored of under reinforced repaired beams previously damaged under levels of constant shearing forces = (0, 1.95Vc) was (110%, 80%) respectively.

– Using GFRP sheets for repairing damaged high strength beams almost restores the section torsion resistance and its overall behavior along with the control cracking propagation and widening.

– The proposed wrapping scheme and the overlap of 15 cm used in retrofitting of under reinforced beams high strength were sufficient to allow the GFRP contribution and finally its rupture.

– The mode of failure of retrofitted beams indicates the importance of the extension of anchored length of the sheets beyond the box section zone and to the supports.

– As for high strength over reinforced beams, in spite of the use of denser (GFRP) than used in repairing normal strength concrete, the wrapping efficiency did not exceed 75% in restoring torsion resistance.

– In retrofitting high strength concrete, with over reinforced steel, the use of carbon fibers sheets (CFRP) as a lateral stiff confinement is a must to contribute in restoring the section strength.

REFERENCES

Anwar, H. & El-Kafrawy, W. (2003). *Retrofitting of Beams Subjected to Combined Shear & Torsion Using GFRP Sheets*, 1st International Conference on Concrete Repair St-Malo, Brittany , 15th – 17th July.

Hammad, Y.H., Shabaan, I.G., Bazzan, I.M. & Awwad, W.H. 2002. *Shear Behavior of High Strength Concrete Beams Reinforced By GFRP And Steel Bars*. Third Middle East Symposium on structural composites for infrastructure application, Aswan, Egypt.

Shaheen, H., El Kafrawy, M. & El Kafrawy, W. (2001). *Behavior of R.C. Box Section Under Combined Shear & Torsion for High & Normal Strength Concrete*, Ph.D. thesis, Faculty of Engineering, Cairo University.

28. Concrete properties and technology

Three-dimensional quantitative simulation of flowable concrete and its applications

M.A. Noor
Department of Civil Engineering, Bangladesh University of Engineering and Technology, Dhaka, Bangladesh

T. Uomoto
Department of Civil Engineering, The University of Tokyo, Japan

ABSTRACT: Three-dimensional distinct element method (DEM) was used, as a new tool, to simulate behaviors of flowable concrete. Effort has been made to correlate and verify the DEM parameters of the simulation to the Bingham coefficients of high flow mortar and concrete. Finally, effort has also been made to propose generalized equation for mortar rheology from the mortar DEM parameters, for quantitative analysis. These equations were then verified by simulating the mortar. After mortar verification the mortar model was combined with aggregate model to simulate the flowable concrete. These simulations were then verified with the experimental results of flowable concrete. Here, some applications based on the developed model have also been simulated to show that this model can be applied all practical cases. It is found that, the DEM model proposed in this research can be used for simulation of flowable concrete.

1 INTRODUCTION

Flowable concrete, requiring no consolidation work in site, developed to improve the reliability of concrete and concrete structures. A large amount of experimental effort can be avoided if a numerical approach can predict the behavior of flowable concrete with reasonable preciseness. To perform the simulation of this type of concrete some method should be there to relate simulation parameter with the concrete parameters, such as Bingham parameters. Three-dimensional distinct element model (henceforth DEM), a particle simulation method (Cundall and Strack 1979), was used to simulate behaviors of this type of concrete under various states. In this paper, first effort has been made to correlate and verify the Distinct Element Model parameters of the mortar and concrete simulation to the Bingham coefficients of high flow mortar and concrete. Here, Bingham coefficients mean yield value (τ_0) and viscosity (η). Why Bingham coefficients were selected instead of mix ratio? Because there is no direct relation between fresh concrete flow properties and mix ratio. DEM parameters for mortar and concrete mean viscosity between particle contact $(V_c = \eta_d^n)$, overlap between particle during contact (O), friction coefficient (f_s) and bond parameter (e_n) in normal direction between particles. Also the normal

to shear viscosity ratio $(V_r = \eta_d^n/\eta_d^s)$ and stiffness ratio of particles $(S_r = K^n/k_s)$. The relationship between Bingham coefficients to DEM parameters, has been plotted and described in detail. Finally, effort has also been made to propose generalized equation for mortar rheology from the DEM parameters, for quantitative analysis. These equations were then verified by simulating the mortar. After mortar verification the mortar model was combined with aggregate parameter to simulate the flowable concrete. These simulations were then verified with the experimental results of flowable concrete.

Some applications based on the developed model have been simulated to show that the model can be applied in all practical cases. These applications have been chosen carefully – so that it can cover a broad range of problems. Several applications could be performed but, if the critical applications can be simulated using the proposed model, any application can be simulated. It is beyond the scope of the current research, to cover all possible applications currently available. Two classes applications were selected. One has relation to the flow of concrete and another has relation to the blocking and segregation of concrete. Both of these are important in practical problems regarding HFC. To demonstrate the flow related problem, slump flow test has been selected. A frame packed with reinforcement

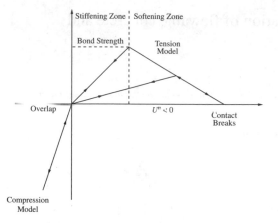

Figure 1. Modified tension constitutive law.

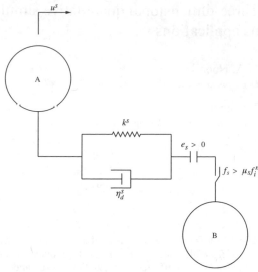

Figure 2. Modified constitutive model for contact between particles (shear direction).

was selected, to show both flow and blocking related phenomena.

2 DEM MODEL

Two-phase model, has been proposed to model the fresh concrete. One particle was used as aggregate and another particle as mortar. The constitutive model of DEM, used during the quantitative simulation was different from the constitutive model, that authors had used in the qualitative simulation (Noor and Uomoto 1999b) earlier. Detail description of the latter constitutive model can be found in (Noor and Uomoto 1999a). In the modified constitutive model, strain softening was included in the tension model. In this model, tensile force carried by mortar, increases to a peak value and then drops down to zero. The increasing and decreasing rate of force after peak depends on loading and unloading stiffness, respectively. If the tensile force becomes zero the contact between the particle breaks. The schematic view of the modified constitutive model is shown in Figure 1. In this figure U^n represents the overlap between the particles. It is positive when one particle overlaps another. Allowance of tension was determined from the experiment. Linear friction model, which was used during qualitative simulation, has been changed and linear viscous model for both normal and shear direction, has been introduced. This shear direction model (same in normal direction) is shown in Figure 2.

3 METHODOLOGY

Lifting Sphere Viscometer (LSV) (Chu and Machida 1998) test has been employed to calculate and see the

variation of Bingham coefficients of model mortar and concrete due to DEM parameters. Different sets of DEM parameters were taken for numerical simulation with the same choices of lifting sphere ball speed (v). Therefore, the numerical simulation was time consuming and exhaustive.

The determination of Bingham coefficients from numerical simulation result was kept similar to that of experiment. The resistance F to pulling ball was calculated by averaging the dragging force within the whole displacement range after the peak, for each lifting speed. A set of points was plotted with $v/2r$ as ordinate and $F/12\pi r^2$ as abscissa, for each set of parameters. Here, r was the radius of lifting sphere. Least square fitting method was applied to obtain linear approximation of Bingham coefficients with the slope of approximated line representing viscosity η and intercept with abscissa representing yielding value τ_0. By treating DEM parameters, including overlap parameter, viscous coefficient, bond and friction, as independent variables and Bingham coefficients as dependent variables different equations have been developed to cater the variation of these parameters.

The main objective here, was to get quantitative parameters for mortar simulation. Then, these parameters were combined with the aggregate DEM parameters to perform the concrete simulation. After completing the whole process it was possible directly, to simulate any high flow concrete using this model from the fresh concrete parameters – such as mix proportion and Bingham coefficients.

4 DEM MODEL FOR MORTAR

A considerable simulation, of lifting sphere viscometer test, has been performed in order to investigate the relationship between DEM parameters and Bingham coefficients of mortar. After Bingham coefficients for each set of parameters are determined (Noor 2000), the relationship between DEM parameters and Bingham coefficients of mortar can be given in equation format. To relate DEM parameters with the yield stress and viscosity of the mortar following two model equations have been proposed after several parametric studies and sensitivity analyses. The yield stress model equation is as follows

$$\tau_f = 512 + 7112.6V_c(1.014 - 0.014V_r)$$
$$+4002.2e^{-O/0.288}(0.76 + 0.57e^{-S_r/11.45}$$
$$+11106.1e_n - 979.5e^{-f_s/0.1798} \qquad (1)$$

The viscosity model equation is as follows

$$\eta_f = -21.1 + 17.68e^{V_c/0.05}(0.58$$
$$+1.04e^{-V_r/1.1}) + 76.8e^{-O/0.213}(0.96$$
$$+0.5e^{-S_r/4.43}) \qquad (2)$$

Interested readers should read Noor, M. A. (Noor 2000) to know how these models have been established step by step. Here, detail description is avoided due to brevity.

5 DETERMINATION OF AGGREGATE ELEMENT PARAMETER

Although DEM is self-valid to simulate the behavior of aggregate from the successful experiences of simulating rock mechanics problems, an attempt has been made to show the agreement between simulation result and lifting sphere viscometer test through aggregate mixture. In brief the parameters used in aggregate mixture simulation are listed in Table 1. These parameters obtained after several trial-and error methods. These parameters were used whenever aggregate parameter was needed for simulation during this research. The comparison between the force-displacement curve generated when lifting sphere ball is dragged upward

through aggregate mixture at the speed of 30 mm/s and that computed by the simulation. The acceptable agreement between experimental result and simulation verifies the sufficiency of the parameter values used for the simulation.

6 DEM MODEL FOR CONCRETE

Concrete model is the combination of mortar model and aggregate model. In (Noor and Uomoto 2003) empirical equations were summarized as to predict yield value τ_f and viscosity η from the mix proportion in case of mortar. In this paper, correlation between DEM parameters and Bingham Coefficients already established. Based on the above available results an attempt was made to predict DEM parameters from mix proportion step by step. This was vitally important because it provides the basis for the numerical simulation of mechanical behavior of fresh concrete by DEM approach. The whole procedure of prediction is explained as follows:

1. A mix proportion was selected and its yield stress and viscosity were determined from the viscosity and yield stress equation given in (Noor and Uomoto 2003).
2. Above calculated yield stress and viscosity were then used in the Equation 1 and Equation 2 to design the DEM parameters. In doing so first overlap (O) and viscosity (V_c) parameter were calculated from Equation 2. Then these parameters put into Equation 1 and bond and friction parameters were determined.
3. After finishing above two steps the DEM parameters should be available for numerical analysis. For selecting the DEM parameters several trials may be required, this resembles the mix design of concrete from raw material.
4. After finding the DEM model parameters for mortar these are combined with the aggregate parameters (Table 1) to get concrete model, and simulation of concrete should be done with these parameters.

In later sections, the verification of the quantitative equations and model parameters were performed. To perform verification, some mix proportions were selected at the beginning. Two different kinds of mix proportions were selected. From these mix proportions first DEM parameters were calculated using above method and then lifting sphere simulations were done for both numerical and real mortar and concrete. Force-displacement curve of numerical mortar and concrete were then compared experiment on mortar and concrete. Later sections give detail description of the comparative study. To verify the quantitative analysis the selected mix proportions given in Table 2.

Table 1. DEM parameters for aggregate simulation.

Stiffness parameter (N/m)		Viscous parameter (N-s/m)		Bond (N)	Friction coefficient
Normal	Shear	Normal	Shear		
1×10^5	5×10^4	0.01	0.01	0	0.01

Table 2. Selected mix proportion.

Mix no.	w/p (v)	S (S_{lim})	G (G_{lim})	ts	Water (kg/m³)	Powder (kg/m³)	Sand (kg/m³)	Gravel (kg/m³)
1	0.90	0.44	0.50	0.94	182	630	746	797
2	0.80	0.60	0.36	0.96	162	635	1009	556

Table 3. Selected parameter values for mortar simulation.

Mix No.	Bond (N)	Friction coefficient	Viscosity (N-s/m)	Overlap percent	Stiffness ratio	Viscosity ratio
1	0.008	0.03	0.3	1	10	1
2	0.01	0.03	0.13	2	50	1

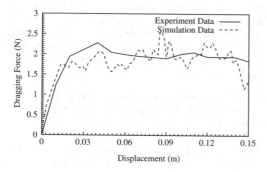

Figure 3. Verification for mortar mix-2 and 12.3 mm/s velocity.

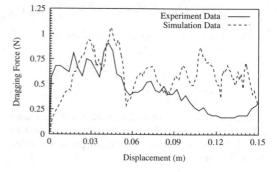

Figure 4. Verification for concrete mix-1 and 8.4 mm/s velocity.

6.1 Mortar simulation

Using the yield stress and viscosity model, yield stress and viscosity for the mortar were first calculated from the model given in Noor and Uomoto (Noor and Uomoto 2003). Using these values DEM parameters for mortar were calculated. First the viscosity was calculated, by trial error. Then with these values and again using trial error method rest of the parameters were calculated. Final selected values for the parameters are given in Table 3. The number of mortar particles in the simulation were calculated directly using the mix proportion. Then with these parameters lifting sphere viscometer simulation was performed, for different velocities. It was found that low yield stress mortar experiment cannot be simulated well using DEM. These can be explained that these mortars are very fluid. And with distinct element method it is very difficult to get such a fluid behavior. This is the inherent limitation of the distinct element method. Figure 3 shows the high yield stress case for mix two. It can be seen from this figure that this data simulate the force-displacement curve very well. It reveals the fact that the more concrete with less fluid behavior,

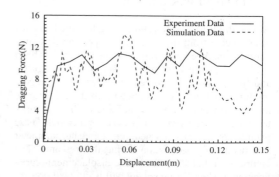

Figure 5. Verification for concrete mix-2 and 8.1 mm/s velocity.

distinct element method can be able to simulate this type of mortar.

6.2 Concrete simulation

For concrete simulation the mortar parameter and aggregate parameter were combined. The comparison is shown in Figure 4. The parameters for mortar mix

Table 4. Mix data and slump flow value.

Mix no.	W/P (vol^m)	Water (kg/m^3)	Powder (kg/m^3)	Sand (kg/m^3)	Gravel (kg/m^3)	SP*	Slump flow (mm)
1	0.90	182	625	746	770	1.3	748.5
2	0.80	162	625	1009	556	1.3	360.0

*Superplasticizer.

Table 5. Parameter values and simulation results.

Mix no.	Bond (N)	Friction coefficient	Viscosity (N-s/m)	Overlap percent	Stiffness ratio	Viscosity ratio	Slump flow (mm)
1	0.007	0.01	0.028	1	10	1	680
2	0.03	0.03	0.13	2	10	50	370

one were combined with aggregate parameter and concrete simulation was performed. With these parameters several lifting sphere viscometer simulations were performed on concrete. One example with very low velocity Figure 4 is shown here. The experiments were conducted with the same velocities to compare with the simulation results. It can be seen from Figure 4 that, the experiment and simulation force–displacement curve coincide well. So, it can be concluded that distinct element method can simulate this type of concrete very well.

The same method, which was performed for Mix No. 1, was also performed for Mix No. 2 and shown in Figure 5. It can be seen from Figure 5 that, the experiment and simulation force–displacement curve coincide well. So, it can be concluded that distinct element method can simulate this type of concrete very well.

7 SLUMP FLOW SIMULATION

Slump flow test method was selected, to verify that proposed model can simulate the flow behavior of concrete. For slump flow simulation two kinds of concrete mixes were selected. Detailed mix proportion and slump flow values, of these concrete, are shown in Table 4. These mix proportions were same mix which were used to verify DEM model earlier. Parameters used for these simulations and the result obtained from the simulation are given in Table 5.

Plan and elevation of the simulation results, for mix one, are shown in Figures 6 and 7, respectively. It was understood that, the slump flow value obtained in the simulation was lower than the experiment. This was due to the fact that, using DEM, it would be very difficult to obtain very high flow behavior.

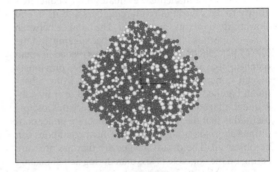

Figure 6. Slump flow simulation for concrete one. (Plan, slump flow = 680 mm.)

Figure 7. Slump flow simulation for concrete one. (Elevation, slump flow = 680 mm.)

The simulation results of mix two, not shown here, is given in Table 5. It was observed that, the slump flow value obtained in the simulation was more or less equal to the experiment (370 mm). As slump flow became less it was possible to simulate the concrete using DEM.

8 FILLING TEST SIMULATION

The test shown in Figure 8 is the one employed during the development stage of self-compacting concrete (Okamura and Ozawa 1994). If a concrete is observed to perfectly fill this formwork packed with a lot of

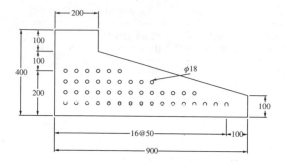

Figure 8. Filling test in formwork packed with reinforced bars (Dimensions are in mm).

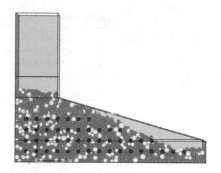

Figure 9. Frame packed with reinforcement simulation (final stage, low yield stress).

reinforcing bars, it is judged to be consolidation free self-compacting concrete. This method does not give the quantitative evaluation of the self-compacting concrete.

This test simultaneously evaluates both a concrete's ability to pass through reinforcing bars and its lateral flowability under its own weight. The balance between these properties changes with the bar spacing and the rate of placement. Since the concrete is placed in small batches, the operators can observe how previously placed concrete moves under pressure from subsequent batches or how it halts and passed over by subsequent placed concrete. The characteristics of this method is that the behavior of the concrete under conditions approximating actual placement conditions can be observed. The disadvantages are that the pressure acting on the concrete is less than during actual placing and that such tests require a relatively large amount of concrete and much labor.

Two different types of concrete were used, one with low yield stress and viscosity and another with high yield stress and viscosity. Former concrete was selected to show the filling behavior of the concrete with packed reinforcement and latter was selected to show the blocking of the concrete in packed reinforcement formwork. The mix proportions used for this simulation are given in Table 4.

Mix one has low yield stress and low viscosity. It has high slump flow. This concrete should fill the formwork. Figure 9 shows the final stage of the particle. It can be observed from these figures that the some part of the formwork could not be filled properly. Same conclusion, which was discussed for mortar, can also be made from this observation. Figures 10 and 11 show the contour line of the different stages of the simulation and experiment, respectively. These show that contour lines of experiment were comparable with the simulation performed with the numerical concrete. This verifies the numerical simulation result performed with DEM model.

Mix two concrete has high yield stress and viscosity. Figure 12 shows the final flow stage with this concrete.

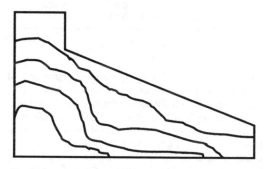

Figure 10. Self-compactability simulation.

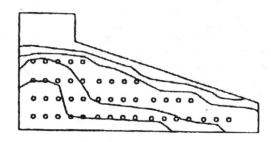

Figure 11. Self-compactability experiment.

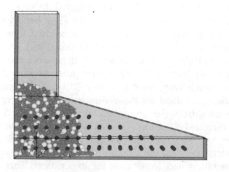

Figure 12. Frame packed with reinforcement simulation (final stage, high yield stress).

Here several phases of concrete were generated and no flow was observed. The flow was blocked and path was blocked for further flow. Which shows the blocking simulation capability of the model.

9 CONCLUSIONS

In this paper, attempt has been made to observe the variation of Bingham coefficients with the variation of DEM parameters. A model was established to get the DEM parameters from the mortar as well as concrete mix proportion. Then with this model verification was performed with different mortar and concrete mix using lifting sphere viscometer test. A comparison between the force–displacement curve generated when ball dragged upward through the aggregate mixture for different mixes at several speed and that computed by DEM simulation using the proposed model showed fairly good agreement which proved the applicability of the proposed model. Also for different concrete mixes the simulation gave the good prediction. The developed model was not sufficient to simulate very high fluid mortar. This was not the shortcoming of the model but of the DEM itself. Two flow simulations such as, slump flow and frame packed with reinforcement were performed successfully. The validity of prediction of DEM parameters from mix proportion was verified by comparing the experimental force displacement curves and those of numerical simulation. Blocking

and segregation were also successfully observed in the simulation by the present model. Finally, it is found that, the DEM model proposed in this research can be used for simulation of flowable concrete.

REFERENCES

Chu, H. & Machida, A. 1998. Experimental & Theoretical Simulation of Self-compacting Concrete by Modified Distinct Element Method (MDEM). *Recent Advances in Concrete Technology*: 691–714.
Cundall, P. A. & Strack, O. D. L. 1979. A Discrete Numerical Model for Granular Assemblies. *Geotechnique 29*(1): 47–65.
Noor, M. A. 2000. *Three Dimensional Discrete Element Simulation of Flowable Concrete*. Ph. D. thesis, The University of Tokyo.
Noor, M. A. & Uomoto, T. 1999a. Three-dimensional Discrete Element Model for Fresh Concrete. *IIS Journal 51*(4): 25–28.
Noor, M. A. & Uomoto, T. 1999b. Three-dimensional Discrete Element Simulation of Lifting Sphere Viscometer Test for Fresh Concrete. *Proceedings of JCI 21*(2): 565–570.
Noor, M. A. & Uomoto, T. 2003. Analytical Modelling of Rheology on High Flowing Concrete. *2nd International Structural Engineering & Construction Conference ISEC-02 Proceedings*.
Okamura, H. & Ozawa, K. 1994. Self-compactable High Performance Concrete in Japan. *Proceedings of ACI seminar, Bangkok*.

Rheological properties of fresh concrete before and after pumping

K. Watanabe, M. Takeuchi & H. Ono
Chubu University, Aichi, Japan

Y. Tanigawa
Nagoya University, Aichi, Japan

ABSTRACT: High rise and long distance pumping experiments were carried out to evaluate the pumpability of high-strength high-fluidity concrete. Rheological properties of fresh concrete before and after pumping were estimated by slump flow value and inverse slump falling time. Normal pressure of concrete in pipe was measured and pressure distribution was discussed. Changing of rheological properties before and after pumping was discussed and a simple model was proposed. It was clarified that the pressure loss of pumping is dependent on the consistency of fresh concrZete and the slip behavior. Slump loss by pumping can be estimated by using the proposed model before pumping.

1   INTRODUCTION

High rise and long distance pumping of concrete which designed strength of 60 N/mm^2 or over has been applied to concrete filled steel tubular column (CFT) system. To insure the compactability of concrete from the bottom of column, good pumpability and enough workability after pumping are required. However, the slump loss due to pumping is high, and the fluidity after pumping becomes worse because high strength concrete has high cement content and the viscosity is high.

In this study, the pumpability of high-strength high-fluidity concrete is discussed. Properties of fresh concrete before and after pumping, and normal pressure of concrete in pipe were measured to discuss theoretically from the rheological point of view.

2   PRESSURE DISTRIBUTION

2.1   *Theoretical approach*

The flow of fresh concrete in conveying pipe is illustrated in Figure 1. Fresh concrete flows by shear deformation and slipping at the wall surface of pipe. Therefore, the flow volume per unit time Q is represented by the sum of the volume per unit time by slip flow (Q_s) and that by shear deformation (Q_v). The value of Q is the same at every position of pipe, as given in Equation 1. The value of Q_v is shown in Equation 2 by using Buckingham's equation if concrete can be regarded as a Bingham's fluid.

$$Q = Q_v + Q_s \tag{1}$$

$$Q_v = \frac{\pi R^4 \Delta P}{8\eta}\left\{1 - \frac{4}{3}\frac{r_0}{R} + \frac{1}{3}\left(\frac{r_0}{R}\right)^4\right\} \tag{2}$$

where, $r_0 = 2\tau_y/\Delta P$: radius of plug flow [m], τ_y: yield value (Pa), ΔP: pressure loss [Pa/m], R: radius of pipe [m].

Slipping resistance σ_h can be expressed by Equation 3 using slipping speed β_s and normal stress σ_n. The quantity by slipping Q_s can be expressed by Equation 4 using slipping speed β_s and area of pipe section. The ratio of Q_s to Q is dependent on the normal stress σ_n and the pressure loss ΔP. Therefore, a model which represents slipping and shear

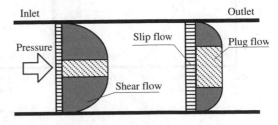

Figure 1.   Outline of slip and flow of fresh concrete in pipe.

deformation behavior can be used to simulate the flow behavior of fresh concrete in the pipe.

$$\sigma_h = S_1\sigma_n\beta_s + S_2\beta_s + S_3\sigma_n + S_4 \qquad (3)$$

$$Q_s = \pi R^2\beta_s \qquad (4)$$

2.2 Outline of pumping experiment

Three kinds of cementitious materials were used in this study. Belite rich cement (HF), low heat cement with 10% fly ash (LF), and three component cement mixed with ordinary portland cement, slag gypsum and silica fume (TC) were used in this experiment. The water-cement ratio was 0.30 and high range water reducing AE agent was used to ensure high fluidity. Designed slump flow value was 65 cm, and water content and volumetric ratio of coarse aggregate were controlled to ensure enough segregation resistance.

Outline of arrangement of concrete pipe is shown in Figure 2. The maximum pressure of pumping machine was 121 MPa. Pressure sensors were set in the pipe, and inner pressure was measured.

2.3 Estimation method of rheological properties

Slump flow test and inverse slump test were carried out to evaluate the properties of fresh concrete. Conventional slump cone was used in up side down style for inverse slump test. Inverse slump cone filled with fresh concrete was gently pulled up and the falling time of concrete was measured. The unit weight of fresh concrete and the yield value affect the shape after slumping. Therefore, the yield value was estimated by Equation 5 using the slump flow value and the unit weight. To apply Buckingham's equation, the inverse slump test was regarded as a narrow funnel test in this study. Equations 6 and 7 were used to estimate the plastic viscosity. The detail of this theory is shown in earlier papers.

$$\tau_y = \frac{GV\rho}{\sqrt{3}\pi r^2} = \frac{396\rho}{S_f^2} \qquad (5)$$

$$\beta = 1 - 3.55\frac{\tau_y}{\rho} + 16.7\left(\frac{\tau_y}{\rho}\right)^4 \qquad (6)$$

$$\eta = 9.59\beta\rho t_i \times 10^{-3} \qquad (7)$$

where, τ_y: yield value (Pa), G: gravity [9.8 m/s²], V: volume of concrete [5.50×10^{-3} m³], ρ: unit weight of concrete [kg/m³], S_f: slump flow value [cm], r = S_f/200 [m], β: parameter of plug flow, η: plastic viscosity [Pa · s], t_i: falling time in inverse slump test [s].

2.4 Experimental and analytical results

The relationship between pressure and distance is shown in Figure 3 in which experimental and analytical

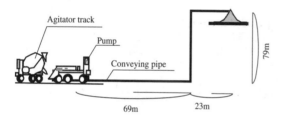

Figure 2.   Pipe arrangement.

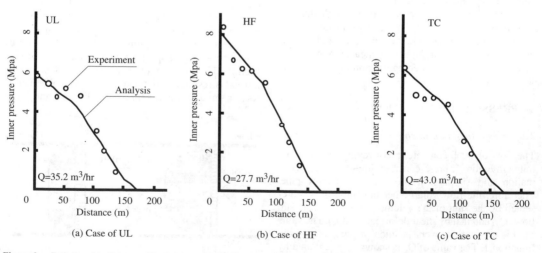

Figure 3.   Relationship between inner pressure and distance (Experimental and analytical results).

Table 1. Parameters for analysis.

Series	Rheological constants		Slipping parameters				
	τ_y	η	S_2 ($\times 10^3$)	S_3 ($\times 10^{-6}$)	P_{max}	Q	Q_s/Q
TC	413	30	1.2	0.8	6.38	43.0	0.8
HF	203	72	3.2	4.0	8.10	27.7	0.64
UL	204	39	1.5	1.0	5.95	35.2	0.71

Notes: τ_y: Yield value (Pa), η: Plastic viscosity (Pa·s), S_1–S_4: Slipping parameters ($S_1 = S_4 = 0$), P_{max}: Maximum of pressure (MPa), Q: Flow volume per unit time (m³/hr), Q_s: Flow volume per unit time by slipping (m³/hr).

results are shown. Input parameters for analysis are shown in Table 1. The rheological constants were estimated by using slump flow test and inverse slump test. The slipping parameters are difficult to estimate by experiment, therefore S_1 and S_4 was ignored, and S_2 and S_3 were calculated by computer to make the difference between experimental and analytical results smaller.

The pressure of fresh concrete became lower from inlet to outlet of pipe, and the pressure loss varied from horizontal portion to vertical portion. It can be said that the analytical results can well simulate the experimental ones. The volumetric ratio by slipping flow against total flow (Q_s/Q) is almost same from inlet to outlet as shown in Table 1. The volume by slipping flow is higher than that by shear deformation flow in high-fluidity high-strength concrete.

3 CHANGE OF RHEOLOGICAL PROPERTIES BEFORE AND AFTER PUMPING

3.1 Experimental results

The properties of fresh concrete was examined before and after pumping. And rheological properties (yield value τ_y and plastic viscosity η) were estimated by using slump flow test and inverse slump test shown above. The relationship between rheological properties of fresh concrete before and after pumping is shown in Figures 4 and 5. In the Figure 4, the yield value of concrete before pumping is about 200 to 400 Pa and the yield value becomes high after pumping in every concrete. On the other hand, the plastic viscosity of concrete before pumping is about 30 to 70 Pa · s, but it becomes low after pumping in every concrete. It can be said that the slump loss by pumping yields not only high yield value but also low plastic viscosity.

3.2 Theoretical approach

It can be considered that the mechanism of slump loss by pumping is different from that is by passing time.

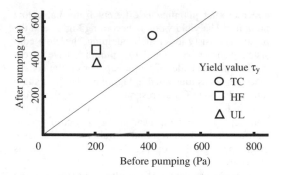

Figure 4. Yield value before and after pumping.

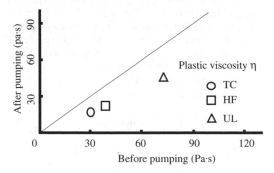

Figure 5. Plastic viscosity before and after pumping.

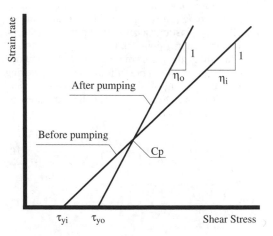

Figure 6. Idealization of consistency before and after pumping.

The slump loss by pumping occurs while the pumping time is very short. In this study, the slump loss by pumping was assumed to be caused by high shear stress in the conveying pipe.

The change of rheological properties by pumping is illustrated in Figure 6. Rheological properties of fresh concrete after pumping is different from that before pumping. The yield value is becoming higher and the plastic viscosisty lower after pumping, therefore consistency curves before and after pumping have a cross point, which is called Cp in this paper on the consistency curves as illustrated in Figure 6. The shear strain rate at the point Cp corresponds to that of fresh concrete in the conveying pipe. The quantity by deformation flow Q_v is same at near inlet or outlet of pipe as shown in Table 1, therefore the strain rate is also same even rheological properties of fresh concrete is changed.

Actually, the shape of the plug flow and also the shear strain rate are changed from inlet of pipe to outlet because the rheological properties in each portion change. However, the difference of the shear strain rate near inlet and outlet may be very small, so that the shear strain rate at the point Cp is treated as a constant value in this paper. The shear stress and the strain rate at the point Cp can be expressed by using rheological properties before and after pumping as shown in Equations 8 and 9, respectively.

$$\tau_p = \tau_{yi} + \eta_i \dot{\gamma}_p, \ \tau_p = \tau_{yo} + \eta_o \dot{\gamma}_p \quad (8)$$

$$\dot{\gamma}_p = \frac{\tau_{yo} - \tau_{yi}}{\eta_i - \eta_o} \quad (9)$$

where, τ_p: shear stress at the point Cp (Pa), τ_{yi}: yield value before pumping (Pa), η_i: plastic viscosity before pumping (Pa · s), τ_{yo}: yield value after pumping (Pa), η_o: plastic viscosity after pumping (Pa · s), $\dot{\gamma}_p$: strain rate at the point Cp (1/s).

In this study, the radius of plug flow (r_o) is assumed to be enough small against the radius of conveying pipe R because high-fluidity concrete has low yield value. Therefore, Equation 2 can be expressed by Equations 10 and 11.

$$Q_v = \frac{\pi R^4 \Delta P}{8 \eta} \left\{ 1 - \frac{4}{3} \frac{r_0}{R} + 0 \right\} = \frac{\pi R^2}{\eta} \left\{ \frac{\Delta P R^2}{8} - \frac{\tau_y R}{3} \right\} \quad (10)$$

$$\Delta P = \frac{8}{R^2} \left\{ \frac{\eta Q_v}{\pi R^2} + \frac{\tau_y R}{3} \right\} \quad (11)$$

The value of Q_v is constant because Q_s/Q is constant as shown in Table 1. Both of rheological properties before and after pumping can be applied to Equations 10 and 11, and Equation 12 can be obtained. Equation 12 can be rewritten by using Equation 13, and Equation 14 can be obtained by using Equations 13 and 9.

The Equation 15 means that the strain rate at the point Cp depends on the quantity of flow (Q_v) and the

radius of pipe (R), and the rheological properties of concrete (τ_y and η) have no effect on the strain rate at the point Cp.

$$Q_v = \frac{\pi R^2}{\eta_o} \left\{ \frac{8R^2}{8R^2} \left(\frac{\eta_i Q_v}{\pi R^2} + \frac{\tau_{yi} R}{3} \right) - \frac{\tau_{yo} R}{3} \right\} \quad (12)$$

$$Q_v = \frac{\pi R^2}{\eta_o} \left\{ \frac{\eta_i Q_v}{\pi R^2} - \frac{(\tau_{yo} - \tau_{yi}) R}{3} \right\} \quad (13)$$

$$\frac{Q_v (\eta_i - \eta_o)}{\pi R^2} = \frac{(\tau_{yo} - \tau_{yi}) R}{3} \quad (14)$$

$$\frac{3Q_v}{\pi R^3} = \frac{\tau_{yo} - \tau_{yi}}{\eta_i - \eta_o} = \dot{\gamma}_p \quad (15)$$

On the other hand, Equation 16 can be written by using Equation 10 and rheological properties before and after pumping because Q_v is constant at every portion of conveying pipe. Equations 17 and 18 are represented by using Equation 16. Equation 18 is expressed by the pressure loss (ΔP), the radius of pipe (R) and the rheological properties before and after pumping. This equation indicates that the rheological properties after pumping can be estimated before pumping if the properties of fresh concrete before pumping and the pumping condition, such as pressure and radius of pipe, are known.

$$\frac{\pi R^2}{\eta_i} \left\{ \frac{\Delta P R^2}{8} - \frac{\tau_{yi} R}{3} \right\} = \frac{\pi R^2}{\eta_j} \left\{ \frac{\Delta P R^2}{8} - \frac{\tau_{yj} R}{3} \right\} \quad (16)$$

$$\frac{\Delta P R^2}{8} (\eta_j - \eta_i) = \frac{R}{3} (\tau_{yi} \eta_j - \tau_{yj} \eta_i) \quad (17)$$

$$\Delta P = \frac{8}{3R} \frac{(\tau_{yi} \eta_j - \tau_{yj} \eta_i)}{(\eta_j - \eta_i)} \quad (18)$$

3.3 Analytical results

Experimental results of pumping as shown above were discussed again to confirm the theory of pumping. The value of Q_v is calculated by using Q and Q_s/Q shown in Table 1. The strain rate at wall surface of conveying pipe is estimated by using Equation 15 and the shear stress at wall surface τ_q is estimated by using Equation 8. The shear stress τ_p at the point Cp is also calculated by using Equation 9. Figure 7 shows that the relationship between shear stress at the point Cp and the shear stress at wall surface of conveying

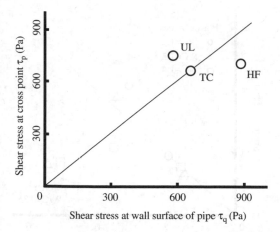

Figure 7. Relationship between shear stress at the cross point Cp and wall surface of pipe.

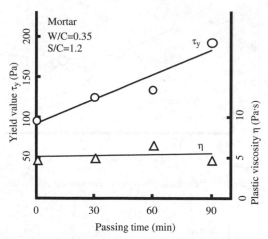

Figure 8. Relationship between rheological properties and passing time.

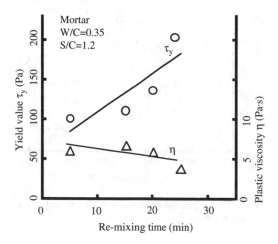

Figure 9. Relationship between rheological properties and re-mixing time.

pipe. Three points illustrate indivisual case of pumping as shown in Figure 3, and a correlation can be observed. It can be said that the theory shown above can be applied to obtain the slump loss at pumping and evaluate the slump loss by pumping before casting into pumping machine.

4 CHANGE OF RHEOLOGICAL PROPERTIES BY RE-MIXING

4.1 Outline of mortar experiment

Indoor experiment was carried out to examine the relationship between shear stress and properties of fresh concrete.

A mortar mixer was used to yield shear stress in fresh mortar after 30 min from mixing, and it is called re-mixing in this paper. Ordinary portland cement was used and the water-cement ratio (W/C) was 0.35 with high-range water reducing AE agent. Three types of sand-cement ratio (S/C), 1.2 to 1.4, were prepared.

Yield value was estimated by using mortar flow test which is regulated in JIS (Japan Industrial Standard) code, but with no impact. Mortar flow value with no impact is 210 mm immediately after mortar was mixed. Plastic viscosity was estimated by using J14 funnel falling test which is regulated in JSCE (Japanese Society of Civil Engineers) code.

4.2 Results and discussion

An example of the relationship between rheological properties of mortar and passing time is shown in Figure 8. It can be observed that the yield value increases when passing time increases, and the plastic viscosity also increases. An example of the relationship

between rheological properties of mortar and re-mixing time is shown in Figure 9. The yield value increases when re-mixing time increase, but the plastic viscosity decreases even re-mixing time increases. The change of plastic viscosity in case of re-mixing is different from that in case of passing time. The behavior of rheological properties in case of re-mixing shown in Figure 9 is similar to that of concrete before and after pumping shown in Figure 5.

The cross points Cp in consistency curves in each mixture are calculated, and it is clarified that these points concentrate in a certain point. The relationship between shear stress at the cross point Cp and sand-cement ratio (S/C) is shown in Figure 10. It can be observed that the shear stress at the cross point Cp

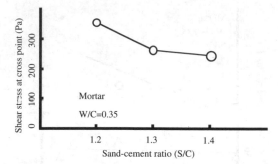

Figure 10. Relationship between shear stress at cross point Cp and sand-cement ratio.

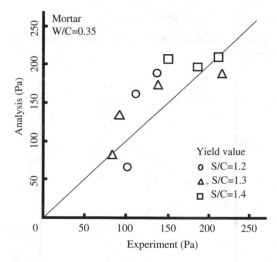

Figure 11. Relationship between analytical and experimental yield value.

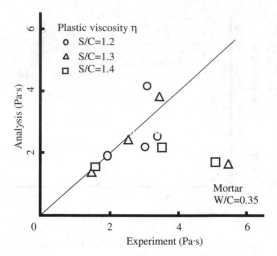

Figure 12. Relationship between analytical and experimental plastic viscosity.

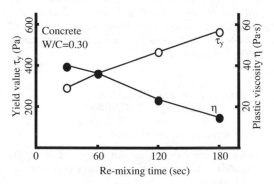

Figure 13. Relationship between rheological properties and re-mixing time.

decrease when the sand-cement ratio increases, and the stress state at the cross point Cp is dependend on the mix proportion of material.

4.3 Change of rheological properties

A simple model is proposed to simulate the change of rheological properties under shear stress. In this model, the plastic viscosity is becoming lower as shown in Equation 19, but the yield value higher because the cross point does not move. The consistency curves move around the cross point anti-clock wise direction.

Analytical and experimental results are shown in Figures 11 and 12 respectively. Analytical results well simulate the experimental ones and good correlation can be observed in these figures.

$$\eta_t = \frac{\eta_0}{1 + a \times \tau_p \times t_m} \qquad (19)$$

where, η_0: initial plastic viscosity (Pa · s), η_t: plastic viscosity after t second mixing (Pa · s), a: experimental constants (1/Pa · s), τ_p: shear stress of cross point (Pa), t_m: mixing time (sec).

4.4 Outline of concrete experiment

Indoor experiment with concrete was also carried out the method of re-mixing was almost the same as in the mortar experiment. Ordinary portland cement was used and the water-cement ratio (W/C) was 0.30 with high-range water reducing AE agent to make the slump flow value 60 cm.

The rheological properties of concrete and the remixing time are shown in Figure 13. According to Figure 13, the yield value increases but the plastic viscosity decreases when the re-mixing time increases as

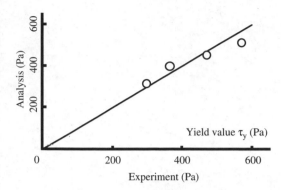

Figure 14. Relationship between analytical and experiment yield value.

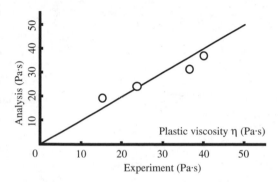

Figure 15. Relationship between analysis and experiment plastic viscosity.

same in the mortar experiment. The rheological properties obtained by analysis and experiment are shown in Figures 14 and 15. It can be said that the analytical results are consistent with the experimental ones.

5 CONCLUSION

In this paper, the pumpability of high-strength high-fluidity concrete was discussed. Inner pressure in conveying pipe was measured and the analytical results were compared. A theoretical approach was carried out to explain the change of rheological properties before and after pumping. Re-mixing experiment were carried out to simulate the pumping of concrete. It was clarified that the behavior of fresh concrete under shear stress is similar to that of concrete under pumping.

ACKNOWLEDGEMENT

The authors wish to express their gratitude to Mr. S. Kuroiwa (Taisei corporation) for his cooperation. The part financial support by JSPS and the Nitto Foundation are gratefully acknowledged.

REFERENCES

Watanabe, K. et al. 2001. Evaluation of pumpability of high-strength high-fluidity concrete, Concrete Under Severe Conditions – Environment and Loading-, Proc. of the Third International Conference on Concrete, Vol. 2, pp. 1642–1649.

Kuroiwa, S. et al. 1998. Application of High-Strength Concrete to Fill Tubular Steel Columns: Fourth CANMET/ACI/JCI International Symposium on Advances in Concrete Technology, pp. 365–387.

Watanabe, K. et al. 1994. Study on Adhesion and Slipping Properties of Fresh Concrete, Transactions of the Japan Concrete Institute, Vol. 16, pp. 33–40.

Tanigawa, Y. et al. 1991. Theoretical Study on Pumping of Fresh Concrete, Transactions of the Japan Concrete Institute, Vol. 13, pp. 25–32.

System-based Vision for Strategic and Creative Design, Bontempi (ed.)
© *2003 Swets & Zeitlinger, Lisse, ISBN 90 5809 599 1*

The properties of lightweight aggregate concrete for pre-cast bridge slab

H. Tanaka, I. Yoshitake, Y. Yamaguchi & S. Hamada
Department of Civil Engineering, Yamaguchi University, Yamaguchi, Japan

H. Tsuda & M. Kushida
Kurimoto, Co. Ltd. Osaka-Rinkai Factory, Osaka, Japan

ABSTRACT: This paper presents fundamental properties of lightweight aggregate concrete to be applied for pre-casting and pre-stressed concrete slab. The properties of volume change which influences the effective pre-stress are discussed based on instantaneous and long-term tests. The tensile and shear strengths by various tests are also investigated and these properties affect the punching shear strength of slab.

1 INTRODUCTION

Double main girder bridges need lighter slab in order to decrease the dead weight of concrete slabs. There are two methods for decreasing the weight of concrete slabs; one is to employ thinner slabs strengthened by pre-stressing and the other is to apply the lightweight concrete to the slabs. The advanced technology may be a combination of these two methods of the pre-stressed concrete and lightweight aggregate concrete (Hamada et al. 2002). Especially, pre-stressed concrete structures become an effective method for strengthening the member, whereas lightweight aggregate concrete has a low strength in the tensile and shear stress fields.

On the other side, prefabrication for concrete slab panels is required in construction of bridge slab in order to build the bridge in short period and with high quality. In prefabrication, concrete slabs are generally produced by the pre-stress of pre-tension method at young age for the factory production system.

This paper presents some experimental data of fundamental properties for effective pre-stress and strength of slab with lightweight aggregate concrete. When the pre-stressing system is applied to the slab with lightweight concrete, the fundamental properties of lightweight aggregate concrete is needed for rational design of these concrete slab. The present study shows the compressive creep due to pre-stressing in long-term experiments. In addition, a mix proportion is attempted in the present research, which is a combination of 2 lightweight aggregates as coarse aggregate for concrete in order to improve the Young's modulus. The Young's modulus obtained from the present experiment has been evaluated from the composite structural

models. A mix design method of lightweight aggregate concrete with high Young's modulus is presented in this paper.

The failure of slab is mainly caused by the punching shear. In order to obtain the fundamental properties concerned with failure of slab, tensile strengths by the splitting test and direct tension test are discussed in the present study. The pure shearing strength by using new shearing machine is also presented in this paper.

2 COMPRESSIVE CREEP

2.1 *Purpose of compressive creep test*

Compressive creep and shrinkage gradually causes a decrease of pre-stressing force. In order to obtain pre-stress loss, the property of compressive creep must be evaluated. However, little research especially loading at early age has focused on the PC slab with the high-strength lightweight concrete. Thus the present study mainly focused on the compressive creep caused by pre-stress.

2.2 *Experimental method*

In the present study, the long-term loading test was conducted in order to investigate the compressive creep of lightweight concrete. The concrete in this experiment was made of artificial lightweight aggregate with low permeability and low density of 5% at the maximum and $0.85\,g/cm^3$ respectively. Water-cement ratio (W/C) of the concrete was determined as 30% so that high strength can be obtained early. Additionally, the lower W/C ratio provides high

Table 1. Materials and mix proportion.

Water Cement	High-early strength Portland cement (3.13 g/cm³)	157 kg/m³ 527 kg/m³
Sand Gravity (G_{LI})	Sea Sand (2.60 g/cm³) Artificial lightweight aggregate of pearl stone (0.85 g/cm³)	786 kg/m³ 290 kg/m³
Admixture	High performance water reducing agent with air-entraining	6.5 kg/m³

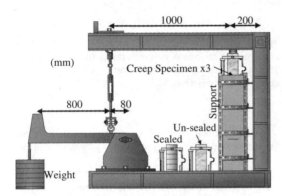

Figure 1. Compressive creep machine.

viscosity to the lightweight aggregates. Table 1 gives the detail of materials and the mix proportion of concrete.

Specimens for creep test were $100 \times 100 \times 380$ mm, and each edge was cut out of $100 \times 100 \times 400$ mm. An initial load of 69 kN [7.0tf] applied to the specimen at 1 day. The load and loading age were determined to simulate the pre-stressing age of pre-fabrication systems.

The testing machine has a two-lever system, 3 specimens were subjected to compression at the same time as shown in Figure 1. Creep strains were evaluated by subtracting instantaneous strain and strain of un-loaded specimen in the common condition from the total strain, where the temperature is 20 Celcius degree and relative humidity is 60 ~ 90%. The strain of un-loaded specimen means the shrinkage and thermal strains. Such strains are usually included in the strain by applied loads.

All strains of concrete specimen were obtained by a revised commpressometer shown in Figure 2. This equipment has the electric displacement sensor with accuracy of 1/1000 mm. That is, the equipment can measure the strain as small as approximately 1.4×10^{-6} of strain for a length of 360 mm.

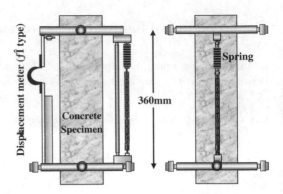

Figure 2. New compressometer for creep test.

Table 2. Loading history.

	Initial	Second	Third
Age (day)	0 1 >>	500 >>	700 >>
Stress (N/mm²)	– 6.9 >>	5.4 >>	4.1 >>

2.3 Experimental program

The present study employed 2 specimens with and without the waterproofing alumni-seal. Therefore, a drying creep by Picket's effect was obtained from the difference between the strain of sealed and un-sealed specimen.

Table 2 gives a loading program for the creep test. As mentioned above, loads were applied to the specimen under an initial stress of 6.9 N/mm² [69 kN/(100 × 100 mm²)] at an age of 1day. The stress was kept until loading period of 500 days. After 500 days a load equivalent to 1.5 N/mm² was removed, the second stress of 5.4 N/mm² was applied during 200 days. A load to the stress of 1.5 N/mm² was removed after loading period of 700 days, and the remained stress was thereafter kept loading.

2.4 Results and discussions

Figure 3 shows the strains, which is strain of the un-loaded specimen subtracted from that of the loaded specimens. This figure indicates that the elastic strains were approximately 500 ~ 600 micro and creep strains under drying condition increased in an early age. Each strain under initial loading becomes an almost constant value after 210 days.

Recovering strains at 500 days and 700 days in this figure are the elastic strains by removing the load of 1.5 N/mm². After the removing the load, the strain-time curve for the sealed specimen became a constant. However, creep strains under drying condition increased slightly after first removing, and decreased gradually after second removing the load.

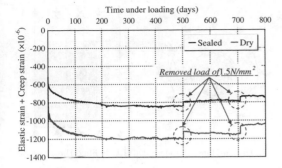

Figure 3. Strains due to loading.

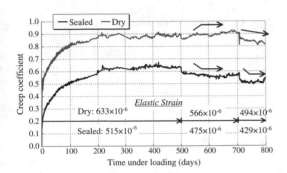

Figure 4. Creep coefficients.

Table 3. Mix proportions for Young's modulus test. (volume ratio %)

Weight (kg/m³)	Unit amount (kg/m³)				
	W	C	S	G_{LI}	G_{LII}
1. 1695				294 (100)	0 (0)
2. 1782	1	5	7	220 (75)	160 (25)
3. 1868	6	3	0	147 (50)	319 (50)
4. 1954	0	3	6	73 (25)	479 (75)
5. 2041				0 (0)	639 (100)

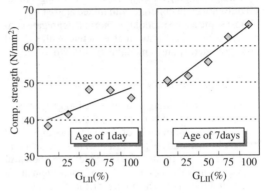

Figure 5. Compressive strength of mixed concrete.

Based on the above results, the creep coefficients are evaluated as shown in Figure 4. This diagram indicates that the drying and basic creep coefficients at initial stresses were 0.9 and 0.6, respectively. Such creep coefficients were considerably smaller than that of normal concrete, which are normally 2.0 ~ 3.0. This is due to the large elastic strain of lightweight aggregate concrete, and the concrete has lower Young's moduli than the normal concrete.

The creep coefficient became smaller immediately after removing the load and became gradually almost constant. This phenomenon was resulted from the influence of recovering creep of prior load. As consequence, the creep coefficients were smaller than the initial coefficient. Such influence on the pre-stress loss seems to be disregarded so that such effects may be however considerably smaller than the influence of elastic strain caused by lower Young's modulus.

3 IMPROVEMENT OF YOUNG'S MODULUS

3.1 Mixed aggregate proportions

As mentioned in the previous chapter, the lightweight aggregate concrete has low Young's modulus, thus the loss of pre-stress tends to be larger at pre-stressing compared with creep and shrinkage. The present study employed the concrete mixed with two different lightweights aggregate in order to obtain the higher Young's modulus concrete.

Mixtures of newly developed concrete are given in Table 3. Fly-ash aggregate G_{LII} was herein employed for additional material to the previous lightweight aggregate G_{LI}. This aggregate has a density of 1.85 g/cm³, and absorption of 3%. The maximum size is 15 mm same as the peal stone aggregate.

Concretes with mixed aggregate were made from 5 mix proportions, changing volume ratio of aggregate. Water cement ratio (W/C) was 30% and the high early-strength Portland cement was employed for this experiment.

Specimens were a cylinder with a diameter of 100 mm and height of 200 mm. The compression tests were conducted at ages of 1 and 7days, and over 3 specimens were used in a test. Concrete strains were measured by the normal compressometer and Young's moduli were evaluated by the strain at 1/3 of the compressive strength of concrete.

3.2 Compressive strength and young's modulus

Figure 5 shows relationships between the compressive strength and volume ratio of the lightweight aggregate G_{LII}. Here, each plotted point indicates the average compressive strength of 3 specimens.

This figure indicates that the compressive strength was higher with increase of G_{LII} and its relation was almost liner. The figure also represents that a compressive strength of 30 N/mm² can be obtained in all the concrete even at an age of 1 day. This result means that the pre-stressing can be given at 1 day according to Japanese specification (JSCE 2002).

Young's moduli of mixed aggregate concrete are shown in Figure 6. This figure indicates that Young's moduli are proportional to the volume ratio of the lightweight aggregate G_{LII}. Especially, Young's moduli of lightweight concrete with mixed aggregate were almost same as the modulus of normal aggregate concrete, so that the lightweight concrete with G_{LII} of 50% provided Young's modulus of 30 kN/mm².

3.3 Estimation by composite structural model

In order to apply more widely mixed aggregate concrete, the present study employed the composite structural models estimation of Young's modulus. The models in this study were Hashin-Hansen model and modified Hirsch model. These models are shown in Figure 7, and Young's modulus of each element is given in Table 4. Such moduli of the element were obtained from the specification of the artificial aggregate or experiments for the mortar with W/C of 30%.

Figure 8 shows relationships of Young's modulus between the experiments and estimations obtained from each model. These figures indicate that each model had relatively high accuracy within 15% for estimation of Young's modulus. Especially, the average error of estimation by the modified Hirsch model was 4.80%, its effectiveness was verified from the error of prediction of Young's modulus. That is, these models can predict Young's modulus when lightweight aggregate concrete are employed as mixed aggregate.

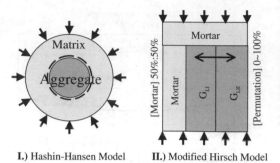

I.) Hashin-Hansen Model   **II.)** Modified Hirsch Model

Figure 7.   Composite structural models.

Table 4.   Young's modulus of element of concrete.

Mortar (W/C = 30%)	32.3 kN/mm²
Coarse Aggregate G_{LI}	9.7 kN/mm²
Coarse Aggregate G_{LII}	42.1 kN/mm²

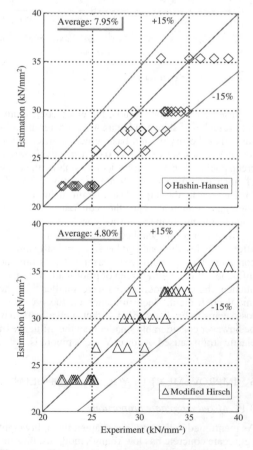

Figure 8.   Experiments and estimations by models.

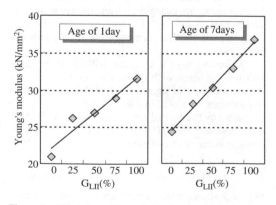

Figure 6.   Young's modulus of mixed concrete.

4 FRACTURE STRENGTH TEST

4.1 Purpose of fracture strength test

Tensile strength of lightweight aggregate concrete tends to be lower than that of normal concrete due to pores in aggregate. Resultantly, the slab with lightweight concrete has lower strength than normal weight slab. The bridge slab is failed generally by punching shear, and concrete needs higher tensile and shear strength. Therefore the present study confirmed the fracture strength of lightweight concrete for precast slab experimentally.

4.2 Tensile strength and young's modulus

Table 5 gives results of tensile strength obtained from the splitting tension test and direct tension tests. W/C of concrete ranged from 30 to 60%, and these aggregates were two different lightweight aggregates and crushed andesite.

Table 5 shows that the direct tension strengths were lower than the strength by splitting test by 50 ~ 20%. This reason is caused from the fact that the local yielding influenced on the direct tension strength (Zheng, W. et al. 2001). However, the ratio of these strengths of the lightweight concrete was almost similar to that of normal concrete or mortar concrete. As shown in Table 5, Young's moduli obtained from the direct tension test were averagely higher than compressive moduli. The concrete is generally failed by tensile stress rather than compressive stress, and the tensile max strain is fairly smaller than the compressive strain. The tensile Young's moduli were higher than compressive moduli, which were caused from the tensile strain capacity and fracture stress of concrete. However, the tensile Young's modulus of lightweight concrete may be equivalent to the compressive modulus.

4.3 Pure shearing strength

Concrete under shearing fails normally in the principal tensile stress that is angle of 45 degree to direction of shear stress. Since shearing strength cannot be evaluated for concrete, the pure shearing strength is generally evaluated by the tensile strength. However, to investigate the concrete behavior under pure shearing stress field is important for understanding the fracture mechanism due to shear force, which is related to the shear strength of beam and punching shear strength of slabs. Thus, the present study is intended to obtain experimental pure shearing strength of lightweight concrete.

A machine for pure shearing strength is developed and illustrated in Figure 9. This machine can subject the pure shearing force to concrete-plate element by changing the direction of the uni-axial force. The specimen for this machine is $170 \times 170 \times 100$ mm, and it was attached with the loading plate by epoxy-adhesion and 11 bolts.

Picture 1 shows an example of fracture pattern by pure shearing test. All specimens were fractured by vertical crack due to principal stress in the present loading method as shown in the picture. This result indicates that a concrete under pure shearing stress fractures in the direction of shearing force, and that the strength represents almost the same strength as tensile strength.

4.4 Correlation of fracture strength

The relationships between the pure shearing strength and compressive strength are shown in Figure 10. As

Table 5. Tensile strength, Young's modulus and pure shearing strength.

Type	WC (%)	Permutation G_{LI}	G_{LII}	Comp. Strength (N/mm^2)	Tens.Strength (N/mm^2) A splitting	B Direct	B/A	Young's Modulus (kN/mm^2) C Comp.	D Tens.	D/C	Pure Shearing Strength (N/mm^2) E	E/A	E/B
Normal	60	—	—	27.4	2.94	2.25	77%	34.2	42.8	125%	2.37	81%	95%
Aggregate	45	—	—	47.9	3.25	2.25	69%	41.6	40.4	97%	2.89	89%	78%
Concrete	30	—	—	66.0	4.72	3.03	64%	44.7	45.7	102%	3.50	74%	87%
Mortar	45	—	—	38.7	2.32	1.01	44%	26.5	34.0	129%	2.14	92%	47%
	30	—	—	64.5	3.25	—	—	29.2	—	—	2.67	82%	—
Light weight	60	100	0	20.1	1.42	1.09	77%	17.8	18.9	106%	1.53	108%	71%
Aggregate	45	100	0	30.9	1.67	1.14	68%	19.4	16.8	87%	2.38	143%	48%
Concrete	30	100	0	42.4	3.23	1.61	50%	23.9	24.7	103%	3.30	102%	49%
		75	25	51.6	3.12	2.32	74%	28.1	31.3	111%	—	—	—
		50	50	55.4	3.87	2.65	68%	30.3	37.8	125%	—	—	—
		25	75	62.2	3.38	2.33	69%	32.9	33.7	102%	—	—	—
		0	100	66.0	4.18	2.94	70%	37.2	37.9	102%	—	—	—

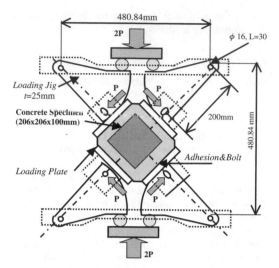

Figure 9.   Pure shearing machine.

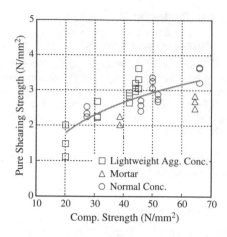

Figure 10.   Pure shearing strength – compressive strength.

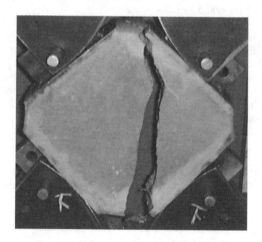

Picture 1.   Fracture pattern by pure shearing test.

shown in this figure, the pure shearing strength became higher with increase of the compressive strength.

Figure 11 shows the relationships between the pure shearing strength and 2 tensile strengths, i.e. splitting-tension and direct-tension. Figure 11a indicates that the pure shearing strength is averagely lower than the splitting tensile strength of normal concrete. This result was caused by the effect of tensile-zone area. On the other hand, the lightweight aggregate concrete had higher strength in the pure shearing test than that of the splitting tensile test. In addition, the pure shearing strengths were higher than the direct tensile strength in both concrete as shown in Figure 11b. It may be due to

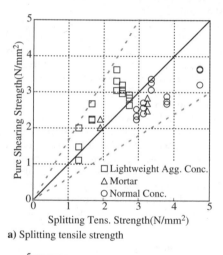

a) Splitting tensile strength

b) Direct tensile strength

Figure 11.   Pure shearing strength – tensile strength.

the influence of size effects, breeding of direct test, and/or stress field in splitting test. However, when the compressive strength of lightweight concrete almost equals to the compressive strength of normal concrete. Such strength of lightweight concrete tends to be lower than the strength of normal concrete.

5 SUMMARY

This study presented some experiments related to the mechanical properties of lightweight aggregate concrete for pre-cast bridge slab. Especially, the volume change and the failure strength of the concrete were studied experimentally. The fundamental properties for the design were discussed in this paper. The conclusions of this study are summarized as follows:

(1) Compressive creep coefficients of lightweight aggregate concrete were under 1.0. It was significantly lower than that of normal concrete. This is resulted from the lower Young's modulus of the lightweight aggregate concrete.

(2) Young's modulus of concrete with mixed aggregate was related to the volume of lightweight aggregate with high strength. Composite structural models are able to estimate Young's modulus with high accuracy.

(3) Pure shearing strengths of lightweight aggregate concrete were higher than the strength of direct tension or splitting tension tests. The fracture strength of such concrete, however, than that of normal concrete.

REFERENCES

Hamada, S., Mao, M., Yoshitake, I. & Tanaka, H. 2002. Slabs with lightweight aggregate concrete of high strength, Proc. of the first fib Congress 2002: 267–272.

Japan Society for Civil Engineers, 2002. Standard specification for concrete structures 2002, materials and construction.

Zheng, W., Kwan, A.K.H. & Lee, P.K.K. 2001. Direct tension test of concrete, ACI Materials Journal, Vol. 98, No.1: 63–71.

System-based Vision for Strategic and Creative Design, Bontempi (ed.)
© *2003 Swets & Zeitlinger, Lisse, ISBN 90 5809 599 1*

Bond behaviour of reinforcement in self-compacting concrete (SCC)

K. Holschemacher, Y. Klug & D. Weiße
Leipzig University of Applied Sciences (HTWK Leipzig), Germany

G. König & F. Dehn
Institute for Structural Concrete and Building Materials, University of Leipzig, Germany

ABSTRACT: Because of its favourable fresh concrete properties and its workability efficiency, self-compacting concrete (SCC) has a high production engineering, economic and architectural-design potential, so it is of high interest for practical construction work. Self-compacting concrete is undoubtedly an engineered high-tech material whose potential has not yet been fully exploited. So, besides some concrete technological also several design aspects are not cleared definitely, among others the time development of the bond properties and the influence of cyclic loading on the bond performance. This paper outlines the experimental programme and its results on the bond behaviour of conventional rebars in self-compacting concrete. In this study, the hardened concrete properties and the bond behaviour were experimentally investigated at 3, 7, 28 and 56 days.

1 INTRODUCTION

Numerous innovations concerning building materials have led the concrete construction industry to new chances and possibilities. One of the most promising developments is SCC. It has the ability to compact itself only under its self weight without additional internal or external vibration energy and deaerates thereby almost completely while flowing in the formwork. SCC fills all recesses, reinforcement spacings and voids even in highly reinforced concrete members and flows without segregation like "honey" nearly to level balance (Skarendahl & Petersson 2000). The elimination of the need to vibrate the concrete is the main advantage in the production process.

The main components of SCC are the same as used in normal concrete, but for SCC the application of combinations of numerous constituent materials to achieve a reduction of the liquid limit and to improve the workability are necessary. SCC normally contains one or more cementitious materials like fly ash or silica fume, inert fine fillers like quartz or lime stone powder and admixtures in particular superplasticizer. Furthermore, a higher content of fine material is necessary to provide lubrication for the coarse aggregates as well as a possible utilization of viscosity agents.

Self-compaction leads to positive dissimilarities in the microstructure in comparison to normal vibrated concrete (NC). Especially the interfacial transition zone between aggregates and matrix is less porous, the distribution of the pores is more homogeneous.

On the basis of the stated differences it is necessary to prove existing design rules based on years of experience on normal concrete. Within the scope of a common research project between the University of Leipzig, Institute for Structural Concrete and Building Materials, and the Leipzig University of Applied Sciences (HTWK Leipzig) experimental investigations on the bond behaviour of self-compacting concrete (SCC) were carried out. This paper reports the results of a study aimed on evaluating the bond performance of reinforcement and the material properties.

2 BOND IN CONCRETE UNDER CYCLIC LOADING

Whereas in the CEB-FIP Model Code 90 (Structural Concrete 1999) a bond law is included, in other design codes like Eurocode 2 (1992) the design rules based on bond are hidden in design tables. In the case of Eurocode 2 it is difficult for the user to perceive the background of the bond design specifications. The following bond law for NC is integrated in the CEB-FIP Model Code 90 (Structural Concrete 1999) (Fig. 1,

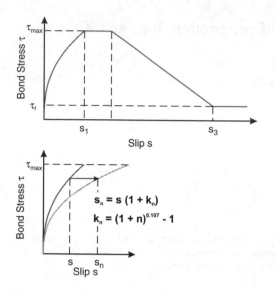

Figure 1. Bond law according to Model Code 90, Equations 2 & 3 (Structural Concrete 1999).

Equation 1), also the relevant parameters depending on the bond conditions in this bond law.

$$\tau = \begin{cases} \tau_{max} \cdot \left(s / s_1\right)^{\alpha} & \text{for } 0 \leq s \leq s_1 \\ \tau_{max} & \text{for } s_1 \leq s \leq s_2 \\ \tau_{max} + \dfrac{\tau_f - \tau_{max}}{s_3 - s_2} \cdot \left(s - s_2\right) & \text{for } s_2 \leq s \leq s_3 \\ \tau_f & \text{for } s > s_3 \end{cases}$$

(1)

Out of numerous investigations on pull-out specimens made of NC it is well known that cyclic loading with an increasing number of load cycles leads to a decrease in the bond strength accompanied with an increasing displacement (Koch & Balázs 1998; Rehm & Eligehausen 1977). This influence can be recognised in the area of the increasing branch in the bond law according to Figure 1, the displacement after first loading is multiplied by the bond creep coefficient. Therefore it is possible to generate a bond law affine to static loading for cyclic loading, Equations 2 & 3 according to Model Code 90 (Structural Concrete 1999):

$$s_n = s_0 \cdot \left(1 + k_n\right)$$

(2)

$$k_n = \left(1 + n\right)^{0.107} - 1$$

(3)

where n = number of load cycles; k_n = bond creep coefficient.

A reduction of the bond strength depending on the number of load cycles is not included in the Model Code MC 90. It seems reasonable, because for design rules in the serviceability limit states, like crack width control, the height of the loading action is relatively low.

Considerably not cleared is nowadays, whether this bond law is applicable also for SCC. Aiming to an increased and wider application of SCC in e.g. bridge construction, with the described test series concerning the bond behaviour under cyclic loading it will be possible to determine the bond creep of reinforcement.

3 INITIAL SITUATION OF THE BOND BEHAVIOUR IN SCC

The bond behaviour of deformed reinforcing bars in concrete is influenced by many parameters. These factors can be divided into three main groups:

- *concrete properties* such as compressive strength, tensile strength,
- *reinforcement* like bar diameter, relative rib area,
- *loading regime* e.g. monotonic or cyclic loading, loading rate and history.

For the first mentioned group not only hardened concrete properties play an important role, but also the properties in the fresh state like the consistency or the mix design, especially the grading curve as well as the fine grain content. In the literature only a few reports are present where these values and their influence on the bond behaviour were investigated. Tests performed by Martin (1982) show, that both the bond strength and the bond stiffness decline with an increasing liquid consistency of the fresh concrete. The differences were up to 100%. The same effect was monitored with an increasing fine grain proportion in the concrete mix. Both influences have to take into account for SCC. Khayat et al. (1997) report of investigations on wall elements cast with self-consolidating concrete where due to the application of superplasticizer and stabilizer an increase of the bond strength was reached. In the literature various tests on bond of SCC are presented. The results out of these papers are partly contradictory. Nevertheless, they can be summarized as follows:

- The influence of the rebar position is less distinct in SCC than in NC (Khayat et al. 1997; Khayat 1998; Schießl & Zilch 2001);
- The bond strength related to the concrete compressive strength of the SCC is partly lower (Schießl & Zilch 2001), in other papers slightly higher than to comparable NC (Sonebi & Bartos 2001) or on the same level (Lorrain & Daoud 2002).

From these statements there can be followed, that it is not possible at the moment to formulate final conclusions about the bond behaviour of reinforcement bars in

SCC, additional tests for clarification of unknown aspects are necessary. Furthermore it should be borne in mind that self-compacting concrete can be produced in different compressive strength classes, with a large range of mix designs using a large number of various aggregates, cementitious materials and admixtures. So there exists not only one SCC, but also many kinds of SCC, so that it is difficult to compare the results with normal vibrated concrete.

4 TESTS FOR THE DETERMINATION OF THE BOND BEHAVIOUR

4.1 Experimental programme

The programme was divided into 2 main series, one for monotonic and the other one for cyclic loading. In order to assess the time development of the bond properties in relation to the strength development of the concrete, the tests were performed at a concrete age of 1, 3, 7, 28 and 56 days in the first series. Three monotonic loaded pull-out tests were carried out for each rebar orientation per day. The bond tests under cyclic loading (series 2) were only performed after 28 days et seq. In order to obtain the loading level for each cyclic one, also three monotonic loaded pull-out tests were implemented. The schedule of 28, 33 and 36 days was necessary because of the duration of the dynamic pull-out tests of about 2½ days. Altogether 60 monotonic and cyclic pull-out specimens were tested in both series as well as associated concrete properties like Young's modulus, the compressive and the splitting tensile strength.

4.2 Pull-out specimen and tests

The bond behaviour under monotonic and cyclic loading was tested on RILEM-specimens (RILEM 1970). To avoid an unplanned force transfer between the reinforcement and the concrete in the unbonded area, the rebars were encased with a plastic tube and sealed with a silicone material (Fig. 2). During the test series the orientation of the rebar with respect to the casting direction was varied, i.e. horizontal and vertical rebar position. The vertical rebar orientation has to be distinguished between loading in and against concrete placing direction (Fig. 3).

Within both series, the pull-out specimens under monotonic loading were stressed path-controlled. In the second series the average of the maximum bond stresses τ_{bu} of 3 specimens out of the measured monotonic bond stress-slip-relationships was determined to set the lower and upper level of the bond stress for the cyclic loading. The constant bond-stress amplitude $\Delta \tau_b$ was 25% and the upper level τ_{bu} was 60% of the maximum bond stress, whereby the frequency was at 5 Hz up to a maximum number of 1 000 000 load cycles.

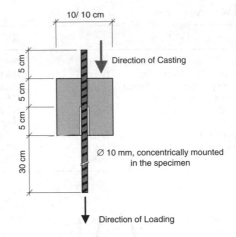

Figure 2.   RILEM pull-out specimen.

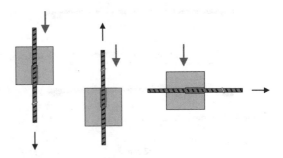

Figure 3.   Different rebar orientations (large arrow: casting direction, small arrow: loading direction).

The slip between rebar and concrete was measured both at the loaded and unloaded end of the specimen with three LVDT's each. For the analysis only the values from the unloaded end were used. As reinforcing steel BSt 500 S (Ø 10) according to the German Standard DIN 488 (1984) was applied (Fig. 2).

4.3 Materials

For the SCC a powder-type was chosen. Because of the high ultrafine material content compared with other SCC-concepts, the most significant differences in the bearing behaviour to normal vibrated concrete were expected. For the mix design usual aggregates out of local resources were used (Table 1).

5 TEST RESULTS

5.1 Fresh concrete properties

To evaluate the consistency, the slump flow was measured with the slump cone according to EN 12350

Table 1. Mix design.

		SCC
Sand (0–2 mm)	[kg/m³]	769.0
Gravel (2/8, 8/16 mm)	[kg/m³]	788.0
Quartz powder (0.1–0.5 mm)	[kg/m³]	66.0
Cement CEM I 42,5 R	[kg/m³]	316.0
Fly ash	[kg/m³]	211.0
Water	[kg/m³]	168.0
Superplasticizer	[kg/m³]	4.4
Retarder	[kg/m³]	2.2
W/b-ratio	–	0.31

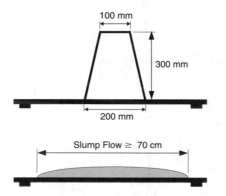

Figure 4. Slump cone according to EN 12350 (2000).

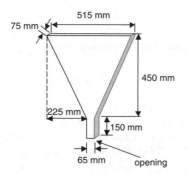

Figure 5. V-Funnel Test.

(2000) (Fig.4). The slump flow was 72 cm for the SCC and fulfils therefore the criterion of 70 cm ± 5 cm. Furthermore, the viscosity of the SCC was evaluated with the V-Funnel Test (Fig. 5). The measured flow time was 14 sec (acceptance criterion flow time between 10 and 20 sec).

5.2 Fresh concrete properties

The compressive strength is the main parameter used in the design of concrete structures. It is determined

Table 2. Hardened concrete properties after 28 days.

	SCC
Cylinder compressive strength $f_{c,cyl}$ [N/mm²]	52
Cube compressive strength $f_{c,cube}$ [N/mm²]	55
Splitting tensile strength $f_{ct,sp}$ [N/mm²]	3.7
Modulus of elasticity E_c [N/mm²]	29,600

mainly either on cylinders (Ø 150/300 mm) or on cubes (150 ×150 ×150 mm). For normal vibrated concrete, the cylinder compressive strength is about 85% of the cube compressive strength. This proportion is embodied in standards like Model Code 90 (Structural Concrete 1999) in the term "compressive strength class", e.g. C 50/60. With SCC this relationship is smeared with an increasing fine material content to 0.9–1.0 (Table 2). This means, the compressive strength is less related to the slenderness of the specimens. Furthermore due to the improved interfacial transition zone the tensile strength is higher compared with normal vibrated concrete of the same compressive strength. The higher content of ultrafine materials plays in this context also an important role. The modulus of elasticity is about in the same range than for NC. Table 2 shows the measured hardened concrete properties as a mean out of 3 values, whereas all specimens were cured under water until the test.

5.3 Pull-out tests – monotonic loading

In the first test series all of the pull-out specimens failed by pulling out of the rebar. Therefore it was also possible to determine the bond behaviour after reaching the maximum bond stress, namely the descending branch. In the beginning of the bond stress–slip relationship the curve of SCC shows a stiff bond, which has a positive influence on the development of crack widths or tension stiffening. The post-peak bond performance of ripped rebars in SCC is characterised by a very ductile behaviour. After a distinctive plateau at maximum bond stress the descending branch of the bond stress–slip relationship has a low decrease, the bond stresses reduce very low and the slip increases continuously (Fig. 6). Even the rebar orientation has no influence on the bond behaviour in contrast to results on normal vibrated concrete out of the literature.

5.4 Pull-out tests – cyclic loading

The results of these tests can be presented in Wöhler-Diagrams indicating the relationship between the slip s versus the number of load cycles n. Figure 7 shows this relationship for the performed tests (only one specimen failed before 1 000 000 load cycles due to splitting of the concrete cover). The results displayed in this figure show that after about 250 000 load

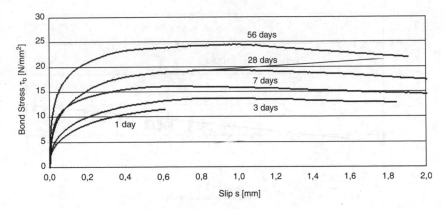

Figure 6.   Time development of bond stress (vertical rebar position during casting).

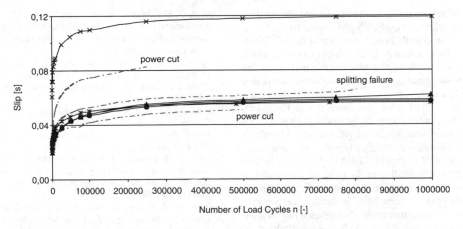

Figure 7.   Slip increase during cyclic loading as a function of load reversals.

cycles almost no further slip increase takes place. Furthermore the scatter of the different curves is quite low, with one exception of the top curve. This also confirms the more homogeneous structure of the SCC concrete matrix, that is nearly independent from the casting direction.

5.5 Residual bond strength after 106 load cycles

The specimens, that did not fail during cyclic loading (series 2), were finally tested monotonic. Tests on NC performed by Rehm & Eligehausen (1977) could be verified for SCC that previous cyclic loading does not negatively influence the bond stress versus slip behaviour near ultimate load compared with the monotonic loading behaviour. The maximum bond stress τ_{bu} was about 25% higher than of specimens without cyclic loading. Furthermore a steeper increase of the bond stress was observed due to the anticipation of the initial slip.

6   BOND CREEP COEFFICIENT UNDER REPEATED LOADING

The measured relationships between the slip and the number of load cycles are shown in a double-logarithmic scale (Fig. 8). It is evident from this image, that there is a nearly linear connection between lg s and lg n. Through the application of regression analysis the bond creep coefficient k_n could be determined. For a bond length of 5 d_s and loading levels below the fatigue strength of the bond, the bond creep coefficient of the tested SCC is:

$$k_n = (1+n)^{0,087} - 1 \qquad (4)$$

7   CONCLUSIONS AND OUTLOOK

It could be shown with the performed test series, that SCC has a high bond stiffness for the design relevant

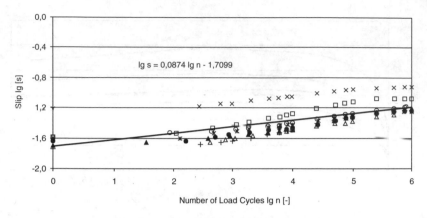

$$\lg s = 0{,}0874 \lg n - 1{,}7099$$

Figure 8. Slip-number of load cycles-relationship in a double-logarithmic scale.

area of the serviceability limit state. In the ultimate limit state, the relative bond strength seems to be lower. Positive is the ductile bond behaviour of the SCC characterised by the nearly linear sloping branch which means after reaching the maximum bond stress the slip increases gradually but the bond stress decreases extremely low.

For the cyclic loaded specimens it was found out, that the bond creep due to cyclic loading is less distinctive compared with results of NC out of the literature (Rehm & Eligehausen 1977). Therefore, concrete structures made of SCC should show a slightly better behaviour under cyclic loading in the ultimate limit states as well as in the serviceability limit states. From the bond properties point of view SCC is suitable for concrete members and structures under dynamic loading so there is no problem to use SCC e.g. in bridge constructions or towers.

Again it has be taken account of the fact that the tested SCC-mix does not represent the whole spectrum of self-compacting concretes. Furthermore, the reached data are restricted to one specimen geometry and at the moment one loading level. On the other hand the complex investigation of the bond behaviour under repeated loading is very time and cost intensive, so it is unlikely to expect short-dated results with all possible parameters varied. At present it can be recommended both for NC and for SCC to use the same bond creep coefficient (Equation 1) after Model Code 90 (Structural Concrete 1999) within the design.

REFERENCES

Skarendahl, Å. & Petersson, Ö. 2000. Self-Compacting Concrete. *RILEM Report 23*.

Martin, H. 1982. Bond Performance of Ribbed Bars (Pull-out-Tests) – Influence of Concrete Composition and Consistency. *Proceedings of the International Conference on Bond in Concrete*, Paisley, Scotland 1982, pp. 289–299.

Khayat, K.H., Manai, K. & Trudel, A. 1997. In Situ Mechanical Properties of Wall Elements Cast Using Self-Consolidating Concrete. *ACI Materials Journal*, Nov–Dec. 1997, pp. 491–500.

Khayat, K.H. 1998. Use of Viscosity-Modifying Admixture to Reduce Top-Bar Effect of Anchored Bars Cast with Fluid Concrete. *ACI Materials Journal*, March–April 1998, pp. 158–167.

Schießl, A. & Zilch, K. 2001. The Effects of the Modified Composition of SCC on Shear and Bond Behaviour. *Proceedings of the Second International Symposium on Self-Compacting Concrete*, Tokyo 2001, pp. 501–506.

Sonebi, M. & Bartos, P.J.M. 1999. Hardened SCC and its bond with reinforcement. *Proceedings of the First International RILEM Symposium on Self-Compacting Concrete*, Stockholm 1999, pp. 275–289.

Lorrain, M. & Daoud, A. 2002. Bond in Self-Compacting Concrete. *Proceedings of the 3rd International Symposium on Bond in Concrete – from research to standards*. Budapest, 2002, pp. 529–536.

RILEM. 1970. Technical Recommendations for the Testing and Use of Construction Materials: RC 6, Bond Test for Reinforcement Steel. 2. Pull-out Test, 1970.

DIN 488, Part 2. 1984. Betonstahl; Betonstabstahl; Maße und Gewichte. (Reinforcing steel; reinforcing bars; dimensions and masses.) (in German).

EN 12350. 2000. Part 2: Prüfung von Frischbeton, Setzmaß. (Testing fresh concrete, Slump test.) (in German).

Structural Concrete. 1999. Textbook on Behaviour, Design and Performance. Updated knowledge of the CEB/FIP Model Code 1990. Volume 1: Introduction – Design process – Materials. *fib-Bulletin No. 1*, July 1999.

Eurocode 2. 1992. Design of concrete structures. 1992.

Rehm, G. & Eligehausen, R. 1977. Influence of a non-monotonic loading on the bond behaviour of ribbed rebars. *Betonwerk + Fertigteil-Technik 1977*, No. 6, pp. 295–299 (in German).

Perry, E.S. & Jundi, N. 1969. Pullout bond stress distribution under static and dynamic repeated loadings. *ACI-Journal*, May 1969, pp. 377–380.

Koch, R. & Balázs, G. 1998. Bond under non-static loading. *Beton- und Stahlbetonbau 93*, 1998, No. 7, pp. 220–223 (in German).

Self-compacting concrete – hardened material properties and structural behaviour

K. Holschemacher, Y. Klug & D. Weiße
Leipzig University of Applied Sciences (HTWK Leipzig), Germany

ABSTRACT: Self-compacting concrete (SCC) is an innovative construction material with a favourable rheological behaviour, which is caused in the modified concrete composition. Based on this fact SCC offers improved properties and therefore many advantages regarding the productivity and the design potential compared with normal vibrated concrete. Consequently, the amount of SCC, used for structural purposes has strongly increased. In this context it has to be clarified, if it is possible to apply the current design rules, e.g. Model Code 90 and Eurocode 2, based on years of experience on normal vibrated concrete, to structural members made of SCC as well. This paper represents the analysis of own and internationally published test results of the compressive strength, tensile strength, modulus of elasticity, bond behaviour as well as the time-dependent deformations of SCC in comparison with conventional concrete, in order to give a general statement regarding the agreements and differences between the hardened material properties of these concretes.

1 INTRODUCTION

SCC, primarily invented in Japan in the late 1980's, has developed more intensive only in the last decade. In this time the application of SCC has increased and many investigations all over the world were carried out to find optimal and economical mix compositions which guarantee the typical fresh concrete behaviour of SCC. Meanwhile there are various concepts for the production of SCC-mixes, which vary mainly in the amount and kind of used additives and admixtures.

Due to the optimised combination of the individual components SCC is capable to compact itself only under its own weight without the requirement of internal or external vibration energy and deaerates itself almost completely while flowing in the formwork. Furthermore, SCC is able to fill all recesses and reinforcement spaces, even in high reinforced concrete members and flows free of segregation near to level balance. These specific material properties were achieved by the excellent coordination of deformability and segregation resistance.

Based on these properties SCC may contribute to a significant improvement of the quality of concrete structures and opens up new fields for the application of concrete, Skarendahl & Petersson (2000).

The designation "self-compacting" is based on the fresh concrete properties of this material and therefore the degree of compactability, the deformability and the viscosity in connection with different mix compositions were investigated very frequently. However, it is also to verify, to what extent the modifications of the mix composition of SCC effect the hardened concrete properties as well as the durability. This fact formed the basis of the creation of a data base with currently known data of own and internationally published test results of hardened concrete properties of SCC. Thus the data base represents a first step in the analysis and generalisation of the numerous investigations of the individual researchers.

2 SCOPE OF THE INVESTIGATION

2.1 Initial situation

A good starting point to discuss the hardened material properties of self-compacting concrete is the mix composition of this material.

Independent of the fact that SCC consists basically of the same components as normal vibrated concrete, there exist clear differences regarding the concrete composition in order to achieve the desired "self-compacting properties". On the one hand SCC has to reach a high segregation resistance and on the other hand a high deformability. Therefore the content of

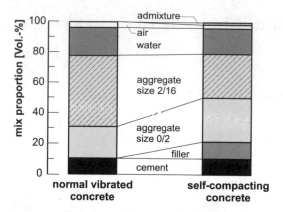

Figure 1. Comparison of mix composition of SCC and normal vibrated concrete.

ultrafine materials at SCC is essentially higher. For this purpose various fillers, e.g. limestone powder, fly ash, blast furnace slag, quartzite powder and silica fume, are given to the mixture or the content of cement will be increased. Furthermore a larger quantity of superplasticiser has to be added and stabilizer is used, if required. Figure 1 shows a mix composition of SCC compared with normal vibrated concrete in principle.

On the basis of the stated differences between the mix composition of SCC and conventional concrete it is necessary to analyse the effects of these modifications on the hardened concrete properties. So, referring to this, the applicability of the currently existing design rules based on years of experience on normal vibrated concrete has to be examined carefully.

Reasons for possible differences between the hardened properties of SCC and conventional concrete may be the modified mix composition as mentioned before, the better microstructure and homogeneity and the absence of vibration.

1 *Modified mix composition*: The higher content of ultrafine materials and the accordingly lower content of coarse aggregates change the granular skeleton. This could influence the strength as well as the modulus of elasticity, since the modulus of elasticity of the several components defines that one of the concrete. Moreover, the addition of fillers leads to a different reaction with water and therefore in connection with the use of superplasticiser the water demand changes.
2 *Better microstructure and homogeneity*: Due to the increased content of ultrafines (cement, filler) at SCC the grain-size distribution and packing density will be improved and therefore the bulk of the cement paste is stable, coherent and flowable and the porosity of the interfacial transition zone (ITZ) between aggregate and cement paste is decreased.

Consequently the tensile strength of SCC could be increased compared with conventional concrete, because of the fact that the transfer of tensile loads is supplied by the adhesion of the cement matrix or the bond within the ITZ, by friction and by aggregate interlock between the crack flanks.
3 *Absence of vibration*: Regarding this fact on the one hand gross defects by vibration can't arise but on the other hand the self-deaeration while flowing in the formwork has to be realized surely in order to avoid new sources of error.

2.2 Aim of the investigation

The aim of our investigations was to compare the hardened material properties of SCC with that ones of normal vibrated concrete, ultimately to give a general estimation regarding the application of the current design codes or calculation methods respectively in case of the usage of SCC.

Thereupon a data base with results of own experimental investigations and a large number of internationally published data of design relevant hardened material properties of several self-compacting concretes was created, Holschemacher & Klug (2002). The data of these properties such as the compressive strength, the tensile strength, the modulus of elasticity, the bond strength and the time-dependent deformations were documented and analysed particular with regard to the given values and limits of the European design code "CEB-FIB Model Code".

Reports regarding the hardened properties of SCC are quite frequently in literature. However, the import of the test results into a data base is frequently problematic owing to the following facts:

– there are often insufficient statements concerning the exact mix compositions, curing conditions and dimensions of used specimens,
– there exists a wide spectrum of different mix compositions,
– the initial parameters of diverse investigations differ strongly from each other.

Systematic investigations with respect to the influence of individual components of the concrete mix, like type and quantity of the filler or superplasticiser have rarely been carried out.

Nevertheless, by the interpretation of the created data base it is possible to recognise the basic relations and dependencies of the hardened properties of self-compacting concrete and to compare them with the well-known rules, valid for normal concrete. For this purpose all utilised data of considered concrete properties are represented in suitable diagrams.

Based on these realities mentioned above the following sections try to demonstrate the present level of

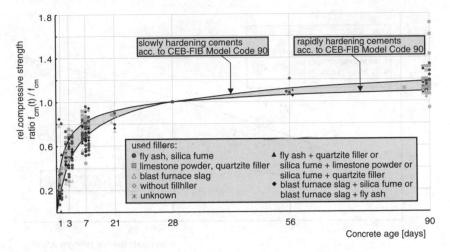

Figure 2. Development of compressive strength of SCC with time in comparison with the regulations of Model Code 90.

knowledge regarding the most important design-relevant hardened concrete characteristics of SCC.

3 PARAMETER STUDY OF THE HARDENED MATERIAL PROPERTIES OF SCC

3.1 *Compressive strength*

In general in national and international codes concrete is classified on the basis of its compressive strength, because the compressive strength is the most important mechanical property of concrete for the most applications. Since the compressive strength depends on the mechanical properties of the hardened cement paste and the adhesion within the ITZ, it is of interest whether the differences in the concrete composition and the positive changes in the microstructure, as mentioned before, affect the short and long term load-bearing behaviour. Moreover, clarification is still necessary to determine whether the hardening process and the ultimate strengths of SCC and conventional concrete differ.

Corresponding to the characteristic compressive strength f_{ck} of cylinders and cubes concrete is classified in concrete grades. As is known there exists a certain dependence on the specimen geometry with conventional concrete (Comité Euro-International du Béton 1991):

$$\frac{f_{c,cyl}\,(150/300)}{f_{c,cube}\,(150)} = 0.8\ldots0.85 \tag{1}$$

However, own tests carried out in Leipzig have shown, that this well-known relation between cylinders

and cubes could not be confirmed with SCC in the expected magnitude. A clearly lower dependency was ascertained:

$$\frac{f_{c,cyl}\,(150/300)}{f_{c,cube}\,(150)} = 0.9\ldots1.00 \tag{2}$$

This fact should be more investigated, especially regarding the classification of SCC.

Mostly the compressive strength of SCC and normal vibrated concrete at the age of 28 days of similar composition does not differ drastically, but in a few cases higher values were observed. Compared with the majority of the published test results the tendency becomes obvious that at the same water cement ratios higher compressive strengths were reached for SCC. This fact gets along with the decreasing water-binder-ratio corresponding to the rising amount of fillers. However, an explicit research program regarding this topic does not exist so far and therefore a generalised conclusion can not be drawn.

The strength development of SCC is subject to similar dependencies like conventional concrete in general (Fig. 2). Some of the published test results show that an increase of the cement content and a reduction of filler content at the same time increases the concrete strength. For young SCC aged up to 7 days the relative compressive strength spreads to a greater extend as given in the Model Code 90, whereas higher values as well as lower values are reached. Especially if limestone powder is used higher compressive strengths are noticeable at the beginning of the hardening process. At higher concrete ages SCC often exceeds the valid range according to the given limits by Model Code 90. By the use of fly

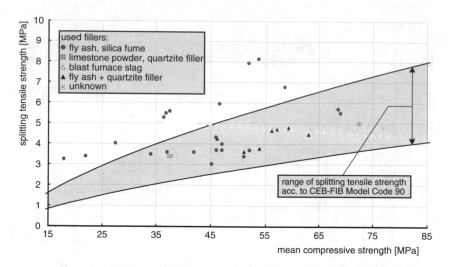

Figure 3. Data base of the splitting strength of SCC with reference to the corresponding compressive strength in comparison with the regulations of Model Code 90.

ash or silica fume this will be caused by the pozzolanic effect of these fillers.

3.2 Tensile strength

All parameters which influence the characteristics of the microstructure of the cement matrix and of the ITZ are of decisive importance in regard to the tensile load bearing behaviour. Figure 3 shows the aranged data of the splitting tensile strength of several self-compacting concretes.

By evaluating the created data base it can be shown, that the most of the measured values are within the valid range of current regulations for normal vibrated concrete. However, in about 30% of all data points clearly higher splitting tensile strengths were reached. Therefore, the tendency of higher tensile strengths of self-compacting concretes becomes obvious.

Presumable the reason for this fact is given by the better microstructure, especially due to the lower and more evenly distributed porosity within the interfacial transition zone with SCC, which is caused by the higher content of ultrafines. Further on the denser cement matrix of SCC enables a better load transfer.

The time development of the tensile strength of SCC and normal vibrated concrete is subject to a similar dependence. Only few publications about SCC refer to a more rapid increase of the tensile strength in comparison to the compressive strength.

3.3 Modulus of elasticity

As it is known, the modulus of elasticity of concrete depends on the proportion of the Young's modules of the individual components and their percentages by volume. Thus, the modulus of elasticity of concrete increases with high contents of aggregates of high rigidity, whereas it decreases with increasing hardened cement paste content and increasing porosity. For this reason lower values of modulus of elasticity can be expected, because of the higher content of ultrafines and additives as dominating factors and the accordingly lower content of coarse, stiff aggregates with SCC.

The evaluation of the data really shows the fact that the modulus of elasticity of SCC is within the lower half of the scattering range according to the Model Code 90. More exactly described, the average value valid for conventional concrete represents the upper limit for SCC, whereas all values were always referred to the mean compressive strength (Fig. 4).

3.4 Time-dependent deformations

Shrinkage and creep are very complex processes regarding the restructuring several components of the concrete structure caused by changes in the humidity balance. Furthermore these time-dependent investigations require time and high technological expenditure. Due to this fact only few data of the plastic shrinkage and the autogenous shrinkage of SCC as well as the time-dependent deformation behaviour under load are published in literature and very different conclusions about these material properties are stated. The drying shrinkage of SCC, however, is examined several times.

A general agreement exists to the fact that SCC is influenced in the same way by the water-cement ratio as well as the kind of specimen curing as normal vibrated concrete. The modified aggregate combination, especially the relation of coarse and fine aggregates as well

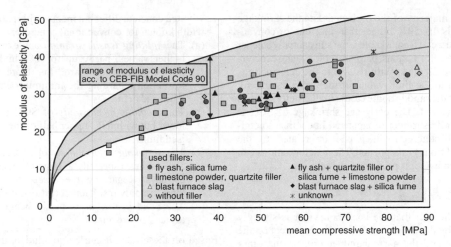

Figure 4. Data base of the modulus of elasticity of SCC with reference to the corresponding compressive strength in comparison with the regulations of Model Code 90.

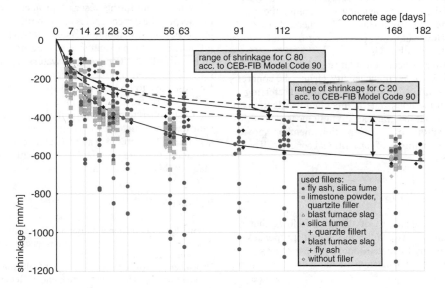

Figure 5. Data base of the shrinking deformations of SCC at different concrete ages in comparison with the regulations of Model Code 90.

as fineness (Blaine) and content of ultrafines seems to influence the shrinking deformations. Thus the shrinking deformations of SCC can increase due to the lower content of coarse aggregates and the minimum paste volume which must be present in order to ensure the optimal self-compaction of SCC without segregation. As a result the conclusion could be drawn that the shrinking deformations of SCC can achieve clearly higher values than that ones of comparable normal vibrated concretes. However, a denser microstructure of the cement paste can be achieved by addition of fillers with a fineness larger than that

of cement, whereby the shrinkage can be affected positively. Thus, it is possible to modify the SCC composition in such way that smaller shrinking deformations appear, similar to those of normal vibrated concrete.

Figure 5 shows the relationship between the shrinkage and the concrete age. The identified areas mark the limits for shrinking deformations of C20 up to C80 with a relative humidity of 60% and notional member size of 50 according to the Model Code 90, since all considered self-compacting concretes of the data base meet these conditions.

In the majority of the evaluated data the shrinkage of SCC is 10–50% higher than that one of conventional concrete. Remarkable is the substantially steeper rise of the deformations, particularly for young concrete aged up to 28 days. With rising age the deformations approach to the limit values of the current standard. Partly similar and in some cases smaller deformations were observed. The early age shrinkage of SCC is substantially stronger pronounced in contrast to conventional concrete, which can be related to the increased flour grain portion. These deformations can be limited by appropriate curing methods.

3.5 Other aspects

Due to the fact that the bond behaviour is strongly affected by the properties of the reinforcement on the one hand and the surrounding matrix on the other hand the bond strength of SCC is different from that one of conventional concrete. In own tests an improvement of the bond strength in the serviceability limit state and a ductile bond behaviour after reaching the maximum load was ascertained, Holschemacher et al. (2003). However, this fact is controversial, since opposed test results exist.

A positive effect of the high segregation resistance of SCC is the better homogeneity. Hence, the concrete strength could be more evenly distributed referring to the overall member. Indeed, in some investigations it was found that the concrete strength measured at different locations of a construction member spreads less than in members of conventional concrete, Khayat et al. (1997).

4 CONCLUSIONS

From the comparisons of the data follows that an exact identity between the mechanical properties of SCC and normal vibrated concrete does not exist. The results of the interpretation of the data base can be summarised as follows:

- The concrete strength of SCC and conventional concrete are similar under comparable conditions, whereas the tendency is obvious that SCC shows higher strengths with same water-cement ratios. The definite relation, however, has still to be clarified.
- The development of concrete strength with time is similar. Deviations are to recognise depending on the type of filler.
- The dependence of the compressive strength on the specimen geometry of SCC is only inarticulately pronounced compared with the well-known relation of conventional concrete. However, this fact is subject to review.
- Splitting tensile strength, modulus of elasticity, shrinkage and bond behaviour of SCC and normal

vibrated concrete differ but mostly within the scatter width known for conventional concrete:

(a) The *splitting tensile strength* achieves clearly higher values, partly up to 40% higher than in the current standard. I.e. there is need for action regarding the minimum amount of reinforcement.

(b) The *modulus of elasticity* of SCC is slightly lower but within the upper half of the standardised limits.

(c) The *shrinking deformations* of SCC are up to 50% higher, especially at concretes aged up to 28 days. Regarding the *creep deformations* there are only insufficient test results known.

(d) The *bond behaviour* is partly better but this topic is controversial.

Based on these facts it can be concluded, that extra design rules for SCC are not necessary, however, it should be take into consideration that restrictions can be added to the current standards.

In this regard further research projects are required to interpret the dependencies of the hardened material properties of SCC more precise. Referring to this the influence of any parameter, e.g. type of cement and filler as well as their portion, water-binder/-cement ratio, proportion of fine and coarse aggregates and fineness have to be investigated specifically.

To benefit from the advantages of SCC not only the fresh concrete properties but also the design relevant hardened material properties have to be known accurately. Only in this way SCC can be used convenient and a realistic calculation of structural members made of this innovative material is possible.

REFERENCES

Comité Euro-International du Béton 1991. CEB-FIB Model Code 1990. Design Code. Lausanne (Switzerland): Thomas Telford Services Ltd.

Deutscher Ausschuss für Stahlbeton 2002. Sachstandsbericht Selbstverdichtender Beton (SVB) – State of the art report "self compacting concrete". Berlin (Germany): Beuth Verlag.

Eurocode 2 2001. European Standard prEN 1992-1. Design of concrete structures. Part 1: General rules. 2nd draft.

Fédération internationale du béton (fib) 1999. Structural Concrete. Textbook on Behaviour, Design and Performance. Updated knowledge of the CEB/FIP Model Code 1990. fib Bulletin N°1: Introduction – Design process – Materials. Lausanne (Switzerland): fib.

Holschemacher, K. & Klug, Y. 2002. A data base for the evaluation of hardened properties of SCC. Leipzig Annual Civil Engineering Report No. 7. Leipzig (Germany): University of Leipzig. Institute for Structural Concrete and Building Materials.

Holschemacher, K., Klug, Y., Weiße, D., König, G & Dehn, F. 2003, Bond Behaviour of Reinforcement in Self-Compacting Concrete (SCC). *Proceedings of the 2nd*

International Structural Engineering and Construction Conference. Rome (Italy).

Skarendahl, A. & Petersson, Ö. 2000. Self-Compacting Concrete. *Proceedings of the First International RILEM Symposium.* RILEM Proceedings PRO 7. Cachan Cedex (France): RILEM Publications sarl.

Khayat, K.H., Manai, K. & Trudel, A. 1997. In Situ Mechanical Properties of Wall Elements Cast Using Self-Consolidating Concrete. *ACI Materials Journal* 96, No. 6: 491–500.

Kochi University of Technology. 1998. *Proceedings of the International Workshop on Self-Compacting Concrete.* Kochi (Japan): CD-Rom

König, G., Holschemacher, K., Dehn, F. & Weiße, D. 2001. Self-Compacting Concrete – Time Development of Material Properties and Bond Behaviour. *Proceedings of the Second International Symposium on Self-Compacting Concrete*: 507–516. Tokyo.

Shrinkage cracking and crack control in fully-restrained reinforced concrete members

R.I. Gilbert
School of Civil and Environmental Engineering, The University of New South Wales, Sydney, Australia

ABSTRACT: Cracking caused by shrinkage in restrained reinforced concrete members is considered both experimentally and analytically. A total of eight fully restrained slab specimens with different reinforcement layouts were monitored for up to 150 days to measure the effects of shrinkage on the time-dependent development of direct tension cracking. Strains in both the reinforcement and the concrete were monitored throughout the tests. The age of the concrete when each crack developed, the crack locations and the gradual change in crack widths with time were also recorded. An analytical model developed previously to study the problem (Gilbert 1992) is modified and recalibrated and the experimental results and analytical predictions are compared. The effects of varying the quantity of reinforcing steel, the bar diameter and the bar spacing are studied in order to gain a clearer understanding of the mechanism of direct tension cracking caused by restrained shrinkage and the factors affecting it.

1 INTRODUCTION

Shrinkage of concrete is the reduction in volume caused by the loss of water during the drying process (drying shrinkage) and also by chemical reactions in the hydrating cement paste (endogenous shrinkage). If shrinkage is unrestrained and the concrete is free to deform, shrinkage is of little consequence. However, in concrete structures, this is rarely the case. The bonded reinforcement in every reinforced concrete element provides restraint to shrinkage, with the concrete compressing the reinforcement as it shrinks and the reinforcement imposing an equal and opposite tensile force on the concrete at the level of the steel. This internal restraining force is often significant enough to cause cracking.

In addition, the connections of a concrete member to other parts of the structure or to the foundations also provide restraint to shrinkage. The tensile restraining force that develops rapidly with time at the restrained ends of the member usually leads to cracking, often within days of the commencement of drying. Thin floor slabs and walls in buildings are particularly prone to significant cracking resulting from restrained shrinkage and temperature changes. It is this problem that forms the focus of this paper.

Direct tension cracks due to restrained shrinkage and temperature changes frequently lead to serviceability problems, particularly in regions of low moment. Such cracks usually extend completely through the thickness of the restrained member and are more parallel-sided than flexural cracks. If uncontrolled, these cracks can become very wide and lead to waterproofing and corrosion problems and may even compromise the integrity of the member.

The onset of cracking reduces the stiffness of a reinforced concrete member and hence reduces the internal actions caused by imposed deformation. In the case of statically indeterminate members, it also redistributes the actions caused by external loads. The width of a crack depends on the quantity, orientation and distribution of the reinforcing steel crossing the crack. It also depends on the deformation characteristics of the concrete and the bond between the concrete and the reinforcement bars at and in the vicinity of each crack. A local breakdown in bond at and immediately adjacent to a crack complicates the modelling, as does the time-dependent change in the bond characteristics caused by drying shrinkage.

The study of cracking is complex. Cracks occur at discrete locations in a concrete member, often under the day-to-day service loads. Great variability exists in observed crack spacing and crack widths and accurate predictions of behaviour are possible only at the statistical level.

In this paper, the problem of cracking caused by shrinkage in restrained reinforced concrete members is considered, both experimentally and analytically. A total of eight fully restrained slab specimens with different reinforcement layouts were monitored for up to 150 days to measure the effects of shrinkage on the time-dependent development of direct tension cracking. Of particular interest are the final crack spacing and crack widths. An analytical model developed previously to study the problem (Gilbert 1992) has been modified and recalibrated and the experimental results and analytical predictions are compared.

2 ANALYTICAL MODEL FOR CRACKING IN RESTRAINED MEMBERS

Consider the fully restrained member shown in Figure 1a (Gilbert 1992). As the concrete shrinks, the restraining force $N(t)$ gradually increases until the first crack occurs when $N(t) = A_c f_t(t)$ (usually within a week of the commencement of shrinkage).

Immediately after first cracking, the restraining force reduces to N_{cr}, and the concrete stress away from the crack is less than the tensile strength of the concrete, $f_t(t)$. The concrete on either side of the crack shortens elastically and the crack opens to a width w, as shown in Figure 1b. At the crack, the steel carries the entire force N_{cr} and the stress in the concrete is obviously zero. In the region adjacent to the crack, the concrete and steel stresses vary considerably, and there exist a region of high bond stress. At the crack, a region of partial bond breakdown exists. At some distance s_o on each side of the crack, the concrete and steel stresses are no longer influenced directly by the presence of the crack, as shown in Figures 1c and 1d.

The distance s_o in which stresses vary on either side of a crack was taken to be (Gilbert 1992)

$$s_o = \frac{d_b}{10\rho} \quad (1)$$

where d_b is the bar diameter and ρ is the reinforcement ratio (A_s/A_c). This expression was earlier proposed by Favre et al. (1983) for a member containing deformed bars or welded wire mesh. Base & Murray (1982) used a similar expression.

Referring to Figure 1, Gilbert (1992) derived the following expressions for N_{cr}, σ_{c1}, σ_{s1}, and σ_{s2}:

$$N_{cr} = \frac{n\rho \, f_t A_c}{C_1 + n\rho(1 + C_1)} \quad (2)$$

$$\sigma_{c1} = \frac{N_{cr}(1 + C_1)}{A_c}; \quad \sigma_{s1} = -C_1 \sigma_{s2}; \quad \text{and} \quad \sigma_{s2} = \frac{N_{cr}}{A_s} \quad (3)$$

where n is the modular ratio (E_s/E_c); f_t is the direct tensile strength of the concrete at the time of first cracking; and

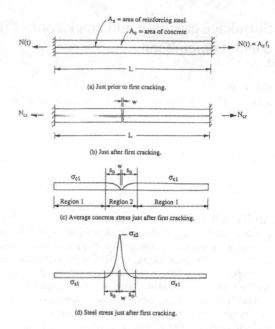

Figure 1. First cracking in a restrained direct tension member.

$$C_1 = \frac{2s_o}{3L - 2s_o} \quad (4)$$

The final number of cracks and the final average crack width depend on the length of the member, the quantity and distribution of reinforcement, the quality of bond between the concrete and steel, the amount of shrinkage, and the concrete properties.

In Figure 2a, a portion of a fully-restrained direct tension member is shown after all shrinkage has taken place and the final crack pattern is established. The average concrete and steel stresses caused by shrinkage are illustrated in Figures 2b and 2c.

By enforcing the requirements of compatibility and equilibrium, Gilbert (1992) derived expressions for the final average crack spacing (s) and width (w) in a fully-restrained member, the final restraining force in the member, and the final concrete and steel stresses (as defined in Figs. 2b and 2c). In these derivations, s_o was assumed to remain constant with time (and given by Equation 1) and the supports of the member were assumed to be immovable.

Recent experimental results (Nejadi & Gilbert 2003) suggest that shrinkage causes a deterioration in bond at the steel-concrete interface and a gradual increase in s_o with time. It is suggested here that for calculations at first cracking (see Fig. 1), s_o may be taken from Equation 1, and for final or long-term calculations (see Fig. 2), the value of s_o given by Equation 1 should be multiplied by 1.33.

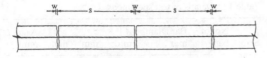

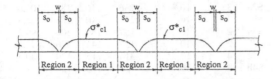

(a)Portion of a restrained member after all cracking.

(b) Average concrete stress after all shrinkage.

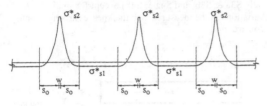

(c) Steel stress after all shrinkage cracking.

Figure 2. Final concrete and steel stresses after direct tension cracking.

In many practical situations, the supports of a reinforced concrete member, which provide the restraint to shrinkage, are not immovable, but are adjacent parts of the structure that are themselves prone to shrinkage and other movements.

If the supports of the restrained member (see Fig. 1a) suffer a relative movement Δu with time, such that the final length of the member is $(L + \Delta u)$, the final restraining force $N(\infty)$ changes and this affects both the crack spacing and the crack width. If Δu increases, $N(\infty)$ increases and so too does the number and width of the cracks. For a member containing m cracks and providing the reinforcing steel has not yielded, the following expressions are obtained by equating the overall elongation of the steel reinforcement to Δu:

$$\frac{\sigma_{s1}^*}{E_s}L + m\frac{\sigma_{s2}^* - \sigma_{s1}^*}{E_s}\left(\frac{2}{3}s_o + w\right) = \Delta u \qquad (5)$$

and, as w is much less than s_o, rearranging gives

$$\sigma_{s1}^* = \frac{-2\,s_o\,m}{3L - 2s_o m}\sigma_{s2}^* + \frac{3\,\Delta u\,E_s}{3L - 2s_o m} \qquad (6)$$

At each crack,

$$\sigma_{s2}^* = N(\infty)/A_s \qquad (7)$$

In Region 1, s_o away from each crack (with s_o taken as 1.33 times the value given by Equation 1), the concrete

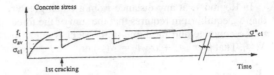

Figure 3. Concrete stress history in uncracked Region 1.

stress history is shown diagrammatically in Figure 3. The concrete tensile stress increases gradually with time and approaches the direct tensile strength of the concrete f_t. When cracking occurs elsewhere in the member, the tensile stress in the in the uncracked regions drops suddenly as shown. Although the concrete stress history is continuously changing, the average concrete stress at any time after the start of drying, σ_{av}, is somewhere between σ_{c1} and f_t, as shown in Figure 5, and the final creep strain in Region 1 may be approximated by

$$\varepsilon_c^* = \frac{\sigma_{av}}{E_c}\,\varphi^* \qquad (8)$$

where φ^* is the final creep coefficient (defined as the ratio of the final creep strain to elastic strain under the average sustained stress σ_{av}). Gilbert (1992) assumed that

$$\sigma_{av} = \frac{\sigma_{c1} + f_t}{2} \qquad (9)$$

The final concrete strain in Region 1 is the sum of the elastic, creep, and shrinkage components and may be approximated by

$$\varepsilon_1^* = \varepsilon_e + \varepsilon_c^* + \varepsilon_{sh}^* = \frac{\sigma_{av}}{E_c} + \frac{\sigma_{av}}{E_c}\varphi^* + \varepsilon_{sh}^* \qquad (10)$$

The magnitude of the final creep coefficient φ^* is usually between 2 and 4, depending on the age at the commencement of drying and the quality of the concrete. ε_{sh} is the final shrinkage strain and depends on the relative humidity, the size and shape of the member, and the characteristics of the concrete mix. Numerical estimates of φ^* and ε_{sh} can be obtained from Standards, such as Eurocode 2, AS3600-2001 and elsewhere. Equation 10 may be expressed as

$$\varepsilon_1^* = \frac{\sigma_{av}}{E_e^*} + \varepsilon_{sh}^* \qquad (11)$$

where E_e^* is final effective modulus for concrete and is given by

$$E_e^* = \frac{E_c}{1 + \phi^*} \qquad (12)$$

In Region 1, at any distance from a crack greater than s_o, equilibrium requires that the sum of the force in the concrete and the force in the steel is equal to $N(\infty)$. That is $\sigma_{c1}A_c + \sigma_{s1}A_s = N(\infty)$ or

$$\sigma_{c1}^{*} = \frac{N(\infty) - \sigma_{s1}^{*}A_s}{A_c} \tag{13}$$

The compatibility requirement is that the concrete and steel strains are identical ($\varepsilon_{s1} = \varepsilon_1$) and using Equation 11, this becomes

$$\frac{\sigma_{s1}^{*}}{E_s} = \frac{\sigma_{av}}{E_e^{*}} + \varepsilon_{sh}^{*} \tag{14}$$

Substituting Equations 6 and 7 into Equation 14 and rearranging gives

$$N(\infty) = \frac{3A_s E_s \Delta u}{2s_o m} - \frac{(3L - 2s_o m)n^{*}A_s}{2s_o m}(\sigma_{av} + \varepsilon_{sh}^{*}E_e^{*}) \tag{15}$$

With the restraining force $N(\infty)$ and the steel stress in Region 1 obtained from Equations 15 and 6, respectively, the final concrete stress in Region 1 is

$$\sigma_{c1}^{*} = \frac{N(\infty) - \sigma_{s1}^{*}A_s}{A_c} \tag{16}$$

The number of cracks m is the lowest integer value of m that satisfies Equation 17.

$$\sigma_{c1}^{*} \leq f_t \tag{17}$$

The direct tensile strength f_t in Equation 17 should be taken as the 28 day value. The final average crack spacing $s = L/m$.

The overall shortening of the concrete is an estimate of the sum of the crack widths. The final concrete strain at any point in Region 1 of Figure 2 is given by Equation 11, and in Region 2, the final concrete strain is

$$\varepsilon_2^{*} = \frac{fn\sigma_{c1}^{*}}{E_e^{*}} + \varepsilon_{sh}^{*} \tag{18}$$

where fn varies between zero at a crack and unity at s_o from a crack. If a parabolic variation of stress is assumed in Region 2, the following expression for the average crack width w is obtained by integrating the concrete strain over the length of the member:

$$w = -\left[\frac{\sigma_{c1}^{*}}{E_e^{*}}\left(s - \frac{2}{3}s_o\right) + \varepsilon_{sh}^{*}s\right] \tag{19}$$

Table 1. Details of test specimens.

Slab	No. of bars	Bar Diam. mm	A_s mm²	c_s mm	s_s mm	Avge, D mm
S1a	3	12	339	109	185	102.2
S1b	3	12	339	109	185	99.8
S2a	3	10	236	110	185	101.6
S2b	3	10	236	110	185	98.3
S3a	2	10	157	145	300	99.2
S3b	2	10	157	145	300	99.3
S4a	4	10	314	115	120	100.5
S4b	4	10	314	115	120	101.1

Note: Slabs S1a & S1b were identical (so too were S2a & S2b, S3a & S3b, and S4a & S4b) except for small variations in the measured slab thicknesses arising during construction.

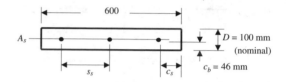

Figure 4. Cross section details of slabs.

The preceding analysis is valid, provided the assumption of linear-elastic behaviour in the steel is valid, i.e. provided the steel has not yielded.

3 EXPERIMENTAL PROGRAM

A total of 8 fully-restrained slab specimens with four different reinforcement layouts were monitored for up to 150 days to measure the effects of shrinkage on the time-dependent development of direct tension cracking due to restrained deformation. Steel strains in the vicinity of the first crack, steel strains along the full length of the reinforcement, concrete surface strains, crack widths and crack spacing were recorded throughout the period of testing. The compressive and tensile strengths of the concrete at various times were measured on companion specimens (concrete cylinders and prisms), together with the elastic modulus and the creep coefficient. In addition, the magnitude and rate of development of free shrinkage in the concrete were measured on unrestrained and unreinforced companion specimens with the same cross-section as the test slabs.

Details of the parameters varied in the tests are given in Table 1. Each specimen consisted of a 2000 mm long prismatic portion (with section dimensions shown in Figure 4), which was effectively anchored at its ends by casting each end monolithically with a 1 m × 1 m × 0.6 m concrete block. The block at each end of the specimen was rigidly clamped to the reaction floor

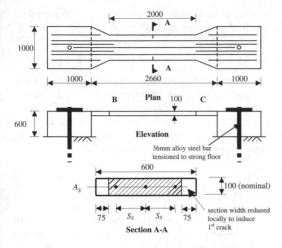

Figure 5. Dimensions and reinforcement details (typical).

Table 2. Final measured elongation Δu (in mm).

	S1a	S1b	S2a	S2b	S3a	S3b	S4a	S4b
Δu	0.305	0.383	0.309	0.315	0.402	0.419	0.245	0.162

After 150 days, the finest final crack width ($w = 0.18$ mm) and the lowest final steel stress ($\sigma_{s2}^* = 190$ MPa) measured during the experimental program were in slab S1b (the slab containing the most longitudinal reinforcement). The widest final crack ($w = 0.84$ mm) and highest final steel stress ($\sigma_{s2}^* = 532$ MPa) were measured in slab S3a (the slab containing the least longitudinal reinforcement). The thickness of each slab was measured at several points across the slab's width in the region of first cracking to determine the average value (see Table 1) and to account for any small variations between specimens arising during construction.

Using the analytical model discussed in Section 2, the final average crack width (w) and crack spacing (s), the final steel stress at the crack (σ_{s2}^*), and the final steel and concrete stresses away from the crack (σ_{s1}^* and σ_{c1}^*) were calculated for each slab, together with the values immediately after first cracking. These quantities have also been determined from the test data and comparisons are made below between theoretical and experimental results. The measured final elongations Δu (see Table 2) were used in the calculations.

4.1 Material properties

Two different concrete batches were used, namely concrete batch I (for slabs S1a, S1b, S2a, S3a, S3b, S4a and S4b) and concrete batch II (for slab S2b). Measured material properties are in Tables 3 and 4.

using 36 mm diameter high strength alloy steel rods tensioned sufficiently to ensure that the ends of the specimen were effectively held in position via friction between the end block and the laboratory floor (see Fig. 5).

A plan view of a typical slab is shown in Figure 5, together with the typical reinforcement layout. The 600 mm wide by 100 mm deep slab specimens were gradually splayed at each end, as shown, to ensure cracking occurred within the specimen length and not at the restrained ends. At the mid-length of each specimen the section was locally reduced in width by 150 mm using thin 75 mm wide plates attached to the side forms, so that first cracking always occurred at this location. The bottom surface of each specimen was supported on smooth supports to ensure negligible bending in the specimen and no longitudinal restraint, except at the specimens' ends.

4 TEST RESULTS

Each slab was moist cured for 3 days after casting and then drying began. All slabs were anchored to the laboratory floor at age 3 days. First cracking, due to direct longitudinal tension caused by restraint to shrinkage, occurred within the first week and subsequent cracks developed over the next 60 days. Crack widths gradually increased with time, with the change being relatively small after about 90 days.

Due to shrinkage of the 1 m × 1 m × 0.6 m anchorage blocks at each end of the specimens, the ends of each 2000 mm long slab specimen (B & C in Fig. 5) suffered a relative longitudinal movement Δu. This elongation was measured throughout each test and the final values are given in Table 2.

4.2 Crack width

For all slabs, first cracking occurred within the first week, except for slab S1b, where first cracking occurred at age 10 days. The development, extent and width of cracks were observed and measured using a microscope. The magnitude of the crack width depends primarily on the amount of bonded reinforcement crossing the crack. In addition, the width of a crack in a restrained member depends on the degree of restraint, the quality of bond between concrete and steel, the size and distribution of the reinforcement bars and the concrete quality. Table 5 provides a comparison between the calculated final crack width (obtained using the analytical model described in Section 2) and average final crack width measured during the laboratory experiments.

4.3 Crack spacing

The distance between cracks was measured every 50 mm across the slab width and averaged to obtain the average crack spacing for each slab (s in Fig. 2a). In Table 6, the observed average crack spacing is compared with the predicted average crack spacing.

In general, an increase in the steel area causes a reduction in the average crack spacing.

4.4 Steel stress

At each crack, the steel carries the full restraining force (N) and the stress in the concrete is zero. Using the measured final steel strains in the vicinity of the first crack, the maximum stress in each steel bar crossing the crack was determined (σ_{s2}^* in Fig. 2). The restraining force was thus determined from the experiments

Table 3. Material properties for concrete batches I and II.

Age (Days)	3	7	14	21	28
Batch I:					
Compressive strength*	8.17	13.7	20.7	22.9	24.3
Flexural tensile strength*	1.91	3.15	3.43	3.77	3.98
Indirect tensile strength*		1.55			1.97
Modulus of elasticity*	13240	17130	21080	22150	22810
Batch II:					
Compressive strength*	10.7	17.4	25.0	27.5	28.4
Flexural tensile strength*	2.47	3.10	3.77	3.97	4.04
Indirect tensile strength*		1.60			2.1
Modulus of elasticity*	16130	18940	21750	22840	23210

*All data are in MPa.

Table 6. Average crack spacing s (in mm).

	S1a	S1b	S2a	S2b	S3a	S3b	S4a	S4b
Theoretical	667	500	1000	667	–	–	667	667
Experimental	670	403	674	700	–	997	783	995

– Single crack only.

Table 7. Theoretical and experimental values for σ_{s2}^* (in MPa).

	S1a	S1b	S2a	S2b	S3a	S3b	S4a	S4b
Theoretical	246	196	337	235	553	562	261	227
Experimental	273	190	250	290	532	467	270	276

Table 4. Creep coefficient and shrinkage strain for concrete batches I and II.

Age (days)	7	14	21	28	36	43	53	77	100	122
Batch I:										
φ_{cc}	.38	.60	.68	.69	.73	.84	.86	.93	.97	.98
ε_{sh} ($\times 10^{-6}$)*	66	115	154	208	244	313	327	342	421	457
Batch II:										
φ_{cc}	.40	.60	.82	.87	.85	1.04	1.00	1.15	1.07	1.16
ε_{sh} ($\times 10^{-6}$)*	72	183	277	258	331	381	457	463	469	495

*in microstrain.

Table 5. Final average crack widths (in mm).

	S1a	S1b	S2a	S2b	S3a	S3b	S4a	S4b
Theoretical	0.19	0.16	0.28	0.26	0.51	0.50	0.18	0.20
Experimental	0.21	0.18	0.30	0.31	0.84	0.50	0.23	0.25
No. of cracks*	4	4	3	3	1	2	3	3

*Observed during test.

Table 8. Theoretical and experimental values for σ_{c1}^{*} (in MPa).

	S1a	S1b	S2a	S2b	S3a	S3b	S4a	S4b
Theoretical	1.68	1.43	1.54	1.22	1.63	1.65	1.66	1.47
Experimental	1.77	1.41	1.13	1.46	1.45	1.31	1.64	1.71

$(N = \sigma_{s2}^{*} A_s)$. Theoretical and experimental results for the final steel stress at each crack (σ_{s2}^{*}) are compared in Table 7.

4.5 Concrete stress

The concrete and steel stresses vary with distance from the crack within Region 2 in Figure 2. At some distance $(>s_o$ in Fig. 2) on each side of the crack at any time instant, the concrete and steel stresses $(\sigma_{c1}$ and σ_{s1}, respectively) remain constant until the next crack is approached (Region 1 in Fig. 2). From the measured steel strain in Region 1 and obtaining the average value along the bar, the steel stress σ_{s1} was calculated for each test. By enforcing equilibrium $(N = \sigma_{c1} A_c + \sigma_{s1} A_s)$, the concrete stress away from the crack σ_{c1} was determined. Comparisons between theoretical and experimental results for the final concrete stress (σ_{c1}^{*}) are made in Table 8.

5 DISCUSSION OF RESULTS

From the results presented in Section 4, the final crack width, the crack spacing and the steel stress at the crack are dependent on the steel area (or more precisely, the reinforcement ratio $\rho = A_s/A_c$). An increase in the steel area (or ρ) causes a reduction in the final crack width and, with more cracks developing, a reduction in the crack spacing.

With an increase in the steel area, the loss of stiffness at first cracking reduces and, hence the restraining force after cracking is greater, but the steel stress at each crack decreases (see Table 7). With a larger restraining force, the concrete stress in regions remote from a crack tends to be higher and hence further cracking is more likely.

Comparing the behaviour after 150 days of drying for slabs S3b $(A_s = 157\,mm^2, \rho = 0.00262)$ and S1a $(A_s = 339\,mm^2, \rho = 0.00565)$, for example, the final measured average crack widths are 0.50 mm and 0.21 mm, respectively. The measured average crack spacings are 997 mm and 670 mm, respectively; the final measured steel stress at the first crack are 467 MPa and 273 MPa, respectively; and the concrete stress away from each crack is 1.31 MPa and 1.77 MPa, respectively.

For a maximum design crack width of 0.3 mm (as is commonly specified in codes of practice), it appears that for the restrained slabs tested in this study a reinforcement area of greater than about 270 mm^2 $(\rho = 0.0045)$ would be satisfactory.

6 CONCLUDING REMARKS

A simple and rational analytical procedure for the determination of the stresses and deformation after cracking in a restrained direct tension member is discussed and long-term tests on eight restrained reinforced concrete slab specimens have been described. The measured width and spacing of shrinkage cracks, steel stresses and concrete stresses, agree well with the results of the analytical model.

ACKNOWLEDGEMENT

The funding for this research was provided by the Australian Research Council (ARC). The author would like to thank Mr Shami Nejadi for assistance with the experimental work.

REFERENCES

AS3600-2001. "Australian Standard for Concrete Structures", Standards Australia, Sydney.
Base, G.D. & Murray, M.H. 1982. "New Look at Shrinkage Cracking", Civil Engineering Transactions, Institution of Engineers Australia, V. CE24 No. 2, 171 pp.
Base, G.D. & Murray, M.H. 1978. "Controlling Shrinkage Cracking in Restrained Reinforced Concrete", Proc. Aust. Road Research Board, Vol. 9 Part 4.
Faver, R. et al. 1983. "Fissuration et Deformations". Manual du Comite Euro-International du Beton (CEB), Ecole Polytechnique Federale de Lausanne, Switzerland: 249 pp.
Gilbert, R.I. 1992. "Shrinkage Cracking in Fully Restrained Concrete Members", ACI Structural Journal, Vol. 89, No. 2: pp. 141–149.
Gilbert, R.I. 1988. "Time Effects in Concrete Structures", Elsevier Science Publishers: 321 pp.
Nejadi, S. & Gilbert R.I. 2003. "Shrinkage Cracking in Fully-Restrained Reinforced Concrete Members", UNICIV Report. School of Civil and Environmental Engineering, The University of New South Wales, Sydney, Australia.

Properties of a self-compacting mortar with lightweight aggregate

L. Bertolini, M. Carsana & M. Gastaldi
Politecnico di Milano, Dipartimento di Chimica, Materiali e Ingegneria Chimica, Milan, Italy

ABSTRACT: The paper shows the results of a work aimed at developing a lightweight mortar with self-compacting properties in the fresh state and with low permeability after hardening. The use of expanded clay aggregates reduced the density of the mortar to about $1600\,kg/m^3$, while the addition of a high content of pulverized coal fly ash provided the high content of fines required for self-compactability and allowed the development of a dense and impermeable cement paste. Consequently the repair material also showed a high resistance to penetration of chlorides and sulphates.

1 INTRODUCTION

Because of the ageing of reinforced concrete structures and infrastructures, as well as the low quality of concrete made in the past, the rehabilitation of reinforced concrete structures is often required. Corrosion of embedded steel is the most frequent cause of degradation (Bertolini et al. 2003a). A repair work should be able to remedy to the present damage and to prevent future degradation of the structure, which continues to be subjected to the action of the aggressive environment. Although several repair techniques have been developed for the rehabilitation of structures damaged by steel corrosion, the conventional repair is the most frequently applied method (Rilem 1993; ENV 1504-9 1997). When this technique is properly applied, it is aimed at restoring protection to the reinforcement by replacing the non-protective concrete with a cementitious material. The durability of the repair work is due to the achievement and maintenance of passivity on the reinforcement by the contact with the protective repair material. A suitable repair material should be selected and a proper cover thickness should be designed to guarantee that, after the repair, carbonation or chloride penetration from the environment will not damage the structure within the required residual life.

The repair material, to be able to protect the reinforcement for a long time, should comply with several requirements (Cambell-Hallen & Ropr 1991, Mailvaganam 1992). It should be a cement based mortar or concrete (i.e., of cementitiuos nature), resistant to carbonation, to chloride penetration (if the structure is exposed to chloride environments) and to other types of attack that could occur in the specific environment (sulphate, freeze-thaw, etc.). As far as the mechanical properties are concerned, the compressive strength of repair materials is often high, even if this is not required for structural reasons, because it is a consequence of the low water/cement ratio required for durability reasons. A low modulus of elasticity could reduce the stresses induced by deformations due to moisture or temperature changes. The repair materials should have a good bond to the concrete sustrate.

The properties required of the repair material in the fresh state depend on the thickness of the mortar layer and the procedure used for the application. If the thickness is high (e.g. higher than 50 mm) formworks may be used and the repair material should have a flowable consistency and should be able to fill the space inside the form without segregating.

Self-compacting concrete (SCC) is a recent development in the concrete technology (Okamura & Ouchi 1999, Umoto & Ozawa 1999). It is able to fill the formwork simply by means of its own weight and it does not require any vibration, even in the presence of a complex geometry of the structure or the reinforcing bars. The fluidity of SCC is so high that the usual tests for measuring workability cannot be used, and specific tests have been developed (Collepardi 2002). Mixes are designed to provide an extremely high flowability and self-levelling properties, to prevent blockage of coarse aggregate between the reinforcing bars and, in the meantime, to avoid segregation. A great amount of paste phase is required to provide a combination of high fluidity and good cohesion to the fresh concrete. High dosage of superplasticizer is combined to large quantity (usually more than

$500 \, \mathrm{kg/m^3}$) of fine particles (<100–$150 \, \mu\mathrm{m}$). These include cement and addition of pozzolanic materials (fly ash or silica fume), ground granulated blastfurnace slag, or ground limestone. Specific admixtures (viscosity modifying admixtures, VMA) have also been developed to further improve the cohesion of the mix.

Self-compacting properties could be also useful for the cementitious materials used in the repair or strengthening of existing structures (Bertolini et al. 2003b). A SCC repair mortar would be easy to cast even in the presence of a complex geometry or when thin layers have to be cast, e.g. when only the concrete cover has to be replaced. It would also be useful to reduce the density of the repair material and thus limit the increase in weight of the repaired structure.

This paper reports the results of a research aimed at developing a repair mortar with a self-compacting behaviour made with lightweight expanded clay aggregates. The protection that such a material can provide to the embedded steel was also investigated. For comparison, an ordinary portland cement concrete and a normalweight self-compacting concrete were also studied.

2 EXPERIMENTAL PROCEDURE

A lightweight mortar with self-compacting properties (LW-SCC) was designed during the initial phase of the work, by using lightweight expanded clay aggregates. For comparison purpose, a self-compacting concrete with normal weight (NW-SCC) and an ordinary concrete with class of consistence S4 (NW-S4) according to European standards (EN 206-1 2002) were also cast.

The properties of LW-SCC and NW-SCC were studied in the fresh state by means of different tests normally used to characterize self-compacting concrete: the slump flow test (that measures the diameter of the concrete spread after the removal of the slump cone), the V-funnel test (that measures the time required to empty a V-shaped funnel), the U-box test (that measures the difference in height between two chambers of a U-shaped apparatus where concrete flows from one chamber to the other one, separated by steel bars), the L-box test (that measures the length reached by concrete after it passes through some steel bars) and the fill-box test (that measures the degree by which concrete can fill a box with congested bars).

The hardened materials were tested to measure the compressive strength, the dynamic elasticmodulus, the water absorption, the capillary suction, the electrical resistivity and the resistance to penetration of chloride ions. Resistance to chloride penetration was tested on cubes exposed to one-day wetting with 3.5% NaCl solution alternated with two-days drying.

Chloride penetration was allowed only from one face and cores were taken after different times of exposure to measure the total chloride content profiles. Cylindrical specimens with a reinforcing bar were also exposed to the wetting-drying cycles; corrosion potential and corrosion rate of steel were monitored. Resistance to sulphate attack was also tested by measuring the linear expansion of slender prism specimens which were immersed in 5% $\mathrm{Na_2SO_4}$ solution at 23°C.

Details of the experimental procedure have been described elsewhere (Bertolini et al. 2003b).

3 RESULTS

3.1 Mix proportions

Table 1 shows the mix proportions of the tested materials. In the lightweight mortar (LW-SCC), the high dosage of fine particles required to achieve the self-compacting properties was obtained by adding $300 \, \mathrm{kg/m^3}$ of portland-limestone cement CEMI I/A-L 42.5R and $300 \, \mathrm{kg/m^3}$ of pulverized coal fly ash (PFA). The water content was $233 \, \mathrm{kg/m^3}$ and a superplasticizer associated with a viscosity-modifying agent were also used. The water/cement ratio was 0.78, while the water/(cement + fly ash) ratio was 0.39.

The density of the material was decreased by using $283 \, \mathrm{kg/m^3}$ of expanded clay particles with maximum size of 3 mm, combined with $591 \, \mathrm{kg/m^3}$ of river sand (maximum size 3 mm). The expanded clay had bulk density of $535 \, \mathrm{kg/m^3}$ and apparent density of $900 \, \mathrm{kg/m^3}$. By assuming an absolute density of $2600 \, \mathrm{kg/m^3}$, a porosity of about 65% could be calculated. Water absorption tests, however, showed that only about 11% of the aggregate volume is occupied by accessible pores and can be filled by water. The lightweight mortar had a density of $1645 \, \mathrm{kg/m^3}$ in the water-saturated conditions and $1560 \, \mathrm{kg/m^3}$ after drying.

For comparison, a normalweight self-compacting concrete (NW-SCC) was also cast, with a water/(cement + fly ash) ratio similar to that of LW-SCC (Table 1). The use of crushed coarse aggregates with

Table 1. Mix proportions of the tested materials ($\mathrm{kg/m^3}$).

	LW-SCC	NW-SCC	NW-S4
CEMI I A/L 42,5	300	190	320
Fly ash	300	250	–
Sand	591	940	1140
Expanded clay (1–3 mm)	283	–	–
Coarse aggregate	–	770	670
Water	233	176	208
Superplasticiser	11.4	6.6	1.6
VMA	3	2.2	–

Table 2. Results of the workability tests on the fresh materials.

Type of test	LW-SCC	NW-SCC
Slump-flow:		
– spread diameter (cm)	75	66
V-funnel:		
– flowing time (s)	9.5	9
U-box:		
– difference in height (cm)	0.5	3.5
L-box:		
– maximum length (cm)	150	119
– time (s)	31	39
Fill-box:		
– degree of filling (%)	91%	88%

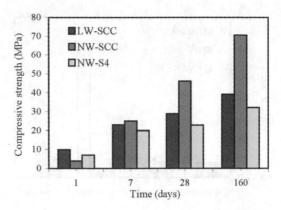

Figure 1. Average values of compressive strength in time.

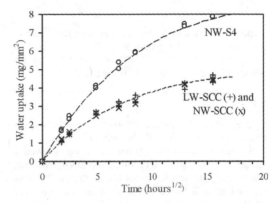

Figure 2. Results of capillary suction tests.

maximum size of 12 mm allowed a reduction in the water content to 176 kg/m^3 and thus also in the cement and fly ash contents.

A normalweight ordinary concrete (NW-S4) with water/cement ratio of 0.65 was also cast, to simulate a low-quality concrete that is often found in the structures that need a repair work, and thus is representative of the substrate where the repair materials have to be applied. The water and cement contents were selected in order to achieve a consistence class S4 according to European standard EN 206-1 (i.e. slump in the range 15–20 cm). The density of the two normalweight concretes was around 2350 kg/m^3 in the water-saturated conditions and 2250 kg/m^3 after drying.

3.2 Properties in the fresh state

Table 2 summarizes the properties of the self-compacting materials in the fresh state. The ordinary concrete NW-S4 had a slump of 20 cm. The workability of the self-compacting materials was studied with the tests developed to study the rheological behaviour of SCC. The slump flow was 66 cm for NW-SCC and 75 cm for LW-SCC, while the flowing time through the V-funnel was about 9 s for both the materials. The lightweight mortar showed a higher fluidity; during the L-box test the material flowed for 150 cm in 31 s, while NW-SCC reached only 119 cm in 39 s (Table 2). The fill box test showed that both LW-SCC and NW-SCC are able to guarantee a high degree of filling even in an element with congested reinforcing bars.

3.3 Properties of the hardened materials

Figure 1 shows the compressive strength of the materials as a function of the curing time. The ordinary concrete NW-S4 had a compressive strength measured on cubes of 25 MPa after 28 days of curing. The compressive strength of the lightweight self-compacting

mortar, that was 10 MPa after only 1 day, reached 29 MPa after 28 days and 40 MPa after about six months. The normalweight self-compacting concrete showed a slower development of the strength during the first day, but in time it reached the highest values.

In fact, the compressive strength was only 4 MPa and 25 MPa respectively after one day and one week, but it was higher than 45 after 28 days and 70 MPa after six months. The dynamic elasticmodulus, which was evaluated after 28 days of curing, was about 24 GPa for NW-S4, 35 GPa for NW-SCC, and 17 GPa for LW-SCC.

Water absorption and capillary absorption were also studied. Water absorption, i.e. the amount of water absorbed by the materials under saturation, was 5.2% by mass for NW-S4, 3.2% for NW-SCC and 5.6 for LW-SCC. Figure 2 shows the results of capillary absorption tests and plots the water uptake as a function of the square root of time. The two self-compacting materials showed a similar behaviour and a water uptake lower than that of the ordinary concrete. Since the two curves in Figure 2 do not show a linear tend, a

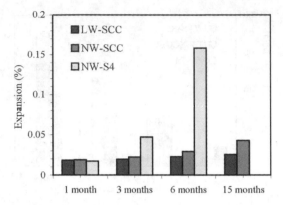

Figure 3. Expansion after immersion in 5% Na$_2$SO$_4$ solution.

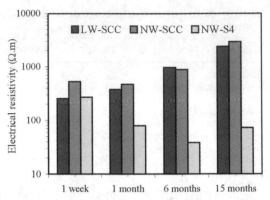

Figure 4. Electrical resistivity as a function of time of exposure to drying/wetting cycles with 3.5% NaCl solution (exposure started after 28 days of wet curing).

sorption coefficient S was calculated as the slope of the intercept of the graph at 24 hours; it resulted 13.8 g/(m^2 × s$^{1/2}$) for NW-S4, 8.7 g/(m^2 × s$^{1/2}$) for NW-SCC and 9.0 g/(m^2 × s$^{1/2}$) for LW-SCC.

3.4 Resistance to sulphate attack

Figure 3 shows the expansion in time of slender specimens immersed in the 5% Na$_2$SO$_4$ solution.

Specimens made of ordinary concrete showed a remarkable expansion since after 2 months of exposure due to sulphate attack; values higher than 0.2% were reached after 6 months and cracking was observed on the surface of the concrete. Expansion of self-compacting materials was negligible even after 15 months of testing. The lightweight mortar LW-SCC, that had an expansion slightly lower than that of NW-SCC, reached values of only 0.025–0.03% after 15 months.

3.5 Corrosion tests

Corrosion tests were carried out in the specimens subjected to cycles of wetting with 3.5% NaCl solution alternated to drying. Figure 4 shows the evolution of the electrical resistivity of the materials as a function of time. At the beginning of the tests, i.e. after 28 days of curing, the three materials had an electrical resistivity of 250–500 Ω·m.

After a few cycles of wetting with the chloride solution, the ordinary concrete showed a sharp decrease in the electrical resistivity, due to the penetration of the chloride ions.

A further slight decrease was observed in time and values of 40–70 Ω·m were measured after 15 months of exposure. Conversely, both self-compacting materials (NW-SCC and LW-SCC) showed a progressive increase in time of the resistivity, that reached values of 2500–3000 Ω·m after 15 months.

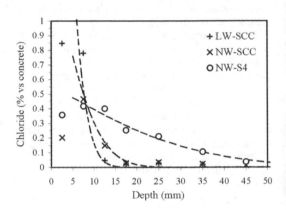

Figure 5. Example of chloride profiles measured after nine months of drying/wetting cycles with 3.5% NaCl solution.

Figure 5 shows, as an example, chloride profiles measured after nine months of exposure (acid soluble chlorides). A roughly constant content of chloride around 0.4% by mass of the dry concrete was measured in the first 15 mm of the depth of specimens made of the ordinary concrete (this value is equivalent to about 3% by mass of cement). Even at the depth of 35 mm the chloride content was 0.1% by mass of concrete (about 0.7% by mass of cement). The self-compacting materials showed remarkable chloride contents on the outermost layer of concrete, but both of them showed a negligible chloride content at depths higher than 15 mm.

Chloride profiles were interpolated with the "erf function" obtained from Fick's second law (Collepardi et al. 1972):

$$C_x = C_s \left(1 - \text{erf} \frac{x}{2\sqrt{D_{app} t}} \right) \qquad (1)$$

where C_x = chloride content (% by mass of cement) at time t (s) and depth x (m); D_{app} = apparent diffusion coefficient (m²/s); C_s = surface content (%). Results obtained from different specimens of the same concrete where fitted together to calculate values of D_{app} and C_s (Table 3); the first analysis, i.e. the outermost point, was not considered in the fitting procedure (Frederiksen 1996).

All the experimental profiles were accurately fitted; thus, even though wetting and drying cycles lead to conditions different from those of pure diffusion, the penetration profiles can be described by means of values of D_{app} and C_s of the fitting curve.

Similarly to results normally obtained on real structures, however, values of D_{app} and C_s changed in time. The apparent diffusion coefficient decreased in time in the self-compacting materials and reached low values (around $0.3 \cdot 10^{-12}$ m²/s) after 15 months, i.e. values more than an order of magnitude lower than the apparent diffusion coefficient measured in the ordinary concrete (Table 3).

The reduced chloride penetration in the self-compacting materials was also confirmed by corrosion tests. Table 4 shows that steel embedded in LW-SCC and NW-SCC remained passive after more than 15 months of exposure. In fact, corrosion potential had values higher than −200 m V vs SCE, typical of

passive steel, and the corrosion rate had negligible values, of the order of 0.1 m A/m² (which is equivalent to about 0.1 µm/year). After 9 months of tests some of the specimens were cracked up to the depth of the reinforcement. The average corrosion rate increased to values of 1-3 m A/m².

Corrosion rate initiated very soon in the specimens made with the ordinary concrete, where a drop in the corrosion potential and an increase in the corrosion rate, which are typically associated with the corrosion onset, were observed after only 2–3 months of testing (Table 4). Corrosion rate reached very high values of 20 mA/m² (about 20 µm/year) even in the absence of cracking.

4 DISCUSSION

The lightweight self-compacting mortar was obtained by the combination of high volume of fines and lightweight expanded clay aggregates (Table 1). The high dosage of fines (300 kg/m³ of cement and 300 kg/m³ of pulverized coal fly ash) and the use of superplasticizer and viscosity modifying admixtures led to a rheological behaviour typical of self-compacting concrete. Different types of tests used for SCC (slump-flow, V-funnel, U-box, L-box and fill-box tests) have shown that the lightweight mortar (LW-SCC) has a good self-compacting behaviour, even better than that of the self-compacting concrete NW-SCC (Table 2). The density of LW-SCC was 1560 kg/m³ in the dry condition, i.e. about two thirds of the density of the normalweight concrete.

The pulverized coal fly ash (PFA), added to LW-SCC and NW-SCC to guarantee a high content of fines required to achieve the self-compacting behaviour, played a major role also in the development of the strength (Fig. 1). The compressive strength of the normalweight self-compacting concrete (NW-SCC) was 45 MPa after 28 days of curing and reached 70 MPa after 6 months. This material has a very high water/cement ratio of 0.93, which cannot account for such a high strength (Table 1). However the water/(cement + PFA) ratio is 0.4 and thus, even though it is not possible to assess the amount of PFA that has actually reacted, it is clear that the contribution of the hydration of the PFA was remarkable. A higher dosage of PFA was also added to the lightweight self-compacting mortar. In this case, however, the beneficial role of hydration of PFA on the compressive strength was reduced because of the lower strength of the lightweight aggregates (about 30 MPa were reached after 28 days and 40 after 6 months). The elasticmodulus was also reduced; this is however beneficial with regards to the reduction of internal stresses in the repair material that can be generated by constrained dimensional changes.

Table 3. Apparent diffusion coefficient (D_{app}) and surface content (C_s) obtained from chloride profiles in time.

Time months	LW-SCC		NW-SCC		NW-S4	
	D_{app} m²/s·10^{12}	C_s %	D_{app} m²/s·10^{12}	C_s %	D_{app} m²/s·10^{12}	C_s %
1.5	2.3	1.6	5.1	0.45	15.1	0.38
3	1.8	0.92	0.70	1.7	17.0	0.37
6	0.78	5.5	1.3	1.5	19.7	0.41
9	0.47	4.7	1.4	1.4	11.7	0.49
15	0.30	3.0	0.35	1.4	13.3	0.36

Table 4. Steel potential (E_{corr}, mV vs SCE) and corrosion rate (i_{corr}, mA/m²) in specimens subjected to chloride penetration.

Time months	LW-SCC		NW-SCC		NW-S4	
	E_{corr}	i_{corr}	E_{corr}	i_{corr}	E_{corr}	i_{corr}
1.5	−208	0.30	−134	0.10	−170	0.11
3	−189	0.15	−102	0.08	−386	1.3
6	−126	0.10	−104	0.06	−510	5.7
9	−222	0.33	−165	0.10	−567	20.4
15	−75	0.14	−210	0.17	−519	13.2
15 cracked	−332	0.9	−411	2.7	−475	15.5

1933

The high amount of PFA also extended its beneficial effects to the durability properties of the materials. Several tests showed that, thanks to the progressive hydration of PFA, the cement paste of LW-SCC reached low porosity and low permeability. Consequently, the lightweight material behaved similarly as NW-SCC, which had no lightweight aggregates. For instance, the two self compacting materials, compared to the ordinary concrete, showed a lower capillary suction (Fig. 2), a progressive increase in time of the electrical resistivity (Fig. 4) and a higher resistance to the penetration of chlorides (Fig. 5) and sulphates (Fig. 3). The increase in time of the electrical resistivity observed in LW-SCC and NW-SCC exposed to chloride ponding (Figure 4) suggests that the pozzolanic reaction of PFA continued for several months after casting. In fact, the increase in resistivity in concrete at early ages is a consequence of the changes that occur in the microstructure of cement paste due to hydration that leads to a finer pores (Polder 2000).

Similarities in behaviour between LW-SCC and NW-SCC show that the high porosity of the expanded clay aggregates did not have any appreciable influence on the transport properties of the lightweight self-compacting mortar. Therefore, the decrease in density obtained by means of the porous aggregate did not affect the resistance to the penetration of aggressive species. The permeability of the cementitious material is essentially governed by the hydrated cement matrix. The porosity of expanded clay particles, that is essentially not accessible, seems only to affect the mechanical properties of the mortar, but not its permeability.

The low permeability of the lightweight self-compacting mortar can provide a good protection to the embedded steel reinforcement when this material is used in the repair of reinforced concrete structures exposed to chloride environments. The penetration of chlorides in specimens exposed to alternate drying and wetting with 3.5% NaCl solution (simulating the seawater), was much slower in the self-compacting materials compared to the ordinary concrete (Fig. 5), as shown by the low value of the apparent diffusion coefficient (Table 3). Corrosion tests on specimens with concrete cover of 25 mm (Table 4) showed that, while corrosion initiated after only two months of exposure on steel embedded in the ordinary concrete NW-S4, no corrosion was detected in the specimens with LW-SCC and NW-SCC after more than 15 months. Corrosion initiated only when the self-compacting materials were cracked up to the depth of the reinforcement; the measured corrosion rate was significantly lower than that measured on the specimen NW-S4, suggesting that corrosion took place only in the neighbourhoods of the crack.

The lightweight self-compacting mortar also showed a good resistance to sulphate attack; expansion was negligible even after 15 months of immersion in a solution with a high concentration of sulphate ions (5% Na_2SO_4), and it was only slightly higher than that of NW-SCC (Fig. 3). This is again a beneficial effect of the hydration of PFA, which decreases the permeability of the cement paste to the sulphate ions and lowers the lime content in the hydration products (Neville 1995).

5 CONCLUSIONS

By means of a combination of pulverised coal fly ash (PFA), used as mineral addition, and expanded clay aggregate, used as lightweight aggregate, a mortar suitable for the repair of reinforced concrete structures damaged by corrosion was obtained. The addition of a high amount of PFA, which lead to rheological properties typical of self-compacting concrete in the fresh state, could also reduce the permeability of the hardened cement paste and increase the resistance to the penetration of aggressive species, such as chlorides and sulphates. The use of expanded clay fine aggregate with bulk density of $535 \, kg/m^3$ reduced the density of the mortar to about $1600 \, kg/m^3$, but did not affect the resistance to chloride penetration of the hardened material.

The lightweight self-compacting mortar can be advantageous in the repair of corrosion damaged structures. The self-compacting behaviour in the fresh state allows casting operations under unfavourable geometrical conditions. The low density allows to increase the thickness of the concrete cover with a lower increase in the additional weight applied to the repaired structure. Finally, the high resistance to aggressive species guarantees protection of the repaired area from further environmental aggression.

ACKNOWLEDGEMENTS

Specimens tested in this work were cast in the laboratories of MAC-MBT of Treviso (Italy).

REFERENCES

Bertolini, L., Elsener, B., Polder, R.B., Pedeferri, P. 2003a. *Corrosion and protection of steel in concrete*, Wiley, in print.

Bertolini, L., Carsana, M., Gastaldi, M. 2003b. Lightweight self-compacting mortar for the repair of reinforced concrete structures. *Int. conf. Management of Durability in the Building Process*, Milan, 25–26 June.

Cambell-Allen, D., Ropr, H. 1991. *Concrete structures: materials, maintenance and repair*, Longman Scientific and Technical, London.

Collepardi, M., Marcialis, A., Turriziani, R. 1972. Penetration of chloride ions into cement pastes and concretes, *Journal of American Ceramic Society*, 55 (10), 534.

Collepardi, M. 2002. *The new concrete* (in Italian), Tintoretto, Villorba (Italy).

ENV 1504-9:1997, Products and systems for the protection and repair of concrete structures - definitions, requirements, quality control and evaluation of conformity – Part 9: General principles for the use of products and systems, European Committee for Standardization.

EN 206-1:2000, Concrete – Part 1: Specification, performance, production and conformity, European Committee for Standardization.

Frederiksen, J.M. (Ed.) 1996. HETEK – Chloride penetration into concrete. State of the art. Transport processes, corrosion initiation, test methods and prediction models, The Road Directorate, Report No. 53, Copenhagen.

Mailvaganam, N.P. 1992. *Repair and protection of concrete structures*, CRC Press Inc., Boca Raton, Florida.

Neville, A.M. 1995. *Properties of concrete*, 4th Ed., Longman Group Limited, Harlow.

Okamura, H. & Ouchi, M. 1999. Self-compacting concrete. Development, present use and future, *Int. Conf. Self-Compacting Concrete*, Stockholm, 13–14 Sept., 3.

Polder, R.B. 2000. Simulated de-icing salt exposure of blended cement concrete – chloride penetration, *2nd Int. RILEM Workshop Testing and Modelling the Chloride Ingress into Concrete*, C. Andrade, J. Kropp, Eds., 189–202.

Rilem T.C. 124-SRC, 1994. Draft recommendation for repair strategies for concrete structures damaged by reinforcement corrosion, *Materials and Structures*, 27, 415.

Umoto, T. & Ozawa, K. 1999. Recommendation for Self-Compacting Concrete, *Japan Society of Civil Engineers*, Tokyo, August 1999.

Durability of cement and cement plus resin stabilized earth blocks

A. Guettala
Civil Engineering Dept., Biskra University, Algeria

H. Houari
Civil Engineering Dept., Constantine University, Algeria

A. Abibsi
Mechanical Engineering Dept., Biskra University, Algeria

ABSTRACT: The main drawback of earth construction is the rapid deterioration of the material under severe weather conditions. The objective of this work is to improve the behavior of stabilized blocks of earth blocks against water attacks. The blocks manufactured with one type of earth were tested in compressive strength as dry blocks and after immersion, in intensive sprinkling and in absorption. Test of wetting-drying. The tests of freeze-thaw were also carried out. The results show the influence of the different manufacturing parameters: compacting intensity, sand, cement and cement plus resin content on the mechanical strength in the dry state as well as in the wet state, water resistance coefficient, weight loss and absorption.

1 INTRODUCTION

The use of earth as a building material dates back to the period of the ancient Mesopotamia (5000–4000 BC). For economical reasons and by studying what already has been done until now, scientists and builders consider that it is judicious to try to improve the life span of construction materials Ghoumari (1989). Large research programs have been undertaken all over the world into the durability of earth walled buildings Gregory & Kevan (2002) and Heathcote (1985). The durability prevision of stabilized earth blocks is still a controversial matter amongst construction actors. In order to know the limits of this kind of material destined to construction, it is intended to find solutions that can improve its life span by the know how of its general use as well as its mass treatment (additions of binders, compacting energies, ...). Obtaining a durable material would need a treatment which would result in sufficient mechanical strength as well as low sensitivity to water attacks Guettala et al (2002).

These two main conditions should be preceded by a very precise study of parameters related to the grading and mineralogy of these materials. The type and the content of the binders, aggregates grading, compacting stresses and water content would be adapted as conditions of making of these materials, Guettala & Guenfoud (1998) and (1997). The durability can be improved by other additions such as lime Guettala et al (2002), cement and lime or cement and microsilica, Keraali (2001). In this present work, we have tried to improve the durability of earth blocks by several methods: by the additions of cement (5, and 8%), cement and resin (5% cement + 50% resin; 8% cement + 50% resin[*]), sand content (0, 10, 20, 30 and 40%) and the compacting stresses (5, 7.5, 10, 12.5, 15, 17.5 and 20 MPa).

2 SOIL PHYSICAL CHARACTERISTICS

Soil samples of the region of Biskra (south east of Algeria) have been taken as reference samples and subjected to several laboratory tests as specified by ASTM standards (1993).

2.1 Atterberg limits

According to Michel (1976), the best earth soils for stabilization are those with low plasticity index (PI) and the product (PI $\times$ M) in the vicinity of 500 to 800, where M is the percentage of mortar, in this case PI $\times$ M = 644, see Table 1.

---

[*] The resin content is relative to compacting water.

Table 1. Atterberg's limits.

WL	WP	PI	Ws	Wa	Ca	PI × M
31	17	14	10	9.5	0.77	644
PZ[(*)]	PZ	PZ	PZ		A.A[(**)]	

[(*)] Preference Zone.
[(**)] Average Activity.

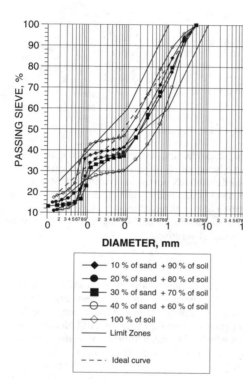

Figure 1. Grading curves aggregate analysis of used soil, corrected soil and the recommended limit zone of stabilized earth concrete.

2.2 Grading aggregate analysis

In Figure 1, the grading curves of the soils as well as the corrected soils with sand and limits of the recommended zone for compressed earth blocks are represented, Rigassi and CRATerre (1995). It is noted through these curves that soil and corrected soil with contents of 10, 20 and 30% of sand are very close to the lower limit of recommended zone; whereas corrected soil with content of 40% of sand is out of the recommended limit zone.

2.3 Chemical analysis

Clay analysis has been accomplished in the cement factory of Hamma Bouziane (Constantine, east of Algeria) using Fluorescence X ray, in accordance to

Table 2. Chemical composition of the soil.

Content, %					
SiO_2	AL_2O_3	Fe_2O_3	MgO	CaO	SO_3
32.22	2.24	0.53	0.03	31.8	5.81
K_2O	Na_2O	Cl	TiO_2	MnO	F W[(*)]
0.15	0.03	0.005	0.2	0.02	26.9

[(*)] Weight loss due to fire.

Table 3. Soil mineralogical constituents.

Clayey minerals, %			Non clayey minerals, %	
Kaolin	Illites	IM[(*)]	Quartz	Calcite
45	40	15	5	10

[(*)] Interstratifiers.

Table 4. Proctor test.

	Optimal (Wc), %	Max. dry density (γ), kg/m³
	11.75	1877
Appreciation	Excellent	Satisfactory

NF6 P 15-467 (1984). The obtained results showing the constituents of the soil are presented in Table 2.

2.4 Mineralogical analysis

To differentiate the clay soils, a mineralogical analysis by X rays is important. The analyses have been carried out in the geology laboratory of Boumerdes (Algiers, Algeria) using a diffractometer SIEMENS 500, interfaced to a computer for data collecting. Tests have been conducted on aggregates passed on sieves of 80 microns. The obtained results see Table 3 show that the soil is composed mainly of kaolin (non-expansive and non-absorbent) and illites.

2.5 Organic matter

During the treatment of soil with oxygenated water, it was noticed that the soil-water reaction is very slow and the organic mater is essentially free vegetable fragment (0.15%).

2.6 Measure of pH

The analysis show that the tested sample was almost neutral, pH = 7.1.

2.7 Mechanical characteristics (Proctor test)

The results are shown in Table 4. The Proctor test shows that the water content (Wc) of the studied sample is excellent and the dry density is satisfactory.

3 PHYSICAL CHARACTERISTICS OF SAND

Using AFNOR (1984) regulations, the sand samples have been tested and found the following results;

- Disturbed apparent density (ρo) = 1520 kg/m^3
- Specific mass (γ) = 2640 kg/m^3
- Fineness modulus (FM) = 2.33
- Sand equivalence value by sight (SE) = 70
- Sand equivalence value by test (SE$_t$) = 64

4 CEMENT

The used cement is manufactured in Algeria, under the commercial label CPJ 45 and has been tested following the AFNOR (1984) regulations in order to determine its real class. Tests carried out on mortar cubes have shown that the strength at 28 days is 46 MPa.

5 RESIN

The resin used for this work has a commercial name of "Medalatex"; supplied by Granitex; private Algerian company of additives making. Medalatex is an aqueous dispersion of resin of white colour. It's compatible with most of cement as well as lime.

In general, the latex content varies between 10 and 20% in respect to the cement mass. The latex addition gives a good adherence to the support. It gives also the impermeability, the durability and the improvement in protection of the reinforcement, thus resistance to chemical attacks.

6 INFLUENCE OF SAND CONTENT

In order to determine the influence of sand content on the mechanical strength, durability and the optimal quantity of soil-sand mix, several blends have been used (0–40%) with cement content of 5% and 5% + 50% resin and a compacting stress of 10 MPa. Samples have been stored in a humid environment.

6.1 Compressive strength

These tests were carried out according to AFNOR (2001). Figure 2 shows that the mechanical compressive strength of dry and humid sand-soil samples increases with increasing the sand content. However, in percentage terms, the compressive strength evolution is 27.5% for dry samples and 30% for humid samples, when the concentration of sand is 30%. For the same sand content, the addition of the resin has resulted in the increase in strength of the order of 11% in the dry state and 29% in the wet state.

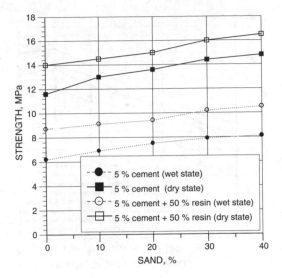

Figure 2. Sand content effect on compressive strength with 5% cement and 5% cement + 50% resin stabilizer, using 10 MPa compacting stress.

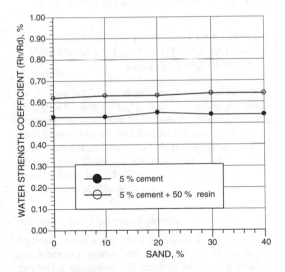

Figure 3. Sand content effect on water strength coefficient with 5% cement and 5% cement + 50% resin stabilizer, using 10 MPa compacting stress.

6.2 Water strength coefficient

The water strength coefficient is determined from the compressive strength ratio for dry and humid states. Figure 3 shows that the sand content does not affect the water strength coefficient which varies between 0.53 and 0.54 when the sand content varies between 0 and

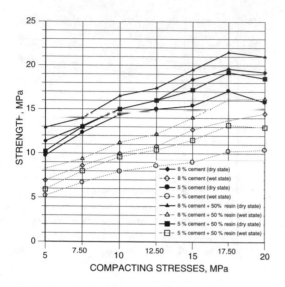

Figure 4.   Influence of compacting stresses and cement and cement plus resin content on the mechanical strength.

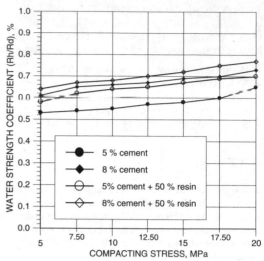

Figure 5.   Influence of compacting stresses and cement and cement plus resin content on the water strength coefficient.

40%. However, the addition of the resin has resulted in an increase of the coefficient of the order of 17%.

7   INFLUENCE OF THE COMPACTING STRESS AND THE CEMENT AND CEMENT PLUS RESIN CONTENT

In the following section, the effect of the compacting stresses (5, 7.5, 10, 12.5, 15, 17.5 and 20 MPa), the cement content (5 and 8%) and cement plus resin (5% + 50% and 8% + 50%) on the mechanical compressive strength on dry and humid sand samples is studied. Also the durability: wetting and drying, freeze-thaw, water absorption (total and capillary) tests with the optimal sand content of 30% are carried out.

7.1   *Mechanical compressive strength in dry samples*

Figure 4 shows clearly that the compressive strength evolution is the same for the different cement and cement plus resin content: the compressive strength increases with increasing the compacting stress until 17.5 MPa which is the optimal compacting stress. Again, the addition of the resin had a great effect on the strength of the samples in the dry state. As can be seen in Figure 4, the compressive strength increases with the addition of resin content for both cement content cases.

7.2   *Mechanical compressive strength in humid samples*

The mechanical strength of humid soil sample increases with increasing the compacting stresses,

Figure 4. The mechanical compressive strength also increases with the addition of resin. Like in the case of dry samples, the compressive strength evolution is not regular. The effect of the resin addition is more important in the case of the 5% sample.

7.3   *Water strength coefficient*

The water strength coefficient evolution depends on the cement and cement plus resin percentage and on the compacting stresses, Figure 5. It increases with increase of cement content, the compacting stresses as well as the addition of resin. For example, the addition of the resin by 50% as is the case in this work had resulted in variation of the water strength coefficient by 15% and 7.5% for both cases of 5% and 8% cement respectively, with a compacting stress of 15 MPa.

7.4   *Water absorption*

The absorption capacity of earth stabilized blocks gives a general idea on the presence and importance of voids. When a volume of soil subjected to the action of a stress, the material is compressed and the voids ratio decreases. As the density of soil is increased, its porosity is reduced and less water can penetrate it, Houben & Guillaud (1984).

7.4.1   *Total absorption*
The present test consists of immersing the soil samples in water and measuring the increase in weight during 24 hours. The absorption is evaluated in dry

Table 5. Influence of the compacting stresses and the cement and cement plus resin on the total absorption.

Total absorption, %

Compacting stress (MPa)	5% cement	5% cement + 50% resin	8% cement	8% cement + 50% resin
5	10.12	8.2	9.17	7.5
7.5	9.78	7.5	9.23	6.6
10	9.12	6.1	8.26	5.5
12.5	8.33	6.0	7.84	5.6
15	8.27	5.9	7.35	5.3
17.5	7.54	5.8	7.25	5.2
20	8.71	5.8	8.59	5.2

Table 7. Influence of the compacting stresses and the cement and cement plus resin on the weight loss.

Weight loss, %

Compacting stress (MPa)	5% cement	5% cement + 50% resin	8% cement	8% cement + 50% resin
5	2.65	2.1	2.07	1.8
7.5	2.28	1.8	1.86	1.7
10	1.9	1.5	1.4	1.2
12.5	1.34	1.2	1.35	1.3
15	1.27	0.9	1.17	0.9
17.5	0.31	0.25	0.25	0.22
20	0.13	0.10	0.05	0.05

Table 6. Influence of the compacting stresses and the cement and cement plus resin on the capillary absorption.

Capillary absorption, %

Compacting stress (MPa)	5% cement	5% cement + 50% resin	8% cement	8% cement + 50% resin
5	3.99	3.8	3.52	3.1
7.5	3.19	3.7	2.48	2.2
10	3.90	3.1	2.30	2.1
12.5	2.74	2.5	2.24	1.9
15	2.35	2.3	2.23	1.8
17.5	2.15	2.1	2.02	1.2
20	1.17	2.0	1.19	1.1

weight percentage. Table 5 shows that the absorption decreases when increasing the compacting stresses. Up to a certain value of 15 MPa and above; it has a minor effect. We also notice that the increase in cement content decreases the water absorption factor. The addition of the resin decreases considerably the water absorption factor for both cement contents. The total absorption varies between 18 to 33% and between 18 to 39% for 5 and 8% cement respectively; for compacting stresses varying from 5 to 20 MPa.

7.4.2 Capillary absorption

Capillary absorption test consists of placing the soil sample on a humid surface with voids, constantly water saturated, and measuring its weight after 7 days. Absorption is evaluated in percentage of dry weight. Table 6 shows that the capillary absorption decreases when increasing the compacting stresses and the cement as well as cement plus resin content. For instance, it varies from 2.15 to 2.02% when the cement content varies from 5 to 8% with a compacting stress of 17.5 MPa. Again, the addition of the resin decreases considerably the water capillary absorption.

7.5 Wetting and drying test

This test is carried out according to the ASTM D 559-57 (1993); it consists of immersing soil samples in water for a period of 5 hours and then removed to be dried in an oven at 71°C for a period of 42 hours. The procedure is repeated for 12 cycles, samples are brushed every cycle to remove the fragment of the material affected by the wetting and drying cycles. For every sample, the variation in weight is computed after the 12 cycles, Houben & Guillaud (1984). Table 7 shows that the loss in weight diminishes when increasing the compacting stress and the cement and cement plus resin content. For the cases of 5 and 8% cement, the effect of cement on the weight loss is important for the compacting stresses up to 12.5 MPa. Above this value the addition of cement is less significant. The resin addition has a slight effect on the weight loss.

7.6 Freeze-thaw

Following the procedure described by ASTM D560 (1993), the freeze-thaw test consists of placing a soil sample on an absorbent water saturated material in a refrigerator at a temperature of −23°C for a period of 24 hours and then removed. The sample is then thawed in a moist environment at a temperature of 21°C for a period of 23 hours and then removed and brushed. The test is repeated for 12 freeze-thaw cycles and dried in an oven to obtain a constant weight, Houben & Guillaud (1984). Table 8 shows that the weight loss diminishes when increasing the compacting stress and the cement content as in the case of wetting and drying test discussed previously. For the 5% cement sample, the weight loss changes from 17 to 3.3% when the compacting stress varies from 5 to 20 MPa. And the weight loss is very important with low compacting stresses. The effect of the resin is more pronounced at lower compacting stresses.

Table 8. Influence of the compacting stresses and the cement and cement plus resin on the weight loss.

Weight loss, %

Compacting stress (MPa)	5% cement	5% cement + 50% resin	8% cement	8% cement + 50% resin
5.0	17.0	3.8	5.28	4.1
7.5	12.0	3.7	4.67	3.2
10.0	4.5	3.1	2.4	2.1
12.5	4.2	2.5	1.3	1.1
15.0	3.9	2.3	1.02	1.0
17.5	3.6	2.1	0.24	0.2
20.0	3.3	2.0	0.14	0.1

8 CONCLUSION

The main objective of this work was to investigate the different factors affecting the durability of cement stabilized earth blocks as well as the importance of the resin addition.

The work showed the importance of the sand content, the compacting stress and the cement and the cement plus resin contents on the behaviour of stabilized earth blocks with respect to water attacks as well as elucidating certain points:

- The principle effect of the stabilization with the cement is to prevent water attacks. We would achieve then a good stabilization if we could obtain a durable material with a limited loss in mechanical strength in a wet state, Guettala et al (2000).
- The sand content does not affect considerably the compressive strength and the water strength coefficient.
- Increasing the compacting stress from 5 to 20 MPa and the cement content from 5 to 8% improve the compressive strength in dry as well as wet state and the water strength coefficient. We notice also that the increase of these two parameters decrease the weight loss and the water absorption.
- The latex addition is shown also to be beneficial concerning the durability in general. This is due to the fact that such additions consolidate the

cementery matrix and play a role of a co-marix. However, when considering the cost factor mainly the resin price (4 to 6 times more expensive than cement), this addition is not economical.

REFERENCES

Ghoumari, F. 1989. Matériau en Terre Crue Compactée: Amélioration de sa Durabilité à l'Eau; Thèse de Doctorat, INSA de Lyon.

Gregory, M. & Kevan, H. 2002. Earth Building in Australia – Durability Research. *Proceedings of Modern Earth Building*, Berlin, 19–21 April 2002: 129–139.

Heathcote, K. A. 1985. Durability earthwall buildings. *Construction and Building Materials* Volume 9, Number 3: 185–189.

Guettala, A., Houari, H., Mezghiche, B. & Chebili, R. 2002. Durability of Lime stabilized Earth Blocks. *Proceedings of international conference University of Dundee*, Scotland UK 9–11 Sep: 145–654.

Guettala, A. & Guenfoud, M. 1998. Influence des Types d'Argiles sur les Propriétés Physico-mécaniques du Béton de Terre Stabilisée au Ciment, *Annales du Bâtiment et des Travaux Publics*: 1 : 15–25.

Guettala, A. & Guenfoud, M. 1987. Béton de Terre Stabilisée Propriétés Physico-Mécaniques et Influence des Types d'Argiles, *La technique moderne* 1-2: 21–26.

Keraali, A. G. 2001. Durability of compressed and cement-stabilized building blocks. PhD Thesis, University of Warwick, School of Engineering, UK, September.

American Society for Testing and Materials. 1993. Annual Book of ASTM Standards, Vol. 4.01, Philadelphia.

Rigassi, V. & CRATerre-EAG (Hoehel-Druck). 1995. Blocs de terre comprimée: Manuel de production (1). Germany.

AFNOR XP P 13-901, 2001. Blocs de terre comprimée pour murs et cloisons.

Michel, J. 1976. Etude sur la Stabilisation et la Compression des Terres Pour leur Utilisation dans la Construction. *Annales de l'Institut Technique de Bâtiment et des Travaux Publics*, *Série Matériaux* 339: 22–35.

AFNOR, 1984. Recueil de Norme Françaises. Bâtiment Béton et Constituants du Béton. Paris.

Houben, H. & Guillaud, H., (CRATerre), 1984 Earth Construction, Primer Brussels, CRATerre/PGC/CRA/ UNCHS/AGCD.

Guettala, A., Mezghiche, B., Chebili, R. & Houari, H. 2000. *Durability of Blocks of Earth Concrete*. Proceedings of II International Symposium Cement and Concrete Technology, September 6–10 Istanbul, Turkey: 273–281.

Experimental investigation into the effect of fly ash on the fresh properties of self-compacting concrete (SCC)

M. Dignan & C.Q. Li
University of Dundee, Scotland, UK

ABSTRACT: This paper presents an initial examination of the direct effect that two types of fly ash (both compliant with BS EN 450) has on the properties of fresh self-compacting concrete. The use of cement replacement materials is thought to be extremely advantageous in the production of SCC due to a number of factors. These factors include; a reduction in yield strength of fresh concrete, a reduction in workability loss, a reduction in heat development which is high in SCC production due to the cement content and reduction in initial concrete cost. However, although fly ash has been used in SCC research, its direct effect on SCC has yet to be evaluated. It is in this regard that the investigation is undertaken in order to accumulate more data. The investigation found that the use of coarse fly ash improves the flowability and cohesion of SCC compared with fine fly ash.

1 INTRODUCTION

Self-Compacting Concrete (SCC) is a relatively new and specialised concrete which possesses a very low yield strength and an adequate viscosity. The low yield strength is required to ensure that the concrete is very flowable and can fill any type of formwork into which it is placed. This is advantageous in instances where there is heavy reinforcement or a complicated form shape. However, the concrete must remain homogenous during placing and at rest. Homogeneity is achieved by ensuring that the viscosity of the concrete is such that the downward movement of aggregate particles is reduced to a minimum. Delicate proportioning of the concrete mix constituents can ensure that there is negligible inter-particle interference and therefore less opportunity for the concrete to segregate.

Interest in SCC begun in earnest in the mid 1980's when Japanese engineers found encountered problems with a reduced skilled labour workforce and poor on-site workmanship which culminated in poor quality concrete (Okamura & Ouchi 1999). Collepardi (1976) first wrote about flowable non-segregating concrete in the mid 1970's while Banfill (1980) used the term "self-compacting concrete" in a study of flowable concrete in 1980. Since these beginnings, SCC has become more popular, although it is still considered to be a niche product.

Fly ash, as a constituent material in concrete, has been studied since the mid 1930s (Joshi & Lohita 1997). Its effects on concrete properties have been established for several years and studies are taking place on concretes with high volumes of fly ash

(HVFA) (Bouzoubaa et al. 2000, Malhotra 2002) However, even though fly ash has been used in SCC studies, there has been little activity in researching the real effects of this material on SCC properties (Khayat 1999, Kim et al. 1998). One study has, indeed, examined the effects of Class F HVFA SCC and has found that this material has beneficial effects (Bouzoubaa 2001). This study looked at fly contents of 40%, 50% and 60%. Interestingly, it was stated that an increase in fly ash volume resulted in a reduction in segregation. This implies that the viscosity of high volume fly ash concrete is greater than normal concrete.

1.3 *Research significance*

Previous research into SCC has often included fly ash as a constituent material but not on the real influence of fly ash on its properties. The inclusion of this material has basically been as a fine filler which increases concrete workability. The following paper outlines an experimental study on the effects of varying volumes of two different types of fly ash on the properties of SCC. The properties will be tested with the accepted test procedures as outlined in section 2.

2 EXPERIMENTAL DETAILS

2.1 *Materials*

Details of the materials used in the tests are as follows:
Portland cement of grade 42.5 N complying with BS EN 197: Part 1 (2000) was used for all concrete mixes. Table 1 gives the cement properties. Two types

Table 1. Chemical composition of cementitious materials.

Bulk Oxide composition, %	PC	Fine ash	Coarse ash
SiO$_2$	21.0	47.91	1.63
Al$_2$O$_3$	4.9	25.74	45.63
Fe$_2$O$_3$	2.6	10.?	22.69
CaO	64.6	1.94	8.14
MgO	1.2	0.98	1.524
P$_2$O$_5$		1.30	0.044
TiO$_2$		1.94	1.013
SO$_3$	3.3	0.82	3.988
K$_2$O	0.7	3.74	4.560
Na$_2$O	0.1	1.41	0.216
MnO		0.12	0.016
Loss on ignition (LOI), %	1.4	3.6	4.0
% Retained on 45 μm Sieve	3.4	8.6	28

Table 2. Mixture proportions (kg/m^3).

Mix Type	PC	FA	Sand	Coarse agg	Water	HRWR (bwc) %	VMA (bwc) %
PC	500		930	800	180	0.8	0.1
70/30	350	150	885	800	180	0.8	0.1
50/50	250	250	855	800	180	0.8	0.1
30/70	150	350	825	800	180	0.8	0.1

strength (Nagataki & Fujiwara 1995). The diameter of the slump flow is measured as well as the time (in seconds) the concrete takes to reach a diameter of 500 mm (T$_{500}$ value) (Bartos 2000).

The j-ring test is used in conjunction with the slump flow test. It demonstrates the concrete's passing ability (ability to flow between bars without segregating) in combination with its flowability. The diameter of the concrete spread is measured and note any blocking and/or segregation (EFNARC 2002).

The l-box is used to give the operator an indication of the SCC's passing and filling ability. The total length travelled by the concrete and time taken to reach 500 mm (T$_{500}$ value) is recorded (Bartos 2000).

This test involves pouring about 12 litres of concrete into a v-shaped funnel. The concrete is left for 2 minutes to allow for internal settlement then a trapdoor at the bottom of the apparatus is opened to allow the concrete to flow out. The time taken for the complete discharge of concrete is recorded (EFNARC 2002).

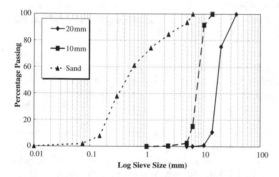

Figure 1. Total aggregate grading curve.

of fly ash, one fine and one coarse, conforming to BS EN 450 (1995), see Table 1, were used in this project. Natural gravel in two sizes (20-10 mm and 10-5 mm) conforming to BS 882 (1992) were used, in an air dry condition, in all concrete mixes. Sand conforming to Zone F of the standard was also used. The aggregate grading curve is shown in Figure 1. Mains tap water and a commercially available HRWR based on a modified synthetic carboxylated polymer conforming to BS EN 934-2 (2001) was used.

2.2 Mix proportions

The mix proportions used in the test programme are set out in Table 2

2.3 Test methods

The slump flow test is an indication of the concrete's flowability. Therefore, it is indicative of its yield

3 RESULTS AND DISCUSSION

3.1 Slump flow and slump loss

The following table (Tab. 3) presents the results from the SCC slump flow tests. The second slump flow test was carried out after all other testing was complete (generally around twenty minutes after the first slump) in order to assess slump loss.

As can be seen from the table the control mix used in the test has a slump flow about half the diameter of the other concretes. This shows that, at a given HRWR dosage, the presence of fly ash, even at a small replacement level, improves the flowability substantially, i.e. by over 100%. Fly ash improves workability of fresh concrete due to its physical properties, i.e. the particles act as small ball bearings. Therefore, workability is increased due to reduced inter-particle friction of the aggregates.

Figure 2 shows the slump flow results for the two types of fly ash used. Firstly it can be noted that, contrary to published knowledge of fly ash in normal concrete (Joshi & Lohita 1997), SCC containing coarse fly ash shows superior flowability than the fine ash SCC at all replacement levels. The reasons for

1944

Table 3. Slump flow results.

Mix type	1st diameter (mm)	1st T500	2nd diameter (mm)	2nd T500	% Slump loss
Control (100% PC)	350	–	–	–	–
70/30 Fine ash	745	2.5	525	6	30
50/50 Fine ash	770	2	700	2.5	9
30/70 Fine ash	740	2	635	3	14
70/30 Coarse ash	755	2.5	675	4	11
50/50 Coarse ash	775	2	735	2.5	5
30/70 Coarse ash	765	2.5	750	3	2

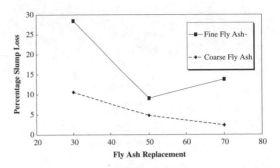

Figure 3. Percentage slump loss for SCC with both types of fly ash.

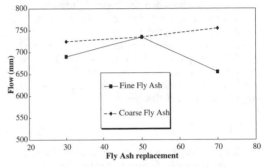

Figure 4. J-ring results for both types of fly ash.

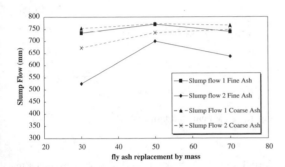

Figure 2. Graph of 1st and 2nd slump flow for SCC with both fly ash types.

this are unclear as it is generally accepted that fine ash, with its greater void filling capability, has a greater effect on workability than coarse ash. It is also clear from the graph that slump flow values decrease when fly ash replacement reaches a certain level. It can be taken from this study that the optimum fly ash level in SCC is 50%. However, due to the small number of tests undertaken, this may be an area where more detailed work can be carried out. The reason for this may be that the mix has "choked" because there is too much fly ash in the concrete. Neville (1995) states that very fine fly ash particles are absorbed onto the surface of cement particles due to electrostatic effects thereby preventing them from flocculating and increasing workability. However, once all of the cement particles have been covered there are no further beneficial effects and the trend is actually reversed. It is suggested that the upper limit of fly ash replacement is 20% (Neville 1995). From this study, however, it is observed that the upper limit is perhaps

as much as 50%. The reduced slump flow effect is much less pronounced for the coarse fly ash SCC. This may be due to the fact that there is a lower volume of fine particles in the coarse ash and therefore it takes a greater mass of coarse ash to reach the point where the fly ash effect reverses.

The slump loss of SCC is greater when fine fly ash is used (Fig. 3). The reason for this result may be that coarse particles hydrate more slowly, thereby reducing slump loss.

3.2 J-ring results

The results for SCC containing coarse fly ash shows the expected trend of an increase in flow with an increase in fly ash level (Fig. 4). However, the fine ash SCC provides results which do not show the expected trend. Fly ash replacement up to 50% shows the expected trend then there is a large reduction in flow at 70% replacement. This may provide further evidence that replacing PC with fly ash is detrimental after a particular level. The "choking" effect will reduce the concrete's ability to carry aggregate particles between reinforcing bars and, therefore, reduce the resulting flow. The presence of too much fine ash

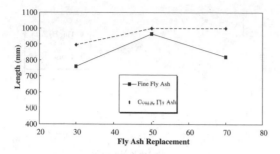

Figure 5.   L-Box final length results.

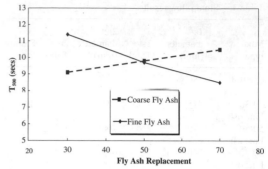

Figure 6.   L-Box T$_{500}$.

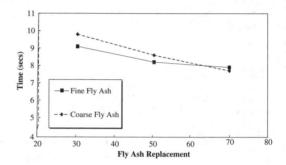

Figure 7.   V-Funnel test time (secs).

would appear to increase friction between the powder (cementitious materials) particles meaning that there may be problems when the concrete must negotiate obstacles such as reinforcing bars. There was no visible blocking during the tests.

The coarse ash SCC appears to improve even when the 50% replacement level is reached. The results show that the use of this particular type of fly ash improves the cohesion of the concrete whilst maintaining its excellent flow properties.

3.3   L-box results

The lengths that the tested concrete flowed along the L-box are shown in Figure 5. Again, there is further evidence here that too much fly ash is detrimental to fresh SCC properties. There is a negative effect between 50% and 70% fly ash content when fine ash is used. This confirms the suggestion that there is an optimum fly ash content. It is also noticeable that the coarse fly ash outperforms the fine fly ash, i.e. it is more flowable and has a greater ability to carry coarse particles between reinforcing bars. Only minor blocking was noted during the tests.

However, after a visual inspection of the point of blocking it was observed that the blocking was caused by oversized coarse aggregate particles. It can be seen from Figure 1 that there are some particles larger than 20 mm contained within the coarse aggregate. The presence of particles of this size will increase the chances of blocking occurring during the test. Therefore, it ought to be stated that quality of materials must be tightly controlled when producing SCC.

The time for the concrete to travel 500 mm along the trough was taken and compared between all mixes. Opposite trends can be noted between the concretes containing fine and coarse fly ashes. The results provided by SCC with coarse ash suggest that the concrete is more viscous as the level of fly ash increases while SCC with fine ash appears to be less viscous. Viscosity is defined as the property of a material to resist deformation caused by internal friction (Tattersall & Banfill 1983). Therefore, the physical action of fly ash

particles ought to make it easier for other particles in the mix to pass over and around each other, thereby reducing the material resistance to movement. Previous work carried out by this author would suggest that the use of fly ash does reduce the viscosity of SCC. Therefore, Figure 6 suggests that an increasing content of coarse fly ash induces greater viscosity of SCC probably due to higher internal friction brought about by larger ash particles.

3.4   V-funnel results

Figure 7 shows the v-funnel times (in seconds) for all concretes. The results show that SCC viscosity decreases as fly ash content increases, probably due to the physical nature of the ash particles resulting the increased ease of particles to move around and by each other.

The figure may also provide further validation that the use of coarse fly ash will increase SCC viscosity. The graph shows that SCC with coarse fly ash gives a longer v-funnel time than SCC with fine fly ash. This may be due to the larger ash particles causing slightly more inter-particle friction than when fine ash is used. However, there is a point on the graph, close to

70% fly ash content, where the lines cross suggesting that the fine ash SCC becomes more viscous. The result may indicate that when there are too many fine ash particles there is increased inter-particle friction resulting in a more viscous concrete.

4 CONCLUSION

It has been observed from this simple series of tests that, contrary to accepted fly ash knowledge, the use of coarse fly ash results in a more flowable, more cohesive concrete than when fine ash is used in SCC. It has also been noted that there appears to be an optimum fly ash replacement level when the material is used in SCC. From the test results shown the optimum fly ash content can tentatively be put at 50% replacement by mass. However, the small number of tests and wide range of fly ash contents tested means that further work is required to be carried out in order to define acceptable ranges of ash replacement levels.

REFERENCES

Banfill, P.F.G. 1980. Workability of Flowing Concrete, *Magazine of Concrete Research*, Vol. 32, No. 110, March, pp. 17–27

Bartos, P.J.M. 2000. Measurement of Key Properties of Fresh Self-Compacting Concrete, *CEN/STAR Workshop, Measurement, Testing and Standardisation: Future Needs in the Field of Construction Materials*, Paris 5–6 June, pp. 6

Bouzoubaa, N. & Lachemi, M. 2001. Self-Compacting Concrete Incorporating High Volumes of Class F Fly Ash – Preliminary Results, *Cement and Concrete Research*, Vol. 31, pp. 413–420

Bouzoubaa, N., Zhang, M.H. & Malhotra, V.M. 2000. Laboratory-Produced High Volume Fly Ash Blended Cements: Compressive Strength and Resistance to the Chloride-Ion Penetration of Concrete, *Cement and Concrete Research*, Vol. 30, pp. 1037–1046

British Standards Institution, BS 882, Specification for Aggregates from Natural Sources for Concrete, 1992

British Standards Institution, BS EN 197: Part 1: Cement – Composition, Specification and Conformity Criteria for Common Cements, 2000

British Standards Institution, BS EN 450, Fly Ash for Concrete – Definitions, Requirements and Quality Control, 1995

British Standards Institution, BS EN 934 – 2 Admixtures for Concrete, Mortar and Grout, 2001

Collepardi, M. 1976. Assessment of the Rheolplasticity of Concretes, *Cement and Concrete Research*, Vol. 6, pp. 401–408

EFNARC, *Specification and Guidelines for Self-Compacting Concrete*, pp. 32

Joshi, R.C. & Lohita, R.P. 1997. *Fly Ash in Concrete; Production, Properties and Uses*, Advances in Concrete Technology, Vol. 2, Gordon and Breach Science Publishers, pp. 269

Khayat, K.H. 1999. Workability, Testing and Performance of Self-Consolidating Concrete, *ACI Materials Journal*, Vol. 96, No. 3, pp. 346–353

Kim, J.K., Han, S.H., Park, Y.D. & Noh, J.H. 1998. Material Properties of Self-Flowing Concrete, *Journal of Materials in Civil Engineering*, Vol. 10, No. 4, pp. 244–249

Malhotra, V.M. 2002. High-Performance HVFA Concrete: A Solution to the Infrastructural Needs of India, *The Indian Concrete Journal*, February, pp. 103–107

Nagataki, S. & Fujiwara, H. 1995. Self-Compacting Property of Highly Flowable Concrete, *Special Publication SP* 154-16, ACI Publications, pp. 13

Neville, A.M. 1995. *Properties of Concrete*, 4th edition, Longman, pp. 844

Okamura, H. & Ouchi, M. 1999. Self-Compacting Concrete: Development, Present Use and Future, *Proceedings of the First International RILEM Symposium*, Edited by A. Skarendahl and O. Petersson, September 13–14, pp. 3–14

Tattersall, G.H. & Banfill, P.F.G. 1983. *The Rheology of Fresh Concrete*, Pitman, pp. 356

Potentialities of extrusion molded cement composites

K. Yamada & S. Ishiyama
Akita Prefecture University, Honjo, Akita, Japan

S. Tanaka
Taiheiyou Cement Co. Ltd., Tokyo, Japan

A. Tokuoka
Kawada Construction Co. Ltd., Tokyo, Japan

ABSTRACT: This paper describes a study about an application of extruded composite material for permanent form. The extruded cement composite material has many advantages in productivity, strength and durability, but the fracture behavior is brittle in general. The authors revealed the potentialities of the material
experimentally and analytically by showing the way to improve the ductility of the material with the use of steel
fiber and continuous fiber rebars.

1   INTRODUCTION

It is well known that extruded cement composite panels have many advantages; such as high productivity,
high strength, high durability including freezing and
thawing resistance and possible light weight by making longitudinal hollows inside. Although the composite is reinforced by short fiber as usual, the
fracture behavior of the panel is brittle (Hirato 1995).
As a result, the allowable stress is low and also
the unreliability of the material leads to the limited
application. There would be great possibility for
broad application, if the reliability of the material is
enhanced.

The best way for achieving high ductility is reinforcing by continuous rebars that can disperse cracks
along rebars. Other than that, there are many ductile
cementitious materials reinforced by short fiber such
as a cement composite reinforced by steel fiber (SF)
and ECC (engineered cementitious composites) (Li
2002) reinforced by polyvinyl alcohol (PVA) fiber.

But under the restriction of the commercialized
process including extrusion followed by autoclave
curing, above mentioned ways for enhancing ductility
was believed to be impossible.

In this study, the authors show the way to improve
the ductility and allowable stress of the extrusion
molded material with the use of steel fiber or fiber
rebars, and try to reveal the potentialities of the
material.

2   CONSIDERATIONS ON PROCESS

The conventional extrusion process has two bottle-
necks in process.

One bottleneck is related to the depressurizing
process. Because the composite material should
squeeze through small slots of a depressurized chamber to compose a material-seal, the fiber should be pli-
able and have limited length. So, vulnerable fiber such
as carbon fiber (CF) and glass fiber (GF) tends to be
easily broken, and stiff fiber such as SF was considered to suffocate the material-seal.

Another bottleneck is related to autoclave curing, as
autoclave curing is indispensable for high-productive
manufacturing process like extrusion process. Because
the temperature during autoclave curing is about
150°C~180°C, almost all synthetic fibers would melt.
So, polyvinyl alcohol (PVA) fiber, which is well known
as advantageous fiber for ECC, is not appropriate.
Only polypropylene (PP) fiber can be durable at the
temperature (Yamamoto 2002) among many synthetic
fibers except for expensive super fibers.

3   EXPERIMENT

3.1   *Reinforcement by continuous rebar*

3.1.1   *Extruder*
Cement composites have now become to be simultaneously reinforced by rebars with the use of a special

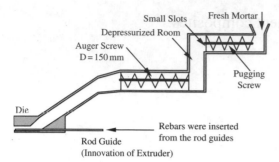

Figure 1. Typical extruder provided with special equipment.

Table 1. Mix proportion of R matrix.

Mix	Cement Vol. %	Silica Vol. %	PP fiber Vol. %	Pulp Vol. %	Additive and water Vol. %
R	32.8	28.7	1.4	0.8	36.3

Table 2. Mechanical properties of CFRP reinforcement bar.

Diameter mm	Strength GPa	Modulus GPa	Area total mm^2	CF mm^2	Bond strength MPa
3.7	1.258	147.6	10.74	4.53	15.1

divaricated pipe and a special die attached to an extruder (Yamada 1995). Figure 1 shows a typical extruder provided with such special equipment.

Utilizing the extruder, a composite material that consists of PP short fiber reinforced cement material and the continuous reinforcement made of carbon fiber plastics reinforcing bar (CFRP rebar) was manufactured. The diameter of auger screw was 150 mm.

3.1.2 Materials
Mix proportion of short fiber reinforced cement material is shown in Table 1. All powdery materials are very fine; the average grain diameter is equal to 13 μm. As fibrous materials, PP fiber and pulp (made from waste paper) were mixed. Methylcellulose (MC) was admixed as an additive to give plasticity to the composite to make extrusion successful.

A special type of CFRP rebar was developed. The outermost layer of the rebar was braided with a special synthetic fiber to achieve excellent bond strength (above 15 MPa). The mechanical properties of the CFRP rebar are shown in Table 2.

Table 3. Two types of specimen for flexural test.

Specimen	Width (mm)	Thickness (mm)	Reinforcement ratio (%)
R-B	67.42	52.35	0.00
R-22	208.11	54.89	0.11

3.1.3 Specimens
After the rebar and the mortar were extruded simultaneously, the composite was cut. In reinforced cases, the CFRP rebar was arranged only at the tensile side. After six hours of hot wet (60°C, 95% RH) cure, they were autoclaved for six hours at 150°C.

All flexural test specimens are summarized in Table 3, which reports that two types of specimens were prepared. One type was reinforced with CFRP rebar (R-22) and the other was without rebar (R-B).

3.1.4 Bending behavior
Test conditions and the measured load-deflection relation are shown in Figure 2. The blank specimen (R-B) showed good toughness. The measured toughness value described by ACI's toughness index I10 is 6.80. The reinforced specimens (R-22) showed tough bi-linear behavior. The final fracture of both reinforced specimens occurred by sudden fracture of the CFRP rebar indicating the maximum load.

3.2 Reinforcement by steel fiber

3.2.1 Extruder and materials
Utilizing the laboratory-size (the barrel diameter was 75 mm) extruder equipped with an improved material-seal, a composite reinforced by SF was manufactured.

The SF was a cut product of steel chord used for steel radial tires, which was 0.2 mm for its diameter, 9 mm for its length, over 1 GPa for the strength and 200 GPa for the modulus of elasticity. It was plated with copper for drawn processing.

Mix proportion of short fiber reinforced cement material is shown in Table 4. As binder, premixed powder for reactive powder concrete material (RPCM) on the market was used. The additive was MC. The green mortar behaves as though it flows like self-compacting concrete without MC, whereas it behaves like clay with MC.

3.2.2 Specimen
After the material was mixed, two types of specimens were extrusion molded by adding or removing the end die (die B) in Figure 3; one was square (40 mm in each side) specimen molded without end die and the other was rectangular (40 mm by 15 mm) specimen molded with end die.

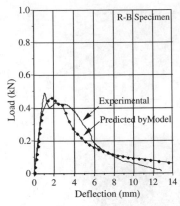

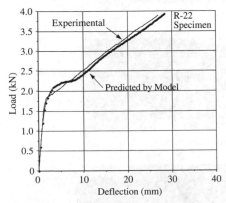

(a) Bending behavior of R-B specimen without rebar

(b) Bending behavior of R-22 specimen with rebars

Figure 2.   Bending behavior of specimens.

Table 4.   Mix proportion of D matrix.

Mix	Premixed powder Vol. %	SF fiber Vol. %	Additive Vol. %	Water Vol. %
D-B	76.14		1.47	22.39
D-S	73.52	1.00	1.48	24.00

After 24 hours of wet (20°C, 95% RH) cure, the specimens were autoclaved for 5 hours at 160°C. All flexural test specimens are summarized in Table 5.

3.2.3   SF in specimen

The distribution and the orientation of fiber were investigated by observing the bare surfaces, the polished surfaces and cut sections of specimens (Fig. 4(a–g)). The surface of the specimen right after AC curing indicated that both square and rectangular specimens have little SF on the surfaces. Figure 4(f, g)

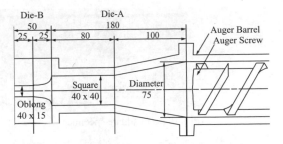

Figure 3.   Die-A and Die-B attached at the end of extruder.

Table 5.   Two types of specimen for flexural test.

Specimen	Width (mm)	Thickness (mm)	Volume fraction of SF (%)
D-B1	38.58	13.51	0.0
D-S1	38.53	13.51	1.0

shows the section of the specimens, which indicates that SF tends to distribute on the fringe of sides.

Figure 5 indicates the distribution of SF showing volume fraction of each layer that was divided into the thickness of about 3.5 mm. The average fraction for the square specimen was 1.07% whereas the fraction of each layer scatters from 0.53% to 2.00%. It is very conspicuous that the volume fraction of SF at the perimeter is greater than that in the heart. The same feature is observed in the plate-type specimen.

3.2.4   Advantage of SF distribution

It is well known that SF projects from the surface of cast concrete, which is dangerous for workers and becomes unsightly after it gets rusted. But it is advantageous that a thin layer of mortar concealed SF after extrusion molded avoiding such problems. Also it is advantageous that the major quantity of SF is located near the fringe enhancing the effectiveness of reinforcement to bending load.

3.2.5   Bending behavior of the composite

Figure 6 shows the bending diagram of specimens. Some mechanical properties related to bending behavior were listed in Table 6.

There was a drop of load after initial crack, because tensile strength of the matrix is too high. To avoid such discontinuity of load, the volume fraction (1%) and the length (9 mm) of SF should be increased.

4   ANALYSIS

4.1   Prediction of bending behaviors

There are many analytical methods of continuously predicting bending behavior of beam before and after

(a) Surface of square specimen

(b) Surface of plate specimen

(c) Polished top surface of square specimen (0.7 mm in the depth)

(d) Polished top surface of plate specimen (0.7 mm in the depth)

(e) Polished side surface of plate specimen (0.7 mm in the depth)

(f) Cross section of square specimen   (g) Cross section of plate specimen

Figure 4.   Orientation and distribution of extrusion molded SF composite material.

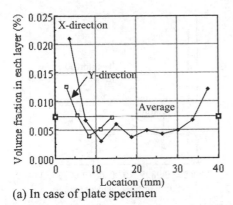

(a) In case of plate specimen

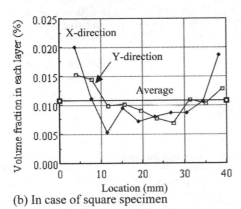

(b) In case of square specimen

Figure 5.   Distribution of SF appeared in section of specimen.

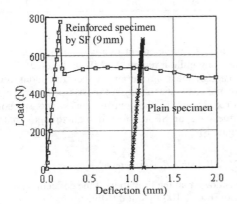

Figure 6.   Bending behavior of D-B1 and D-S1 specimen.

initial crack. When initial cracking, some type of specimen has a special behavior called snap back. It means the method should have a special consideration about the treatment of unloading by cracking. For example, it is resolved by arc-length method or by

Table 6.   Mechanical properties.

| Specimen | Flexural strength | | Flexural modulus (GPa) |
	At initial crack (MPa)	At fiber pullout (MPa)	
D-B1	20.5	–	36.8
D-S1	23.4	16.3	37.6

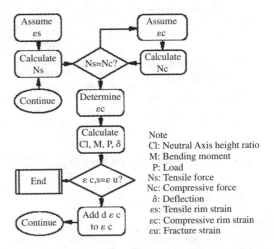

Figure 7.   Flow chart of program for fiber element method.

inclination of the coordinate system (Yoshikawa 1995) in finite element method.

The authors have been investigating fiber element method as a tool for prediction of the bending behavior of beams including snap back behavior by a localized crack (Yamada 1992, 1999, 2000, 2001). The fiber element method is a traditional analytical method that has been successfully used (Lim 1987, Craig 1987) in predictions of concrete structures and members.

In fiber element method, following three major assumptions are made. The first is that the member is assumed to be layers of thin element called fiber element. The second is that stress-strain relation of fiber element follows measured tensile and compressive behaviors of the material. After cracking, the strain of fiber element is assumed to be the average strain achieved by dividing crack width by total length of the element. The third is an assumption of the equilibrium of compressive force and tensile force in the section.

The calculation of the bending behavior progress in the way described in Figure 7, which is the flow chart of the program developed by the author Yamada (1992).

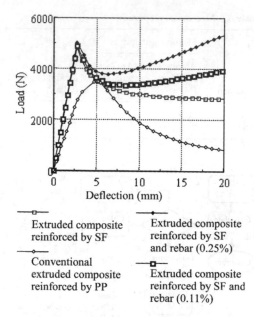

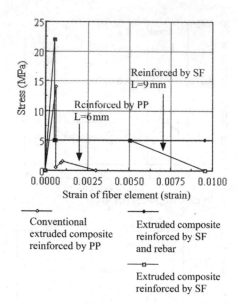

Extruded composite
reinforced by SF

Extruded composite
reinforced by SF
and rebar (0.25%)

Conventional
extruded composite
reinforced by PP

Extruded composite
reinforced by SF and
rebar (0.11%)

Figure 8. Predicted behavior of typical permanent form.

Conventional
extruded composite
reinforced by PP

Extruded composite
reinforced by SF
and rebar

Extruded composite
reinforced by SF

Figure 9. Stress–strain relation of fiber element for prediction.

4.2 *Predicted bending behaviors*

In Figure 2, predicted bending behavior is depicted together with experimental one, which was calculated by fiber element method. The diagram indicates that the prediction was successful.

Utilizing the same fiber element method for prediction, bending behaviors of the panel made of D-S matrix were calculated. The result is displayed in Figure 8 with including the one of conventional extrusion molded material (Watabe 1999). The Bending configuration for prediction is 10 mm for the thickness, 1000 mm for the width and 500 mm for the span in a condition of four points loading. The tensile stress of the modeled fiber element is depicted in Figure 9.

5 DISCUSSION

5.1 *Efficiency of continuous reinforcement*

It can be said from Figure 2 that the small amount of continuous reinforcement greatly improved the bending behavior by dispersing cracks, which add tensile force by short fiber into the equilibrium of the moment in the section.

Figure 8 also shows the predicted bending behavior of panels reinforced by CFRP in which the dimensions are the same as the one without CFRP. The composite is made of D-S matrix reinforced by CFRP in Table 2. In this case, the cracks were expected to disperse at spacing of 100 mm, so the stress–strain

Table 7. Allowable load and stress.

Case	Reinforcement ratio (%)	Allowable load (N)	Allowable stress (MPa)
A	0.25	5000	24.9
B	0.11	4000	19.9
C	0.0	2700	13.4
D	0.0	1156	5.8

relation of fiber element in a calculation model has quite long pullout region at 5 MPa in Figure 9.

Though the contribution of the continuous reinforcement is outstanding in the case of low-strength matrix (Figure 2), the contribution is inconspicuous in the case of SF reinforced high-strength matrix (Figure 8).

5.2 *Allowable flexural stress*

It is known that the limitation of the deflection is ordinarily from 1/100 to 1/400 (Ministry of Construction 1994) for permanent forms. If 1/100 is adapted, the allowable load would be 5000 N in the case-A in Figure 8. In the case-B, the allowable load would be about 4000 N, because load capacity drops after the crack. For the same reason, the allowable load would be about 2700 N in case-C.

The results are shown in Table 7, which indicates the allowable stress is more than twice (case-C) to four times (case-A) compared to the one for conventional composite (case-D) for permanent form (Yamada 2002).

6 CONCLUSION

Two types of extruder were employed to manufacture new type extrusion molded cement composites. One is an extruder that has a special divaricated pipe and die, which can mold the cement composite material reinforced by CFRP rebar. The other is a small extruder that has a special material seal, which can mold the cement composite reinforced by SF with the length of 9 mm. The major quantity of SF distributed on the fringe of the sides without appearing on the surface.

The bending behaviors of a type of permanent form were predicted employing the fiber element method and based on the measured material properties. The simulated results indicate that the SF reinforced composites would have two to four times greater allowable stress than the conventional materials as well as the improved ductility, which shows a great potentiality for permanent form.

REFERENCES

Craig, R.J. Decker, J. & Dombrowsky, L. 1987. Inelastic behavior of reinforced fibrous concrete. *Journal of the structural engineering ASCE 113(ST4)*: 802–817.

Hirato, Y. Yamada, K. & et al. 1995. Characteristics of extrusion molded cement panels reinforced with CFRP rebars. *Summaries of Technical Papers of Annual meeting AIJ (Sect. A)*: 1009–1014.

Li, V. C. Wu, C. Wang, S. Ogawa, A. & Saito, T. 2002. Interface tailoring for strain-hardening polyvinyl alcohol engineered cementitious composite (PVA-ECC). *ACI materials journal 99(5)*: 463–472.

Lim, T.Y. Paramasivam, P. & Lee, S.L. 1987. Analytical model for tensile behavior of steel-fiber concrete. *ACI Materials Journal / July-August*: 286–298.

Ministry of Construction. 1994. Evaluation sheets for NALC formwork method. Tokyo: Ministry of Construction of Japan.

Watabe, H. Tokuoka, A. Ryu, S. & Ohta, S. 1999. Rationalized construction method of highway bridge guard wall employing permanent form, *Proceedings of 9th symposium on prestressed concrete structure*: 223–228.

Yamada, K. & Hirato, Y. 1992. A discussion on method for prediction of load and deflection of thin plate made of cement composite reinforced by carbon fiber mesh. *Journal of Structure and Construction Engineering AIJ (440)*: 1–7.

Yamada, K. Suenaga, T. & Mihashi, H. 1995. Safe bending behaviour of extruded mortar reinforced with CFRP. *Proceedings of Non-Metallic Reinforcement for Concrete Structure*: 259–266.

Yamada, K. & Mihashi, H. 1999. A mechanism leading to high ductility for extruded cementitious composite reinforced with polypropylene short fibre. *Journal of Structure and Construction Engineering AIJ (520)*: 1–8.

Yamada, K. & Mihashi, H. 2000. A study on enhancement of flexural ductility for short fiber reinforced cementitious composite beams reinforced by high strength fiber rebars. *Journal of Structure and Construction Engineering AIJ (537)*: 1–6.

Yamada, K. & Ishiyama, S. 2001. Prediction of snap back behavior in beam specimen made of fiber reinforced composites employing fiber element method. *Proceedings of the JCI 23(1)*: 193–198.

Yamada, K. & Ishiyama, S. 2002. Extruded cement composite panel for formwork, In R. K. Dhir, P. C. Hewlett and L. J. Csetenyi (eds), *Innovations and developments in concrete materials and construction*: 741–750.

Yamamoto, M. & Yamada, K. 2002. Bending behavior of polypropylene fiber reinforced cementitious composites cured in autoclave. *Proceedings of the JCI 24*: 231–236.

Yoshikawa, H. & Nishioka, M. 1995. Condition on stable and unstable behavior of one-directional concrete member yielding localized region of strain. *Concrete research and technology JCI 6(1)*: 89–101.

Analytical modelling of rheology on high flowing mortar and concrete

M.A. Noor
Department of Civil Engineering, Bangladesh University of Engineering and Technology, Dhaka, Bangladesh

T. Uomoto
Department of Civil Engineering, The University of Tokyo, Japan

ABSTRACT: The detail rheology analysis high flowing mortar and concrete were studied in this paper. The rheology of mortar and concrete was calculated using coaxial cylinder rheometer. Efforts have also been made to propose generalized equations for viscosity and yield stress of mortar and concrete. The model selected for calculating viscosity in this research was Farris model. Because no model was available that predicts the yield stress of Bingham fluid, this paper proposed one. It was found that these equations gave promising result.

1 INTRODUCTION

The detail rheology study of high flowing concrete (HFC) is not available. It is expensive to prepare this kind of concrete. If the final behavior of concrete before casting could be predicted it would be economical. There has been no attempt to relate the mix proportion ratio to the rheology property so that if a designer knows the mix proption he can predict the rheology of the final mortar and concrete. Effort has been made to propose generalized equations of rheology from the mix proportion, assuming the mortar and concrete behaves as Bingham fluid.

The concrete was referred as HFC not the SCC (Okamura and Ozawa 1994); as the variation of mix proportion used in this study does not fall into the small range of SCC mix proportion. This research has been divided into two main parts. First – tests on mortar, and second – tests on concrete. The mortar has been prepared to achieve the same mortar property of previously tested concretes. Wet screening has been avoided to obtain the mortar. In the process of wet screening time was required from the time of mix and vibration was required to separate the mortar from the coarse aggregate. These two factors affect the mortar property, which were there during the mixing time. The main objective of mortar preparation was to keep the same mortar inside the concrete.

The materials used to create the concrete were coarse aggregate, fine aggregate, Ordinary Portland Cement (OPC), Ground Granulated Blast Furnace Slag (GGBS), a liquid superplasticizer, and water.

2 MATERIALS USED

In this research OPC, GGBS, Sand, Gravel, and Superplasticizer were used. OPC complying with JIS A 6206 was used. The specific gravity was 3.16. The Bogue equations (Mehta and Monteiro 1993) was used for estimating the theoretical or potential compound composition of OPC, from the chemical composition, which is given in Table 1.

In this study, only one type of cement replacement material was used. This was, GGBS, complying with JIS A6206. The superplasticizer used was SP-8Sx$_2$. The specific gravity of the admixture used was 1.05. The amount of superplasticizer used in the concrete mixes, was calculated from the mortar tests and the selected amount was kept constant for all other mixes. Tap water was used in all the mixes, and the temperature of water was normally about 20°C.

River sand (produced in Fujigawa) – with specific gravity 2.63, fineness modulus (FM) 3.21 and solid

Table 1. Composition of Portland Cement.

Compound	Abbreviation	Mineral composition
3CaO · SiO$_2$	C$_3$S	63.3
2CaO · SiO$_2$	C$_2$S	11.8
3CaO · Al$_2$O$_3$	C$_3$A	8.2
4CaO · Al$_2$O$_3$ · Fe$_2$O$_3$	C$_4$AF	8.3
Impurities	—	Others

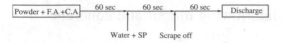

Figure 1. Mixing procedure for concrete.

volume content 64.3 percent was used as fine aggregate. Crushed stone (produced in Chichibu), with specific gravity 2.71, FM 6.45 and solid volume content 57.5 percent, was used as coarse aggregate. Maximum size of gravel was limited to 13 mm to avoid using larger size of rheometer and a large amount of concrete for tests. The sieve analysis measurement for all aggregates (coarse and fine) was conducted according to, ASTM C 136-84a, Standard Test Method for Sieve Analysis for Fine and Coarse Aggregates.

3 MIXING PROCEDURE

A basic assumption of this research, was that the mortar may be considered as the fluid phase of concrete, and as such it may be evaluated independently. The mortar phase was rheologically evaluated in the same manner as the concrete. The mixing procedure has been kept identical for mortar.

For concrete – first OPC, GGBS and all aggregate were mixed for 60 seconds then water and superplasticizer were added and mixed for another 60 seconds then the mixing was stopped and again the concrete was further mixed for 60 seconds before it was discharged from the mixture, this procedure is shown in Figure 1. For concrete, pan type mixture was used whose capacity was 100-liter. For thorough mixing, a concrete volume of 50 liters was used. Same mixing procedure was followed for mortar also. Two type of mortar mixers were used. One two-liter mixer was used, on the way to decide the final mix proportion of concrete. Another 60 liters mixer was used to perform the mortar test.

4 RHEOLOGY TEST

The rheological properties of mortar were measured by a coaxial cylinder rheometer (Fig. 4) with inner and outer diameters of the cylinder were 120 mm and 220 mm, respectively. The height for the cylinder was 250 mm. The rheological properties of concrete were measured by a coaxial rheometer with inner and outer diameters of the cylinder were 100 mm and 250 mm, respectively. The height for the cylinder was 250 mm. The inner side of the outer cylinder (Fig. 2) and outer side of the inner cylinder (Fig. 3) were modified to give the roughness on the surface to reduce the slippage on the surface.

Figure 2. Modified coaxial cylinder rheometer used for concrete. Outer cylinder.

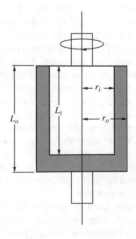

Figure 3. Modified coaxial cylinder rheometer used for concrete. Inner cylinder.

Figure 4. Coaxial cylinder rheometer.

For such instruments, the means of inducing the flow are two-fold: one can either drive one member and measure the resulting couple or else apply a couple and measure the subsequent rotation rate. There are two ways that the rotation can be applied and the couple measured: the first is to drive one member and measure the couple on the same member, whistle the other method is to drive one member and measure the couple on the other. In this research the second method is exercised. In this method, the speed of the rheometer was gradually increased and then gradually decreased. The rheology test and flow test were completed with 15 minutes after mixing of mortar and concrete. If the gap between two concentric cylinders is small enough and the cylinders are in relative rotation, the test liquid enclosed in the gap experiences an almost constant shear rate. The shear rate in the liquid at the inner cylinder is then given by:

$$\dot{\gamma} = \frac{2\Omega_1}{n(1 - b^{2/n})} \tag{1}$$

where b is the ratio of the inner to outer radius (i.e. $b = r_i/r_o$). The shear stress in the liquid at the inner cylinder is given by

$$\tau = \frac{T}{2\pi r_i^2 L_i} \tag{2}$$

the value of n can be determined by plotting T versus on a double-logarithmic basis and taking the slope at the value of Ω_1 under consideration (Barnes et al. 1989). This method was employed to estimate the rheology parameters in this research.

5 DESIGN OF THE BEST MIX

The experimental plan was so designed as to make a reasonable survey of the full range of workable mixtures of the chosen components. It was expected that segregation problems would be encountered for some compositions. The main objective was to make HFC with very little segregation and bleeding in the extreme range of the mixes. First it was attempted to prepare a segregation and bleeding free HFC. Then this mixture became the central mix and from that other mixtures were derived.

This mix proportion was designed following the mix design proposed by H. Okamura (Okamura and Ozawa 1986; Okamura et al. 1993). Only element was considered with this design method was blocking criteria. As the coarse aggregate size was limited to 13 mm, the blocking and no blocking zone was developed for this size of the aggregate for different gravel to total aggregate ratio using a developed model, based on the theory

proposed by Ozawa and Tangtermsirikul et al. (Ozawa et al. 1992) (Fig. 5). There should not be any blocking if the aggregate to total concrete ratio remains below the line for a particular gravel to aggregate volume ratio.

To design the central mix first superplasticizer amount was selected for the fine aggregate selected for the experimental plan. For this purpose several mortar tests have been done with 30% replacement and is shown in Figure 6. From this figure water to powder ratio and superplasticizer amount, for relative flow (Γ) (Okamura and Ozawa 1986) 5 and relative V-funnel speed (R) (Okamura and Ozawa 1986) 1, suggested by Okamura (Okamura et al. 1993),were calculated. The amount of superplasticizer was kept constant over all the ranges during the experiment. The selected values for w/p ratio and superplasticizer were 0.24 (by weight) and 1.3 (weight fraction of total powder), respectively.

After selecting the water-powder ratio and super-plasticizer, amount of coarse and fine aggregate volume was selected so that its ratio remain in the no-blocking zone. This required several trial and error. After selecting this mix proportion, sand aggregate ratio was varied on both side, but the powder volume was varied only on

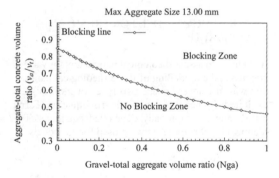

Figure 5. Blocking and no blocking zone.

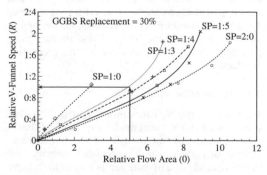

Figure 6. Relation between relative flow area and relative V-funnel speed (SP constant).

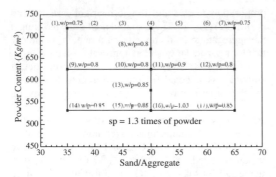

Figure 7.   Mix plan for the present research.

one side, because for the best mix the powder content was very high. It was highly impractical to select powder content above that value. The final design values of all mixes are given in Figure 7. The number four mix in the figure, is the best mix. The mortar mix design was done to get the same mortar property inside the concrete considered. So, for mortar, the coarse aggregate was removed from the corresponding concrete mix data and then mix proportion volume was made one.

6   RHEOLOGY MODEL FOR FRESH CONCRETE

In addition to assess rheological parameters for various concretes, the modelling of those rheological parameters was also investigated. Two types of models were exploited: 1) theoretical, using Farris Equation (Farris 1968), and 2) empirically developed equations. The identical input parameters were used for all models.

6.1   Model input parameters

The aggregate parameters are: sand volume fraction (S), and maximum solid volume (S_{lim}), gravel volume fraction (G) and maximum gravel solid volume (G_{lim}) for viscosity, and water to powder ratio (w/p), and total apparent aggregate volume (ts) for yield stress. The total apparent aggregate volume was defined as summation of S/S_{lim} and G/G_{lim}. These are the unique parameters which can be obtained if a mix proportion is selected.

6.2   Viscosity modelling

For viscosity model, Farris model (Farris 1968) has been employed. The Farris model, Equation 3, is based on the concept that "the viscosity of a multimodal suspension of particles can be calculated from the unimodal viscosity data of each size as long as the relative sizes in question are sufficient to have this condition of zero interaction" (Farris 1968). The modes Farris

alludes to are the different sizes of particles within the suspension.

$$\eta = \eta_s \left(1 - \frac{\phi_1}{\phi_m^1}\right)^{-[\eta_1]\phi_m^1} \left(1 - \frac{\phi_2}{\phi_m^2}\right)^{-[\eta_2]\phi_m^2} \quad (3)$$

where the first particle fraction is denoted with the subscript 1 and the second particle fraction is denoted with the subscript 2.

Since the Farris model is based on the theory that the particles within the solid phase of the suspension can be divided into two or more specific size fractions, and concrete is typically divided into coarse and fine aggregate, this model seemed applicable. Therefore, within the bounds of this research, the Farris model was used in following form:

$$\eta_c = \eta_p \left(1 - \frac{S}{S_{lim}}\right)^{-[\eta^{FA}]S_{lim}} \left(1 - \frac{G}{G_{lim}}\right)^{-[\eta^{CA}]G_{lim}} \quad (4)$$

FA and CA are two constants. The fine aggregate fraction is denoted with FA, while the coarse aggregate fraction is denoted with CA. η_p is supposedly the viscosity of the paste.

6.3   Yield stress modelling

As no model was available that predicts the yield stress of a material with Bingham characteristics, this research attempted to do so. The yield stress of the concrete was assumed to be a function of yield stress of mortar and the volume fraction of aggregate. The yield stress of the concrete must be equal to the yield stress of mortar if the gravel content is zero. With these model conditions in place, the development of a yield stress model was explored. This model is completely empirical one based on experimental data on mortar and concrete. With these in mind, the proposed empirical equation is as follows:

$$\tau_c = \tau_m + f(ts) \quad (5)$$

Total apparent solid volume is denoted by $f(ts)$. τ_c and τ_m is the yield stress of mortar and concrete, respectively.

7   DISCUSSION ON RESULTS

Following sections discussed the results obtained from the experiment plan. The result was grouped according to yield stress and viscosity. Inside each group mortar and concrete results are discussed separately.

8   YIELD STRESS

To control yield stress, one should know the effect of the variables that controls it. In this paper, an effort has

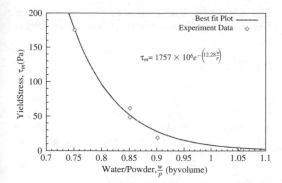

Figure 8. Relation between yield stress of mortar and w/p.

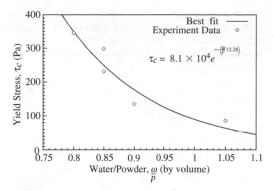

Figure 9. Relation between yield stress concrete and w/p.

been made to perceive the effect of various factors affecting yield stress and compare this yield stress with traditional testing methods for high flowing concrete.

8.1 Water powder ratio

Water powder ratio was a major variable to investigate. From Figure 8 it can be said that there was a clear relationship between the yield stress and the w/p ratio. An increase in the w/p produced a decrease in the yield stress for mortar. In above figure it can be seen that yield stress and w/p conform well to an exponential decay equation for mortar as $\tau_m = ae^{b\,w/p}$ where, a and b are two constants. After fitting the equation from the experiment data following equation was obtained for mortar. Equation 6 forms the first part of the empirical yield stress model, which shown in Equation 5.

$$\tau_m = 1.757 \times 10^6 e^{12.28\frac{w}{p}} \qquad (6)$$

As mortar, water to powder ratio has same effect on concrete. From Figure 9 it can be said that same relationship as mortar was also observed for concrete.

8.2 Volume fraction of aggregate

This was one of the key factors affecting the yield stress of concrete. From Figure 10 the effect of apparent total solid volume on yield stress can be seen. It was found that the greater the apparent total solid content, the higher the yield stress. It can be seen from the Figure 10 that yield stress of concrete follow the power law as follows.

$$\tau_c = 242.57 ts^{7.41} \qquad (7)$$

Equation 7 has been derived from the best fit curve of the Figure 10, which form the second part of the empirical yield equation given in Equation 5.

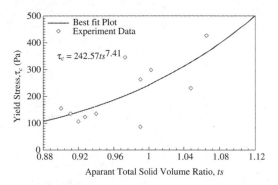

Figure 10. Relation between yield stress and apparent total solid volume.

9 VISCOSITY

Once the yield stress overcomes, the viscosity decides how fast the material will flow. The same variables those affect the yield stress also affect the viscosity.

9.1 Volume fraction of aggregate

As the viscosity model described earlier was based on the S/S_{lim} the relation between these with the viscosity is plotted in Figure 11. It can be seen that if the sand content increases the viscosity decreases. Following equation has been found by best fitting the experiment data,

$$\eta_m = 0.74 \left(1 - \frac{S}{S_{lim}}\right)^{[1.9]S_{lim}} \qquad (8)$$

η_m is the viscosity of the mortar. The value of S_{lim} can be obtained from the particular fine aggregate. The value of S_{lim} for this research was 0.643.

As the viscosity model described earlier was based on the G/G_{lim} the relation between these with the viscosity is plotted in Figure 12. It can be seen that if the gravel content increases the viscosity increases. This is because in a total volume of aggregate if the sand content increases the gravel content decreases. Following equation has been found by best fitting the experiment data,

$$\eta_c = 118.1 \left(1 - \frac{G}{G_{lim}}\right)^{[3.2]G_{lim}} \qquad (9)$$

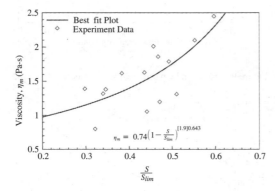

Figure 11. Relation between mortar viscosity and S/S_{lim}.

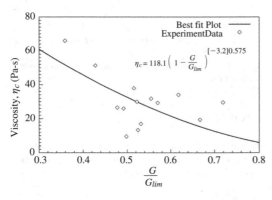

Figure 12. Relation between concrete viscosity and G/G_{lim}.

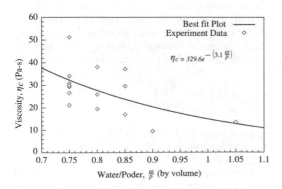

Figure 13. Viscosity Vs. w/p ratio.

The value of G_{lim} can be obtained from the particular fine aggregate. The value of G_{lim} for this research was 0.575.

9.2 Water powder ratio

The w/p ratio affects the plastic viscosity directly; an increase in w/p produces a decrease in the plastic viscosity (Fig. 13). As discussed with yield stress an attempt was made to describe the relationship between plastic viscosity and w/p with an exponential decay equation.

10 PROPOSED MODEL EQUATIONS

After analyzing the viscosity data given in Figure 11 and Figure 12 the following empirical modified version of the Farris model was developed.

$$\eta_c = C \left(1 - \frac{S}{S_{lim}}\right)^{-[2.0]S_{lim}} \left(1 - \frac{G}{G_{lim}}\right)^{[3.0]G_{lim}} \quad S > 0 \text{ and } G > 0 \quad (10)$$

In this formula constant C (60 for this research) was decided to satisfy some specific concretes values by few experiments. Because no model was available for the prediction of yield stress, it was decided to explore empirical model and analyzing the Figure 9 and Figure 10 the following equation was proposed.

$$\tau = 1.75 * 10^6 e^{-w/p*12.28} + D * ts^{7.4} \ ts > 0 \quad (11)$$

Again in this formula constant D (200 for this research) was decided to satisfy some specific concretes, by few experiments. Figures 14 and 15 show the comparison between calculated result and measured result. The range of validity, of these equations, was the range of parameters, which was used in the mix proportion.

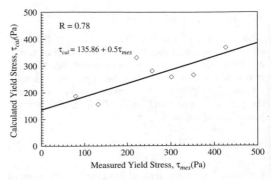

Figure 14. Comparison between calculated and measured values.

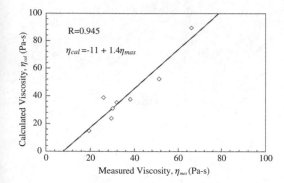

$$R=0.945$$
$$\eta_{cal} = -11 + 1.4\eta_{mas}$$

Figure 15. Comparison between calculated and measured values.

Outside the range these equation should be used with extreme care. This equations are valid for the range of w/p ratio of 0.24 to 0.28 by weight.

11 CONCLUSIONS

It was possible to predict the plastic viscosity of the concrete with the Farris equation. It was also possible to predict the yield stress of concrete based on a model developed from the experimental results of this research.

REFERENCES

Barnes, H. A., Hutton, J. F. & Walters, K. 1989. *An Introduction to Rheology* (First ed.). Elsevier Science.

Farris, R. J. 1968. Prediction of the Viscosity of Multimodal Suspensions from Unimodal Viscosity Data. *Transactions of the Soc. of Rheology 12*(2), 281–301.

Mehta, P. K. & Monteiro, P. J. M. 1993. *Concrete Structure, Properties, & Materials* (Second ed.). New Jersey: Prentice Hall, Inc.

Okamura, H., Maekawa, K. & Ozawa, K. 1993. *High Performance Concrete (In Japanese)*. Gihoudou Pub., Tokyo.

Okamura, H. & Ozawa, K. 1986. Mix Design for Self-Compacting Concrete. *Concrete Library of JSCE* (25), 107–120.

Okamura, H. & Ozawa, K. 1994. Self-compactable High Performance Concrete in Japan. *Proceedings of ACI seminar, Bangkok*.

Ozawa, K., Tangtermsirikul, S. & Maekawa, K. 1992. Role of Powder Materials on the Filling Capacity of Fresh Concrete. *The 4th Canmet/ACI International Conference on Fly Ash, Silica Fume, Slag & Natural Pozzolans in Concrete*, 121–130.

29. Innovative tools for structural design

System-based Vision for Strategic and Creative Design, Bontempi (ed.)
© 2003 Swets & Zeitlinger, Lisse, ISBN 90 5809 599 1

A web-based control and management environment for interoperation of structural programs

J. Wang & C.-Mo Huang
Department of Civil Engineering, Tamkang University, Tamsui, Taipei county, Taiwan, ROC

ABSTRACT: The procedure of traditional structural analysis and design usually consists of several individual programs and software that may be distributed among different computers or workstations. These programs are usually executed in sequence, and their inputs and outputs are more or less dependent. Moreover, the outdated systems or programs in business enterprises or engineering consultants are still in use because of their proven capabilities and user custom. Hence, integrating the classic programs and software scattered on different computers is an important issue. The objective of the work reported is to apply the concept of distributed computing to build a control and management interface for structural design and computing using the latest network technology. The emphasis is to solve the problems of data transfer, software management and execution control. The system development concept, system architecture and program interface are discussed in this paper. A short review of the technical details is presented as well.

1 INTRODUCTION

Usually a lot of programs and application software are involved in a typical structural engineering analysis and design project. These programs are always carried out in a particular order, and their input and output files more often than not have certain correlation. Moreover, these programs and software may be installed on different computers, which may even running different operating systems, in a design office. In order to effectively run a modern design office, the problems of program interoperation (Howie et al. 2000), which involve program remote execution, procedure control, data transfer, information exchange and software management, have to be solved.

The developments of computer networks and distributed computing provide new light of solving our problem. However, re-writing old programs (old in terms of operating system and/or language used) is not easy for enterprises, which is a time consuming task and requires plenty of resources. In transition, integrating the classic programs and software scattered on different computers in structural design offices by means of network links and standardized data management can from a high-powered network computing mechanism.

The objective of our research is to apply Internet and web-based technologies to build a networked control and operation interface for structural design and computing using the latest network technology. This can largely increase the effectiveness of the overall design-computing environment.

After a technology survey and prototype test, Java Server Page (JSP) was selected as the core technology used to develop the web-based control and management environment in this research. The system is based on intranet architecture. However, with an Internet connection, it can be operated on extranet. The major operating elements are built of JSP, JavaScript, Perl and MySQL database along with basic HTML and DHTML format web pages. All modules can be installed and operated on different operating systems, which accomplishes the goal of cross platform integration.

The following section introduces the core technology, JSP, used to implement the web-based program control and management environment.

2 INTERACTIVE WEB PAGE TECHNOLOGY: JSP

Internet has caused a revolutionary whirlwind in recent years. Enterprises and individuals swamped the Internet with their own web sites because of this trend of weblization and e-commerce. Information provided by static web sites can no longer satisfy the needs of WWW users. Interactive web sites offering dynamic contents are the most popular.

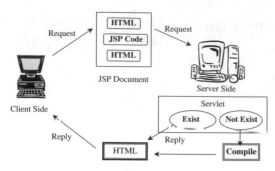

Figure 1. JSP operation procedure.

So the demand of server software and web page interpreting programs as well as the back-end database software for building interactive web sites has increased significantly.

JSP (Malk et al. 2001) is a server-side scripting language based on Java for developing dynamic web sites. JSP and JDBC together provide excellent server side scripting support for developing database driven web application. JSP enables the developers to embed Java code in HTML pages. Web pages requested by the client browsers are processed by a web server the embedded scripts are compiled and executed, and then results are sent back to the client browser as shown in Figure 1. JSP is Java technology's answer to CGI programming. It is platform independent and is Sun's answer to Microsoft's ASP (Active Server Page).

JSP files are finally compiled into a servlet in the form of "byte code" by the JSP engine, JVM (Java Virtual Machine). Compiled servlet is used by JVM to serve the client requests. JSP is efficient. A page is compiled and loaded into a web server's memory on receiving the request the very first time and the subsequent calls to the same page are served in a very short period of time without compiling again (see Figure 1).

3 SYSTEM DEVELOPMENT CONCEPT

It is not easy to replace the classic systems and programs in engineering consultants. These programs are usually in-house programs developed using Fortran. Rewriting them to cooperate with the current networked computing environment is not an easy task, which requires the domain knowledge of a particular program and the knowledge of network programming. Therefore, it is highly desirable to integrate all the analysis and design programs of a design office within an easy to setup and use environment under the premise that no early structural programs need to be rewritten. The interface of the environment should be developed utilizing the current Internet and web technology, and support the access either via intranet or Internet with

standard web browsers. Users no longer need to install special software to accomplish the interoperation of structural engineering programs.

On the other hand, for license control reasons, commercial software cannot be installed on every engineer's computer. Some programs are hardware demanding and can only run on certain workstations in a design office. There are a lot of reasons, i.e., program maintenance, version control and personnel competence, etc., prevent the installation of every software and program on all the computers in a enterprise. Therefore, how to share the computing resources is an important issue. The web-based program control and management environment should server as a convenient networked workspace for engineers working together on projects. Computer and program resources can be shared, and data can be transferred easily within the workspace.

3.1 System analysis

Take the aforementioned requirements into consideration, the web-based program control and management environment should display the following characteristics to come up to expectations.

3.1.1 Networking of old programs
The main objective is to add network ability to early structural engineering programs without altering the original program codes. That is, within the environment, input data can be prepared on a workstation, a program installed on a remote computer can be activated with the input data, and then outputs can be transmitted back to the workstation for examination.

3.1.2 Interoperation of programs on different platforms
In a series of the procedures of a structural computing project, a lot of different programs and software may be involved. These programs may not be all confined to the same computer or the same operating system. So, the program control and management environment is also targeted at interoperation of programs on different computing platforms.

3.1.3 Program process control and data sharing
A team of engineers working on a design project, each of them is responsible for a portion of the structural analysis and design tasks and they each has some program to run. This is a familiar scenario in structural design offices. Therefore, project based program process control and data sharing are very important functions that the web-based program interoperation environment should support. The knowledge of what programs are involved in a particular project and how these programs are interrelated, both procedurewise and datawise, should be stored in the system within easy reach of engineers instantly through web browsers on the network.

3.1.4 Project and program management

From the program management point of view, different type of users of the system should have different rights. Only the system administrators and project managers are allowed to use the project and program management features, including adding, deleting and modifying project contents, program information, file structures, project members, workstation information and user information. These functions should also be provided by the web-based interface to let the project managers enter the project management knowledge into the database without the help of MIS engineers. Changes can also be made in time for quick response to program revisions and procedure modifications.

3.1.5 Automatic data exchange among programs

Traditional structural analytic software such as ETABS and SAP utilizes a pre-defined text file as a medium of data input. The users fill in the data in the input file and execute the program. The outcome is written into an output file. In a series of calculation using different programs, data are sometimes passed from on program to another. In order to make all the programs involved in a project work together to solve the design problem, the interoperation environment must be able to read output files of upstream programs and generate input files for the subsequence programs. If the data exchange operation is hard-coded, adding another program to the process will require a lot of efforts. This approach lacks flexibility and expansibility.

A better approach is to establish a file format definition and transformation language (Wang & Chao 1999). In order to run a program, information sources and format that needs to be passed to the program is predefined in the environment using the language, and data is automatically collected and transferred to the input file before the program is called. The results of the program execution can also be retrieved from the output files in the same fashion. Details of the text file data exchange idea can be found in Chou (1999).

3.2 System architecture

The design of the system architecture of the web-based program interoperation environment is targeted at typical enterprise intranet environment. However, with an Internet connection, it can be operated on extranet. Figure 2 is an overview of its architecture.

According to the cross-platform principle, all the elements building up the interoperation environment including the network module are implemented using portable language, such as JSP, JavaScript, Perl, etc. Therefore, It is not limited to one operating system. The whole architecture can be easily ported to other hardware platforms.

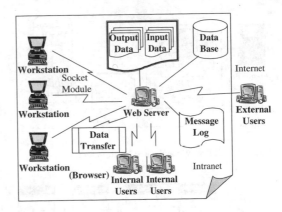

Figure 2. Architecture of the web-based program interoperation environment.

4 IMPLEMENTATION

Based on the system development concept, the program control and management environment is actually implemented as explained in the following three subsections.

4.1 Environment setup

The program interoperation environment consists of a server, several workstations and thin clients. The server computer is the heart of the program interoperation environment, which hosts a web server and a database. Database can also be installed on a separate server. The workstations are where the structural analysis and design programs actually installed. A Socket network module needs to be installed and activated on each workstation in order to link the workstation to the server to join the program interoperation environment. Thin clients are used to access the web-based program interoperation environment. No particular hardware and software requirement is to be declared. Just a web browser and a network connection are considered necessary.

In our test environment, the server is running on a Pentium 4 1.5 GHz computer with 256MB of memory. The software setup is as follows:

– Operating system: Redhat Linux 7.2
– JSP compiler: JDK (Java Development Kit) 1.3.1
– Web server: Tomcat 3.2.4
– Database: MySQL 3.23.41.
– Database tool: phpMyAdmin.2.2.2
– JDBC interface: mm.mysql.jdbc-1.2c

The way they interact with each other is shown in Figure 3.

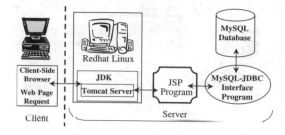

Figure 3.   JSP server and client environment.

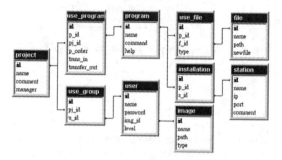

Figure 4.   Database schema.

4.2 Database design

A relational database model is used to store the required information for program interoperation. MySQL database (Dubois & Widenius 2000) was selected because of its low cost and high efficiency as well as its JDBC and SQL (Date 1995) support.

According to system analysis and schema normalization, ten tables and forty-two fields are arranged. The database schema and their relations are shown in Figure 4. With the help of phpMyAdmin, a web-based interface to MySQL, initial database setup and data entering are made easier for the system administrator.

4.3 Socket network module

Socket is a kind of application programming interface used to implement the TCP/IP communication protocol. It is also the interface between the application programs and TCP/IP. Most current operating systems offer TCP/IP API functions that provide a standard interface for programming.

There are many ways to transfer files using TCP/IP communication protocol, such as FTP or WWW. However, this research needs a low-level operation that is flexible and efficient. General application software is not suitable. Custom-built Socket program was used instead. Perl was chosen to be the Socket programming language because of its portability. Users only need to assign the parameters that the Socket program needs, and then the Socket program delivers the file to the

remote computer, activates the remote program, and retrieves the results.

The Socket program has two parts. One is the server module, which is installed on the server computer running Linux. It is called when the program interoperation environment gets a request to execute a structural analysts program on a remote computer. This Socket program demands the workstation Socket module (specified in the calling parameters) for connection and wait for the results.

The other Socket program is the workstation module, which is installed with the remote program on a workstation computer (running Microsoft Windows in our case), waiting for requests from the server module. After a successful connection, the program starts running the requested remote program. The output files of the remote program are sent back to the server side after the remote program finishes execution. Then the workstation Socket module restarts to wait for the next connection.

Although only two operating systems are tested so far, according to the cross-platform characteristic of Perl, the Socket programs mentioned above can be easily implant in any operating system without any modification. Of course Perl interpreter has to be installed first. One alternative is to use PerlApp in Perl Dev Kit (ActiveState 2003) to compile the Perl Socket program to DOS or Windows executables.

5 WEB-BASED INTERFACE

The web-based control and management environment for interoperation of structural program has two major functions: program control and program management. This section describes the web interface for these two tasks.

5.1 Program controlled interface

A representative session using the interoperation environment for program remote execution is described below:

– Enter the welcome display of the program interoperation environment and type in the user name and password to enter the main page. Click the [Project] item on the menu bar and select [Open Project] to get a list of the projects you belong to.
– Click on the desired project icon to enter the program flow page (Fig. 5).
– Select the programs for execution one by one.
– Click the [Open] button on the left after a program is selected to show the input file icons (Fig. 6). Click an icon to modify the input file (Fig. 7).
– Click the [Execute] button on the left after all the input files are prepared to show all the workstations

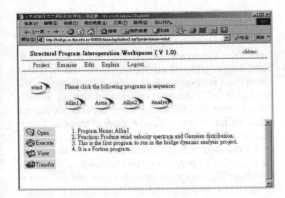

Figure 5.   Select a program to run.

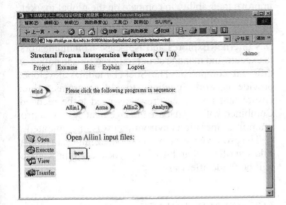

Figure 6.   Select an input file.

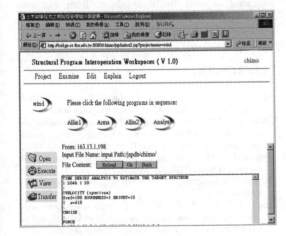

Figure 7.   Edit the input file.

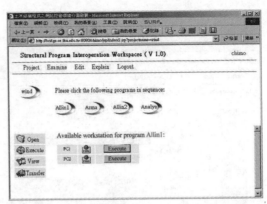

Figure 8.   Select a computer to execute the current program.

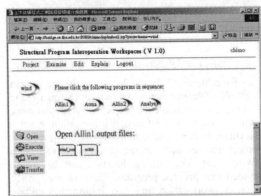

Figure 9.   View output files.

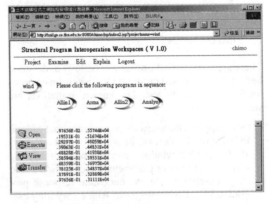

Figure 10.   Contents of an output file.

that have this program installed (Fig. 8). Select one for execution.

– Select [View] from the left after the message "Remote job completed" is shown. Then you can select the output files (Fig. 9) one at a time to examine the results (Fig. 10).

– Go to the next program in the sequence and repeat the procedure again until all the programs are finished.

Figure 11.   Edit workstation information.

5.2   *Project and program management interface*

The Examine and Edit on the menu bar are the two functions of the management interface. The query function is used to examine the project, program and workstation contents as well as user information. A set of interfaces is designed to help the users to locate his project and resources in the program interoperation environment. The editing function can be further divided into project, program, file structure, project member, workstation and user information editing. Each has a set of interfaces designed to perform information adding, deleting and modifying. See Figure 11 for an example.

6   LIMITATIONS

There are some limitations have to be fully understood before the web-based program interoperation environment can be put to test. The known limitations are stated as follows:

- Only supports the software or programs that can run in batch mode. That is, the software must not operate by graphic window interface without offering script commands or parameters, such as SAP2000.
- Only incorporates fairly immediate response programs. The remote programs with a very long execution time are not suitably. This is restricted by the response time of HTTP protocol and web browsers.
- Only supports the software or programs that use text I/O. The software or programs operated in the environment must use text files as their primary I/O media. Display of special output format, such as AutoCad DXF files, is not supported now.

7   CONCLUSION AND FUTURE DEVELOPMENT

How to utilize the latest information technology to build a high efficient computing environment within enterprises is always a goal to strive for. However, the resistance of industrial informationlization usually comes from the following three aspects: economic effect, technical support and personnel training. Therefore, previous efforts on program integration in enterprises and engineering consultants have not been totally successful. The purpose of this research is the same, searching for a program interoperation and integration model that can remove the abovementioned resistance. This can accelerate the process of informationlization and networklization in engineering enterprises and improve their ability to compete on the global market.

The major contribution of the research is that engineers can make use of standard WWW browsers (e.g., MS Internet Explores and Netscape Communicators) to execute and manage structural programs installed on different computers. The Socket modules developed provide network functions to old programs, which ensure that no classic programs need to be recoded. Combined with a database management system, a program interoperation environment was implemented.

The prototype system is now under full evaluation at a design office of architecture associates. Improvement will be made afterward.

ACKNOWLEDGEMENTS

Financial support for this project from National Science Council of Taiwan, ROC under grant NSC-89-2211-E-032-023 is gratefully acknowledged.

REFERENCES

ActiveState 2003. ActiveState – Products – Productivity Tools – Perl Dev Kit. URL: http://www.activeperl.com/.

Chou, Y.-C. 1999. *The study of file format transformation for associated structural design programs*, master thesis, Tamkang University, Tamsui, Taipei County, Taiwan, ROC.

Date, C.J. 1995. *An introduction to database systems*, 6th Edition. Boston: Addison-Wesley.

Dubois, P. & Widenius, M. 1999. *MySQL*. Indianapolis: New Riders.

Howie, C.T., Law, K.H. & Kunz, J.C. 2000. *A model for software interoperation engineering enterprise integration*, CIFE Working Paper #61, Stanford University, Palo Alto.

Malks, D. et al. 2001. *Professional JSP 2nd edition*. Chicago: Wrox Press.

Wang, J. & Chao, C.-W. 1999. Cooperative networked workspaces for structural design. In B.H.V. Topping & B. Kummar (eds), *Novel design and information technology applications for civil and structural engineering*: 33–38. Civil-Comp.

System-based Vision for Strategic and Creative Design, Bontempi (ed.)
© 2003 Swets & Zeitlinger, Lisse, ISBN 90 5809 599 1

Attitudes to 3D and 4D CAD systems in the Australian construction industry

S. Saha, M. Hardie & A. Jeary
School of Construction, Property & Planning, University of Western Sydney, N.S.W., Australia

ABSTRACT: New technology in 3D and 4D CAD systems offers a myriad of possibilities in terms of efficiency gain, coordination, error detection and forward planning. These benefits are far from being fully utilised at the moment and there appears to be some resistance to their greater incorporation in the general construction process. This study seeks to analyse the extent and the sources of such resistance. It further intends to look at the most readily accepted forms of computer documentation and the most useful extensions to these systems. The results point to areas for future study in terms of the usefulness of 3D and 4D CAD as training devices and "dry-run" facilities in the construction industry.

1 INTRODUCTION

Although the residential construction industry in Australia is a significant slice of the national economy in dollar terms, it has been slow to adopt new technologies. This research aims to show that productivity gains and improvements in end-user satisfaction can be made by the adoption of integrated computer based training and documentation systems. Although the home building process has tended to remain resolutely "low-tech", this has not led to a lack of defects or a low rate of necessary reworking. In fact, consumer dissatisfaction with the products of the home building industry is at a high level, and complaints about shoddy workmanship and the cost of repairs are rapidly increasing. Of course, shoddy workmanship can have other causes apart from failure to properly communicate the task at hand to the construction worker. Inadequate time allocation, pressure of ongoing work schedules, lack of adequate supervision and even deliberate sabotage may all contribute to poor quality results (Tilley 1997). The use of training and documentation based on animated CAD modelling may be expected to reduce or eliminate at least those defects that are due to discrepancies, omissions and simple mistakes in the drawings and specifications. In addition, the effect of animating the construction process is hypothesized to lead to a better level of understanding in the building worker, and therefore result in less reworking.

2 BACKGROUND

The CAD systems that were first introduced into the construction industry were strictly two-dimensional in scope. They were, in fact, merely an automated version of traditional drawing systems. It soon became apparent that computer systems offered the option of designing in three dimensions, rather than in plan and elevation separately, and that 3D work enabled better visualization of the project and resulted in fewer discrepancies and anomalies in the documentation (Dorsey & McMillan 1998). 3D CAD systems developed inherent checking processes, so that they proved to be superior in this respect even to the use of scale models in conjunction with drawings. Anadol and Akin (1994) found a significant reduction in "change orders" or reworking, necessary in projects designed with CAD systems compared to conventional media. In addition, CAD documentation resulted in readily shared information between the different specialist groups and therefore in better co-ordination between them.

As well as eliminating certain kinds of errors, computer visualization lends itself to the incorporation of many additional functions into the design process. According to John Mitchell, architect and director of the International Alliance for Interoperability Australian Chapter (IAI AC) "The 3D visualization of the plan shows the client what the building will look like and how it will function. The next step... is to include building codes, standards, specifications, reports and automatic procurement. This will automate the process of documentation, schedules and quantities, and will be capable of checking planning accuracy, such as whether the fire exits are in the right place to conform with the code" (Practicing e.Construction 2000).

The introduction of the time element to 3D documentation represents a further leap in both complexity and possibility. 4D CAD documentation may be used in construction planning to identify problems (McKinney, Fischer & Kunz 1998). In addition, H.-Y. Kang at the Texas A & M University has conducted a comparison study using civil engineering students of the relative effectiveness of 4D visualization software relative with 2D web drawings. The study concludes that the 4D group "detected more logical errors, made fewer mistakes, detected logical errors faster, and communicated less frequently than the 2D drawing groups"(Kang 2001).

The development of 4D animation models to describe construction operations is proceeding apace in several locations. Kamat and Martinez (2000) have developed a simulation model of the block-laying process that can provide the user with insights into the consequences of different techniques and strategies. It can help define choices in the construction process that are otherwise not easily quantifiable. The system can be developed as an effective training tool or as a dry run simulation for a complex operation. Xu and AbouRizk (1999) have used a case study on CAD simulation of earthmoving to suggest that the simulation of the whole construction process on 4D CAD is not beyond the realms of possibility.

3 AIMS AND OBJECTIVES

This study aims to verify the demand for integrated information systems in the residential construction industry. It also seeks to identify areas of resistance to computerisation of the construction documentation process and to analyse and address the reasons for such resistance. It will try to examine perceptions about whether 3D CAD systems can ultimately lead to fewer errors and reworking in construction.

4 METHODOLOGY

A survey was conducted among various construction industry personnel to gauge attitudes to, and interest in the various forms of CAD documentation. The survey sought responses from architects, building designers, quantity surveyors, building estimators, construction managers, project managers, builders and building surveyors.

Both email and personal interviews were used to seek responses. 87 surveys were sent out and 29 responses were received with a completed questionnaire (a response rate of 33%). The survey sought to ascertain whether the general response to the use of new technology in construction was positive or negative, and whether any pattern emerged in the responses.

5 DATA RESULTS

5.1 Overall response

Generally, the response to many questions was significantly more negative than anticipated. There maybe several reasons for this. Most of the questionnaires were responses to unsolicited emails. It could be that those wishing to lodge a protest about CAD systems would make the time to respond. The percentage standard error for the individual questions ranged from 9% to 18% (Siegel 1988). It is noteworthy that on a number of the questions asked, the largest response was neutral. Comments indicate that this may, in fact, mean simply that judgement is reserved on the matter and no conclusions can yet be drawn. In other words, there is a considerable constituency waiting to be convinced of the value of new systems and ways of working.

The response to proposition 1 indicated that a clear majority of those surveyed see 3D CAD as "the way of the future". However, proposition 2, which sought to determine the level of current usage of 3D CAD systems in the respondent group had roughly 40% preferred users to 60% who prefer other options. It is clear, however, that a positive attitude to CAD usage was significantly more common among architects and building designers than among other professionals surveyed. Overall there was some scepticism that 3D CAD would lead to fewer discrepancy-type errors (proposition 3). Further research is necessary to determine whether this response correlated with negative answer to the second question.

Slightly less than half of the respondents agreed that 3D CAD was easier for builders to understand and comments indicated that this was due to a widely held belief that it is the quality of the construction knowledge of the documenter that is crucial rather than the system used. This theme was also reflected in the responses about whether fewer construction mistakes or more accurate costing would result from 3D CAD usage (proposition 5). Architects as a group were negative on this proposition in contrast to their general attitude.

The response was more positive from all groups on the use of 3D CAD for construction scheduling (proposition 7) and comments indicated that some of the extensions related to CAD systems could lead to the automation of some of the repetitive areas of construction planning. There was a definite positive response to the proposed use of 3D and 4D CAD for training purposes (propositions 8, 9 & 10). This is an area that seems to have a ready market for well-developed reality-based construction modelling. Skepticism reasserted

Table 1. General survey response (29 responses).

Proposition	Total positive	Total neutral	Total negative
1. 3D CAD is the way of the future	61%	32%	7%
2. 3D CAD is my preferred form	42%	29%	29%
3. 3D CAD leads to fewer drafting errors	39%	32%	29%
4. 3D CAD is easier for builders	42%	33%	25%
5. 3D CAD leads to fewer construction mistakes	36%	28%	36%
6. 3D CAD makes costing more accurate	29%	42%	29%
7. 3D CAD is useful for construction scheduling	50%	36%	14%
8. 3D CAD is useful for training design office staff	61%	32%	7%
9. 3D CAD is useful for training management staff	57%	32%	11%
10. Animated (4D) documentation would be useful for training workers	61%	22%	17%
11. 4D CAD would be useful in construction planning	54%	36%	10%
12. Accurate 4D CAD would make estimating more precise	29%	39%	32%
13. Residential building is resistant to increased computer use	29%	39%	32%
14. My current work method will not change in the next ten years	25%	18%	57%
15. The next ten years will see increased use of computers in residential building	86%	14%	0%

Table 2. Response of architects and building designers (12 responses).

Proposition	Total positive	Total neutral	Total negative
1. 3D CAD is the way of the future	75%	17%	8%
2. 3D CAD is my preferred form	67%	17%	16%
3. 3D CAD leads to fewer drafting errors	50%	25%	25%
4. 3D CAD is easier for builders	67%	17%	16%
5. 3D CAD leads to fewer construction mistakes	42%	8%	50%
6. 3D CAD makes costing more accurate	42%	33%	25%
7. 3D CAD is useful for construction scheduling	67%	17%	16%
8. 3D CAD is useful for training design office staff	82%	9%	9%
9. 3D CAD is useful for training management staff	67%	25%	8%
10. Animated (4D) documentation would be useful for training workers	58%	25%	17%
11. 4D CAD would be useful in construction planning	67%	17%	17%
12. Accurate 4D CAD would make estimating more precise	33%	25%	42%
13. Residential building is resistant to increased computer use	16%	42%	42%
14. My current work method will not change in the next ten years	16%	17%	67%
15. The next ten years will see increased use of computers in residential building	100%	0%	0%

itself, however, on the question of whether 4D CAD could make estimating more accurate (proposition 12). Once again the industry remains to be convinced that such systems will work in practice. The survey shows widespread acknowledgment of the likelihood that the use of computer technology in residential building will increase in the coming decade (proposition 14). Furthermore there is considerable willingness to respond to this change, as a large majority of respondents perceive that their own working methods will change in the next ten years (proposition 15). The standard error percentages for the overall group ranged from 9% to 19%.

5.2 Response broken down by employment category

Since survey participants came from a broad range of roles within the construction and design industry it was considered useful to separate out responses according to employment category. The most significant divergence of opinion, as previously mentioned, was that between those primarily concerned with design and the rest of the industry.

Architects and building designers as a separate group were significantly more positive than the overall response as shown in Table 2. This group was considerably more enthusiastic at the prospect of using animated CAD for training purposes but also more enthusiastic about all the extensions that can be attached to CAD systems. Two thirds of this group were current users of 3D CAD and greater familiarity may at least partly explain their enthusiasm. The standard error percentages for this sub-group ranged from 12% to 21%.

6 DISCUSSION OF COMMENTS

It is clear that despite some cynicism there are demonstrable advantages to be gained from the use of integrated IT systems in the residential construction industry. In particular, Object Oriented CAD documentation systems can have a number of positive effects for the industry (Anadol & Akin 1994). For example, the ready exchange of information on computer file between the various design and construction consultants can be expected to lead to fewer discrepancies in design documentation. Land surveyors, architects, structural and mechanical engineers are able to produce documents in the same format, so that conflicts of interest may be resolved at the documentation stage rather than on site at greater expense. Quantity surveyors can also have input at this stage with resultant better control over cost outcomes. This appears to be well recognised in the industry.

Nevertheless there remains considerable resistance to the wholesale transfer of the industry to IT based work systems. There are a number of sources for this disquiet. Firstly, several industry sources in our survey stressed that it is the skill of the operator rather than the operating system that determines the quality of the documentation. "Garbage in–Garbage out" was the refrain of several architects and quantity surveyors interviewed for this study. Jim Peet, senior principal of Widnall, Quantity Surveyors, stated that:

"In my experience (now over 45 years) and having developed computer systems for our own practice over the last twenty years, any form of drawing is only as good as the level of knowledge of construction of the person doing the drawing work by whatever method."

In addition, some respondents believed that newer CAD systems could (temporarily) mask the skill level of the operator, and lead to documentation that appeared to be adequate when this was not in fact the case.

The second area of objection to the computerisation of the documentation process related to the constant updating of platforms and systems. This leads to an ongoing capital outlay in order to keep up with the latest technology and small business particularly finds this prohibitive. In addition, incompatibility of old and new systems and between platforms is a further difficulty that plagues small business. Some architects have chosen to remain outside the CAD revolution and continue to produce traditional documentation for this reason. This is an issue that needs to be addressed before further integration of computer technology into the construction industry can proceed.

Thirdly, there is some concern among designers that many CAD systems fall short of their promise in terms of producing usable construction documents in 2D form. The kind of diagrammatic sections so easily produced by 3D CAD systems may not contain sufficient information to make them useful for structural and engineering services purposes (Dakan 2001). One of our survey respondents, Vanovac Associates confirmed this opinion stating that although "We use 3D Studiomax for 3D modelling and animation. 3D documentation is not, in our opinion, very effective."

Finally, our contractual and legal system still makes it necessary to produce a 2D paper document than can be signed off as the "Contract Document". This tends to mean that some of the benefits of integrated transfer of information are lost when the legal system ties all parties to the initial document produced to establish the project.

On the other hand some of the survey respondents did feel that resistance to implementing integrated IT throughout the whole construction industry was largely due to the inertia of established businesses wishing to go on operating in the manner they had first learnt when they started out in the industry.

There was broad agreement among our surveyed group that 4D animated documentation of construction processes could be a useful training tool. Furthermore it was believed that it might be useful in construction scheduling. Independent research has confirmed this (Kang 2001). However, the view was strongly expressed that the use of animated documentation for actual construction projects was likely to meet with firm resistance. Respondents suggested that experienced construction workers were certain to feel condescended to by such documentation and to believe that it impinges on their exclusive territory. There is a great deal of existing expertise in turning paper documents into finished buildings and this would be somewhat threatened by animated documentation that purported to tell a builder how to do his job. To borrow from the terminology used in the Building Code of Australia, it could be said that builders prefer "performance based" documentation that specifies the outcome required, rather than "deemed to satisfy" documentation that tells them precisely how to do it.

7 CONCLUSION

Our indicative survey showed quite a wide diversity of opinion on the subject of CAD documentation as it currently operates in the construction industry. The survey included both enthusiastic proponents of 3D CAD systems and sceptics about its overall value. It also included some who have eschewed the use of CAD systems altogether. Future studies should seek to separate out negative attitudes due to failures and difficulties experienced with CAD systems and negative attitudes due to innate resistance to change.

Enthusiasm for the use of animated 4D CAD as a training tool was widespread. Also widespread was scepticism about the use of 4D CAD as documentation. Of course, this is speculative as such systems are not widely available as yet. Development of the technology to demonstrate construction systems in realistic animated form may convince some of the doubters. However the contractual and legal barriers to its use as construction documentation have yet to be addressed.

In future developments, this study area will be extended by practical experimentation in the efficacy of 4D construction documentation using a concrete slab as an example. This will develop work done previously on the rate of error detection in concrete slab construction (Saha, Greville & Mullins 1999).

REFERENCE LIST

Anadol, Z. & Akin, O. 1994. Determining the impact of CAdrafting tools on the building delivery process, *International Journal of Construction Information Technology*, Vol 2 no 1, pp. 1–8

CSIRO Built Environment Innovation & Construction Technology, *Practising e.Construction*, Number 16, December 2000. http://www.dbce.csiro.au/innovation/2000_12/e_construct.htm Accessed 1/10/02

Dorsey, J. & McMillan, L. 1998. *Compute graphics and architecture: state of the art and outlook for the future*, ACM Press, New York http://doi.acm.org//10.1145/279389.279449 Accessed 17/9/02

Kamat, V.R. & Martinez, J.C. 2001. Visualizing simulated construction operations in 3D, *Journal of Computing in Civil Engineering*, October 2001, American Society of Civil Engineers, New York

Kang, H-Y. 2001. *Web-based four-dimensional visualization for construction scheduling*, Texas A&M Accessed 24/9/02

McKinney, K., Fischer, M. & Kunz, J. 1998. Visualization of Construction Planning Information, *Proceedings of the 3rd International Conference on Intelligent user interfaces*, January 1997. http://doi.acm.org//10.1145/268389.268414 Accessed 24/9/02

Saha, S.K., Greville, C. & Mullins, T. 1999. Simulation Experiment: The effects of experience and interruption in predicting error rate for a construction inspection task", *Modelling and Simulation Society 1999 Conference*, December 1999, New Zealand

Siegel, A. 1988. *Statistics and Data Analysis, An Introduction* John Wiley & Sons, New York

Singh, A.K. 2002. *Electronic simulation in construction*, University of Cincinnati. Abstract available at http://wwwlib.umi.com/dissertations/fullcit/1408476 Accessed 24/9/02

Tilley, P.A. 1997. Causes, Effects and Indicators of Design and Documentation Deficiency" *Proceedings 1st International Conference on Construction Industry Development: Building the Future Together*, Singapore, 9–11 November, 1997,2, 388–95. http://www.dbcc.csiro.au/research/papers/abstract.cfm/148 Accessed 1/10/02

Xu, J. & AbouRizk S. 1999. Product-based model representation for integrating 3D CAD with computer simulation, *Proceedings of the 1999 Winter Simulation Conference*, December 1999, pages 971–7. http://doi.acm.org//10.1145/324898.324975 Accessed 24/9/02

System-based Vision for Strategic and Creative Design, Bontempi (ed.)
© 2003 Swets & Zeitlinger, Lisse, ISBN 90 5809 599 1

The 4D-CAD: a powerful tool to visualize the future

M. Barcala, S.M. Ahmed & A. Caballero
Department of Construction Management, Florida International University, Miami, Florida, USA

S. Azhar
Department of Civil and Environmental Engineering, Florida International University, Miami, Florida, USA

ABSTRACT: Visualization is becoming increasingly important for construction projects because it can help to visually understand the implications of any design or construction technique without performing actual construction. Visualization tools such as the 4D model (3D graphical model plus time) gives planners the ability to improve construction sequences, identify and resolve construction conflicts, and track and manage labor and resources. The objective of this paper is to analyze the effectiveness of 4D models for construction. Case studies of a number of construction companies using these kinds of models were analyzed. Most companies have the opinion that the 4D models are helping them to deliver highly quality projects with reduced cost and time. However, some professionals argued that the 4D models are still expensive for small-to-medium size construction projects. This paper will explain the advantages and disadvantages of 4D models for different type of projects and their possible uses into the construction arena.

1   INTRODUCTION

The AEC industry has been witnessing a steady increase in the use of 3D and 4D (graphical illustration of 3D models with time as the 4th dimension) tools for project planning. The idea to link 3D-CAD models to construction schedules was conceived in 1986–87 when Bechtel (a leading, international, engineering and construction company) collaborated with Hitachi Ltd., to develop CAE/4D Planner software (Smith 2001). A 4D model involves linking the CPM schedule to the 3D-CAD model to visualize the construction schedule; actually showing which pieces of the project will be constructed in what sequence (Rischmoller et al. 2001).

4D models display the progression of construction overtime, sometimes dramatically improving the quality of construction plans and schedules (Rischmoller et al. 2001). Several documented studies have shown 4D-CAD as a good visualization and schedule review tool. More project stakeholders can understand a construction schedule more quickly and completely with 4D visualization than with the traditional construction management tools (Fischer 2000).

4D models allow simulating and interacting with construction sequences (schedules) through graphical display devices. If the sequence is not just right, schedulers adjust the schedule and rerun the 4D simulation to verify it. It gives planners the ability to improve construction sequences, identify and resolve construction conflicts, track and manage labor and resources such as formwork, scaffolding, and cranes to make sure they all are applied effectively. It lets planners formulate a tighter and more finely tuned construction plan (Rischmoller et al. 2001).

4D models can also help to develop contingency plans to handle delays in material deliveries or address the unavailability of resources. To avoid rework, important decisions concerning deadlines, sequences, and resource utilization, which would ordinarily be made at the jobsite, are better made ahead of time. Feedback from construction to the design teams resulting from 4D model reviews can often lead to a more readily constructible, operable and maintainable project (Rischmoller & Alarcon 2002).

1.1   The need

The need for new tools that can contribute to the development of the construction industry is indispensable. Construction professionals must be capable to use and adapt IT solutions into their projects in order to optimize and save resources. Because of that, construction professionals must be aware that traditional design and construction planning tools, such as 2D drawings and network diagrams, are not enough to maintain competitiveness and quality anymore (Teicholz 2000). These tools do not support and bring the required information to move projects forward as quickly as is needed in today's world. These tools do not have the capability to provide visualization, analysis, and modelling to support decision-making. Because of that, professionals have been asking and working to get computerized system that can keep all the data, interconnect users, and help to manage projects at the speed that is required in today's world.

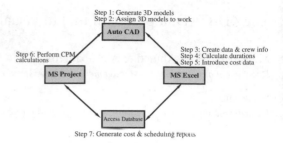

Step 1: Generate 3D models
Step 2: Assign 3D models to work
Auto CAD
Step 6: Perform CPM calculations
Step 3: Create data & crew info
Step 4: Calculate durations
Step 5: Introduce cost data
MS Project
MS Excel
Access Database
Step 7: Generate cost & scheduling reports

Figure 1. Components of 3D-CAD.

2 EVOLUTION OF 4D MODELS

2.1 Computer integrated construction (CIC)

A computer integrated construction (CIC) system is a flexible and advantageous way to expedite construction operations. It helps to integrate the management, planning, design, construction, and operation of constructed facilities. Computer integrated construction systems automate many of the labor-intensive tasks associated with construction management of new facilities. This means that knowing applications, functions, and benefits of computer, professionals can put together all of the activities involved in a construction projects. Through the integration of the 3D models, database management systems, and expert systems, CIC systems allow users to automatically calculate material quantities from CAD models, test the constructability of the design before actual construction, report construction progress graphically, and improve collaboration between project members.

The main function of CIC system is to interconnect and share data with all project participants. Due to this integration, CIC systems increase the effectiveness of communication and the management processes through the idea of managing all the data relative to projects. This data is available anytime to anyone who needs it and is authorized to get it.

2.2 The 3D-CAD model

The whole core of the CIC is the 3D-CAD model. This program allows all project participants to visualize the construction process at the same time and the way they built. Additionally, this technique could be used as planning and management tool to promote interaction and collaboration between the parties involved. As a result, professionals can get quick solutions that save time and cost for all entities involved in the construction projects.

The 3D-CAD model is a very powerful tool to help and improve visualization. This model represents clearly the entire design information of projects, and help to avoid misunderstanding during the construction process in the field. The database incorporated to the 3D-CAD model has a lot of information to determine the 3D properties of true solids, draw full size objects to establish possible interferences, and simulate construction process in order to familiarize participants with the scope of a specific work.

According to Fisher (1999), the 3D Construction CAD was designed to integrate all tasks involved in construction projects, and facilitate the management of them. This was accomplished by the integration of the different applications as shown in Figure 1:

1. AutoCAD to create all 3D models.
2. Microsoft Access Database to store all the data relative to crews, units costs, and productivity.
3. Microsoft Project Scheduling System to develop CPM charts, and
4. Microsoft Excel to report quantities related to calculations.

Through these applications we can create "intelligent" objects. The information is stored in different layers in order to create accurate cost estimates, quantities, and durations for a specific element or any part of the project.

3 THE 4D MODEL

The 4D model, which combines the 3D model with scheduling software, is quickly gaining popularity in the construction industry after years of laboratory studies. This tool represents a great technological advantage to construction professionals. They can create construction projects digitally, and get all the information related together in just one database. An example is illustrated in Figure 2 (Goldstein 2001).

4D-CAD is more than just a visualization tool. The power of 4D-CAD is not only the visual simulation of the projects. It additionally provides a common view for designers to produce proposals, and to explore the real knowledge, and the relationship between different dimensions of the construction elements.

1980

Figure 2.　A 4D model for a pumping plant.

4　CASE STUDIES OF SELECTED COMPANIES

4.1　*Walt Disney Imagineering Research and Development Inc.*

Walt Disney Imagineering Research and Development Inc, who is responsible for the design and construction of Disney's theme parks, was among the first companies to use 4D models for their projects. Through the 4D model, they got the opportunity to increase productivity and reduce waste on job sites. Their executive director, *Mr. Ben Schwegler*, says that any activity which can produce huge amount of waste has to have room for productivity improvements, if a comparison is made with manufacturing industry. 4D models provide an opportunity to improve productivity in construction in the same way as lean manufacturing techniques helped the manufacturing industry (Goldstein 2001).

4.2　*DPR Construction Inc.*

The main value for 4D model comes from using it strongly to visualize the construction sequence of work. Everyone would see what will go first, where you can store materials on site, and fit equipments in specific places of the project site. Additionally, 4D tools have the ability to run different scenarios in order to determine optimum scheduling and resource management. Also, by linking 3D drawings to a project schedule, all entities involved in a project can see how the project is going and/or supposed to go.

As another example, Peter Allen, project manager for DPR Construction Inc., used 4D-CAD on the $72 million Bay Street Entertainment and Retail Complex in Emeryville, California as shown in Figure 3. DPR won this project because they could cut some weeks

Figure 3.　Actual construction and 4D model of the Bay Street Entertainment and Retail Complex.

off the project schedule by speeding up some activities in a specific area of the complex.

"By running a 4D simulation using InviznOne software developed by Walt Disney Imagineering, we found we could accelerate the steel works in the theater area and save three weeks, Allen says. (Teicholz 2000)"

The creation of virtual construction scenarios can help project managers and contractors to visualize specific situations. Therefore, construction professionals could plan alternative ways to attempt the construction activities such as concrete formwork, crane operations, and materials lay down by avoiding delays and conflicts on site. In addition, professionals could identify crew sizes, space availabilities, and possible obstructions that would save time on the general schedule and money to owners.

4.3　*Intel Corporation*

Another example of using 4D-CAD into the construction world is Computer Chipmaker Intel Corporation. The 4D model has been used by them in order to plan a new $400-million-plus fob facility close to Portland, Oregon. The results of using 4D modelling has eliminated significantly design conflicts and redundancies on this project.

4.4 *Drawbacks of 4D models*

Even though 4D-CAD represents a great improvement into the construction industry, some professionals believe that this tool is expensive and slow to use it. 4D models are very time-consuming and too complicated to generate manually. Therefore, the difficulty and cost of creating and using such models is currently blocking their widespread adoption. Prof. Paul Teicholz of Department of Civil and Environmental Engineering at Stanford University said that research and work are needed to redefine 4D-CAD tools in order to make them easier and faster to use into the construction industry. He also recommended that these tools must generate information and plans that would be analyzed directly by the computer (Teicholz 2000).

4 CONCLUSIONS

Construction professionals must be aware of the advantages that IT solutions can bring into the construction industry. Professionals have to be able to introduce different technological tools in projects in order to save resources and increase productivity. Even though 4D-CAD is expensive and complicated tool, it is still an important step to make complex projects easier to accomplish by having the idea of where and when professionals can get design conflicts. Therefore, they can focus on those aspects, and save enormous amounts of time and cost. Technologies such as the 3D and 4D-CAD model represent a time consuming process at the beginning of every project; however, the earlier the professionals can identify and visualize problems and conflicts, the easier would be to find out solutions and alternatives to improve productivity.

REFERENCES

Burchard, B. & Pitzer, D. 1999. *Inside AutoCAD 2000*. Indianapolis: New Riders.

Fischer, M. 2000. Benefits of 4D Models for Facility Owners and AEC Service Providers, *6th Construction Congress, ASCE*, Orlando, Florida, February 20–22, pp. 990–995.

Fisher, M. 2001. *Introduction to 4D Research*. Available: http://www.stanford.edu/group/4D/index.shtml

Goldstein, H. 2001. *4D: Science Fiction or Virtually Reality*. Available: http://new.construction.com/NewsCenter/it/features/01-20010416.jsp

Rischmoller, L., Fischer, M., Fox, R. & Alarcon, L. 2000. 4D Planning and Scheduling (4D-PS): Grounding Construction IT Research in Industry Practice. Available at: http://fourd.stanford.edu/papers/kathleen/asce-96.pdf

Rischmoller, L. & Alarcon, L.F. 2002. 4D-PS: Putting an IT New Work Process into Effect. *Proceedings of the International Council for Research and Innovation in Building and Construction*, CIB W78 Conference, Aarhus School of Architecture, June 12–14, 2002, pp. 1–6.

Smith, S. 2001. *4D-CAD Goes Beyond Mere Representation*. Available at: http://www.aecvision.com/October2001/1_feature_full.pdf

Teicholz, P. 2000. *Vision of Future Practice*. Available: http://www.ce.berkeley.edu/~tommelein/CEMworkshop/Teicholz.pdf

Visual Engineering. 2001. *Visual Management of Data*. Available: http://www.visual-engineering.com/cadstuff. html

Computer aided design of composite beams

V.K. Gupta & R. Kumar
Deptt. of Civil Engg., I.I.T. Roorkee, India

ABSTRACT: Steel and concrete both have some distinct advantages over each other. Composite construction is the technique, which utilizes advantages of both materials in the optimum manner. On the basis of efficiency of connection, composite beam may be designed for full interaction or partial interaction. Due to complicity in analysis and unavailability of set rules of design for partial interaction, generally composite beams are designed for full interaction. Design being an iterative procedure, is a laborious and time taking process. Computer Aided Design is the tool, which saves a lot of designer's time and provide more accurate results. The present paper deals with the development of the software 'COMPBEAM-2001' for the design of steel-concrete composite beams. The program prepared in 'C' language is interactive and quite flexible in nature. Design is based on limit state approach. Recommendations of Indian Standard, IS: 11384-1985 are built in.

1 INTRODUCTION

Steel and concrete are widely used building materials. Steel is suitable for high rise and large span construction using specialised equipments. Reinforced cement concrete has ease in construction limiting the beam span and ductility. Composite construction technique utilises advantages of both the materials and is effectively used for purposes. Composite construction leads to the advantages of reduction in depth of beam, reduction in weight, less construction time, less weight of foundation of structure and has high earthquake resistance and economy. Figure 1 shows a typical section of composite steel-concrete construction.

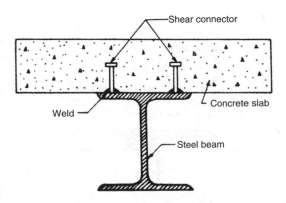

Figure 1.   Typical cross-section of composite beam.

This paper is an attempt towards the analysis and design of steel concrete composite beams with shear connectors through the computer programme.

2 ANALYSIS AND DESIGN

Depending upon amount of shear connectors at the interface in concrete, steel concrete composite beam may have no interaction (non composite), full interaction (fully composite), partial interaction (partial composite) sections. In case the slip occurs at the interface, it induces appreciable strain differences at the interface. The equilibrium of the section takes into account this strain difference. Therefore, strain difference cannot be determined from bending moment as in case of non composite and fully composite beams. Due to complexity in analysis of partially composite beams, normally composite beams are designed for full interaction.

2.1 *Design for fully composite action*

A steel concrete composite beam for building structures is designed as per recommendations of IS: 11384-1985, by limit state method. Partial safety factors (y_f for load and y_m for material strength) are taken as per IS: 456-2000. Analysis for ultimate limit state is done taking account of inelastic properties of concrete and steel as given in section. Analysis for serviceability limit states is done by elastic theory assuming the values of young's modulus for concrete and steel as given in IS: 456-2000 and neglecting the tensile stresses in concrete.

For the design of floor slab, usual methods for one way and two way slab design are applicable when the slab serves as the flange of one or more composite beams. Design steps to be followed are listed below:

(i) With guidance from recommended span-depth ratios for composite beams (generally taken 16 for design purpose) depth of steel section h_s is guessed.
(ii) Assuming lever arm to be $(h_c + h_s)/2$ approximately c/s area of steel required A_s, is calculated from $0.87A_sf_y (h_c + h_s)/2 = M_c$.
(iii) For the c/s area required A_s, a steel section is selected.
(iv) Section is checked for shear. It is assumed that the web, of area Aw, carries the whole of shear and it is checked that ultimate shear V_c does not exceed shear capacity of section $(0.45A_wf_y)$.
(v) Moment of resistance M_p, is calculated and checked that it exceeds ultimate moment M_c.

In a section of homogeneous material, the plastic neutral axis coincides with the equal area axis of section, that is, the axis which divides the section into two equal areas on either side. The same concept can be used in the case of composite beams also, provided the steel area is converted into equivalent concrete area by multiplying it with the stress ratio a = (0.87 f_y)/(0.36f_{ck}), where f_y: Yield strength of steel section and f_{ck}: Characteristic strength of concrete. Plastic moment carrying capacity for different cases can be given as below.

2.1.1 Plastic N.A. within concrete slab (Fig. 2)
This is valid when, $0.36f_{ck}bh_c \geqslant 0.87A_sf_y$
Compressive force in concrete C = $0.36f_{ck}bx_u$
Therefore, $x_u = (0.87A_sf_y)/(0.36f_{ck}b)$
Moment of resistance $M_p = 0.87A_sf_y (d - 0.42x_u)$

2.1.2 Plastic N.A. within top flange of steel (Fig. 3)
$M_p = 0.87f_y[A_s(d - 0.42h_c) - b_f(x_u - h_c)(x_u + 0.16h_c)]$

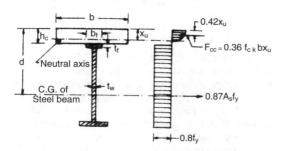

Figure 2. Stress distribution in a composite beam with neutral axis within concrete slab.

2.1.3 Plastic N.A. within web of steel beam (Fig. 4)
$M_p = 0.87A_sf_y (d - 0.42h_c) - 2 \times 0.87A_sf_y (0.58h_c + t_f/2) - 2 \times 0.87f_yt_w (x_u - h_c - t_f) (0.5 x_u + 0.08h_c + 0.5t_f)$

2.2 Effective width of composite section

Effective width for composite T beam is taken as minimum value obtained by (Fig. 5).

$b = L/6 + t_w + 6h_c$ or,

$b = 12h_c + b_f$

where, L = distance between points of zero moment in beam,
a = stress ratio.

Serviceability analysis is carried out to check that deflection at service loads does not exceed the

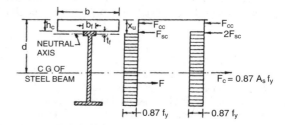

Figure 3. Stress distribution in a composite beam with neutral axis within the flange of steel beam.

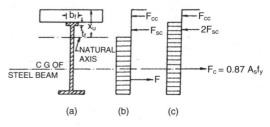

Figure 4. Stress distribution in a composite beam with neutral axis within the web of steel beam.

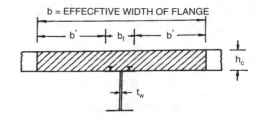

Figure 5. Effective width of composite section.

permissible limits and the stresses are within the elastic range. Using elastic analysis with transformed section method and assuming full penetration. Modular ratio m may have two values.

$m_1 = E_s/E_c$ for imposed or live load.
$m_2 = E_s/k_c E_c$ for dead load with creep coefficient k_c.

IS: 11384 1985 recommends value of m_1 and m_2 as 15 and 30 respectively.

M. I. of composite section is given by

$I = bx^3/3\,m + I_s + A_s\,(d-x)^2$ when $X < h_c$
$I = bh_c^3/12\,m + (bh_c/m)\,(x - h_c/2)^2 + I_s + A_s\,(d-x)^2$

Maximum compressive stress in concrete,

$fc_{max} = Mx/mI$, and,

Maximum tensile stress in steel,

$fst_{max} = m\,(h_c + h_s - x)/I$

Taking the method of construction into account, the max^m stress f_c and f_{st} can be written as follows:

(i) For propped construction, where whole load is carried by composite section.

$fc_{max} = M_L x_1/m_1 I_1 + M_D x_2/m_2 I_2$
$fst_{max} = M_L\,(h_c + h_s - x_1)/I_1 + M_D$
$\qquad (h_c + h_s - x_2)/I_2$

where, x_1 and x_2 = depth of NA for L.L. and D.L. cases respectively.

M_L = moment due to live load,
M_D = moment due to dead load
I_1 and I_2 = second moment of area of composite section for LL. and D.L. respectively.

(ii) For unsupported construction where D.L is caused by steel section alone,

$fc_{max} = M_L\, x_1/m_1 I_1 \leqslant 1/3\, f_{ck}$
$fst_{max} = M_L\,(h_c + h_s - x_1)/I_1 + M_D h_s/2 I_s \leqslant 0.87\, f_y$

These are as per IS: 11384-1985.

3 SHEAR CONNECTORS

These are used essentially for effective shear transfer and consequently, ensure full interaction for complete composite action of beam. Figure 1 shows a vertical type of shear connector. These have been shaped to ensure (i) shear transfer from one material to another and (ii) resistance against uplift.

Cost of shear connectors plays an important role in the efficient utilisation of composite beams. The shear connector may be either continuous or intermittent. Continuous connections may be achieved by (i) taking advantage of bond between steel and concrete and (ii) by gluing steel beam and concrete slab by means of epoxy resin adhesive.

Intermittent connectors be anchored by welding shear connectors to the top of steel beam, known as mechanical shear connectors.

The strength of shear connectors (P) is determined by the use of standard tests.

Strength $P = kd_s^2(f_{cu}\,E_c)^{1/2}/1.25$

where E_c = short term elastic modulus of concrete
$\qquad k$ = empirical constant
$\qquad = 0.32$ for $h/d_s \leqslant 4.2$
$\qquad = 0.25$ for $h/d_s = 3.0$

Design shear capacity $Q_p = 0.8Q_k$ in areas of sagging moment,

$Q_p = 0.6Q_k$ in areas of hogging moment.

where Q_k is characteristic strength of connector. Design strength of various shear connectors are given in Table 1 as per IS: 11384-1985.

Table 1. Design strength of shear connectors for different concrete strength.

		Decision strength of connector for concrete of grade		
Type of connector		M20	M30	M40
Headed stud				
Diameter (mm)	*Height (mm)*	*Load per stud (P_c) (kN)*		
25	100	86	101	113
22	100	70	85	94
20	100	57	68	75
20	75	49	58	64
16	75	47	49	54
12	62	23	28	31
Bar connector		*Load perbar (P_c) (kN)*		
50 mm × 38 mm × 200 mm		318	477	645
Channel connector		*Loar per channel (P_c) (kN)*		
125 mm × 65 mm × 12.7 kg × 150 mm		184	219	243
100 mm × 50 mm × 9.2 kg × 150 mm		169	204	228
75 mm × 40 mm × 6.8 kg × 150 mm		159	193	218
Tee connector		*Load per connector (P_c) (kN)*		
100 mm × 100 mm × 10 mm × 50 mm		163	193	211
Helical connector				
Bar diameter (mm)	*Pitch circle diameter (mm)*	*Load per pitch (P_c) (kN)*		
20	125	131	154	167
16	125	100	118	96
12	100	70	83	90
10	75	40	48	52

Number of elastic connectors can be evaluated by (i) elastic (ii) limit state approaches.

3.1 Elastic approach

Horizontal S.F. at interface $F = VA\tilde{y}/I$

No. of connectors $N = F/q_e$

where,

V = shear force in Newton
A = transformed area of composite beam
$\tilde{y}$ = distance between cetre of area A and elastic NA
I = moment of inertia of composite section
F = horizontal S.F. at interface
$q_e = 19.14 (f_{cu})^{1/2}$
$q_u = 54.18 (f_{cu})^{1/2}$
f_{cu} = cube crushing strength of concrete after 28 days
qe = permissible load carrying capacity of stud
qu = ultimate load carrying capacity of stud.

3.2 Limit state approach

Minimum three tests should be conducted and design value P_c should be taken as 67% of the lowest ultimate capacity.

Longitudinal force in steel and concrete at mid span,

$$F = 0.87 A_s f_y$$

Number of shear connector in half span, $N = F/P_c$ where, P_c = design strength of shear connector.

3.3 Recommendations for shear connectors

Following limitations are recommended for detailing of shear connectors.

(i) Maximum spacing between connectors should not be greater than four times the thickness of slab or 600 mm, whichever is less.
(ii) Clear distance between edge of the beam and the edge of shear connectors should not be less than 25 mm.
(iii) Clear depth of concrete cover over the top of shear connector should not be less than 25 mm.
(iv) The connection should extend at least 50 mm above the bottom of the main body of the slab.
(v) Minimum transverse spacing between the heads of the two adjacent connectors should not be less than 12 mm.
(vi) Minimum longitudinal spacing of studs should not be less than 6 times diameter of stud.

4 COMPUTER PROGRAMME

A software, "COMPBEAM-2001", for the design of simply supported steel concrete composite beam, of type as shown in Figure 1 has been developed. It is based on limit state approach. Program is prepared in "C" language under DOS environment at CAD centre, Civil Engineering Department, IIT Roorkee (Erestwhile University of Roorkee, Roorkee). It follows recommendations of Indian standards IS: 11384-1985, IS: 456-2000 and IS: 800-1984.

Whole program is divided in mainly three parts: (i) General, (ii) Design of composite beam (iii) Design of shear connectors. Many functions have been used in this program to conduct some specific tasks. Some important functions are discussed below:

(i) input_data ( ) – Input data may be given through keyboard or file.
(ii) eff_width ( ) – This function calculates effective width of beam.
(iii) showtable ( ) – These are the functions to show tables during execution of program. Showtable1 ( ) display of table of c/s area and designation of Indian standard steel beam section to facilitate the section for user. Showtable2 ( ) and showtable3 ( ) displays tables of stud and channel shear connectors.
(iv) Mor ( ) – This function calculate moment of resistance for the section.
(v) bilt_up ( ) – This function is to ask the user, properties of built up section chosen.
(vi) service_check ( ) – This function deals with the serviceability analysis of beam. Function calculates second moment of area of beam section, neutral axis for dead and live loads separately. Values of modular ratio m have been taken as per IS: 11384-1985. Maximum stress in steel and concrete and maximum deflection in beam are calculated in this function.
(vii) spacing ( ) – This is the function to calculate longitudinal spacing of shear connectors to be provided.
(viii) trans_rein ( ) – Function calculates quantity of minimum transverse reinforcement required in the bottom of slab to prevent horizontal shear failure of concrete.

This program is prepared for design of simply supported steel concrete composite beam. It includes analysis for both, propped and unpropped methods of construction. Properties of mostly used Indian standards steel beam sections and those of shear connectors are in built in this program. The program has been developed and built on IBM PC/Pentium III running under Windows Millennium 4.90.3000 version with EGA/VGA colour monitor. Salient features of the program includes data input, correction to input data, flexibility checks during analysis, option of built up section, codal provisions and function output file.

Due to paucity of space, flow chart of the program is not included.

Testing of software developed has been done by solving two problems of propped method of construction using hand calculations and it is found that the results obtained by software and through manual calculations are quite comparable. It is also observed that the requirement of steel considerably increases when opted for un-propped method of construction.

5 VALIDATION OF SOFTWARE

To test the working of software, few test problems have been solved manually as well as using the software developed. It is observed that results obtained manually match with computer results.

5.1 Test problem no. 1 (Naithani et al. 1997)

Design a simply supported composite T-beam for the following data by limit state method and apply various checks. Span = 9 m, spacing of beams = 3.5 m, method of construction = propped, concrete grade = M-20, yield strength of steel = 250 MPa, yield strength of reinforcement = 250 MPa, dead load = 15 kN/m, live load = 17.5 kN/m. Computer results are given in Table 2.

Table 2. Output file for test problem no. 1.

OUTPUT FOR COMPOSITE BEAM DESIGN

PROBLEM TITLE	: PROB. 1
***** INPUT DATA *****	
METHOD OF CONSTRUCTION	: PROPPED
TYPE OF BEAM	: T-BEAM
GRADE OF CONCRETE	: M-20
YIELD STRENGTH OF STEEL SECTION	: 250 N/sq. mm
YIELD STRENGTH OF REINF.	: 250 N/sq. mm
SPAN OF BEAM	: 9 m
DISTANCE BETWEEN TWO CONSECUTIVE BEAM	: 3.5 m
DEPTH OF SLAB	: 120 mm
DEAD LOAD	: 15.0 kN/m
LIVE LOAD	: 17.5 kN/m
***** DESIGN RESULTS *****	
PROVIDE BEAM SECTION	: ISMB 500
DEPTH OF SLAB	: 120 mm
TYPE OF CONNECTOR	: STUD TYPE
DIAMETER OF CONNECTOR	: 20 mm
HEIGHT OF CONNECTOR	: 75 mm
PROVIDE CONNECTORS IN DOUBLE ROW AT DISTANCE	: 185 mm
TRANSVERSE REINF. NOT LESS THAN/m RUN OF BEAM	: 529.73 sq. mm
***** SERVICE CHECKS *****	
MAX. COMP. STRESS IN CONCRETE	: 4.86 N/sq. mm
MAX. TENSILE STRESS IN STEEL	: 199.13 N/sq. mm
MAX. DEFLECTION IN BEAM	: 20.50 mm

5.2 Test problem no. 2 (Naithani et al. 1999)

Design a simply supported steel concrete composite T-beam of span 7.25 m spaced at 3.8 m, with slab depth 120 mm and dead load of 20.7 kN/m and live load of 4.58 kN/m. Concrete grade is M-20 and yield strength of steel section is 250 MPa. Yield strength of reinforcement is 415 MPa. Design the beam for propped case and check it for un-propped case also. Computer results are given in Tables 3–4.

6 CONCLUSIONS

Time of the designer can be saved and accuracy can be achieved by using CAD systems. In this paper a software entitled "COMPBEAM–2001" has been developed in C language for design of composite beam. It incorporates following features.

(i) It is quite flexible, interactive and user friendly.
(ii) Software has been supported with the required provisions of relevant Indian codes of practices.
(iii) Properties of steel sections and connectors are in-built in this program. So users need not to refer these tables while using this software.

Table 3. Output file for test problem no. 2 (Propped case).

OUTPUT FOR COMPOSITE BEAM DESIGN

PROBLEM TITLE	: PROB. 2.1
***** INPUT DATA *****	
METHOD OF CONSTRUCTION	: PROPPED
TYPE OF BEAM	: T-BEAM
GRADE OF CONCRETE	: M-20
YIELD STRENGTH OF STEEL SECTION	: 250 N/sq. mm
YIELD STRENGTH OF REINF.	: 415 N/sq. mm
SPAN OF BEAM	: 7.25 m
DISTANCE BETWEEN TWO CONSECUTIVE BEAM	: 3.8 m
DEPTH OF SLAB	: 120 mm
DEAD LOAD	: 20.7 kN/m
LIVE LOAD	: 4.58 kN/m
***** DESIGN RESULTS *****	
PROVIDE BEAM SECTION	: ISMB 350
DEPTH OF SLAB	: 120 mm
TYPE OF CONNECTOR	: STUD TYPE
DIAMETER OF CONNECTOR	: 20 mm
HEIGHT OF CONNECTOR	: 75 mm
PROVIDE CONNECTORS IN DOUBLE ROW AT DISTANCE	: 255 mm
TRANSVERSE REINF. NOT LESS THAN/m RUN OF BEAM	: 231.51 sq. mm
***** SERVICE CHECKS *****	
MAX. COMP. STRESS IN CONCRETE	: 4.48 N/sq. mm
MAX. TENSILE STRESS IN STEEL	: 217.45 N/sq. mm
MAX. DEFLECTION IN BEAM	: 19.62 mm

Table 4. Output file for test problem no. 2 (Un-propped case).

OUTPUT FOR COMPOSITE BEAM DESIGN

PROBLEM TITLE	: PROB. 2.2

***** INPUT DATA *****

METHOD OF CONSTRUCTION	: UN-PROPPED
TYPE OF BEAM	: T-BEAM
GRADE OF CONCRETE	: M-20
YIELD STRENGTH OF STEEL SECTION	: 250 N/sq. mm
YIELD STRENGTH OF REINF.	: 415 N/sq. mm
SPAN OF BEAM	: 7.25 m
DISTANCE BETWEEN TWO CONSECUTIVE BEAM	: 3.8 m
DEPTH OF SLAB	: 120 mm
DEAD LOAD	: 20.7 kN/m
LIVE LOAD	: 4.58 kN/m

***** DESIGN RESULTS *****

PROVIDE BEAM SECTION	: ISMB 450
DEPTH OF SLAB	: 120 mm
TYPE OF CONNECTOR	: STUD TYPE
DIAMETER OF CONNECTOR	: 20 mm
HEIGHT OF CONNECTOR	: 75 mm
PROVIDE CONNECTORS IN DOUBLE ROW AT DISTANCE	: 180 mm
TRANSVERSE REINF. NOT LESS THAN/m RUN OF BEAM	: 327.98 sq. mm

***** SERVICE CHECKS *****

MAX. COMP. STRESS IN CONCRETE	: 2.98 N/sq. mm
MAX. TENSILE STRESS IN STEEL	: 174.16 N/sq. mm
MAX. DEFLECTION IN BEAM	: 19.96 mm

(iv) Provision of built-up section is in-built in this software.

(v) Beam can be designed for both propped and un-propped methods of construction using this software.

(vi) It has been observed that the requirement of steel considerably increases if design is done for un-propped construction than that for propped construction.

REFERENCES

Byron, S.G. 1991. *Theory and problems of programming with C*, Tata McGraw Hill Publishing Limited, New Delhi.

IS: 11384-1985. *Indian standard code of practice for composite construction in structural steel and concrete*, BIS, New Delhi.

Johnson, R.P. 1971. Transverse reinforcement in composite beams for buildings, *The Structural Engineer*, No. 5, Vol. 49, May.

Johnson, R.P. 1975. *Composite structures of steel and concrete*, Vol.1, Crosby Luckwood Staples, London.

Naithani, K.C., Kumar, S., Gupta, V.K. & Mittal, M.K. 1997. *Handbook on composite steel concrete construction*, C.B.R.I., Roorkee.

Naithani, K.C., Gupta, V.K., Kumar, S., Mittal, M.K., Ghosh, S.K., Karmakar, D. & Maini, P.K. 1999. Cost economic study of composite construction and its comparison with traditional R.C.C. construction, *Civil Engg. & Construction Review*, March.

Viest, I.M., Fountain, R.S. & Singleton, R.C. 1958. *Composite construction in steel and concrete*, McGraw Hill Book Company, Inc.

30. Issues in the theory of elasticity

Variable pressure arresting opening of internal and external cracks weakening a plate

R.R. Bhargava & S. Hasan
Department of Mathematics, Indian Institute of Technology, Roorkee, India

P.K. Bansal
M.S. College, Saharanpur, India

ABSTRACT: A crack arrest model is proposed for unbounded elastic perfectly-plastic plate weakened by an internal and two external cracks. These internal and external cracks are hairline and collinear. The faces of the cracks open due to remotely applied uniform unidirectional tension. The tension is applied in a direction perpendicular to the rims of the cracks. Due to the opening of the rims a plastic zone is developed ahead of each tip of the cracks. Applying at their rims normal cohesive quadratically varying yield point stress arrests developed plastic zones. Consequently cracks are arrested from further opening. The solution of the problem is obtained from that of an Auxiliary Problem. The Auxiliary Problem is appropriately derived from the original problem. Analytic solution is obtained for the Auxiliary Problem using complex variable technique. A case study is carried out to study the qualitative behavior of load required to arrest the developed plastic zones, thus arresting the crack with respect to affecting parameters. Results obtained are reported graphically and analyzed.

1 INTRODUCTION

Stress intensity factors at the tip of two semi-infinite plate using linear elastic fracture mechanics is given by Paris (1965). Employing integral transform method, the solution for the same problem was obtained by Lowengurb (1968). Tada (1973) using Green function obtained the stress intensity factor for a single finite and two semi-infinite cracks weakening a body under different loading conditions. A semi-infinite crack model, for determining Mode I type stress intensity factors using crack surface displacements, is proposed by Jing & Ng (1997). Sadowsky (1956) obtained elastic solution for two external and an internal crack weakening an elastic plate when a pair of concentrated forces act at the center of section between the cracks. The Dugdale model solution for two semi-infinite and an internal crack weakening an unbounded plate is obtained by Bhargava & Hasan (2001).

In the present paper a generalized Dugdale model solution is obtained for an elastic perfectly-plastic plate weakened by two external and an internal collinear hairline cracks. The plastic zones developed, due to Mode I type deformation of the crack faces, are closed by applying normal cohesive quadratically varying yield point stress along their rims.

2 FUNDAMENTAL FORMULATION REQUIRED

As is well-known, in complex potential formulation P_{ij} and displacement components u_{ij} (i, j = x, y) may be expressed as

$$P_{yy} - iP_{xy} = \Phi(z) + \Omega(\bar{z}) + (z - \bar{z})\overline{\Phi'(z)} \qquad (1)$$

$$2\mu(u_x + iu_y) = \kappa\phi(z) - \omega(\bar{z}) - (z - \bar{z})\overline{\phi'(z)} + const \qquad (2)$$

where $\omega(z) = \int \Omega(z)\,dz$ and μ is shear modulus and $\kappa = 3 - 4\gamma$ for plain strain case and $\kappa = (3 - \gamma)/(1 + \gamma)$ for plane stress case. The Poisson's ratio is denoted by γ.

An infinite plate (occupying z-plane) cut along n straight collinear hairline cracks L_i (i = 1, n). The end points of crack L_i are denoted by a_i and b_i. The configuration is subjected to following boundaries conditions

(a) The rims of the cracks are subjected to known P_{yy} and P_{xy}
(b) Infinite boundary is subjected to $P_{yy} = \sigma_\infty$ and
(c) Displacements are single-valued around cracks. Boundary condition (a) together with Equation 1 yields following Hilbert problems:

$$\Phi^+(t) + \Omega^-(t) = P_{yy}^+ - iP_{xy}^+ \; ; \; \Phi^-(t) + \Omega^+(t) = P_{yy}^- - iP_{xy}^- \qquad (3)$$

superscripts '+' and '−' denote the value of the function as t on any crack is approached from $y < 0$ and $y > 0$ plane, respectively. The desired potential $\Phi(z)$ is obtained from solution of the Equation 3, refer Muskhelishvili (1954), as

$$\Phi(z) = \Phi_0(z) + \frac{P_n(z)}{X(z)} - \frac{\sigma_\infty}{4} \tag{4}$$

$$\Omega(z) = \Omega_0(z) + \frac{P_n(z)}{X(z)} + \frac{\sigma_\infty}{4} \tag{5}$$

where

$$\Phi_0(z) = \frac{1}{2\pi i X(z)} \int_L \frac{X(t)\,p(t)}{t-z}dt + \frac{1}{2\pi i}\int_L \frac{q(t)}{t-z}dt$$

$$\Omega_0(z) = \frac{1}{2\pi i X(z)} \int_L \frac{X(t)\,p(t)}{t-z}dt - \frac{1}{2\pi i}\int_L \frac{q(t)}{t-z}dt$$

$$p(t) = \frac{1}{2}\left[P_{yy}^+ + P_{yy}^-\right] - \frac{i}{2}\left[P_{xy}^+ + P_{xy}^-\right]$$

$$q(t) = \frac{1}{2}\left[P_{yy}^+ - P_{yy}^-\right] - \frac{i}{2}\left[P_{xy}^+ - P_{xy}^-\right]$$

$$X_0(z) = \prod_{j=1}^{n}(z - a_j)^{-1/2}(z - b_j)^{-1/2}$$

$$P_n(z) = C_0 z^n + C_1 z^{n-1} + C_2 z^{n-2} + ... + C_n$$

$$\text{and } X(t) = X^+(t) \tag{6}$$

Arbitrary constants C_i ($i = 0,1...n$) determined using conditions (b) and (c).
Opening mode stress intensity factor at the tip $z = a$ is defined as

$$K_1 = \sqrt{8\pi}\,\underset{t\to a^+}{\text{Lim}}\left\{\sqrt{(a-t)}\Phi(t)\right\} \text{ or } K_1 = \sqrt{8\pi}\,\underset{t\to a^-}{\text{Lim}}\left\{\sqrt{(t-a)}\Phi(t)\right\}$$

$$\tag{7}$$

3 THE PROBLEM

An unbounded homogeneous, isotropic elastic perfectly-plastic plate is weakened by collinear, hairline straight cracks L_i ($i = 1, 2, 3$). Crack L_2 is an internal crack and two external cracks are L_1 and L_3. Cracks L_1 and L_3 occupy the intervals $(-\infty, -b_1]$ and $[b_1, \infty)$, respectively, on ox-axis and internal crack L_2 lies symmetrically between L_1 and L_3 and occupying the interval $[-c_1, c_1]$ on ox-axis. Unidirectional tension, $P_{yy} = \sigma_\infty$, applied at infinity boundary of the plate opens the faces of the cracks. Consequently a plastic

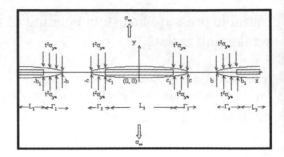

Figure 1. Configuration of the problem.

zone is formed ahead of each finitely distant tip of the cracks. To arrest the crack from further opening each rim of the developed plastic zones is subjected to cohesive normal stress distribution $P_{yy} = t^2\sigma_{ye}$ and $P_{xy} = 0$, where t is any point on any of the rim of the plastic zones and σ_{ye}, is the yield point stress the plate. Schematically the configuration is depicted in Figure 1.

4 ALGORITHM OF THE PROBLEM

The solution of the above problem is obtained from that of an *Auxiliary problem*. This *Auxiliary problem* is appropriately derived from the original problem.

4.1 *Auxiliary problem and solution*

An unbounded elastic perfectly-plastic plate is cut along three finite hairline collinear straight cracks C_i ($i = 1, 2, 3$). The cracks C_1, C_2 and C_3 lie on ox-axis and each occupies each of the intervals $[-a_1, -b_1]$, $[-c_1, c_1]$ and $[b_1, a_1]$, respectively. Unidirectional uniform constant tension applied at remote boundary of the plate and the faces of the cracks C_i ($i = 1, 2, 3$) are opened forming a plastic zone is developed ahead each tip $-a_1, -b_1, -c_1, c_1, b_1$ and a_1, these are denoted by $\Gamma_6, \Gamma_1, \Gamma_2, \Gamma_3, \Gamma_4$ and Γ_5. All the plastic zones developed occupy the intervals on ox-axis. Plastic zone Γ_6 occupies the interval $[-a, -a_1]$; Γ_1 occupies the interval $[-b_1, -b]$; Γ_2 occupies the interval $[-c, -c_1]$; Γ_3 occupies the interval $[c_1, c]$; Γ_4 occupies the interval $[b, b_1]$ and plastic zone Γ_5 occupies the interval $[a_1, a]$. Union of developed plastic zone is denoted by $\Gamma = \bigcup_{i=1}^{6}\Gamma_i$.

Boundary conditions of the problem are

(i) Each rim of the developed plastic zone Γ_i ($i = 1$ to 6) is subjected to stress distribution $P_{yy} = t^2\sigma_{ye}$ and $P_{xy} = 0$.
(ii) Rims of the cracks C_i ($i = 1, 2, 3$) are stress free.

(iii) Infinite boundary of the plate is subjected to stress distribution $P_{yy} = \sigma_\infty$ and $P_{xy} = 0$.

(iv) Displacements are single-valued around the closed counter enclosing each crack C_i ($i = 1, 2, 3$) and corresponding plastic zones Γ_i ($i = 1, 2, \ldots, 6$). Desired potential $\Phi(z)$ is written using Equations 4–6 in conjugation with the boundary conditions (i) to (iii)

$$\Phi(z) = \frac{\sigma_{yc}}{\pi i X(z)} \int_\Gamma \frac{t^2 X(t)}{t - z} dt + \frac{2P_3(z)}{X(z)} - \frac{\sigma_\infty}{2} \qquad (8)$$

where

$$X(z) = \sqrt{z^2 - a^2} \sqrt{z^2 - b^2} \sqrt{z^2 - c^2}$$

$$P_3(z) = \frac{\sigma_\infty}{2} z^3 + C_1 z^2 + C_2 z + C_3 \qquad (9)$$

Constants C_i ($i = 0, 1, 2, 3$) are determined using boundary condition (iv) in conjugation with Equation 2. Evaluating the integral in Equation 8 and simplifying, one obtains the complex potential $\Phi(z)$ which solves the Auxiliary problem completely. The potential $\Phi(z)$ can be written as

$$\Phi(z) = \frac{-g\sigma_{yc}}{2\pi X(z)} + z^2 X^2(z) \left\{ \begin{array}{l} \left[\begin{array}{l} -(S_2 V_1 + L_0 V_2 + V_3) - \\ -z^2(S_2 V_1 + L_0 V_2 + V_3) + z^4 S_2 + z^6 L_0 + \end{array} \right] \\ \dfrac{\Pi\left(\varphi_a, \alpha_a^2, k\right)}{a^2 - z^2} - \\ -\dfrac{\left(b^2 - c^2\right)\Pi\left(\varphi_b, \alpha_b^2, k\right)}{\left(z^2 - b^2\right)\left(z^2 - c^2\right)} - \\ -\dfrac{\left(b^2 - c^2\right)\Pi\left(\varphi_c, \alpha_c^2, k\right)}{\left(z^2 - b^2\right)\left(c^2 - z^2\right)} \end{array} \right\} + \frac{\sigma_\infty z^3}{2X(z)} - \frac{\sigma_\infty z \alpha_3}{2X(z)\alpha_1} - \frac{\sigma_{ye}}{4}$$

where

$$S_1 = (L_1 - M_1 - N_1) - A(L_2 - M_2 - N_2) + a^2 b^2 (L_3 - N_3) - a^2 c^2 (M_3 + N_3) + c^2 b^2 (L_3 - M_3 - N_3)$$

$$S_2 = (L_2 - M_2 - N_2) - b^2 L_3 + c^2 M_3 + \left(a^2 + b^2 + c^2\right) N_3,$$

$$L_1 = b^4 D_0 + 2b^2\left(c^2 - b^2\right) D_2 + \left(c^2 - b^2\right)^2 D_4$$

$$L_2 = b^2 D_0 + \left(c^2 - b^2\right) D_2$$

$$L_3 = D_0 = L_0$$

$$D_0 = F(\varphi_c, k),$$

$$D_2 = F(\varphi_c, k) - \frac{E(\varphi_c, k)}{k'^2} + \frac{\sin\varphi_c \sqrt{1 - k^2 \sin^2 \varphi_c}}{k'^2 \cos\varphi_c}$$

$$D_4 = \frac{1}{3k'^4} \left[\begin{array}{l} k'^2\left(2k'^2 - k^2\right) F(\varphi_c, k) + 2\left(2k^2 - 1\right) E(\varphi_c, k) \\ + \left(2 - 4k^2 + k'^2 \sec^2 \varphi_c\right) \\ \dfrac{\sin\varphi_c \sqrt{1 - k^2 \sin^2 \varphi_c}}{\cos\varphi_c} \end{array} \right]$$

$$g = \frac{2}{\sqrt{a^2 - c^2}}, k = \frac{a^2 - b^2}{a^2 - c^2}, \varphi_c = \sin^{-1}\left(\sqrt{\frac{c^2 - c_1^2}{b^2 - c_1^2}}\right)$$

$$, \alpha_c^2 = \frac{z^2 - b^2}{z^2 - c^2}, k'^2 = 1 - k^2$$

$$M_1 = c^4 I_0 + 2c^2\left(b^2 - c^2\right) I_2 + \left(b^2 - c^2\right)^2 I_4$$

$$M_2 = c^2 I_0 + \left(b^2 - c^2\right) I_2$$

$$M_3 = I_0$$

$$I_0 = F(\varphi_b, k), I_2 = \frac{E(\varphi_b, k)}{k'^2} - \frac{k^2 \sin\varphi_b \cos\varphi_b}{k'^2 \sqrt{1 - k^2 \sin^2 \varphi_b}}$$

$$I_4 = \frac{1}{3k'^4} \left[\begin{array}{l} \left(2 - k^2\right) E(\varphi_b, k) - k'^2 F(\varphi_b, k) \\ -\dfrac{k^2 \sin\varphi_b \cos\varphi_b}{\sqrt{1 - k^2 \sin^2 \varphi_b}}\left(4 - 2k^2 + \dfrac{k'^2}{\sqrt{1 - k^2 \sin^2 \varphi_b}}\right) \end{array} \right]$$

$$g = \frac{2}{\sqrt{a^2 - c^2}}, k = \frac{a^2 - b^2}{a^2 - c^2},$$

$$\varphi_b = \sin^{-1}\left(\sqrt{\frac{\left(a^2 - c^2\right)\left(b_1^2 - b^2\right)}{\left(a^2 - b^2\right)\left(b_1^2 - c^2\right)}}\right), \alpha_b^2 = \frac{\left(a^2 - b^2\right)\left(c^2 - z^2\right)}{\left(a^2 - c^2\right)\left(b^2 - z^2\right)},$$

$$k'^2 = 1 - k^2$$

$$N_1 = \left[\begin{array}{l} v_1 F(\varphi_a, k) + v_2 E(\varphi_a, k) + \\ \left(a^2 - b^2\right)\left(a^2 - c^2\right)\sin\varphi_a \cos\varphi_a \sqrt{1 - k^2 \sin^2 \varphi_a} \end{array} \right]$$

$$v_1 = a^4 - 2a^2\left(a^2 - c^2\right) + \frac{2}{3}\left(a^2 - c^2\right)^2 + \frac{\left(a^2 - c^2\right)\left(a^2 - b^2\right)}{3}$$

$$v_2 = 6a^2\left(a^2 - c^2\right) - \left(a^2 - c^2\right)^2 - \left(a^2 - b^2\right)\left(a^2 - c^2\right)$$

$$N_2 = \left[c^2 F(\varphi_a, k) + \left(a^2 - c^2\right) E(\varphi_a, k)\right]$$

$$N_0 = F(\varphi_a, k)$$

and

$$g = \frac{2}{\sqrt{a^2 - c^2}}, \quad k = \frac{n^2}{a^2 - c^2} \frac{b^2}{}$$

$$\varphi_a = \sin^{-1}\left(\sqrt{\frac{\left(a_1^2 - a^2\right)}{\left(a^2 - b^2\right)}}\right), \quad \alpha_a^2 = \frac{\left(a^2 - b^2\right)}{\left(a^2 - z^2\right)} \qquad (10)$$

where F, E and Π are the normal elliptic integrals of first, second and third kind respectively.

4.2 *Solution of original problem*

Required complex potential $\Phi_R(z)$, for the problems stated in the section 3 is obtained from the Equation 8–10 by

$$\Phi_R(z) = \underset{\substack{a = a_1 \to \infty \\ \sigma_\infty \to 0 \\ a\sigma_\infty \text{ is finite}}}{\text{Lim}} \Phi(z) \qquad (11)$$

5 PLASTIC ZONE LENGTH

Length of each developed plastic zone at the finitely distant tips $\pm b_1$ {of the external cracks L_i ($i = 1, 3$)} and $\pm c_1$ {of the internal crack L_2} is determined using the fact that stresses remain finite at every point of the plate. This yields following two equations to determine b and c

$$b\sigma_\infty = -2\sigma_{ye}\left(b^2 - c^2\right)\left[T_1 + b^2 T_2 + b^2 P_1\right] \qquad (12)$$

and

$$c\sigma_\infty = -2\sigma_{ye}\left(b^2 - c^2\right)\left[T_1 + c^2 T_2 + c^2 P_2\right] \qquad (13)$$

where

$$p_1 = \text{Ln}\left(\frac{\sqrt{c^2 - c_1^2} + \sqrt{b^2 - c_1^2}}{\sqrt{b^2 - c^2}}\right)$$

$$p_2 = \text{Ln}\left(\frac{\sqrt{b_1^2 - b^2} + \sqrt{b_1^2 - c^2}}{\sqrt{b^2 - c^2}}\right)$$

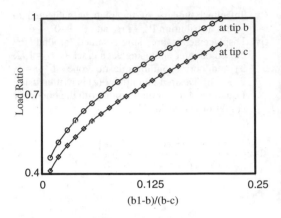

Figure 2. Variation between load required to close the plastic zone length/inter crack distance at the finitely distant tips of external and internal crack.

$$T_1 = p_1 W_1 - \frac{W_2}{\left(b^2 - c^2\right)}$$

$$T_2 = \frac{1}{2}(p_2 - \Pi(\phi_c, 1, 1) + \Pi(\phi_b, 1, 1))$$

$$W_1 = \frac{b^2}{2}\left[1 - \frac{E(k_1)}{F(k_1)}\right], \quad k_1 = \frac{c^2}{b^2}$$

$$W_3 = \underset{a = a_1 \to \infty}{\text{Lim}} \{S_1 V_1 + L_0 V_2 + V_3\}$$

Plastic zone length is then calculated using $|b - b_1|$ and $|c - c_1|$ by substituting b and c determined from above two equations and substituting corresponding b_1 and c_1.

6 CASE STUDIES AND CONCLUSION

To study the qualitative behavior of load required to arrest the crack opening a case study was carried out. Figure 2 shows the variation of load ratio (load applied at infinity/yield point stress of the plate) required to arrest the crack opening at the adjacent tips of internal and external cracks. It is observed from figure that more load is required to close the opening of an external crack. The presence of an external crack in neighborhood of a small finite crack affects the opening of the small crack. The study is carried for a fixed plastic zone length at each of the tip of internal and external cracks

Figure 3 depicts the load required to arrest the plastic zone developed at the finitely distant tip of external crack with respect to ratio $(b_1 - b)/(b - c)$ {plastic zone length at tip of an external crack/distance between

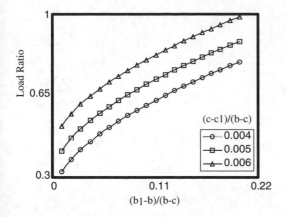

Figure 3. Variation between the loads required to close the plastic zone ratio (plastic zone length at the tip of external crack/inter plastic zone distance between internal and external crack) at the finite tip of external crack.

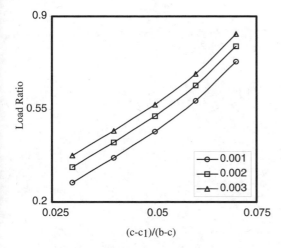

Figure 4. Variation between the load required to close the plastic zone ratio (plastic zone length at the tip of internal crack/inter plastic zone distance between internal and external crack) at the tip of internal crack.

plastic zone length of internal and external cracks}. It may be noted that as the ratio $(b_1 - b)/(b - c)$ increases more load is required to arrest. Also if the size of plastic zone at the tip of internal crack is increased than more load is required to arrest.

Variation of load required to arrest the opening of small crack, when an external crack exit in its neighborhood is plotted in Figure 4. It may be noted that although with increase in plastic zone at the tip of internal crack more load is required but behavior in Figure 3 and 4 is opposite nature, as expected.

ACKNOWLEDGEMENT

Authors acknowledge their gratitude to Prof. R. D. Bhargava (Senior Prof. and Head, retd.) Indian Institute of Technology, Bombay for his continous advise during course of this work.

REFERENCES

Bhargava, R.R. & Hasan, S. 2002. Crack arrest model for a internal and external cracks weakening a plate-a Dugdale model approach. *Proc.14th U.S. National Congress of Theoritical and Applied Mechanics, June 23–28, 2002.* In R.C. Batra, E.G. Hennekw (eds).

Jing, L.K. & Ng, S.W. 1996. A semi-infinite crack model for determining Mode I stress intensity factors using crack surface displacements. *International Journal of Fracture* 76: 355–371.

Lowengrub, M. & Srivastva, K.N. 1968. On two copla nar cracks in an infinite medium. *International Journal of Engineering Science*.6: 359–371.

Muskhelisvili, N.I. 1954. Some basic problems in mathematical theory of elasticity, In J.R.M. Radok(Ed.), *3rd. P. Noordhoff Ltd, The Netherlands.*: 506–508.

Paris, P.C. & Sih, G. 1965. Stress analysis of cracks in fracture toughness testing and its applications. *Special Technical publication: American Society for Testing and Materials, Philadelphia* 381: 30–46.

Sadowsky, M. 1956. Stress concentration caused by multiple punches and cracks. *Journal of Applied Mechanics* 23(1): 80–84.

Tada, H., Paris P.C. & Erwin, G.R. 1973. The stress analysis of cracks. *Handbook DEl. Carporation.*

System-based Vision for Strategic and Creative Design, Bontempi (ed.)
© 2003 Swets & Zeitlinger, Lisse, ISBN 90 5809 599 1

Vertical stress formulas for surface loads

R.V. Jarquio
New York City Transit, USA

ABSTRACT: This paper illustrates the total vertical stress formulas for surface loads using the Boussinesq's elastic equation for a point load. Previous investigators from Steinbrenner to Terzaghi and other modern analysts were unable to integrate this elastic equation. Instead, graphical methods such as the Newmark's chart and finite-element techniques were employed. The paper shows the derived formulas to determine the value of "Δp" in the standard soil settlement formula as well as the dispersion of surface loads through soil. The current method approximates "Δp" by using the published Steinbrenner's formula at the middle of a soil layer subject to settlement. The AASHTO's frustum method assumes the total surface load is transmitted within the area of the frustum at any depth. The derived formulas will provide accurate and reliable results. Comparison between methods indicates greater values of "Δp" are obtained by use of derived formulas than the current graphical methods of calculations.

1 INTRODUCTION

For more than half a century, the integration of the Boussinesq's elastic equation for a rectangular and trapezoidal surface loading has eluded investigators on this subject. W. Steinbrenner first presented the integration of the elastic equation for rectangular or uniform loading. His solution, however, was only halfway done and hence, no better than the Newmark charts since mechanical integration has to be performed to get the total vertical stress. His formulas for the vertical stress at any depth appeared in some textbooks without his named mentioned. He admitted in his paper that the integration for total vertical stress became difficult. Terzaghi probably read his paper and concluded in his book that the Boussinesq's equation cannot be simplified.

This paper will show that Terzaghi's conclusion was premature as evidenced from the results of the integration of the elastic equation for vertical stress, using standard integration formulas. Three levels of substitutions have been employed to solve the total vertical stress under a uniform load. Intermediate steps in the derivations are not shown.

2 DERIVATION

2.1 *Rectangular loading*

The vertical stress under a point load through a soil medium is given by the Boussinesq's elastic equation as

$$\sigma_z = (3Q/2\pi) \{ z^3 / (z^2 + r^2)^{5/2} \} \tag{1}$$

Figure 1 shows a rectangular area of dimensions "a" and "b", loaded with a uniform surface load, "q".

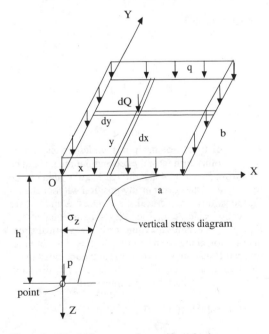

Figure 1. Vertical stress under a uniform load.

Integrate Equation 1 to yield the expression for the total vertical stress "h" deep under any corner. This involves integration of limits in the X, Y and Z axis indicated by a, b and h respectively. Start with the following substitution as follows:

Let $z = h$, $r^2 = x^2 + y^2$, $dQ = qdxdy$, $\sigma_z = p$ and

$$dp = (3dQ / 2\pi) \{ h^3 / (h^2 + x^2 + y^2)^{5/2} \} \text{ or}$$

$$p = (3qh^3/2\pi) \int_0^b \int_0^a \{(dx\,dy)/ (h^2 + x^2 + y^2)^{5/2}\} \quad (2)$$

Integrate Equation 2 from zero to "a" with respect to "x" to obtain the expression for the vertical stress at a point "h" deep below the end of a finite load, i.e.,

$$p = (qah^3/2\pi)[\{(2a^2 + 3h^2) + 3y^2\}/(h^2 + y^2)^2 (h^2 + a^2 + y^2)^{3/2}] \quad (3)$$

Using algebraic and trigonometric substitutions, integrate Equation 3 with respect to "y" to obtain the expression for the vertical stress under the corner of a rectangular area. Evaluating limits from zero to "b" yields

$$p = (q/2\pi)[abh \{(A^2 + B^2)/A^2 B^2 C\} + (\pi/2) - \arctan (hC/ab)] \quad (4)$$

in which,

$$A = (h^2 + a^2)^{1/2}, \ B = (h^2 + b^2)^{1/2}, \text{ and}$$

$$C = (h^2 + a^2 + b^2)^{1/2}$$

are the distances of the point under consideration from the other corners of the rectangular area. It can be seen from Equation 4 that the rectangular area can be rotated about the corner in any position without changing the value of the vertical stress, i.e., "a" and "b" are interchangeable. Also, when "h" is made equal to "z", Equation 4 becomes the equation of the vertical stress distribution along the depth or "Z" axis. Steinbrenner reduced Equation 4 to two terms and concluded in his paper that the integration from here on is difficult. His formula is given by the expression

$$p = (q/2\pi)[abh \{(A^2 + B^2)/A^2 B^2 C\} + \arctan (ab/hC)] \quad (5)$$

and appeared in the book by Poulous and Davis with no reference that it was first derived by Steinbrenner. Equation 5 is used as an approximation for the "Δp" value at mid-height of soil layer.

Using integration by parts and partial fractions, Equation 4 can in fact be integrated to the expression

$$p_1 = (q/2\pi)[(\pi h/2) + b \, Ln\{(D + a)(C - a)/(D - a)(C + a)\} + a \, Ln\{(D + b)(C - b)/(D - b)(C + b)\} - h \arctan(hC/ab)] \quad (6)$$

in which,

$D = (a^2 + b^2)^{1/2}$ = diagonal distance of the rectangular area. The integration is lengthy and cannot be shown in the limited number of pages allowed.

Equation 6 is the total vertical stress at any point "h" under a rectangular area loaded with a uniform load, "q". To get the average vertical stress at this point, divide Equation 6 by "h" i.e.

$$\Delta p = (q/2\pi h)[(\pi h/2) + b \, Ln\{(D + a)(C - a)/(D - a)(C + a)\} + a \, Ln\{(D + b)(C - b)/(D - b)(C + b)\} - h \arctan(hC/ab)] \quad (7)$$

2.2 Trapezoidal loading

Trapezoidal loading can be derived by superposition using triangles. In Figure 2, using Equation 3 the vertical stress under the zero load corner for a triangular loading is given by the expression

$$p_o = (qah^3/2\pi b) \int_0^b [\{(2a^2 + 3h^2)y + 3y\}/(h^2 + y^2)^2 (h^2 + a^2 + y^2)^{3/2}] \, dy \quad (8)$$

Integrate Equation 8 using algebraic and trigonometric substitutions to obtain

$$p_o = (qah^3/2\pi b)[(1/h^2 A) - (1/B^2 C)] \quad (9)$$

The vertical stress formula under the "q" corner is given by the expression

$$p_q = p - p_o \quad (10)$$

The total vertical stress under the zero load corner is obtained by integrating Equation 8 by letting $h = z$

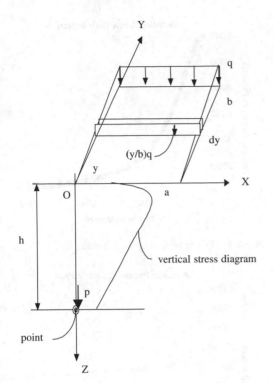

Figure 2.   Vertical stress under a triangular load.

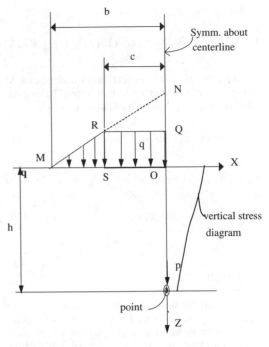

Figure 3.   Vertical stress under a trapezoidal load.

and dh = dz. Using integration by parts and evaluating limits yields

$$p_{oT} = (qa/2\pi)[(a/b)(A-a) - b\ Ln\ \{(b/B)(c - a)/(D - a)\}] \tag{11}$$

The average vertical stress, Δp for this case is also obtained by dividing Equation 11 by "h", i. e.

$$\Delta p_o = (qa/2\pi h)[(a/b)(A-a) - b\ Ln\ \{(b/B)(c - a)/(D - a)\}] \tag{12}$$

The total vertical stress at the "q" corner is given by the expression

$$p_{qT} = p_T - p_{oT} \quad \text{or,}$$

$$p_{qT} = (q/2\pi)[(\pi h/2) + b\ Ln\{(D + a)(C - a)/(D - a)(C + a)\} + a\ Ln\{(D + b)(C - b)/(D - b)(C+ b)\} - h\ arctan(hC/ab)] - (qa/2\pi)[(a/b)(A - a) - b\ Ln\ \{(b/B)(c - a)/(D - a)\}] \tag{13}$$

The average vertical stress, Δp is again obtained simply by dividing Equation 13 by "h", i.e.

$$\Delta p_q = (q/2\pi h)[(\pi h/2) + b\ Ln\{(D + a)(C - a)/(D - a)(C + a)\} + a\ Ln\{(D + b)(C - b)/(D - b)(C+ b)\} - h\ arctan(hC/ab)] - (qa/2\pi h)[(a/b)(A-a) - b\ Ln\ \{(b/B)(c - a)/(D - a)\}] \tag{14}$$

Equations 9 to 14 supersede equations published in the ASCE Journal of Geotechnical Engineering January 1984.

Figure 3 shows a trapezoidal loading common to roadway embankment. Using the principle of super-position, the vertical stress under the center of the embankment at a point "h" is derived as follows:

Due to MON = Eq. (13) in which "q" is replaced by

"bq/(b - c)" (15)

Due to RQN = Eq. (13) in which "q" is replaced by

"cq/(b - c)" (16)

Hence,

$$p_T = 2(MON - RQN) = \{Eq. (15) - Eq. (16)\} \quad (17)$$

Similarly, the total vertical stress under point M can be derived by superposition of three (3) triangular loads ($q_1 - q_2 - q$) using Equation 13 in which,

$$q_1 = 2bq/(b - c) \text{ with } b_1 = 2b \quad (18)$$

$$q_2 = \{q(b + c)/(b - c)\} \text{ with } b_2 = b + c \quad (19)$$

and q with $b_3 = (b - c)$ \quad (20)

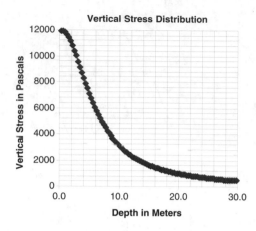

Figure 4.   Plot of vertical stress @ corner.

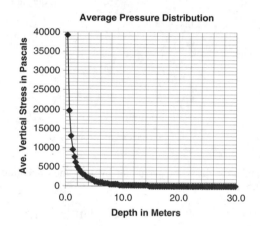

Figure 5.   Plot of average pressure @ corner.

3   APPLICATION

The derived formulas for total vertical stress are applicable in calculating magnitude of soil settlements due to surface loads and magnitude of surface loads transmitted to underground structures such as tunnels, bridges and culverts. The principle of superposition must be utilized in conjunction with the derived formulas to obtain total vertical stresses at corners, center and at any other point of a rectangular area under these surface loads. From these values the total vertical stress on this rectangular area can be calculated to any desired degree of accuracy and hence, an accurate prediction on the soil settlement and accurate estimate of the transmitted vertical stress on the roof of underground structures.

Illustrative Example 1: An area 3.05 m × 6.10 m is loaded with a uniform load of 47.88 kPa. Plot the variation of vertical stress and the average vertical stress through depth at the corner of the rectangular area, using the previously derived formulas for vertical stresses due to a rectangular surface loading.

Solution: Figures 4 & 5 are the results from the Excel program when the derived formulas are used.

Illustrative Example 2: In Example 1 the rectangular area is loaded with a triangular load of 47.88 kPa. Plot the variation of vertical stress and average stress through depth at the "q" load corner using the derived formulas of Figure 2.

Solution: Figures 6 & 7 are the results from the Excel program of the derived formulas.

Illustrative Example 3: In Figure 3 an embankment 180 m. long has the values b = 19.15 m, c = 10 m and q = 86 kPa. Determine the vertical stress distributions under the center of the embankment.

Solution: Using the Excel Program for uniform and triangular loading Figures 8 & 9 are the required

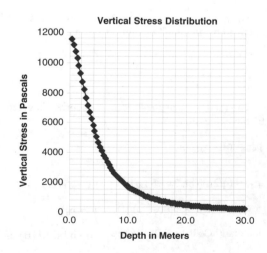

Figure 6.   Plot of vertical stress @ "q" corner.

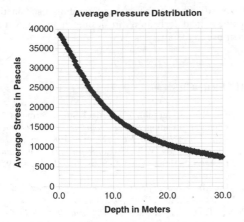

Figure 7.   Plot of average pressure @ "q" corner.

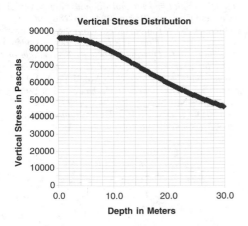

Figure 8.   Plot of vertical stress @ center.

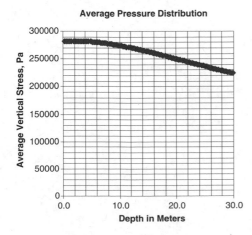

Figure 9.   Plot of average pressure @ center.

Table 1.

Method	Δp = Average Pressure in kPa	
	Corner	Center
Newmark	14.36	32.56
Steinbrenner	14.85	32.80
AASHTO	22.98	22.98
EXACT	14.85	33.09

solutions to this problem. The superpositioning of triangular loadings are as follows:

For triangle MNO the parameters are: a =90 m, b = 19.15 m and q = 180 kPa and for triangle RQN the parameters are a = 90 m, b = 10 m and q = 94 kPa. The vertical stresses under the center is obtained from 4(MNO – RQN) loadings.

Hence, two sets of Excel program for triangular loading has to be set up before the final tabulation and plots of vertical stress distribution can be obtained.

4   COMPARISON OF METHODS

For this purpose an example from Foundation Engineering book by Peck, Hanson and Thornburn will be used with SI units. A clay soil layer 4.27 m in thickness is overlain by 10.06 m of sand on which a raft foundation 12.20 m × 18.30 m sits with a net base load of 77 kPa. Calculate and compare results of average pressure, Δp at the mid-height of the clay layer using the Newmark's chart, Steinbrenner's formula, AASHTO's method and the exact method of Equation 7.

Solution: The Newmark's method is already indicated in the reference book. For Steinbrenner, the average pressure, Δp is assumed to equal the value from Equation 5 with q = 77 kPa and a = 12.2 m., b = 18.30 m at corner and a = 6.10, b = 9.15 m at center. For the new exact method, use Equation 6 at the top and bottom of the clay layer. Get the difference and divide by 4.27 m to obtain Δp. Do this for the corner and center of the raft foundation. For AASHTO the calculation for the average pressure at the mid-height of the clay layer is

$$\Delta p = 77(12.2)(18.3)/(24.4)(30.50) = 22.98 \text{ kPa.}$$

Table 1 shows the comparison of results. It indicates that the exact method predicts a higher value at the center of the raft foundation.

5   CONCLUSIONS

1. The derived formulas showed that the Boussinesq's equation could be simplified using standard

integration formulas in calculus to solve the total vertical stress under a rectangular and trapezoidal surface loading.

2. These derived formulas along with the principle of superposition predict greater values of the average vertical pressure, Δp underneath center of surface loads.

3. This method eliminates the use of graphical and finite-element method of calculations for vertical stresses due to surface loads.

6 NOTATIONS

σ_z = the vertical stress at any point under a concentrated load, Q

h = depth below surface load

p = the vertical stress at a depth "h" under a rectangular area with triangular, uniform or trapezoidal load

p_T = the total vertical stress at a depth "h" under a rectangular area loaded with triangular, uniform or trapezoidal load

q = loading intensity

p_o and p_q = the vertical stresses at a depth "h" due to a triangular load, "q"

p_{oT} and p_{qT} = the total vertical stresses at a depth, "h" under a rectangular area loaded with a triangular load, "q"

Δp = average pressure for settlement calculations

Note: All other alphabets and symbols used in the mathematical derivations are defined in the context of their use in the analysis.

REFERENCES

Jarquio, R.V. & Jarquio, V., Vertical Stress formulas for Triangular Loading, *ASCE Journal of Geotechnical Engineering*, Vol. 110, No. 1, January, 1984, pp. 73–78.

Peck, R.B., Hanson, W.E. & Thornburn, T.H, *Foundation Engineering*, John Wiley & Sons, Inc., London, 1953, pp. 276–278.

Bowles, J.E., *Physical and Geotechnical Properties of Soils*, McGraw-Hill Inc., 1979, pp. 286–291.

Holtz, R.D., & Kovacs, W.D., *An Introduction to Geotechnical Engineering*, Prentice Hall Inc., Englewoods Cliffs, N. J., 07632, 1981, pp. 346–366.

Smith, P.F., Longley, W.R. & Granville, W.A., *Elements of the Differential and Integral Calculus*, Revised Edition, Ginn and Company, Boston, 1941, pp. 286–306.

Steinbrenner, W., A Rational Method for the Determination of the Vertical Normal Stresses under Foundations, *Section E: Stress Distribution in Soils, Proceedings of the First International Conference on Soil Mechanics and Foundation Engineering*, 1936, Vol. II, pp. 142–143.

Terzaghi, K., *Theoretical Soil Mechanics*, John Wiley and Sons, Inc., New York, N. Y., 1943, pp. 373–384.

Poulous, H.G. & Davis, E.H., *Elastic Solutions for Soil and Rock Mechanics*, John Wiley & Sons, 1974, p. 40.

Stress fields for a semi-infinite plane containing a circular hole subjected to stresses from infinite points

T. Tsutsumi
Kagoshima National College of Technology, Kagoshima, Japan

K. Hirashima
Yamanashi University, Kofu, Japan

ABSTRACT: The problem of semi-infinite plane containing a circular hole has been treated using the stress functions in bi-polar coordinate. In these analyses, the boundary conditions have been given by only stresses or displacements on the hole or the straight edge. The aim of this study is that the solutions are presented for the semi-infinite plane containing a circular hole subjected to stresses from infinite points. For introducing the solutions, the solution of semi-infinite plane and that of infinite plane with a circular hole are superposed to converge to both boundary conditions. Some numerical results are shown by graphical representation.

1 INTRODUCTION

The problem of an elastic semi-infinite plane with one circular hole is very important for the strength of materials (Jeffery 1921) or driving tunnels (Mindlin 1939). Solutions for these problems have been induced using stress functions on bipolar coordinate system. In most of studies dealing these problems, the boundary conditions are estimated by stresses or displacements on the circular hole and straight boundary for obtaining the solution of the problem. Therefore, very few studies have dealt the problem of this shape subjected to stresses from infinite points.

On the other hand, solutions have been induced for doubly connected elastic problem by superposing the two kinds of elastic solution for simply connected problems until stresses on both boundaries converge to the boundary conditions (Tsutsumi 2002). This procedure allows us to obtain the final solution simply by solving the doubly connected elastic problems.

In this paper, this method is called the Constraint-release technique and is used to verify the solution to the problem of a semi-infinite plane with one circular hole subjected to stresses from infinite points.

2 FUNDAMENTAL EQUATIONS

Consider a two-dimensional semi-infinite plane with one circular hole, as presented in Figure 1. The stress components σ_x, σ_y and τ_{xy} and displacement components u_x and u_y are represented by the following equations:

$$
\begin{aligned}
\sigma_x &= 2\,\mathrm{Re}\left[\varphi'(z)\right] \\
&\quad - \mathrm{Re}\left[\bar{z}\varphi''(z) + \psi''(z)\right], \\
\sigma_y &= 2\,\mathrm{Re}\left[\varphi'(z)\right] \\
&\quad + \mathrm{Re}\left[\bar{z}\varphi''(z) + \psi''(z)\right], \\
\tau_{xy} &= \mathrm{Im}\left[\bar{z}\varphi''(z) + \psi''(z)\right].
\end{aligned}
\tag{1}
$$

$$
u_x - iu_y = \frac{1}{2G}\left[\kappa\overline{\varphi(z)} - \left\{\bar{z}\varphi'(z) + \psi'(z)\right\}\right].
\tag{2}
$$

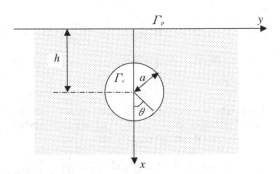

Figure1. Semi-infinite plane with one circular hole.

$$\kappa = \begin{cases} \dfrac{3-v}{1+v} & : plane\ stress \\ 3-4v & : plane\ strain \end{cases} \tag{3}$$

where $z = x + iy$, i is the imaginary unit; $\varphi(z)$ and $\psi(z)$ are the stress functions introduced by Kolosov and Muskhelishvili (Muskhelishvili 1963); Re and Im represent the real part and the imaginary part of the complex functions; the upper bars of the terms represent complex conjugates; and v and G represent Poisson ratio and shear modulus, respectively.

The formulae which map stress and displacement components into curvilinear coordinates $(\xi\eta)$ given by

$$\left. \begin{aligned} \sigma_\xi + \sigma_\eta &= \sigma_x + \sigma_y, \\ \sigma_\eta - \sigma_\xi + 2i\tau_{\xi\eta} &= e^{2i\theta}\left(\sigma_y - \sigma_x + 2i\tau_{xy}\right). \end{aligned} \right\} \tag{4}$$

$$u_\xi - iu_\eta = e^{i\theta}\left(u_x - iu_y\right). \tag{5}$$

3 FORMULATION OF THE PROBLEM

The purpose of this paper is to obtain the solution for a semi-infinite plane containing a circular hole subjected to stresses from infinite points. The stress functions for the infinite plane containing a circular hole as shown in Figure 2 is given by

$$\left. \begin{aligned} \varphi_{c,0}(z) &= A_0\left(z - h\right) + \frac{A_{0,-1}}{z - h}, \\ \psi_{c,0}(z) &= B_0\left(z - h\right)^2 \\ &\quad + K_0 \log\left(z - h\right) + \frac{B_{0,-2}}{\left(z - h\right)^2}. \end{aligned} \right\} \tag{6}$$

where

$$\left. \begin{aligned} A_0 &= \frac{\sigma_{x0} + \sigma_{y0}}{4}, \\ B_0 &= \frac{1}{2}\left(\frac{\sigma_{y0} - \sigma_{x0}}{2} + i\tau_{xy0}\right), \\ A_{0,-1} &= -2\overline{B}_0 a^2, \quad B_{0,-2} = \overline{B}_0 a^2, \\ K_0 &= -2A_0 a^2. \end{aligned} \right\} \tag{7}$$

The normal stress $\sigma_{x,0}^*$ and the shear stress $\tau_{xy,0}^*$ on the virtual straight boundary $(x = 0)$ are represented using the stress functions:

$$\left. \begin{aligned} \sigma_{x,0}^* &= 2\,\mathrm{Re}\left[\varphi_{c,0}'(z)\right] - \mathrm{Re}\left[\overline{z}\varphi_{c,0}''(z) + \psi_{c,0}''(z)\right], \\ \tau_{xy,0}^* &= \mathrm{Im}\left[\overline{z}\varphi_{c,0}''(z) + \psi_{c,0}''(z)\right]. \end{aligned} \right\} \tag{8}$$

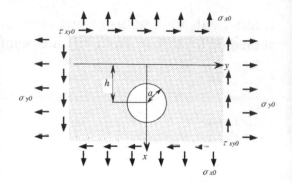

Figure 2.   Infinite plane under the stresses from infinite point.

The terms in the above equations are expanded as in the following equations, where $F_{0,n}$, $\hat{F}_{0,n}$, $H_{0,n}$, $\hat{H}_{0,n}$, $H_{0,n}^*$, $\hat{H}_{0,n}^*$, $J_{0,n}$, $\hat{J}_{0,n}$, $J_{0,n}^*$, $\hat{J}_{0,n}^*$ are complex coefficients which are determined by M_0, N_0, K_0, $A_{0,-m}$, and $B_{0,-m}$.

$$\left. \begin{aligned} \mathrm{Re}\left[\varphi_{c,0}'(z)\right] &= \sum_{k=1}^{K}\frac{F_{0,k}}{\left(y^2 + h^2\right)^k} + \sum_{k=1}^{K}\frac{y\hat{F}_{0,k}}{\left(y^2 + h^2\right)^k}, \\ \mathrm{Re}\left[\overline{z}\varphi_{c,0}''(z)\right] &= \sum_{k=1}^{K}\frac{H_{0,k}}{\left(y^2 + h^2\right)^k} + \sum_{k=1}^{K}\frac{y\hat{H}_{0,k}}{\left(y^2 + h^2\right)^k}, \\ \mathrm{Im}\left[\overline{z}\varphi_{c,0}''(z)\right] &= \sum_{k=1}^{K}\frac{yH_{0,k}^*}{\left(y^2 + h^2\right)^k} + \sum_{k=1}^{K}\frac{\hat{H}_{0,k}^*}{\left(y^2 + h^2\right)^k}, \\ \mathrm{Re}\left[\psi_{c,0}''(z)\right] &= \sum_{k=1}^{K}\frac{J_{0,k}}{\left(y^2 + h^2\right)^k} + \sum_{k=1}^{K}\frac{y\hat{J}_{0,k}}{\left(y^2 + h^2\right)^k}, \\ \mathrm{Im}\left[\psi_{c,0}''(z)\right] &= \sum_{k=1}^{K}\frac{yJ_{0,k}^*}{\left(y^2 + h^2\right)^k} + \sum_{k=1}^{K}\frac{\hat{J}_{0,k}^*}{\left(y^2 + h^2\right)^k}. \end{aligned} \right\} \tag{9}$$

Therefore, $\sigma_{x,0}^*$ and $\tau_{xy,0}^*$ are represented as:

$$\left. \begin{aligned} \sigma_{x,0}^* &= \sum_{k=1}^{K}\left\{\widetilde{\sigma}_{x,k}^* \frac{1}{\left(y^2 + h^2\right)^k} + \widetilde{\sigma}_{x,k}' \frac{y}{\left(y^2 + h^2\right)^k}\right\}, \\ \tau_{xy,0}^* &= \sum_{k=1}^{K}\left\{\widetilde{\tau}_{x,k}^* \frac{1}{\left(y^2 + h^2\right)^k} + \widetilde{\tau}_{x,k}' \frac{y}{\left(y^2 + h^2\right)^k}\right\}. \end{aligned} \right\} \tag{10}$$

where

$$\left. \begin{aligned} \widetilde{\sigma}_{x,k}^* &= 2F_{0,k} - H_{0,k} - J_{0,k}, \\ \widetilde{\sigma}_{y,k}' &= 2\hat{F}_{0k} - \hat{H}_{0,k} - \hat{J}_{0,k}, \\ \widetilde{\tau}_{xy,k}^* &= H_{0,k}^* + J_{0,k}^*, \\ \widetilde{\tau}_{xy,k}' &= \hat{H}_{0,k}^* + \hat{J}_{0,k}^*. \end{aligned} \right\} \tag{11}$$

2004

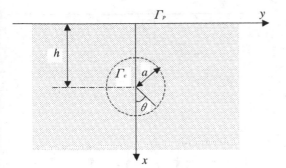

Figure 3. Semi-infinite plane without holes.

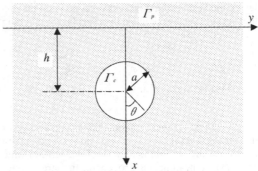

Figure 4. Infinite plane containing one circular hole.

Then, in order to cancel out the stresses represented in equation 8, the negative values of these stresses are loaded on semi-infinite plane. The stress functions for a semi-infinite plane are represented as:

$$
\left.
\begin{aligned}
\varphi'_{p,1}(z) &= \frac{a_1(0)}{2}\frac{1}{z} \\
&\quad + \int_0^\infty e^{-zt}\frac{a_1(t)-a_1(0)+ib_1(t)}{2}\,dt, \\
\psi''_{p,1}(z) &= \frac{\overline{a_1(z)}}{2}\frac{1}{z} \\
&\quad + \int_0^\infty e^{-zt}\left\{\frac{\overline{a_1(t)}-\overline{a_1(0)}+i\overline{b_1(t)}}{2}\right. \\
&\quad + \left.\frac{a_1(t)+ib_1(t)}{2}+zt\frac{a_1(t)+ib_1(t)}{2}\right\}dt.
\end{aligned}
\right\} \quad (12)
$$

where

$$
\left.
\begin{aligned}
a_1(0) &= -\frac{1}{\pi}\left(P_{y,1}+iP_{x,1}\right), \\
a_1(t) &= -\frac{2}{\pi}\int_0^\infty\left[\sigma^*_{x,0}+i\tau^*_{xy,0}\right]\cos(ty)dy, \\
b_1(t) &= -\frac{2}{\pi}\int_0^\infty\left[\sigma^*_{x,0}+i\tau^*_{xy,0}\right]\sin(ty)dy.
\end{aligned}
\right\} \quad (13)
$$

In the above equations, $P_{y,1}$ and $P_{x,1}$ represent the resultant forces in the x-direction and y-direction on the virtual straight boundary, respectively. Considering the x-axial symmetry of this problem, $a_1(t)$ and $b_1(t)$ can be expanded as following series:

$$
\left.
\begin{aligned}
a_1(t) &= -\frac{2}{\pi}\int_0^\infty\sum_{k=1}^K\frac{\left(\widetilde{\sigma}^*_{k,n}+i\widetilde{\tau}'_{xy,k}\right)}{\left(y^2+h^2\right)^k}\cos ty\,dy, \\
b_1(t) &= -\frac{2}{\pi}\int_0^\infty\sum_{k=1}^K\frac{\left(\widetilde{\sigma}'_{x,k}+i\widetilde{\tau}^*_{xy,k}\right)}{\left(y^2+h^2\right)^k}y\sin ty\,dy.
\end{aligned}
\right\} \quad (14)
$$

Furthermore, the above equations are rewritten as follows:

$$
a_1(t)=\sum_{k=1}^K\widetilde{a}_{1,k}t^{k-1}e^{-th}, \quad b_1(t)=\sum_{k=1}^K\widetilde{b}_{1,k}t^{k-1}e^{-th}. \quad (15)
$$

By using the above equations and the Laplace transformation, equation 12 may be represented as follows:

$$
\left.
\begin{aligned}
\varphi'_{p,1}(z) &= \sum_{k=1}^K\frac{(k-1)!\left(\widetilde{a}_{1,k}-\widetilde{b}_{1,k}\right)}{2(z+h)^k}, \\
\psi''_{p,1}(z) &= -\sum_{k=1}^K\left\{\frac{(k-1)!\widetilde{b}_{1,k}}{(z+h)^k}+\frac{k!\left(\widetilde{a}_{1,k}-\widetilde{b}_{1,k}\right)z}{2(z+h)^{k+1}}\right\}.
\end{aligned}
\right\} \quad (16)
$$

On using these stress functions, the normal stress, $\sigma_{\xi,1}$, and the shear stress, $\tau_{\xi\eta,1}$, arising on the virtual circular hole boundary in the semi-infinite plane without holes are represented as:

$$
\begin{aligned}
\sigma_{\xi,1}-i\tau_{\xi\eta,1} &= \overline{c}_{1,0} \\
&\quad + \sum_{m=1}^M\left(\overline{c}_{1,m}\cos m\theta+\overline{d}_{1,m}\sin m\theta\right).
\end{aligned} \quad (17)
$$

For canceling out these stresses, the negative values of the stresses represented in equation 17 are again loaded on the circular hole of the infinite plane with one circular hole. The stress functions for the infinite plane acting the arbitrary loads on the circular hole are given by

$$
\left.
\begin{aligned}
\varphi_{c,1}(z) &= M_1\log(z-h)+\sum_{m=1}^M A_{1,-m}(z-h)^{-m}, \\
\psi_{c,1}(z) &= N_1(z-h)\log(z-h) \\
&\quad + K_1\log(z-h)+\sum_{m=1}^M B_{1,-m}(z-h)^{-m}.
\end{aligned}
\right\} \quad (18)
$$

Using these stress functions, stresses arise on the virtual straight boundary in the infinite plane containing a circular hole as follows.

$$\left.\begin{array}{l} \sigma_{x,1}^{*} = 2\,\mathrm{Re}\!\left[\varphi_{c,1}'(z)\right] - \mathrm{Re}\!\left[\bar{z}\,\varphi_{c,1}''(z) + \psi_{c,1}''(z)\right], \\[2mm] \tau_{xy,1}^{*} = \mathrm{Im}\!\left[\bar{z}\,\varphi_{c,1}''(z) + \psi_{c,1}''(z)\right] \end{array}\right\} \tag{19}$$

For canceling out these stresses, the negative value of the stresses represented in equation 19 are again loaded on the straight boundary of the semi-infinite plane. By repeating this procedure, the stress functions for this problem are obtained using the following equations:

$$\left.\begin{array}{l} \varphi(z) = \displaystyle\sum_{n=0}^{N} \varphi_{c,n}(z) + \sum_{n=1}^{N} \varphi_{p,n}(z), \\[4mm] \psi(z) = \displaystyle\sum_{n=0}^{N} \psi_{c,n}(z) + \sum_{n=1}^{N} \psi_{p,n}(z). \end{array}\right\} \tag{20}$$

where

$$\left.\begin{array}{l} \varphi_{c,n}(z) = M_n \log(z - h) \\[2mm] \qquad + \displaystyle\sum_{k=1}^{K} A_{n,-k}(z - h)^{-k}, \\[4mm] \psi_{c,n}(z) = N_n z \log(z - h) + K_n \log(z - h) \\[2mm] \qquad + \displaystyle\sum_{k=1}^{K} B_{n,-k}(z - h)^{-k} \quad (n \ge 1), \\[4mm] \varphi_{p,n}(z) = \dfrac{\overline{a_n(0)}}{2} \log z \\[2mm] \qquad - \displaystyle\int_0^{\infty} e^{-zt}\, \frac{\overline{a_n(t)} - \overline{a_n(0)} + i \overline{b_n(t)}}{2t}\, dt, \\[4mm] \psi_{p,n}(z) = -\dfrac{\overline{a_n(0)}}{2} \log z \\[2mm] \qquad + \displaystyle\int_0^{\infty} e^{-zt}\{- \frac{\overline{a_n(t)} - \overline{a_n(0)} + i \overline{b_n(t)}}{2t} \\[2mm] \qquad - z\, \frac{\overline{a_n(t)} + i \overline{b_n(t)}}{2}\} dt. \end{array}\right\} \tag{21}$$

4 RESULTS AND DISCUSSION

Figure 6 shows the tangential stress around the circular hole at $h/a = 2.0$ and 5.0 when the stress parallel to x axis is loaded from the infinite point as shown in Figure 5. In this graph, the inside region of initial shape represents compression, the outside region represents tension. The number of calculation repetition

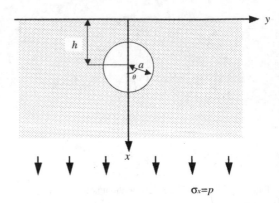

Figure 5. Stress parallel to x axis from infinite point.

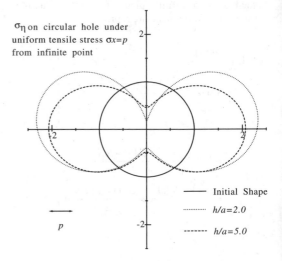

σ_η on circular hole under uniform tensile stress $\sigma x = p$ from infinite point

—— Initial Shape

·········· $h/a = 2.0$

------ $h/a = 5.0$

Figure 6. Tangential stress around hole subjected to stress $\sigma_x = p$ from infinite point.

N is 9 at $h/a = 2.0$ and 6 at $h/a = 5.0$. Tension appears on both side of the hole and compression appears on the top and bottom of the hole. The value of the compression appearing on the top of the hole is same as that appearing on the bottom of the hole when the hole is deep (i.e. $h/a = 5.0$). However, when the hole is shallow (i.e. $h/a = 2.0$), the compression appearing on the top of the hole is about 1.3 times larger than that appearing on the bottom of the hole. Also, the tension when the hole is shallow is larger than that when the hole is deep.

Figure 8 shows the tangential stress around the circular hole at $h/a = 2.0$ and 5.0 when the stress parallel to y axis is loaded from the infinite point as shown in

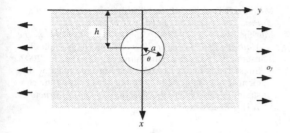

Figure 7. Stress parallel to y axis from infinite point.

σ_η on circular hole under
uniform tensile stress σ_y=p
from infinite point

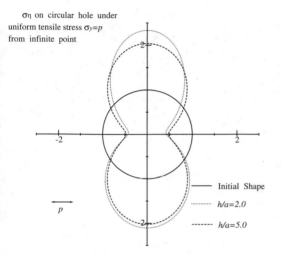

— Initial Shape
········ h/a=2.0
------ h/a=5.0

Figure 8. Tangential stress around hole subjected to stress $\sigma_y = p$ from infinite point.

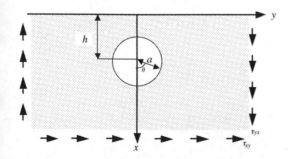

Figure 9. Shear stresses from infinite point.

in Figure 7. Tension appears on the top and bottom of the hole and compression appears on both side of the hole. The value of the tension appearing on the top of the hole is same as that appearing on the bottom of the hole when the hole is deep (i.e. $h/a = 5.0$). However,

σ_η on circular hole under
uniform shear stress τ_xy=p
from infinite point

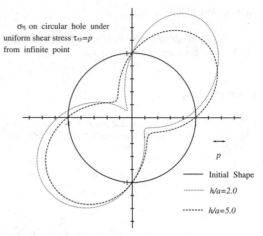

— Initial Shape
········ h/a=2.0
------ h/a=5.0

Figure 10. Tangential stress around hole subjected to shear stress $\tau_{xy} = p$ from infinite point.

when the hole is shallow (i.e. $h/a = 2.0$), the tension appearing on the top of the hole is about 1.1 times larger than that appearing on the bottom of the hole. It is observed that the straight boundary influences more to the tangential stress on the top of the hole subjected to σ_x than that subjected to σ_y. Also, the compression when the hole is shallow is a little larger than that when the hole is deep.

Figure 10 shows the tangential stress around the circular hole at $h/a = 2.0$ and 5.0 when the shear stress τ_{xy} is loaded from the infinite point as shown in Figure 9. The maximum value of compression or tension appears on the diagonal direction. Both values of tension and compression arising on the nearest side to straight boundary is same as those arising on the opposite side when the hole is deep. However, both values of tension and compression arising on the nearest side to the straight boundary is larger than those arising on the opposite side when the hole is shallow.

5 CONCLUSION

In this paper, solutions are proposed for an isotropic elastic semi-infinite plane subjected to stresses from infinite points by superposing the solutions for an isotropic elastic infinite plane containing one circular hole and solutions for an isotropic elastic semi-infinite plane until the required boundary conditions are met.

Some numerical results are presented and it is shown that the semi-infinite plane containing a circular hole can be regarded as an infinite plane containing a circular hole when hole is deep. However, in these results the hole is influenced by straight boundary when hole is shallow.

ACKNOWLEDGEMENT

The authors wish to thank Professor M. Otsuki and Mr. H. Arai for much helpful advice.

REFERENCES

Jeffery, G.B. 1921. Plane stress and plane strain in bipolar co-ordinates, *Transaction of Royal Society* A(221): 265–293.

Mindlin, R.D. 1939. Stress distribution around a tunnel, *Proceedings of A.S.C.E.* 65(4): 619–642.

Muskhelishvili, N.I. 1963. *Some Basic Problem of the Mathematical Theory of Elasticity*, 4th ed., Noordhoff.

Tsutsumi, T., Sato, K., Hirashima, K. & Arai, H. 2002. Stress fields on an isotropic semi-infinite plane with a circular hole subjected to arbitrary loads using the constraint-release technique, *Steel and Composite Structures* 2(4): 209–220.

System-based Vision for Strategic and Creative Design, Bontempi (ed.)
© *2003 Swets & Zeitlinger, Lisse, ISBN 90 5809 599 1*

Study of two-dimensional elasticity on Functionally Graded Materials

T. Seto, M. Ueda & T. Nishimura
Department of Mechanical Engineering, College of Science and Technology, Nihon University, Tokyo, Japan

ABSTRACT: Recently, composite materials called Functionally Graded Materials (FGMs) have been developing as future materials. One of the FGM is made by mixing non-metal fine particles and metal particles, and sintered together by spark plasma sintering method (SPS method). Changing their mixing ratio, FGMs can possess any required character of mechanical properties and chemical resistance for the actual use. Therefore, FGMs are quite useful on designing an optimum structure. Although the classical theory of elasticity is available for the body of uniform elastic property, this theory cannot be applied for the structure of FGMs, because their mechanical property varies over the body. Hence, two-dimensional problem for the body of non-uniform elastic modulus will be discussed in this paper.

1 INTRODUCTION

Functionally Graded Materials (FGMs) have been developing as future materials. One of the FGMs is realized by mixing non-metal particles with metal particles and sintering these together. Selecting the compound substances and changing their mixing ratio, any character of materials, i.e. the new progressive material which has a required property of strength or flexibility and a chemical resistance according to actual applications can be obtained. The elastic coefficients can also be varied throughout the body of FGM, while the ordinary materials have uniform properties.

However, there is no theory to evaluate the mechanical rigidity of a structure composed of FGM which has a non-uniform elastic modulus. The classical theory of elasticity is only applicable for the ordinary materials with constant elastic modulus. It is required to evaluate theoretically the mechanical rigidity of the FGM in order to design the adaptive structure. Hence, the new theory applicable for the body with a non-uniform elastic modulus is demanded.

In this paper, two-dimensional elastic problem is discussed for the material which has a non-uniform modulus of elasticity over the body. Then, the problem of edge-bonded dissimilar materials, which can be considered as one of FGM, is cleared from the view of two-dimensional elastic theory.

2 TWO-DIMENSIONAL PROBLEM

2.1 *Government equation of classical theory*

In the classical theory, the differential equation of equilibriums, the compatibility equation and Hooke's law must be satisfied together with the boundary conditions to determine the stress distribution of two-dimensional problems. Using stress function ϕ, compatibility equation, namely the government equation, can be expressed as follows.

$$\frac{\partial^4 \phi}{\partial x^4} + 2\frac{\partial^4 \phi}{\partial x \partial y} + \frac{\partial^4 \phi}{\partial^4 y} = 0 \tag{1}$$

Here the body forces are ignored. This equation holds both for the case of plane stress state and for the case of plane strain state. Thus the solution of a two-dimensional problem reduces to finding a solution of Equation 1, which also satisfies the boundary conditions.

2.2 *Government equation for FGMs*

In the case of FGMs, the Hooke's law includes the Young's modulus represented by the function of x and y, because Young's modulus varies over the cross section. The equations of equilibrium and compatibility equation are also identical with those of FGM.

Therefore, using the same procedure to the classical theory, the government equation can be derived as follows:

$$a\left(\frac{\partial^4\phi}{\partial x^4}+2\frac{\partial^4\phi}{\partial x^2\partial y^2}+\frac{\partial^4\phi}{\partial y^4}\right)$$

$$+2\frac{\partial a}{\partial x}\left(\frac{\partial^3\phi}{\partial x^3}+\frac{\partial^3\phi}{\partial x\partial y^2}\right)+2\frac{\partial a}{\partial y}\left(\frac{\partial^3\phi}{\partial y^3}+\frac{\partial^3\phi}{\partial x^2\partial y}\right)$$

$$+\frac{\partial^2 a}{\partial x^2}\left(\frac{\partial^2\phi}{\partial x^2}-v\frac{\partial^2\phi}{\partial y^2}\right)+\frac{\partial^2 a}{\partial y^2}\left(\frac{\partial^2\phi}{\partial y^2}-v\frac{\partial^2\phi}{\partial x^2}\right)$$

$$+2(1+v)\frac{\partial^2 a}{\partial x\partial y}\frac{\partial^2\phi}{\partial x\partial y}=0 \qquad (2)$$

Where $a(x) = 1/E(x, y)$ and Poisson's ratio v is constant. The body force is also ignored here for simplicity. The main theme of this study is to clear the problems on the stress analysis of FGM. This equation holds only for the case of plane stress state, because Equation 2 includes the Young's modulus and Poisson's ratio. However, for the plane strain problem, Equation 2 can be used by replacing the Young's modulus $1/E(x, y)$ by $(1 - v^2)/E(x, y)$ and Poisson's ratio v by $v/(1 - v)$ as usual procedure. Therefore two-dimensional problem with non-uniform Young's modulus is reduced to find the stress function, which satisfies Equation 2 under the given boundary conditions on each problem. It is clear from government Equation 2 that stress distribution is affected by the derivative of inverse function of Young's modulus, in other words, the manner of distribution $a(x, y)$. The government equation for FGM is too difficult to solve theoretically. The uni-axial tension of rectangular plate of FGM is discussed in the next section.

2.3 Uni-axial tension of a rectangular plate

As the government Equation 2 is difficult to solve theoretically, let consider a uni-axial tensile problem of rectangular plate which has a Young's modulus varying over x direction (Fig. 1). In this case, Equation 2 can be reduced to the next equation.

$$a\left(\frac{\partial^4\phi}{\partial x^4}+2\frac{\partial^4\phi}{\partial x^2\partial y^2}+\frac{\partial^4\phi}{\partial y^4}\right)$$

$$+2\frac{\partial a}{\partial x}\left(\frac{\partial^3\phi}{\partial x^3}+\frac{\partial^3\phi}{\partial x\partial y}\right)+\frac{\partial^2 a}{\partial x^2}\left(\frac{\partial^2\phi}{\partial x^2}-v\frac{\partial^2\phi}{\partial y^2}\right)=0 \qquad (3)$$

At first, let assume the stress function as,

$$\phi=\frac{1}{2}Py^2 \qquad (4)$$

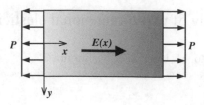

Figure 1. Uni-axial tension of rectangular plate.

Then the stresses are given by

$$\sigma_x=\frac{\partial^2\phi}{\partial y^2}=P \ , \ \sigma_y=\frac{\partial^2\phi}{\partial x^2}=0 \ , \ \tau_{xy}=\frac{\partial^2\phi}{\partial x\partial y}=0 \qquad (5)$$

Above equations explain the uniform stress state and satisfy the boundary conditions along the periphery and equilibrium equation in the plate. Then, substituting Equation 4 into Equation 3, the next equation must be satisfied in order to hold the government equation.

$$\frac{\partial^2 a}{\partial x^2}(0-vP)=0 \qquad (6)$$

Here, since vP is not zero, Equation 3 is satisfied only if the second derivative of inverse function of Young's modulus is zero.

$$\frac{\partial^2 a}{\partial x^2}=0 \qquad (7)$$

In other words, only if the distribution of the inverse function of Young's modulus is linear or constant, Equation 3 is consistence to the uniform stress distribution, that is,

$$a=Ax+B=\frac{1}{E(x)} \qquad (8)$$

Eequation 3 is hold if the Young's modulus satisfies the above condition. Hence, in the case of uni-axial tension or compression of rectangular plate of FGM with Young's modulus varying over the x direction, the constant stress distribution never exist unless the above conditions hold.

However, apart from a simple tension or compression, it is found from Equation 6 that if the load vP is applied to another pair of sides of rectangular plate, the constant stress distribution may exist for arbitrary distributed Young's modulus along the x direction (Fig. 2). Then, it is clear from the discussion that the stress distribution may be complex generally for FGMs.

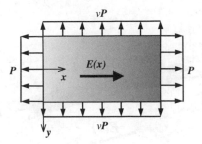

Figure 2. Bi-axial tension of rectangular plate.

2.4 Calculated result of stress distribution for uni-axial tension

It is very difficult to solve the government equation for FGM directly even if the loading condition is supposed as uni-axial tension. Then, the finite difference method is adopted to solve Equation 3 in this study. In the numerical calculation, the distribution of Young's modulus is assumed to be a linear distribution.

$$E(x) = 100x + 10 \qquad (9)$$

The value of Young's modulus changes 10 to 110 from $x = 0$ to 1 continuously. The external forces $P = 200$ is applied to both ends $x = 0$, and 1. The numerical calculation is performed about the plate of following three aspect ratios.

$$Length(x) : Width(y) = 1:1, 1:4, 4:1 \qquad (10)$$

The calculated results of stress distribution are shown in Figures 3–5. It is interesting to note that not only tensile stress σ_x but also lateral stress σ_y and shearing stress τ_{xy} are appeared in the plate. The tensile stress σ_x becomes larger than the mean stress around the mid portion of free sides and smaller around the central portion of the plate (Fig. 3-a), while the lateral stress σ_y yields to be larger around the loading sides and smaller around the center of the plate (Fig. 3-b). The shearing stress distribution τ_{xy} is indicated by so-called chess mode (Fig. 3-c).

In the case of the aspect ratio 1:4, the large disturbance of the tensile stress σ_x appears at the limited portion along the free sides of plate, while the stress disturbance becomes small in the remainder central portion (Fig. 4-a), because the stress state at the center portion tends to become the plane stress state. Therefore the lateral stress σ_y becomes uniform along the x axis around the center portion of the plate (Fig. 4-b), and the shearing stress τ_{xy} approaches to zero (Fig. 4-c). In the case of the aspect ratio 4:1, the

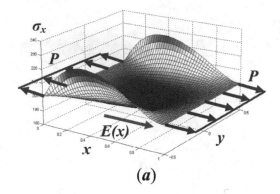

(a)

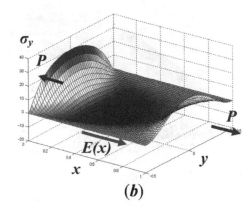

(b)

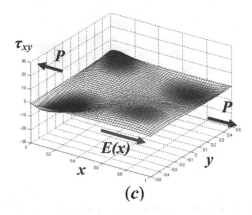

(c)

Figure 3. Stress distribution of the plate with $L:W = 1:1$.

tensile stress σ_x takes maximum values around the edge of the plate where Young's modulus is small (Fig. 5-a), because the second derivative of $a(x)$ takes maximum values at $x = 0$, and decrease suddenly with increasing of x (Fig. 6). It is concluded that the stress disturbance is strongly affected by the second derivative of $a(x)$.

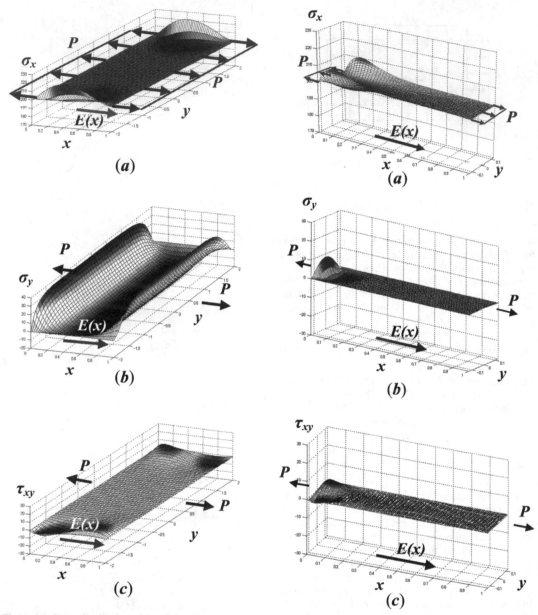

Figure 4. Stress distribution of the plate with $L:W = 1:4$.

Figure 5. Stress distribution of the plate with $L:W = 4:1$.

3 APPLICATION TO THE STRESS ANALYSIS OF THE EDGE-BONDED DISSIMILAR MATERIALS

The government Equation 3 is also applicable to the problem of edge-bonded dissimilar materials (Fig. 7). The distribution of inverse function of Young's modulus is assumed to be expressed by so called sigmoid function.

$$a(x) = \frac{1}{E(x)} = a_A - \frac{a_A - a_B}{1 + \exp(-\lambda x)} \quad (11)$$

Where, a_A and a_B are inverse of Young's modulus of material A and B. Assuming the parameter of λ takes larger to infinity, the slope in Young's modulus distribution becomes larger. Namely, Young's modulus changes E_A to E_B more steeply. The change of distribution of

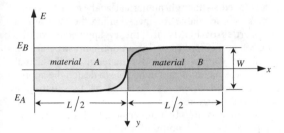

Figure 6. Distribution of derivative $a(x)$.

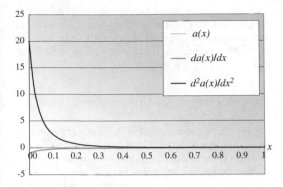

Figure 7. Distribution of Young's modulus of Edge-bonded dissimilar materials.

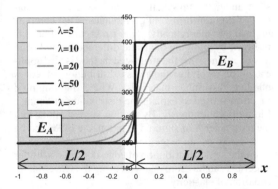

Figure 8. Distribution of Young's modulus.

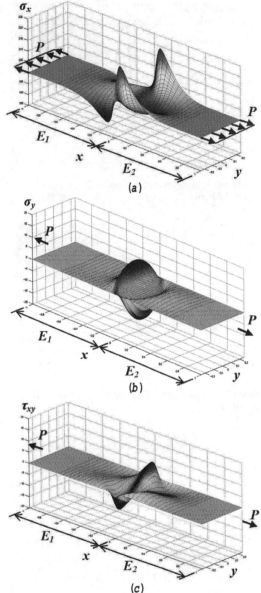

Figure 9. Stress distribution of the edge-bonded dissimilar materials.

Young's modulus with the parameter λ is shown in Figure 8.

Figure 9 shows the calculated result in the case of $\lambda = 50$. Here the length $L = 2$ and width $W = 0.5$. The young's modulus of material A and B are 200 and 400 respectively. The external force P is 200. The tensile stress σ_x around the joint portion at the free sides takes maximum value in the material A with smaller Young's modulus and minimum value in the material B with larger Young's modulus. Here, the distribution of derivative $a(x)$ is shown in Figure 10. In this figure, the distribution of second derivative of $a(x)$ is similar to the disturbance distribution of tensile stress σ_x around the joint portion of the free sides and has much larger value than in the remainder portion. That is, it is clear from this figure that the disturbance of stress distribution σ_x on the side of the plate is strongly affected by the second derivative of $a(x)$.

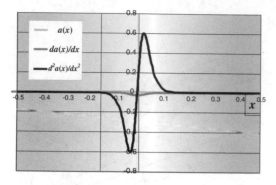

Figure 10. Distribution of differentiating $a(x)$ on inverse of sigmoid function ($\lambda = 50$).

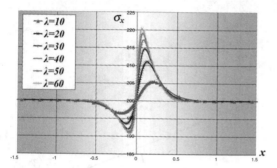

Figure 11. Calculation results of σ_x at the side of the plate ($y = 0.25$).

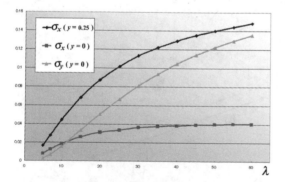

Figure 12. Stress disturbance rate.

Figure 11 shows the comparison of calculated results σ_x when λ is ranged from 5 to 60. If the parameter λ becomes larger, the stress disturbance becomes larger. And Figure 12 indicates the calculation result of difference between maximum and minimum stresses which are divided by external force P, namely

stress disturbance rate of tensile and lateral stresses. It is clear from this figure that these stress disturbances may converge to some values when the large numerical number is put into parameter λ of sigmoid function.

4 CONCLUSION AND DISCUSSION

FGMs are advanced materials, the properties of which are possible to adjust corresponding to the demands on actual applications. With development of production technique, it has been promoted rapidly. However, since the classical theory of elasticity is not directly applicable to FGMs, the theoretical study on non-uniform mechanical properties has not been developed. Although numerical analysis and experiments are available for investigating the mechanical properties of FGMs, the theory applicable for the FGMs is required to design an optimum structure. In order to supply an adaptive material for the actual demand, it is important to be able to design the appropriate arrangement of FGM so as to satisfy the required functions.

In this study, two-dimensional elastic problem applicable for the non-uniform Young's modulus was discussed. The uniform stress distribution is no more exist for the FGM except for the condition that an inverse of Young's modulus is a linear function or constant under the uni-axial tension. Government equation of non-uniform Young's modulus is too complex to analyze theoretically. The stress distribution is resolved by applying the finite different method. And it is mentioned that the suggested method is also applicable for resolving the stress distribution of edge-bonded dissimilar materials. Where the sigmoid function is used to explain the distribution of Young's Modulus continuously. By putting large number into the parameter λ of sigmoid function, the FGM can be considered as an edge-bonded dissimilar material. Then the stress distribution of edge-bonded dissimilar materials can be estimated with using the Government equation of elasticity.

REFERENCES

Timoshenko, S. & Goodier, J.N. 1951. *Theory of Elasticity (second edition)*. McGraw-hill book company, inc.
Kuranishi, M. 1980. *Applied elasticity*. Kyoritsu-zensho.
Omori, M. et al. 1994. *Proceedings of the 3rd International Symposium on structural and Functionally Gradient Materials*.

System-based Vision for Strategic and Creative Design, Bontempi (ed.)
© *2003 Swets & Zeitlinger, Lisse, ISBN 90 5809 599 1*

Equivalent shear modulus of Functionally Graded Materials

T. Sakate, M. Ueda, T. Seto & T. Nishimura
Department of Mechanical Engineering College of Science and Technology, Nihon University

ABSTRACT: Recently, composite materials such as Functionally Graded Materials (FGMs) have been developing as future materials. One of the FGMs is realized by mixing non-metal particles with the metal and sintering these together. Selecting the compound substances and changing their mixing ratio, we can obtain any properties of material. Especially the mixing ratio can be varied over the cross-section so as to get the required function for actual use. However there is no theoretical method to evaluate the mechanical rigidity of a cross section over which the mechanical property varies. The general theory of torsion cannot be applied to a FGM shaft. In this paper, the torsion theory applicable for the FGM is discussed and the concept of equivalent shear modulus is introduced for simple evaluation of torsional rigidity. It is mentioned that the ordinal torsion theory is developed by using the equivalent shear modulus suggested in this paper.

1 INTRODUCTION

Functionally Graded Materials (FGMs) have been developing as a future material. One of the FGMs is realized by mixing non-metal fine particles with metal particles and sintering these together. By selecting the compound materials and changing their mixing ratio, any character of material i.e. the new progressive material, which has a required character of strength or flexibility and the chemical resistance corresponding to the actual applications, can be obtained. This new material may possess the properties that vary over its cross-section while ordinary materials have the uniform properties. In this paper, the authors discuss an elastic problem for the FGM, especially the stress distribution by the effect of the non-uniform elastic modulus over the cross-section.

FGM is provided by changing the mixing ratio of compound materials over the cross-section of a body for satisfying the required function on an actual use. However, there is no general theory to evaluate the mechanical rigidity of a structure composed of FGMs, over the cross-section of which the mechanical properties vary. The torsion theory of a prismatic bar was developed by using the analogy of the soap film inflation (membrane analogy), introduced by L. Prandtl. However it is only applicable for a shaft having a uniform shear modulus, because the membrane theory is correct for homogeneous materials. Therefore, it is impossible to adopt the membrane analogy for the torsion problems of a shaft composed of FGM.

Since the mechanical property of FGMs has been estimated experimentally up to the present, it is necessary to develop the theoretical resolution of the property of new materials for adaptability evaluation on the actual application. A new theory applicable to materials with a non-uniform shear modulus is required. In this research, the general theory of torsion of a bar with constant shear modulus is expanded for a bar with non-uniform shear modulus.

2 TORSION OF A PRISMATIC BAR WITH NON-UNIFORM SHEAR MODULUS

2.1 *Differential equation*

The classical torsion theory is only applicable for a shaft having a uniform shear modulus. Therefore, it is impossible to adopt the classical theory for a shaft composed of FGM. In this paper, a new method, which is applicable for FGM, is developed.

Taking the origin of coordinates at an end cross-section (Fig. 1), the displacements corresponding to the rotation in the cross-section and warping of cross-section are defined as ordinary by

$$u = -\theta zy, v = \theta zx \tag{1}$$

$$w = \theta \psi(x, y) \tag{2}$$

where θz = angle of the rotation of cross-section at a distance z from the origin, and ψ = warping function.

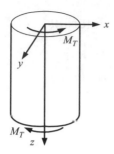

Figure 1. Decision of coordinate.

Then, the equilibrium equations are reduced to the next formulae.

$$\frac{\partial \tau_{xz}}{\partial z}=0, \quad \frac{\partial \tau_{yz}}{\partial z}=0, \quad \frac{\partial \tau_{xz}}{\partial x}+\frac{\partial \tau_{yz}}{\partial y}=0 \qquad (3)$$

And the relations between displacements and strains are identical with the usual theory. Consequently, the stresses are related to the warping function in the usual manner as shown below.

$$\sigma_x = \sigma_y = \sigma_z = \tau_{xy} = 0$$

$$\tau_{yz} = G(x,y)\,\gamma_{yz} = G(x,y)\,\theta\left(\frac{\partial \psi}{\partial y}+x\right) \qquad (4)$$

$$\tau_{xz} = G(x,y)\,\gamma_{xz} = G(x,y)\,\theta\left(\frac{\partial \psi}{\partial x}-y\right)$$

where $G(x,y)$ = non-uniform shear modulus. The shear modulus is not constant over the cross-section but is a function of x and y. Substituting Equation 4 into the equilibrium Equation 3, the following equation is obtained.

$$\frac{\partial G(x,y)}{\partial x}\theta\left(\frac{\partial \psi}{\partial x}-y\right)+\frac{\partial G(x,y)}{\partial y}\theta\left(\frac{\partial \psi}{\partial y}+x\right)$$

$$+G(x,y)\,\theta\left(\frac{\partial^2 \psi}{\partial x^2}+\frac{\partial^2 \psi}{\partial y^2}\right)=0 \qquad (5)$$

Since the first and second terms of the left side of the equation vanish for usual materials because of constant shear modulus, only the last term is discussed. In generally, it is very hard to solve Equation 5 directly for arbitrary non-uniform shear modulus function. However, if the sum of the first two terms is equal to zero regardless of the existence of derivatives of shear modulus, Equation 5 may be solved. Now, instead of the stress function introduced by the classical theory, a new concept, the strain function, is introduced. The strain function is defined by the following equations.

$$\gamma_{yz} = -\theta\frac{\partial \eta}{\partial x}=\theta\left(\frac{\partial \psi}{\partial y}+x\right)$$

$$\gamma_{xz} = \theta\frac{\partial \eta}{\partial y}=\theta\left(\frac{\partial \psi}{\partial x}-y\right) \qquad (6)$$

where η = strain function. Strain function η is the function of x and y, and independent of z. If the shear modulus is constant, the configuration of the strain function is identical with the stress function. By using the strain function, the first and two terms of Equation 5 can be rewritten as follows.

$$\frac{\partial G(x,y)}{\partial x}\theta\left(\frac{\partial \psi}{\partial x}-y\right)+\frac{\partial G(x,y)}{\partial y}\theta\left(\frac{\partial \psi}{\partial y}+x\right)$$

$$=\left[\frac{\partial G(x,y)}{\partial x}\frac{\partial \eta}{\partial y}-\frac{\partial G(x,y)}{\partial y}\frac{\partial \eta}{\partial x}\right]\theta$$

$$=\left[\frac{\partial G(x,y)}{\partial x}\quad \frac{\partial G(x,y)}{\partial y}\right]\begin{bmatrix}\dfrac{\partial \eta}{\partial y}\\[2mm]-\dfrac{\partial \eta}{\partial x}\end{bmatrix}\theta \qquad (7)$$

In the last side of Equation 7, the factor of the first bracket indicates the normal direction of the curve $G(x, y)$ = constant, and the second bracket indicates the tangential direction of the curve $\eta(x, y)$ = constant. Here, although η is not yet determined, if the normal direction of the contour curve $G(x,y)$ = constant in the cross-section is perpendicular to the tangential direction on the curve $\eta(x, y)$ = constant (Fig. 2), Equation 7 is equal to zero. In other words, when the shear modulus $G(x, y)$ is a function of η, and is uniquely determined by η, this condition is satisfied. Under this restriction, the distribution of the shear modulus $G(x, y)$ is defined by the strain function η, Equation 5 is reduced to

$$G(x,y)\,\theta\left(\frac{\partial^2 \psi}{\partial x^2}+\frac{\partial^2 \psi}{\partial y^2}\right)$$

$$=G(\eta)\,\theta\left(\frac{\partial^2 \psi}{\partial x^2}+\frac{\partial^2 \psi}{\partial y^2}\right)=0 \qquad (8)$$

It is noted that Equation 8 is the Equilibrium equation defined by the warping function. Differentiating Equation 6 firstly with respect to y, and secondly with x, and adding them together, it is self-evident that Equation 8 is satisfied. Equation 8 is the same Equilibrium equation as that of constant shear modulus by the warping function. Therefore, the configuration of the warping of non-uniform shear modulus is identical with that of constant shear modulus.

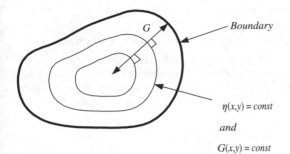

Figure 2. The contour lines of strain function and non-uniform shear modulus in the cross section.

$\eta(x,y) = const$

and

$G(x,y) = const$

The strain function satisfies equilibrium equations. Eliminating ψ from Equations 6 by differentiating firstly with respect to x, secondly with respect to y, and subtracting the first from the second, the compatibility equation represented by the strain function is derived.

$$\frac{\partial^2 \eta}{\partial x^2} + \frac{\partial^2 \eta}{\partial y^2} = -2 \qquad (9)$$

Since the strain function satisfies the equilibrium equation self-evidently as mentioned above, the strain function must satisfy Equation 9 together with the boundary conditions.

The boundary conditions of a prismatic shaft are now discussed. As the lateral surface of a shaft is free from external forces, the resultant shearing stress on the periphery of the cross section must be directed along the boundary. Hence the following equation is derived.

$$\tau_{xz}l + \tau_{yz}m = 0 \qquad (10)$$

where l and m = direction cosines of the normal vector of the boundary. Substituting Equation 4 and 6 into Equation 10, the following boundary condition can be derived.

$$G(\eta)\theta\frac{\partial \eta}{\partial y}\frac{dy}{ds} + G(\eta)\theta\frac{\partial \eta}{\partial x}\frac{dx}{ds} = G(\eta)\theta\frac{\partial \eta}{\partial s} = 0 \qquad (11)$$

Consequently, the strain function must be constant along the boundary (see Fig. 2). This constant can be chosen arbitrarily, and then in the following discussion, it is taken to be zero. It is known that, in the case of uniform shear modulus, differential equation and boundary condition are expressed by

$$\frac{\partial^2 \phi}{\partial x^2} + \frac{\partial^2 \phi}{\partial y^2} = -2, \quad \frac{\partial \phi}{\partial s} = 0 \qquad (12)$$

Then the differential Equation 9 and the boundary condition Equation 11 for the problem of non-uniform

shear modulus are basically identical to those of the constant shear modulus. Therefore the same procedure can be applied to this problem. If the problem of the constant shear modulus can be solved and obtained the stress function, the strain function of a bar with non-uniform shear modulus can be obtained by using the same procedure. The torsion problem of the constant shear modulus is possible to solve by theoretical analysis or the method of membrane analogy. Then as the strain function is known, it is possible to produce a FGM so that the distribution of shear modulus depends only on the strain function. Although the suggested analytical method is restricted to the distribution of shear modulus, in actual application of FGM, this assumption is not useless, since the most important requirement is to adjust the rigidity or strength of the shaft of FGM for the actual use.

2.2 Derivation of twisting moment

Here, the verification that the resultant of shearing stress, that is shearing force, in the end cross-section becomes to zero is omitted. Now the twisting moment caused by the shearing stress is introduced. At first, the new function is introduced for convenience, that is,

$$g(\eta) = \int G(x,y)d\eta = \int G(\eta)d\eta \qquad (13)$$

As the strain function is already known, $g(\eta)$ is able to derive and is constant along the boundary. With using $g(\eta)$, the twisting moment can be explained by the following equation.

$$
\begin{aligned}
M_T &= \iint \left(\tau_{yz}x - \tau_{xz}y\right)dxdy \\
&= \iint \left[G(x,y)\,\gamma_{yz}x - G(x,y)\,\gamma_{xz}y\right]dxdy \\
&= -\iint \left[G(x,y)\,\theta\frac{\partial \eta}{\partial x}x + G(x,y)\,\theta\frac{\partial \eta}{\partial y}y\right]dxdy \\
&= -\theta\iint \left[\frac{dg}{d\eta}\frac{\partial \eta}{\partial x}x + \frac{dg}{d\eta}\frac{\partial \eta}{\partial y}y\right]dxdy \\
&= -\theta\iint \left[\frac{\partial g}{\partial x}x + \frac{\partial g}{\partial y}y\right]dxdy \\
&= \theta\left[-\int gxdy + \iint gdxdy - \int gydx + \iint gdxdy\right] \\
&= -\theta g\big|_{boundary}\left[\int xdy + \int ydx\right] + 2\theta\iint gdxdy \\
&= -2A\theta g\big|_{boundary} + 2\theta\iint gdxdy \qquad (14)
\end{aligned}
$$

where A = area of cross-section. The problem of estimating the torsional rigidity of a prismatic bar reduces to the problem of resolving Equation 14. While the twisting moment of a bar with constant shear modulus is obtained from the classical theory as follows.

$$M_T = 2 \int \int \phi dx dy \qquad (15)$$

It is apparent that Equation 14 and 15 give the same result if shear modulus is constant. Therefore, Equation 14 is also applicable to the torsion problem of constant shear modulus. This equation will contribute to the arrangement design of FGM.

2.3 Equivalent shear modulus

In this section, the equivalent shear modulus is suggested for FGMs instead of the constant shear modulus as follows.

$$G_{eq} = \frac{\int \int g dx dy - A g|_{boundary}}{\int \int \eta dx dy} \qquad (16)$$

Since the strain function is already determined for the special problem under consideration, the function g in Equation 16 is able to derive. Hence, with using Equation 16, the twisting moment for a shaft of FGM can be easily explained by the following equation.

$$M_T = 2 G_{eq} \theta \int \int \eta dx dy \qquad (17)$$

It is on more to say that above equation is also applicable for a shaft of the uniform material property.

Next, it will be represented that warping in the case of non-uniform shear modulus can be easily obtained from the warping function of constant shear modulus by replacing the constant shear modulus with the equivalent shear modulus. In other words, the classical torsion theory is applicable for the FGM by using the equivalent shear modulus instead of the uniform shear modulus.

The warping function ψ is as same as that of constant shear modulus, because it must satisfy Equation 8. Therefore, the warping in the case of non-uniform shear modulus is obtained by using Equation 17.

$$w_{eq} = \theta \psi = \frac{M_T}{2 \int \int \eta dx dy \cdot G_{eq}} \psi \qquad (18)$$

It can be seen from the equation that the difference of warping between the uniform and non-uniform shear modulus is only the term of shear modulus. Hence, the warping can be generally obtained from the warping function of the constant shear modulus by replacing the constant shear modulus with the equivalent shear modulus.

3 SIMPLE EXAMPLE OF TORSION

3.1 Determination of the twisting moment

Let's consider a shaft with elliptical cross-section twisted by couples applied at the ends. In this case, Equation 9 and Equation 11 as the boundary condition are satisfied, if the strain function is assumed as

$$\eta = m \left(\frac{x^2}{a^2} + \frac{y^2}{b^2} - 1 \right) \qquad (19)$$

where m = unknown constant; a = major axis; and b = minor axis. Unknown constant m is determined by the substitution of Equation 19 into Equation 9, and then the strain function is given by

$$\eta = \frac{-a^2 b^2}{a^2 + b^2} \left(\frac{x^2}{a^2} + \frac{y^2}{b^2} - 1 \right) \qquad (20)$$

As an example, the shear modulus is assumed as

$$G(\eta) = \beta + \alpha \eta \qquad (21)$$

where α = constant; β = constant. Thus substituting Equation 20 into 21, the non-uniform shear modulus can be expressed as a function of x and y by

$$G(x, y) = \beta + \alpha \left[\frac{-a^2 b^2}{a^2 + b^2} \left(\frac{x^2}{a^2} + \frac{y^2}{b^2} - 1 \right) \right] \qquad (22)$$

As Equation 21 can be replaced by any function of η, an appropriate distribution of shear modulus can be chosen. The distribution of non-uniform shear modulus given by Equation 22 is parabolic in the cross-section, and is shown in Figure 3.

Substituting Equation 22 into 13, we obtain the function $g(\eta)$.

$$g(\eta) = -\beta \frac{a^2 b^2}{(a^2 + b^2)} \left(\frac{x^2}{a^2} + \frac{y^2}{b^2} - 1 \right)$$
$$+ \frac{\alpha a^4 b^4}{2(a^2 + b^2)^2} \left(\frac{x^2}{a^2} + \frac{y^2}{b^2} - 1 \right)^2 \qquad (23)$$

Finally, equivalent shear modulus is obtained from Equation 16.

$$G_{eq} = \beta + \frac{1}{3} \alpha \frac{a^2 b^2}{a^2 + b^2} \qquad (24)$$

Twisting moment of a shaft with elliptical cross-section that has the constant shear modulus is expressed by

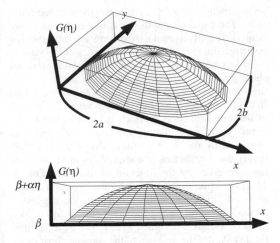

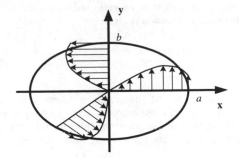

Figure 4. Distribution of shearing stress.

$$\beta + \alpha\eta$$

$$\beta$$

Figure 3. Distribution of non-uniform shear modulus $G(\eta)$ in the cross section.

$$M_T = \frac{a^3 b^3 \pi}{a^2 + b^2} G\theta \qquad (25)$$

Using the equivalent shear modulus instead of the uniform shear modulus, the twisting moment in the case of non-uniform shear modulus can be written as follows.

$$M_T = \frac{a^3 b^3 \pi}{a^2 + b^2} G_{eq}\theta$$

$$= \frac{a^3 b^3 \pi}{a^2 + b^2} \left(\beta + \frac{1}{3}\alpha \frac{a^2 b^2}{a^2 + b^2} \right)\theta \qquad (26)$$

It is no more to say that the same result is obtained from the direct calculation of Equation 14. If $\alpha = 0$ and $\beta = G$ (constant), this means the condition of constant shear modulus, are substituted in Equation 26, it is reduced to Equation 25 which expresses the twisting moment with the constant shear modulus.

3.2 Distribution of shearing stress

In the case of non-uniform shear modulus, the stresses are expressed using warping function, as follows.

$$\tau_{yz} = G(\eta)\theta\left(\frac{\partial\psi}{\partial y} + x\right) , \quad \tau_{zx} = G(\eta)\theta\left(\frac{\partial\psi}{\partial x} - y\right) \qquad (27)$$

In the case of uniform shear modulus, stresses can be written as

$$\tau_{yz} = G\theta\left(\frac{\partial\psi}{\partial y} + x\right) , \quad \tau_{zx} = G\theta\left(\frac{\partial\psi}{\partial x} - y\right) \qquad (28)$$

It is clear that, since warping function is same for each problem, the stresses for the non-uniform shear modulus are obtained from the stress in the case of constant shear modulus by replacing the constant shear modulus with the non-uniform shear modulus.

In this example, if shear modulus is supposed as constant, shearing stresses are distributed linearly and expressed as well-known equation.

$$\tau_{yz} = \frac{2b^2\theta}{a^2 + b^2} xG , \quad \tau_{zx} = -\frac{2a^2\theta}{a^2 + b^2} yG \qquad (29)$$

The shearing stresses of non-uniform shear modulus can be obtained from Equation 29 by replacing the constant shear modulus with the non-uniform shear modulus.

$$\tau_{yz} = \frac{2b^2\theta}{a^2 + b^2} xG(x, y)$$

$$= \frac{2b^2\theta}{a^2 + b^2} x\left[\beta - \alpha\frac{a^2 b^2}{a^2 + b^2}\left(\frac{x^2}{a^2} + \frac{y^2}{b^2} - 1\right) \right]$$

$$\tau_{zx} = -\frac{2a^2\theta}{a^2 + b^2} yG(x, y) \qquad (30)$$

$$= -\frac{2a^2\theta}{a^2 + b^2} y\left[\beta - \alpha\frac{a^2 b^2}{a^2 + b^2}\left(\frac{x^2}{a^2} + \frac{y^2}{b^2} - 1\right) \right]$$

This shearing stress is also obtained from the Equation 4 by substituting Equations 20 and 22. Thus, it is clear from Equation 30 that the stress distribution is expressed by a cubic function in the cross section (Fig. 4). This is a feature of shearing stress of FGM which has a non-uniform shear modulus $G = \beta + \alpha\eta$ over its elliptical cross section.

3.3 Determination of warping

The warping function ψ is same for each problem because it satisfies the same Equation 8. Here, the warping for a shaft of constant shear modulus with elliptical cross section is given by

$$w = -\frac{\left(a^2 - b^2\right) M_T}{a^3 b^3 \pi \, G} xy \qquad (31)$$

In the case of non-uniform shear modulus $G = \beta + \eta$, since the warping can be obtained by replacing the constant shear modulus with the equivalent shear modulus, the warping can be written as

$$\begin{aligned} w &= -\frac{\left(a^2 - b^2\right) M_T}{a^3 b^3 \pi \, G_{eq}} xy \\ &= -\frac{\left(a^2 - b^2\right) M_T}{a^3 b^3 \pi \left(\beta + \frac{1}{3} \alpha \frac{a^2 b^2}{a^2 + b^2} \right)} xy \end{aligned} \qquad (32)$$

Hence, the warping can be easily derived from that of a constant shear modulus by replacing the constant shear modulus G with the equivalent shear modulus G_{eq}.

4 DISCUSSION AND CONCLUSION

FGMs are advanced materials, the properties of which are possible to adjust in accordance with the demands of actual applications. With development of production technique, it has been promoted rapidly. However, since the classical theory of elasticity is not directly applicable to FGM, the theoretical study on non-uniform mechanical properties has not been developed. In order to supply an adaptive material for the actual demand, it is important to be able to design the appropriate arrangement of FGM so as to satisfy the required functions. The usual elastic theory is not useful, and only numerical analysis and experiments are available for investigating the mechanical properties of FGMs.

In this paper, the theory of torsion with constant shear modulus is modified so as to be applicable for the torsion problem with non-uniform shear modulus. In this theory, the strain function η is suggested instead of the stress function ϕ which is usually used in the ordinary theory. Although the suggested method is restricted by the distribution of the shear modulus, that is, the distribution of non-uniform shear modulus must be given by the function of strain function η. This restriction is no obstacle for the structural design of FGM. In spite of the restriction, the suggested method is useful, when the arrangement of components of a shaft is designed so as to realize a required rigidity. And if the problem of the constant shear modulus can be resolved, the problem of a bar with non-uniform shear modulus can also be resolved because of the same form of differential equation and boundary condition. This advantage allows wide applications to designers.

Finally, the concept of equivalent shear modulus is introduced and a simple example of torsion problem of a prismatic shaft with elliptical cross-section is mentioned. This concept gives the powerful method for the calculation of torsional rigidity.

REFERENCES

Omori, M. et al. 1994. *Proceeding of the 3rd International Symposium on Structural and Functionally Gradient Materials.*

Timoshenko & Goodier 1951. *Theory of Elasticity* (second edition), McGraw-hill book company, Inc.

31. Special session on "Structural monitoring and control"

System-based Vision for Strategic and Creative Design, Bontempi (ed.)
© 2003 Swets & Zeitlinger, Lisse, ISBN 90 5809 599 1

Structural health monitoring based on dynamic measurements: a standard and a novel approach

F. Vestroni, F. dell'Isola & S. Vidoli
Dipartimento di Ingegneria Strutturale e Geotecnica, Università di Roma La Sapienza, Italy

M.N. Cerri
Istituto di Scienza e Tecnica delle Costruzioni, Università di Ancona, Italy

ABSTRACT: Standard approaches to the health monitoring of structures are discussed in some detail; these techniques are mainly based on indirect measures of the structural response to external loads, are less effective than the direct investigations but reveal to be extremely advantageous from a practical point of view. An advanced technique, accounting for the peculiar aspects of damage detection problems and based on measures of the structural eigenfrequencies, is presented. In some specific problems this technique can however manifest a low sensitivity to small local damages. To this end, a novel technique is also introduced and discussed: it is based on purely electric measures of the state variables of an auxiliary electric system coupled to the main structure through a distributed set of piezoelectric patches. The topology of this auxiliary circuit and its constitutive parameters are optimized to increase the sensitivity of global measures, as eigenfrequencies, with respect to local variations of structural mechanical properties. Also due to the high experimental sensitivity of electric measures, the proposed method allows for accurate results in the identification and localization of damages.

1 INTRODUCTION

Health monitoring of existing structures is a subject receiving an increasing interest in the last decades; this is due to the urgent need of a continuous check of the structural reliability and safety, and to the convenience of a scheduled maintenance program with respect to unexpected out-of-service-conditions.

Among the wide variety of health monitoring techniques, an important position is occupied by those making use of indirect measurements to ascertain the structural characteristics related to damage. Even though these techniques cannot be effective as the direct investigations, they are less expensive. Moreover, the direct on-site testing result often to be not practicable; recourse to it is suggested when the results of the indirect analyses show some anomalies in the structural behavior.

Within this framework, measurements of structural response to dynamic loads is an attractive possibility in order to monitor the evolution of the system integrity. Dynamic tests are easy to carry out and great improvements have been recently registered in processing dynamic data. Ambient excitations can be used to excite the structure; for instance, in the case of bridges, the oscillations induced by traffic loads do not require to limit the serviceability of the structure.

Since damage modifies the mechanical characteristics of a structure, the modification in the dynamic response can be used to identify structural damage (Salawu 1997, Doebling et al. 1998). Notwithstanding the simplicity of the idea, the correct solution of this inverse problem is not straightforward. As in most inverse problems, accuracy depends on many factors; some are connected specifically to the problem under consideration, others are related to the completeness and quality of data, others to the interpretative model used to represent the mechanical system in the undamaged and damaged conditions (Cawley 1979, Davini et al. 1995, Vestroni et al. 1996, Vestroni & Capecchi 1996, Friswell et al. 1997, Vestroni & Capecchi 2000).

The case of concentrated damage in beam structures is considered, and linear behavior is assumed before and after the damage; therefore the damage is represented by a localized decrease of stiffness. According to the usual techniques of structural monitoring, a small amount of damage should be expected between two checks. Thus, in the detection problem, the unknown quantities are generally very limited in number, as are the location and stiffness reductions of damaged sections. Although this circumstance was known to a certain extent for many years (Cawley 1979, Liang et al. 1990, Casas et al. 1994), a great number of papers on damage detection has been published practically ignoring this specific peculiarity of the problem. The desired solution should be such that the stiffness is known throughout to be equal to the undamaged value except in the few damaged zones, as first illustrated in (Vestroni & Capecchi 1996) and, recently, partially found in (Friswell et al. 1997); consequently, a small number of measured data should be sufficient to obtain the solution.

Taking into account the peculiarity of the damage detection problem with respect to the general model identification (Gladwell 1984), an inverse procedure has been proposed, based on the analysis of all possible damage scenarios for a given number of damages. Notwithstanding it is shown that damage identification is not an underdetermined problem and a few data can be strictly sufficient to damage identification – that is it is much better than frequently expected – the problem remains strongly ill-conditioned and the possibility to reach good results is notably influenced by the accuracy of the experimental data.

Indeed, both the eigenfrequencies and the external work represent global quantities and the information concerning the local changes of structural characteristics is deeply hidden and not directly available. This is why global measures are less sensitive to local moderate variations of structural characteristics. To avoid this difficulty, the basic problem, to be solved, is how to increase the sensitivity of selected global measures up to a level enabling an effective damage detection. Thus a novel approach to damage identification is here proposed through the use of an auxiliary system coupled to the main structure; the coupling, between the mechanical (main structure) and the electric (auxiliary electric system) sides, is guaranteed by piezoelectric patches glued along the structure.

The classical problem of damage detection due to an open crack in a beam, that has received great attention in the literature, is tackled by means the classical and novel approaches. The structure is dealt with as a continuous model and the expected damage is described by two parameters only, location and magnitude. The optimal estimate of damage parameters is obtained by minimization of a suitable objective functional using experimental data in the standard approach, whereas pseudo-experimental data are used in the novel approach because no experimental results are yet available. It is shown how purely electric measurements of voltages in the auxiliary system allow for detection of mechanical local damages.

2 DAMAGE IDENTIFICATION: A STANDARD APPROACH

Reference is made to continuous models, which allows a more explicit mathematical treatment; the results can be extended to discrete models, which are more important in practice (Vestroni & Capecchi 1996 and 2000).

2.1 Frequency modification due to damage: direct problem

A concentrated damage in a supported beam, shown in Figure 1, is dealt with to illustrate the identification technique. In this case, a rotational spring can accurately model the dynamic behavior of the damaged beam. Damage is described by two parameters, location S_c and magnitude; the stiffness value K of the spring is related to damage by means of different relationships, depending on the kind of damage.

In the case of one crack only, as in Figure 1, the characteristic equation for the eigenvalue problem furnishes a relationship among the square frequency λ, and S_c and K, which is explicit with respect to K and can be written synthetically as

$$k\, g_1(\bar{\lambda}) + g_2(\bar{\lambda}, s_c) = 0, \qquad (1)$$

where the non-dimensional quantities $s_c = S_c/(L/2)$ and $k = K/(EI/L)$ are introduced, and $\bar{\lambda} = \lambda(\rho A)^2/EI]^{1/4}$ is the wave vector. For given k and s_c, eqn. (1) is satisfied for each eigenvalue $\lambda = \lambda_r$. On the contrary, when $\bar{\lambda}_r$ is known, a 1-dimensional manifold $k_r(s)$ satisfies eqn. (1), i.e., for any location s there is a value of damage intensity k, for which the r-th eigenvalue is equal to the given $\bar{\lambda}_r$. Thus, two values of $\bar{\lambda}_r$ are, in principle, sufficient to determine s_c and k. Due to the nonlinear character of eqn. (1), not all couples define a unique solution. However, it is clear that the addition of a new frequency is sufficient to eliminate the multiplicity. After determination of s_c, eqn. (1) furnishes the value of k for any $\bar{\lambda}_r$.

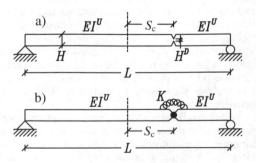

Figure 1.   (a) Beam with a crack ; (b) Model with rotational spring.

These considerations can be extended to the case of r cracks; it is worth to remark that it is possible, Vestroni & Capecchi (1996), to determine magnitudes and locations of r cracks with a number m of frequencies comparable with r, that implies that the identification damage is not in general an underdetermined problem as usually stated.

2.2 Frequency modification due to damage: inverse problem

The direct problem is simple and well-posed; when the entity and location of damage are given, eqn. (1) furnishes the frequencies of the system. The inverse problem, which consists of determining the damage parameters when a certain number of $\bar{\lambda}_r$ is known, is usually more complex; in absence of errors it is equivalent to the solution of the algebraic nonlinear problem:

$$\bar{\lambda} - h(k, s_c) = 0, \qquad k, s_c \in \mathbb{R}^r, \bar{\lambda} \in \mathbb{R}^m, \qquad (2)$$

where the m-dimensional vector $\bar{\lambda}$ collects the experimental eigenvalues (the square of the frequencies) in the damaged state and $h(k, s_c)$ is the function that furnishes the corresponding analytical quantities, associated with the r-dimensional vectors s_c and k collecting the r positions of the cracks and the corresponding r values of the rotational stiffness respectively.

Based on the above expression, a classical procedures for damage identification, denoted as the response quantity procedure, can be implemented. In the real world a noise term appears in eqn. (2), due to experimental and modelling errors; so, the problem to find the correct values of k and s_c can be formulated as a minimum problem for the objective function:

$$l(k, s) = \sum \|\bar{\lambda} - h(k, s)\|^2, \qquad (3)$$

defined by the difference between experimental and analytical frequencies of the damaged beam. The minimization of the functional (3) can be divided in two phases and has an attractive mechanical meaning, which is useful in case of discrete systems. For each scenario of damage location, defined by s, the objective function:

$$\tilde{l}(s) = min_k l(k, s), \qquad (4)$$

as a function of parameter s only, is initially determined by the minimization of $l(k, s)$ with respect to parameters k. For each s, $\tilde{l}(s)$ gives the best value of k to minimize the error between experimentally and analytically observed quantities. The solution to the inverse problem is then given by the minimum of $\tilde{l}(s)$.

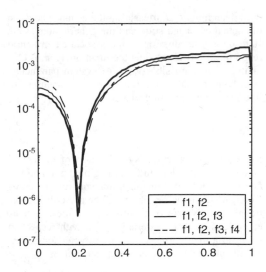

Figure 2. Functional $\tilde{l}(s)$ for different measured frequencies.

In the case of one crack $(r = 1)$, for a number of frequencies equal or greater than 2, it is possible to determine one single global minimum of $\tilde{l}(s)$, that is a global minimum of $l(k, s)$ as well.

2.3 Application: experimental tests on a free beam

In a free-free steel beam $(60 \times 40\,mm$, $950\,mm$ length) damage is inflicted through a crack without removal of material, defined by residual $H^D = 0.75$ $H = 30\,mm$ and position $S_c = 0.19$ L. Four frequencies are measured in the undamaged and damaged conditions; in Figure 2 frequencies in undamaged (f_n^U) and damaged (f_n^D) conditions and their percentual variations (Δf_n) are reported; the objective functionals $\tilde{l}(s)$, obtained with different number of measured frequencies, are also drawn. In all cases the results are very satisfactory; using only the first two frequencies the exact location $s_c = 0.19$ is obtained while using more than two frequencies the value $s_c = 0.20$ is reached; for what concerns the spring constants a unique value $k = 36$ is found coincident with the expected one.

	1	2	3	4
f^U (Hz)	228.6	623.1	1202.5	1951
f^D (Hz)	221.5	615.1	1196.0	1897
Δf (%)	3.1	1.29	0.54	2.77

3 DAMAGE IDENTIFICATION THROUGH AUXILIARY SYSTEMS

As already mentioned the damage identification process can be usefully regarded as the minimization of a given functional measuring the distance between the actual damaged state and the generic admissible state through evaluation of the associated structural responses. Indeed, consider the functional $l(\pi; I_*, O_*)$, mapping an admissible value of system parameters $\pi \in \Pi$ the actual inputs I_* and system responses O_*, into real positive numbers such that:

$$l(\pi_*; I_*, O_*) = 0, \qquad l(\pi; I_*, O_*) \succ 0, \qquad (5)$$

for $\pi \neq \pi_*$. Clearly the identification of the actual value π_* is equivalent to find the global minimum of $l(\cdot; I_*, O_*)$ in Π. Within this framework, the main difficulty concern the sensitivity of the prescribed functional, usually based on global measures such as eigenfrequencies or external works, with respect to local variations of the structural characteristics; from a mathematical viewpoint, it may happen that

$$|l(\pi; I_*, O_*) - l(\pi_*; I_*, O_*)| < \varepsilon, \qquad (6)$$

for $\pi \in \Pi$, being ε a positive number measuring the experimental sensitivity.

To overcome this drawback, it is here proposed to couple the main structure (MS), whose mechanical properties have to be detected, with an auxiliary electric circuit (AEC). The coupling between the two subsystems (MS and AEC) is here ensured by distributing along the structure an array of piezoelectric transducers but in general can be provided by other coupling devices. Therefore, for a suitable choice of the auxiliary electric circuit the electric response to any kind of forcing inputs is influenced by the mechanical constitutive properties, and one can detect structural damages through purely electric inputs and measurements. The proposed augmented system sensibly reduces the value ε of the experimental sensitivity, the right hand term in (6). Indeed the electric voltages (or currents) can be measured within tolerances much lower than the typical tolerances of structural quantities. Moreover the constitutive parameters of the auxiliary circuit can be tuned to enhance the sensitivity of the chosen functional l, namely the left hand term in (6). Indeed suppose to partition Π into

$$\Pi = \Pi_1 \cup \Pi_2, \qquad \Pi_1 \cap \Pi_2 = \varnothing, \qquad (7)$$

i.e. into the set Π_1 of parameters to be identified and the set Π_2 of parameters perfectly known (for instance the electric ones); these last can be then used to enhance the sensitivity of the functional with respect to the parameters in Π_1. If a value $\bar{\pi}_2$ can be found for which the restricted functional $l(\cdot, \bar{\pi}_2; I_*, O_*)$ resolves more sharply the actual value π_{*1} then $\bar{\pi}_2$ has to be

chosen in the experiment to increase the detection sensitivity. The entire process for the identification of the actual unknown parameters would consist into a sequence of maximization and minimization problems according to the following scheme:

$$\pi_{2,k+1} = \pi_2 \text{ to achieve } \max_{\pi_2 \in \Pi_2} l(\pi_{1k}, \pi_2; I_*, O_*),$$

$$\pi_{1,k+1} = \pi_1 \text{ to achieve } \min_{\pi_1 \in \Pi_1} l(\pi_1, \pi_{2,k+1}; I_*, O_*).$$

3.1 Description of the overall system

The proposed method relies on the coupling of the main structure with an auxiliary electric network. The energy is transformed from the mechanical to the electric form by means of a set of piezoelectric patches distributed along the structure. These kind of electromechanical devices has been proposed by dell'Isola et al. (2002) to control structural vibrations; they guarantee a multi-modal control and lead to the most efficient coupling between the mechanical and electric subsystems, a very effective characteristic in order to electrically sense the structural damage.

In Figure 3 a sketch of such an electromechanical structure is shown; the auxiliary electric circuit, i.e. the electric network connecting the piezoelectric patches has been chosen as the fourth-order transmission line synthesized by Andreaus et al. (2003). Two moduli of this connection are shown; note that an inductance and a transformer are needed in each module.

Once an homogenization procedure has been carried out, the governing equations for the proposed augmented electromechanical system can be written as:

$$\begin{cases} (\alpha u'')'' + \ddot{u} - \gamma \dot{\varphi}'' = 0, \\ \beta \varphi^{IV} + \ddot{\varphi} + \gamma \ddot{u}'' = 0, \end{cases} \qquad (8)$$

for each regular interval in [0, 1]. Here, and in what follows, a superposed prime means the space derivative, a superposed dot the time derivative and all the quantities are dimensionless. While u represents the transverse displacement in the beam, the scalar field φ physically represents the electric flux-linkage i.e. the time primitive of the electric potential.

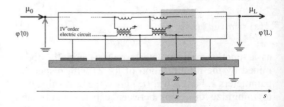

Figure 3. Euler beam coupled to its electric analog circuit.

3.2 Choice of a suitable functional $l(\pi)$

In the system identification a crucial role is played by the choice of the functional $l(\pi; I_*, O_*)$ to be minimized over the space Π of admissible parameters. A possible choice is given in eqn. (3) where the difference between the squared eigenfrequencies are accounted for. Since, from a practical point of view, the measurement of the frequency response function of the electric dof is very easy, we adopt here the functional introduced in (Liu & Chen 1996) based on the comparison of the frequencies response functions in several sampled frequencies. Thus, if the equation

$$D(\omega, \pi_*) O_*(\omega) = I_*(\omega), \qquad (9)$$

describes the relation between the forcing vector $I_*(\omega)$ at frequency ω and the resulting response vector $O_*(\omega)$ in a given experiment, the functional l to be minimized to identify the actual values π_*, is chosen as follows:

$$l(\pi) = \sum_{k=1}^{K} |D(\omega_k, \pi) O_*(\omega_k) - I_*(\omega_k)|^2. \qquad (10)$$

The sum over a set of frequencies ω_k is important to include, in the functional, informations over a large frequency bandwidth. Clearly, due to eqn. (9), the functional (10) vanishes when $\pi = \pi_*$; moreover l is continuous with respect to the parameter vector π since involves continuous functions of π. When $\pi \neq \pi_*$, the balance eqns. (9) are not satisfied and the functional weighs these unbalanced generalized forces.

3.3 Application: damage detection in Euler beams

The discussed procedure is applied to the relevant case of a simply supported Euler-Bernouilli beam. This choice has been also suggested by the relative simplicity in the assembly of a forthcoming experimental set up. The system described in Figure 3, is supposed subjected to the following boundary conditions:

$$u''(0, t) = u''(L, t) = 0; \\ \varphi''(0, t) = \mu_0(t)/\beta; \quad \varphi''(L, t) = \mu_1(t)/\beta. \qquad (11)$$

These correspond to a simply supported beam and to an electrically grounded transmission line that is subjected to a non-vanishing value of the electric bending moment at its edges, μ_0 and μ_1 (Figure 3).

We identify the damaged dimensionless bending stiffness $\alpha(s)$ in eqn. (8), by means of purely electric measurements and forcing inputs in the auxiliary network. Indeed, only the following quantities

$$O_* = \{\varphi_0', \varphi_L'\}^\top, \qquad I_* = \{\mu_0, \mu_L\}^\top, \qquad (12)$$

are measured and contribute to the functional l to be minimized, namely the functional chosen in eqn. (10). If a lumped damage is to be identified, the space of parameters Π is reduced to a finite dimensional space, assuming for the unknown bending stiffness function $\alpha(s)$, the following form:

$$\alpha(s) = \begin{cases} \alpha_0, & \text{for} & s \leq x - \epsilon, \\ \alpha_0 d, & \text{for} & x - \epsilon < s < x + \epsilon, \\ \alpha_0, & \text{for} & x + \epsilon \leq s. \end{cases} \qquad (13)$$

Here $d \in [0, 1]$ is the percentage loss with respect to the undamaged state, $x \in (0, 1)$ the abscissa where the damaged zone is centered, and 2ε the fixed range of the damaged zone, refer to the gray shaded zone in Figure 3. Thus, setting $\Pi_1 = \text{span}\{d_*, x_*\}$, $\Pi_2 = \text{span}\{\beta\}$ means to seek for the correct value $\pi_{1*} = \{d_*, x_*\}$ of the unknown constitutive mechanical parameters, using the electric parameter β supposed constant along the beam, to increase the functional sensitivity.

The dimensionless quantities, used in the numerical simulations, correspond to technically feasible conditions and materials and are assumed as follows:

$$\alpha_0 = 1, \qquad \gamma = 1/20, \qquad \epsilon = 1/20, \\ d_* = 0.5, \qquad x_* = 0.8, \qquad \mu_0 = (\omega) = \mu_L(\omega) = 1.$$

In Figure 4 for a fixed value of the parameter $\beta = 1$ the contour plot of the functional $l(d, x, \bar{\beta})$ over the parameters space $\Pi_1 = [0.1, 1] \times [0, 1]$ is drawn. There are several local minima but only one global minimum correspondent to the actual values of state parameters.

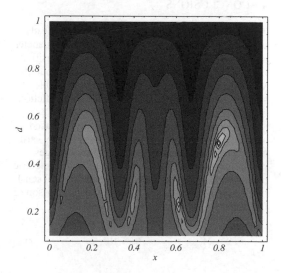

Figure 4. Level plot of $\log l(d, x, \bar{\beta})$ on the space Π_1. The point $d = 0.5, x = 0.8$ is the global minimum.

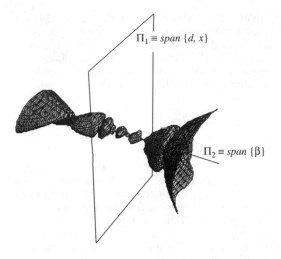

$\Pi_1 \equiv span\ \{d, x\}$

$\Pi_2 \equiv span\ \{\beta\}$

Figure 5. Points in Π satisfying the condition $l(d, x, \beta) = \varepsilon$. The parameter β ranges from 0.8 to 1.2.

To understand how the electric Π_2-parameter β can affect the sensitivity of the functional with respect to the Π_1-parameters, d and x, in Figure 5 is drawn the locus of all the point satisfying $l(d, x, \beta) = \varepsilon$, with the experimental sensitivity fixed to the value $\varepsilon = 1\%$. Note that the tightened regions in the plot correspond to values of $\beta \simeq 1 = \alpha_0$; this condition, in case of an undamaged structure, would lead to a complete modal coupling of all the mechanical and electrical modes, since all the mechanical and electrical eigen-frequencies of eqns. (8) would match.

4 CONCLUSIONS

The chosen damage ($d_* = 0.5$, and $x_* = 0.8$) leads to variations Δf of the first three natural frequencies ranging from 2% to 4%, the experimental sensitivity ε_f for natural frequencies measures being about 1%. When accurate experimental results are available, the standard technique produces satisfactory predictions on both locations and damage values.

On the other side, in the novel approach based on an auxiliary coupled system, the percentage varia-tions $\Delta\varphi$ of the measured electric flux-linkages can range from 0 to 100%, depending on the electric parameter β, while the associated experimental sensi-tivity ε_φ is at least 0.1%. Therefore the comparison of the ratios

$$\frac{\Delta f}{\varepsilon_f} \approx 2 \to 4, \qquad \frac{\Delta\varphi}{\varepsilon_\varphi} \approx 0 \xrightarrow{\beta} 10^3, \qquad (14)$$

meaning the experimental confidences, shows an evident advantage of the proposed technique with respect to the standard method of damage detection, where only the variation of frequencies are measured.

ACKNOWLEDGEMENTS

This work has been partially supported by the Italian Ministry of Education, under the Grant COFIN 2001-02, *http://www.disg.uniroma1.it/fendis/*.

REFERENCES

Cawley, P. & Adams, R.D. 1979. The location of defects in structures from measurements of natural frequencies. *J. of Strain Analysis* 14(3): 49–57.

Liang, R.Y., Choy, F. & Hu, J. 1990. Identification of cracks in beam structure using measurements if natural frequen-cies. *J. of Franklin Institute* 324(4): 505–518.

Casas, J.R. & Aparicio, A.C. 1994. Structural Damage Identification from Dynamic-Test Data, *J. Structural Eng.* 120(8): 2437–50.

Davini, C., Morassi, A. & Rovere, N. 1995. Modal analysis of notched bars: tests and comments on the sensitivity of an identification technique, *J. Sound and Vibration* 179(3), 513–27.

Vestroni, F., Cerri, M.N. & Antonacci, E. 1996. The problem of damage detection in vibrating beams, *Proc. Eurodyn'96 Conference*, Firenze, Augusti, Borri & Spinelli (Eds.), Balkema, Rotterdam: 41–50.

Vestroni, F. & Capecchi, D. 1996. Damage evaluation in cracked vibrating beams using experimental frequencies and FE models, *J. Vibration and Control* 2: 69–86.

Friswell, M.I., Penny, J.I.T. & Garvey, S.D. 1997. Parameter subset selection in damage location, *Inverse Problems Eng.* 5: 189–215.

Salawu, O.S. 1997. Detection of structural damage through changes in frequency: a review, Eng. Structures 19(9), 718–723.

Doebling, S.W., Ferrar, C.R. & Prime, M.B. 1998. A sum-mary review of vibration based damage identification methods, *Shock and Vibration Digest* 30: 91–105.

Vestroni, F. & Capecchi, D. 2000. Damage detection in beam structures based on measurements of natural frequencies, *J. of Eng. Mech.*, ASCE, 126(7): 761–768.

Gladwell, G.M.L. 1984. The inverse problem for the vibrat-ing beam, *Proc. Royal Society of London*, 393 Series A: 277–295.

Vidoli, S. & dell'Isola, F. 2001. Vibration control in plates by uniformly distributed actuators interconnected via electric networks, *Eur. J. of Mech. A/Solids*, 20 (3): 435–456.

dell'Isola, F., Vestroni, F. & Vidoli, S. 2002. A class of electro-mechanical systems: linear and nonlinear mechanics, *J. of Theoretical and Applied Mech.*, 40 (1): 47–71.

Andreaus, U., dell'Isola, F. & Porfiri, M. 2003. Piezoelectric passive distributed controllers for beam flexural vibra-tions to appear in *J. of Vibration and Control*.

Liu, P.L. & Chen, C.C., 1996. Parametric identification of truss structures by using transient response, *J. Sound and Vibration*, 191 (2): 273–287.

System-based Vision for Strategic and Creative Design, Bontempi (ed.)
© 2003 Swets & Zeitlinger, Lisse, ISBN 90 5809 599 1

Control of vortex-induced oscillations of a suspension footbridge by Tuned Mass Dampers

E. Sibilio
Universities of Florence, Italy, and Braunschweig, Germany

M. Ciampoli
University of Rome "La Sapienza", Department of Structural and Geotechnical Engineering, Italy

V. Sepe
University of Chieti-Pescara "G. D'Annunzio", Department of Design, Rehabilitation and Control of Architectonic Structures, Italy

ABSTRACT: The across-wind oscillations due to vortex-shedding from the deck are often of concern in the design of suspension and cable-stayed bridges. This kind of oscillations may cause cumulated damage to the structural elements due to fatigue and discomfort to the users. The vortex-shedding can arise for wind velocities usually lower, and therefore more likely to occur, than those corresponding to other aeroelastic phenomena, like flutter or torsional divergence. In this paper, the control of vortex-induced oscillations of a suspension bridge by Tuned Mass Dampers (TMDs) is dealt with; namely, a procedure for the design and location of the control devices is implemented and applied, as a case example, to a suspension footbridge with a span of 252 m. The evaluation of the structural response is obtained in time domain by a finite element model, that allows the optimal location of the TMDs, according to the modal shapes involved in the vortex-shedding excitation, as well as the calibration of their characteristics. The dynamic parameters of the TMDs are optimised through a parametric analysis, that considers the main aspects of the dynamic response of the coupled system. The efficiency of the control system is proved by comparison of the responses of the bridge deck with and without TMDs.

1 INTRODUCTION

It is well known that slender decks of suspension and cable-stayed bridges can be very sensitive to wind action. Phenomena as flutter, buffeting and vortex-shedding may induce in these structures problems of higher relevance than other ordinary variable loads related to serviceability; therefore, structural design may be strongly influenced by technical solutions adopted to safeguard the bridge from the effects of wind.

In the last decades, many studies on control of deck oscillations due to wind action have been developed, and useful results obtained. An efficient solution is represented by Tuned Mass Dampers (TMDs); these control devices are usually placed in correspondence of the maxima of the modal shapes that significantly contribute to the overall dynamic response, and can be designed according to a procedure, that is well defined at least from a theoretical point of view.

Other mechanical and aerodynamical control devices, of the *passive, active* or *semi-active* types, have been proposed; the relevant state-of-the-art is briefly summarized in the following.

This paper discusses in some detail the optimal location of the TMDs and the calibration of their dynamic parameters, with reference to an example of vortex-shedding induced oscillations. The adopted procedure is based on a dynamic analysis in time domain, that is able to follow the evolution of the oscillations also in presence of second order effects.

2 CONTROL SYSTEMS

The control systems proposed in technical literature to limit the oscillations of bridge decks due to wind action, can be of three different types, i.e., *passive, active* or *semi-active*. They usually differ according to their ability in modifying the fluid-structure interaction and/or the dynamic characteristics of the structure itself, depending on the specific phenomenon (e.g., flutter, buffeting or vortex-shedding).

2.1 Passive control systems

A well known solution is represented by TMDs or by Multiple Tuned Mass Dampers (MTMDs); these latter are used in order to operate on a wide range of natural frequencies. These control systems have many advantages: the dynamic behaviour of the controlled structural system can be easily interpreted, and the whole system is robust and reliable.

The dynamic response of bridge decks equipped with TMDs or MTMDs is usually evaluated in frequency domain, and on a simplified model of the structure. In (Gu et al. 1998), two TMDs are used to increase the critical flutter speed, whose value has been derived on a section model. In (Lin et al. 1999, 2000), a MTMD and a TMD are respectively used to increase the flutter speed or to reduce the amplitude of the buffeting response. In (Gu et al. 2001), a MTMD is proposed to reduce the buffeting response, evaluated on a section model. In (Know et al. 2000), two TMDs are used to activate plates that modify the flow around the deck, thus increasing the flutter speed.

The passive control of deck oscillations in suspension and cable-stayed bridges can be also obtained by means of aerodynamic devices (*flaps*) mechanically connected to the deck. In (Wilde et al. 1999), two flaps are placed below the deck, while in (Wilde et al. 2000) two flaps are placed in line with the deck. The flaps may move relatively to the deck when the wind acts on the bridge, as a direct consequence of the motion of the deck itself: their rotation modifies the aerodynamic load on the deck, and leads to an increase of the critical flutter speed.

2.2 Active control systems

Typical examples are feedback active control systems formed by actuators and sensors, operated by a computer. In (Achkire et al. 1998) and (Yang & Giannopoulos 1979), active stays are considered, whose tension can be varied, thus modifying the stiffness of the system. The use of an active eccentric mass placed inside the deck has been proposed in (Wilde et al. 1996): the mass may move transversally, thus modifying the dynamic characteristics of the deck section.

Other solutions are obtained by using flaps placed above the deck (Kobayashi & Nagaoka 1992), below the deck (Wilde et al. 1998) or in line with it (Wilde et al. 2001); the control system determines the rotation of the flaps, and modifies the characteristics of the aerodynamic load. An increase of the critical flutter speed is always obtained. The efficiency of these active systems has been always assessed on section models, taking into account only the first vertical and the first torsional natural modes.

2.3 Semi-active control system

A semi-active control system has been proposed in (Gu et al. 2002): it consists of a series of semi-active TMDs with variable stiffnesses, that are able to mitigate the buffeting oscillations, either vertical and torsional. The investigation performed on a section model in frequency domain shows also an increase of the critical flutter speed.

3 OSCILLATIONS DUE TO VORTEX-SHEDDING

The oscillations due to vortex-shedding can be caused, in road bridges, by wind at a relatively low velocity, and have consequences both in terms of discomfort to the users and of fatigue damage to structural components. The solutions currently used to reduce their effects consists in elements, known as *guide vanes*, that modify the characteristics of the flow around the deck; an example is the case of the Great Bealt Bridge (Larsen et al. 2000).

To reduce this kind of oscillations, it has been proposed, also recently, the use of TMDs. Two such systems have been proposed for the Osterøy suspension bridge (Strommen & Hjorth-Hansen 2001) and the Kessoch Bridge (Owen et al. 1996), while an Active Tuned Vibration Absorber (ATVA), composed by a series of tuned masses activated by an actuator, has been proposed for the Rio Niteroi bridge (Batista & Pfeil 2000).

The analyses of the efficiency of the control systems illustrated in (Strommen & Hjorth-Hansen 2001) and (Batista & Pfeil 2000), are based on the following simplified assumptions: the structure is modelled as an equivalent SDOF system, e.g., for a given deformed shape, only the vertical displacement is assumed to characterize the dynamic response; the analysis is carried out in frequency domain, considering only the steady-state response, or in time domain, considering just one modal shape.

4 A PROCEDURE FOR THE DESIGN OF SYSTEMS CONTROLLED BY TMDS

In the following, a procedure for the design and the optimal location of TMDs aimed at controlling the oscillations due to vortex-shedding is applied: the analysis is carried out in time domain on a complete model of a bridge that, in this study, is proposed to be equipped with TMDs.

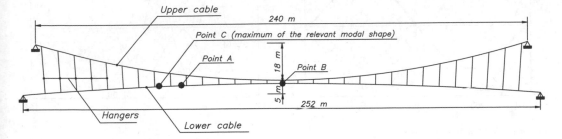

Figure 1. Schematic diagram of the footbridge.

4.1 The case example

The considered example, whose geometrical, aerodynamical and mechanical characteristics are illustrated in (Pirner 1994), is a suspension footbridge built in Moravia, with a simply-supported deck sustained by two pairs of parabolic cables with opposite concavity (Fig. 1). The deck is made by precast concrete elements, and is supported by the upper cables and prestressed by the lower cables. The sag of the upper cables is 18 m, and the sag of the lower cables is 5 m; the span of the bridge is 252 m.

In order to analyse the sensitivity of the bridge to wind action, Pirner (1994) carried out an experimental research on a complete model (in scale 1/130) of the whole structure in the wind tunnel of the Aircraft Research and Testing Institute of Prague. The displacements of two points (see Fig. 1) of the model were continuously monitored, namely the vertical displacement of point A at about one-quarter of the span, and the horizontal displacement of point B at midspan. In experiments, the vertical displacement of A strongly increased (up to a value that in the real structure can be assumed to be approximately equal to 0.11 m) for a wind velocity of 30 m/s, that corresponds to a full-scale value of about 15–17 m/s; in this range, the dynamic response to wind action is dominated by just one modal shape, more specifically the first vertical modal shape.

The phenomenon was interpreted by Pirner (1994) as the result of oscillations due to vortex-shedding from the footbridge deck in lock-in conditions, i.e. for vortex-shedding frequency f_{she} equal to one of the natural frequencies (in this case, the frequency of the first vertical mode). f_{she} is related to the width of the deck D and to the non-dimensional Strouhal number St (in the example case, $St = 0.12$) by the relation:

$$St = (f_{she} D)/U \tag{1}$$

4.2 Footbridge FE model

To reproduce the experimental response of the footbridge in time domain, a FE model implemented in ADINA 7.4 has been analysed. The FE model is composed by 1164 3D cable elements (the upper cables, the lower cables and the hangers) with 4 nodes for each element and 3 degrees of freedom for each node, and by 204 frame elements (the deck) with 2 nodes for each element and with 6 degrees of freedom for each node. The concrete and the steel have a linear elastic behaviour.

The TMDs have been modelled as spring elements, with a lumped mass, an axial stiffness and a linear damping; the efficiency of this model has been confirmed by validation tests on simple structural schemes.

A second order analysis has been carried out, evaluating the stiffnesses of the cables according to their deformations. In the equilibrium configuration under dead loads, tension is equal, respectively, to $H_u = 4625$ kN in the upper cables, to $H_l = 662$ kN in the lower cables and to $H_h = 180$ kN in the hangers. A Rayleigh's damping matrix has been introduced, with coefficients $a = 0.011$ and $b = 0.002$ (corresponding to a damping ratio in the two first natural modes equal to 0.005).

Analysis in time domain has many advantages, that is, it allows to take into account the influence of all relevant modes of vibrations, the geometrical nonlinearities, a model of loading more representative of the actual action, and the possible local effects due to interaction between the principal structure and the control devices.

The modal shapes of the footbridge are illustrated in Figure 2, where the first vertical mode is represented in bold line; the corresponding natural frequencies are reported in Table 1. Both modal shapes and frequencies evaluated with the adopted FE model are in good agreement with the results reported by Pirner (1994).

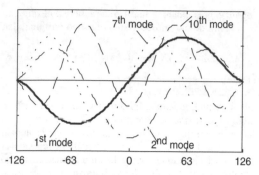
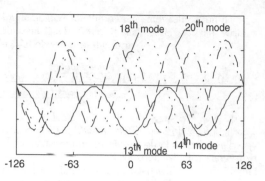

Figure 2. Vertical modal shapes.

Table 1. Modes and natural frequencies.

Mode	f [Hz]	Modal shape
1	0.304	1st vertical
2	0.424	2nd vertical
3	0.438	1st torsional-lateral
4	0.496	2nd torsional-lateral
5	0.528	1st lateral upper cables
6	0.581	3rd torsional-lateral
7	0.607	3rd vertical
8	0.692	4th torsional-lateral
9	0.693	2nd lateral upper cables
10	0.730	4th vertical
11	0.732	5th torsional-lateral
12	0.891	6th torsional-lateral
13	0.912	5th vertical
14	0.932	6th vertical
15	1.000	7th torsional-lateral
16	1.007	3rd lateral upper cables
17	1.013	8th torsional-lateral
18	1.117	7th vertical
19	1.167	9th torsional-lateral
20	1.278	8th vertical

4.3 Mathematical model of the loading due to vortex-shedding

In general, the effects of vortex-shedding may be evaluated by applying a time-varying distributed load to the FE model of the structure.

Different load patterns have been proposed for cylindrical structures (e.g. chimneys) or for bridge decks; their parameters are usually calibrated in wind tunnel tests, and can be adopted for representation in frequency or time domains.

Two of these models appear more suitable to analyses in time domain, the first proposed by Scanlan & Eshan (1990), the second by D'Asdia & Noè (1997).

Both models assume that the structure has a sinusoidal deformed shape, and that the lock-in interval varies in time at the same frequency as the vortex-shedding. According to these assumptions, confirmed by the results of wind tunnel tests (Pirner 1994), in the considered example the vortex-shedding occurs with the same frequency of the first vertical mode of vibration, which has an unsymmetrical shape with two half-waves (Fig. 2). Thus, if f_1 is the first natural frequency, and f_{she} is the vortex-shedding frequency, it happens that $f_{she} = f_1 (= 0.304\,\text{Hz})$. The wind speed which causes the vortex-shedding can be obtained by Strouhal equation, as

$$U = \frac{St \cdot D}{f_{she}} = 17.45\,m/s \qquad (2)$$

where $D = 6.9\,\text{m}$ is the deck width.

The distributed vertical loading due to vortex-shedding is thus equal to

$$F_L = \frac{1}{2}\rho U^2 H C_L \sin(\omega_{she}t) \qquad (3)$$

where: $\rho = 1.25\,\text{kg/m}^3$; $H = 0.45\,\text{m}$ is the deck thickness; C_L is the lift coefficient; $\omega_{she} = 2\pi f_1 = 1.907\,\text{rad/s}$ is the circular frequency of the vortex-shedding. The lift coefficient can be determined once the load distribution along the deck is assigned. According to Ruschweyh (1988), the load is assumed to act on the deck around the maximum and the minimum of the considered modal shape (Fig. 3). The extension L_e of the loading zone (the *effective length of correlation*) is given by Equation 4:

$$L_e/D = \begin{cases} 6 & \text{if} & s_{max}/D < 0.1 \\ 4,8 + 12\,s_{max}/D & \text{if} & 0.1 \leq s_{max}/D \leq 0.6 \\ 12 & \text{if} & s_{max}/D > 0.6 \end{cases}$$

$$(4)$$

where s_{max} is the maximum vertical displacement of the deck. In the considered case, $s_{max} = 0.11\,\text{m}$, $L_e/D = 6$, $L_e = 41.4\,\text{m}$.

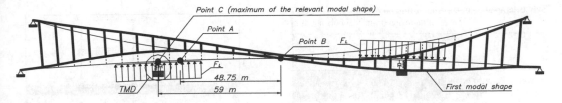

Figure 3. Loading pattern, TMD location and first modal shape of the footbridge.

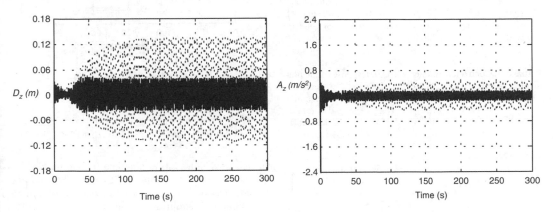

Figure 4. Controlled and uncontrolled response of the point C ($\beta = 1$, $\alpha = 0.99$, $\mu = 0.01$, $\zeta_{TMD} = 6.09\%$); D_z vertical displacement, A_z vertical acceleration.

4.4 Analysis of the dynamic response and efficiency of the TMDs

The response of the footbridge to wind action has been studied in time domain by solving the equations of motion by Newmark's implicit integration method (with parameters: $\gamma = 0.5$, $\beta = 0.25$); the integration step has been taken as $\Delta t = 0.025$ s, which is about 1/130 of the first natural period of the structure ($T_1 = 3.295$ s).

To evaluate the dynamic response, the value of the lift coefficient C_L must be carefully calibrated. According to the results of wind tunnel tests (Pirner 1994), that is, to reproduce by numerical simulation a full scale maximum displacement of 0.11 m, a value $C_L = 0.55$ was required.

In the considered example, the design of the control system has been carried out by placing the TMDs in the points of relative maxima of the modal shape characterizing the motion (Fig. 3), and assuming that the first mode of vertical vibration is the basis for the calibration of the dynamic parameters of TMDs.

After some numerical calibration, the following parameters have been assumed to characterise the coupled system:

$m_{TMD} = 2.52 \times 10^3$ kg (total mass of TMDs);
$m(x) = 2.1 \times 10^3$ kg/m (distributed mass of the deck);

$M^* = 252 \times 10^3$ kg (generalized mass associated to the first modal shape); $\mu = m_{TMD}/M^* = 0.01$ (mass ratio); $\alpha = \omega_{TMD}/\omega_1$ (tuning parameter); $\omega_1 = 2\pi f_1 = 1.907$ rad/s (first circular frequency); ζ_{TMD}: damping ratio of the TMDs; ω_{she} = circular frequency of loading; $\beta = \omega_{she}/\omega_1$.

The assessment of the optimal values of the dynamic parameters α and ζ_{TMD} of the TMDs can be carried out iteratively. As a first step, the relationships proposed by Warburton (1980) for an undamped structure subject to a sinusoidal loading are considered, thus obtaining:

$$\alpha = \frac{1}{1+\mu} = 0.99 \qquad \zeta_{TMD} = \sqrt{\frac{3\mu}{8(1+\mu)}} = 0.061 \qquad (5)$$

The dynamic response of the deck is reported in Figures 4 and 5, where the displacements D_z and the accelerations A_z of points C and A (see Fig. 3) evaluated for the controlled and the uncontrolled systems are compared. With the dynamic parameters given by Equation 5, the maximum displacements and the maximum accelerations of points C and A are reduced of about 60% and 65%, respectively.

Previous calculations have been carried out with reference to the first natural frequency f_1 of the system.

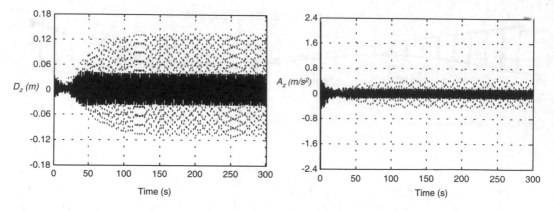

Figure 5. Controlled and uncontrolled response of the point A ($\beta = 1$, $\alpha = 0.99$, $\mu = 0.01$, $\zeta_{TMD} = 6.09$ %); D_z vertical displacement, A_z vertical acceleration.

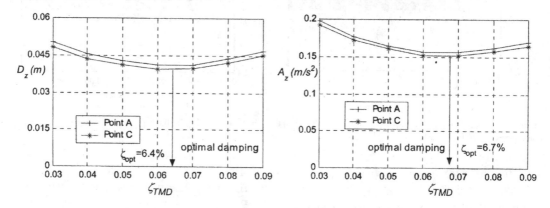

Figure 6. Optimal damping of the TMDs.

However, for the coupled system, two frequencies, slightly different from f_1, are of interest. In order to take into account the possibility of a lock-in with one of these two frequencies, the damping of TMDs has been considered as a variable parameter, and its optimal value evaluated as the one that minimises the structural response in a range of frequency that includes both values.

The two modal frequencies of the coupled system (the SDOF corresponding to the bridge + the TMDs), if $\mu = 0.01$, are equal to $f_A = 0.287$ Hz and $f_B = 0.317$ Hz, thus giving: $\beta_A = 0.947$ and $\beta_B = 1.045$. Assuming β in the range β_A to β_B, the optimal damping of the TMDs, that is the value corresponding to the minimum amplitude of the displacements or of the accelerations of the point C, becomes respectively equal to $\zeta_{TMD} = 6.4$% and $\zeta_{TMD} = 6.7$% (Fig. 6). These two values are in good agreement with the approximate value given by Equation 5; this can be explained by the fact, verified through a parametric analysis, that the maximum displacement of the

example footbridge is relatively small. However, this behaviour cannot be expected in general for more slender and flexible structures, characterized by larger displacements; for these cases, only a time-domain analysis (similar to the one implemented in this work) can lead to a proper design of the TMDs.

5 CONCLUSIONS

The control of vortex-induced oscillations of a suspension bridge by Tuned Mass Dampers (TMDs) has been dealt with; namely, a procedure for the check of the efficiency of the control devices has been applied, as an example, to a suspension footbridge with a span of 252 m. The evaluation of the structural response has been obtained in time domain by a finite element model, that allows to derive the optimal location of the TMDs, according to the modal shapes involved in the vortex-shedding excitation. The dynamic parameters of the TMDs have been optimised through a parametrical analysis, that considers the main aspects of the dynamic response of the coupled system. The efficiency of the control system is proved by comparison of the dynamic responses of the footbridge with and without TMDs.

ACKNOWLEDGEMENTS

The research illustrated in this paper has been carried out in the framework of the "Research Programme of National Interest" WINDERFUL (2001–2003), co-financed by the Italian Ministry for University and Research.

REFERENCES

Achkire, Y., Bossens, F. & Preumont, A. 1998. Active damping and flutter control of cable-stayed bridges. *Jour. of Wind Engin. and Indu. Aerody.* 74–76: 913–921.

Batista, R. C. & Pfeil, M. S. 2000. Reduction of vortex-induced oscillations of Rio-Niterói bridge by dynamic control devices. *Jour. of Wind Engin. and Indu. Aerody.* 84: 273–288.

D'Asdia, P. & Noè, S. 1998. Vortex induced vibration of reinforced concrete chimneys: in situ experimentation and numerical previsions. *Jour. of Wind Engin. and Indu. Aerody.* 74–76: 765–776.

Gu, M., Chang, C. C., Wu, W. & Xiang, H. F. 1998. Increase of critical flutter speed of long-span bridges using tuned mass dampers. *Jour. of Wind Engin. and Indu. Aerody.* 73: 111–123.

Gu, M., Chen, S. R. & Chang, C. C. 2001. Parametric study on multiple tuned mass dampers for buffeting control of Yangpu Bridge. *Jour. of Wind Engin. and Indu. Aerody.* 89: 987–1000.

Gu, M., Chen, S. R. & Chang, C. C. 2002. Control of wind-induced vibrations of long-span bridges by semi-active lever-type TMD. *Jour. of Wind Engin. and Indu. Aerody.* 90: 111–126.

Kobayashi, H. & Nagaoka, H. 1992. Active control of flutter of a suspension bridge. *Jour. of Wind Engin. and Indu. Aerody.* 41–44: 143–151.

Kwon, S. & Jung, M. S. & Chang, S. 2000. A new passive aerodynamics control method for bridge flutter. *Jour. of Wind Engin. and Indu. Aerody.* 86: 187–202.

Larsen, A., Esdahl, S. & Andersen, J. E. & Vejrum, T. 2000. Storebælt suspension bridge-vortex shedding excitation and mitigation by guide vanes. *Jour. of Wind Engin. and Indu. Aerody.* 88: 283–296.

Lin, Y., Cheng, C. & Lee, C. 1999. Multiple tuned mass dampers for controlling coupled buffeting and flutter of long-span bridges. *Wind and Structures* 2: 267–284.

Lin, Y., Cheng; C. & Lee, C. 2000. A tuned mass damper for suppressing the coupled flexural and torsional buffeting response of long-span bridges. *Engineering Structures* 22: 1185–1204.

Owen, J. S., Vann, A. M., Davies, J. P. & Blakeborough, A. 1996. The prototype testing of Kessock Bridge: response to vortex shedding. *Jour. of Wind Eng. and Ind. Aerody.* 60: 91–98.

Pirner, M. 1994. Aeroelastic characteristics of a stressed ribbon pedestrian bridge spanning 252 m. *Jour. of Wind Engin. and Indu. Aerody.* 53: 301–314.

Ruscheweyh, H. & Sedlacek, G. 1988. Crosswind vibrations of steel stacks, critical comparison between some recently proposed codes. *Jour. of Wind Engin. and Indu. Aerody.* 30: 173–183.

Scanlan, R. H. & Eshan, F. 1990. Vortex-induced vibrations of flexible bridges. *Jour. Engine. Mechan. ASCE* 116-I: 1392–1411.

Strommen, E. & Hjorth-Hansen, E. 2001. On the use of tuned mass dampers to suppress vortex shedding vibrations. *Wind and Structures* 4: 18–30.

Warburton, G. B. & Ayorinde, E. O. 1980. Optimum absorber parameters for simple systems. *Earth. Engin. and Struc. Dynam.* 8: 187–217.

Wilde, K., Fujino, Y. & Prabis, V. 1996. Suppresion of bridge deck flutter by variable eccentricity. *Proc. 2nd Int. Workshop of structural control*: 564–574.

Wilde, K., & Fujino, Y. 1998. Aerodynamic control of bridge deck flutter by active surfaces. *Jour. Engine. Mechan. ASCE* 124: 718–727.

Wilde, K., Fujino, Y. & Kawakami, T. 1999. Analytical and experimental study on passive aerodynamic control of flutter of a bridge deck. *Jour. of Wind Engin. and Indu. Aerody.* 80: 105–118.

Wilde, K., Fujino, Y. & Omenzetter, P. 2000. Suppression of wind-induced instabilities of a long span bridge by a passive deck-flaps control system. *Jour. of Wind Engin. and Indu. Aerody.* 87: 61–79.

Wilde, K., Fujino, Y. & Omenzetter, P. 2001. Suppression of bridge flutter by active deck-flaps control system. *Jour. Engine. Mechan. ASCE* 127: 80–89.

Yang, J. N. & Giannopoulos, F. 1979. Active control and stability of cable-stayed bridge. *Jour. Engine. Mechan. ASCE* 105: 677–694.

System-based Vision for Strategic and Creative Design, Bontempi (ed.)
© *2003 Swets & Zeitlinger, Lisse, ISBN 90 5809 599 1*

Active control of cable-stayed bridges: large scale mock-up experimental analysis

G. Magonette & F. Marazzi
E.C., Joint Research Centre, ELSA Lab, Ispra (VA), Italy

H. Försterling
Bosch Rexroth, Jahnstrasse, Lohr, Germany

ABSTRACT: A large-scale cable-stayed bridge mock-up has been constructed and tested at the ELSA Laboratory of the JRC-Ispra. This mock-up has been equipped with two types of active tendon actuators to increase the structural damping in order to mitigate vibrations. The results of several damping experiences are presented; they clearly demonstrate the efficiency of the active damping system. The proposed technology is directly applicable to real structures by scaling up the devices. This paper is associated with (Dumoulin 2003) likewise submitted to this conference.

1 INTRODUCTION

1.1 Background

As modern cable-stayed bridges move toward taller and more flexible design the problems related to vibrations have become increasingly apparent. Wind and traffic can induce dangerous oscillations both in the deck and in the cables. To prevent such vibrations, passive measures like cross-ties interconnecting the stay cables or dampers installation at the bridge deck have been widely used. But some problems have occurred with these systems. The initial tension of the cross-ties must be selected with care in order to avoid de-tensioning and shock effects in the cable system. Moreover, viscous dampers located near the cable anchorage at bridge deck have a limited damping effect, in particular in the case of parametric excitation.

Traditional counteract measures are reaching their limits and new vibration mitigation technologies are currently studied and tested. In particular active and semi-active techniques have received increasing attention due to their effectiveness.

1.2 Necessity of a large-scale mock-up

During the last decade researchers have devoted an increasing attention to active and semi-active control technologies. Many numerical benchmark studies were carried out to compare different control strategies and to validate concepts imported from the automatic control field. The effectiveness of the active damping control was experimentally confirmed on a small-scale cable-stays model (Ackire 1997). This model was in scale 1/100 and used piezoelectric actuators.

The mock-up used for the experimental analysis presented here is a unique large-scale cable-stayed bridge especially designed to investigate in details the controlled system dynamics. The overall performances have been verified on a realistic structure equipped with industrial components.

The controlled structure consists of a number of important components such as sensors, controllers, actuators, and power generators that must be part of an integrated system. Moreover, a number of implementation-aspects must be addressed such as intermittent and fail-safe operations, integrated safety, reliability and maintenance. These issues require experimental verification under realistic conditions. Furthermore the experimental validation on a large-scale prototype gives an enhanced knowledge of the non-linear dynamic behavior of the cables and of the real loads in the anchorages.

After the description of the mock-up and of the control actuators, details are given on the measurement systems and on the main results of the testing campaign to assess the performances of the active tendon system with two types of actuators (of first and second generation).

Figure 1. Large scale cable-stayed bridge mock-up at the ELSA laboratory of the JRC-Ispra.

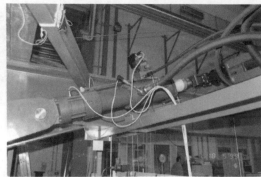

Figure 2. The right hydraulic actuators acting on the stay cable.

2 MOCK-UP DESCRIPTION

2.1 The deck

The bridge mock-up has been designed by Bouygues. It is a cable stayed cantilever beam that will basically represent a cable-stayed bridge under construction. The deck is 30 m long and is mainly composed of two H-beams whose axis are spaced 3.0 m apart (Fig. 1). They are appropriately linked each 3.5 m with transversal H-beams to provide to the whole structure sufficient transverse and torsion stiffness. Each H-beam is fixed to the Reaction Wall. The vibration excitation source is anchored at the free end of the deck.

2.2 The stay cables

Four pairs of parallel stay cables support the deck. Each stay-cable is composed of one T13 strand with a slope of 1/3. This slope is very close to that of the longest stay cable of modern bridges. The static tension being about 70 kN, the first free frequency of the longest stay cables (29.5 m long) is higher than 5 Hz, which can be rarely met on modern cable stayed bridges. To give the stay cables enough sag and consequently reduce their free vibration frequencies, they are heavily overloaded with split lead cylinders. This increases their average mass to an amount of about 15 kg/m. In this way, the sag of the longest stay cables is about 0.8% of their length and their first free frequency in the vertical plane is close to 1.2 Hz. With this arrangement the first vertical free frequency of the complete structure (flexion) is very close to that of the longest stay cable. By positioning an intermediate support under the deck, the vibration frequencies of the whole structure can be varied and eventually adjusted to be very near to the double value of the first vertical free frequency of the longest stay cable.

Similarly by modifying very slightly the length of the stay cables or by reducing the amount of additional masses gripped on the stay cables, it is possible to merge or to separate modal frequencies and to create critical situations for the structural behavior.

As designed, the mock-up allows the complete analysis of numerous particular situations. The mock-up is a demonstrator allowing the verification of the efficiency of the active control system in the worst conditions that can be faced by real structures. Without the intermediate support, the first modal frequencies of the structure were 1.12 Hz for the first flexional vertical mode, 1.16 Hz for the first torsional mode and 3 Hz for the first flexional horizontal mode.

2.3 The basic hydraulic actuators (first generation)

The main function of the hydraulic actuator (Fig. 2) is to track the displacement command required by the damping law. As the actuator is mounted between the bridge deck and a stay cable has to carry both the static and the quasi-static loads (bridge deck and live loads). Only the two longest stays, which give the best controllability of deck vibrations, are actively controlled.

Safety and energy-saving concerns led to the design of an actuator integrating two functions within one device (Aupérin 2001): an asymmetric large cylinder (pressurized by an external accumulator) compensates the static load while a smaller double rod cylinder provides the displacement necessary to satisfy the damping law (Fig. 3). Fails safe is guaranteed by the static pressure part of the actuator. In case of troubles in the control chain the control can be disabled and the actuator works as a passive device sustaining the static tension of the cable.

Energy consumption is a major concern. This has two implications: from one side there is a need for a permanent pumping system able to supply pressurized oil to the valve controlled actuators, from the other side this pumping system need large amount of energy to function. The first problem can be neglected if a

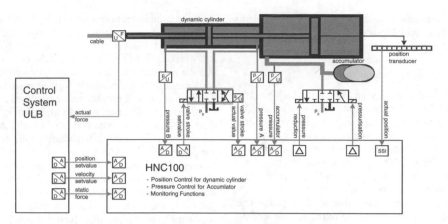

Figure 3. Sketch of the first generator actuator.

pumping system is already present, as for example in the construction phase of the bridge. In field application, energy consumption may be reduced by imposing that the active system is normally in stand-by and becomes operational only if the excitation level is greater than a fixed trigger level.

The good way to reduce the energy consumption is to replace the valve controlled actuator by a semi-active system allowing the recovery of induced vibration energy.

2.4   The optimized hydraulic actuators (second generation)

A very promising concept in structural control sets to work energy-transformation systems, in which the vibration energy is transformed in re-useable energy. A transformation of the vibration energy into mechanical or electrical form avoids the heat problems inherent to conventional energy dissipation and allows recovering energy. This concept is especially appealing where a large amount of induced vibration energy is acting over a long time.

Following this idea, Bosch-Rexroth has developed a new hydraulic device. Its functional scheme is illustrated in Figure 4. The connection pattern is the opposite of the usual one encountered in hydraulic applications, the backside being attached to the front and vice versa. This device is designed for cable-stayed bridges so the larger area of the rod has to carry big static loads, whereas the small area compensates the dynamic forces (in the order of magnitude of 10% of the static load).

The static chamber of the actuator shown in Figure 4 is pressurized at about 20 MPa whereas the average pressure of the dynamic chamber is 10 MPa only. This allows a "pressure stroke" of 20 MPa (from 10 to 0 MPa and from 10 to 20 MPa during control). Two pumps are used: one feeds oil from the tank to the dynamic chamber at an average differential pressure of 10 MPa; this

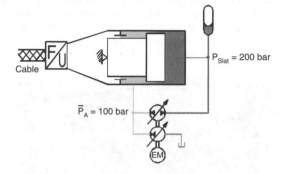

Figure 4.   Scheme of the second generation optimized actuator.

pump needs external energy. The other pump is placed between the dynamic chamber and the static pressure accumulator, feeding oil at 20 MPa to the dynamic level at 10 MPa. Also in this case, there is a 10 MPa differential pressure acting, but now in opposite direction. This pump is working as hydraulic motor. This means that it generates mechanical energy. For this quasi-stationary state there is – in principle – no resulting torque on the electric motor.

If vibration forces act on the hydraulic damper, the additional energy input minus the frictional losses is acting on the electric motor which works as a generator: the recovered energy can be fed into the electrical net.

The electrical motor, pumping system and hydraulic actuator are embedded into a single assembly. This integrated system, shown in Figure 5, is compact, robust and easy to install. The piston moves on low friction bearings. This improves energetic efficiency and allows using a differential pressure signal instead of a load force signal for the feedback of the IFF controller. This leads to a sensible reduction of instrumentation cost and improves the global reliability. Similarly to the first generation actuator, in case of problems in the control chain the device can be switched off and then work as a passive element.

2.5 Exciter and instrumentation

An electro-hydraulic inertial exciter was located at the free end of the deck to excite the mock-up. This shaker produced a vertical or horizontal unidirectional force with continuously adjustable amplitude in a frequency range from 0.01 to 50 Hz. The total mass of the structure being about 10,000 kg, the necessary exciting force was estimated in the range of 1.5 to 2.5 kN. The objective was to induce a vertical vibration of the uncontrolled deck with amplitude close to 25 cm peak-peak.

A special care was assigned to the selection of the most appropriate transducers and conditioning electronics. Measurement equipment included inductive, temposonic and laser displacement transducers as well as accelerometers, velocity, strain gauges, and force transducers. To measure tendon vibrations other techniques were considered such as CCD line-scan camera and laser scanning systems. Accelerometers were also

fastened in different positions on the cables. Depending of its frequency spectrum, each signal was acquired with a constant sample frequency selected between 50 to 500 Hz. Each test had duration from 3 to 10 minutes. All the measured values were recorded continuously on hard disc to be processed off-line subsequently. Figure 6 shows the measurement positions.

A dedicated data acquisition and control system was developed at ELSA (Marazzi 2002) with a recording capacity of 100 channels; 30 channels collected signals from the controllers and from the electro-hydraulic system. The remaining 70 channels were used to gather enough data to fully characterize the behavior of deck and cables. It was possible to record the values of each internal control variable: this option was very useful especially during the debugging or tuning phase. To help the users to interact with the experiment, some measures such as the cable tension or the deck displacement were displayed in real-time.

Note that from the point of view of the control system, only four signals are necessary: the two forces at the anchorage points of the two active stays and the corresponding actuator displacements.

3 CONTROL OF DECK AND CABLES

3.1 The control strategy

It is widely accepted that the active damping of linear structures is much simplified if one uses collocated actuator-sensor pairs (Bossens 2001); for nonlinear systems, this configuration is still quite attractive, because there exists control laws that are guaranteed to remove energy from the structure. The direct velocity feedback is an example of such "energy absorbing" control.

When using a displacement actuator (active tendon) and a force sensor, the (positive) Integral Force Feedback

$$u = g \int T dt \tag{1}$$

(refer to Figure 7 for notations) also belongs to this class, because the power flow from the control system is

$$W = -T\dot{u} = -gT^2 \tag{2}$$

Figure 5. The second generation optimized actuator placed at the anchorage point of the cable-stayed bridge mock-up.

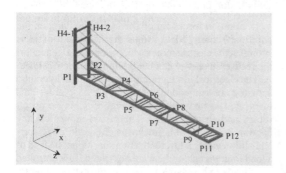

Figure 6. Scheme of the controlled structure.

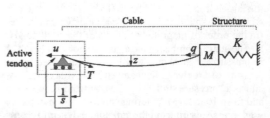

Figure 7. Scheme of the controlled structure.

This control law applies to nonlinear structures; all the states that are controllable and observable are asymptotically stable for any value of g (infinite gain margin).

3.2 The decentralized control

The foregoing approach can readily be extended to the decentralized control of a structure with several active cables, each tendon working independently with a local control law following Equation 1.

An important advantage of this control strategy is that only the measure of the dynamic component of the tension force inside the cables is needed. This signal permits to calculate the displacement to be imposed to the actuator to maximize the overall damping (stays and deck) of the bridge.

The force into the cable can be measured in different ways: by a load cell inserted in series with the cable and the actuator; recovered from calculation using the differential pressure value inside each piston; with stain-gauges or optic fibers located on the cable. During this testing campaign the direct measure of the forces via load-cells was in general considered, the second possibility was successfully implemented only with the second-generation actuators and the last option was disregarded. The direct force measurement provides the best control but the load-cells must be appropriately strong to bear the static tension of the stays and in the same time must be sensible enough to sense the dynamic component of the forces.

4 TESTING CAMPAIGN

The specific objectives of the testing campaign were:

- to improve the understanding of induced vibrations;
- to validate the numerical tools for prediction of dynamic behavior of cables;
- to verify the capability of the two active systems to mitigate induced vibrations;
- to evaluate in detail the performances and the reliability of the whole implementation with both types of actuators;

An exhaustive tests campaign including more than 200 tests was undertaken. As a preliminary step, modal analysis of the structure was performed. The vertical and horizontal bending modes, the torsional modes and the modal damping coefficients were accurately determined by achieving frequency sweep excitation and free vibration tests. These experimental data were used to validate the numerical simulations and to size the lead masses fitted on the longest cables to tune down their vibration frequencies in order to insure correct similitude ratio (Fig. 8). Afterwards, the structure equipped with the over-loaded cables was

Figure 8. Scheme of the controlled structure.

Table 1. Modal parameters of the cable-stayed bridge.

1st	Deck mode *(vertical bending)*	1.122 Hz
2nd	Deck mode *(torsion)*	1.158 Hz
3rd	Deck mode *(horizontal bending)*	1.647 Hz
4th	Deck mode	2.220 Hz
1st	Long cable mode *(vertical bending)*	1.188 Hz
1st	Long cable mode *(horizontal bending)*	1.094 Hz
2nd	Long cable mode	2.188 Hz
1st	Short cable mode *(vertical bending)*	1.594 Hz
1st	Short cable mode *(horizontal bending)*	1.531 Hz
2nd	Short cable mode	3.094 Hz

characterized. Some modal parameters are presented in Table 1. The first bending mode of the longest cables is very close to that of the deck.

The active control performance achieved by using the basic and the optimized actuators are summarized in the next paragraphs.

4.1 First generation actuators

The following results were obtained in the framework of the EC-ACE project (ACE 1997).

Thanks to the excellent characteristics of the electro-hydraulic shaker, it was possible to generate random excitations within a limited frequency bandwidth and to reproduce quite well realistic excitation conditions. In order to evaluate the real damping induced by the active control, random excitations were generated across the first resonance frequency of the bridge. The frequency spectrum of this limited bandwidth random excitation was contained in the range from 0.6 to 1.3 Hz.

The effectiveness of the active control approach appears clearly in Figures 9 and 10 as regards to displacements attenuation and in Figures 11 and 12 as regards to forces reduction in the cables. The reduction of fatigue at the cable anchorage is very significant.

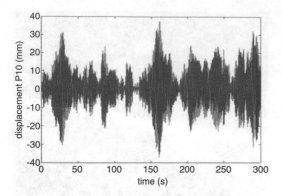

Figure 9. Displacement of the free bridge edge in response to a random excitation: without control (first generation).

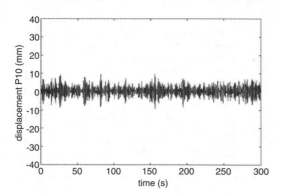

Figure 10. Displacement of the free bridge edge in response to a random excitation: with control (first generation).

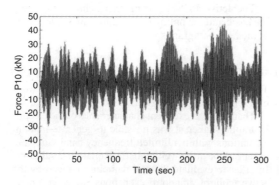

Figure 11. Dynamic tension of the longest stays in response to a random excitation: without control (first generation).

Thanks to an appropriate instrumentation based on a laser scanning system, cables motions were recorded. Figure 13 shows the trajectory of a long cable in the transversal plane crossing its mid length. The cable vibration mitigation is quite apparent.

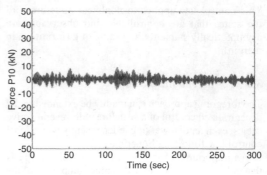

Figure 12. Dynamic tension of the longest stays in response to a random excitation: with control (first generation).

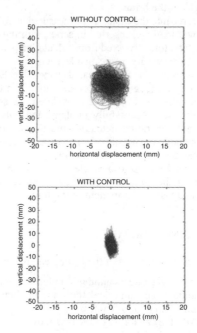

Figure 13. Cable response of the longest stays in response to a random excitation with and without control (first generation).

4.2 Second generation actuators

The following results were obtained in the framework of the EC-CaSCo project (CaSCo 2000).

The excitation signal generated for the ACE project was re-used here in order to easily compare the performances of the optimized (new) and basic actuators. Figures 14 and 15 show the force acting on the longest cables without and with control inserted. Figures 16 and 17 are relative to the displacement measured at the free edge of the bridge. The vibration damping in the controlled structure is once again evident.

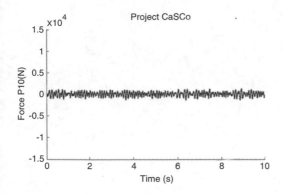

Figure 14. Force at active cable anchorage in response to a random excitation: with control (optimized actuators).

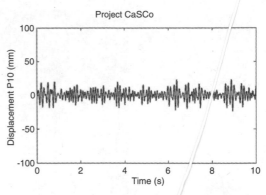

Figure 16. Displacement of the free bridge edge in response to a random excitation: with control (optimized actuators).

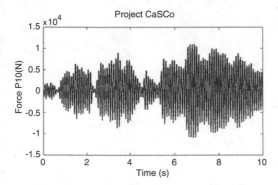

Figure 15. Force at active cable anchorage in response to a random excitation: without control (optimized actuators).

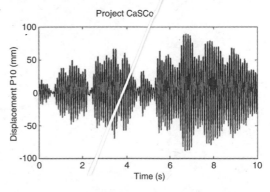

Figure 17. Displacement of the free bridge edge in response to a random excitation: without control (optimized actuators).

5 CONCLUSIONS

For the first time, a large-scale cable-stayed bridge mock-up equipped with active tendons was constructed and tested at ELSA. The purpose was to increase the structural damping of the whole structure. The tested mock-up was equipped with two different types of electro-hydraulic actuators.

The actuator of first generation gave excellent results in terms of performance. The damping of the bridge, less than 1% of critical damping in ordinary conditions, was increased to more than 10% by the active system. This produces an evident reduction of the deck displacement and of the dynamic tension acting on the stays with consequent mitigation of fatigue effects.

Starting from these successful results, a second generation of actuators was developed in order to reduce the energy consumption. The testing campaign shows that the goal was achieved: the external pumping system is no longer needed, but the vibration

reduction remains very good and comparable to the previous case.

REFERENCES

ACE 1997, Active Control in Civil Engineering, EC Brite-Euram Contract BRPR-CT97-0402.

Achkire, Y. 1997. Active Tendon Control of Cable-Stayed Bridges, Ph. D. dissertation, Active Structural Laboratory, Université Libre de Bruxelles, Belgium.

Aupérin, M. et al. 2001, Active Control in Civil Engineering: From Conception to Full Scale Applications, Journal of Structural Control, Vol. 8, Number 2.

Bossens, F. et al. 2001, Active Tendon Control of Civil Structures: Theoretical and Experimental Study, XIX International Modal Analysis Conference IMAC, Orlando, Florida.

CaSCO 2000, Consistent Semi-Active System Control, ECGrowth Contract G1RD-CT1999-00085.

Marazzi, F. 2002. Semi-Active Control of Civil Structures: Implementation Aspects, Ph. D. dissertation, Department of Structural Mechanics, University of Pavia, Italy.

The implementation of a cable-stayed bridge model and its passive seismic response control

V. Barberi & M. Salerno
Structural Engineer, Rome, Italy

M. Giudici
Department of Structural Mechanics, University of Pavia, Italy

ABSTRACT: The objective of the present study is to realize and critically evaluate a 3D model of the cable-stayed CAPE GIRARDEAU BRIDGE and successively to define a passive control system against seismic oscillations. By finite element commercial program SAP 2000 various models are realized more and more sophisticated using monodimensional and bidimensional finite elements. Static, modal and dynamic analyses are developed. After a critical comparison of these models, the more realistic one is used to study a passive control system. Elastoplastic and viscoelastic devices with different mechanical properties are tested, modelled using non linear elements. The controlled structure is subjected to a non linear dynamic analysis to determine the control system's benefits. A comparison between the best passive control system and an active one, tested in Benchmark control problem for seismic response of cable stayed bridge study by Washington University of St. Louis, is also conducted. Passive and active devices are positioned in the same place at pier and deck's connection, to evaluate differences between these control systems and to choose the best one.

1 INTRODUCTION

Seismic action together with wind and other environmental actions either natural or artificial acting on a structure, causes vibrations that, with critical intensity, can prejudice the comfort of users or, in more serious cases, the total stability of the construction. In order to attenuate effects caused by these dangerous vibrations, a structural control on the structure was attempted to exercise.

The considered structure is a cable-stayed bridge. A cable-stayed bridge has become a popular type of bridges throughout the world because of its aesthetic shape, structural efficiency, and economical construction. However, such a structure might be vulnerable to strong earthquake excitations due to its large flexibility and low damping ratios.

The cable-stayed bridge considered in this study is the Missouri 74–Illinois 146 bridge spanning the Mississippi River near Cape Girardeau, Missouri, USA.

This bridge was the object of Benchmark control problem for cable stayed bridge study realized by Prof. Dyke's staff of Washington University of St. Louis; more details can be found at http://wusceel.cive. wustl.edu/quake/.

Seismic problems were tacked into consideration in the design of this bridge due to the location of the bridge (in the New Madrid seismic zone) and its critical role as a principal crossing of the Mississippi River.

Just in early stages of the design process, the loading case governing the design was determined to be due to seismic effects. Various design configurations were considered, including full longitudinal restraint at the tower piers, no longitudinal restraint. Temperature effects were also considered.

Deck restraining in the longitudinal direction was found to result in unacceptably large stresses. Thus, force transfer would provide the most efficient design.

With this regard, bearings at bent 1 and pier 4 are designed to permit longitudinal displacements and rotations about the transverse and vertical axes.

Soil-structure interaction is not an issue with this bridge as the foundations of the cable-stayed portion are attached to bedrock.

As shown in Figure 1, the bridge has a total length of 1205.8 m, with two towers each supporting 64 cables. Furthermore there are 12 additional piers in the approach bridge from the Illinois side. The main span is 350.6 m in length, the side spans are 142.7 m in length, and the approach on the Illinois side is 570 m.

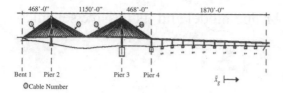

Figure 1. Cape Girardeau Bridge global configuration.

The bridge has four lanes plus two narrower bicycle lanes, for a total width of 29.3 m. The deck is composed from the structural point of view of steel beams and pre-stressed concrete slabs.

The H-shaped towers have a height of 100 m at pier 2 and 105 m at pier 3. Each tower supports a total 64 cables.

The towers are constructed by reinforced concrete.

The cross section of each tower varies five times over the height of the tower. As aforementioned, the approach bridge from the Illinois side is supported by 11 piers, but it has not be considered for its negligible effect on the dynamics of the cable-stayed portion of the bridge.

2 STRUCTURAL MODELLING

From the recalled Benchmark study data regarding geometrical and inertial structural properties were collected. By these data, to study the effective behavior of the bridge a 3D model, using those elements that commercial finite element program SAP 2000 names "general sections" was realized.

These sections, defined only by stiffness and inertial properties, gives the model named "USA" and defined as the reference model in the structural modelling.

Two different model's series were created to simulate in the best way the structural behavior of the bridge. These models were respectively named "FRAME" and "SHELL" from the name of the finite elements used. Another model named "FRAME CONTRO-VENTATO" was created to simulate concrete slab's contribution to deck's stiffness. These two last models are illustrated in Figure 2.

These models were obtained using elements with specific geometrical characteristics extracted from technical drawings: for these models deck's inertial properties were calculated and compared with "USA".

Evaluations about inertial properties convergence were therefore made to obtain a refined model. This convergence was obtained changing geometrical properties of deck's elements. Models created in this

way were defined "NUOVO" and "NUOVISSIMI", so the total number of models was 7 or 8 considering also "FRAME CONTROVENTATO" (this model was studied only in the transverse analysis), as in the table.

Cables were modelled by truss elements, towers and deck by beam elements in the SHELL model plate elements were used to simulate the bidimensional behaviour of concrete slabs. The stiffening cross used on the deck in "FRAME CONTROVENTATO" were modelled by truss element.

Considering static load cases were DEAD load self calculated by the program by material density and geometrical characteristics of the sections. LIVE case load was assigned only to the central span to obtain the worst load case for the vertical deck's deformation.

Another load case was cable PRESTRESS force, this is calculated to cancel deck's vertical deformation when permanent load and 20% of accidental load one considered together. Deck's masses allocation was made by Wilson and Gravelle method. By this method, the total mass of the deck was modelled by two lumped masses, having only traslational properties, connected by rigid links to the central beam element that has on its joint only mass' rotational properties.

The connection between piers and soil was modelled by embedded joints; abutment's behavior was simulated by constrained joints in transverse, vertical and rotational direction. The connection between pier and deck was modelled by a rigid link element acting only transversally.

2.1 Analysis

The models were subjected to several analyses. Consequently in order to point out the most realistic model static and dynamic behavior of the bridge, together with its modal shapes were estimated.

Static analysis was conducted to estimate bridge's deformation under DEAD and LIVE load cases, considering also cable prestress contribution.

Modal analysis was a way to evaluate any differences in models' behavior regarding modal participation mass ratios, natural periods and modal deformed shapes. At this regard, the structural response's accuracy was evaluated considering excited mass as the reference parameter; so modes number was so increased to excite at least 85% of the total mass in every direction. This result was obtained by considering 300 natural modes.

Dynamic analysis was developed to simulate bridge's behaviour under a real seismic action. This time-history analysis was carried out using EL CENTRO, MEXICO CITY and GEBZE seismic history, Fig. 6 that, for their characteristics, concur to estimate a complete structural behavior. These three earthquakes

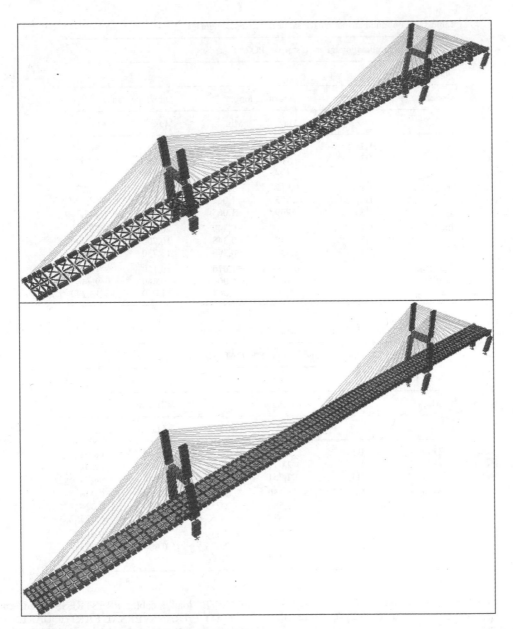

Figure 2. FRAME CONTROVENTATO & SHELL models.

are each at or below the design PGA level of 0.36 g for the bridge.

Analyses were made considering earthquake attack, at the beginning, in the only longitudinal direction. To consider 3D structural behavior, the bridge was subjected to a transverse earthquake too. Eventually to obtain a more realistic evaluation of structural responce, 3 time history cases were applied with different incidence directions, going from 0 to 90°.

2.2 Modelling results

To evaluate the behavior differences of the models three parameters were considered:

1. deck's displacements (in deck's central nodes),
2. base piers' shear (in the only longitudinal direction),
3. base piers' moment (in the only longitudinal direction).

Table 1. Longitudinal and transversal deck's displacements.

Displacement combination transv. history + PO2AT

Unit [cm]

Model	History	Direzione long.		Direzione transv.	
		Max	Min	Max	Min
USA	H1	1,182	−0,153	16,610	−16,040
	H2	1,181	−0,152	6,080	−5,057
	H3	1,182	−0,153	10,860	−15,800
Frame	H1	0,917	−1,073	38,780	−31,830
	H2	0,905	−1,062	43,730	−58,690
	H3	0,909	−1,065	181,690	−144,200
Shell	H1	1,541	−0,349	20,930	−23,500
	H2	1,539	−0,348	10,700	−10,400
	H3	1,539	−0,348	13,730	−18,840
Frame contr.	H1	0,910	−1,070	19,150	−17,750
	H2	0,903	−1,061	9,660	−9,440
	H3	0,909	−1,064	11,060	−13,220

Table 2. Longitudinal and transversal deck's displacements.

Displacement combination long. history + PO2AT

Unit [cm]

Model	History	Long.		Transv.	
		Max	Min	Max	Min
USA	H1	3,391	−2,156	0,006	−0,008
	H2	2,071	−0,962	0,002	−0,002
	H3	2,751	−1,742	0,003	−0,003
Frame	H1	15,350	−14,939	0,039	−0,046
	H2	5,580	−3,759	0,026	−0,024
	H3	9,620	−11,639	0,023	−0,025
Shell	H1	7,768	−7,539	0,013	−0,012
	H2	3,488	−2,676	0,004	−0,005
	H3	4,078	−3,806	0,007	−0,007

Those characteristics were calculated in the static analysis, while in dynamic analysis were also considered:

1. effective base piers' strains (shear and moment in every direction and their vectorial combination,
2. earthquake's input energy.

In Table 1, longitudinal and transversal deck's displacements are shown under a load combination composed by: DEAD + 20% of LIVE load + cable PRESTRESS load case (PO2AT) + LONGITUDINAL EARTHQUAKE. In Table 2 are illustrated longitudinal and transversal deck's displacements, they are shown under a load combination including: DEAD + 20% of LIVE load + cable PRESTRESS load case (PO2AT) + TRANSVERSAL EARTHQUAKE. This analysis is made on every model for the third iteration's models (NUOVISSIMI).

Table 3 portrays dynamic analysis results: in particular, there is vectorial combination of transversal and longitudinal shear, calculated at piers' base, considered unit is kg × cm, under a load combination composed by: DEAD + 20% of LIVE load + cable PRESTRESS load case (PO2AT) + TRANSVERSAL EARTHQUAKE and there is a vectorial combination of transversal and longitudinal moments, calculated at the base of the piers, considered unit is kg, under the same combination.

Table 3. Vectorial combination of moment and shear.

	Vectorial combination history + PO2AT					
	Moment [kg × cm]			Shear [kg]		
Model	H1	H2	H3	H1	H2	H3
USA	7,68E + 09	4,09E + 09	6,48E + 09	3,68E + 06	1,60E + 06	3,11E + 06
Frame	3,03E + 09	9,63E + 08	2,75E + 09	1,33E + 06	4,24E + 05	1,11E + 06
Shell	5,61E + 09	2,85E + 09	4,88E + 09	2,81E + 06	1,26E + 06	2,33E + 06
Frame contr.	3,61E + 09	1,67E + 09	2,08E + 09	1,63E + 06	5,78E + 05	1,39E + 06

Table 4. Shear and moment.

	History + PO2AT longitudinal					
	Moment [kg × cm]			Shear [kg]		
Model	H1	H2	H3	H1	H2	H3
USA	1,345E + 10	8,271E + 09	1,160E + 10	4,908E + 06	2,197E + 06	2,893E + 06
Frame	7,267E + 09	2,431E + 09	4,515E + 09	2,810E + 06	9,317E + 05	1,766E + 06
Shell	1,097E + 10	1,325E + 10	2,120E + 10	5,238E + 06	1,839E + 06	2,394E + 06

Table 5. Excited mass percentage.

% Excited mass				
Model	n° modes	Long.	Transv.	Vert.
USA	20	2,52	19,24	11,42
	100	14,38	68,58	60,6
	300	84,52	90,59	80,07
	500	94,32	97	93,02
Frame	20	2,48	66,87	9,3
	100	10,45	70,4	32,01
	300	81,04	83,6	58,38
	500	94,09	94,16	69,2
Shell	20	3,64	25,55	14,07
	100	89,11	85,38	48,38
	300	93,92	92,55	78,47
	500	98,89	98,66	97,09

In Table 4 there is shear, calculated at piers' base (considered unit is kg × cm), under a load combination composed by: DEAD + 20% of LIVE load + cable PRESTRESS load case (PO2AT) + LONGITUDINAL EARTHQUAKE and there is moment, calculated at piers' base (considered unit is kg), under the same combination.

Table 5 is shows the excited masses' percentage in three directions for the third iteration's models (NUOVISSIMI) and for USA model at various modes' number.

As conclusion of these analyses was determined that model named "SHELL NUOVISSIMO" has the same behavior of USA but, better then USA, simulates tridimensional structural bridge's behavior under a seismic attack acting out of the longitudinal plane of the bridge.

For this reason SHELL NUOVISSIMO model was used to test structural control devices' efficiency.

3 PASSIVE CONTROL: MODELIZATION AND POSITIONING

Cape Girardeau bridge is situated in a seismic region.

On the structure was thus placed a structural control system: this control was effected by passive dissipating devices.

New structural behavior may be represented by the following equations:

$$[M]\{\ddot{x}\} + [C]\{\dot{x}\} + [K]\{x\} + [\Gamma]\{x\}$$
$$= -[M + \overline{M}]\ddot{x}_g\{r\}$$
$$[\Gamma] = [\overline{M}]\{\ddot{x}\} + [\overline{C}]\{x\} + [\overline{K}]\{x\}$$

where Γ is a linear function of acceleration, velocity and displacement. Mechanic properties of devices were implemented refering to industrial production's indications. Evaluation parameters were:

– deck and piers displacement,
– base shear and moment,
– energy considerations.

NON LINEAR LINK ELEMENTS (NLL) were used to model control devices this is the only way for the used FE program to simulate material non linearity. These devices are ELASTOPLASTIC and VISCOELASTIC dampers. The purpose of devices' inserting is to reduce both stress properties and seismic input energy on the structure.

Finite element program SAP 2000 has 6 different NLL kinds, each one of these imply caratteristics' definition of non linear springs in each DOF. In particular in the bridge DAMPER and PLASTIC 1 link elements were used.

The first NLL property is based on a viscoelasticity model, proposed by Maxwell. It's made by a non linear damper in series with an elastic spring; the consequent non linear relationship is then:

$$f = k * d_k = c * (\delta d_c / \delta t)^\wedge \; cexp \quad cexp = 0.2 \div 1.2$$

The second element is based on an hysteretic behavior proposed by Wen: all internal deformation are independent among themselves, therefore a single DOF's yielding doesn't condition other deformation's behavior. This model is based on the following relationship:

$$f = ratio \; k * d + (1 - ratio)yield * z$$

where z is an hysteretic constant including between 0 and 1.

The analysis considered out is a NON LINEAR TIME HISTORY ANALYSIS. In particular, FE program make use of an exstension of FAST NON LINEAR ANALYSIS developed by Wilson: this method is extremely efficient and was developed to be applied to all structural systems with a basic linear behavior, and a limited number of predefined non linear elements. Traditional bearings were substitute by any devices both to absorb seismic energy and to reduce seismic response amplitudes. Devices were positioned at connection between deck and piers. In particular there are 8 devices corresponding to each of central

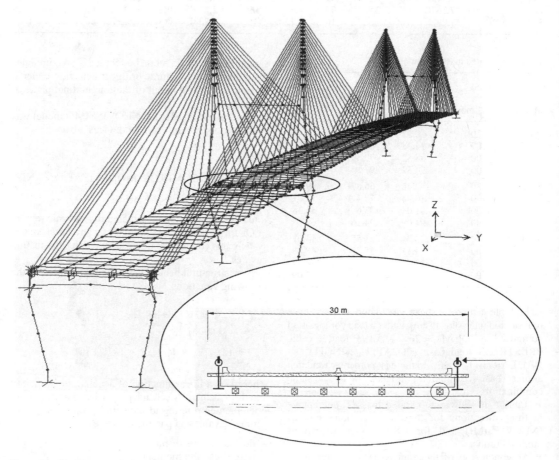

Figure 3.   Deck's controlled model view.

towers and 4 at the connection between abutement and deck-end (Fig. 4). Executed analyses and tested devices are listed in Figure 5.

Listed devices' mechanical properties (k1 = linear elastic constant; k2 = postelastic constant; Fy = yielding value; C = viscous constant), are those required by SAP 2000. Those properties are obtained by producers catalogues.

3.1 Results for the controlled model

A First kind of analysis is characterized by longitudinal seismic attack and by non linear devices' properties in the same direction.

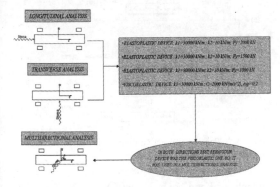

Figure 4. Executed analyses and tested devices.

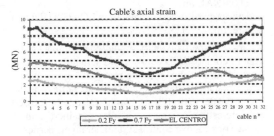

Figure 5. Axial strain state for each cable.

Devices' introduction transfers loads to cables. This wasn't a problem for their resistence, being axial strain under dead, live and time history loads always between 20 and 70% of ultimate load.

In the follow diagrams, Figure 6, are portraied, as example, only for EL CENTRO earthquake, axial strain state for each cable (were considered only 32 of 128 cables, infact a double symmetry is in the bridge).

To represent and summarize in the best way results obtained by dynamic analyses under EL CENTRO earthquake in the only longitudinal direction, all the properties by which the comparison was made, Table 6. As shown in the upper table there are extermely high reductions in base shear and moment, instead longitudinal deck displacement increases, because deck-pier connection is less rigid then in traditional connection.

Energy dissipation entity was deducted by hysteretic diagrams of devices, were considered only 2 of many studied solutions.

Producers assign a maximum displacement for elastoplastic devices equal to 15 cm and equal to 30 cm for viscoelastic ones.

In the following diagram Figure 7 is portrayed the reduction of deck's acceleration obtained by introduction of control devices in the structure.

A second kind of analysis is characterized by considering seismic attack in the only transverse direction and also non linear devices' properties in the same direction.

Devices were efficient also in the transverse analysis; the only negative result is the exceeding of maximum admissible displacement for elastoplastic devices, in particular this problem was evident during GEBZE earthquake.

This limitation was only a mechanical problem (material's yielding) and not geometrical, infact between various devices' position there is much more 50 cm.

After devices' efficiency checking in longitudinal and transversal direction, was effected a "MULTIDIRECTIONAL" analysis, submitting the structure to a seismic attack from different directions to evaluate the real behaviour of the bridge.

Considered devices are viscoelastic ones, the best working in previous analyses. Non linear characteristics

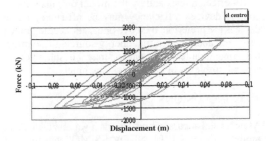

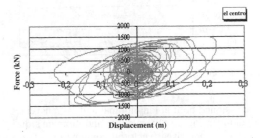

Figure 6. Energy dissipation.

Table 6. Controlled vs Uncontrolled model.

	Uncontrolled	Elastoplastic			Viscoelastic
		K1 = 50000 kN/m K2 = 10 kN/m Fy = 2000 kN	K1 = 50000 kN/m K2 = 10 kN/m Fy = 1500 kN	K1 = 80000 kN/m K2 = 10 kN/m Fy = 1000 kN	K1 = 50000 kN/m C = 2000 kN/ms exp = 0.2
Displacement (m)	2,093E − 01	2,130E − 01	1,970E − 01	1,960E − 01	2,120E − 01
Shear (kN)	2,770E + 04	1,320E + 04	1,440E + 04	1,280E + 04	1,330E + 04
Shear % Reduction	–	52,35	48,01	53,79	51,99
Moment (kN × m)	5,180E + 05	3,080E + 05	2,720E + 05	2,380E + 05	2,460E + 05
Moment % Reduction	–	40,54	47,49	54,05	52,51
Dissipated Energy %	–	14,91	33,24	33,69	37,46

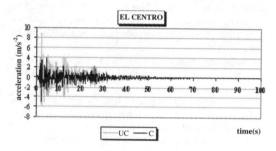

Figure 7. Reduction of deck's acceleration.

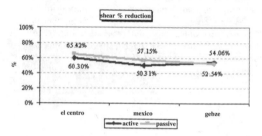

Figure 9. Passive and active control system: % reduction of base shear.

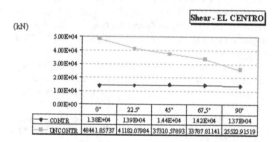

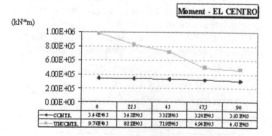

Figure 8. Shear and moment reduction.

were assigned both in longitudinal and in transversal direction. In Figure 8 base shear and moment's diagrams were portrayed (vectorial combination of two components). It is shown the comparison between uncontrolled and controlled structure. By these diagrams looks that:

– the worst direction of seismic attack is the LONGITUDINAL one;
– devices give the best behaviour in LONGITUDINAL direction.

Notre Dame University modelled an active control system composed by sensors and actuators placed on the structure as shown in the previous figure; even in our modelation devices are in the same position because, after various tests, that revealed to be the best. It's interesting comparize results obtained using a passive or an active control system: strain properties' % reduction are confrontable for the three time histories, as shown in Figure 9.

By effected tests and by the model outputs passive control devices' efficiency may be affirmed. These devices are preferible to active ones, not needing much servicing and any external energy source, these infact for considerable earthquakes, may be subjected to typical problems of electronics machinery.

4 CONCLUSIONS

After several accurate modelations it was evaluated that model named "SHELL NUOVISSIMO" was the best simulator of Cape Girardeau Bridge's real behavior.

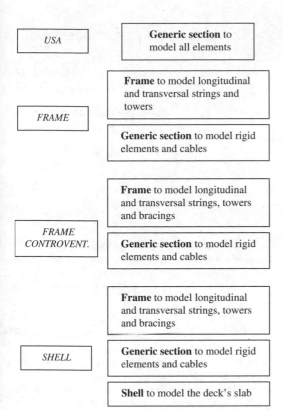

Figure 10. Models.

On this model were tested and got some different passive structural control devices. Best were *Viscoelastic Dampers* ($\mathbf{f} = \mathbf{k} * \mathbf{d_k} = c * (\mathbf{\delta d_c/\delta}t)^\wedge$ **cexp**) having: $K = 50000$ kN/m, $C = 2000$ kN $*$ s/m and cexp $= 0,2$ as mechanical properties. These devices,

opportunely positioned between piers and deck, reduced strain properties even about 60%, and about 35% seismic input energy. Tested the equality between passive and active control system's results, passive control is preferable to the active one that is more vulnerable during seismic attacks.

APPENDIX

Figure 10 shows a synthesis of the different models developed with their specific sections.

ACKNOWLEDGMENTS

The authors thank Prof. S.J. Dyke and J.M. Caicedo for information and data on Cape Girardeau Bridge.

REFERENCES

Dyke, S.J., Caicedo, J.M., Turan, G., Bergman, L.A. & Hague, S. 2000. *Benchmark Control Problem for Seismic Response of Cable-Stayed Bridges*.

Bontempi, F., Casciati, F. & Giudici, M. *Risposta Sismica di un Ponte Strallato: sistemi di Controllo Attivo e Passivo* (Bridge Benchmark Problem), (in italian) .

Wei-Xin Ren & Makoto Obata. August 1999. *Elastic-Plastic Seismic Behavior of Long Span Cable-Stayed Bridge*. Journal of Bridge Engineering.

Gimsing, N.J. *Cable Supported Bridge*. John Wiley & Sons, West Sussex 1998.

Active control in civil engineering: a way to industrial application

M. Aupérin & C. Dumoulin
Bouygues Travaux Publics, St Quentin en Yvelines, France

ABSTRACT: The present document provides information about an active tendon control system suitable for the large cable-supported structures used in civil engineering. The active tendon control system has been widely tested on a large-scale mock-up during a research project partly funded by the European Commission. This paper is a follow-up to previous documents describing the experimental results; it mainly focuses on the industrial application aspects.

1 INTRODUCTION

Improvements in materials and computational technology have led progressively to structurally more efficient and more slender bridges, particularly as regards cable-supported structures. Consequently structures increasingly flexible. The natural damping ratio of such structures is often lower than one percent. In service as well as in construction state, vibrations with a more or less large amplitude can be observed. Even if these vibrations do not lead to the quick collapse of the structure, they increase anyway the fatigue of the materials and the discomfort of the end-users. During construction, excessive vibrations can be particularly dangerous for security and can lead to expensive interruptions of the civil works on site. To reduce the impact of these undesirable effects, a solution consists in increasing the damping of the structure. This can be achieved by an active damping strategy able to upgrade structural damping in a large frequency bandwidth. With the proposed active control, damping ratios up to 5–10 percent are obtained.

2 CONTROL STRATEGY

The active control system is based on an active tendon including an actuator collocated with a force sensor. The control law is the so-called Integral Force Feedback (IFF). Detailed information on the IFF control strategy are presented in (Achkire 1997, Preumont 2000, Bossens 2001).

The command provided by the IFF law with collocation is equal to the stroke imposed to a linear viscous damper installed at the location of the actuator. If the controlled element is a cable, it is not able to cope with a compressive force. A permanent tensile force is therefore required which cannot be balanced by a viscous damper. It is necessary to install a spring mounted in parallel with the damper. This spring reduces the static apparent stiffness of the controlled element, what is not wished in most cases. Consequently the additional spring must have a large stiffness, and this reduces drastically the provided damping (see Aupérin 2001). An active device allows to retrieve the complete damping capability of the system, even in case of a permanent tensile force in the controlled element. It is just necessary that the force taken into account by the controller is the total dynamic force in the controlled element.

3 PROJECT OBJECTIVES

The aim of the project was to bring the proposed active tendon control system from a laboratory development to an industrial level. Specific hydraulic actuators were developed. To demonstrate and to validate the proposed system, a very large-scale mock-up was built using industrial components and the hydraulic actuators. More than two hundreds of tests were carried out. More information about the experimental testing campaign will be find in a sister paper presented during this conference.

4 THEORETICAL MODEL

4.1 *Theoretical model description*

To be able to analyze the test results, a theoretical system is needed. This system should be as simple as possible, so that the main characteristics of the control can be easily pointed out. In most of the tests, mainly the first bending mode of the structure has been excited. In spite of the strong non-linearity of the cables, an approximate linear design method has been developed and validated.

The theoretical model is a SDOF linear mass-spring-damper model where a second so-called controlled spring has been added (Fig. 1). The end of the second spring is controlled by an actuator able to impose a displacement. The control law follows the IFF law: $du/dt = g\, k\, (z - u)$ where k, z and u are defined in Figure 1 and g is the called the gain.

The system is characterized by four parameters:

- The natural circular frequency Ω when the actuator is fixed,
- the natural circular frequency ω when the actuator is completely free,
- the natural damping ratio ξ_0 often very small,
- the dimensionless gain $q = g\, k/\Omega$ (also called reduced gain)

This is a reliable system for describing the oscillations of a mode of the mock-up assuming that the control modifies the mode-shape only slightly.

With the help of the theoretical model, a design criterion for the active control system for an "optimal" control is established.

Some major experimental results are described hereafter and compared to experimental results.

4.2 Response to stationary random excitation

Wind is a key factor in the design of civil cable-supported structures, especially regarding vibrations. With the help of the theoretical model, a design criterion for the active control system as regards an "optimal" control is established.

Knowing the excitation spectrum, for each value of the gain, the parameters (natural circular frequency ω_e and damping ratio ξ_e) of a mass-spring-viscous damper

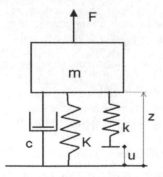

Figure 1.   SDOF linear system.

system equivalent to the SDOF linear system can be computed.

This equivalence must be defined: subjected to the same random excitation, the controlled system and its equivalent have the same variance of the displacements and the same central frequency of vibration. If the excitation is a stationary white noise with a very large bandwidth, the optimal value of the dimensionless gain is q_1:

$$q = q_1 = \frac{\omega}{\Omega} - 2\,\xi_0 \quad \Rightarrow \quad \omega_e = \frac{\Omega + \omega}{2} \quad ; \xi_e = \frac{\Omega - \omega}{2\Omega} + \xi_0$$

As long as the spectral bandwidth of the excitation contains the range $[\omega, \Omega]$, the optimal value of the dimensionless gain is close to q_1 and the damping ratio is not very different of the value ξ_e given in the previous formula. The variance of the actuator displacement and the mean power consumed to move the actuator increase with the gain. If a value higher than the optimum is chosen for the gain, compared to the optimal value, an efficiency loss and an increase of the stroke and of the power consumed by the actuators will be obtained.

The actuators developed in the project are hydraulic ones, well suited for civil engineering structures. The energy required by the actuator motion is provided by an external source. In these conditions, the mean power to be provided increases with the gain of the control law. The economic optimum does not necessary fit with the maximum damping.

4.3 Bending moments and strokes

Among the two hundreds of tests carried out on the mock-up, a lot of experimental results have been compared to theoretical predictions based on the SDOF linear system. An interesting comparison concerns eight tests carried out on the same structure, with the same white noise excitation. The only parameter that changes is the gain value.

Figure 2 gives two pairs of theoretical curves deduced from the system and the real spectrum of the excitation:

- the variance and the central frequency of the vertical motion of the mass;
- the reduced variance and the central frequency of the stroke u.

The abscissa represents the dimensionless gain q. The experimental points are shown in Figure 2: squares for the bending motion and triangles for the stroke. Regarding the bending variance, the experimental minimum is equal to the minimum predicted by the theory. Moreover it is important to notice that around the minimum value the curve are really flat. A precise tuning of the gain is not necessary to be close to the minimum value.

4.4 Relation between strokes and damping

Figure 2 suggests another representation of the results, eliminating the variable gain. The abscissa axis represents the reduced variance of the stroke u and the ordinate axis represents the reduced variance of the bending motion z. This leads to Figure 3. Here again the experimental points lie on the theoretical curve. The non-linear behavior of the cables induces a modification of their apparent stiffness. Despite this effect, the relation between the variances of the stroke and the bending motion is similar to that deduced from a system with a constant stiffness.

The dotted line in Figure 3 is a line with a slope of fi. It can be noticed that it cuts the theoretical curve at the minimum value of that curve. When the gain is tuned to provide a damping as large as possible, the variance of the stroke is equal to half of that of the projection along the chord of the imposed motion of the anchorage.

4.5 Power consumed by the actuators and damping

Because of the dimensions of important civil engineering structures, designers look for effective damping methods with low-energy demand. This aspect justifies

the establishment of a criterion to optimize energy consumption rates of the active control unit compared with the additional damping ratio that will be obtained for the structure. The results of the theoretical system are described hereafter and related to the experimental results.

The hydraulic actuators are provided with oil from a tank at constant pressure. The role of that accumulator is to avoid shocks. The hydraulic pump should only provide the oil mean flow. In output of an actuator, the oil returns to the tank supplying the pump. It results that the mean power P_m provided by the pump is equal to the product of the pump over-pressure and the oil mean flow. At each time, the oil flow is proportional to the absolute value of the stroke velocity. The overpressure being constant, the power P_m is proportional to the mean value of the absolute value of the stroke velocity.

Figure 4 shows the theoretical curve obtained in this way and the experimental points. Once more again, the experimental points lie on the theoretical curve. The figure exhibits that there is an optimal power value giving the maximal damping.

5 INDUSTRIAL ASPECTS

5.1 Structure and actuator design requirements

In order to apply this technology to real structures, engineers need rules to be able to quickly design structures equipped with an active control system.

The location of the controlled elements is driven by the formula of section 4.2. The expected damping ratio is driven by the frequency shift $\Omega - \omega$. Two dynamic analyses have to be carried out to compute the natural frequencies of the structure: one with the controlled elements and one without.

The additional damping can be estimated with the formula of section 4.2. Then the structure can be analyzed with the classical tools available in the design office (linear dynamic analysis software packages).

Based on the structural analysis, the engineer can define the allowable displacement variance of the structure and consequently estimate the actuator stroke and

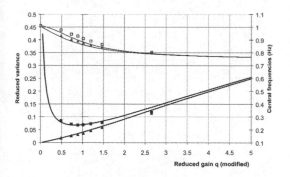

Figure 2. Reduced variances versus reduced gain q.

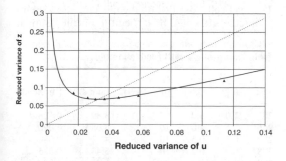

Figure 3. Reduced variance of z versus reduced variance of u.

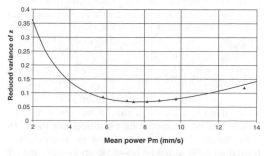

Figure 4. Reduced variance of z versus mean power P_m.

the required mean power by using results of Figures 3 and 4.

Structure and active control system design does not require sophisticated dynamic step by step calculations.

5.2 Command saturation

The command saturation aspect has been tested. During test 203, the standard deviation of the strokes is a little bit lower than 1 mm regarding each actuator and the maximal value is 3.8 mm. During test 204, we have repeated the test but limiting the command amplitude to 2 mm, which allows to simulate a saturation. In practice, the command is computed by integration of the dynamic force then its absolute value is limited to 2 mm each time it over passes this value.

With the basic IFF control law, the power provided by the dynamic force is proportional to the squared value of this force. On the contrary, the power necessary for the stroke is proportional to the absolute value of the stroke velocity, so to that of the dynamic force. During a cycle, the saturation of the command occurs when the dynamic force is weak. It should not induce unfavourable effect on the efficiency. The representative point of test 204 lies on the theoretical curve, a little bit on the left side of the point associated to test 203. As long as it remains occasional and limited, the command saturation has no much effect on the efficiency. Without loss of efficiency, it is possible to limit the actuator stroke to a value equal to two times the standard deviation of the stroke computed without saturation (for more information, see Aupérin 2001).

5.3 Actuator failure

Thanks to the IFF control law, the efficiency of the actuator is only based on the dynamic force measured by the collocated sensor. If an actuator fails, the other ones continue to work and an additional damping is always provided, with of course a lower value than the foreseen optimal value. Tests have been carried out successfully to verify this assumption.

5.4 Power saturation

In control theory, power saturation is generally a difficult problem. When it happens, the phenomena exhibits a frequency which is equal twice times to the controlled frequency. This leads to high frequencies and risk of spillover. The curve plotted on Figure 4 shows that in a wide range around the optimum value, power can be reduced without modifying the optimum value. Consequently, in case of power saturation, the problem can be solved if the power demand is reduced, for instance by decreasing the gain.

5.5 Command law with variable gain

The command law has been modified by introducing a gain g variable in relation with the dynamic force, according to a non linear law.

$$u = \int_0^t g(\Theta) F(t) \, dt$$

with

$$\Theta = \frac{\max \{|F(t)| - A \, ; \, 0\}}{B}$$

and

$$g(\Theta) = \frac{\Theta^n}{1 + \Theta^n} G$$

F(t) represents the dynamic force at time t. The variable Θ is equal to zero as long as the absolute value of the dynamic force is lower than the value A, which allows to suppress the control when the excitation is to weak or equal to zero. B is a parameter homogeneous to a force. The parameter n allows tuning of the shape of the gain law. The value $n = 2$ has been selected. According to the absolute value of the dynamic force, the gain g has a value contained between zero and a asymptotic value G. It has been checked experimentally that this law suppresses the self excitation modes at medium frequency, by providing a command equal to zero. Five complete tests have been carried out using this law. The data and the results (power P_m and efficiency, i.e. the reduced variance of the bending motions) are summarised in Table 1 hereafter.

Regarding tests 205 to 207, the experimental points are located under the theoretical curve: for a same efficiency, the modified law needs less hydraulic power than the basic law. The experimental points related to the two last tests lie on the theoretical curve (for more information, see Aupérin 2001).

5.6 Scale filter, non frequential

The load cells installed on the mock-up are extremely precise. It would be difficult to obtain a similar

Table 1. Law parameters and test results.

Test	A (kN)	B (kN)	G (mm/s/kN)	P_m (mm/s)	Efficiency
205	1.0	1.0	10	3.39	0.126
206	0.5	1.0	10	4.08	0.098
207	0.5	0.5	10	4.71	0.084
208	0.5	0.2	10	5.44	0.075
209	0.5	0.2	20	5.49	0.080

precision (better than 0.01% of full scale) with a load cell coping with some thousands of kN and to be installed on a Civil Engineering structure. In order to check the influence of load cell precision, a set of tests have been carried out including a modification of the force measurement. The signal of each load cell has been filtered according to the following formula:

$$F^* = C \times E\left[\frac{F}{C}\right] \quad with \quad E[X] = Integer\ Part\ of\ X$$

C represents a positive constant, homogeneous to a force. This "scale" filter transforms a signal F(t) with continuous variations into a signal F^*(t) varying by equidistant steps. If the amplitude of the variations of F(t) is too small, F^*(t) could even be a constant value. This filter is sensitive to the signal amplitude and non sensitive to its frequency. The output signal F^*(t) is used as input signal of the computation chain of the stroke command. The command law is the basic law with a constant gain (10 mm/s/kN for all the tests) and without saturation. Table 2 summarises the value of the filter parameter C and the results (power P_m and efficiency).

The experimental point of test 210 lies on the theoretical curve and that related to test 211 is a little bit above this curve. The two last points are clearly above the curve: a significant part of the hydraulic power is dissipated through low frequency strokes (lower than 0.7 Hz) which deliver no damping.

A load cell with a medium precision provides a signal largely better than the filter with a high value of constant C. The results of the tests show that we can be satisfied by a load cell of good precision, certainly more robust and really cheaper than a load cell of exceptional precision.

We have also noticed that even with a constant C equal to 0.2 kN, the "scale" filter eliminates all the self excitations, what is particularly interesting.

5.7 Structures with multiple degrees of freedom

Regarding the single-degree-of-freedom model, all the equations are linear. Considering a structure with more than one component, the same theory could be extended. Thus modal analysis is applicable. This means that each degree of freedom of the structure is associated to a single-degree-of-freedom system. Even if the simplified theory shows that the gain g is mode dependent, experiments have exhibited that in fact efficiency is not g dependent in a wide area

around the optimum. Consequently, the same actuator could be used to control several modes, separately or simultaneously.

To damp a mode with efficiency, adding an element between a node and an anti-node of the mode shape is a very good choice. For instance, regarding the first vertical mode of a suspension bridge, the addition of a tie between the node where the deck is connected to the tower and a point on the main cable located at 15% of span length from the tower, having a cross section of about 3% of the cross section of the main cable can provide a damping ratio ξ close to 10%. The same elements can also damp the modes 2 to 7 (bending and torsion). With such a system, it is clear that the Tacoma Narrows bridge would have hardly vibrated during the wind story that induced its collapse.

6 REPRESENTATIVE STRUCTURES

6.1 Suspension bridge

The application of active damping elements has been studied on several civil structures like a suspension bridge with a main span length of 850 m as shown in Figure 5. The bridge is supported by two main cables whose cross section is equivalent to 1000 strands .5".

Vertical vibrations, caused by vortex shedding, large enough to discomfort the traffic have been observed on this kind of structure. The initial solution was to install spoilers. The aim of the study was to carry out an alternative solution based on active control. A first study trying to apply active control directly on some hangers led to unsatisfactory results. This should have been expected, because the removal of some hangers does not really modify the first mode frequencies. A frequency shift is a prerequisite for active damping. Consequently the use of additional elements has been foreseen. Their location was based on the analysis of the mode shapes of the structure. In order to control the displacements of the first five modes of the structure, controlled elements linked between the pylon and the quarter-span of the main cable should be efficient. (Fig. 6)

Calculations were carried out using 4 stay-cables with 31 strands · 5" (to be compared to the main cable with equivalent 1000 strands · 5"), one can obtain between 5 and 11% additional damping. In case of

Table 2. Values of the filter parameters and test results.

Test	210	211	212	213
C (kN)	0.2	1.0	2.0	4.0
P (mm/s)	6.90	6.93	5.70	2.87
Efficiency	0.061	0.072	0.107	0.266

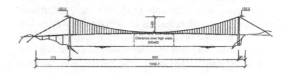

Figure 5. Suspension bridge.

Figure 6. Sketch of the bridge with additional controlled cables.

Table 3. Main results of the analysis of the suspension bridge.

| Mode | Frequency (Hz) | | Expected damping ratio (%) |
	Without controlled elements	With controlled elements	
1 antisymmetric	0.151	0.188	10.9
1 symmetric	0.196	0.216	4.9
2 symmetric	0.279	0.305	4.5
2 antisymmetric	0.309	0.388	11.3
3 symmetric	0.414	0.491	8.5
3 antisymmetric	0.522	0.573	4.7
4 symmetric	0.651	0.671	1.5

Table 4. Main results of the analysis of the cable-stayed bridge.

| Mode | Frequency (Hz) | | Damping ratio (%) |
	Without control	With control	
1 vertical	0.284	0.268	2.9
2 vertical	0.354	0.335	2.8
3 vertical	0.490	0.458	3.4
4 vertical	0.598	0.582	1.4
5 vertical	0.702	0.689	0.9

vortex shedding cables with only 12 strands · 5″ would certainly be sufficient.

Active control could therefore be an excellent method for retrofitting this kind of bridges and could be economically competitive compared with spoilers.

6.2 Cable-stayed bridge

Considering a cable-stayed bridge with a 530 span length, more than 3% structural damping on the first modes has been obtained by controlling only 2 * 3 pairs of the 4 * 26 pairs of cables of the complete bridge.

Expected damping values could be found in table 4. Here again, active control could offer easily additional damping, but with actuators located on stay cables. But applied to large bridges, the solution faces the size of the devices. The required actuators become big, the oil accumulators balancing the static force become very big, and it looks difficult to locate them in the bridge deck.

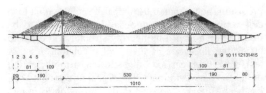

Figure 7. Cable stayed bridge.

6.3 Guyed mast

Considering a guyed mast composed of a high tower supporting a cantilever bridge at the end of the construction stage, the structure behavior could be easily controlled with only six cables composed of 31 strands · 5″ and equipped with the active control system. This solution was foreseen for the construction of the high towers of Grand Viaduc de Millau. Compared to the basic solution using static guys, the proposed solution was competitive and required less guys. It will offer more security and more comfort during construction. The expected damping should be higher than 10% for the first bending mode.

7 CONCLUSION

The proposed active control system is a technology applicable to Civil Engineering and able to increase largely structural damping of a complete structure. It is directly applicable by scaling up the devices.

A simplified theory for design is proposed. The expected additional damping ratio obtained by the theoretical system is in accordance with the experimental values and provides adequate parameters for the design of the structure and the active control unit.

The theoretical system takes into account economic aspects to optimize the dimensioning if the actuators, which is very important to carry out real structures.

Analyses have been carried out in order to take into account actuator failure, command saturation, load cell precision ...

Moreover the information provided by the force sensors used by the system could offer a continuous monitoring of the structure.

ACKNOWLEDGMENTS

The work reported is part of a research partly funded by the European Commission under the BriteEuram program (contract BRPR-CT97-0402). The authors gratefully acknowledge the contributions of the other partners of the consortium, Defense Evaluation Research

Agency (GB), Johs.Holt (NO), Joint Research Center of EC – Ispra (EC), Mannesmann Rexroth (D), Newlands Technology (GB), Technische Universität Dresden (D), Université Libre de Bruxelles (B) and VSL (F).

REFERENCES

ACE 1997, Active Control in civil Engineering, EC Brite-Euram Contract BRPR-CT97-0402.

Achkire Y. 1997, Active Tendon Control of Cable-Stayed Bridges, Ph.D. dissertation, Active Structures Laboratory, Université Libre de Bruxelles, Belgium.

Aupérin M. et al. 2001, Active Control in Civil Engineering: from Conception to Full Scale Applications, Journal of Structural Control, Vol. 8 N.2.

Aupérin M. et al. 2000, Structural Control: Point of View of a civil engineering company in the field of cable-supported structures, 3rd International Workshop on Structural Control, Paris.

Bossens F. et al. 2001, Active Tendon Control of Civil Structures: Theoretical and Experimental Study, XIX International Modal Analysis Conference IMAC, Orlando, Florida.

Preumont A. et al. 2000, Active Tendon Control of Vibration of Truss Structures, Theory and Experiments, Journal of Intelligent Material Systems and Structures, 11(2), pp 91–99.

System-based Vision for Strategic and Creative Design, Bontempi (ed.)
© *2003 Swets & Zeitlinger, Lisse, ISBN 90 5809 599 1*

Modal control for seismic structures using MR dampers

S.W. Cho & K.S. Park
Korea Advanced Institute of Science & Technology, Daejon, Korea

M.G. Ko
Kongju National University, Kongju, Korea

I.W. Lee
Korea Advanced Institute of Science & Technology, Daejon, Korea

ABSTRACT: This paper proposes the implementation of modal control for seismic structures using magnetorheological (MR) dampers. Numerous control algorithms have been adopted for semi-active systems including MR dampers. In spite of some advantages of each algorithm, there are not many differences in the performance. Among many control algorithms, modal control represents one control class, in which the motion of a structure is reshaped by merely controlling some selected vibration modes. Although modal control has been investigated for several decades, their potential for semi-active control, especially for MR dampers, has not yet been exploited. To improve the performance, modal control scheme is applied together with modal state observer and low-pass filter. Through the numerical example, the performance is compared with other control algorithms that are previously proposed. The numerical example considers a six-story structure controlled with MR dampers on the lower two floors. For the modal control, three cases of the structural measurements are considered; displacement, velocity and acceleration feedback. Using each structural measurement, Kalman filter estimates modal states. In simulation, an El Centro earthquake is used to excite the system, and the reduction in the drifts, accelerations, and relative displacements throughout the structure is examined. Simulation results indicate that the performance is improved by modal control scheme.

1 INTRODUCTION

Magnetorheological (MR) dampers are one of semi-active control devices, which use MR fluids to provide controllable damping forces. A number of control algorithms have been adopted for semi-active systems including the MR damper. Jansen and Dyke (2000) discussed recently proposed semi-active control algorithms, formulated these algorithms for use with MR dampers, and evaluated and compared the performance of each algorithm.

Modal control is especially desirable for the vibration control of civil engineering structure may involve hundred or even thousand degrees of freedom, its vibration is usually dominated by the first few modes. The purpose of this study is to implement modal control for seismically excited structures that use MR dampers and to compare the performance of the proposed method with that of other control algorithms previously studied. A modal control scheme

with a Kalman filter and a low-pass filter is applied. A Kalman filter is included in a control scheme to estimate modal states from measurements by sensors. Three cases of the structural measurement are considered by a Kalman filter to verify the effect of each measurement; displacement, velocity, and acceleration, respectively. Moreover, a low-pass filter is applied to eliminate the spillover problem.

2 MODAL CONTROL

Consider a seismically excited structure controlled with m MR dampers. In modal control, only a limited number of lower modes are controlled. Hence, l controlled modes can be selected with $l < n$ and the displacement may be partitioned into controlled and uncontrolled parts as $x(t) = x_C(t) + x_R(t)$, where x_C and x_R represent the controlled and uncontrolled displacement vector, respectively. We refer to the uncontrolled modes as residual. Considering orthogonal condition between eigenvectors, we obtain

$$\ddot{\eta}_r + 2\zeta_r \omega_r \dot{\eta}_r + \omega_r^2 \eta_r = \phi_r^T \Lambda f - \phi_r^T \Gamma \ddot{x}_g, \quad r = 1,2,...,l \tag{1}$$

where $\eta_r(t)$ is the r th modal displacement; ϕ_r is the rth eigenvector; Φ is a eigenvector set; η is a modal displacement vector; ζ_r are modal damping ratios; ω_r is a natural frequency; $f = [f_1, f_2, ..., f_m]^T$ is the vector of measured control forces generated by m MR dampers; x_g is ground acceleration; Γ is the column vector of ones; and Λ is the matrix determined by the placement of MR dampers in the structure. Then, Equation 1 can be rewritten in state-space form such as

$$\dot{w}_C(t) = A_C w_C(t) + B_C f(t) + E_C \ddot{x}_g$$
$$y_C(t) = C_C w_C(t) \tag{2}$$

where w_c is a $2l$-dimensional modal state vector by the controlled modes and

$$A_C = \begin{bmatrix} 0 & I_C \\ -\Omega_C^2 & -\Delta_C \end{bmatrix}, B_C = \begin{bmatrix} 0 \\ B'_C \end{bmatrix}, E_C = \begin{bmatrix} 0 \\ E'_C \end{bmatrix} \tag{3}$$

are the $2l \times 2l$, $2l \times m$ matrixes and a $2l \times 1$ vector, respectively, and Δ_C is the diagonal matrix listing $2\omega_r\zeta_r$; Ω_C^2 is the diagonal matrix listing $\omega_1^2, ..., \omega_n^2$; $B'_C = \Phi^T \Lambda$ and $E'_C = \Phi^T \Gamma$. For feedback control, the control vector is related to the modal state vector according to

$$f(t) = -K_C w_C(t) \tag{4}$$

where K_C is an $m \times 2l$ control gain matrix. Because the force generated in the i th MR damper depends on the responses of the structural system, the MR damper cannot always produce the desired optimal control force f_{Ci}. Thus, the strategy of a clipped-optimal control (Dyke et al. 1996) is used. Control gain matrix K_C should be decided. Although a variety of approaches may be used to design the optimal controller, $H2/LQG$ (Linear Quadratic Gaussian) methods are advocated because of their successful application in previous studies (Dyke et al. 1996).

3 MODAL STATE ESTIMATION

To estimate the modal state vector $w_C(t)$ from the measured output $y(t)$, we consider a Kalman-Bucy filter as an observer (Meirovitch, 1990). Not only the state feedback including velocities or displacements is considered, but also the acceleration feedback is implemented for the modal state estimation using a Kalman-Buch filter. In any case, we can write a modal observer in the form

$$\dot{\hat{w}}_C(t) = (A_C - B_C K_C)\hat{w}_C(t) + LC_C(w_C(t) - \hat{w}_C(t))$$
$$+ LC_R w_R(t) + E_C \ddot{x}_g \tag{5}$$

where $w_C(t)$ is the estimated controlled modal state and L is the optimally chosen observer gain matrix by solving a matrix Riccati equation, which assumes that the noise intensities associated with earthquake and sensors are known. C_C is changeable according to the signals which are used for the feedback and D_C is generally zero except the acceleration feedback. The error vector is defined such as $e_C(t) = w_C(t) - w_C(t)$. Then the equations can be written in the matrix form

$$\begin{bmatrix} \dot{w}_C(t) \\ \dot{w}_R(t) \\ \dot{e}_C(t) \end{bmatrix} = \begin{bmatrix} A_C - B_C K_C & 0 & -B_C K_C \\ -B_R K_C & A_R & -B_R K_C \\ 0 & LC_R & A_C - LC_C \end{bmatrix} \begin{bmatrix} w_C(t) \\ w_R(t) \\ e_C(t) \end{bmatrix}$$
$$+ \begin{bmatrix} E_C \\ E_R \\ E_C \end{bmatrix} \ddot{x}_g \tag{6}$$

Note that the term $-B_R K_C$ in Equation 6 is responsible for the excitation of the residual modes by the control forces and is known as control spilloveralas. If C_R is zeros, which means the sensor signal only include controlled modes, the term $-B_R K_C$ has no effect on the eigenvalues of the closed-loop system. Hence, we conclude that control spillover cannot destabilize the system, although it can cause some degradation in the system performance. Normally, however, the above system can not satisfy the separate principle because the term LC_R affects eigenvalues of the controlled system by the observer. This effect is known as observation spillover and can produce instability in the residual modes. However, a small amount of damping inherent in the structure is often sufficient to overcome the observation spillover effect. At any rate, observation spillover can be eliminated if the sensor signals are prefiltered so as to screen out the contribution of the uncontrolled modes (Meirovitch 1990).

4 NUMERICAL EXAMPLE

To evaluate the proposed modal control scheme for use with the MR damper, a numerical example is considered in which a model of a six-story building is controlled with four MR dampers (Fig. 1).

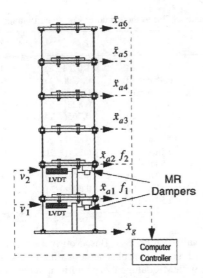

Figure 1. Schematic diagram (Jansen & Dyke 2000).

Table 1.* Normalized controlled maximum responses due to the scaled El Centro earthquake.

Control strategy	J_1	J_2	J_3	J_4
Passive-off	0.862	0.801	0.904	0.00292
Passive-on	0.506	0.696	1.41	0.0178
Lyapunov controller A	0.686 (+35)	0.788 (+13)	0.756 (−16)	0.0178
Lyapunov controller A	0.326 (−35)	0.548 (−21)	1.39 (+53)	0.0178
Decentralized bang-bang	0.449 (−11)	0.791 (+13)	1.00 (+11)	0.0178
Maximum energy dissipation	0.548 (+8)	0.620 (−11)	1.06 (+17)	0.0121
Clipped-optimal A	0.631 (+24)	0.640 (−8)	0.636 (−29)	0.01095
Clipped-optimal B	0.405 (−20)	0.547 (−21)	1.25 (+38)	0.0178
Modified homogeneous friction	0.421 (−17)	0.599 (−20)	1.06 (+17)	0.0178

*(Jansen and Dyke 2000)

This numerical example is the same with that of Jansen and Dyke (2000) and is adopted for direct comparisons between the proposed modal control scheme and other control algorithms. In simulation, the model of the structure is subjected to the NS component of the 1940 El Centro earthquake. Because the building system considered is a scaled model, the amplitude of the earthquake was scaled to ten percent of the full-scale earthquake. The various control algorithms were evaluated using a set of evaluation criteria based on those used in the second generation linear control problem for buildings (Spencer & Sain 1997) such as

$$J_1 = \max_{t,i}\left(\frac{|x_i(t)|}{x^{max}}\right), \quad J_2 = \max_{t,i}\left(\frac{|d_i(t)/h_i|}{d_n^{max}}\right)$$
$$J_3 = \max_{t,i}\left(\frac{|\ddot{x}_{ai}(t)|}{\ddot{x}_a^{max}}\right), \quad J_4 = \max_{t,i}\left(\frac{|f_i(t)|}{W}\right) \quad (7)$$

The resulting evaluation criteria are presented in Table 1 for the control algorithms previously studied (Jansen & Dyke 2000). The numbers in parentheses indicate the percent reduction as compared to the best passive case. To compare the performance of the semi-active system to that of comparable passive systems, two cases are considered in which MR dampers are used in a passive mode by maintaining a constant voltage to the devices. The results of passive-off (0 V) and passive-on (5 V) configurations are included.

For modal control, three cases of the structural measurements are considered; displacements, velocities and accelerations. Using each structural measurement, a Kalman filter estimates the modal states.

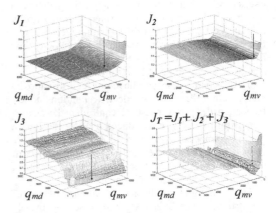

Figure 2. Variations of evaluation criteria with weighting parameters for the acceleration feedback.

Figure 2 represents the variations of each evaluation criteria for increasing weighting parameters in a 3-dimensional plot. J_T is the summation of evaluation criteria, J_1, J_2 and J_3. We can find the weighting for reduction of overall structural responses from the variations of J_T, whereas we can find the weighting for reduction of related responses from J_1, J_2 and J_3, Designer can decide which to use according to control objectives. By using the controller (H2/LQG) with designed weighting matrices from (Fig. 2), we can get the results in Table 2. Similarly, for the displacement and velocity feedback cases, Table 2 summarize

Table 2. Normalized controlled maximum responses of the various feedback due to the scaled El Centro earthquake.

Control strategy	Weighting parameters	J_1	J_2	J_3	J_4
Modal control A_{J1}	qmd = 400, qmv = 1500	0.310(−39)	0.529(−24)	1.07(+18)	0.0178
Modal control A_{J2}	qmd = 1, qmv = 500	0.398(−21)	0.485(−30)	0.870(−4)	0.0178
Modal control A_{J3}	qmd = 2200, qmv = 100	0.549(+8)	0.618(−11)	0.697(−23)	0.0176
Modal control A_{JT}	qmd = 500, qmv = 600	0.380(−25)	0.488(−30)	0.823(−9)	0.0178
Modal control D_{J1}	qmd = 100, qmv = 4900	0.403(−20)	0.560(−20)	0.765(−15)	0.0177
Modal control D_{J2}	qmd = 100, qmv = 4900	0.403(−20)	0.560(−20)	0.769(−15)	0.0178
Modal control D_{J3}	qmd = 200, qmv = 4900	0.702(+39)	0.728(+5)	0.671(−26)	0.0178
Modal control D_{JT}	qmd = 3300, qmv = 4700	0.408(−19)	0.566(−19)	0.721(−20)	0.0178
Modal control V_{J1}	qmd = 700, qmv = 800	0.327(−35)	0.554(−20)	1.06(+17)	0.0178
Modal control V_{J2}	qmd = 1, qmv = 400	0.383(−24)	0.487(−30)	0.874(−3)	0.0177
Modal control V_{J3}	qmd = 1300, qmv = 100	0.541(+7)	0.611(−12)	0.632(−3)	0.0178
Modal control V_{JT}	qmd = 600, qmv = 500	0.354(−30)	0.502(−28)	0.825(−9)	0.0176

the results for each minimum evaluation criteria of the designed weighting matrices.

5 CONCLUSIONS

In this paper, modal control was implemented to seismically excited structures using MR dampers. To this end, a modal control scheme was applied together with a Kalman filter and a low-pass filter. A Kalman filter considered three cases of the structural measurement to estimate modal states: displacement, velocity, and acceleration, respectively. Moreover, a low-pass filter was used to eliminate spillover problem.

Modal control reshapes the motion of a structure by merely controlling a few selected vibration modes. Hence, in designing phase of controller, the size of weighting matrix Q was reduced because the lowest one or two modes were controlled. Therefore, it is more convenient to design the smaller weighting matrix of modal control. This is one of the important benefits of the proposed modal control scheme.

The numerical results show that the motion of the structure was effectively suppressed by merely controlling a few lowest modes, although resulting responses varied greatly depending on the choice of measurements available and weightings. The modal controller A and V achieved significant reductions in the responses. The modal controller A_{J1}, A_{J2} and V_{J3} achieve reductions (39%, 30%, 30%) in evaluation criteria J_1, J_2 and J_3, respectively, resulting in the lowest values of all cases considered here. The modal controller A_{JT} and V_{JT} fail to achieve any lowest value of evaluation criteria, but have competitive performance

in all evaluation criteria. Based on these results, the proposed modal control scheme is found to be suited for use with MR dampers in a multi-input control system. Further studies are underway to examine the influence of the number of controlled modes on the control performance.

ACKNOWLEDGEMENTS

This research was supported by the National Research Laboratory Grant (No.: 2000-N-NL-01-C-251) in Korea. The financial support is gratefully acknowledged.

REFERENCES

Dyke, S.J., Spencer, Jr., B.F., Sain, M.K. & Carlson, J.D. 1996. Modeling and Control of Magnetorheological Dampers for Seismic Response Reduction, *Smart Materials and Structures*, Vol. 5, pp. 565–75.

Jansen, L.M. & Dyke, S.J. 2000. Semi-active Control Strategies for MR Dampers: Comparative Study, *Journal of Engineering Mechanics*, Vol. 126, No. 8, pp. 795–803.

Meirovitch, L. 1990. *Dynamics and Control of Structures*, John Wiley & Sons.

Sack, R.L., Kuo, C.C., Wu, H.C., Liu, L. & Patten, W.N. 1994. Seismic Motion Control via Semiactive Hydraulic Actuators. *Proc. of the U.S. Fifth National Conference on Earthquake Engineering, Chicago*, Illinois, Vol. 2, pp. 311–20.

Spencer, Jr., B.F., Dyke, S.J., Sain, M.K., & Carlson, J.D. 1997. Phenomenological Model off Magnetorheological damper, *Journal of Engrg. Mech., ASCE*, 123(3), pp. 230–238.

New development of semi-active control with variable stiffness

F. Amini & F. Danesh
Civil Eng. Dept., IUST, Tehran, Iran

ABSTRACT: In the last two decades active variable stiffness method has been successfully used to control the structures. In recent years semi-active control method has mostly been studied because of its low cost and reliability. The main objective of this study is to obtain required practical stiffness for any selected short period of time to control the behavior of structure so that displacements and accelerations do not exceed the allowable values. The semi-active system proposed herein is based on the orderly iteration method (OIM) by changing stiffness of structures. On the other hand it is assumed that the number of bracings which can be active automatically is limited. The feasibility of the method is shown by a ten-storey frame structure.

1 INTRODUCTION

In recent years, semi-active control techniques and methods have relentlessly developed in relation to those of active control, because it is more reliable and low cost. In semi-active control systems variation of structure dynamic characters such as damping or stiffness are obtained with no external forces to control the behavior of structures.

Previously several methods of semi-active control such as "Pole Assignment" (Amini 1996, Amini 2002) and "Optimal Control" (Amini 1996) have been presented, the main objective of this paper is to obtain a simple and precise method to determine a practical stiffness matrix in any short period of time and to control responses of structure such as displacements and accelerations not to exceed allowable values.

In this study practical changing of stiffness matrix is considered by orderly iteration method to control responses of structures. The procedure is capable of adding and removing limited number of bracings on stories automatically.

2 ORDERLY ITERATION METHOD (OIM)

Dynamic equation of a structure is shown as follows:

$$M.X^{\circ\circ} + C.X^{\circ} + K.X = f(t) \tag{1}$$

in which M, C, and K are mass, damping, and stiffness matrices respectively and f(t) is external force vector exerted on the stories. Another form of this equation in the state space takes form as follows:

$$q^{\circ} = A.q + B.f(t) \tag{2}$$

where

$$A = \begin{bmatrix} 0 & I_n \\ -M^{-1}K & -M^{-1}C \end{bmatrix} \qquad B = \begin{bmatrix} 0 \\ M \end{bmatrix}$$

$$q = \begin{Bmatrix} X \\ X^{\circ} \end{Bmatrix} \qquad q^{\circ} = \begin{Bmatrix} X^{\circ} \\ X^{\circ\circ} \end{Bmatrix}$$

State equation is a linear differential equilibrium of type I with 2n degree of freedom.

Since control process is applied for any short period of time during an earthquake, accelerograph is divided into time steps (ΔT) and control procedure runs over any of the short time steps. It is assumed that the forces of stories are stationary in any of the time steps.

$$P_0 = \begin{Bmatrix} P_1 \\ \vdots \\ P_n \end{Bmatrix}$$

Now if unknown stiffness matrix that is K^*, fulfils all the allowable conditions, the state equation will be changed through the time step as follows:

$$q^\circ_{al} = A^* . q_{al} + B.P_0 \tag{3}$$

where

$$A^* = \left[\begin{array}{c|c} 0 & I_n \\ \hline -M^{-1}K^* & -M^{-1}C \end{array} \right]$$

$$q_{al} = \left\{ \begin{array}{c} X_{al} \\ X^\circ_{al} \end{array} \right\} \qquad q^\circ_{al} = \left\{ \begin{array}{c} X^\circ_{al} \\ X^{\circ\circ}_{al} \end{array} \right\}$$

where X_{al}, X°_{al} and $X^{\circ\circ}_{al}$ are allowable displacements, velocities, and accelerations respectively.

The state equation before the control procedure in the time step is shown in Equation 4 as follows:

$$q^\circ = A^* . q + B.P_0 \tag{4}$$

Since velocities are not controlled, they are assumed to be equal to the allowable values.

$$\left\{ X^\circ_{al} \right\} = \left\{ X^\circ \right\} \tag{5}$$

For shear dominant structures, K^* will be

$$K^* = \begin{bmatrix} K_1^* & -K_1^* & & & 0 \\ -K_1^* & K_1^*+K_2^* & \ddots & & \\ & & \ddots & \ddots & \\ & & & \ddots & -K_{n-1}^* \\ 0 & & & -K_{n-1}^* & K_{n-1}^*+K_n^* \end{bmatrix} \tag{6}$$

where $K_1,...,K_n$ are stiffness components. It is assumed

$$\left[K^* \right]\left\{ X_{al} \right\} = \left[X_{al} \right]\left\{ K^* \right\} \tag{7}$$

where $X_{al1},...,X_{aln}$ are allowable displacements of stories.

Combination of Equations 3 and 4 is simplified by the Equation 5 and results in:

$$\left[X_{al} \right]\left\{ K^* \right\} = \left[K \right]\left\{ X \right\} - \left[M \right]\left\{ \left\{ X^{\circ\circ}_{al} \right\} - \left\{ X^{\circ\circ} \right\} \right\} \tag{8}$$

$$\left[X_{al} \right] = \begin{bmatrix} X_{al_2} - X_{al_1} & & & 0 \\ X_{al_2} - X_{al_1} & X_{al_2} - X_{al_3} & & \\ & X_{al_3} - X_{al_2} & X_{al_3} - X_{al_4} & \\ & & \ddots & \\ 0 & & X_{al_{n-1}} - X_{al_{n-2}} & X_{al_{n-1}} - X_{al_n} \end{bmatrix}$$

This equilibrium is effectively solved by the transformation 6 and result will be:

$$\left\{ K^* \right\} = \left[X_{al} \right]^{-1}\left[K \right]\left\{ X \right\} - \left[X_{al} \right]^{-1}\left[M \right]\left\{ \left\{ X^{\circ\circ}_{al} \right\} - \left\{ X^{\circ\circ} \right\} \right\} \tag{9}$$

Stiffness components are imbedded in Equation 6 to reach the required stiffness matrix.

A semi-active control which uses this method, requires discrete values of stiffness to add or remove bracings, therefore it is necessary that the stiffness values get rounded to the closest existing and actual value. The control method used throughout this paper uses an iteration with stiffness matrix feedback to eliminate errors of discrete stiffness values and also the assumption of stationary forces on stories. This procedure is a new semi-active control algorithm with variable stiffness which is called "OIM".

3 NUMERICAL EXAMPLE

Two span braced frame structure with ten stories are designed which are capable of removing or adding up to two bracings in another span. In this example accelerograph of "El Centro" simulates the earthquake ground acceleration, and if accelerations or displacements of frames exceed the ultimate values in any time step, control procedure calculates required stiffness matrix based on allowable values.

The ten-storey braced frame is illustrated in Figure 1 and Table 1. Maximum absolute accelerations and displacements are shown in Table 2.

Table 3 shows maximum absolute values after exerting the following control conditions.

1. Ultimate acceleration value: $17\,\text{m/s}^2$
2. Allowable acceleration value: $15\,\text{m/s}^2$
3. Ultimate displacement value: Storey Elevation/ $100\,\text{m}$
4. Allowable displacement value: Storey Elevation/ $500\,\text{m}$
5. Time step: $\Delta T = 0.65$ sec

Pattern of bracings during earthquake is illustrated in Table 4 and Figure 2.

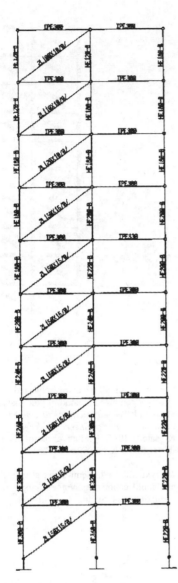

Figure 1. Ten-storey braced frame.

Table 1. Bracings of the ten-storey frame.

	Bracing
1st storey	2L100 × 10
2nd storey	2L110 × 10
3rd storey	2L120 × 10
4th storey	2L150 × 10
5th storey	2L150 × 10
6th storey	2L150 × 10
7th storey	2L150 × 10
8th storey	2L150 × 10
9th storey	2L150 × 10
10th storey	2L150 × 10

Table 2. Maximum absolute accelerations and displacements.

	Maximum absolute accelerations (m/s²)	Maximum absolute displacements (m)
1st storey	19.3	0.110
2nd storey	16.1	0.105
3rd storey	13.9	0.095
4th storey	13.9	0.080
5th storey	13.2	0.075
6th storey	12.0	0.065
7th storey	10.8	0.050
8th storey	9.7	0.040
9th storey	8.3	0.025
10th storey	7.2	0.015

Table 3. Maximum absolute accelerations and displacements after control.

	Maximum absolute accelerations (m/s²)	Maximum absolute displacements (m)
1st storey	17.2	0.075
2nd storey	16.5	0.065
3rd storey	14.8	0.058
4th storey	13.2	0.053
5th storey	13.0	0.044
6th storey	12.9	0.041
7th storey	11.7	0.032
8th storey	8.4	0.026
9th storey	7.9	0.017
10th storey	6.1	0.008

Table 4. Pattern of bracings during earthquake.

	0 S < t < 1.44 S	1.44 S < t < 2.52 S	2.52 S < t
1st storey	2L100 × 10	L100 × 10	L100 × 10
2nd storey	2L110 × 10	2L110 × 10 + L100 × 10	2L110 × 10 + L100 × 10
3rd storey	2L120 × 10	2L120 × 10 + 2L100 × 10	2L120 × 10 + 2L100 × 10
4th storey	2L150 × 10	2L150 × 10 + L100 × 10	2L150 × 10 + L100 × 10
5th storey	2L150 × 10	2L150 × 10 + 2L100 × 10	2L150 × 10 + L100 × 10
6th storey	2L150 × 10	2L150 × 10 + 2L100 × 10	2L150 × 10 + 2L100 × 10
7th storey	2L150 × 10	2L150 × 10 + 2L100 × 10	2L150 × 10 + L100 × 10
8th storey	2L150 × 10	2L150 × 10 + 2L100 × 10	2L150 × 10 + 2L100 × 10
9th storey	2L150 × 10	2L150 × 10 + 2L100 × 10	2L150 × 10 + 2L100 × 10
10th storey	2L150 × 10	2L150 × 10 + 2L100 × 10	2L150 × 10 + 2L100 × 10

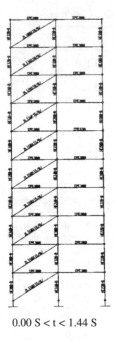 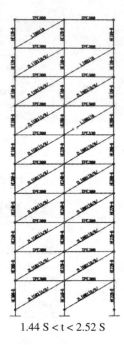 

| 0.00 S < t < 1.44 S | 1.44 S < t < 2.52 S | 2.52 S < t |

Figure 2. Pattern of braced frame during earthquake.

4 CONCLUSION

The proposed algorithm is very effective in controlling both maximum accelerations and displacements through changing the stiffness matrix according to the earthquake. It is more probable to control the responses of structure with one single loop of iteration than two or more.

The OIM is an applicable method which is more precise and straightforward rather than the other similar methods.

REFERENCES

Amini F. 2002. Optimal active control of structures by pole assignment method, *Proc. high performance structures and composites*: 419–425. Spain.

Amini F. 1996. Response of tall structures subjected to earthquake using a new combined method of pole assignment and optimal control, *Second international conference on seismology and earthquake engineering (SEE-2)*. Iran.

Nagaraiah S. & Mate D. 1996. Semi-ative control of continuously variable stiffness system, *Struct. Control*: 397–405. Spain.

Spencer B.F. & Dyke S.F. 1996. Semi-active structural control, System Identification for Synthesis and Analysis, *Struct. Control*: 568–584. Spain.

System-based Vision for Strategic and Creative Design, Bontempi (ed.)
© *2003 Swets & Zeitlinger, Lisse, ISBN 90 5809 599 1*

Reducing damages in structures during earthquake using energy dissipation devices

K.K.F. Wong & D.F. Zhao
School of Civil and Environmental Engineering, Nanyang Technological University, Singapore

ABSTRACT: A computational algorithm of inelastic structural behavior is used to quantitatively calculate the plastic energy of moment-resisting framed structures. This algorithm uses the force analogy method to evaluate the structural response and energy in the inelastic domain subjected to earthquake excitations. The advantage of using force analogy method is that it directly gives a closed-form analytical equation to evaluate plastic energy due to inelastic deformation in the structure. Energy dissipation devices such as viscous dampers, viscoelastic dampers, and active control are then used to improve the structural performance by providing an alternative path of energy dissipation through the devices and therefore the plastic energy dissipation required in the structure will be reduced. Results show that these energy dissipation devices are very effective in dissipating control energy, and thereby reducing the plastic energy dissipation and damage in the structure.

1 INTRODUCTION

Earthquake hazard mitigation is one of the most important issues facing structural engineers today. Recent destructive seismic events have demonstrated the importance of mitigating earthquake hazards in both the design of new structures and retrofit of existing structures. Since 1970s, the concept of structural control has been advancing rapidly with the need for adaptability, increased flexibility and safety levels, better utilization of material, and lower cost (Housner et al. 1997). Structural control offers a promising alternative to protect structures by applying passive and/or active control force while maintaining desirable structural dynamic properties. Structural control can be classified as four types: (1) passive control, (2) active control, (3) semi-active control, and (4) hybrid control. Much research works have been done on analyzing the structural responses using different control devices, and these devices include viscous dampers (Makris et al. 1991; Soong & Dargush 1997), viscoelastic dampers (Zhang et al. 1989; Zhang & Soong 1992), active control (Yang et al. 1987; Soong 1990; Wong & Yang 2001) and etc.

It is well known that using structural control devices will dissipate energy during the vibration process. However, none of the published works has ever quantitatively calculated the actual amount of energy dissipation in these devices and in the structure as a whole. Therefore, the objective of this research is to propose a simple analytical method of calculating the seismic energy dissipation of inelastic structures when control devices are installed. This algorithm uses the force analogy method (FAM) to perform inelastic structural analysis, and this makes the evaluation of plastic energy due to inelastic deformation in the structure very simple, as the accumulation of plastic energy over the entire duration of the earthquake can be derived in a closed analytical form. Energy dissipation devices such as viscous dampers, viscoelastic dampers, and active control are then used to improve the structural performance. Results show that these energy dissipation devices are very effective in dissipating control energy, and thereby reducing the plastic energy dissipation and damage in the structure.

2 FORCE ANALOGY METHOD

The concept of FAM has been discussed in Wong & Yang (1999). The important equations for an n-degree of freedom system with m plastic hinges in a moment-resisting frame are summarized as follows.

2.1 *Total displacement equation*

Total displacement is equal to elastic displacement plus inelastic displacement, i.e.,

$$\mathbf{x}(t) = \mathbf{x}'(t) + \mathbf{x}''(t) \tag{1}$$

where $\mathbf{x}(t)$ is the total displacement vector, $\mathbf{x}'(t)$ is the elastic displacement vector, and $\mathbf{x}''(t)$ is the inelastic vector.

2.2 Elastic moment–displacement relationship

Elastic moment is related to the elastic displacement, and the relationship is:

$$\mathbf{M}'(t) = \mathbf{K}_P^T \mathbf{x}'(t) \qquad (2)$$

where $\mathbf{M}'(t)$ is the elastic moment vector and $\mathbf{K}_P$ is the $n \times m$ matrix that relates the plastic rotation vector with the restoring forces at the structural degrees of freedom (DOFs).

2.3 Governing equations

The governing equations of the FAM are:

$$\mathbf{M}(t) + \mathbf{K}_R \Theta''(t) = \mathbf{K}_P^T \mathbf{x}(t) \qquad (3)$$

$$\mathbf{x}''(t) = \mathbf{K}^{-1} \mathbf{K}_P \Theta''(t) \qquad (4)$$

where $\mathbf{M}(t)$ is the total moment vector, $\Theta''(t)$ is the plastic rotation vector, $\mathbf{K}$ is the $n \times n$ global stiffness matrix, and $\mathbf{K}_R$ is the $m \times m$ matrix that relates the plastic rotation vector with the restoring moments at the plastic hinge locations (PHLs).

3 STATE SPACE RESPONSE

Consider the dynamic equilibrium equation, where the elastic force is the stiffness multiplied by the elastic displacement:

$$\mathbf{M}\ddot{\mathbf{x}}(t) + \mathbf{C}\dot{\mathbf{x}}(t) + \mathbf{K}\mathbf{x}'(t) = -\mathbf{M}\ddot{\mathbf{x}}_e(t) + \mathbf{D}\mathbf{f}_c(t) \qquad (5)$$

where $\mathbf{M}$ is the mass matrix, $\mathbf{C}$ is the damping matrix, $\dot{\mathbf{x}}(t)$ is the relative velocity, $\ddot{\mathbf{x}}(t)$ is the relative acceleration, $\mathbf{f}_c(t)$ is the $p \times 1$ lateral control force vector, $\mathbf{D}$ is the $n \times p$ distribution matrix that relates the effect of control force with each DOF, $\ddot{\mathbf{x}}_e(t)$ is the earthquake ground acceleration and p is the number of dampers or actuators in the system. Solving for the elastic displacement in Equation 1 and substituting the result into Equation 5 gives

$$\mathbf{M}\ddot{\mathbf{x}}(t) + \mathbf{C}\dot{\mathbf{x}}(t) + \mathbf{K}\mathbf{x}(t) = -\mathbf{M}\ddot{\mathbf{x}}_e + \mathbf{D}\mathbf{f}_c + \mathbf{K}\mathbf{x}''(t) \qquad (6)$$

Representing Equation 6 in state space form gives

$$\dot{\mathbf{z}}(t) = \mathbf{A}\mathbf{z}(t) + \mathbf{H}\mathbf{a}(t) + \mathbf{B}\mathbf{f}_c(t) + \mathbf{F}_p^c \mathbf{x}''(t) \qquad (7)$$

where

$$\mathbf{z}(t) = \begin{Bmatrix} \mathbf{x}(t) \\ \dot{\mathbf{x}}(t) \end{Bmatrix} , \quad \mathbf{A} = \begin{bmatrix} \mathbf{0} & \mathbf{I} \\ -\mathbf{M}^{-1}\mathbf{K} & -\mathbf{M}^{-1}\mathbf{C} \end{bmatrix}$$

$$\mathbf{H} = \begin{bmatrix} \mathbf{0} \\ -\mathbf{1} \end{bmatrix} , \quad \mathbf{B} = \begin{bmatrix} \mathbf{0} \\ \mathbf{M}^{-1}\mathbf{D} \end{bmatrix} , \quad \mathbf{F}_p^c = \begin{bmatrix} \mathbf{0} \\ \mathbf{M}^{-1}\mathbf{K} \end{bmatrix}$$

and $\mathbf{0}$ is the zero matrix, $\mathbf{I}$ is the identity matrix, and $-\mathbf{1}$ is the matrix with either 0s or -1s in all entries. The matrix $\mathbf{H}$ contains 6 columns corresponding to the ground acceleration in 6 different directions, and $\mathbf{a}(t)$ is the 6×1 earthquake ground acceleration vector related to $\ddot{\mathbf{x}}_e(t)$. The solution to Equation 7 is

$$\mathbf{z}(t) = e^{\mathbf{A}(t - t_o)} \mathbf{z}(t_o)$$
$$+ e^{\mathbf{A}t} \int_{t_o}^{t} e^{-\mathbf{A}s} \left[\mathbf{H}\mathbf{a}(s) + \mathbf{B}\mathbf{f}_c(s) + \mathbf{F}_p^c \mathbf{x}''(s) \right] ds \qquad (8)$$

Let $t_{k+1} = t$, $t_k = t_0$ and $\Delta t = t - t_0$, and represent the ground acceleration and inelastic displacement using pulses within the small time step Δt. Performing the integration in Equation 8 gives

$$\mathbf{z}_{k+1} = \mathbf{F}_s \mathbf{z}_k + \mathbf{H}_d \mathbf{a}_k + \mathbf{G}\mathbf{f}_{ck} + \mathbf{F}_p \mathbf{x}_k'' \qquad (9)$$

where

$$\mathbf{F}_s = e^{\mathbf{A}\Delta t} , \quad \mathbf{H}_d = e^{\mathbf{A}\Delta t} \mathbf{H}\Delta t$$
$$\mathbf{G} = \mathbf{A}^{-1}\left(e^{\mathbf{A}\Delta t} - \mathbf{I}\right)\mathbf{B} , \quad \mathbf{F}_p = e^{\mathbf{A}\Delta t} \mathbf{F}_p^c \Delta t \qquad (10)$$

and $\mathbf{a}_k$, $\mathbf{f}_{ck}$, and $\mathbf{x}_k''$ are the discretized forms of $\mathbf{a}(t)$, $\mathbf{f}_c(t)$, and $\mathbf{x}''(t)$, respectively. Equation 9 is a recursive equation that will be used in the current time history analysis.

The control force $\mathbf{f}_{ck}$ in Equation 9 represents the force exerted on the structure from the structural control devices. When linear passive viscous dampers are used, this control force follows the relationship:

$$\mathbf{f}_{ck} = \mathbf{C}_o \dot{\mathbf{x}} \qquad (11)$$

where C_0 is the viscous damping matrix which depends on the property and location of the dampers. When linear passive viscoelastic dampers are used, then the control force can be written as

$$\mathbf{f}_{ck} = \mathbf{C}_1 \dot{\mathbf{x}} + \mathbf{C}_2 \mathbf{x} \qquad (12)$$

where $\mathbf{C}_1$ and $\mathbf{C}_2$ are the viscoelastic damping matrices which again depend on the property and location of the dampers. Finally when active control with optimal linear control algorithm is used, then the control force can be obtained by first defining a cost function J to be

$$J = \frac{1}{2} \sum_{k=0}^{N} \left(\mathbf{z}_k^T \mathbf{Q}\mathbf{z}_k + \mathbf{f}_{ck}^T \mathbf{R}\mathbf{f}_{ck} \right) \qquad (13)$$

where $\mathbf{Q}$ and $\mathbf{R}$ are the weighting matrices. Minimizing this cost function subject to the constraint given in

Equation 9, it follows that

$$\mathbf{f}_{ck} = -\left(\mathbf{G}^T\mathbf{P}\mathbf{G} + \mathbf{R}\right)^{-1}\mathbf{G}^T\mathbf{P}\mathbf{F}_s\mathbf{z}_k \qquad (14)$$

where

$$\mathbf{P} = \mathbf{Q} + \mathbf{F}_s^T\mathbf{P}\left(\mathbf{I} + \mathbf{G}\mathbf{R}^{-1}\mathbf{G}^T\mathbf{P}\right)^{-1}\mathbf{F}_s \qquad (15)$$

4 ENERGY BALANCE

Once the response time histories are known, the variation of energy with time can be calculated. The energy equation can be derived based on Equation 5. Define the absolute acceleration $\ddot{\mathbf{y}}(t)$ to be $\ddot{\mathbf{y}}(t) = \ddot{\mathbf{x}}(t) + \ddot{\mathbf{x}}_e(t)$. It follows from Equation 5 that

$$\mathbf{M}\ddot{\mathbf{y}}(t) + \mathbf{C}\dot{\mathbf{x}}(t) + \mathbf{K}\mathbf{x}'(t) - \mathbf{D}\mathbf{f}_c(t) = 0 \qquad (16)$$

Integrating both sides of Equation 16 over the path of structural response from time zero to t_k gives

$$\int_{t=0}^{t=t_k} \ddot{\mathbf{y}}^T\mathbf{M}d\mathbf{x} + \int_{t=0}^{t=t_k} \dot{\mathbf{x}}^T\mathbf{C}d\mathbf{x}$$
$$+ \int_{t=0}^{t=t_k} \mathbf{x}'^T\mathbf{K}d\mathbf{x} - \int_{t=0}^{t=t_k} \mathbf{f}_c^T\mathbf{D}^T d\mathbf{x} = 0 \qquad (17)$$

Since $d\mathbf{x}(t) = d\mathbf{y}(t) - d\mathbf{x}_e(t)$ where $\mathbf{y}(t)$ is the absolute displacement vector and $\mathbf{x}_e(t)$ is the ground displacement vector, substituting this result into the first term of Equation 17 gives

$$\int_0^{t_k} \ddot{\mathbf{y}}^T\mathbf{M}d\mathbf{y} + \int_0^{t_k} \dot{\mathbf{x}}^T\mathbf{C}d\mathbf{x}$$
$$+ \int_0^{t_k} \mathbf{x}'^T\mathbf{K}d\mathbf{x} - \int_0^{t_k} \mathbf{f}_c^T\mathbf{D}^T d\mathbf{x} = \int_0^{t_k} \ddot{\mathbf{y}}^T\mathbf{M}d\mathbf{x}_e \qquad (18)$$

In addition, since $d\mathbf{x}(t) = d\mathbf{x}'(t) + d\mathbf{x}''(t)$ from Equation 1, substituting this result into the third term of Equation 18 gives

$$\int_0^{t_k} \ddot{\mathbf{y}}^T\mathbf{M}d\mathbf{y} + \int_0^{t_k} \dot{\mathbf{x}}^T\mathbf{C}d\mathbf{x} + \int_0^{t_k} \mathbf{x}'^T\mathbf{K}d\mathbf{x}'$$
$$+ \int_0^{t_k} \mathbf{x}'^T\mathbf{K}d\mathbf{x}'' - \int_0^{t_k} \mathbf{f}_c^T\mathbf{D}^T d\mathbf{x} = \int_0^{t_k} \ddot{\mathbf{y}}^T\mathbf{M}d\mathbf{x}_e \qquad (19)$$

The first three terms on the left side of Equation 19 are the kinetic energy (KE), damping energy (DE), and strain energy (SE) of the system, respectively. The fourth term on the left side of Equation 19 represents the plastic energy (PE) of the structure, and it can be simplified by using FAM equations. Substituting Equations 2 and 4 into this term give

$$PE = \int_{t=0}^{t=t_k} \mathbf{x}'^T\mathbf{K}d\mathbf{x}'' = \int_{t=0}^{t=t_k} \mathbf{M}'^T d\Theta'' = \sum_{i=1}^m PE_i \qquad (20)$$

where PE_i represents the plastic energy dissipated at the ith plastic hinge, i.e.,

$$PE_i = \int_{t=0}^{t=t_k} M_i' d\theta_i'' \qquad (21)$$

The fifth term on the left side of Equation 19 represents the control energy (CE), and the term on the right side represents the input energy (IE) due to earthquake excitation. In summary, Equation 17 can be written as

$$KE + DE + SE + PE + CE = IE \qquad (22)$$

Further discussion on each of these energy forms can be found in Wong and Yang (2002).

5 NUMERICAL EXAMPLE

A six-story moment-resisting steel frame is used to study the effect of active control on the response of multi-degree of freedom structure subjected to earthquake excitations. The masses of each floor are assumed to be 300,000 kg, while 3% damping is used for all six modes of vibration. Figure 1 shows the structural model with a total of 40 potential plastic hinges (labeled from PHL #1 to PHL #40). The plastic hinges in the beams and columns are assumed to deform in an elastic–plastic behavior with yield moments computed based on the plastic section modulus of the members multiplied by the yield stress of steel of 248.2 MPa. Assume that all beams are subjected to a gravity load of 21.89 kN/m, and the interaction between axial force and moment in the columns is ignored.

First consider the case where no control is used. Subjected to the 1995 Kobe earthquake ground motion, Table 1 summarizes the maximum global responses

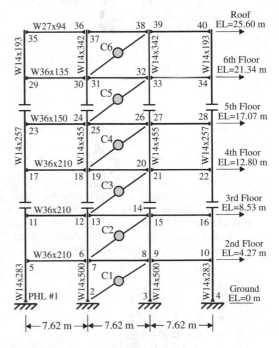

Figure 1. Six story moment-resisting frame.

Table 1. Maximum global responses due to Kobe earthquake.

Floor/Type	UC	VD	VE	AC
Displacement (cm)				
2nd Floor	10.17	9.31	8.09	8.38
3rd Floor	18.19	17.68	15.97	15.94
4th Floor	22.36	22.56	21.36	20.21
5th Floor	25.32	25.16	25.20	22.73
6th Floor	27.63	27.68	28.59	24.69
Roof	27.50	29.89	30.72	26.26
Velocity (cm/s)				
2nd Floor	51.1	34.8	38.9	36.7
3rd Floor	82.2	73.9	80.6	77.7
4th Floor	113.0	109.0	119.0	114.2
5th Floor	142.8	137.5	149.6	142.0
6th Floor	167.2	163.8	178.0	165.2
Roof	185.4	179.9	195.7	178.2
Acceleration (g)				
2nd Floor	0.790	0.699	0.580	0.732
3rd Floor	0.800	0.563	0.602	0.647
4th Floor	0.619	0.600	0.662	0.675
5th Floor	0.841	0.599	0.636	0.796
6th Floor	0.782	0.690	0.753	0.738
Roof	1.003	0.897	1.012	1.041
Energy (kJ)				
SE	409	396	396	407
KE	782	666	688	660
DE	1602	1014	1063	1027
PE	3348	1808	1916	1788
CE	–	2711	2711	2711
IE	5058	5538	5693	5531
Control energy in each actuator (kJ)				
C1	–	471	510	449
C2	–	695	685	696
C3	–	473	476	475
C4	–	415	408	397
C5	–	449	430	445
C6	–	210	201	248

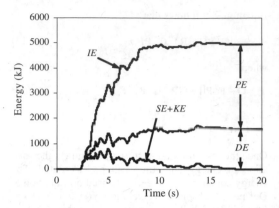

Figure 2. Energy distribution in UC model.

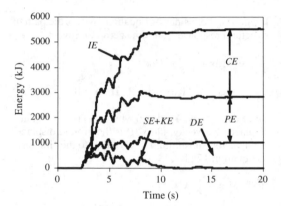

Figure 3. Energy distribution in VD model.

of the uncontrolled (UC) model. Figure 2 shows the energy distribution time history where the input energy is decomposed into storage energy (i.e., *SE* and *KE*) and dissipative energy (i.e., *DE* and *PE*).

Now consider the use of either linear passive dampers or optimal linear control to reduce the structural responses. Six actuators (labeled from C1 to C6 as shown in Figure 1) are used with one installed in every floor. Based on this setup, it follows that the distribution matrix **D** in Equation 5 is defined as:

$$\mathbf{D} = \begin{bmatrix} 1 & -1 & 0 & 0 & 0 & 0 \\ 0 & 1 & -1 & 0 & 0 & 0 \\ 0 & 0 & 1 & -1 & 0 & 0 \\ 0 & 0 & 0 & 1 & -1 & 0 \\ 0 & 0 & 0 & 0 & 1 & -1 \\ 0 & 0 & 0 & 0 & 0 & 1 \end{bmatrix} \quad (23)$$

The parameters used in the actuators are adjusted to give the same level of control energy dissipation. Based on an iterative procedure, the following parameters are selected:

- Passive viscous damper (VD) model:

$$\mathbf{C}_o = -3651 \times \mathbf{D}^T \quad (24)$$

- Passive viscoelastic damper (VE) model:

$$\mathbf{C}_1 = -3500 \times \mathbf{D}^T \ , \quad \mathbf{C}_2 = -14000 \times \mathbf{D}^T \quad (25)$$

- Active control (AC) model:

$$\mathbf{Q} = \begin{bmatrix} \mathbf{K} & 0 \\ 0 & \mathbf{M} \end{bmatrix} \ , \quad \mathbf{R} = 3.46 \times 10^{-4} \times \mathbf{I} \quad (26)$$

The maximum global responses subjected to the 1995 Kobe earthquake and the maximum control forces in each actuator are summarized in Table 1.

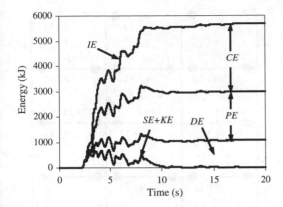

Figure 4.   Energy distribution in VE model.

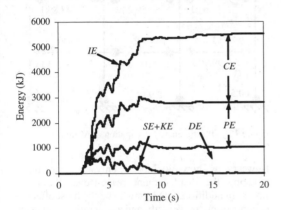

Figure 5.   Energy distribution in AC model.

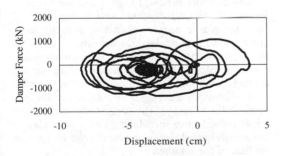

Figure 6.   Viscous damper C2 force–displacement response.

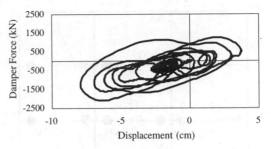

Figure 7.   Viscoelastic damper C2 force–displacement response.

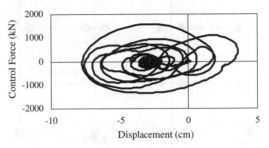

Figure 8.   Active control C2 force–displacement response.

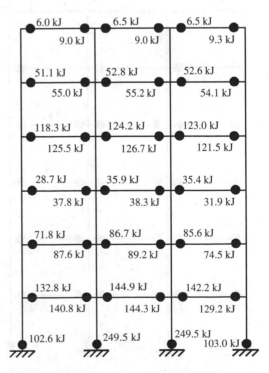

Figure 9.   Plastic energy dissipation in UC model.

Figures 3–5 show the energy distribution time histories of VD, VE, and AC models, respectively. According to Table 1, the largest control force occurs in C2 for all three cases. Figures 6–8 show the force-displacement response for C2 in these three cases.

Comparing the results obtained in Figures 3–5 for VD, VE, and AC models shows that these three control strategies give similar energy response time histories.

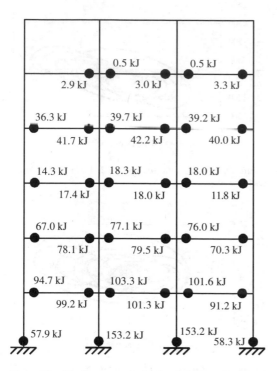

Figure 10. Plastic energy dissipation in VD model.

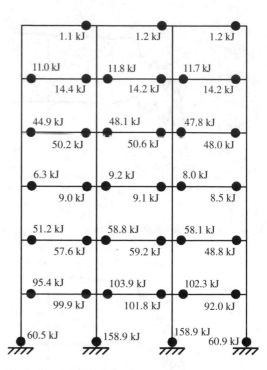

Figure 12. Plastic energy dissipation in AC model.

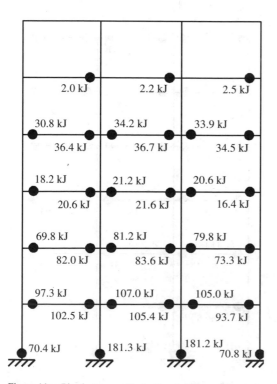

Figure 11. Plastic energy dissipation in VE model.

However, as shown in Table 1, only AC is capable of reducing the displacement response at the upper floors. In addition, VD seems to be the most effective in controlling the velocity and acceleration responses. Finally, VE seems to give the poorest performance among the three control strategies.

Now consider the local response of the structure for the UC, VD, VE, and AC models. Figures 9–12 show the plastic energy dissipation at each plastic hinge for the four models, respectively. As shown in these figures, PHLs #2 and #3 are the locations where maximum plastic energy dissipation occurs. Among the three control strategies used, VE seems to be the most effective in reducing the plastic energy dissipation in PHLs #2 and #3. This suggests that even though VE is not effective in reducing global responses, it has the advantage of reducing the damage occurred at the base of the columns.

6 CONCLUSIONS

In this paper, energy dissipation during structural vibration using different control devices is studied. While input energy entering the structure due to earthquake ground motion induces structural damage when plastic energy is dissipated, using energy dissipation devices can reduce the damage by dissipating control energy and thereby reducing the necessary amount of plastic energy dissipation. Therefore, using energy dissipation devices certainly has the advantage of protecting structures from suffering severe damage.

Numerical simulations have been performed to study the energy flows in structure using control devices including viscous dampers, viscoelastic dampers, and active control. Comparison of results show that active control is very effective in reducing the global responses in general, while viscous dampers are particularly effective in reducing the velocity and acceleration responses. Although viscoelastic dampers are the least effective in reducing the overall structural response, it is very effective in reducing the displacement at the lower stories and thereby reducing the amount of plastic energy dissipation at the base of the columns.

REFERENCES

Housner, G.W., Bergman, L.A., Caughey, T.K., Chassiakos, A.G., Claus, R.O., Masri, S.F.R., Skelton, E., Soong, T.T., Spencer, B.F., & Pao, J.T.P. 1997. Structural Control: Past, Present, and Future. *Journal of Engineering Mechanics ASCE*, 123(9): 897–971.

Makris, N. & Constantinou, M.C. 1991. Fractional-derivative Maxwell model for viscous dampers. *Journal of Structural Engineering ASCE* 117(9): 2708–2724.

Soong, T.T. 1990. *Active Structural Control, Theory & Practice*. Longman Scientific & Technical, U.K.

Soong, T.T. & Dargush, G.F. 1997. *Passive Energy Dissipation Systems in Structural Engineering*. John Wiley and Sons Inc., New York.

Wong, K.K.F. & Yang, R. 1999. Inelastic dynamic response of structures using force analogy method. *Journal of Engineering Mechanics ASCE* 125(10): 1190–1199.

Wong, K.K.F. & Yang, R. 2001. Effectiveness of structural control based on control energy perspectives. *Earthquake Engineering and Structural Dynamics* 30(12): 1747–1768.

Wong, K.K.F. & Yang, R. 2002. Earthquake response and energy evaluation of inelastic structures. *Journal of Engineering Mechanics ASCE* 128(3): 308–317.

Yang, J.N., Akbarpour, A., & Ghaemmaghmi, P. 1987. New optimal control algorithms for structural control. *Journal of Engineering Mechanics ASCE* 113(9): 1369–1386.

Zhang, R.H., Soong, T.T., & Mahmood, P. 1989. Seismic response of steel structures with added viscoelastic dampers. *Earthquake Engineering and Structural Dynamics* 18(3): 389–396.

Zhang, R.H. & Soong, T.T. 1992. Seismic design of viscoelastic dampers for structural applications. *Journal of Structural Engineering ASCE* 118(5): 1375–1392.

Economical seismic retrofitting of bridges in regions of low to moderate risk of seismic activity

M. Dicleli
Department of Civil Engineering and Construction, Bradley University, Peoria, Illinois, USA

M. Mansour
Department of Civil Engineering, University of British Columbia, Vancouver, BC, Canada

A. Mokha & V. Zayas
Earthquake Protection Systems Inc., Richmond, California, USA

ABSTRACT: The economical efficiency of friction pendulum bearings (FPB) for seismic retrofitting of typical Illinois bridges is studied. A typical bridge was carefully selected by Illinois Department of Transportation (IDOT) for the purpose of this study. The seismic analysis of the bridge revealed that the bearings, and substructures of the bridge are vulnerable and need to be retrofitted. A conventional retrofitting strategy is developed for the bridge and the cost of retrofit is estimated. Next, the bridge is further studied to develop appropriate techniques for upgrading its seismic capacity using FPB and the seismic analysis is repeated. It is observed that the FPB effectively mitigated the seismic forces and eliminated the need for retrofitting of the substructures of the bridge.

1 INTRODUCTION

In early 1990's IDOT made a statewide assessment of the seismic vulnerabilities of its highway bridges. The bearings, and substructures of these bridges were found vulnerable. Conventional retrofitting methods based on strengthening and enhancing the ductility of substructure members may be used to mitigate the seismic damage risk for such bridges (FHWA 1995). Yet, most of these methods are expensive and difficult to implement. Thus, seismic retrofitting methods based on response modification may provide a more efficient solution. This may be achieved by replacing the existing bearings by FPB to mitigate the seismic forces and eliminate the need for costly retrofitting of substructures. To assess the economical and structural efficiency of FPB for seismic retrofitting of the bridges in Illinois, a representative bridge was carefully selected by IDOT and studied.

2 FRICTION PENDULUM BEARINGS

FPB are sliding-based seismic isolators. The main components of FPB are; a stainless steel concave spherical plate, an articulated slider and a housing plate as shown in Figure 1. FPB use the characteristic of a pendulum motion to lengthen the natural period of the isolated structure, thus, deflect the earthquake input energy to mitigate the seismic forces. A pendulum motion for the supported structure is achieved as the articulated slider rides on the concave surface as illustrated in Figure 1. The movement of the slider generates friction that results in hysteretic energy dissipation.

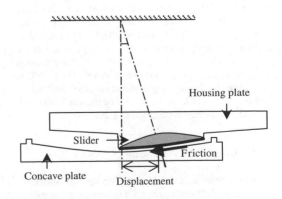

Figure 1. Friction pendulum bearing.

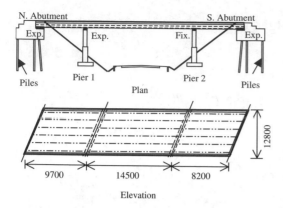

Figure 2. Bridge geometry.

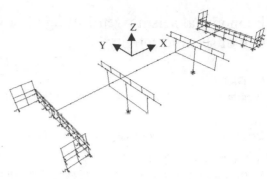

Figure 3. Structural model of the bridge.

3 TYPICAL BRIDGE SELECTED FOR STUDY

The bridge is located in southern Illinois in a region of low to moderate risk of seismic activity. It has three continuous spans carrying two traffic lanes and is supported by two multiple-column piers as shown in Figure 2. Elastomeric bearings are provided underneath each girder at both abutments. Steel rocker and fixed bearings are provided at Piers 1 and 2 respectively.

Both abutments are seat type and are nearly identical. Each abutment is supported by four straight and three battered HP200x54 piles. The length of the piles at the north and south abutments are 4.2 and 6.2 m respectively

The piers are reinforced concrete multiple column frames supported on a crush wall as typically found in Illinois bridges. The piers are supported on spread footings. The bridge site has stiff soil conditions.

4 SEISMIC ANALYSIS OF THE BRIDGE

Iterative multi-mode response spectrum (MMRS) analyses of the bridge are conducted to simulate; cracking of pier columns, flexural yielding of the wingwalls and nonlinear soil-bridge interaction effects. A detailed description of the iterative analysis procedure is presented elsewhere (Dicleli & Mansour 2002). The normalized acceleration response spectrum of AASHTO (1998) is used in the analyses. For the bridge site, the zonal acceleration ratio is obtained as 0.14. For the site's stiff soil, the site coefficient is obtained as 1.20.

5 MODELLING FOR SEISMIC ANALYSIS

A 3-D structural model of the bridge illustrated in Figure 3 is built and analyzed. The model is capable of simulating the nonlinear behavior of bridge components

and soil-bridge interaction effects when used in combination with iterative MMRS analyses. In the model, equivalent linear stiffness (ELS) properties are used for the bridge components exhibiting nonlinear behavior.

The bridge superstructure is modelled using 3-D beam elements. The 13.40 tons/m superstructure mass is lumped at each nodal point. The bridge deck which has a large in-plane translational stiffness is modelled as a transverse rigid bar at the abutments and piers as shown in Figure 3.

All the bearings are idealized as 3-D beam elements connected between the superstructure and substructures as shown in Figure 3. Pin connection is assumed at the joints linking the bearings to the substructures. As the stiffness of elastomeric bearings is temperature dependent, an effective shear stiffness, $K_b = 1.35x(K_{min})$, is used for the elastomeric bearings, where K_{min} is the shear stiffness of the bearing at 20°C.

The pier is modelled using a set of 3-D beam elements as shown in Figure 3. The columns are connected to a horizontal rigid bar at their lower end to simulate the interaction between their axial deformation and wall's flexural rotation. The abutments are modelled using a grid of beam elements as shown in Figure 3. The tributary masses of the abutment and wingwalls are lumped to the nodes.

5.1 Bridge foundation-soil interaction modelling

At the abutments, three translational springs are connected at each pile location to model the foundation flexibility. The horizontal boundary springs connected at the pile-substructure interface nodes are assigned an ELS obtained from the piles' non-linear lateral-force displacement (P-Y) curves as the slope of the line connecting the origin to the point representing pile's seismic lateral force on the curve. The average ELS for the abutment piles is obtained as 2900 kN/m.

The soil-footing interaction effect at the piers is included in the model using three translational and three rotational uncoupled boundary springs connected at the interface nodes of soil and rectangular spread footings. The methods proposed by Dorby and Gazetas (1986) and FHWA (1986) are employed in the calculation of the stiffness of the boundary springs for the vertical, horizontal, rocking, and torsional modes. The stiffness of the pier footings in the vertical, longitudinal and transverse directions of the bridge are obtained as 10.28×10^6, 8.62×10^6 and 10.28×10^6 kN/m respectively. The rotational spring constants about the global X and Y-axis are obtained as 292×10^6 and 40.4×10^6 kN.m/rad respectively and the torsional spring constant is calculated as 729×10^6 kN.m/rad.

5.2 Bridge abutment-soil interaction modelling

In the longitudinal direction, translational springs are attached to the nodes of only one of the abutments to simulate backfill-abutment interaction effects as illustrated in Figure 3. In the transverse direction, translational springs are attached to the nodes of only one of the wingwalls at each abutment to simulate backfill-wingwall interaction effects. Translational springs are also attached at each node of the abutment to model the shear stiffness of the backfill. The shear stiffness of the backfill is calculated assuming that only the portion of the backfill between the wing-walls will deform in a shearing mode.

Using the relationship defined by Clough & Duncan (1991) for the variation of the earth pressure coefficient as a function of the abutment movement, the horizontal subgrade constant for the backfill is obtained as a function of the depth from the abutment top. The stiffness of the boundary springs connected at the abutment-backfill interface nodes at the abutment and wingwalls are then calculated by multiplying subgrade reaction constant by the area tributary to the node.

6 ANALYSIS RESULTS

The fundamental period of the bridge is calculated as 0.584 sec. and corresponds to its modal vibration in the longitudinal direction. The second, third and fourth modal periods of the bridge are 0.541, 0.395 and 0.308 respectively.

The bearings at Pier 2 attract the largest seismically induced forces (106 kN) due to their larger lateral stiffness. The bearing relative displacements at all substructure locations are almost identical due to larger in-plane stiffness of the deck in the longitudinal direction and equal to 24.1 mm. The displacements are less than the width of the expansion joint. Therefore, the superstructure is not anticipated to impact the abutment back-wall.

The total seismic force acting on the structure is 1594 kN in the longitudinal and 1776 kN in the transverse direction. The seismic force in the transverse direction includes the contribution of friction and passive resistance of the backfill and the embankment soil. The one in the longitudinal direction includes the effect of static and dynamic backfill pressures. The piers carry 67% of the total seismic force and are the most vulnerable bridge substructures. At the abutments, the maximum longitudinal direction lateral force is 99 kN for the battered piles and 28 kN for the vertical piles. The battered piles helped to reduce the lateral seismic force and displacement demand on all other abutment piles.

7 CAPACITY-DEMAND RATIOS FOR VULNERABLE COMPONENTS

The factored elastic seismic demands for each bridge component are calculated following the procedures outlined by AASHTO (1998). AASHTO allows the design and evaluation of properly detailed ductile structural members using a seismic force smaller than that obtained from elastic seismic analysis of the bridge. However, for the members of the bridge considered in this study, the calculated elastic seismic demands are directly used to evaluate their seismic vulnerability since a ductile behavior is not expected due to poor reinforcement detailing. The capacity demand (C/D) ratios for the vulnerable bridge components are presented below.

The displacement capacity of elastomeric bearings is limited to 50% of the bearing height per AASHTO (1998) and is calculated as 33 mm for the bearings at both abutments and Pier 1 and as 12.5 mm for those at Pier 2. The seismically induced relative displacements at Pier 2 bearings are found to be larger than their 12.5 mm displacement capacity. The C/D ratio for the displacement of the bearings at Pier 2 is calculated as 0.82.

The ultimate shear and flexural capacities of abutment components are calculated based on AASHTO (1998). The ultimate flexural capacities of the wingwall and back-wall are calculated as 57 kN.m/m and 89 kN.m/m, respectively, and their shear capacities are calculated as 180 kN/m and 307 kN/m, respectively. Under transverse direction earthquake loading, the passive resistance of the backfill is found to induce transverse seismic moments in the wingwalls in excess of their ultimate flexural capacities. The C/D ratio for the wingwalls is calculated as 0.52.

Careful examination of the structural drawings of the bridge revealed that the reinforcement in the footings of the piers is not detailed in compliance with AASHTO (1998). No reinforcement is provided at the footing top to resist the tensile stresses introduced by the reversed seismic loading. Thus, ductile behavior of the pier footings is not expected. Based on the provided reinforcement detail, the ultimate shear and flexural capacities of the footings are calculated as 423 kN/m and 120 kN.m/m The ultimate bearing resistances of the soil at the foundations of Pier 1 and 2 are calculated as 1200 and 1100 kPa, respectively AASHTO (1998) limits the eccentricity of the applied vertical loads to one forth of the dimension in the direction of interest to prevent overturning of the footing and/or bearing failure of the foundation soil. The eccentricity limits for the pier foundations in the longitudinal and transverse directions of the bridge are thus calculated as 571 and 2400 mm respectively. The sliding resistance of the pier foundations is calculated as 1500 kN following the procedure recommended by AASHTO (1998). The seismically induced moments at the foundations of both piers resulted in vertical load eccentricities in excess of the maximum limits specified by AASHTO (1998). The C/D ratios for the eccentricities at Pier 1 and 2 foundations are calculated as 0.62 and 0.39 respectively. Thus, overturning of the piers and/or bearing failure of their foundation is anticipated.

8 CONVENTIONAL SEISMIC RETROFITTING

For the bearings at Pier 2, although the seismically induced displacement exceeds their displacement capacity, it is anticipated that the failure of such shallow elastomeric bearings may not cause a major damage. Thus, any repair may be done after the occurrence of a major earthquake.

The retrofitting of wingwalls at both abutments includes the addition of cleats at the joints between the wingwalls and the abutments to enhance their flexural capacities in the transverse direction.

The widths of the footings at piers 1 and 2 need to be increased from 2286 to 3800 and 4800 mm respectively to obtain a vertical load eccentricity smaller than 25% of the footing width as specified by AASHTO (1998). This will ensure the stability of the piers. Increasing the footing width results in larger seismically induced flexural moments in the footing at the face of the crush walls. Therefore, the flexural capacities of the footings at both piers need to be increased. For this purpose, the depths of the footings at Pier 1 and 2 are increased from 610 mm to 1100 and 1200 mm respectively. This involves casting of an overlay of reinforced concrete doweled to the existing footing.

8.1 Seismic retrofitting cost

The strengthening of the pier foundations and the wingwalls requires, excavation of the soil to expose the footings and the wingwalls completely, concrete removal, placing new concrete, reinforcement and shear dowels. The summary of the cost estimate is presented

Table 1. Conventional retrofitting cost.

Item	Quantity	Total cost range
Excavation	311 m^3	$7,775–$9,330
Concrete removal	5 m^3	$5,000–$7,500
New concrete	92 m^3	$92,000–$138,000
Shear dowels	245 kg	$513–$540
Reinforcing bars	2860 kg	$5,985–6,300

in Table 1. Mobilization and traffic control costs are not included in the cost estimate. IDOT provided ranges of unit prices for the purpose of cost estimation. Based on this unit prices the total cost of retrofitting the bridge is found to range from $111,273 to $161,670.

9 SEISMIC RETROFITTING USING FPB

FPB with a spherical concave surface is recommended over the substructures to control and mitigate the seismic forces acting on the substructures.

9.1 Structural model with FPB

The structural model used for the detailed seismic analysis of the bridge is slightly modified to incorporate the FPB instead of the existing bearings. The FPB are also modelled using 3-D vertical beam elements. An ELS is used for the FPB to estimate the stiffness properties of the beam elements. Hysteretic damping of the bearing is also considered in the analyses following the procedures defined by AASHTO (1999). An iterative analysis procedure is performed to calculate the seismic response of the bridge as the ELS and hysteretic damping are functions of the bearing displacement.

9.2 Analysis results with FPB

The periods of vibration for the first and second modes of the seismically isolated bridge are 1.42 and 1.39 sec. respectively.

The bearing displacements at all substructure locations are almost identical and equal to 50 mm due to the large in-plane stiffness of the bridge deck relative to the equivalent stiffness of the FPB. This displacement is larger than the width of the expansion joints at both abutments. Thus, the superstructure is expected to impact the abutment back-walls. For the back-wall to fail in shear, the magnitude of the impact force must be at least equal to the shear capacity of the back-wall plus the force required for mobilizing the passive resistance of the soil behind the back-wall only. This force is calculated as 3970 kN. Thus, a force equal to 3970 kN may potentially be transferred to the breast wall and the piles. The sum of the lateral capacities of the abutment piles and passive resistance of the portion of the backfill behind the breast wall is 1407 kN. This is smaller than the anticipated 3970 kN force transferred by the backwall. Therefore, damage to the abutment piles may occur. This damage may be prevented if the back-wall is modified to fail in shear, for instance by providing a knock-off device. Essentially, in seismic design of bridges, it is recommended to use such a mode of damage as a "fuse" as it is much easier to repair the upper portion of the abutment after a seismic event. Therefore, either the abutment may need to be modified to fail in shear by providing a knock-off device or the expansion joints may be widened to accommodate the required displacements if the back-wall damage is not desired.

The analysis results revealed that FPB eliminated the in-plane torsional rotation of the bridge and resulted in a more uniform distribution of seismic forces among substructures. The forces are generally reduced by a factor ranging between 3 and 4 due to the presence of FPB when compared to those of the same bridge with elastomeric bearings. This is a result of energy dissipation and lower equivalent linear stiffness of FPB compared to that of the elastomeric bearings and even other rubber-based seismic isolation systems for this particular application. The lower equivalent linear stiffness mainly results from the relatively smaller weight of the only 32.4 m long bridge superstructure transferred to the FPB.

9.3 Capacity-demand ratios with FPB

The C/D ratios for the previously determined vulnerable structural members are now all larger than 1.0 after replacing the existing bearings with FPB. The FPB effectively mitigated the seismically induced forces such that seismic retrofitting of the pier foundations and wingwalls is no more required.

9.4 Retrofitting of the bridge using FPB and cost estimate

The isolated bridge must be free to move in any horizontal direction for the seismic isolation system to perform as desired. Considering a maximum thermal movement of 10 mm in each end of the bridge, the minimum required expansion joint width at the abutments is calculated as 60 mm. Thus, the width of the expansion joint at the north and south abutments needs to be increased by 22 and 35 mm, respectively. Furthermore, the minimum required displacement capacity of the bearings including a tolerance for thermal movements needs to be 60 mm. It is noteworthy that the low profile of FPB required only minor modifications to adjust the bearing seat elevations for their installment.

The estimated range of cost for the seismic retrofitting of the bridge using FPB is calculated as

Table 2. Retrofitting cost with FPB.

Item	Quantity	Total cost range
Concrete removal	5 m³	$5,000–$7,500
New concrete	5 m³	$5,000–$7,500
Reinforcing bars	410 kg	$855–$900
Steel plates	680 kg	$3,000–$6,000
Preformed joint seal	24.6 m	$4,860–$5,670
Jack & remove bearings	24 bearings	$19,200–$28,800
FPB erection	24 bearings	$4,800–$12,000
FPB cost	24 bearings	$64800

$107,515 to $133,170 using the retrofitting cost ranges provided by IDOT. The details for the cost estimation are presented in Table 2. The retrofitting cost mainly includes jacking the bridge and removing the existing bearings, widening the expansion joints, adjusting the elevation of bearing pedestals, and erection and cost of FPB. The average cost of retrofitting using FPB is found to be only 88% of the average conventional retrofitting cost. Retrofitting with FPB requires shorter construction time than that required for conventional retrofitting. Therefore, retrofitting with FPB may become much more economical when the additional cost of construction time and cost of traffic mobilization is considered. Thus, FPB may effectively be used for seismic retrofitting of typical bridges in the state of Illinois or in regions of low to moderate risk of seismic activity.

10 CONCLUSIONS

The economical and structural efficiency of FPB for retrofitting of typical bridges in the state of Illinois is investigated. The following observations are made:

1. The FPB eliminated the in-plane torsional rotation of the bridge and resulted in a more uniform distribution of seismic forces among substructures.
2. A three to four times reduction in the seismically induced forces in the structure components is achieved with the FPB. This is a result of; (i) energy dissipation and (ii) lower equivalent linear stiffness of FPB compared to that of the elastomeric bearings and even other rubber-based seismic isolation bearings for this particular application. The lower equivalent linear stiffness mainly results from the relatively smaller weight of the only 32.4 m long bridge superstructure transferred to the FPB. Thus, FPB effectively mitigated the seismic forces and eliminated the need for costly retrofitting of the bridge substructures.
3. However, FPB resulted in large superstructure displacements in excess of the expansion joint widths. This required either increasing the widths of the

expansion joints or providing knock-off devices at the abutments to eliminate the possibility of the superstructure impacting the abutment back-wall and a potential damage to the abutment piles. Nevertheless, the low profile of FPB required only minor modifications to adjust the bearing seat elevations.

4. An average retrofitting cost using FPB is calculated as only 88% of that using conventional retrofitting method. Retrofitting with FPB requires shorter construction time than that required for conventional retrofitting. Therefore, retrofitting with FPB may become much more economical when the additional cost of construction time and cost of traffic mobilization is considered.

REFERENCES

AASHTO. 1998. LRFD Bridge Design Specifications, Washington, DC.

AASHTO. 1999. Guide Specifications for Seismic Isolation Design, Washington, DC.

Clough, G.W. & Duncan, J.M. 1991. Foundation Engineering Handbook, 2nd Edition, Edited by H. Y. Fang, Van Nostrand Reinhold, New York, NY, 223–235.

Dicleli, M. & Mansour, M.Y. 2002. Seismic Retrofitting of Highway Bridges Using Friction Pendulum Seismic Isolation Bearings, Technical Report: BU-CEC-02-01, Department of Civil Engineering and Construction, Bradley Univ., Peoria IL,

Dobry, R. & Gazetas, G. 1986. Dynamic Response of Arbitrary Shaped Foundations. *ASCE Journal of Geotechnical and Geoenvironmental Engineering* 112(2): 109–135.

FHWA. 1986. Seismic Design of Highway Bridge Foundations – Volume II: Design Procedures and Guidelines. Publication No. FHWA-RD-94-052. Federal Highway Administration, US Department of Transportation, Washington, DC.

FHWA. 1995. Seismic Retrofitting Manual for Highway Bridges. Publication No. FHWA-RD-86-102. Federal Highway Administration, US Department of Transportation, Washington, DC.

Electro-inductive passive and semi-active control devices

M. Battaini
ALGA spa, Milano

F. Casciati & M. Domaneschi
University of Pavia, Pavia

ABSTRACT: Electro-inductive devices were recently proposed for the passive control of buildings and structures. The feasibility of large size devices for long span bridges was already studied in the occasion of the project of a specific bridge. This paper investigates the way to make the device semi-active without loosing its passive control properties.

1 INTRODUCTION

This paper discusses the exploitation of electro-inductive dampers in the design of long span bridges. Electro-inductive devices were introduced in (Marioni et al. 1997) and developed by the research pool working on them (Battaini et al. 1998; Marioni et al. 1999; Battaini et al. 2000 and Battaini et al. 2002).

The basic concept relies in the action of an electro-magnetic field on a moving ferromagnetic body. The simplest implementation is achieved by a linear device, as tested in (Martelli 2002). A better performance can be achieved by rotational devices, with the complication of a mechanical transfer of a linear motion into a rotational one. Apart for its reduced efficiency and the required maintenance, this solution also allows one to introduce multiplicative factors which can help in terms of dissipated energy.

The adoption of energy dissipation devices in long span bridges covers both the longitudinal oscillation (where large device spans are required) and some transversal motion that varies on the basis of the bridge structural scheme. The problem was discussed, among others, in (Battaini et al. 2001) where the different role of ultimate and serviceability limit states is emphasized. The main conclusion is the need of devices for which one can switch from one to another (or several other different) configuration(s).

This concept is reported in literature as the need of semi-active devices. The basic property however is the ability to act as passive devices in case the system making them semi-active fails. This applies perfectly to electro-inductive devices adopting permanent magnets in order to produce the electro-magnetic field.

If this aspect is disregarded, the electro-inductive system can be realized by a simple electric motor, for which special control parameters are adopted to modify the amount of energy dissipated at the end of each cycle.

These motors are the object of this paper which is focused on the ability to make them semi-active, in the sense of showing the capability of varying the dissipated energy from cycle to cycle and within the single cycle according to a control command.

2 THE DEVICE

A market survey led the authors to select an electric machine able to show hysteretic cycles in the torque rotation plane, with the first variable ranging from zero to the maximum value allowed by the available testing machine (150 Nm, Fig. 1). The motor is the SMV58I 6 poles brushless synchronous (Fig. 2) produced by Servo Drive Technologies. The characteristics are summarized in Table 1 and the torque vs velocity diagram is given in Figure 3.

The motor mounts a coaxial epicyclical reducer SUMER NR140 P3. It consents a reduction ratio 1:5 with a torque stiffness of 371000 Nm/rad and a clearance of 3'.

The motor controller which was selected is the SDH 01P shown in Figure 4. This controller jumps from a current of 10 A to a 30 A peak current. It has an anti-noise filter integrated for the power feeding, a two key command, a three code led display integrated to set the main control parameters and a RS 232 interface to realize the pc-connection and the reaction of the encoder located into the motor.

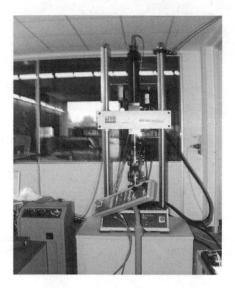

Figure 1. Biaxial universal machine MTS 858: tension – compression and torsion.

Figure 2. Motor SMV 58I.

Table 1. Motor SMV58I characteristics.

	Type	Units	SMV58I
Rated data	Stall torque	Nm	27
	Stall current	A	17.4
	DC link voltage	V	560
	Rated torque	Nm	14
	Rated current	A	9.2
	Rated speed	min^{-1}	3300
	Rated power	W	4838
	Constant voltage	Vkrpm	94
	Constant torque	Nm/A	1.555
	Widing resistance	Ohm	0.32
	Widing inductivity	mH	5.5
Peak value	Peak torque	Nm	104
	Peak current	A	77
	Peak velocity	min^{-1}	6000
Mech. data	Armature inertia	Kgm^{2}10^{-3}	2.47
	Weight	Kg	27

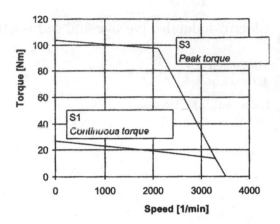

Figure 3. Motor SMV 58I: torque velocity diagram.

Figure 4. Motor controller SDH 01P.

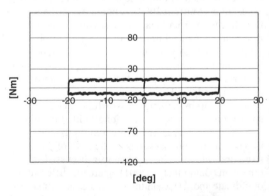

Figure 5. Initial test with the motor without power.

The first experiment was conducted with the motor disconnected from the electrical source.

The rotation ($-40°$, $+40°$) is followed with a very low frequency 0.0125 Hz. The sample size is 0.01 and Figure 5 shows the obtained result in terms of torque-rotation cycle.

Table 2. Test conditions for different frequency experiments.

Intensity [A]	Frequency [Hz]	Amplitude [deg]
1A	0.06	200
1A	0.03	40
1A	1	15
1A	2	7

3 EXPERIMENTAL ENVIRONMENT

The universal testing machine uses its torsion actuator and it is set in rotation control. The controller is programmed in velocity control and torque limit. The control velocity is set to zero, while the torque limit is fixed by assigning a specific value of the current: higher this value is, higher the torque limit results.

The first experiments were conducted maintaining constant the current intensity of the motor for different values of the cycle frequency of the universal machine. The conditions are summarized in Table 2. The results of Figure 6 shows the dissipation cycles for the frequency range considered. A good cycle stability is recorded.

It is remarkable the variation of the cycles shape with the frequency increment: oblique lines appear in the rotation reversal, like a delay in the resistant torque application.

An increment of the motor current intensity induce a larger dissipation cycle like the one represented in Figure 7.

Further experiments were focused on the system transient study: the resistant torque is varied into the same load cycle. In Table 3 the testing conditions are summarized and Figures 8 and 9 show the graphic results. They represent a good outcome for shape regularity and symmetry. In particular Figure 9 was realized on two hysteretic cycles.

It is worth noting that for all the resulting figures it was possible to find an oscillation on the horizontal lines of the hysteretic cycles and the same occurs in Figure 5 with no motor power. This behavior can be attributed to the mechanical part of the motor, the coaxial epicyclical reducer.

Figure 10 is a simple example where the motor operates like a semi-active device. Figure 11 shows repeated increments and decrements of current intensity in the same cycle.

4 MOTOR ACTIVE POWER REMARKS

When operating under an external mechanic excitation with velocity control set to zero and torque limit the motor acts as a current generator.

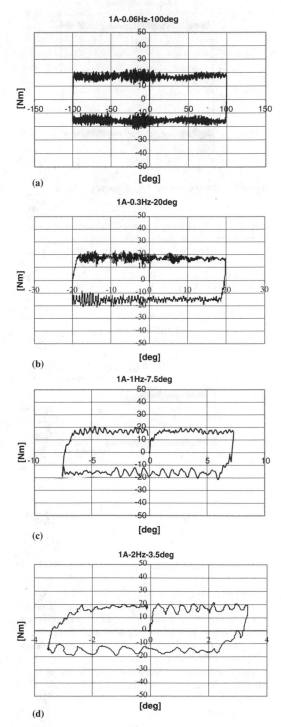

Figure 6. Single cycle tests: (a) driving current 1 A, frequency 0.06 Hz: (b) 0.3 Hz: (c) 1 Hz: (d) 2 Hz.

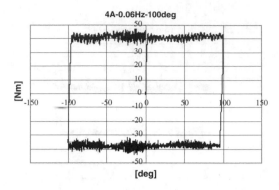

Figure 7. Driving current 4 A, frequency 0.06 Hz.

Table 3. Test conditions for system transient study.

Intensity variation [A]	Frequency [Hz]	Amplitude [deg]
from 4A to 3, 2, 1A	0.06	200
FROM 4A TO 1A	0.3	40

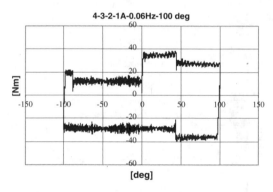

Figure 8. Reducing the driving current from 4 A to 3, 2, 1A in a single cycle, frequency 0.06 Hz.

An opportune electric resistor (fixed to 33 Ω) internal to the controller is provided to dissipate the current intensity generated.

Two samplings on continue cycles of operation are shown in Figure 12 under the condition of having 1 A current intensity, a frequency of 1 and 2 Hz, respectively, with 15 and 7 deg of cycle amplitude (the same conditions of Figure 6 (c) and (d)).

The active power generated during the motor operation under external excitation can be found as:

$$P(t) = I^2(t) R \qquad (1)$$

Where I is the current intensity at the time instant t, R the electric resistor internal to the controller. So

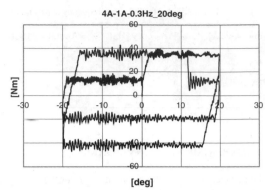

Figure 9. Reducing the driving current from 4 A to 1 A in two cycles, frequency 0.3 Hz.

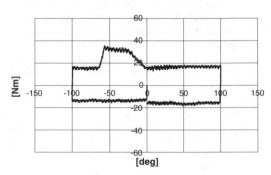

Figure 10. Motor operating as semi-active device.

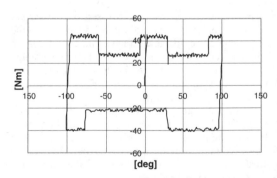

Figure 11. Motor operating as semi-active device. Single cycle and multiple increments and decrements.

for the two frequency considered the active power resulted to be about the same.

When the motor is not subjected to external mechanic excitation and remains with velocity control set to zero and torque limit the generated current intensity is zero, and hence also the active power is null. Of course, one must also mention a reactive power, that in any case is less the 10 W. This is the energy bill one has to pay.

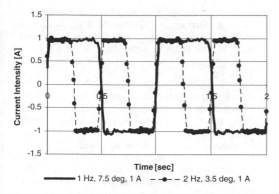

Figure 12. Current samplings on continue cycles.

where: u is the screw end displacement;
ϑ is the motor rotation;
ι is the slope;
M is the torque moment;
F is the force producing the moment, under the assumption of no friction.

Under the last assumption, one can eventually write down the device model in the form

$$F = c(i)\,sign(\dot{u})|\dot{u}|^{0.0001} \qquad (3)$$

where c is the model parameter depending on the driving current i.

For the case of $d = s = 25\,\text{mm}$ and $i = 1$ A, the value of c results $2300\,\text{N/(mm/s)}^{0.0001}$.

Now, further tests with the screw mounted on the motor are required in order to identify the mechanical losses. When these results were available, then the actual dependence of c on i is identified and the full device model is built.

From this last measurements appears in first approximation that the dissipated energy is not related to the operation frequency and the cycle amplitude but only to the current intensity selected and the time duration. The mechanic energy is transferred to the motor and then transformed in heat on the resistor internal to the controller.

5 MODELLING REMARKS

The results summarized in Figure 6 require some further remarks before one can start to built up a model on them. The moment positive constant branch was obtained with a test speed which is one half of the speed used for the moment negative constant branch. This explain the different period of the waves along the two branch. The different speed was adopted in order to emphasize a dependence on the velocity of the limit moment. Indeed there is an apparent asymmetry in the hysteresis, but this turned out to be related to some features of the way the device was fixed on the grips. This was corrected and the results of the subsequent tests (see for instance Figures 7 and 10) show a better symmetry. In conclusion, the limit moment is rather independent of the velocity and hence the standard exponential law adopted for dampers will see a very low value of the exponent, say 0.0001.

The motor is made a damper by coupling it with a screw able to transform a linear motion into a rotational motion. The main parameters are: the diameter of the screw, d, the step, s. Then the following formulae hold:

$$u = \vartheta\,\frac{s}{360°}$$

$$\iota = \tan^{-1}(\frac{\pi d}{s}) \qquad (2)$$

$$F = \frac{M}{d\,\sin(\iota)\cos(\iota)}$$

6 CONCLUSIONS

For those passive control devices the design of which depends on the required span, the technology of construction can have a basic role in the final size of the damper. A required span of 2 m results into a full length of approximately 12 m of the whole damper when standard oil dampers are considered.

This paper considers an electro-inductive solution which could result in a 30% reduction of the total length. Other important aspects of using electro-inductive devices instead of fluid dampers are:

(1) the response behavior is practically independent of the external temperature because the device operating temperature is always higher that the air temperature and it is reached in few seconds;
(2) the device maintenance is reduced because there is no ageing or leakage effects as in the fluid dampers.

The dissipation capacity of the electro-inductive dampers can also be easily adapted by changing the number of dissipation units (magnets plus resistances) inside the devices with a reduced impact of the overall dimensions respect to the fluid dampers.

Such electro-inductive devices work like an electric motor. The object of the present paper is focused on making the device semi-active, with special attention on the possibility of controlling the device response by a suitable input electric signal.

ACKNOWLEDGEMENTS

This paper reports the basic principle of an industrial research program partially supported by MIUR (the

Italian Ministry of Instruction, University and Research) with the leadership of ALGA s.p.a. and with the partnership of the University of Pavia, the Polytechnic of Milan and the ENEA of Bologna.

REFERENCES

Battaini, M., Casciati, F., Marioni, A., Di Gerlando, A., Silvestri, A., Ubaldini, M. 1998. Semi-active Control by Electro Inductive Energy Dissipators. In Kobori et al. (eds.), *Proceedings of the Second World Conference on Structural Control, Kyoto, Japan*, Vol. 1, 437–444.

Battaini, M., Casciati, F., Marioni, A., Di Gerlando, A., Nicolini, G., Silvestri, A., Ubaldini, M. 2000. Experimental Tests on Passive Electro Inductive Dissipators. In *Proceedings of the Second European Conference on Structural Control, Paris, France*.

Battaini, M., Casciati, Faravelli, L. 2001. Ultimate vs. Serviceability Limit State in Designing Bridge Energy Dissipation Devices. In Spencer & Hu (eds.), *Earthquake Engineering in the New Millenium*: 293–297. 2001 Swets & Zeitlinger, ISBN 90 265 1852.

Battaini, M., Marioni, A., Casciati, F., Di Gerlando, A., Perini, R., Silvestri, A., Ubaldini, M. 2002. Semi-Active Electro-Inductive Dissipators: Design and Experimental Tests. In Casciati (ed.), *Proceedings of the Third World Conference on Structural Control*, Vol. 2, 51–58. 2003. Wiley, UK.

Marioni, A., Silvestri, A., Ubaldini, M. 1997. Development of Innovative Energy Dissipation Systems in the EC Countries. In *Proceedings of the International Post-SMiRT Conference Seminar on Seismic Isolation, Passive Energy Dissipation and Active Control of Seismic Vibration of Structures, Taormina, Italy*.

Marioni, A., Battaini, M., Del Carlo, G., Casciati, F., Di Gerlando, A., Perini, R., Silvestri, A., Ubaldini, M. 1999. Development and Application of Electro Inductive Dissipators. In *Proceedings of the International Post-SMiRT Conference Seminar on Seismic Isolation, Passive Energy Dissipation and Active Control of Seismic Vibration of Structures, Cheju, Korea*, Vol. 1, 347–354.

Martelli, A., Forni, M., 2002. Key Issues in the Development and Application of Passive Seismic Vibration Control Techniques. In Casciati (ed.), *Proceedings of the Third World Conference on Structural Control*, Vol. 2, 51–58. 2003. Wiley, UK.

Wireless communication between sensor/device stations

L. Faravelli
Department of Structural Mechanics, University of Pavia, Pavia, Italy

R. Rossi
Department of Electronics, University of Pavia, Pavia, Italy

ABSTRACT: The monitoring/control of bridges is realized by stations where the main components are lodged: sensors, actuators and controller. The communication between these stations in long span bridges can cause significant inconveniences. The wireless technology of communication is the present answer to such kind of problem. The main limitation results from the power harvesting problem. This aspect is discussed in this paper and a system architecture is conceived.

1 INTRODUCTION

Structural monitoring and structural health monitoring are becoming the central aspect of long bridge management (Casciati et al. 2003a). A realization which is widely reported in the literature is the monitoring system covering the bridges in the Hong Kong area, with its early artifact, the Tai Ma bridge, as initiator (Chen et al. 2002).

Nevertheless, in the last years, the limitations resulting from the adopted technology solutions became manifest. On the other side the new products coming from the innovation industry have been more oriented to the bio-medical and environmental areas rather than to structure diagnostics. This situation resulted in the availability of scenarios, operative in other areas, which show the direction along which to proceed. Unfortunately, several technical problems have still to be solved when applications in bridge monitoring are pursued.

Three are the fascinating features offered by present sensor technologies:

1) sensors able to work in a wireless mode (Farrar et al. 2002; Linch et al. 2002);
2) sensors coupled with a microcomputer to form a smart sensor, once the suitable software has been downloaded in the microprocessor (Faravelli 2002; Casciati et al. 2003b);
3) sensor networks which can be made of a redundant number of devices: not all their outputs are continuously monitored, but the array can be modified and made denser where required. This results in a sort of adaptive monitoring scheme.

The system architecture to be implemented, however, is strongly influenced by the power supply of the network, which is the aspect discussed in this paper. The presentation is organized with the goal of emphasizing the dichotomy between target(s) and technological limitations.

2 WIRELESS TECHNOLOGY

Monitoring large structures is usually a challenging task. A number of design aspects has to be faced, each posing a set of pros and cons to system designers. More often than not, power is the most limiting factor, as we will see in next section.

Consider, for the sake of the discussion, a large structure such as a church or a bridge. In order to properly and reliably monitor such a structure, a huge number of sensors should be deployed. This would allow extensive coverage of important structural parameters such as accelerations, strains and so on. Unfortunately, this approach comes with a bulky set of drawbacks that, if not properly addressed, can make the deployment of such a system hardly feasible.

A structural monitoring system usually requires that the following functions be implemented: power supply distribution, data acquisition, local data processing, data transmission and central data processing. Optionally, a connection of the central station to a wide-area network such as the Internet would allow remote monitoring of the structure.

Having the possibility to create wireless sensors networks offers indubitable and well known

Figure 1. Wireless accelerometer station mounted on a steel bridge span to be tested in the laboratory (from (Casciati et al, 2003a).

Figure 2. Details of the wireless accelerometer station assembled at the University of Pavia.

advantages. The dream is to see time and complexity of sensors installation to decrease drastically.

But the characteristic that wireless sensors network needs to replace traditional wired systems are really tightening. Wired network are the best under several points of view. A wired acquisition system grants a huge bandwidth, low disturbances and allows synchronous acquisition. The convenience of a wireless network is mainly in the simplicity of installation.

The first problem experimented using a radio transmitter is the range in which communication is available. This is not due to technical limits but to a legal one. The use of a wide sector of the frequency spectrum is, in fact, ruled by international agreements and subject to the obtainment of a license, or, simpler, to the payment of the service from the license owner. This applies to the use of a GSM, IS-95A, UMTS or, generally speaking, of any telephone standard with, of course, the support of a telephone company.

This choice grants, certainly, a worldwide cover, but it has several drawbacks. Firstly, it has a cost proportional to the use, secondly it requires to renounce to a complete transmission control.

Anyway international agreements fixed up some frequency band free for Industrial, Scientific and Medical (ISM) purpose.

The use of this kind of frequency is subject to limits on the power used in transmission. Evidently this limit conveys in a range limit, being the range proportional to the transmission power.

The power limit is due to consideration of electromagnetic compatibility, that means that measurement equipment does not have to interfere with any other electronic system. Typically, the range is smaller than 100 feet, which makes transmission not functional in several application. Nevertheless, example are available as discussed in (Casciati et al. 2003 a and b), from where Figures 1 and 2 are taken. They mount MEMS accelerometers (Crossbow 2003) and the power is provided by batteries.

3 THE POWER SUPPLY PROBLEM

A large number of sensors spread over a large area means a lot of data that must be transmitted, stored and processed. More importantly, it also means that an adequate power supply network must be deployed for making acquisition and transmission possible.

Data transmission involves different issues from those intrinsic to power transmission over a large area.

In recent years, attention has focused on wireless data transmission, due to its many attractive features: no need for cables means lower costs, higher reliability, easier installation and maintenance. It also makes it possible to deploy sensors in hardly accessible parts of the structure without having to worry about cable connections. The first evident shortcoming is that a wireless connection requires more power than a wired one. In order to compensate for it, the duty cycle of data transmission may be reduced, possibly introducing data processing local to the sensors. In this way, less data need be transmitted, thus lowering requirements in terms of power. The weakest point is that a wireless connection is not truly wireless if power is transported on cables. However, today wireless power supply seems possible only in a very limited number of cases, because it is not possible to transmit power in environments where the presence of people cannot be excluded. Microwave power transmission is a well-known problem, but it poses serious health concerns and, hence, may only be implemented in very particular cases (e.g., satellites). On the other hand, at very low frequencies, the amount of power that can be transported by the electromagnetic field is negligible. Furthermore, national regulations set a hard limit on the maximum electric and magnetic field intensities that electrical and electronic appliances must comply with. For example, at a frequency of 100 Hz, the maximum allowed magnetic field intensity is 100 μT. This requires a current of 100 A on a coil with a radius of 1 m. The voltage induced on a small coil (e.g., on a small sensor board) at 1 m distance is just in the order of 500 µV. Induced voltages are even lower for bigger transmitting coils, due to the expression of the magnetic field in a coil:

$$B = \frac{\mu_0}{2\pi} \cdot \frac{\pi R^2 I}{(R^2 + d^2)^{3/2}} \tag{1}$$

where R is the radius of the power transmitting coil, I is the current and B is the magnetic field intensity on the axis of the coil at a distance d from its center. If $d \ll R$, the magnetic field decreases as $1/R$ does, which means that bigger coils transmit less power.

Since the transmitted power is so low, today the only applications where this kind of solution is viable are smartcard-type applications, where the duty cycle is almost zero. On the contrary, a structural monitoring sensor should be active for most of the time, though transmission could be less frequent.

Nevertheless, wireless power supply is very attractive. For the time being, until much lower processing powers can be achieved, the only means of wireless power supply seems to be solar cells, which require backup batteries.

Since, for the time being, cables cannot be completely eliminated, adopting wireless data transmission does not seem very useful because it requires more power than wire-based communication. A technically viable solution in order to overcome this problem could be the use of the same two wires for both data communication and power supply. For example, a small coaxial cable could be used for this purpose. A single cable would connect all of the sensors and would allow data networking while, at the same time, distributing power over the network of sensors. Of course, this solution is not wireless, but, at least, it is a step in the direction of cable cutting.

In all cases, the power issue must be addressed first, because it is the most limiting one. Once an adequate power distribution policy has been devised, all the other aspects of a monitoring system can be taken care of.

3.1 Available wireless power

Since no electronic component can function with power supplies in the millivolt range, a step-up voltage converter must be used in order to raise the coil voltage to a reasonable value of a few volts (e.g., 3 V or 5 V). Switching regulators are known for their high conversion efficiency. A possibility would be to adopt a boost converter, whose simplified schematic diagram is depicted in Figure 3.

The boost converter acts as a kind of energy pump, whereby energy is first drawn from the power supply into the inductor and then transferred to the capacitor

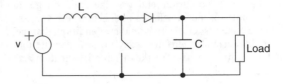

Figure 3.   Boost converter.

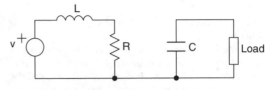

Figure 4.   Equivalent circuit of the boost converter when the switch is closed.

and to the load. Such a converter is capable, in theory, of converting a 1 mV voltage to a voltage of a few volts. The question is about the amount of power available at the load.

To answer this question, the boost converter must be analyzed in more detail. It is clear that, in the case under investigation, the input voltage generator is the coil, which produces a small sinusoidal voltage. If the switch is operated at the same frequency and, more precisely, is kept open for negative voltages and closed for positive voltages, energy can be pumped into the load.

When the switch is closed, the positive part of the input supply voltage charges the inductor. When the switch is open, the energy stored in the inductor is transferred to the load. Actually, only a fraction of the energy is transferred to the load, as a part of it is dissipated in the diode and in the inductor itself due to its Q factor. Therefore, at every cycle, some energy is first loaded into the inductor and then transferred, in part, to the load. This is equivalent to a pulsed flow of power from the coil to the load, where part of this power is lost en route. In order to have an estimate of the amount of power available at the load, one can evaluate the power flowing from the coil to the inductor, keeping losses into account. Then, we will have an upper bound on the amount of power available at the load.

Evaluating the power drawn from the coil into the inductor is quite straightforward. One considers the first half of the cycle, when the switch is closed. The current at the end of this half cycle allows evaluating the amount of energy drawn from the coil into the inductor. The equivalent circuit when the switch is kept closed is given in Figure 4.

One may write the following differential equation:

$$Ri + L\frac{di}{dt} = v \tag{2}$$

where i is the current and v is the coil voltage. R accounts for inductor losses and switch series resistance. One assumes, as a boundary condition, that the initial current is zero. It is easy to show that the following function satisfies the differential equation and the boundary condition:

$$i(t) = \frac{1}{L} e^{\frac{R}{L}t} u(t) * v(t) \tag{3}$$

where $*$ denotes convolution and $u(t)$ is the step function.

In the present case, $v(t)$ is the first half of a sinusoid of period T and amplitude $A = \omega BS$, where B is the magnetic field on the coil and S is the area of the coil. See Figure 5 for a sketch of $v(t)$.

One may write:

$$v(t) = \begin{cases} A\sin\omega t & \text{for } 0 \le t \le T/2 \\ 0 & \text{elsewhere} \end{cases} \tag{4}$$

Carrying out the convolution and taking the final value of the current (i.e., at $t = T/2$), one gets:

$$i_{T/2} = \frac{A\left(1 + e^{-\frac{\pi R}{\omega L}}\right) L\omega}{R^2 + L^2\omega^2} \tag{5}$$

The energy in the inductor is $Li^2/2$. Multiplying this value by the frequency and recalling the value of A one obtains the amount of power drawn from the coil into the inductor:

$$Power = \frac{B^2 S^2 L^3 \omega^5 \left(1 + e^{-\frac{\pi R}{\omega L}}\right)^2}{4\pi\left(R^2 + \omega^2 L^2\right)^2} \tag{6}$$

Given B, R and S, one may say that power increases about linearly with frequency and there exists a value of L for which the amount of power is maximized.

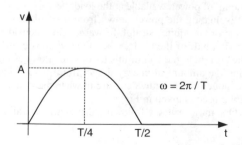

Figure 5.   Coil voltage while the switch is kept closed.

Substitute now some values in order to understand how much power is available. Assume the optimistic value of $0.1\,\Omega$ for R. Also, assume that the coil has an area $S = 25\,\text{cm}^2$, which is a reasonable size for a sensor board.

At $100\,\text{Hz}$, the field limit is $100\,\mu\text{T}$, in Italy; thence the optimum L is $400\,\mu\text{H}$, which means $10\,\text{nW}$ of available power. No device can work with such a low power.

At higher frequencies, things are not much better. On the one hand, power increases with frequency; on the other, national regulations at higher frequencies enforce a stricter limit on the magnetic field B. For example, at $100\,\text{kHz}$, the field limit (in Italy) is $250\,\text{nT}$, the optimum L is $500\,\text{nH}$, which means $60\,\text{nW}$. Again, no device can work with such a low power. With $100\,\mu\text{T}$ one would have $10\,\text{mW}$ available and would be able to operate a device. Of course, this would only be possible for smartcard-like applications, for which people would be exposed to such a high magnetic field for very limited amounts of time.

Beyond $3\,\text{MHz}$, the magnetic field limit is even lower, but, thanks to the higher frequency, one can draw some additional power. However, this is still not sufficient unless the duty cycle is that of a smartcard-like application.

1.2   Solar power

Today, solar power seems the most viable power supply technology for sensors. The most limiting factor is efficiency. For example, a panel of size $15\,\text{cm}$ by $5\,\text{cm}$ can deliver about $500\,\text{mW}$ in full sunlight. However, in cloudy conditions the amount of power generated can drop to a mere 5%–20% of the peak value. Careful sizing of solar panels and backup batteries is therefore necessary. Furthermore, in order to minimize costs, the electronic devices used should be designed for low power operation.

4   SYSTEM ARCHITECTURE

Despite one of the modern trend in sensor technology is to use the same coaxial cable for both power supply and data transmission, the present section conceives the two needs as separate.

4.1   Power supply architecture

Attention is focused on structures of such an importance to justify that power be made available, even if the infrastructure is far from inhabited areas. For instance one has to guarantee lighting in fog conditions or one must mark the presence of the bridge to avoid helicopters or aircraft impact. In the larger cases, power must be supplied because a management office has to be established near-by the artifact.

To have power available means that one can rely on cables along the bridge, where standard reliability considerations apply: two cables are better than one; a network of cables is more reliable than a chain; power supplies, alternative to the main one, have to be planned when no interruption of the monitoring system is accepted.

By these cables or cable network, units distributed along the bridge can be powered. Then each of these units serves a bridge area where sensors are located. They can be reached by:

- a star architecture of the power supply cables, so that failure of one cable results in the loss of a single sensor;
- a series architecture of the power supply cable, with the risk of losing all the sensors in case of cable failure;
- a wireless scheme as the one conceived in the previous section.

The main advantage of the last solution consists in creating a powered volume where sensors can be spread around. The amount of sensors is limited by the global power available, but their location is *de facto* free. This allows the artifact manager a fully adaptive policy of monitoring.

4.2 *Signal acquisition architecture*

One distinguishes here between the links of the powered units which can be realized with or without cables, and the links between units and sensors.
With reference to the three architectures discussed for power supply, one realizes that:

- the star architecture leads one toward a single coaxial cable technology, to be used for both power supply and data recovering;
- the series architecture would suggest a wireless configuration for data acquisition, and this does not pose special difficulties since each sensor is powered by the cable;
- the wireless scheme of power supply strongly relies on a wireless network for data acquisition but here the power requirements represent a strong constraint on the technological solution.

5 CONCLUSIONS

The design of a wireless communication between sensor/device stations strongly depends on the power supply. Three main system architectures are described throughout the paper.

When one is dealing with standard structures, the legal constraints decide the feasibility or not of the system. The availability of sensors which require a battery for power supply could result not convenient when compared with a traditional cabled sensor network.

By contrast, when large and important structures are analyzed, there are not strict cost limitations. Therefore, one can think to acquire the use of a frequency for local transmission to solve the communication problem. Nevertheless, to supply power to the sensors without wires, one still does not have a general solution available.

ACKNOWLEDGEMENTS

This paper is partially funded by the ASI (Italian Space Agency) research program coordinated at a national level by Prof. F. Bernelli-Zazzera of the Polytechnic of Milan. Many of the results were achieved by cooperation made possible by the ESF (European Science Foundation) program CONVIB, with the first author as responsible.

REFERENCES

Casciati, F., Faravelli, L. & Borghetti, F. 2003a. Wireless Links between Sensor-Device Control Stations in Long Span Bridges, *Proceedings SPIE Annual Meeting*, San Diego.

Casciati, F., Faravelli, L. & Fornasari, A. 2003b. Wireless Links between Sensor-Device Control Stations in Buildings, *Proceedings China-US Workshop on Protection of Urban Infrastructure and Public Buildings against Earthquakes and Manmade Disasters*, Beijing.

Chen, Z.Q., Wang, X.Y., Ni, Y.Q. & Ko, J.M. 2002, in Casciati F. (ed.), *Proceeding of 3rd World Conference on Structural Control*, J. Wiley, Chichester, UK, Vol. 3, 393–401.

Crossbow 2003. Smarter Sensors in Silicon.

Faravelli, L. 2002. Innovative Control Technologies for Vibration Sensitive Civil Engineering Structures in Europe, *National Workshop on Future Sensing Systems*, Lake Tahoe, Granlibakken. http://www.ce.berkeley.edu/sensors.

Farrar, C.R., Hemez, F. & Sohn, H. 2002. Developing Damage Prognosis Solution, *Proceeding of Structural Health Monitoring*, Paris, DEStech, 31–45.

Linch, J.P., Kiremidjian, A.S., Law, K.H., Kenny, T. & Carryer E. 2002. Issues in Wireless Structural Damage Monitoring Technologies, in Casciati F. (ed.), *Proceeding of 3rd World Conference on Structural Control*, J. Wiley, Chichester, UK.

System-based Vision for Strategic and Creative Design, Bontempi (ed.)
© 2003 Swets & Zeitlinger, Lisse, ISBN 90 5809 599 1

A benchmark problem for seismic control of cable-stayed bridges: lessons learned

S.R. Williams, S.J. Dyke & J.M. Caicedo
Washington University, St. Louis, USA

L.A. Bergman
University of Illinois, Urbana, USA

ABSTRACT: This paper summarizes the main results of several researchers participating in the benchmark structural control problem for cable-stayed bridges. The benchmark problem is based on the Bill Emerson Memorial Bridge that is currently under construction in Cape Girardeau, Missouri, USA, and expected to be completed late in 2003. The goal of this study is to provide a testbed for the development of strategies for the control of cable-stayed bridges. Based on detailed drawings of the Emerson bridge, a three-dimensional evaluation model has been developed to represent the complex behavior of the full scale benchmark bridge. The linear evaluation model is developed using the equations of motion generated around the deformed equilibrium position. Phase I of this study was first made available in 2001 and considers longitudinal excitations only. Phase II considers more complex structural behavior than Phase I, resulting from multi-support and transverse excitations. Evaluation criteria are presented for the design problem that are consistent with the goals of seismic control of a cable-stayed bridge. Control constraints are also provided to ensure that the benchmark results are representative of a control implementation on the physical structure. Each participant in this benchmark bridge control study is given the task of defining (including devices, sensors and algorithms), evaluating and reporting on their proposed control strategies. Participants also evaluate the robust stability of their designs by including additional mass due to snow loads. These strategies may be either passive, active, semi-active or a combination thereof. Four recent designs addressing the benchmark bridge control problem are summarized herein. The benchmark problem statement and programs are available at: *http://wusceel.cive.wustl.edu/*

1 INTRODUCTION

The control of long-span bridges presents a challenging and unique problem, with many complexities in modelling, control design and implementation. Cable-stayed bridges exhibit complex behavior in which the vertical, translational and torsional motions are often strongly coupled. Clearly, the control of very flexible bridge structures has not been studied to the same extent as buildings have. As a result, little expertise has been accumulated. Thus, the control of seismically excited cable-stayed bridges presents a challenging problem to the structural control community.

The first generation of benchmark problems on cable-stayed bridges focused on the Bill Emerson Memorial Bridge under construction in Cape Girardeau, Missouri, USA (Dyke et al. 2003a, b; also see: *<http://wusceel.cive.wustl.edu/>*). Based on detailed drawings of the Emerson bridge, a three-dimensional evaluation model was developed to represent the complex behavior of the full scale benchmark bridge. A linear evaluation model, using the equations of motion generated around the deformed equilibrium position, was deemed appropriate. Because the structure is attached to bedrock, the effects of soil-structure interaction were neglected. To simplify the problem for phase I, the problem focused on a one dimensional ground acceleration applied in the longitudinal direction and uniformly and simultaneously applied at all supports. Researchers reported their Phase I results during a theme session devoted to this problem held at the *Third World Conference on Structural Control* in April 2002 in Como, Italy (Agrawal et al. 2002; Turan et al. 2002; Moon et al. 2002a; Bakule et al. 2002).

Although a significant amount of expertise was accumulated during phase I, the assumptions made regarding the excitation limited the extent to which this problem modelled a realistic situation. A structure's response to an earthquake is based on the simultaneous

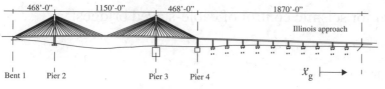

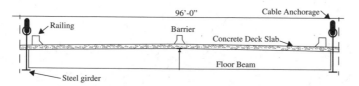

Figure 1a. Cape Girardeau bridge.

Figure 1b. Cross section of bridge deck.

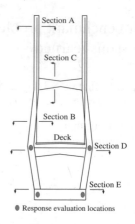

Figure 2.   Diagram of the towers.

action of three translational components of ground motion: two in the horizontal plane, and one in the vertical direction. Structures are typically analyzed for the two horizontal components of ground motion and the response depends on the incidence angle (the angle between the ground motion components and the structural axes). Additionally, the excitation is expected to vary at each of the supports due to the length of these structures. A phase II problem was developed to extend the problem to consider these issues.

This paper presents the second generation of benchmark control problems for cable-stayed bridges. In this problem the ground acceleration may be applied in any arbitrary direction using the two horizontal components of the historical earthquake with a specified incidence angle. Multi-support excitation is also considered in this phase of the study. Here the prescribed ground motion is assumed to be identical at each support, although it is not applied simultaneously. We assume that bent 1 undergoes a specified ground motion, and the motion at the other three supports is identical to this motion but delayed based on the distance between adjacent supports and the speed of the L-wave of a typical earthquake (3 km/sec). The total response of the structure is obtained by superposition of the response due to each independent support input (Chopra 2001; Clough and Penzien 1993).

Summaries of the results of four recent studies addressing this benchmark problem are presented in this paper. This benchmark has been prepared to provide a testbed for the development of effective strategies for the control of long-span bridges. To evaluate the proposed control strategies in terms that are meaningful for cable-stayed bridges, appropriate evaluation criteria and control design constraints are specified within the problem statement. Additionally, an alternate model of the bridge is developed for evaluating the robustness of the designs by including the effects of snow loads on the bridge deck. Those participating

in this benchmark study will define all components of their control system, evaluate them in the context of their proposed control strategies, and report the results. The problems are available for downloading on the benchmark web site in the form of a set of MATLAB® programs *http://wusceel.cive.wustl.edu/quake/>*.

2   BENCHMARK BRIDGE

The cable-stayed bridge used for this benchmark study is the Missouri 74 – Illinois 146 bridge spanning the Mississippi River near Cape Girardeau, Missouri, designed by HNTB Corporation.[1] The bridge is currently under construction and is scheduled to be completed in 2003. Earthquake load combinations in accordance with American Association of State Highway and Transportation Officials (AASHTO) division I-A specifications were used in the design. As shown in Figure 1a, it is composed of two towers, 128 cables, and 12 additional piers in the approach bridge from the Illinois side.

The bridge has a total length of 1205.8 m (3956 ft). The main span is 350.6 m (1150 ft), the side spans are 142.7 m (468 ft), and the approach on the Illinois side is 570 m (1870 ft). A cross section of the deck is shown in Figure 1b. The bridge has four lanes plus two narrower bicycle lanes, for a total width of 29.3 m (96 ft). The deck is composed of steel beams and prestressed concrete slabs. Steel ASTM A709 grade 50W is used, with an f_y of 344 MPa (50 ksi). The concrete slabs are made of prestressed concrete with a f_c' of 41.36 MPa (6 ksi). Additionally, a concrete barrier is located at the center of the bridge, and two railings are located along the edges of the deck.

---

[1] HNTB Corporation, 1201 Walnut Suite 700, Kansas City, Missouri, 64106.

The 128 cables are made of high-strength, low-relaxation steel (ASTM A882 grade 270). Sixteen 6.67 MN (1,500 kip) shock transmission devices are employed in the connection between the tower and deck. These devices are installed in the longitudinal direction to allow for thermal expansion of the deck. Under dynamic loading these devices are extremely stiff. Additionally, in the transverse direction, earthquake restrainers are employed at the connections between the tower and the deck, and the deck is constrained in the vertical direction at the towers. The bearings at bent 1 and pier 4 are designed to permit longitudinal displacement and rotation about the transverse and vertical axes. Soil-structure interaction is not expected to be significant in this site as the foundations of the cable-stayed portion of the bridge are attached to bedrock.

The H-shaped towers have a height of 100 m (336 ft) at pier 2 and 105 m (356 ft) at pier 3. Each tower supports a total 64 cables. The towers are constructed of reinforced concrete with a resistance, f_c', of 37.92 MPa (5.5 ksi). The cross section of each tower changes five times over its height. The deck consists of a rigid diaphragm made of steel, with a slab of concrete at the top.

A three dimensional finite element model of the Cape Girardeau bridge has been developed. The finite element model employs beam elements, cable elements and rigid links. The nonlinear static analysis is performed in ABAQUS® (1998), and the element mass and stiffness matrices at equilibrium are output to MATLAB® for assembly. Subsequently, the constraints are applied, and a dynamic condensation is performed to reduce the size of the model. A linear evaluation model is used in this benchmark study, however, the stiffness matrices used corresponding to the deformed state of the bridge considering dead loads. The first ten natural frequencies of the evaluation model are 0.290, 0.370, 0.468, 0.516, 0.581, 0.649, 0.669, 0.697, 0.710, and 0.720 Hz.

To make it possible for designers/researchers to place devices acting longitudinally between the deck and tower, a modified evaluation model is constructed in which the existing connections between the tower and deck are disconnected. If a designer/researcher specifies devices at these nodes, the modified model will be assembled as the evaluation model, and user-specified control devices must connect the deck to the tower. As expected, the natural frequencies of this modified model are lower than those of the nominal bridge model. The fundamental frequency of this modified model, in the absence of user-specified devices between the tower and deck is 0.162 Hz. Note that the uncontrolled structure used as a basis of comparison for the controlled system corresponds to the original model in which the deck-tower connections are fixed (i.e., the dynamically stiff shock transmission devices are present).

3 BENCHMARK PROBLEM STATEMENT

Designers/researchers are given the task of designing a control system for the benchmark bridge and evaluating its performance based on a specified set of evaluation criteria. The files contained in the MATLAB® problem statement will generate the evaluation models of the bridge. This includes specifying the type and location of each of the control devices and sensors, as well as the control algorithm. The procedure to follow for development and evaluation of a control strategy is described in the flow chart in Figure 3.

Devices may be attached to any active node of the bridge model, including the connection between the deck and the tower. A sample active control design is included to guide benchmark participants through the required constraints and design criteria. The following sections describe the procedure to be followed in using the benchmark bridge problem statement.

3.1 Interfacing with the benchmark bridge evaluation model

To interface with the benchmark files, the researcher/designer will select the components of each proposed control strategy, including: the control device locations; the measured outputs sent to the sensors, $\mathbf{y}_m$; the evaluation outputs $\mathbf{y}_e$; and the connection outputs $\mathbf{y}_c$. These inputs and outputs are specified within an input/output file provided with the benchmark problem statement. A MATLAB® graphical user interface is provided to simplify this procedure. However, this information can be inserted directly into the input/output file in terms of the node numbers, if preferred.

The evaluation model and earthquake inputs are fixed for this benchmark problem. Models of the sensors and control devices are to be defined by the participants. Dynamics of control devices may be neglected. However, dynamic device models may be employed by participants if they so choose.

3.2 Benchmark problem simulation

Each proposed control strategy is evaluated through simulation with the evaluation model. The SIMULINK® program provided has been developed for evaluation of the control strategies. Participants should insert models of their sensors, devices, and algorithms into the simulation to evaluate their proposed control strategies. The simulation uses the analysis tool provided by Ohtori and Spencer (1999).

3.3 Evaluation criteria

The seismic response of cable-stayed bridges is highly coupled. Lateral excitations can generate vertical and

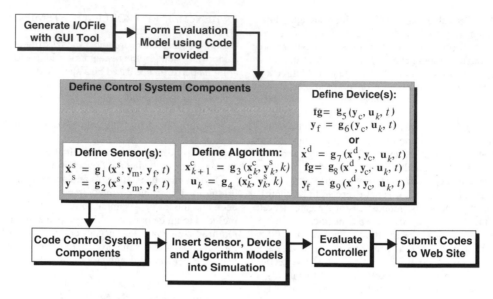

Figure 3. Flow chart of benchmark solution procedure.

longitudinal motions. For cable-stayed bridges subjected to earthquake loading, the relevant responses are related to the structural integrity of the bridge rather than to serviceability issues. Thus, in evaluating the performance of each control algorithm, the shear forces and moments in the towers at key locations (see Figure 2) must be considered. Additionally, the tension in the cables should never approach zero and should remain close to the nominal pretension.

A set of eighteen criteria have been developed to evaluate the capabilities of each control strategy. Because the earthquake is assumed to have two horizontal components at a specified incidence angle, several of these criteria are evaluated in both the X (longitudinal) and Z (transverse) directions. The first six evaluation criteria consider the ability of the controller to reduce peak responses, the second five consider normed responses over the entire time record, and the last seven consider the requirements of the control system itself.

For each control design, the evaluation criteria should be determined for each of three earthquake records provided in the benchmark package: i) *El Centro*. Recorded at the Imperial Valley Irrigation District substation in El Centro, California, during the Imperial Valley, California earthquake of May, 18, 1940; ii) *Mexico City*. Recorded at the Galeta de Campos station with site Geology of Meta-Andesite Breccia in Sep. 19, 1985; iii) *Gebze*. Recorded at the Gebze Tubitak Marmara Arastirma Merkezi on Aug. 17, 1999. These three earthquakes are each at or below the design peak ground acceleration level for the bridge. The evaluation criteria are shown in Table 1.

The values of the uncontrolled responses for the three earthquakes required to calculate the evaluation criteria are provided in the MATLAB® files and in Dyke et al. (2003a, b). All eighteen criteria should be reported for each proposed controller. The Mexico City, El Centro, and Gebze earthquakes should all be considered in determining the evaluation criteria. However, designers/researchers are encouraged to include additional criteria in their results if, through these additional criteria, their results demonstrate an overall desirable quality.

3.4 *Control constraints and procedures*

To allow researchers/designers to compare and contrast various control strategies, each of the controllers must be subjected to a uniform set of constraints and procedures. A set of constraints has been formulated for this benchmark problem to ensure that the proposed control strategies can realistically be implemented on the real structure. For instance, the precision and span of the A/D and D/A converters used for implementation of control algorithms are fixed, and the amount of noise in the sensors is specified. When active systems are implemented, participants should discuss the robust stability of their control systems. Additionally, participants must provide justification for sensors and devices that are employed in their proposed control strategies. Force and stroke limitations of the devices used should be provided, and the controller should meet these limitations upon implementation in the system. Furthermore, tension in the stay cables must remain within a specified range of values to reduce the

2100

Table 1: Evaluation criteria.

PEAK RESPONSES				
Base Shear[*] $$J_1 = \max_{\substack{\text{El Centro}\\\text{Mexico City}\\\text{Gebze}}} \left\{ \frac{\max_{i,t}|F_{bi}(t)|}{F_{0b}^{max}} \right\}$$	**Shear at Deck Level**[*] $$J_2 = \max_{\substack{\text{El Centro}\\\text{Mexico City}\\\text{Gebze}}} \left\{ \frac{\max_{i,t}|F_{di}(t)|}{F_{0d}^{max}} \right\}$$	**Overturning Moment**[*] $$J_3 = \max_{\substack{\text{El Centro}\\\text{Mexico City}\\\text{Gebze}}} \left\{ \frac{\max_{i,t}|M_{bi}(t)|}{M_{0b}^{max}} \right\}$$	**Moment at Deck Level**[*] $$J_4 = \max_{\substack{\text{El Centro}\\\text{Mexico City}\\\text{Gebze}}} \left\{ \frac{\max_{i,t}|M_{di}(t)|}{M_{0d}^{max}} \right\}$$	
Stay Cable Tension $$J_5 = \max_{\substack{\text{El Centro}\\\text{Mexico City}\\\text{Gebze}}} \left\{ \max_{i,t} \left| \frac{T_{ai}(t) - T_{0i}}{T_{0i}} \right| \right\}$$	**Displacement at Abutments** $$J_6 = \max_{\substack{\text{El Centro}\\\text{Mexico City}\\\text{Gebze}}} \left\{ \max_{i,t} \left| \frac{x_{bi}(t)}{x_{0b}} \right| \right\}$$	**Base Shear**[*] $$J_7 = \max_{\substack{\text{El Centro}\\\text{Mexico City}\\\text{Gebze}}} \left\{ \frac{\max_i \|F_{bi}(t)\|}{\|F_{0b}(t)\|} \right\}$$	**Shear at Deck Level**[*] $$J_8 = \max_{\substack{\text{El Centro}\\\text{Mexico City}\\\text{Gebze}}} \left\{ \frac{\max_i \|F_{di}(t)\|}{\|F_{0d}(t)\|} \right\}$$	NORMED RESPONSES
Overturning Moment[*] $$J_9 = \max_{\substack{\text{El Centro}\\\text{Mexico City}\\\text{Gebze}}} \left\{ \frac{\max_i \|M_{bi}(t)\|}{\|M_{0b}(t)\|} \right\}$$		**Moment at Deck Level**[*] $$J_{10} = \max_{\substack{\text{El Centro}\\\text{Mexico City}\\\text{Gebze}}} \left\{ \frac{\max_i \|M_{di}(t)\|}{\|M_{0d}(t)\|} \right\}$$	**Stay Cable Tension** $$J_{11} = \max_{\substack{\text{El Centro}\\\text{Mexico City}\\\text{Gebze}}} \left\{ \max_i \frac{\|T_{ai}(t) - T_{0i}\|}{T_{0i}} \right\}$$	
CONTROLLER				
Control Force[*] $$J_{12} = \max_{\substack{\text{El Centro}\\\text{Mexico City}\\\text{Gebze}}} \left\{ \max_{i,t} \left(\frac{f_i(t)}{W} \right) \right\}$$	**Device Stroke**[*] $$J_{13} = \max_{\substack{\text{El Centro}\\\text{Mexico City}\\\text{Gebze}}} \left\{ \max_{i,t} \left(\frac{|y_i^d(t)|}{x_0^{max}} \right) \right\}$$	**Device Peak Power**[*] $$J_{14} = \max_{\substack{\text{El Centro}\\\text{Mexico City}\\\text{Gebze}}} \left\{ \frac{\max_t \left[\sum_i P_i(t) \right]}{x_0^{max} W} \right\}$$	**Device Total Power**[*] $$J_{15} = \max_{\substack{\text{El Centro}\\\text{Mexico City}\\\text{Gebze}}} \frac{\sum_i \int_0^{t_f} P_i(t)\,dt}{x_0^{max} W}$$	
$J_{16} = \#$ of control devices		$J_{17} = \#$ of sensors	$J_{18} = \dim(\mathbf{x}_k^c)$	

[*]. Twelve evaluation criteria should be evaluated for both the X (longitudinal) and Z (transverse) directions.

possibility of unseating or failure of the cables. These constraints and procedures are described in detail in Dyke et al. (2003a, b).

4 DISCUSSION OF PARTICIPANT RESULTS

Here we summarize the main results obtained by several investigators in phase I and phase II of this problem statement.

The first case to be discussed compares active and passive controllers (Bontempi et al. 2003) for phase I. An LQG active control design and a passive systems are considered. The actuators considered are electro-hydraulic with a capacity of 1000 kN to 5000 kN, and actuator dynamics are not considered as they are assumed to be of a much higher frequency than those of the bridge. As this paper considers Phase I criteria, the excitation is in the longitudinal direction only, with simultaneous excitation of the supports. The passive systems considered include viscous dampers, viscoelastic dampers, and elastoplastic dampers. Also, three schemes of active control are considered. The first employs eight actuators between the deck and supports, the second has eight actuators on the longest stays of the bridge, and the third employs a combination of the two previous schemes (16 actuators). Sensors include both displacement sensors and accelerometers. Their findings show that, for active

control, the combination method results in the best performance followed by the first and second methods. For the passive systems, the viscoelastic systems gave the best performance. Overall, they conclude that the passive system is most suited to the situation due to performance and energy considerations. They also mention that they would need to consider the stipulations of Phase II in order to have an accurate understanding of the efficacy of their systems.

Another group, Yang et al. (2003), considered the Phase I problem using H_2-based controllers. In this control strategy, there are two basic approaches: the first utilizes energy-bounded excitations and the second employs peak-bounded excitations. Both state feedback and dynamic feedback control algorithms are developed in the H_2 strategy, and the performance is compared to the sample controller (Phase I) presented in the benchmark problem. The state feedback controllers are governed by online estimators developed using the Kalman filter while dynamic feedback controllers employ an LMI approach. The authors have chosen to place 24 hydraulic actuators on the structure in eight locations. The first four take the place of the fixed connections between the towers and deck, two are between the bent and deck, and two are between deck and pier 4. These are all oriented to apply forces longitudinally with a force limit of 1000 kN. For the control problem, the same Kalman filter estimator is used as in the sample design, and only the hard

2101

constraints are considered. By varying the parameters of the H_2 controllers, they were able to push the control resources to obtain better performance than was obtained in the sample method. This approach was selected as the preferred design, as dynamic feedback methods require more computation and memory to achieve insignificant gains in performance.

In response to the Phase II problem definition, Iemura et al. (2003) employ a pseudo-negative stiffness (PNS) to obtain a robust controller in the presence of snow loads. The idea of PNS is to improve on the performance of conventional semi-active control strategies. In this case, variable orifice fluid dampers are used to achieve PNS. The damping force developed in this method is inversely proportional to the square of the opening ratio h, which remains between 0.05 and 0.8. The force which is desired is developed by calculating the opening ratio from this relationship. At times, though, the algorithm demands that the force have the same sign as the velocity of the device, which is impossible with semi-active devices, and thus the orifice opening size is then maximized to allow as little force to develop as possible. In conventional semi-active control, the total force developed can exceed the maximum elastic force the towers are capable of withstanding. By using PNS, the authors have been able to dissipate the same amount of energy while keeping the forces within the acceptable range. Displacements are measured and fed back to the control algorithm. Results show that this method absorbs much more energy than traditional passive systems, and even out-performs active control in responses such as the shear and moment developed at tower bases. Under a snow load, which changes the natural frequencies (max of 4.8%) of the bridge, the PNS method performed better than passive control.

Finally, Park et al. (2003) employ two hybrid control systems to maximize efficacy of the control strategy. The two types of hybrid control are as follows: hydraulic actuators (HA's) plus lead rubber bearings (LRB's), or magnetorheological dampers (MRD's) and LRB's. The LRB's constitute the passive element, and the active or semi-active systems supplement the action of the passive system. This type of system is more robust due to redundancy, as well as interaction between control systems lending the benefits of each to the control strategy. While the passive system reduces forces transmitted to towers, it allows large displacements of the deck: the active or semi-active system control these displacements but tend to increase the forces on the towers. In the active system, there are a total of 14 LRB's installed on the bridge, and 24 HA's. They are oriented so that the control force is applied in the longitudinal direction. Fourteen accelerometers and four displacement sensors monitor the movement of the bridge, measuring biaxially. In the semi-active hybrid system, 14 LRB's and 24 MRD's are applied to the structure. In addition to the 18 sensors for structural movement, there are 24 force transducers for the MRD's. The control algorithm from the active case is employed in the semiactive design, but modified to only apply force when it is dissipative, as that is the main limitation of a semi-active device. It was found that the hybrid system outperformed passive, active, and semi-active systems alone in nominal conditions. In the case of a snow load which changes the mass properties of the bridge the active and semi-active failed where the hybrid system performed as expected. In general, their conclusion is that the hybrid system is a robust and efficient control method.

5 SUMMARY

A summary of the benchmark problem for control of a seismically-excited cable-stayed bridge has been presented including four recent studies by problem participants. The problem statement is available at:

http://wusceel.cive.wustl.edu/quake/

If you cannot access the World Wide Web or have questions regarding the benchmark problem, please contact Dr. Shirley J. Dyke via e-mail at: *sdyke@seas.wustl.edu*.

ACKNOWLEDGMENTS

This research is supported in part by National Science Foundation Grant No. CMS 97-33272 (Dr. S.C. Liu, Program Director). The authors thank Shyam Gupta and Bill Strossener from the MODOT for information on the Cape Girardeau Bridge. The helpful advice of Mr. Steven Hague (HNTB Corp.), Prof. Yozo Fujino (Univ. of Tokyo), Prof. Masato Abe (Univ. of Tokyo), Prof. Hirokazu Iemura (Kyoto Univ.), Prof. Joel Conte (UCLA), and Prof. Fabio Biondini (Politecnico di Milano) is gratefully acknowledged. Additional input provided by members of the ASCE Task Group on Benchmark Problems and member-at-large of the structural control community is also recognized.

REFERENCES

ABAQUS® (1996). Hibbitt, Karlsson & Sorensen Inc. Pawtucket, RI.

Bontempi, F., Casciati, F., and Giudici, M. (2003). "Seismic Response of a Cable Stayed Bridge: active and passive control systems," *Journal of Structural Control* (submitted).

Caicedo, J.M., Dyke, S.J., Moon, S.J., Bergman, L.A., Turan, G., and Hague, S. (2003) "Phase II Benchmkar Control Problem for Seismic Response of Cable-Stayed Bridges," *Journal of Structural Control*, (submitted).

Caughey, T. K. (1998). "The Benchmark Problem." *Earthquake Engr. and Structural Dynamics*, Vol. 27, pp. 1125.

Celebi, M. (1998), *Final Proposal for Seismic Instrumentation of the Cable-Stayed Girardeau (MO) Bridge*, U.S. Geological Survey.

Dyke, S.J., Caicedo, J.M., Turan, G., Bergman, L.A., and Hague, S. (2003). "Phase I Benchmark Control Problem for Seismic Response of Cable-Stayed Bridges," *Journal of Structural Engineering*, ASCE, (in press).

Hague, S. (1997). "Composite Design for Long Span Bridges." *Proc. of the XV ASCE Struc. Congress*, Portland, Oregon.

Iemura, H., and Pradono, M.H., "Application of pseudo negative stiffness control to the benchmark cable-stayed bridge," *Journal of Structural Control*, 2003 (submitted).

MATLAB® (1997). The Math Works, Inc. Natick, Massachusetts.

Ohtori, Y. and Spencer, B.F., Jr. (1999). "A MATLAB®-Based Tool for Nonlinear Structural Analysis," *Proceedings of the 13th Engineering Mechanics Conference*, Baltimore, Maryland, June 13–16.

Park, K., Jung, H. Spencer, B.F., Jr., and Lee, I. (2003). "Hybrid Control Systems for Seismic Protections of a Phase II Benchmark Cable-Stayed Bridge," *Journal of Structural Control* (submitted).

Yang, J.N., Lin, S., and Fabbari, F. (2003). "H_2 Based Control Strategies For Civil Engineering Structures," *Journal of Structural Control*, (submitted).

Interval expression of uncertainty for estimated ambient modal parameters

Y.G. Wang
Xi'an Jiaotong University, Chang'an University, Xi'an, China

H. Li & J.H. Zhang
Xi'an Jiaotong University, Xi'an, China

ABSTRACT: The modal parameters play the important role in the structural health monitoring or evaluating performance capacity of the existing constructions. The ambient modal parameters should be employed firstly because the test cost is the cheapest. But a variety of the ignored uncertainties, which rises from the noise dependence, ambient excitation dependence, amplitude dependence, model dependence and the restricted measurement points, instruments, data period and so on, limits the applications of the ambient modal parameters. The interval expression of the uncertainty for the ambient modal parameters is presented in this paper. A comprehensive PSD-based method (CPM) of estimating the ambient modal parameters is described. In order to analyze the uncertainty, the modal parameters are given in the form of interval sets to take the unavoidable uncertainties into account. For estimating the upper and lower limits of the modal frequencies and modal shapes, a probabilistic expression is cited. By using the statistic method, the mean values and standard deviations are estimated and their distributions are verified. A seven-story building has been measured and practiced. The results show that CPM has high resolution and accuracy. For short data series with low Signal-to-Noise Ratio (SNR), MEM (Maximum Entropy Spectral Method) has the excellent performance. It is more reasonable that the interval values of modal parameters can be employed for assessment of the existing constructions.

1 INTRODUCTION

In structural health monitoring, the modal parameters are employed to evaluate structural state, which is studied by many scholars (Vestroni & Capecchi 2000). However, there are two critical factors: the sensitivity to damage and uncertainty of modal parameters, which limit the method in practical application. It is difficult to use the modal frequencies detecting the local damage because of the insensitivity, for example, the local crack. However, the variation of the modal parameters can be employed to assess the global performance of a structure. But it is very important to study the uncertainty of the estimated modal parameters. This paper serves to evaluate the seismic property of an existent building, so the uncertainty of the estimated modal parameters is discussed in this application filed.

To employ the modal parameters for evaluating the assessed property of an existent structure, the forced vibration test or ambient vibration test can be used (Memari 1999, Ivvanonic 2000). The forced vibration test is a developed technology but costs much more than the ambient vibration test. The ambient vibration measurement and modal parameters are only discussed in this paper. The estimated ambient parameters are with big uncertainty. The uncertainty of the ambient modal parameters can be divided into model-dependence, uncompleted information-dependence, measured noise-dependence and estimation method-dependence. The model-dependent uncertainty is mainly from the non-linearity that makes the estimated modal parameters dependant on the amplitude of the excitation. There is micro-vibration in ambient test, the ambient modal parameters are only employed for seismic check of a structure in elastic raging not in plastic raging, so non-linearity is not considered. Additionally, the diurnal variation of temperature gives rise to the model-dependent uncertainty. The uncompleted information-dependent uncertainty comes from the information without excitation force, a bit of measurement points, too short recording period or too less data samples. The ambient vibration signal is with fairly low SNR and the length of recorded stationary data is always limited. So it is necessary to consider the uncertainty from measure noise-dependence. It is better that the ambient vibration test is done in the night. The estimated modal parameters are with different statistic properties by using the different estimation procedures. PSD-based Method employed in this paper will also introduce errors due to its resolution and average number. The spectral estimation procedure needs to be improved.

In modern engineering problem, the spectrum estimation methods can be mainly classified into two groups. non-parameter spectrum estimation and parameter methods. Non-parameter spectrum estimation has been widely used and shows good performance for a long data record period. But when the recorded data is not enough long, its resolution will not be adequate. Parameter spectrum estimation, also named modern spectral analysis, has many great advantages, for example, its resolution is far higher than that of non-parameter method especially in case of very short data series. There are mainly ARMA spectrum analysis, Maximum likelihood method, Entropy spectrum estimation and Eigen-analysis method. ARMA method estimates the PSD of a signal by modelling a stationary linear process to fit a white noise input. Entropy spectrum estimation includes the famous Maximum Entropy Method (MEM) and the Minimum Cross-Entropy Method. Prony Method, MUSIC Method and so on belong to Eigen-analysis method. Compared with classic spectral estimation, the modern spectral analysis methods are fairly suitable for estimating the modal parameters under ambient vibration, not only because the signal of ambient vibration is always non-stationary and its SNR is fairly low, but also the availableness of recorded data is limited (Sim 2000). MEM is adopted to estimate the modal parameters in this paper (Zhu 1999).

In the past, uncertainty in estimating the modal parameters was paid a little attention. Limited research has focused on this subject as follows. Both the noise parameters and the uncertainty in the noise parameters are studied (Briaire 1980) when the noise is used as a diagnostic tool to determine the reliability of a device. Meanwhile, the error is calculated on frequency index for Gaussian 1/f noise. Considering the uncertainty in the updated model parameters, the paper cites a Bayesian probabilistic method for structural health monitoring in theory (Vanik 2000). In view of the complication of uncertain factors, a statistical analysis of modal parameters is studied in their paper and the theoretic distribution is verified. It is very difficult to use the probabilistic approach of modal parameters in structural seismic evaluation, especially, modal shapes. Therefore, in the probability framework an interval expression is presented in this paper.

2 INTERVAL EXPRESSION OF UNCERTAINTY

Suppose that the structural parameters are without uncertain. Let the modal frequency and modal shape sets be respectively

$$\Omega := \{\omega_r, r = 1 \cdots n\} \tag{1}$$

$$\Phi := \{\varphi_r, r = 1 \cdots n, \varphi_r \in C^n\} \tag{2}$$

where n is number of degree of freedom . The statistic properties of the estimated modal parameters come from the ambient vibration test. Let the ambient modal frequency and modal shape sets be respectively

$$\hat{\Omega} := \{\hat{\omega}_r, r = 1 \cdots m; \hat{\omega}_r \in [\hat{\omega}_{rL} \quad \hat{\omega}_{rU}]\} \tag{3}$$

$$\hat{\Phi} := \{\hat{\varphi}_r, r = 1 \cdots m, \hat{\varphi}_r \in C^{n_r}; \hat{\varphi}_r \in [\hat{\varphi}_{rL} \quad \hat{\varphi}_{rU}]\} \tag{4}$$

where m and n_r are number of the measured modes and measurement positions respectively. Generally, $n >> n_r > m$. $\hat{\omega}_{rL}$ and $\hat{\omega}_{rU}$ are lower and upper limits of r th order modal frequency respectively, $\hat{\varphi}_{rL}$ and $\hat{\varphi}_{rU}$ are lower and upper boundaries of r th order modal shape respectively.

To calculate $\hat{\Omega}$ and $\hat{\Phi}$, the CPM estimation procedure with MEM is presented firstly, then the source of the uncertainty is analyzed and the probability expression is quoted to estimate the lower and upper limits of interval expression.

3 MEM ESTIMATION PROCEDURE

The spectral estimation methods have been fully developed. Modern spectral analysis is especially suitable for the low SNR and short data series such as ambient signal. MEM is adopted to estimate the modal parameters of a structure under ambient vibration test in this paper.

It is well known that the Maximum-Entropy Method (MEM) not only has high resolution in spectral analysis, but also suits short data series well. The idea of Burg spectrum estimation is to retain the auto-correlation function (ACF) of the known data series, and extrapolate the ACF of unknown delayed series with satisfying Maximum Entropy criterion at the same time. Since entropy is a measurement of uncertainty and randomicity, MEM is identical to find an auto-correlation sequence to make the signal as stochastic (white) as possible. The high resolution of MEM spectrum estimation is due to the extended information of autocorrelation. In general, MEM fits a high order autoregressive (AR) model to the signal. The spectrum estimation is as follows (Zhang, 1995)

$$P_{MEM}(\omega) = \frac{\sigma^2}{\left|1 + \sum_{i=1}^{p} a_i e^{-j\omega\Delta i}\right|^2} \tag{5}$$

where σ^2 is the variance of the white noise in AR model, a_i is the coefficient of AR model.

Since the recorded data of the ambient vibration is long enough, but non-stationary, the test data is divided into hundreds of short data samples. To estimate the modal parameters more accurately, original recorded data are filtered respectively with the center frequencies, which are estimated from a sample series. Using MEM with 15 orders, the first three frequencies and the corresponding PSD values are estimated for every sample. Since the MEM can estimate only the PSD and the PSD do not contain phase information, signal function derived from the CSD by FFT is necessary.

Let

$$\widetilde{X} = \{\widetilde{X}_i = \{\widetilde{x}_j, j = 1 \cdots N, \widetilde{x}_j \in R^{n_s}\}, i = 1 \cdots n_r\} \quad (6)$$

be the sampled ambient vibration set. There are n_r measurement points and N samples for each measurement point. n_s is the length of a time series. Let P be the PSD set estimated by MEM. For estimated modes, $(n_r \times N) \times n_s$ data are employed in the statistics of the modal frequencies and modal shapes. Figure 1 shows the flow chart of the estimation procedure by MEM method.

Theoretically, the Probability Density Function (PDF) of $\hat{\Omega}$ and $\hat{\Phi}$ can be derived by the statistical properties of $\widetilde{X}$, but it is very complex. As mentioned above, there are great of factors to make the estimated modal parameters be uncertainty, so that Gaussian distribution can be supposed (Vanik 2000).

$$p(\hat{\omega}_r^2) = c_1 \exp\left[-\frac{1}{2}\left(\frac{\hat{\omega}_r^2 - \omega_r^2}{\varepsilon_r}\right)^2\right] \quad r = 1 \cdots m \quad (7)$$

$$p(\hat{\vartheta}_r) = c_2 \exp\left[-\frac{\vartheta_r^T \Gamma^T (I - \hat{\vartheta}_r \hat{\vartheta}_r^T)\Gamma \vartheta_r}{2\delta_r^2 \|\Gamma \vartheta_r\|^2}\right] \quad r = 1 \cdots m$$

$$(8)$$

where c_1 and c_2 are the normalized constants, $\Gamma \in R^{n_r \times n}$ is a position matrix which points the measured degrees of freedom ω_r and ϑ_r are the true modal parameters, but they are unknown.

In fact, the source of uncertainty is always very complicated and it is quite difficult to get an accurate

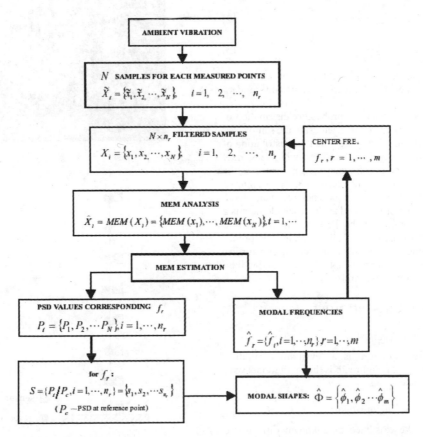

Figure 1.   Flow chart of modal estimation by MEM method.

2107

PDF. Also, the length of recorded data is always not enough. As mentioned above, MEM has a great advantage which especially suits short series. Then the recorded data are divided into N short data samples. Instead of considering only one group of estimated modal parameters, the modal parameters estimated from each sample are expressed in the form of a set to take the uncertainty into account. The statistic characteristics of the modal parameter set, such as mean value μ, standard deviation D and percent standard deviation d, are given.

$$\mu_r^2 = \frac{1}{N}\sum_{i=1}^{N}\hat{\omega}_{r,i}^2 \quad r = 1\cdots m \tag{9}$$

$$D_r^2 = \frac{1}{N}\sum_{i=1}^{N}\left(\hat{\omega}_{ri} - \mu_r\right)^2 \quad r = 1\cdots m \tag{10}$$

$$d_r^2 = \frac{1}{N}\sum_{i=1}^{N}\left(\frac{\hat{\omega}_{ri}^2 - \mu_r^2}{\mu_r^2}\right)^2 \quad r = 1\cdots m \tag{11}$$

Therefore

$$\hat{\omega}_{rL} = \mu_r - a_r D_r \quad r = 1\cdots m \tag{12}$$

$$\hat{\omega}_{rU} = \mu_r + a_r D_r \quad r = 1\cdots m \tag{13}$$

For modal shapes, similarly, the upper and lower boundary are given with the mean value and standard deviation.

Meanwhile, the theoretic probability distribution of modal parameters can be verified using the sample set. From the Equation 7, it is clear that the distribution of the estimated frequencies is consistent with the Normal Distribution, if the true frequency ω_r is treated as the expectation of samples. For Equation 8, the meaning of the PDF of modal shape vectors is not well understood. A stochastic vector is treated as that each element is a stochastic variable and the PDF of the vector is a combination of the PDF of all elements. The probability meaning for the stochastic vector, and further for the stochastic matrix, is not very clear and will be studied in the future research.

4 A PRACTICAL STRUCTURE

4.1 Measured structure

The measured structure is a seven-story building shown in Figure 2a. Figure 2b is the math model with the measured positions.

4.2 Pre-processing of signal

Considering the existence of unknown disturbance and the limited record of the ambient vibration, signal is divided into 138 sequences each containing 1024 data. Sampling rate is 120 HZ. The following examinations are conducted to check if stochastic theory is applicable for the sampling data (Li 1981). The verification of stationary is paid much attention to whether the collecting signal is stationary. While the signal fluctuates greatly, it will be rejected and collects a new one. Figure 3 is an example recorded at the 3rd story that the mean value of signal fluctuates little, peaks and wave shape vary equally and the makeup of frequencies is relatively identical, which indicates that the stochastic sampling signal is fairly stationary.

Figure 2a.

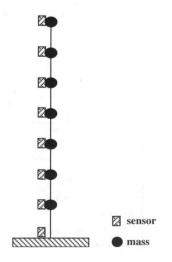

☒ sensor

● mass

Figure 2b.

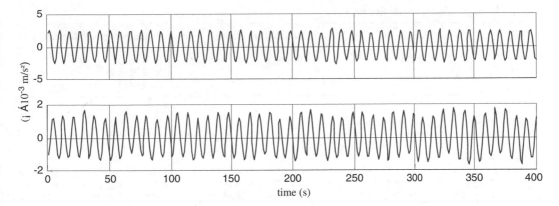

Figure 3. Acceleration response at 3rd floor.

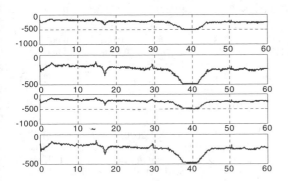

Figure 4. CSD of long data sequence.

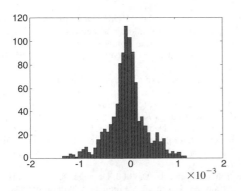

Figure 5. Histogram of sampling data at 4th floor.

Figure 4 is the cross power spectrum density (CSD) of a long data sequence (1024 × 10) at the 1st, 2nd, 5th and 7th floor respecter. Peaks of CSD curves are relatively easy to be identified, which shows that the periodic components containing in the stochastic signal are distinct. Stochastic signal at each measuring story is analyzed by plotting histograms. It can be seen that signal is stationary during the whole band and its histogram shape tallies with that of the theoretic Normal Distribution. It's feasible to treat the sampling signal as a normal one. Figure 5 is the histogram of a sample signal at the 4th story.

On the basis of above examinations, according to obtaining the interval estimation of the modal parameters, the average value of statistic is only needed.

4.3 Statistic results

Table 1 is the statistic results of the estimated frequencies. In order to give a better estimation of the mode shapes, these singular values should be removed from the sets to guarantee the rationality of mode shape. The statistic results of mode shapes in the table 2 are all pretreated.

For the estimated frequency sets, statistical results verify that the first and third frequency sets are consistent with the Normal Distribution but the second frequency is not. For the mode shapes, because the meaning of the PDF for stochastic vector and matrix is not clear, only statistical parameters of the estimated stochastic mode shape vectors are given and the concerning problems will be studied in the future research.

5 CONCLUSION

Modern spectral analysis presents some evident advantages compared with non-parameter spectrum method such as high resolution, excellent suitability for short data series and satisfactory performance for low signal-to-noise (SNR) signal and so on. From the above study, it's clear that peaks of MEM spectrum are sharp and the resolution is quite high even though the length of data series is very short, so the CPM with MEN is very efficient for estimating the ambient modal parameters.

Table 1. Statistics of the first three frequency sets.

	Mean value	Standard deviation	Percent standard deviation (%)
1st Freq	3.2214	0.1565	4.8590
2nd Freq	14.2315	0.4155	2.9194
3rd Freq	29.1151	0.5653	1.9415

Table 2. Statistics of the first three mode shape sets.

Story		1	2	3	4	5	6	7
1st Mode	Mean	0.0246	0.0541	0.1079	0.2279	0.4010	0.6179	1.000
(1 × 123)	SD*	0.0305	0.0408	0.0776	0.1034	0.0750	0.0529	0.000
	SDP(%)	77.3765	58.3747	51.8408	29.4816	15.6361	6.8608	0.000
	Mean	−0.5451	−0.7275	−0.9371	−0.9035	−0.4289	0.4527	1.000
2nd Mode	SD*	0.4329	0.6335	0.9449	0.7857	0.3912	0.3592	0.000
(1 × 106)	SDP(%)	79.4292	87.0838	100.8245	86.9694	91.2212	79.3418	0.000
	Mean	1.0999	−0.6956	−1.1664	−1.1000	−0.5709	−0.0019	1.000
3rd Mode	SD*	0.9317	0.5092	0.4309	0.5278	0.2442	0.0017	0.000
(1 × 123)	SDP(%)	84.7089	73.2001	36.9432	47.9772	42.7805	93.1869	0.000

Note: SD* means standard deviation; SDP means percent standard deviation (%).

The interval expression of the ambient model parameters is very flexible for the seismic evaluation. The statistic procedure can determine the lower and upper limitation of the natural frequencies as well as the boundary of the modal shapes. Each order of estimated frequencies intensively assembles at a narrow band separately. The band becomes wider but the percent standard deviation decreases with the growth of order. For modal shapes, statistical results show that the estimation of the second order mode shape is not as good as the first and the third orders.

Since there is a strong power disturbance whose frequency is adjacent to the second frequency of the structure, the estimation of this modal shape is unavoidable to be influenced to some extent though the disturbance signal has been filtered. Meanwhile the mode shapes are more sensitive in modal identification.

In the further research, rather than consider only point estimates, the probabilistic method will take the uncertainties into account. By using the stiffness derived from the sets of modal parameters, a probabilistic seismic performance and damage measure may be developed.

ACKNOWLEDGEMENT

The financial support of the research project was provided by Chinese Nature Science Fundament No.10176024. The authors would express their thanks to Dr. C.C. Zhu and Dr. C.Y. He for their valuable cooperation and support.

REFERENCES

Briaire, J.L., Vandamme, K.J. 2000. "The Influence of a digital spectrum analyzer on the uncertainty in 1/f noise parameters". *Microelectronics Reliablity*, 40, 1975–1980

Ivanovic, M., Trifunac, D. et al. 2000. Ambient Vibration Tests of a Seven-story Reinforced Concrete Building in van Nuys, California, Damaged by the 1994 Northridge Earthquake. *Soil Dynamics and Earthquake Engineering*, 19, 391–411

Li, L., Chunyu, Wu 1998. Determining Dynamic Characteristics of Building Structures by Detection of Pulsatings at Separate Points One by One. *Journal of Experimental Mechanics*, Vol.13, No.2, Jun., 155–161, (In Chinese)

Memari, A.M., Aghakouchak, A.A. et al. 1999. Full-scale dynamic testing of a steel frame building during construction. *Engineering Structure*, 21, 1115–1127

Sim, C.-M., Chang, K.-O. et al. 2000. Burg Spectrum Estimation of Neutron Noise for the Monitoring of Reactor Internal Vibration with a Hypothesis Test. *International Journal of Pressure Vessels and Piping*, 77, 27–33

Vestroni, F., Capecchi, D. 2000. Damage Detection in Beam Structures Based on Frequency Measurements, in Special Issue: Structural health Monitoring, *Journal of Engineering Mechanics*, Vol.126, No.7, July. 761–8

Vanik, M.W., Beck, J.L. & Au, S.K. 2000. Bayesian Probabilistic Approach to Structural health Monitoring. *Journal of Engineering Mechanics*. July, 738–45

Zhang, Xianda. 1995. Modern Signal Processing, *Tsinghua University Publishing Company*. May (In Chinese)

Zhu, C.C., He, C.Y., Zhang, J.H. & Li, H. 1999. Modal Identification of Building Using Ambient Test. *Journal of Experimental Mechanics*, 14 (2): 243–5 (In Chinese)

32. Special session on "New didactical strategies"

System-based Vision for Strategic and Creative Design, Bontempi (ed.)
© 2003 Swets & Zeitlinger, Lisse, ISBN 90 5809 599 1

Complex of thinking-development in IT-based education

P. Toth & P. Pentelenyi
Budapest Polytechnic, Budapest, Hungary

ABSTRACT: Our study is based on cognitive pedagogy, according to this conception, problem solving as productive thinking may be explained as a complex procedure. In our present study we would like to focus on the ability development based on a special subject, and within that we wish to explore the possibilities of analogous knowledge transfer as a thinking operation within ability development. During the 30s and 40s a lot of experiments were published in Hungary and abroad to explore the possibilities of ability development implicit in the syllabus. Now we would like to emphasize the role of analogous knowledge transfer in teaching IT.

1 THE COMPLEX EXPLANATION OF PRODUCTIVE THINKING

With the help of factor analysis, Guilford organized the intellectual abilities into a unified system (Guilford 1986). According to this, the aspects of categorization are the following: the nature, the operations (cognition, recollection, divergent thinking, convergent thinking and evaluation) of, the contents (figural, symbolic, semantic, behavioural) of, and the products (the representation of information in units, categories, connections, systems, transformations and implications) of the thinking process. According to Guilford, operational factors correspond with definite abilities. In the course of problem solving both convergent and divergent thinking may simultaneously play a defining part, which he regards the components of productive thinking, hereby created the possibility of the complex explanation of productive thinking. According to this, problem solving as productive thinking may be explained as a complex procedure, the determining factors of which are the so-called "tool skills" and the foundations. Guilford terms thinking in different directions and the search for different possibilities of solution divergent thinking, while he calls the search for exclusively correct and optimal solutions convergent thinking (Fig. 1). As the preconditions for all these, mention may be made of quality and quantity, practical knowledge, a firm dedication to problem solving (motivation and dispositions) as well as metacognitive system. (Treffinger et al. 1990)

Lipman talks of superior thought, which includes critical and creative thinking. According to him, the two types of thought mentioned above comprise similar

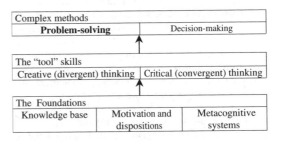

Figure 1. Overall model of productive thinking.

components, however, these elements are organized in different ways in either case. He explains creative thinking as one directed by intellectual connections, trying to transcend itself, sensitive to contradictory criteria, and leading to judgement. It is important to state that no creative thinking exists without being combined with critical judgement and no critical thinking exists without creative judgement to play a role in it, either (Lipman 1991).

According to Lipman superior thinking is under the "wings" of imagination balanced by common sense and reality (practicability). He considers the classroom development of superior thought important. As long as the syllabus is processed with regard to the development of students' problem solving thinking, problems will have to be solved with the active participation of the whole class in which the dialogues between students (argumentation, refutation, debate) have an immense significance. All this may successfully contribute to the development of students' cognitive abilities and, as it has been mentioned earlier, simple cognitive abilities and cognitive components serve as the foundation for the development of superior thought (Lipman 1991).

Algorithms, which decrease the necessity of creative judgements but are still indispensable in the course of problem solving, have a crucial role in the development of students' critical thinking (Lipman 1991). However, in the development of problem solving thinking emphasis has to be laid on the creation of algorithms rather than on their application. Students may be invited to participate – with continuous and appropriate help – in the exploration of the individual steps of algorithm, thus gradualness in education for independence is ensured. Many times we experience that understanding is not necessarily to be considered knowledge. It can easily be checked if we ask our students to reproduce the solution to a problem earlier jointly done and understood. We often see that this kind of self-test fails. The question arises as to what the teacher or the student is to do in this case. It is the review of the correct solution, in the course of which it may transpire that even understanding has not been well-founded at certain points (Pentelenyi 1998).

As we have earlier said, critical judgement plays a decisive part in creative thinking, or the reverse, whereas for discoveries creative thinking is primarily required. According to Polya, modern heuristics is the science, which wishes to explore the process of problem solving together with its essential thought operations.

In summary it can be said that algorithm is a method or a rule which results in the solution of a problem in any case, while heuristic methods apply practical rules that do not guarantee the successful solution of a problem, but once they prove successful, they save a lot of time and energy for the person solving the problem.

2 IMPROVABLE ELEMENTS OF THINKING-DEVELOPMENT STRATEGIES

In order to release from primary and secondary education students who are ready to do and capable of doing independent and self-motivated work, an ever more significant role should be given to teaching ability development methods in teacher training, too. We want to highlight the following methods in half measure:

– emphasis on the method of inductive syllabus processing beside deductive ones,
– acquisition of problem solving algorithms,
– high priority should be given to creative activities beside the reproduction of algorithms (Pentelenyi 1998),
– the presentation of the most possible alternative solutions in the course of problem solving to contribute to the development of the flexibility of thinking (variational teaching),

– problem-posing education means the creation of a chain of problem situations, the process of which students understand before solving the problem,
– having quality knowledge acquired has an outstanding role in problem solving, therefore it is important for our students of teacher training to learn the logical systems, connections, cognitive forms (inductive and deductive methods) implicit in the syllabus to be taught – all of them contribute to the formation of students' network of knowledge,
– emphatic application of thought operations (analysis, synthesis, comparison, relations, completion, analogy) during syllabus processing,
– acquisition of general problem solving procedures. Mention may be made here of the problem solving plan outlined by Polya, which may be universally valid independently of areas of science (Polya 1957),
– methods of forming dedication (motivation) to problem solving in the course of syllabus processing.

Before proceeding to clarify the role of analogous knowledge transfer in teaching informatics, we would like to make a short detour to present the cognitive components of divergent thinking which take part in the process of problem solving:

– problem sensitivity, flexibility and originality of thinking in the preparatory phase
– flexibility of thinking, highlighting the essence, the abilities of analysis and synthesis, posing novel questions and rephrasing in the phase of incubation
– originality of thinking, transformational and adaptable flexibility of thinking in the phase of discretion
– posing novel questions and rephrasing in the phase of evaluation (Landau 1971).

3 THE ROLE OF ANALOGOUS KNOWLEDGE TRANSFER IN PROBLEM SOLVING

One field of the investigation of problem solving thinking throws light upon the role of analogous knowledge transfer.

In analogy earlier experience is used indirectly for the solution of the problem. In this sense we, making use of experience gained in the course of our knowledge acquisition so far, establish analogous mapping (correspondence) between a certain set of thought in the initial stage of knowledge and the stage of knowledge to be achieved. Therefore on expanding our initial stage of knowledge it is important for new knowledge to be organically incorporated, thus creating the so-called "network structure" of knowledge. At the same time, it is also important in the course of problem solving to take the given problem into consideration as widely as possible, from many angles, and to analyze conditions and requirements as deeply as possible. We are informed of instances of such analogous transfer from the descriptions of famous inventors.

Ilona Barkoczi tells of a most instructive experiment. Three groups were formed to solve a particular problem, for the solution of which analogous knowledge transfer should have been recognized. Experiment Group 1 formerly solved several analogous problems. The Control Group never solved any such problems before. In the case of Experiment Group 2 conditions corresponded with those of Group 1, except that before trying to solve the new problem they had been warned to recognize analogies that they had already seen in the solution of problems earlier. The results were the following with regard to successful solutions (analogous relationship recognized): Experimental Group 1: 20% (!), Control Group: 10%, Experimental Group 2: 90% (!). So there is only a very slight difference between those having practiced analogous knowledge transfer and those not having done it (spontaneous analogous knowledge transfer).

She also pointed out that the efficiency of recognition significantly increases in the case of close analogies without or requiring only minor modification. From another investigation it transpired that the experts of a particular field became able to recognize analogies to a much greater extent.

Gick & Holyoak wished to explore the mechanisms that direct analogous transfer. They postulated that two stages of knowledge in analogous relationship can be abstracted onto a general problem scheme. The researchers experimentally justified that spontaneous analogous transfer only occurred if the objective-problem was identical in several respects with the source problem (semantic similarity). Once the appropriate objective stage was evoked, only structural and pragmatic similarity directed joining (Gick & Holyoak 1983).

It was also pointed out that experiments did not always recognize analogy between the so-called "base domain" and "objective domain". In such cases the attention of experiments was drawn to the analogies, which resulted in successful problem solving in most of the cases.

Note was also taken of the fact that the more distant analogy experiments had to recognize, the less successful their problem solving turned out to be. This can be explained by the fact that to recognize distant analogous relations experiments can activate their former experience from declarative memory only with difficulty.

Some of the research considers correspondence between the "base domain" and "objective domain" a pivotal question of analogy, while according to some other research analogous knowledge transfer may occur without correspondence.

Another field of research deals with the role of analogous knowledge transfer with regard to inductive and deductive thinking (Sternberg 1986). Inductive thinking may be approached from the side of problem solving through rule-induction on the one hand (Simon 1974), and through analogy on the other (Polya 1957).

For the measurement of inductive thinking Raven progressive matrix test (Raven 1962) and Csapo complex test (Csapo 1995) are for instance both suitable. In the former, the experiment has to examine the figures in each row and column, then has to select the appropriate missing figure out of the 6–8 given ones on the basis of rule induction. The complex test is suitable for measuring the two most important components of inductive thinking: analogous thinking (number analogy, word analogy) and rule induction (sequence of numbers). The relationship between the two measuring methods may be characterized by the correlation coefficient $R = 0.57$, which may be generalized at even 99% probability.

4 THE ROLE OF ANALOGOUS KNOWLEDGE TRANSFER IN TEACHING IT

On examining problems of informatics we can see that they normally have a single solution. So, in a strict sense, their solution would require convergent thought. However, it is important to note that the single solution may be arrived at by several trains of thought (Fig. 2). Therefore it is important to present a problem of informatics in the phase of processing new information from the most possible points of view (variational teaching).

Gentner says that inter-structural analogous knowledge transfer as a process may be divided into three parts: recognition of analogies, selection of representation and its use. Further, he states that structure mapping may be selectively explained, since it first lifts then maps onto the destination structure those characteristics and relations of the base (source) structure that are common. These recognized common features may of course be unimportant, too, but the consequence of this may be the failure of knowledge transfer (Gentner 1983).

Halford distinguishes four levels of structure mapping: (Halford 1987):

– elementary mapping – an element of the source structure is mapped into another element of the destination structure on the basis of the similarity of the elements,

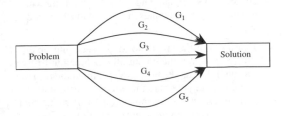

Figure 2. The character of problems of IT.

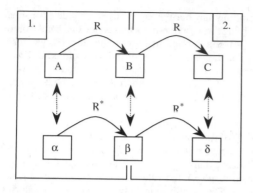

Figure 3. The general explanation of system mapping.

	Configuration of file characteristics with the help of the local operational system	Configuration of rights co-ordinated with network files in Novell environment
1.	start Windows Commander	start menu-commanded NE-TADMIN
2.		select user or group
3.		mark "modification of rights" on the menu
4.	select local drive	select network volume
5.	actualize local directory	actualization of network directory
6.	mark files	mark files
7.	activate characteristics to file	active list of rights
8.	select characteristics to configure	select rights to configure
9.	validate configurations	validate configurations

Figure 4. The explanation of analogous relationship between two operation-executing algorithms.

– relational mapping – the relationship between two elements of the source structure (relation) is mapped onto the relationship between two elements of the destination structure on the basis of the similarity between the relations (independently of the similarity of the elements),
– system mapping – relational mapping recognized in one system is transferred to the relational mapping of another system, where the relationship between the two systems is guaranteed by the mapping of a common element (Fig. 3),
– complex system mapping – system mapping where three relational mappings may be established among systems.

Examples of the latter two can be seen in problems of informatics, but these naturally include the former two as well. Let us see here as a specimen the transfer of a file copying operation in a local operational system environment to the configuration of rights realized in a network environment. According to Figure 4 this transfer may be interpreted as the transfer of an

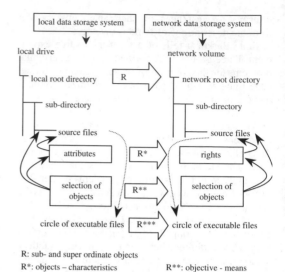

R: sub- and super ordinate objects
R*: objects – characteristics   R**: objective - means
R***: cause – effect

Figure 5. The explanation of complex system mapping onto a problem of IT.

operation-executing algorithm, within which the hierarchical system of local data storage has been mapped onto the hierarchical system of network data storage and NDS (Fig. 5).

On the basis of all these it can be stated that the levels outlined by Halford may be completed by the transfer of operation-executing algorithms (Halford 1987). The source algorithm may be explained on the basis of the examples shown above as a procedure specimen, which can be transmitted to the solution of new, "algorithmable" problems, taking the appropriate elementary, relational and system mappings into consideration. As long as our students acquire these procedure specimens involuntarily and apply them, the development of their ability of thought is not promoted. The acquisition of algorithms develops critical thought, which results in the optimalization (successful execution of operation) of the given series of operations. As it is seen in Chapter 1, the objective of education is not only the acquisition and transfer of algorithms, but we also have to enable our students to create algorithms (Pentelenyi 1998).

5   THE APPLICATION OF ANALOGOUS KNOWLEDGE TRANSFER IN THE TOPIC OF NUMBER CONVERSATION

We carried out my investigations in the field of vocational training. Experiments were members of a class who would receive a software user specialist certificate after the final examinations within the framework of a year-long training. It is a characteristic feature of the public education system in Hungary that almost 40% of final examiners continue their studies, so students of moderate or poor school achievement take part in secondary vocational training. Training focuses besides the acquisition of the basic skills of informatics, operational systems and network skills, on the application level acquisition of different programs with general objectives. In the first half of the training process students learn conversion from one numerical system into another (10 to 2; 2 to 10; 10 to 16; 16 to 10; 2 to 16; 16 to 2).

We were curious to see whether students were able to recognize analogous relations and execute them once they were given conversion into a numerical system they had not studied before. Using Raven progressive matrix test we organized experiment and control groups. In the former group we laid great emphasis on the exploration of analogous relationships implicit in the syllabus, while in the latter one we entrusted their recognition to the students. Let us see, by way of illustration, a couple of examples which may be explained as the algorithm of number conversion, too (where "r" stands for the base figure denoting the optional numerical system):

- $10 \rightarrow r$ (where M stands for the whole of the number to be conversed and N stands for its fraction). It is important to emphasize that divisions are to be continued until remainder is 0. (Fig. 6)
- $r \rightarrow 10$ (where A stands for the real number to be converted into numerical system 10; a_i provides the formal value of digits on individual places, where a_i may assume a value between $0 \ldots [r-1]$)

$$A = \sum_{i=m}^{n} a_i * r^{-i} \qquad (1)$$

- $r_1 \rightarrow r_2$ (conversion from an optional numerical system into another one). It is traced back to the previous two methods with the help of this procedure.

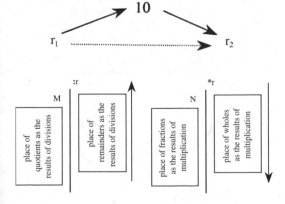

Figure 6.   The process of conversion from 10 to r.

- $r_1 \rightarrow r_2$, where either r_1 or r_2 is 2, while the other one is E {the powers of two}, thus the problem becomes a special case of the previous one. We would like to illustrate the method of conversion by the following example ($2 \leftrightarrow 16$):

$$
\begin{array}{cccc|cccc}
2^3 & 2^2 & 2^1 & 2^0 & 2^3 & 2^2 & 2^1 & 2^0 \\
1 & 0 & 1 & 1 & 0 & 1 & 1 & 0 \\
\end{array}
\quad
\begin{array}{ccccc}
2^3 & 2^2 & 2^1 & 2^0 & 2^3 \\
1 & 1 & 0 & 1 & 1 \\
\end{array}
$$

B   6   D   8

- $2^n = r_1$ or r_2, thus the basis of grouping place values is "n". By becoming conscious of this method, students may acquire a fast and very efficient method.

In the case of the control group we only concentrated on the solution of individual problems ($10 \rightarrow 2$; $2 \rightarrow 10$; $10 \rightarrow 16$; $16 \rightarrow 10$; $2 \rightarrow 16$; $16 \rightarrow 2$), but consciously did not place emphasis on their generalization, therefore its recognition became incidental.

Processing and practicing the syllabus was followed by testing, in the course of which students had to solve, beside the conversions in the main body of their studies ($10 \rightarrow 2$; $2 \rightarrow 10$; $10 \rightarrow 16$; $16 \rightarrow 10$; $2 \rightarrow 16$; $16 \rightarrow 2$), special problems as well, e.g. ($2 \rightarrow 8$; $10 \rightarrow 8$; $10 \rightarrow 4$; $2 \rightarrow 4$; $16 \rightarrow 8$). In these problem situations students had to recognize the relationship between problems solved earlier and the new ones (recognition of similarity), then they had to do the conversion on the basis of the selected conversion procedure (knowledge transfer). For knowledge transfer to take place, of course, both qualitatively and quantitatively adequate acquisition of the skills in the main body is required. Therefore we only included in the evaluation the results of those students who acquired the main body at a minimum 30% level. Thus we separated those students (maximum 5–8%) who had not adequately prepared for the test or had not understood the main body or had not practiced conversion algorithms enough.

As it had been expected, generalized conversion algorithms successfully helped the formation of skills and abilities in the area investigated. In the sense of Figure 7 the near 75% acquisition of the main body was required for knowledge transfer to work in a slight degree, and the near 100% one was required for a reasonable degree.

As a result of the generalization (Fig. 8) a knowledge transfer of some degree takes place with almost every student above the 30% level of knowledge of the main body. The higher level of acquired knowledge it relies on, of course, the higher degree transfer will be. As can be seen from both figures, successful analogous knowledge transfer may take place at a lower level of acquired knowledge, too. It can be explained by the fact that students' insufficient preparation was balanced by their higher general creativity, which was by the way reinforced by the Torrance test of creativity.

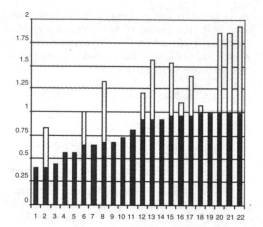

Figure 7. The efficiency of knowledge transfer application in the control group.

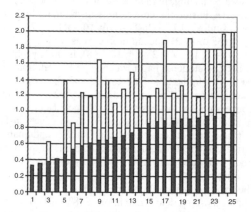

Figure 8. The efficiency of knowledge transfer application in the experiment group.

In the case of the experimental group knowledge acquired during the procession of the topic of number conversions is in a correlation described by a near r 0.5 coefficient with knowledge acquired through knowledge transfer in the problem situation (supposing a 1% margin of error). We measured a correlation of nearly similar degree between knowledge acquired by analogous knowledge transfer and the Raven test. The latter reinforces the relation of analogous knowledge transfer to inductive thought (Csapo 1995).

In order to compare the two teaching methods, we also carried out the χ^2 test, which unambiguously proved the higher efficiency of teaching number conversions through generalized algorithms with regard to analogous knowledge transfer.

6 CONCLUSION

We pointed at the role of thought operations (i.e. knowledge transfer) in the process of syllabus processing, since they enable us to develop our students' ability of thinking through problems that can be solved by a few steps. Our experiments proved, generalized conversion algorithms successfully helped the formation of skills and abilities in the area investigated.

REFERENCES

Csapo, B. 1995. Improving inductive reasoning through the content of teaching materials in primary and secondary schools. In: *Fostering high order skills. Fifth Conference of the International Association of Cognitive Education.* New York.

Gentner, D. 1983. Structure-mapping: a theoretical framework for analogy. *Cognitive Science, 7.*

Gick, M.L., Holyoak, K.J. 1983. Schema induction and analogical transfer. *Cognitive Psychology, 15.*

Guilford, J.P. 1986. *Creative talents. Their Nature, Uses and Development.* Buffalo: Bearly Limited.

Halford, G.S. 1987. *Children's Understanding: The Development of Mental Models.* Hillsdale, N. J.: Erlbaum.

Landau, E. 1971. *Die Psychologie de Kreativität.,* München-Basel: Ernst Reinhardt Verlag.

Lipman, M. 1991. *Thinking in education.* Cambridge: Cambridge University Press.

Pentelenyi, P. 1998. *Development of algorithmic thinking.* Budapest: Ligatura Ltd.

Polya, G. 1957. *How to solve it?* Garden City, New York: Doubleday.

Raven, J.C. 1962. *Advanced progressive matrices, set II.* London.

Simon, H. 1974. Problem solving and rule induction: a unified view. In: Gregg, L.W. (ed.): *Knowledge and cognition.* Potomac, M.L.: Lawrence Earlbaul Associates.

Sternberg, R.J. 1986. Toward a unified theory of human reasoning. *Intelligence, 10. 4.*

Treffinger, D.J., Feldhussen, J.F., Isaksen, S.G. 1990. Organization and structure of productive thinking. *Creative Learning Today.*

System-based Vision for Strategic and Creative Design, Bontempi (ed.)
© 2003 Swets & Zeitlinger, Lisse, ISBN 90 5809 599 1

When less is more: a practical approach to selecting WWW resources for teaching IT hardware

N.N. Berchenko
Institute of Physics, Rzeszow University, Rzeszow, Poland; Lviv Polytechnic National University, Lviv, Ukraine

I.B. Berezovska
Ternopil State Technical University, Ternopil, Ukraine

ABSTRACT: The information needs of educators result from the necessity to stay current with new relevant IT developments and to find answers to student-centered questions. The volume of available information makes the task of rapidly identifying high-quality resources discouraging. Effective sources evaluate the rigor and relevance of information and summarize it in the form of systematic reviews and meta-analysis. These sources have the opposite focus of many other information tools in that they aim at providing less information rather than more. With the development of these sources a new search approach is needed to locate the information for courseware in which speed is the benchmark. The existing IT literature can be considered as a hierarchy, with the most useful information placed at the top. Use of this concept allows educators to navigate through increasing layers until they retrieve information with the highest relevance and validity with the lowest work.

1 INTRODUCTION

Fundamental knowledge of the information technology has become an integral part of higher technical education in each and every engineering field. However very often this means just mastering basic programming skills while the current technical environment requires an in-depth coverage of general concepts forming the operation basis of IT hardware as well as instruction on how to properly use and adjust this hardware to develop an optimum solution for a particular application. It is not an easy task, the more so as the increasingly accelerating development of IT technology significantly decreases the lifetime of many technical solutions which only recently have seemed rather long lasting.

The necessity of keeping pace with this progressing process forces the design of effective educational models and tools, with an Internet-based courseware being one of promising options which immediately involves a new challenge (Berchenko et al. 2002). Permanence is not the Internet's long suit, therefore the evaluation of any instructional materials of this kind, beyond content, is whether the general advice and recommendations hold up long after the uniform resource locators mutate or disappear. Designed as a reference for students who desire to dig deeper into a particular topic such manual can be as reliable as any Internet-related book can be. That is why this paper focuses on a practical approach to selecting WWW resources on a deadline to indicate the moving target of teaching IT hardware courses.

2 EDUCATORS AS INFORMATION CONSUMERS

The information needs of faculty are quite distinct from the needs of researchers, engineers, designers or non-teaching university personnel. Educators usually seek information for two reasons: to stay current with new developments in IT relevant to their practice or to find answers to student-centered questions. Different tools and methods are required for these different information needs. Educators need to be told about new information but also need a tool for quickly finding the information again when they need it.

The type, format, and sources of information in education are undergoing significant and rapid change. Educators, in addition to their role in teaching, are now focusing on providing more effective methods of information retrieval, usually through electronic means.

Due to the time constraints imposed by IT practice, the usefulness of information retrieval systems and the information they provide are determinant to busy teachers.

To meet the growing demand for electronic "just-in-time" information, many educators are designing their own personalized portals and pages (Jaen et al. 2002) for quick access to the resources and services they use the most.

These new retrieval systems also have the potential to provide new types of information, information that synthesizes "raw" information originating from original research findings into summaries and conclusions. Educators are increasingly looking for information that is filtered by scientific rigor and relevance to the teaching practice and then is summarized in the form of synthesized answers to specific questions. Summary sources of information have existed for some time, usually in the form of books, reviews, and materials from professional organizations. This new type of information differs from these older sources in that it strives to provide information that is more useful to faculty in the day-to-day teaching students.

3 INFORMATION AS A COURSEWARE CORNERSTONE

The usefulness of any information source is in proportion to its validity and relevance and in inverse proportion to the amount of time and effort required to locate and implement it into teaching practice.

The validity of information refers to its scientific rigor. A hierarchy exists of research study design, with some methodologies having greater scientific strength.

Information in the IT literature also has various levels of relevance to educators. The most relevant information is research that comprehensively considers the effectiveness of IT solutions in an explicit manner that matters the most to students.

There are many instances in which the early, "makes sense" data did not translate into widely applied technology, devices, equipment, etc. (e.g., high temperature superconductivity). While this preliminary information is necessary to increase our knowledge of a particular issue, it is "not ready for prime time" in the sense that teachers should not integrate it in practice. While this type of pioneer research is crucial to the development of better IT solutions, it is not sufficient, in itself, for educating needs.

The traditional information seeking approach focuses mainly on the critical evaluation of original research and other sources of primary information. However, this approach – the evaluation of the validity of information by individuals using it, – is not as useful as it could be because of the excessive time involved and difficulty of integrating it into educating practice at the point of teaching. Original research, as typically published, is not useful in the education of students until it has been transformed in some manner.

The goal of the alternative new approach to IT-related information is to provide highly valid and relevant information while requiring the least amount of time and effort to locate and apply it to educational practice. To meet this goal, these information sources have the opposite focus of many other information tools in that they strive to provide less information rather than more.

4 ELECTRONIC RESOURCES LANDSCAPE

Even with the development of electronic archiving and searching, the amount of the scientific and technical literature is still so large as to effectively prevent its readily integration into education. Since its inception, the Internet has been attracted educators seeking instructional materials. One of this unprecedented resource's strengths is its size, but this size also makes it more challenging to search, and the responsibility for determining the validity and relevance of its information items is up to users.

Even information that can be rapidly retrieved must be evaluated for validity, and irrelevant information must be removed. Following retrieval and evaluation for relevance and validity, research findings must be compared and combined in ways that can be used to educate the students.

Designing a quality courseware involves planning and considering several structural/functional key factors carefully to ensure its utility (how well it functions) and usability (how effectively students can navigate it). The main stages of the designing process are shown in Figure 1. Courseware should have a core of fundamental knowledge, which forms a pretty unchanged basis of teaching a particular course. Other parts of courseware deal with an increasingly growing range of topics like technology, device design, measurement and testing techniques, specifications of new products and results of industrial benchmarking which form applied knowledge. To be successful in learning the IT hardware students need to learn how to use different educational technology components, including those available on the Internet, i.e. professional forums and lists, benchmarks and tests, glossaries.

Methods have been developed for combining research findings in an explicit and reproducible manner. Systematic review and meta-analysis are two such methods. Research findings are obtained in a comprehensive manner, evaluated for scientific rigor, and combined in a way that makes both instructional and scientific sense. In this way, a vast amount of research literature can be summarized in a single document, "refining" the raw information into a finished product ready to meet educational needs.

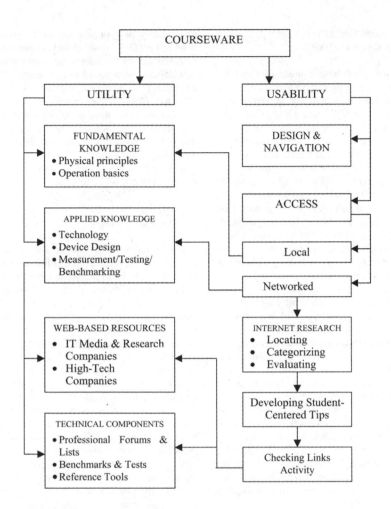

Figure 1. Main stages of designing Internet-based courseware.

Information of this kind is available at web sites of many IT media and research companies, e.g. **PCWorld. com** (*http://www.pcworld.com/*), **Tom's Hardware Guide** (*http://www.tomshardware.com/*), **Anand Tech** (*http://www.anandtech.com*) and **iXBT. com** (*http://www.ixbt.com*).

The aim of such resources is to provide a clearing-house for the best IT - relevant research information. When this information is put all in one spot, educators can quickly access this information.

Each of reviews and how-to articles considers a particular question (e.g., "Established Technology Standards"). If possible, the authors of studies try to combine all of the study results (meta-analysis), trying to treat all of the separate studies as one big study to answer the question. These reviews are updated regularly.

Practice guidelines are also designed to refine IT information into practical ways that can be used by teachers. Not all practice guidelines, though, are designed equal. Guidelines can be categorized as target audience-oriented e.g., **Beginners.co.uk**, (*http:// tutorials.beginners.co.uk/*), skills-oriented e.g., **FindTutorials.com**, (*http://tutorials.findtutorials.com/*), component-oriented e.g., **Michael Karbo's Online Service**, (*http://www.karbosguide.com/*) or universal **PC Technical Guide**, (*http://www.pctechguide.com/*).

The last type is the most useful, because the comprehensive materials are collected, summarized and available at a single access point along with many reference resources and current periodicals. In this way, readers can see for themselves the strength of the materials, rather than relying on the opinion of the authors of the guidelines for interpretation.

5 TOP-DOWN APPROACH TO INFORMATION RETRIEVAL

With the development of these sources of validated and refined information, a new approach is needed to access IT information in which speed is the new benchmark. The existing IT literature, including these new refinement tools, can be conceptualized as a pyramid (Fig. 2), with the most useful information, based on validity and relevance, placed at the vertex. The review resources are placed at the top of the pyramid, because they provide the best utility and usability due to information, synthesized and presented in a highly usable format. At the bottom of the pyramid are sources that are either expert based, and thus difficult to validate, or raw information that has not yet been synthesized into usable forms.

Use of this hierarchy allows searching to begin at the level of information with the highest usefulness. Starting at the top, searchers "drill down" through the progressively enlarging layers, encountering information along the way that is either less valid, less relevant, or harder to use. Rather than focusing on comprehensiveness, which would be the goal when preparing for a research, educators search only until finding the answer to a specific question. The value of the hierarchy is that the best information is searched first, reducing the need for comprehensiveness.

This approach to the IT literature is similar to the tertiary-secondary-primary literature pyramid used by information specialists (Encyclopedia of Library and Information Science 1979). What is different, though, is that searchers more definitely focus on information of greater usefulness (both valid and relevant), rather than treating each level of literature as being essentially equivalent.

To help the faculty efficiently navigate the information pyramid and identify information of high relevance

and validity, two specific tools are needed. Educators need a "first alert" instrument for relevant new student-oriented information as it becomes available. Many newsletters, Web-based systems, and other "current awareness" services attempt to fill this need e.g., **IDG.net**, (*http://www.idg.net*).

Educators also need a tool for rapid retrieval of the information to which they have been alerted but that has not yet been integrated into their daily practice. **TutorGig.com** (*http://www.tutorgig.com/*) provides the search engine to get up to date information on hundreds of online IT tutorials.

6 CONCLUSION

All information in IT is not created equal; most of the currently available research information either is too preliminary to be implemented into educational practice or is otherwise not relevant to the needs of a particular course. The goal of educators is to rapidly identify and use high-quality information in the course of their practice. Unfortunately, the volume of information available to them makes this task daunting without specific tools. Further, information that is presented in its raw (i.e., originally published) form is not useful to students, until teachers or someone else can evaluate and summarize it. A growing number and variety of new tools that are sources of highly filtered, highly relevant information are available. These new tools, placed within a searching framework based on the usefulness criteria, offer the promise that all educators can use resources that retrieve information with the highest relevance and validity with the lowest work.

Selecting instructional resources, as with so much else in education, remains more of an art than a science. Better data and better definitions will certainly help us, but, in a long run, we will continue to rely on the expert judgment of professionals who know the literature and, most important of all, know the needs of people who use it.

REFERENCES

Berchenko, N. et al. 2002. Teacher-Driven, Student-Focused Internet-Based Electronic Textbook on IT Hardware. In Imre J. Rudas (ed.), *Information Technology Based Higher Education and Training*; *Proc. intern. conf., Budapest, 4–6 July 2002*. Budapest: Budapest Polytechnic.

Encyclopedia of Library and Information Science 1979. New York: Marcel Dekker Inc.

Jaen, X. et al. 2002. A web-based educational library. In Imre J. Rudas (ed.), *Information Technology Based Higher Education and Training*; *Proc. intern. conf., Budapest, 4–6 July 2002*. Budapest: Budapest Polytechnic.

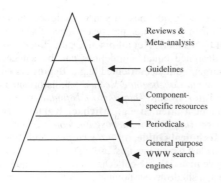

Figure 2. The hierarchy of access levels to IT-related literature.

System-based Vision for Strategic and Creative Design, Bontempi (ed.)
© 2003 Swets & Zeitlinger, Lisse, ISBN 90 5809 599 1

Evaluation of higher educational services by means of value analysis

M. Gyulaffy, R. Meszlényi & F. Nádasdi*
College of Dunaújváros, Dunaújváros, Hungary
* *Licensing and Administration Bureau of the Ministry of Economic Affairs*

ABSTRACT: Higher educational institutes perform not only educational tasks; rather, they also provide a variety of services in the process of education. Therefore, in addition to the most important task (principal function) of the institution, there are additional functions that are indispensable in holding its ground in the market competition. The competitiveness of a given institution is strengthened not only by the field of education it is engaged in, but also by additional services offered. The FAST provides assistance in the orientation, evaluation and also offers the possibility of re-considering the process.

1 A NEW SITUATION IN HUNGARY

1.1 Economic and political transition

In 1990 Hungary voted for economic and political transition. It took a relatively short time for the political transition to take place. However, the economic transition may require several decades to unfold. The reason for this is that several generations were "eliminated" from market economy. It is not possible to simply copy the practice followed by countries with a highly developed market economy. Every possible means must be used to speed up the transition process. The changes have had a strong impact on higher education as well.

Through several examples we intend to present the sorts of problems we have to face, which are most likely characteristic of several other countries that are undergoing the same transition.

1.2 Changes in the higher education

Ten years ago some 10% of the 18–24 age group studied at a higher educational institute. This figure amounts to some 36% at present. A wide range of new opportunities has emerged as a result of the impact that market economy has had on higher education. Several private universities and colleges have been established and foreign universities have also founded their own sections in Hungary. This new situation has become manifest in two ways: a fight for more students, and a fight to keep or attract excellent professors. The management of the College of Dunaújváros has decided to take steps in three different directions:

- the teaching of value analysis on a high international level.
- the diversification of education as a service through the introduction of new training programs.
- the application of value analysis in the elaboration of the strategic plan of the College of Dunaújváros.

2 THE STRATEGIC PLAN OF THE COLLEGE OF DUNAÚJVÁROS

2.1 The diversification of education: the introduction of accredited vocational training in higher education (AIFSZ)

During the 1990s it became clear that there was a need for the diversification of education. People needed more flexible training opportunities, and the analyses of needs clearly showed that there was a strong demand for training programs which offered solid vocational training in fields that could be well utilized in the market, though these programs did not award any higher educational degree. It was also evident that the new training programs should be implemented in the already existing educational system on the border of higher education and public education. This new form of training is called AIFSZ. AIFSZ is a short, two-year training program that does not give any higher educational degree but instead it offers a vocational qualification. Such diversification in higher education implements a flexible form of training, making it possible to uti-lize earlier, often part-time studies. The new training programs of AIFSZ are chosen by a large variety of different groups. One such group comprises students with secondary education who intend to pursue their studies in higher education but for some reason drop out. The new form of education can provide an organizational framework for selection within higher education by making it easier to find a job for those who drop out. By its vocational training program AIFSZ can broaden and enlarge the basic function of higher education. Within the framework of the PHARE program the Hungarian Government announced a competition with the aim to facilitate the introduction of this new vocational training program. In 1998 AIFSZ was launched in 21 institutions. In 1999 the College of Dunaújváros also started its own program. We believe that AIFSZ will become an important form of training in the future. It ensures significant mobility for members of the society, especially for young people. AIFSZ meets two goals: on the one hand it provides general education, on the other, it meets the special requirements of the labor market. In sum: the students enrolled in AIFSZ receive a training that enables them to fill a specific job, but – in case need arises and the required results are achieved – they can pursue their studies in the higher educational institute. After the necessary permits had been issued by the competent government body the training began in four majors in the 2000/2001 academic year: financial administrator, bank executive, accounting executive and tourist industrial manager (Kántor et al. 2000).

2.2 The teaching of value analysis and the provision for the opportunity to obtain an international certificate at the College of Dunaújváros

In the course of the past few years we have seen that in addition to the mandatory subjects the students want to acquire knowledge that can readily be utilized in practice. Another important factor besides utility is whether the acquired knowledge can be officially certified. Value analysis has been taught at the College of Dunaújváros close to 30 years. In 1996 the events took an unexpected turn. The Society of Hungarian Value Engineers become a member of SAVE International (SAVE = Society of American Value Engineers) and adopted its training and qualification system. Making use of this opportunity the College of Dunaújváros expanded the teaching of value analysis. At present it is taught in 2 class hours per week, 15 weeks per semester, in 2 semesters, so altogether in 60 hours. The students meeting the necessary requirements obtain the SAVE International AVS qualification (Associated Value Specialist) that enables them to work efficiently in a value analysis team. Starting in 2002 it became possible for us to teach the full training course of SAVE (Module

I and Module II). We believe it is a task of utmost importance to establish broad cooperation with member states of EU in the field of value analysis.

2.3 The application of value analysis in the elaboration of the strategic plan of the College of Dunaújváros

In 1998 some 100 students started their studies of value analysis, and approximately 50% will obtain the AVS qualification. In addition to the students, six teachers also studied value analysis. This means that the teaching capacity has also increased considerably. The management of the College of Dunaújváros provides significant support for conferences and other programs in the field of value analysis. With the financial and intellectual support of the Hungarian government authorities we organized an international conference on value analysis on April 18–21, 2001. The conference was attended by value engineers from the United States, South Korea and Russia. The conference lectures were followed by a two and half a day workshop. One outstanding topic of this workshop was the support provided by value analysis for the strategic plan of the College of Dunaújváros. Participants in this workshop comprised Hungarian value engineers, experts from other areas, students and foreign experts. Invaluable help was given by Mr. Bruce Lenzer and Mr. Dale Daucher who are outstanding American value engineers. In another section Mr. James Rains, another excellent American expert, presented the method of target costing. The various phases of the procedure were as follows:

1. *Collection of information.* In this phase the needs and the goals of the students and the college were analyzed.

   – Student goals:
     • Strengthening a homely atmosphere
     • Training with an emphasis on practice
     • Healthy team spirit
     • Teacher ready to help
     • Eligibility to other economic schools
     • Available room I student hostels
     • A rich storehouse of traditions
     • Easy transfer of credits between higher educational institutes
     • Flexible training programs
   – Goals of the College:
     • Change the attitude towards organizing programs
     • Attract young talents
     • Modernize the teaching materials
     • More emphasis on quality rather than quantity
     • Greater national prestige

Table 1. Target functions.

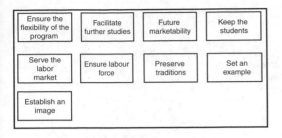

Ensure the flexibility of the program	Facilitate further studies	Future marketability	Keep the students
Serve the labor market	Ensure labour force	Preserve traditions	Set an example
Establish an image			

Table 2. Possible value-increasing functions.

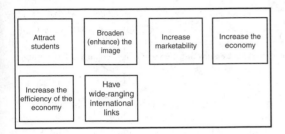

| Attract students | Broaden (enhance) the image | Increase marketability | Increase the economy |
| Increase the efficiency of the economy | Have wide-ranging international links | | |

After defining the goals the target functions (Tab. 1) and the value-increasing functions (Tab. 2) were established. The structure of the FAST diagram is shown in Figure 1. In this case the higher order function, F_o gives a qualification. Essentially, four basic functions cover all the activities:

- F_A: obtains accreditation
- F_B: teaches, educates students
- F_D: provides services for the students
- F_C: runs the College
  - F_C is further divided into two branches:
    - F_E manages the College
    - F_F manages contacts

As an example of detailed function analysis we present the details of function F_D (Fig. 2). In this phase cost were disregarded, for it is the long-term goals and function that must be identified at this stage. A large number of ideas were generated which we will consider when working out the details of the strategic plan.

2.4 Assistance to the value analysis activities of small and medium-sized companies

In Hungary some 99% of the enterprises are small-sized or medium-sized companies. Most of them lack the necessary financial and intellectual resources required for the implementation of a value analysis project. We have started to work out methods which – we hope – will assist the value analysis activities of these companies. The experience we have gained in the past few years in teaching value analysis helped us a lot in working out this new methodology.

3 SUMMARY

In Hungary the higher educational institutes had to face new challenges in the 1990s. The College of Dunaújváros responded to these new challenges by diversifying its activities. Value analysis was used as a tool in working out a long-term strategic plan. Cooperation is given priority while taking into consideration of the mutual interests of the cooperating parties.

REFERENCES

Brzezinski, R. 1999. Company/Supplier Alliances, Tools for 2000 and Beyond. *6th Value Analysis Conference. Society of Hungarian Value Analysts*. Budapest, December 7–9.

Florian, S. 1999. Leadership in Globalization. *6th Value Analysis Conference. Society of Hungarian Value Analysts*. Budapest, December 7–9.

Kántor, Károlyné & Rózsa Meszlényi 2000. Gondolatok az akkreditált rendszeru felsofokú szakképzésrol. (Thoughts on the Accredited Vocational Training System in Higher Education.) In *Dunaújvárosi Foiskola közleményei. XXI. Dunaújváros*. pp. 131–134.

Kelly, R.J. & Male, S. 2000. The application of Value Management to the UK public sector construction supply chain. *SAVE International Conference*.

Kirk, S.J. & Spreckelmeyer, K.F. 1998. *Enhancing Value in Design Decisions*. Kirk Associates, LLC, USA.

Kirk, S.S. & Sherwood, D.R. 2000. Conversation about Establishing a Value Management Program. *SAVE International Conference*.

Kmetty, G. 1999. Project Scoping with Value Analysis methodology. *6th Value Analysis Conference. Society of Hungarian Value Analysts*. Budapest, December 7–9.

Körmendi, L. & Ferenc Nádasdi 1996. Az értékelemzés elmélete és gyakorlata. (Theory and Practice of Value Methodology). Info-Prod. Budapest.

Molnar, P. 1999. Value Management an Active Tool to Improve Quality in Innovation Process. *6th Value Analysis Conference. Society of Hungarian Value Analysts*. Budapest, December 7–9.

Nádasdi, Ferenc & János Lasányi 2000. Új gazdaságpolitika – új elemzési stratégia = értékelemzés. (New Economic Policy – New Analysis Strategy = Value Analysis.) In *Dunaújvárosi Foiskola közleményei. XXI. Dunaújváros*. pp. 97–108.

Nádasdi, Ferenc 2002. The Possibilities of Applying Value Management in Countries in Transition (Based on Hungarian experience). In *SAVE International Annual Conference Proceedings*, Volume XXXVII, May 5–8. USA, Denver, CO.

APPENDIX 1

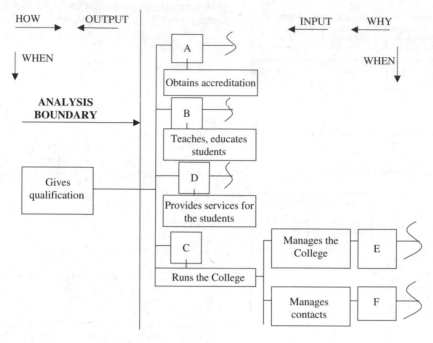

Figure 1.   The structure of the FAST diagram.

APPENDIX 2

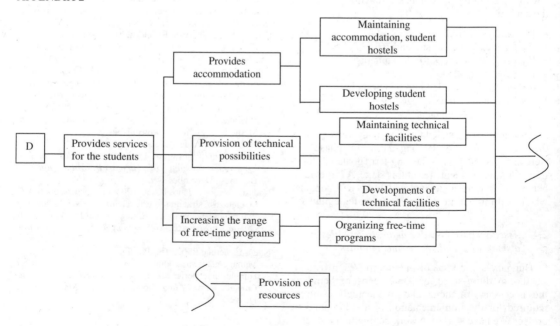

Figure 2.   Provides services for the students.

System-based Vision for Strategic and Creative Design, Bontempi (ed.)
© 2003 Swets & Zeitlinger, Lisse, ISBN 90 5809 599 1

BigBang: a web-learning portal in education

G. Luciani, P. Giunchi & S. Levialdi
University of Rome «La Sapienza» Rome, Italy

ABSTRACT: To be effective, the teaching of English and other foreign languages at university level requires resources not easily available. However a major problem is one of practice: the ability to provide students with opportunities to improve their English language skills. BigBang is a portal aimed at giving an answer to this problem, an attempt to create a virtual linguistic space where English is the only language allowed. It is also a web-drilling application in education which offers students and academics quality access to the Internet within the context of a growing implementation of several e-learning tools.

1 THE TEACHING OF ENGLISH IN ITALIAN SCHOOLS AND UNIVERSITIES

1.1 Teaching and practice

University students don't have much time to study English nor much chance to practice it in their curriculum: English rarely gets more than three or four credits, which is equal to 30 hours teaching at the most and all of their other classes are taught in Italian, although books, as in science, are often in English. Even in labs and at research level, Italian is the language spoken day to day while English is basically used to read and write books and papers and keep in contact with foreign colleagues. If we add to all of this the fact that there's no reliable standard in the way English is taught in secondary schools, it's easy to understand why most students end up knowing English at a very basic level indeed.

It is a matter of teaching: we need both to raise its level in schools and to invest more resources in the language courses offered at university. But in a sense the biggest problem of all is the lack of practice – both because a "sleeping" language is quickly forgotten and because arguably a good deal of learning takes place "incidentally" while a language is being used.

1.2 The priority of acquisition

In the field of language acquisition/learning, scholars have long debated on the interaction between declarative and procedural strategies in "internalising" a second language. In this context, there has been a paradigm shift which originated in the mid-seventies with the radical thesis of the American psycholinguistic S. Krashen, who advocated the priority of acquisition, along with the need to provide learners with comprehensible input, to lower their affective filter and trigger their motivation towards a language different from their own.

Krashen determined a change of paradigm in language teaching that has met with resistance in Italian schools. Some teachers pay only lip service to the 'communicative approach': they keep focusing on the formal aspects of language teaching, i.e. the study of grammar, leaving the use of it outside the classroom. Students come to university thinking that knowing that regular verbs add *ed* while 'irregular' ones have to be learnt by heart is crucial to activate their learning process. When faced with the use of language for authentic communicative purposes they experience a feeling of inadequacy and frustration.

2 A VIRTUAL LINGUISTIC SPACE: BIGBANG

2.1 A way to practice English

BigBang is born as an attempt to rebalance this situation by creating occasions to practice English by taking full advantage of the resources made available by the recent IT technologies. BigBang is a portal for improving English fluency through courses, pronunciation guides, written reports, music and song lyrics, audio channels from English radio broadcasts, video channels from English speaking TV stations, job opportunity databases, interactive activities like forums, discussion lists and ordinary e-mail on top of games (traditional and role-playing). This way, Italian students wishing to test themselves, to improve their English or learn how to write a curriculum vitae or be interviewed for a job, can do it at home whenever they prefer.

2.2 Quality access to Internet

But BigBang is also something else. At an initial stage, the content of BigBang was instrumental to the aim – the practice of English language; but later on, when we actually started creating the portal we soon realized that through BigBang students could do several things at the same time: practise English, but also improve their computer skills, while looking for material which is very valuable in itself. It offers also quality access to topics as varied as literature, astrophysics, travel, music and job finding. It could be seen as a kind of barter: we give students interesting links that many of them might not easily find by themselves, but the service is only provided in English.

3 E-LEARNING AND INTERNET

3.1 MultiCommunity and MultiCom

From this point of view, BigBang could be seen as a web-drilling application in education. At "La Sapienza", we have been involved with a number of projects (MultiCommunity, BigBang, MultiCom) all dealing with distance education in cooperation with face-to-face learning. The rationale of this approach is to combine the positive features of face to face with those allowed by the recent IT technologies, to help augment and personalize the learning process. MultiCommunity is a portal for students that has updated information on exams, seminars, laboratory projects, taxes, deadlines, professors' addresses, etc. that in the near future will be accessible also by cellular phones or new portable/wearable computers. MultiCom is a multi-platform support for the communication between professors, students and administrators involved in the teaching–learning experience. This support, written in Java, can read documents written in html (and next in xtml) which correspond to university courses to be downloaded from home asynchronously but, at the same time, synchronous activity may be performed via chat with other students, professors or administrators. A number of interesting plug-ins are also provided for managing multiple choice tests, annotation activities and more general communication through FAQ channels, discussion lists, forum and standard e-mail.

3.2 Company training and university education

Some people believe that e-learning is the university of the future, possibly an alternative to traditional teaching methods or the other side of the coin, namely in cooperation with face-to-face education, i.e. the solution to some of the problems that badly affect the university system. This seems particularly appropriate for "La Sapienza" with its 150 thousand students and its lack of facilities. However the initial enthusiasm for e-learning has partly evaporated. A university course is considerably more complex than a private company training one, not just because it tends to be more difficult but also because it has an educational element – we teach students a method of study, of researching, of thinking – which a company-training course doesn't need to have. There's a growing consensus on the fact that at least in a European context, e-learning should be seen as a support to traditional teaching rather than an alternative to it. A support for students who miss classes or play a more active role in a virtual class (communicating with other parties, linking external documents to those belonging to distant classes, adding comments to web pages, etc.), but also a support for teachers who can follow their students more closely. One aspect that deserves greater attention, concerns the students curriculum. In the present credit system, a teacher is limited in what he can actually teach within the number of credits given to his/her discipline. However a university does not offer only courses: it offers also educational tools which students can use by themselves according to their own needs and interests.

3.3 Internet and education

Among these tools, the Internet is becoming more and more important. Here the problem is that using the Internet one can find everything and nothing at the same time, for on each topic there are tens of websites whose quality is not checked by anybody. Before an educational institution can use the Internet on a daily basis there's a need for a filter. BigBang could play this role, if adequately supported by the academic community of our university. If our colleagues at "La Sapienza" volunteer to help us in this project so that each finds in the Internet the four or five best websites in his/her discipline, we can build quite a useful tool for students and teachers alike.

In order to encourage colleagues to help us we are trying to build the BigBang portal as an **open source system**: its architecture has been kept as simple and as functional as possible to allow even normal computer users to add links and a news field in the area of their concern. At present the group working on BigBang is made by some 20 people: students, language assistants, lecturers, professors; and what is interesting is the fact that maybe for the first time in the history of "La Sapienza" the group is made up by people coming from both science and the humanities.

An investigation on the students' learning styles in an Advanced Applied Mechanics Course

A. Matrisciano
Campus One Management – Dept. Mechanics and Aeronautics, Rome, Italy

N.P. Belfiore
Dept. Mechanics and Aeronautics, Rome, Italy

ABSTRACT: In this paper, an application of Multiple Intelligence Theory (M.I.) to the analysis of some adult students' learning characteristics is presented. The investigation is conducted over a group of students enrolled at the level of a graduate course in Aerospace Engineering of the University of Rome *La Sapienza*. Results allow students to better understand their own characteristics in learning, and help teachers in tuning up the adopted didactical methods. The investigation offers the possibility of acquiring a comparison *learning profile* to be used during freshmen candidates' orientation.

1 INTRODUCTION

In the last decades, teaching styles have evolved. The new methods, stimulated by the rapid progress in Information Technology, have suggested to teachers new approaches to be used in classrooms.

Most of the newest theories in teaching methods directly originate from the Pedagogy Area and deal, mainly, with elementary and secondary education. In fact, these phases of the individual's life can be reasonably regarded as tremendously important. Young students have to be protected and guided, as best as possible, in some fundamental choices regarding their curricula and careers: a wrong choice at a young age is very effective and can lead to unpredictable bad consequences in one's life. On the other hand, higher education is also a complicated process. Adult individuals are more conscious about what they want, but they are also already shaped in some of their important characteristics. This makes a bit more difficult to plan a teaching style that could suit all the attendants' sorts.

The average age of the students attending curricula in Universities depends on various factors, as for example, the country's traditions, the curriculum's subject and difficulty, and the economic trend.

As for this investigation, the attention is focused on students attending Engineering Courses at the University of Rome *La Sapienza*, whose typical age runs from 19 to 25 years. In practice, people who are able to complete their curricula by the established time terms are only a small part, while most require two or three extra years. This fact has risen the average age of the students attending courses in Engineering.

The mentioned delays in the "*Laurea*" achievement have induced the various Italian Ministries of the Education which have joined up in the last decade, to undertake a significant process of evolution in University. To support this renewal phases, they have sponsored some Courses with well structured National projects, such as *Campus* and *Campus One* (CRUI 2000).

The need of improving the teaching methods has induced the Authors of the present paper, who have been involved in some of the local activities related to the above mentioned projects (Belfiore et al. 2002), to investigate about students' characteristics. In particular, some aspects of the student learning preferences have been investigated in relation to the *Gardner's Multiple Intelligence* (M.I., for short) model, hoping that the results could be useful both to students and teachers, as well as to the course management.

2 LEARNERS' PREFERENCES AND STYLES: A SHORT REVIEW

Undergraduate and graduate students are about to be considered as adult learners. They undertake programs from different starting points and have different natural capabilities and prior instruction. For all these reasons, the success of the educational process is strongly dependent on the teaching strategy, which has to cope with such differences in classrooms. In the field of Engineering education, courses are rather difficult to be planned, maybe because of the subjects' intrinsic difficulty and of the strict and strong prerequisites. Hence, the teaching method becomes even more critical for the course's success, which is measured not only in terms of the students' test scores, but also of their personal satisfaction.

Among the possible activities that can improve Courses' quality, there is the analysis of the students' *learning styles*, which can be useful for many reasons, as it will be better described in paragraph 3. For example, *cognitive* and *learning* styles can be, theoretically, used to predict the best instructional strategy for a given individual at a learning task.

Generally speaking, *cognitive style* is referred to as *the preferred way an individual processes information*, while the *learning style* is regarded as *the preferred way an individual learns*.

Cognitive style describes the individual's typical mode of thinking, remembering or problem solving, and *it is a personality dimension which influences attitudes, values, and social interaction*. It would appear very easy to jump to the conclusion that if there is a preferred manner a student uses *to organize information*, then *the same* manner must be used *to learn*. However, learning is somehow different from *organizing the perceived information*. In fact, the cognitive style represents a *cognitive strategy that can be defined in terms of the desired targets*. Hence, it may vary, for the occasion, from case to case, in the same individual, and it is not necessarily related with *temperament*. Furthermore, some recent investigations have shown only week relationships between learning and cognitive styles, although there are also examples of useful applications of the learning models, to create teacher awareness of individual differences in learning.

Learning and *cognitive styles* have been ordered by Curry (1983) into a three-level stratified scheme, whose innermost, middle, and outermost layers correspond, respectively, to the *cognitive style*, the *information processing style*, and the *instructional preferences*. The fundamental concepts regarding the three models have been extensively discussed in the literature, such as, for only representative examples, in (Reichmann & Grasha 1974), (Witkin et al. 1977), (Kolb 1984), (Gorham 1986), (Riding 1991), (Sadler-Smith 1996) and (Allison & Hayes 1996).

The innermost category, namely the *cognitive personality elements*, refers to particular bipolar dimensions of the *cognitive style*, such as: (i) *Field dependence-Field independence* dimension, which refers to an individual's greater or lesser tendency have

confidence in external or internal references; (ii) *Intuitive-Analytical* dimension; (iii) *Wholist-Analytical* dimension, which refers to individuals' habit of processing information by organizing them into their component parts, or by retaining them from a global or overall view; (iv) *Verbaliser-Imager* dimension, which describes how an individual habitually represents information in memory during thinking.

The middle category, namely the *information processing style* refers to the individual's *learning style* and has been analyzed through special bipolar dimensions such as: (i) *Converger/Diverger*; (ii) *Accomodator/Assimilator*; (iii) *Activist/Reflector*; (iv) *Theorist/Pragmatist*.

Finally, the *instructional preferences* category regards how learners fit with particular instructional approaches which can be classified into three types: (i) *dependent* learners, who prefer structured programs and direct teaching; (ii) *collaborative* learners, who prefer group work, discussion, and interaction; (iii) *independent* learners, who like having a certain control over the contents and the methods.

Among the first specific contributions to the improvement of the teaching methods in Engineering Education, Felder's one deserve a special mention (Felder & Silverman 1988). According to the model therein adopted, the preferred learning styles have been classified into five dimensions, each corresponding to two possible preferred categories, as reported in the following scheme:

Dimension	Dimension's categories	
Perception	Sensory	Intuitive
Input	Visual	Auditory
Organization	Inductive	Deductive
Processing	Active	Reflective
Understanding	Sequential	Global

The proposed dimensions are neither orthogonal nor comprehensive. Hence, there are a total of $2^5 = 32$ possible combinations of learning profiles.

More recently, Sadler-Smith & Riding (1999) addressed the importance of relating the *cognitive styles* to the *instructional preference*, arguing that cognitive styles have an important role to play in determining an individual's instructional preferences and that this may affect the learning performances.

3 THE AIM OF THIS INVESTIGATION

The first goal of this investigation has been the identification of the *preferred learning approach* adopted by the given group of students. The working hypothesis is that the students attending the Course of Aerospace

Engineering at the University of Rome *La Sapienza* are characterized by a *typical learning style* and a *typical intelligence profile*, which are both different from those of the students who attend the other Programs.

The identification of these characteristics could be useful to induce teachers to organize their courses by adopting a method that better fits the students *learning characteristics*, and by suggesting additional and related activities specifically designed for the group. *Learning style* detection, for example, can be a useful tool mainly to help students which have the worst results in terms of progress in the selected curriculum. These students could receive support and counseling about the studying methods, without being considered as simply *obtuse* individuals. As the matter of facts, there are some evidences that some differences in students' success in University can be related to their differences in *learning styles* (Brunas-Wagstaff 1998). However, there are many and complex factors which can affect success in studying, such as *anxiety* and *motivation* (Marton & Saljo 1976).

Should the working hypothesis be true, it would be possible to identify a significant *tendency* that the analyzed class of Engineering students shows. This has suggested the idea of monitoring the characteristics of the younger students coming from the secondary level schools, who want to enroll that particular Course. In this way, younger students' *profiles* can be compared to seniors' ones, or, as another possibility, to the *profiles* presented by Junior or Senior Engineers who have archived success in career. Freshmen candidates would, so, have more information about themselves in relation to the typical characteristics of the best Engineers. Such information can be complementary to the results that candidates might have achieved in the classical attitude tests in Mathematics and Physics. After a reasoned analysis, which could be performed under the guidance of a specialized counselor, they could decide to practice more in some particular aspects of their cognitive and/or learning functions or, in alternative, to avoid to undertake a program which is expected to be not suitable for them. The advantages are quite interesting because, as known, the costs of a wrong choice are heavy, not only for the individual, but also for the whole Society.

4 THE ADOPTED METHOD

The adopted approach has been based on the analysis of the various aspect of *personality*. Concerned by the over-abundant different cognitive and learning models, we have decided to walk along a novel, and unknown, path. In fact, we tried to investigate about a new experimental approach to the problem, by restricting ourselves to a simple question: *which aspects of their intelligence the students have the*

tendency to employ in their activities and, mainly, in their educational process?

This question has led us to apply a quite recent Theory called the *Gardner's Theory of Multiple Intelligence* (Gardner 1983), with the purpose of attempting to identify a student's frame of mind, *as a personal cognitive profile in the educational (University) environment.*

With this aim in mind, a novel test has been set up in order to measure the student's inclinations of using particular capacities. The developed test has been tailored to measure *preferences*, and not *capacities*.

4.1 *Gardner's model of multiple intelligence*

A plethora of different models of the human thinking have been conceived. Among them, the selected one has received increasing attention during the last years, specially in the field of Pedagogy. Accordingly, the human mind relies on seven distinct forms of intelligence: Verbal, Logical-Mathematical, Musical, Kinestetic, Visual, Interpersonal, Intrapersonal (the latest *Naturalistic* form has not been taken into account for the purposes of this investigation).

Gardner's frame has been adopted to define seven corresponding dimensions, although the *Verbal* dimension has been divided into ones, namely, *Verbal* (written) and *Auditory*. For each dimension a raw score has been defined. Finally, the individual mean of the 8 scores has been also considered as variable to be investigated.

4.2 *Characteristics of the adopted test*

The developed test consists in a list of about 130 statements concerning the students' preferences in various cognitive contests, with special reference to learning in their Academic Course. To each statement, the student has been called to can give one among 5 different answers: one neutral, to which correspond a null score, two of agreement (either moderate or strong) and two of disagreement (either moderate or strong), corresponding, respectively to positive and negative scores or vice versa, depending on the nature of the statement. To give an idea of the kind of statements that have been submitted let us consider the following group of three of them, as a representative example.

1. I think that the following diagram (*actually, the one depicted in Figure 1*) is the best way to introduce the concept of stiffness constant of a linear spring.
2. I think that the best way to introduce the concept of stiffness constant is the following:

   "If: x = spring elongation; F = applied force; and k = stiffness constant of the linear spring; then: $F = kx$".

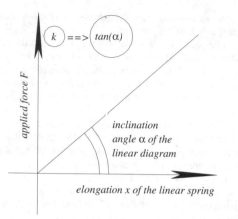

Figure 1. Graphical definition of the linear spring's stiffness constant.

3. I think that the best way to introduce the concept of constant of elasticity is the following:

   "*in a linear spring, the elongation x is always linearly proportional to the applied force F, the constant of proportionality being defined as the stiffness constant k*".

A large grade of agreement with statements like the 1, 2, and/or 3, give, for example, the higher scores in the variables related to the *Visual, Logical–Mathematical*, and/or *Verbal* Intelligences, respectively. From the presented sample statements it will be clear that the scope of the questionnaire is the identification of the *preferences* in studying, rather than the measure of the individual's *capacities*. Of course, the latter are still of a capital importance for the students in order to achieve success, but they are, at the same time, among those characteristics which cannot be easily improved in a simple supporting activity.

5 RESULTS AND VALIDATION

The results presented in this paragraph have been checked by means of simple statistical indicators. Unfortunately, the number of examined individual exceeds 30 only in one case. Some of the presented results must be, therefore, considered as only provisional.

5.1 *Group 1*

The first group consists in a sample of 88 male students enrolled at the 4th year of the Italian Laurea V.O. in Aerospace Engineering, who have an average age of 23.

A first check has been made on the samples by referring to a *null Hypothesis* H_o based on the assumption that an infinite number of individuals had randomly answered to the questions. This allowed to evaluate the means (all null) and the standard deviations of the Populations relative to the considered dimensions, and the means (all null, as well) and the standard deviations of the *sampling distributions of means* which are, respectively

0.622   0.500   0.783   0.603   0.754   0.917
0.839   0.754   0.259

for the nine analyzed dimensions, ordered as follows: (1) Verbal; (2) Auditory; (3) Logical–Mathematical; (4) Musical; (5) Kinestetic; (6) Visual; (7) Interpersonal; (8) Intrapersonal; (9) Individual mean of the eight previous scores.

The latter computations have been performed according the well-known equations:

$$\mu_{sd\_i} = \mu_{o\_i} \tag{1}$$

$$\sigma_{sd\_i} = \frac{\sigma_{o\_i}}{\sqrt{N}} \tag{2}$$

where subscript $i = 1, \dots, 9$ denotes the i-th considered dimension, while subscripts o and sd refer, respectively, to the *original* population and the *sampling distribution* (N is the sample size).

The significance analysis has been firstly based on the evaluation of the probability that the students might have given random answers to the questions. According to Equations 1 and 2, and accepting a level of significance $\alpha = 0.01$ (bilateral), the critical values of the means, for the nine dimensions,

1.276   1.027   1.609   1.238   1.548   1.883
1.724   1.548   0.332

were all lower than the ones measured over the sample:

4.580   0.659   8.989   6.432   11.864   4.330
−2.239   10.898   5.689

with the exception of the one relative to the *auditory* dimension, which could be, however, accepted with a level of significance of $\alpha = 0.05$ (bilateral), since in this case its critical value is equal to $0.640 < 0.659$.

The means, which characterize the *group's typical profile*, can be used as a reference frame in the individual's score interpretation, during counseling.

Finally, by elaborating the *standardized scores*, it has been possible to obtain the correlation matrix, reported in Table 1. There is not strong correlation among dimensions, except for the *Individual mean*, which seems correlated with all the other dimensions, specially with the *Visual* one (see Figure 2), and *Auditory* and *Visual* dimensions, which show a weak positive correlation (see Figure 3).

Table 1. Correlation Matrix relative to the nine analyzed dimensions in the sample *Group 1*.

1	0.311	0.343	0.296	0.314	0.433	0.257	0.319	0.671
	1	0.308	0.110	0.150	0.472	0.186	0.100	0.520
		1	0.185	0.178	0.394	0.015	0.170	0.543
			1	0.193	0.257	0.226	0.335	0.554
				1	0.420	0.292	0.225	0.610
					1	0.152	0.202	0.745
						1	0.039	0.509
							1	0.487
								1

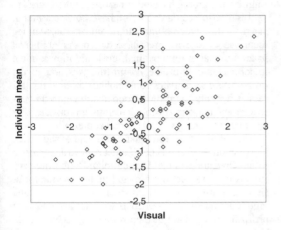

Figure 2. *Visual–Individual mean* dispersion diagram in the sample *Group 1*.

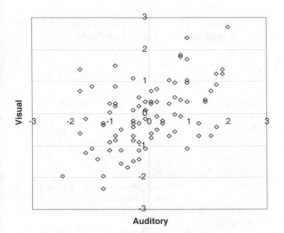

Figure 3. *Visual–Auditory* dispersion diagram in the sample *Group 1*.

5.2 Group 2

The second group consists in a sample of 10 female students enrolled at the 4th year of the Italian Laurea V.O. in Aerospace Engineering. This group is smaller than the previous because the number of female students in the Italian Engineering Faculties is generally relatively small.

The values of the standard deviations of the sampling distributions are for this case:

1.844 1.483 2.324 1.789 2.236 2.720 2.490 2.236 2.174

while the critical values of the means for $\alpha = 0.05$, are the following:

2.363 1.900 2.863 2.293 2.866 3.486 3.191 2.866 2.786

which can be compared to the measured means:

6.100 0.200 13.400 3.900 11.100 6.000 0.600 7.300 6.075

which show that the values are quite significant except the one relative to the Auditory and the Interpersonal dimensions, which do not differ significantly from a random distribution of answers.

5.3 Comparing the class group by gende

A further analysis can be made by comparing the first two samples. In particular the distributions of the differences of the means can be analyzed. Since the second group has only a size of 10 elements, the *Student's t distributions* must be recalled, in order to decide about significance. By using relations

$$t = \frac{\mu_{s1} - \mu_{s1}}{\sigma\sqrt{\frac{1}{N_1} + \frac{1}{N_2}}} \quad \text{with} \quad \sigma = \sqrt{\frac{N_1\sigma_{s1}^2 + N_2\sigma_{s2}^2}{N_1 + N_2 - 2}} \quad (3)$$

the *t scores* of the 9 differences of the means are equal to

−0,567 0,202 −1,280 0,734 0,224 −0,337 −0,643 1,124 −0,192

while by accepting a confidence coefficient $\alpha = 0.05$, the 9 critical values for the *t scores* are all equal to 1.290. Although the value of the confidence coefficient is rather low, the critical value is always higher than the *t scores* reported. Therefore, it must be stated that there are no significant differences due to gender, in the analyzed class.

For the sake of completeness, it can be reported that in the smaller sample (female) *Visual–Logical Intelligence* results are positively correlated by $r = 0.832$, while *Interpersonal–Auditory* are negatively correlated by $r = -0.724$.

5.4 Group 3

Finally, a *check* group of 14 female students enrolled at the graduate programs in *Arts and Humanities* have been analyzed. This allowed to have a test sample to be compared with the ones of interest.

Once again, the values of the standard deviations of the sampling distributions have been obtained, namely:

1.558   1.254   1.964   1.512   1.890   2.299
2.104   1.890   0.650

while the critical values of the means for $\alpha = 0.05$, are the following:

1.997   1.607   2.517   1.938   2.422   2.946
2.696   2.422   0.833

which can be compared to the measured means:

5.143   0.929 -8.929 8.286   3.429   9.286
-3.214 7.000   2.741

The absolute values of the measured means are always greater than the corresponding critical values, which shows that the answers do not come from a random population. The same result can be found by using the *Student's t* distribution.

Since the results obtained for the first two groups do not differ much, the test group can be conveniently compared with group 2, since these samples are characterized by the same gender, and so are more homogeneous. By following the same method of analysis, based on the *Student's t* distributions, the confidence analysis shows that there is one dimension in which there is an enormous difference: by accepting $\alpha = 0.001$ the differences in the *Logical–Mathematical* dimension can be expressed, in terms of *t score* as 3.723, while the corresponding critical value is only 3.213. By accepting a confidence coefficient less restrictive, namely $\alpha = 0.05$, two other dimensions show significant differences, that are *Kinestetic* and *Individual mean*, for which the *t score* is equal, respectively, to 1.585 and 1.381, both greater than the corresponding critical value 1.321. Hence it seems reasonable inferring that the Engineering students show an *M.I. profile* where the *Logical-Mathematical* and *Kinestetic* approaches are preferred in learning, with respect to a given *test group*.

6 CONCLUSIONS

This investigation has been motivated by the unsatisfactory results obtained in terms of share of students which are able to complete their program successfully at the Italian Engineering Courses. This paper is intended to contribute, hopefully, in changing the old philosophy that the *only* thing learners must do is working hard.

The idea is that learners should be conscious that they can reach knowledge and skills, by using one or more methods tailored over their typical capacities.

M.I. Theory has not been well experimented yet at the highest level of education. For this reason, the present research has offered the opportunity to test another application of Gardner's model. According to the adopted model, the examined group of Engineering students has presented a very high score in the *Logical-Mathematical* dimension, while a less strong superiority in the *Kinestetic* dimension, with respect to another test-group. No significant differences, within the Engineering group, have been detected due to gender.

Results (in standard scores) have been communicated to the students with some comments. Among them, the interviewed ones have shown a great interest in the test and in the obtained results. However, some important aspects of the *cognitive styles* and of the *instructional preferences* could not be detected. For this reason, it seems that the M.I. model, although very general, promising, and appealing, cannot be considered as completely satisfying for the given contest.

The future work concerns: (i) the use of the actual test (revisited) in combination to a standard Cognitive Styles Inventory and an Instructional Preference test; (ii) an efficient Web implementation in dynamic form; (iii) the setting up of a counseling service for the enrolled students who need help, which could take profit from the result of the test.

ACKNOWLEDGEMENTS

Miss Amanda Iorio is gratefully acknowledged for her help in data collecting and processing.

REFERENCES

Allison, C.W., Hayes, J., 1996, The Cognitive Style Index: a Measure of Intuition-Analysis for Organizational Research, *Journal of Management Studies*, Vol. 33, No. 1, pp. 119–135.

Belfiore, N.P., et al., 2002, The Development of an Integrated Approach to the Planning of the Higher Techincal Education: past experience and proposals, *Information Technology Based Higher Education and Training: ITHET 2002, Proc. of the 3rd In.1 Conf., Budapest, July 4–6.*

Brunas-Wagstaff, J., 1998, Personality: A Cognitive Approach, *Routledge Publ.*, London.

Curry, L., 1983, Learning Styles in Continuing Medical Education, Ottowa: *Canadian Medical Association.*

Felder, R.M., Silverman, L.K., 1988, Learning and Teaching Styles in Engineering Education, *Journal of Engineering Education*, Vol. 78, No. 7, pp. 674–681.

Gardner, H., 1983, Frames of mind, the Theory of Multiple Intelligence, *Basic Book Inc.*, New York.

Gorham, J., 1986, Assessment Classificaiton and Implication of Learning Styles as Instructional Interactions,

Communication Education, ERIC Reports, Vol. 35, No. 4, pp. 411–417.

Kolb, D.A., 1984, Experiential learning, *Englewood Cliffs, Prentice Hall*, NJ.

Marton, F. & Saljo, R., 1976, On qualitative differences in learning: 1-outcome and process. *British Journal of Educational Psychology*, Vol. 46, pp. 4–11.

Reichmann, S.W. and Grasha, A.F., 1974, A Rational Approach to Developing and Assessing the Construct Validity of a Study Learning Styles Scale Inventory, *Journal of Psychology*, Vol. 87, pp. 213–223.

Riding, R.J., 1991, Cognitive Style Analysis, Birmingham: *Learning and Training Technology.*

Sadler-Smith, E., 1996, Learning style: A holistic approach, *Journal of European Industrial Training*, Vol. 20, No. 7, pp. 29–36.

Sadler-Smith, E., Riding, R., 1999, Cognitive Style and Instructional preferences, *Instructional Science,* Vol. 27, pp. 355–371.

Witkin, H.A., Moore, C.A., Goodenough, D.R., Cox, P.W., 1977, Field-dependent and field-independent cognitive style and their educational implications, *Review of Educational Research,* Vol. 47, pp. 1–64.

System-based Vision for Strategic and Creative Design, Bontempi (ed.)
© 2003 Swets & Zeitlinger, Lisse, ISBN 90 5809 599 1

MIPAA, Multimedia Interactive Platform for an Appropriate Architecture: a prototype of dissemination and education

M. Nardini

I.T.A.C.A., Università degli Studi di Roma "La Sapienza", Rome, Italy

ABSTRACT: The spread of IT&C has changed the face of dissemination, its intrinsic characteristics and requirements. Among other nations all over the world India is substantially involved in this phenomena. This paper describes the Multimedia Interactive Platform for an Appropriate Architecture, a project focused on modalities to increase the sharing of information in a collaborative form, among several sectors of society (corporate, organizations, NGO's) in India. MIPAA has done some basic ground work in looking at the situation of dissemination of information about appropriate architecture, putting some milestones on future courses of action. Keep in touch activity has been elaborated in collaboration with "Architecture et Developpement" (a French NGO) and others local and national organizations from India, with funds of European Commission (Asia-Information Technology and Communication).

1 INTRODUCTION

When we have started to consider a dissemination of appropriate architecture based on interactive multimedia technology, in a country as India, we had few informations regarding cases in such context. Although we understood as this technology has been essential in a large country as India, culturally diversified and with a high scientific degree, wc doesn't be able to imagine the level of growth in which we operated and how much this development of dissemination has been strategic for a diffusion of knowledge based sharing of experiences on the appropriate architecture. Today we understand as that initial perception was true.

2 APPROPRIATE ARCHITECTURE AND IT&C TECHNOLOGIES

2.1 *Appropriate architecture/appropriate technology*

Any technology is able to be appropriate. Opposite to a problem of choice all technologies be set on the starting line, theoretically with equal possibility to be use: only a punctual judgment, case for case, it will be able to define what of this it be better to choose. Not it may speak of "appropriateness" if not are reported to a specific intervention, in a specific context, in a determinate period of time, in an exact social and political situation. In this sense any technology is able to be or not appropriate under some conditions. However beyond different contexts, that it may bring to a kind of cultural relativism, it is possible to give more indications of "preferential" technologies and establish principles and criterions, even if generals and indicative.

Products or technologies that wants answer with effectiveness and efficiency to the diverged needs of the developing countries has to hold account of some general questions about the contexts and to the wanted quality:

- Environmental friendly solution, with saving of the natural factors, of the resources not renewable (finalized and qualified use, elimination of dissipation);
- Global cost reductions (of acquisition or construction, management, maintenance) to output parity or quality of the built;
- Correspondence to the autonomy of the local stakeholders (usable in participative methodology and in self-construction). The self-construction not gate to a decline of the quality if the local consumers they define in the better way (with the advisor support, with the growth of a local technical culture, with a process of conscious sensitiveness on the really housing needs and on the feasible resources) the specific demands and so the type of product that is wants be built;
- Utilization of the local manpower (tall intensity of the manpower to low investment in capital) and

local materials, for implementing and enhancing the local improvement;

- Simplicity of use, of management, maintenance, what however it doesn't have to mean not acceptance of the technological complexity and of increase performances and outputs;
- Development of the amelioration of the quality of life for the possible greatest part of the peoples.

The adoption of the principles of Appropriate Architecture not means the exclusion of more complex technologies. On the contrary the contribution of IT&C technologies should be strategic to make feasible a progressive switch toward a more appropriate use of resurces, intended as natural and human assets. The aims of this adoption are, more specifically, to stimulate different goals:

- Upgrade and enhance the skills of stakeholders through an involving procedure on the IT&C finalized to provide a prototype of an interactive platform on A.A.;
- Promote the exchange of experience and encourage mutual knowledge and recognition on IT&C tools for A.A.;
- Stimulate the creation of basis for future development, including: transfer of technology, bottom-up processes, training, advising and guidance;
- Amplify and enhance the information about IT&C tools in order to increase the attractiveness of A.A. techniques;
- Offer to stakeholders a broader choice based on IT&C opportunities;
- Host knowledge of opportunities in the sector of IT&C and develop link for further co-operation between beneficiaries, stakeholders and organizations.

On the other hand the IT&C technologies are absolutely necessary if it is considers the quantitative data of the need of population in the countries in the development process.

Finally it will be necessary study new approaches, constituted from a mixture among new and old technologies to diffuse and disseminate information on appropriate architecture material, techniques and production as well as challenge new statements on local development, including: bottom-up processes, people participation and local improvement.

2.2 The challenge of virtual community

The challenge of new age communication tools is to develop the digital age toward a knowledge-based economy. The way in which we will able to build and manage this transition will decide quality of life, work opportunity and environmental condition: in one word our common future.

Always greater diffusion of the IT&C tools we make accustomed to an idea of living in a virtual community. The impression it is not only that we live in such community, "… but even that the virtual community lives in our life…" (Howard Rheingold, *Virtual communities*, 1994). It seems that it could be formulate a new evolutionary concept of community identified by groups with common interests (not only such institutional or territorial) and with an individual and collective vision of the world mediate from the virtual. In this new condition we are influenced to modify our behaviours, our form of aggregation, our pattern of production and fruition of goods and services, as well as our use of language.

Starting from this observation also the concept of community need to be better explained, taking in consideration some sides as *glocality* (the relationship between local and global dimension) and *networking*.

The spread of new forms and new notions of community it bring us to outline some thoughts about the meaning of this term. Firstly we have to explain some visible differences among real and virtual communities. Inside the cyberspace we perceive the world as a more extended place. This place is also more open and less bound. Differently from a real community, deeply rooted in a local context, those virtual become established especially with ideas and concerns that pursue. No need to express a judgement if we said that the cyberspace adapt our mind to a notion of conceptual space lived like a real place. A virtual community is therefore a community, moreover heterogeneous of a real one, moving to mutual help. Some differences of place, time and social status are, in part, balanced by the IT&C tools. However this type of community needs a certain degree of acceptance of diversity, also because the mediation performed by technology is on a deep level and isn't able to distil such superficial aspects of difference. A primary tasks of community is to do as filter among his members and the sea of information that flow in the net. Pierre Levy states: "…this type of groups is less sensitive about local factor and less limited by the institutions…" (Pierre Levy, *Virtual*, 1997). It could be said that the fact to exist without a physical presence, allows all to widen the concept of community. A such group, just for its "horizontal" conditions, can apply a certain decisional weight not only on local but also on global degree. The case of consumer-citizens groups is emblematic. This type of community perform a remarkable role of guide in denunciation of concrete episode to damage the consumers, support the requests, also in legal situs, for singles and for the whole group. A primary role is the advice and the guidance about the concerns of citizens (sophistication of foods, emergency, quality of services, etc.). They are joined from a common interest, tied not only to the local field (greatest number of this problems can wide be shared), and every single subject can take advantage and benefits trough the opportunities offered from the net, to supply support and advice from all of community. A characteristic that we can call horizontal subsidiarity, typical of a virtual community. In this way, maybe better than in a real community, it can be possible to establish an open dialog with all the components of group. The consumer-citizen is in effect a complex subject, he can be interested of several topics, as g.m.o., atmospheric pollution, transport, job, health, life of own quarter (city, environment). He can make it in various forms: discussing, elaborating projects, supplying advising, suggesting experiences, without limits of place and time. Trough the community everyone can adapt the information-communication pattern on its own requirements and necessities.

Moreover such pattern can be an effective referee for the global policy, also because its articulation can carry skilled opinions in a given matter. After this observations we can recognize new contents and new forms in a virtual community, unlike a real one. Three points seem to be representative:

- the *horizontal subsidiarity*: the possibility to enable every subject to receive the know-how of community, without hierarchical limits and on equal base;
- the skill to be referee for the *global policy*;
- a *flexible model*, customisable trough singular or collective requirements.

3 THE MULTIMEDIA INTERACTIVE PLATFORM

3.1 *The current situation*

Our first objective is to understand the complex relationship between appropriate architecture, information technology and development. How can information technology be used to accelerate development and increase sustainability.

The uneven distribution of IT within society and the world is resulting in a "digital divide" between those who have access to information resources and those who do not.

The problem of "digital divide" in India, directly results from low levels of literacy, varied languages, a low degree of availability of information about "best practices" and sustainability strategies in general. Even though IT presents an unique opportunity for more sustainable strategies of use for the resources, a fuller participation in policies and an effective communication between communities and outside world, IT itself is not a panacea for surmounting all the obstacles. For this reason we have chosen an approach that held into account two main questions, in the implementation of

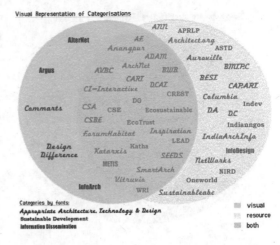

Figure 1. Visual representation of data categories about appropriate architecture.

IT&C tools for popularization of knowledge about appropriate architecture:

- we think the main point is to fill the discrepancy that separates info-rich and info-poor, in the application of this tools, above all for what concerns the teaching or diffusion of literacy skills and education;
- we consider adopting a multimedia methodology approach will be a better way to improve the existing competencies and exchange of information, in constituting a network.

3.2 *Experiences in key means of dissemination*

We focused on the activity of collecting documentation as a first step. The two relevant objectives were: (a) to verify the correspondence of the structure and strategy outlined to the real local needs and possibilities and (b) to collect materials which could be used in the future for the creation of the prototype. The purpose of this activity, therefore, is to actualize the MIPAA such that it is relevant to the correspondent problems and to verify these objectives through action.

At the outset it is important to state that the quality of the collected publications is very high content wise as well as the manner of communication. Having said this, it is important to underline that much more information needs to be made available on appropriate architecture from a growing set of actors: the public authorities and institutions (operating on national, regional and local levels) as well as individuals and groups of individuals. It is also required that the sensitivity and responsibility of such organizations and groups be aroused for far-reaching and efficient protection of environment.

It is hence important to directly involved people to whom information is addressed, in the process and

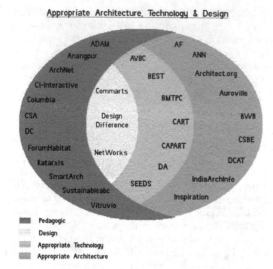

Figure 2. Visual representation of type of data about appropriate architecture.

strategy of diffusion of this topics. To be able to do this a book or a CD Rom cannot be considered effective enough, as these are unidirectional systems of dissemination and communication. But the great advantage of IT&C technologies is that it can even make a "book" in a "writable" form, which allows the stakeholder to "write" some of the contents. This way stakeholders can also contribute in building competencies that trainers must supply.

Besides one of the reasons of the little use of multimedia tools is because the low percentage of customers and users of these services. In India the percentage of internet users to the total population is only 0,2%, even though the actual number of users is around 5.000.000 (e.g. the same number of Slovakia or Turkey).

We have noticed that the documents are frequently not led in relationship with one another. Moreover, there is no clear, access level driven, teaching, to communicate evenly with the local community. As a consequence of this gap, local actors do not adopt IT&C tools or get involved in local strategy.

How to develop the use of these tools, if their adoption – even at the mediator-actor level, is almost unusual as a device of communication that can be more effective? We think there is a work to do on the field of communication at the level of single organizations that produce dissemination materials on appropriate architecture, especially concerning the publication of brochures and of presentation articles, common to all organizations, in terms of developing addressed topics as well as in terms of the means in which they are introduced and addressed. Besides, it is felt important to have a data-base of all the materials that allows to compare the various documents with each other.

3.3 Critical points

It seems that the following critical points emerge, from the evaluation procedure, as sequential steps for an amelioration hypothesis of information dissemination (at the different levels):

- strengthen the level of learning, teaching and diffusion of literacy skills;
- increase the degree of involvement among various organization for increasing the exchange of information;
- adopt effective techniques of communication;
- increase the collaboration and partnership on various subjects in common projects;
- use the technical tools of IT&C;
- learn to communicate through multimedia tools;
- involve a large number of organizations working on the theme of appropriate architecture in a collaborative effort of creating a common information structure (data-bank) on the knowledge in this sector;
- launch initiatives as exhibitions, seminars, and workshop about the dissemination of knowledge in the field of appropriate architecture through IT&C;
- have an open dialogue even with the international organizations in the field of appropriate architecture;
- involve the sector of IT&C in the question of dissemination of knowledge about appropriate architecture.

These steps can be achieved with the implementation of two main actions:

- to upper level (e.g. mediator-actor) to create an interactive multimedia virtual community;
- to lower level (e.g. local actor) to teach and disseminate literacy skills about IT&C.

3.4 Development of MIPAA

Although, we have seen that some of the data are in contradiction and, often, not perfectly comparable amongst themselves, we are able to say that we have recognized a great potential for the use of IT&C tools, considering the high level of know-how of computer science and appropriate architecture (e.g. among the stakeholder). This can also bring the possibility of employing the existing networks in creation and diffusion of a multimedia platform.

In addition to defining more appropriate architectures of communication, the emergence of the environmental topics has led to the search for new models of sharing activities, the development of new strategies, as well as singling out and exploiting technologically advanced tools in appropriate architecture, such as informatic tools, capable of rendering the new schemes formulated usable and effective, in order to make a policy of information and participating in environmental decisions possible.

The home access to a computer and the Internet is infrequent in India. Most persons who use IT&C do so at work. Most of them enter the world of IT only as a tool of production for usual actions. Only few persons use it as a tool of communication for the creation of a network and for the exchange of information. On the side of production of IT materials we have found a good level of knowledge about software design and programming. The e-mail, on the other hand, is much more prolifically used as compared to web services. The main constraint to using web services is the bandwidth available. The most important factor in increasing the use, and the advantages, of IT&C tools is education. This requires interventions at all levels, from literacy through scientific and technological education. Indeed, improving the quality and the accessibility of IT&C tools is progressively essential to moderate the digital divide. Furthermore, access to basic education can allow to prepare people and give them skills to be active in a wide range of roles in IT, as creators, designers and managers. The effort should focus on the increasing of transfer of information through IT&C systems (e.g. like information about appropriate architecture in the MIPAA).

In rural areas the resources and infrastructure are lacking. Connectivity is typically available only in capitals and perhaps secondary cities. Increasing the access to IT involves increasing the availability of infrastructures (e.g. wireless and satellite communications) in rural and semi-urban areas. This extension of infrastructure should develop on the establishment of common use facilities (e.g. tele centers or mobile facility), which are more convenient and accessible to people.

3.5 Policy process for implementation of MIPAA

As last step, we focused on how IT&C could be incorporated and empower several policy processes in education and dissemination of appropriate architecture. We have tried to do this saying with greater precision what it should be the brief of multimedia interactive platform for appropriate architecture.

The MIPAA is a combined initiative in order to offer a reference point for institutions, local organization and communities interested to apply and take advantage in development of multimedia didactic tools about appropriate architecture.

MIPAA provides its own technical support, the matured acquaintances and some instruments in order to increase the productivity, to reduce the costs and to improve the quality of the products.

Its objectives are:

• to stimulate the birth of new plans;
• to create a consultation point where it can be possible to experience and to estimate how much the

Table 1. Policy steps in the MIPAA approach.

Policy step	MIPAA approach
Problem definition	Focus on general information and description of phenomena connected to the employment of IT&C in dissemination of appropriate architecture: – who holds the information – what type of information are there – are information compatible with IT&C tools?
Definition of goals	Statement of MIPAA platform: – how to improve access to knowledge about appropriate architecture; – how to implement a mutual knowledge process; – which tools are compatible with the context where we act?
Formulation of policy option	Policy to increase the sharing of ideas and collaborations about appropriate architecture education and research: – develop methods and instruments – create a consultation point – stimulate the birth of new plans – produce multimedia tools – promote research in multimedia field – develop didactical interactive instruments – organize seminars and workshop
Choice of preferred option	Focus on overall and specific impacts: – create a consultation point – organize seminars and workshop – stimulate the birth of new plans
Implementation of new policy	Develop support from stakeholder, mediator-actor, local actor
Implementation of policy decision	Define implementation modalities and administer process for compliance
Evaluation and monitoring	Based on overall and specific data and goals and qualitative methods of analysis. Decision based on overall and specific impact

information technology provides, in order to improve the didactic effectiveness;
• to develop methods and instruments for the production of multimedia didactic-interactive material;
• to produce multimedia tools for didactic inner and the external ones;
• to promote research in multimedia field.

Moreover MIPAA organizes:

• courses on the production of multimedia didactical interactive instruments;
• seminars and workshops on multimedia and new technologies for didactics and permanent education in national and international field.

We have understood that the question is not only to receive and diffuse information but to submit this data flow to a critical analysis. The aim is to supply reading keys and possible models of efficient communication, not exclusively based on top-down solutions (e.g. pre constituted data-base). For this reason our intention is to listen to local operators to be able to understand what would be the best solution. In order to achieve this we need a collection of different joint experiences and joint knowledge, not only thought as one of one singular interventions. The various knowledge must be sensitive to IT&C instruments which can increase the performances and goals of communication. We think that these instruments can carry not only the knowledge from the worldwide scene to the local contexts, but also to implement an inverse procedure, from the local towards the world.

For these reasons we think that it is strategic to enable the local actors to use IT&C tools.

In the following table we have reassumed the single footsteps in comparison with the implementation of the MIPAA project. They derive from the collecting and analysis of documentation as from the field observation, from the suggestions received by the participants to the project and from those people with we have gotten in touch with.

4 CONCLUSIONS

At last we want to affirm that there is a strong connection among the social organization and the management of information space (e.g. trough IT&C tools). It is necessary to define new relationship among institutions and community, for this aim IT&C tools are able to help us.

We can work on the emergent social group to discover, in one hand, the local knowledge heritage and, in the other, to build nets that activate a bottom-up globalization process. With the aim to create an effective activity of exchange in local fields that represents a type of sociality with deep condition to develop social and productive projects, together with a more complex concept of "local", qualified toward subsidiarity.

We want to underline that the uneven distribution of IT&C within society and the world is resulting in a "digital divide" between those who have access to information resources and those who do not. The problem of low level of literacy, the different languages, a lower degree of information around IT&C in general, determine a marginalisation from the mainstream of the world of a great part of population. But in this field IT&C tools present an unique opportunity for a more sustainable strategies of use for the resources, a fuller participation in policies and an effective communication between communities and outside world.

ACKNOWLEDGMENTS

This paper would not exist without the support of: Ludovic Jonnard (Director of A&D, Architecture et Developement), Claudia Melani (Head of Office, A&D India), Radha Kunke and Sayalee Joshi (Local Staff of A&D India).

REFERENCES

Architecture & Development, http://www.archidev.org/mipaa

HTG, Habitat Technologies Group, http://www.habitatgroup.org/

CART, Centre for Appropriate Rural Technologies, *http://www.oneworld.org/cart*

DA, Development Alternatives, *http://www.devalt.org/*

Auroville Building Centre, *http://www.auroville.org/research/csr*

The internet-enabled Rural Marketplace. Towards the digital village, TARAhaat.com, Development Alternatives, New Delhi, 2001.

Development Alternatives, a magazine for sustainable human development, vol. 12 n°9, September 2002.

Basin News. Building Advisory and Information Network, Changing education for sustainable construction, December 2001, N°22.

Algorithmic approach in engineering educational process

P. Pentelenyi & P. Toth
Budapest Polytechnic, Budapest, Hungary

ABSTRACT: Computer studies do not present the only opportunity for us to form and develop students' algorithmic awareness. Algorithmic thinking can be taught using problems arising in any areas. Problems and tasks arise in each of the various subjects, and it is necessary to introduce the students to the algorithmic solution of problems and require this of them even if they do not have to use a computer to solve the problem.

In our presentation teaching characterised by an algorithmic approach itself and also its significance will come into the centre of the attention. A compact survey of the historical development of teaching programming, which is to be given in our paper, will inevitably and naturally explain the favoured role of algorithm. A comparative analysis of the teaching and programming activities, the effects on the entire educational process and the methodological consequences of the algorithmic approach will be explored and demonstrated.

1 INTRODUCTION

Before studying the possibilities of the formation of algorithmic attitude and thinking, let us take a look at the origins of the term "algorithm".

The word "algorithm" entered our dictionaries in our age as one of the basic technical terms of computer science; in this sense it only made its appearance in the middle of the century. Different theories exist with regard to the development of the word. Donald E. Knuth observes that at first glance it looks as if the first four letters of the word "logarithm" were mixed up. In Webster's New World Dictionary only the old word: "algorism" is to be found up to 1957, and originally it meant arithmetic operations done in Arabic figures. According to historians of mathematics, the word "algorism" derives from the Latin distortion of the name of Abu'Abd Allah Muhammad ibn Musa al-Khwarizmi ("Father of Abdullah, Mohammed, son of Moses, native of Khwarizm"), the famous Arab mathematician (c. 825). The meaning and form of the word "algorism" has changed through time. In its current sense, algorithm is the sequence of such instructions as defines the order of operations yielding the solution for any possible output of the given problem. Examining the definition, one is beginning to feel that algorithm is a universal concept. From the instructions in a public telephone box to "lift the receiver ... wait for the dialling tone ... insert the coin ... dial ... etc. ....", through the (Euclidean) mechanical procedure to define the greatest common divisor of two natural numbers, to the description of the production process of a particular machine part, we are in fact faced with algorithms. The recipe for paprika potatoes or poppy-seed rolls is also an algorithm, just like the migrational instinct of swallows – even if we are not exactly familiar with the individual steps of the latter algorithm. (One may wonder whether the Universe itself, which betrays a constructedness and development according to strictly planned principles in every detail, is not the still ongoing "execution" of an Algorithm, probably having a reaction on itself as well ...)

2 PROBLEM SOLVING

When, in the course of teacher training, we teach trainee teachers the teaching of learning, for example, we actually have to study and elaborate the algorithm of learning. When teaching beginner programmers, we use the outline of the process of computerised problem solving itself in order to give an analogous – probably first – illustration for the concept of algorithm. Of course the main steps: problem – model – algorithm – programme represent a rough algorithm, which is the one for computerised problem solving. It is by no means accidental that the latter conspicuously resembles the algorithm recommended for (traditional) problem solving by György Pólya, the main steps of which are

– understanding the problem
– preparation of a plan

- execution of the plan
- retrospection.

From this introduction naturally follows that computer studies and other subjects of information technology are not the only areas, which can form students' algorithmic thinking. Any algorithmic problem coming up in the practice of teaching (either in mathematics or in other basic subjects) may prove an excellent opportunity to show: how to follow the "problem-model-algorithm" path, one that is unavoidable in computer programming. Unfortunately, when solving a problem without a computer we often do not proceed in this way, even though this problem-solving plan ensures a complex view. Our experiences suggest that the integration of computer studies into other basic subjects produce an interaction that multiplies the possibilities for both sides. Education in algorithmic thinking and efforts to form and develop algorithmic awareness raise the standards of training in any basic subject while they also help programming taught in computer studies. Problems and tasks arise in all of the basic subjects, and it is important to introduce the students to the algorithmic solution of problems and also require this of them even if they do not necessarily have to use a computer to solve the problem. Teaching algorithmic problem solving is to be taught to trainee-teachers, as teacher education itself can prove to be the main opportunity for putting our concept into practice.

3 HISTORICAL DEVELOPMENT

When one sets out to investigate a particular area, he or she will first of all have to start with a historical survey of it. Pedagogical experiences show that a historical analysis of a given subject area and also of its specific professional terminology can give a particular motivation to the student as well as a more understandable explanation of the problem in question. For those dealing with the teaching of algorithmic problem solving, a concrete historical survey of teaching programming and an examination of its development, achievements and shortcomings will help the better understanding of the present as well as a definition as precise as possible of future needs. The adaptation of the requirements of historicity to the teaching of computer studies, information technology and programming can be justified even if the relatively young age of the science in question allows only a modest time span to be investigated. "Development is often not the result of a natural sequence of events, but of entirely different reasons, the ignorance or misinterpretation of which may lead to wrong conclusions. The results of development cannot be understood in the present without an understanding of the past." Studying the history of teaching programming it can be seen that in the course of problem solving the role of algorithm construction, and algorithm independent of a particular programming language or type of computer have gained an increasing significance.

Up to the 1950s programmers merely learnt the use of one or two particular computers. The designers and builders taught the users how to operate a given computer. In practice, obviously, every programmer used a method of his own to construct algorithm in a computer. In the beginning programming mainly meant experimenting with the computers themselves. At that time there was yet no need for generally usable programming methods on account of the limited possibilities and the imperfect reliability of computers.

The 1960s are usually regarded as the age of programming languages. Gradually increasing efforts to unify programming methods lead to the elaboration of high quality programming languages. After several hundreds of (more or less computer independent) programming languages had been created, the question of teaching programming came to the forefront. In the beginning it meant the teaching of programming languages, and the process of the construction of algorithm was neglected in teaching. Tasks concerning teaching objectives were also aimed only at teaching the elements of a given programming language.

The era of algorithmic programming began when the methodology of teaching programming developed in the 1970s. The achievements in the methodology of structured programming are linked with the names of O. J. Dahl, E. W. Dijkstra and C. A. R. Hoare. The teaching of programming started as an "experimental science", since it meant in the beginning a constellation of procedures, operations and methods based on the various areas of science as well as on practice. The title of a book written by the well-known American professor, Donald. E. Knuth in 1968, "The Art of Computer Programming", proves that the book is more about a professional skill than about theoretical science.

Development is well demonstrated by the title of a book by E. W. Dijkstra in 1976: "A Discipline of Programming". Similarly, the title of the work by D. Gries in 1981, "The Science of Programming", shows still another landmark in the way programming has developed.

4 ALGORITHMIC THINKING AND PROBLEM SOLVING

The survey of the historical development of teaching programming inevitably and naturally explains the favoured role of algorithm: it is clear for us that we are to place emphasis on the concept of algorithm instead of programme. It is quite wrong to start writing a problem solving algorithm immediately in a certain programming language, because it means, that instead of the problem it is the programming language which prescribes the details to be dealt with. With knowledge of algorithm, that is the textual system of instructions for problem-solving, it is a dangerous job, especially for beginners, to write a computer programme directly in the case of a somewhat complex algorithm. Searching for and correcting faults in the programme may take a lot more of our time than keeping neat order throughout our work. One of the best means of keeping order is the graphic sketch of algorithm, the flowchart.

To draw a flowchart is principally none other than observing the classical recommendation: to "design in advance" the solution of a problem, that is "program" in Ancient Greek and "pre-draw" in English. Therefore, from a methodological point of view, in the course of solving a certain problem by computer it is absolutely sensible to follow the route of model – algorithm – flowchart – programme. And although, having a given high quality programming language in sight, it is possible for us to draw a so-called programme-oriented flowchart, we have to attach even graver significance during the constant changes of technical development in our era to the preparation of a general flowchart, which at the same time entails algorithm independent of a particular programming language or type of computer.

Although graphic algorithms and flowcharts are drawn for human use and not for the machine, and we therefore have more freedom as regards the symbols and expressions used, it is still advisable to agree in advance on a certain system of symbols. Large flowcharts for more complicated and long-winded problems are often unclear. In such cases it is advisable initially to prepare an outline (a so-called main flowchart) which, being brief, is easy to read and follow. The idea of preparing a main flowchart in outline entails the opportunity of creating modular system. Individual program details and sub-programmes can be used as building blocks to construct edifices of unlimited complexity. What is more, the complex programmes themselves, as modules, can be used as building blocks for a possibly even higher structure. In the formulation of graphic algorithms this corresponds to the programming methodology known as *structured programming*. The essence of structured programming is the methodological application of the principle "divide et impera!" (divide and rule!): we repeatedly deconstruct the algorithms and data objects using specific control structures and data structures. In order to win acceptance for the principle described above, we clearly need to keep a strict order when constructing algorithms and drawing flowcharts. We should only assemble those programme details, which can genuinely be inserted as building blocks into the system. Each such unit (module) should ideally only

have one input and one output. And vice versa: we should be able to "extract" almost any detail of the flowcharts we draw.

It is interesting to see that the concept we today call graphic algorithm, or flowchart, is covered by "programme" in its original meaning of the two words "algorithm" and "programme", as is clear from the facts mentioned above. However, the word "programme" is nowadays used for the coded version of algorithm: when programming, we are in fact describing a particular algorithm in a certain programming language.

Examining the role and "place" of algorithm in problem solving shows that a certain order of problem solving was of course created before the advent of computers. In the following we would like to throw some light upon the extent to which the use of computers modifies this order by a comparative survey of the phases in traditional and computerised problem solving. We begin the full solution of a problem by analysing the problem, constructing a mathematical model on this basis, developing an algorithm for the mathematical model to guide carrying out the solution process either without or with computer.

Thus a certain *problem* is given and we have to set up a mathematical model for it. We are in the same situation as a secondary pupil solving a textual equation: he or she must write down some kind of formula or model on the basis of the problem posed in the text. At such times we omit the irrelevant parts of the text and concentrate only on the relevant sections. On the level of a secondary school example this is extremely simple; in the problems we are presented in real life, however, it is often difficult to construct a correct model, since there is room for argument about what is relevant and what is not. The question therefore is to decide what actual task we need to set ourselves for the real solution of the problem. The *mathematical model* therefore contains the outlines of the task to be set.

Once equipped with the model we can return to the formulation of the *algorithm*. We need to compose a specific sequence of instructions to define the order of those operations providing a solution for any possible output of a problem. The key question for the effective use of the computer is whether the algorithm we are working from is suitable. Once we have a precise and detailed textual or graphic algorithm, writing the source programme (codifying the algorithm into a concrete programming language) is an almost mechanical activity, and it is therefore obvious that during the teaching-learning process we need to concentrate on the creative task of algorithm construction.

For ease of use it is advisable to give the algorithm in graphic form. Graphic algorithms are known as flowcharts or block diagrams. Of course, other methods of writing algorithms can also be used, such as structural diagrams (so-called structograms) or description in sentences. The essential point is that we should be able to compose *a well-arranged algorithm irrespective of the concrete programming language or type of computer.*

It is perhaps worth observing that even if we are using a computer to solve the problem it is still incorrect to start writing a problem-solving algorithm in some concrete programming language, as in this case it would be the programming language, and not the problem itself, which would determine which details of the description we are to deal with. In most cases the actual basic algorithms and objects of programming languages are at a much lower level than those needed to begin working out the algorithm. Thus even when teaching programming we should emphasise the concept of the algorithm, not that of the programme. This fact also justifies our intention of not only using algorithms in computer studies or other subjects connected with information technology. Algorithmic thinking can be taught using problems arising in any of the basic subjects, even if we do not necessarily want to solve them by computer.

The phases in traditional problem solving:

(1) posing and analysing the problem,
(2) formulating or selecting the model of the solution,
(3) preparing a solution plan, working out the algorithm of the solution,
(4) **the actual problem solving according to the instructions of the algorithm**,
(5) analysing and using the results.

The phases in computerised problem solving are the following:

(1) posing and analysing the problem,
(2) formulating or selecting the model of the solution,
(3) preparing a solution plan, working out the algorithm of the solution,
(4) **writing the algorithm into a computer programme, then running the programme on a computer**,
(5) analysing and using the results.

On the face of it, one could say that the only difference between the two is that in the case of the first one the solution plan is implemented by man, whereas in the case of the second it is done by the computer. In reality, of course, the decisive difference between the traditional and computerised processes is to be found in the extent to which algorithm is elaborated as well as in its precision. In the course of the traditional solution, the organisation and implementation of the algorithm may concur, because in the mind of the thinking and calculating person the selection of the successive operations is, in most of the cases, also linked to their implementation. In order to solve lengthier problems, however, it is always sensible to prepare a so-called

solution plan, which, in other words, means just the more or less precise construction of the algorithm.

There is a choice of methods for writing algorithms, e.g. flowcharts, struktograms, and written descriptions. The aim of each one is the same: the preparation of a solution plan for a given type of problem: such a description of the process of problem solving, which is independent of computers and a particular programming language, clearly and unambiguously follows the logical chain of ideas and reflects structural units. These methods for writing algorithms can be used even when a particular problem need not necessarily be solved by computer. During the teaching-learning process we have to concentrate on this creative stage, which is the construction of algorithm within problem solving, since implementation is a purely mechanical activity, whether it be done by man or computer.

5 THE ROLE OF MODEL PROBLEMS

In the course of analysing the teaching-learning process of algorithm, it is becoming evident that we should not primarily teach complete algorithms, rather the way how an algorithm is created. Algorithms to be learnt mainly make sense and are in place in the teaching-learning process only in so far as they appear at the highest level (preferably at that of automatism) (Gyaraki 1976). The "multiplication table" will of course have to be remembered with one's eyes closed, too: $3 \times 8 = 24$ (without thinking! – which of course does not mean that we were not earlier shown and we did not understand that 3×8 stands for $8 + 8 + 8$ and this surely equals 24). The problem in view should be life-like and "tangible". The construction of the algorithm should in every case be preceded by an analysis of the practical problem and model-formation as a result of this. It is to be clarified what kind of data are available (INPUT) and what results are expected (OUTPUT). Algorithm is the precise guide (timetable) which, if followed, leads to the aim, which is the solution of the problem. To be able to set this we ourselves have to be able to solve the given problem. In essence we have to answer the question we have asked ourselves: "How would I go about it?" At the same time, however, in the case of computerised problem solving, we immediately have to think of the fact that the computer interprets and executes more elementary operations than man – the steps have to be broken up for it accordingly. Students, provided with continuous and appropriate help, may also be drawn into the exploration of the individual steps of the algorithm. This way the principle of progressiveness is secured in their training for independence. It is a great advantage that the student tends to "feel ownership" over the problem jointly solved. At the same time it may hide some danger, too: most of the students think that once they have understood something, they also know it. They must be made to realise that understanding is only a necessary but not sufficient prerequisite for knowledge. Let us for the time being take reproductive knowledge only. This is the easiest to check: is the student able to reproduce on a blank piece of paper the solution which has previously been jointly found (and understood)? If students are successfully made to check themselves, they will on any number of occasions experience that understanding is not necessarily the same as knowledge. What is there to be done in this case? The scrutiny of the correct solution again – in the course of which it can turn out that even understanding has not a certain points been fully grounded. After clarifying these students can feel again that they know it already, but this has to be checked again. The "cycle-core" sketched here will have to be executed from time to time until the "condition for exit" (flawless reproduction) is fulfilled. How could one succeed in solving an unfamiliar problem until they have difficulty in reproducing the familiar? Unfortunately, we, teachers are also to be blamed for underrating reproductive knowledge (due to vanity?) and letting this fundamentally important link in the teaching-learning process be missed. It is a serious contradiction, as, at the same time, we tend to require students to know definitions and theorems precisely, which means that in reality we do not deny the necessity of reproductive learning. In the case of teaching-learning a foreign language, the acquisition of words and grammatical rules is unanimously considered important, but what is the case with rote learning? It looks as if we were forgetting nowadays that the word by word learning of set texts (understood, at the same time, of course, consciously even in its details) does not simply help the acquisition of requisites necessary for the knowledge of a language, but through them we store such patterns for the complex manifestations of language which, partly or in full, or even already transformed by the control of the composition patterns, can in a certain sense create something new. If materials are selected well and appropriately for everyday use, motivation will be guaranteed, too. Anyone may see that if they learn 10–20 original foreign language sentences about a certain topic (consciously, understanding it also in its details), by the time they are able to say them fluently (automatically), they will at the same time be able, even if not quite fluently, to interpret the topic in question in the given foreign language practically without a mistake in variations transformed almost at discretion in their own wording. This is a great thing, since implicitly it means that they are able to expound their own ideas about the given topic. In our case the well-selected model problems may play the role of such "rote learning". Nobody would question the inexhaustible intellectual potentials inherent in chess, and still, who would

claim that the acquisition of the steps in chess or the reproductive learning of basic strategies could be omitted or would, by any chance, be shameful?

According to Gestalt psychologists, thinking man "penetrates" the structure of a problem and in the interest of the solution he "restructures" it. The majority of research dealing with expertise examines the how of solving the more or less familiar problems, where, as a matter of fact, schematic knowledge becomes directly applied. Research into creativity, however, seeks the answer to the question of how we are able to produce a solution in such cases when we are not in possession of facts that are related to the particular problem. It is very likely that analogue thinking is the key with the help of which a person dealing with the solution of a problem indirectly still makes use of his former experience. Concerning cognitive psychological aspects of teaching algorithmic problem solving it can be concluded that the cognitive theories of problem solving could be extended to algorithmic problem solving, too. Through analogical problem solving, algorithmic solution contributes to the extension of algorithmic approach in the framework of the entire teaching-learning process.

6 METHODOLOGICAL CONSEQUENCES

With respect to teacher training, it is especially important to see that these ideas are to be taught to trainee-teachers, as it is more often than not that the teachers' way of thinking is reflected in the students' development and so – algorithmic thinking can be passed on. For future teachers – regardless of their chosen special subjects – it is in any way imperative to have adequate programming skills and the capacity for algorithmic thinking.

In what follows we shall make a comparitive analysis of the teaching and programming activities, and examine the effects on the teaching process and the methodological consequences of the algorithmic approach. Let us just consider the fact that whenever we have to solve a problem by computer, the writing of a computer programme in reality means that the programmer "teaches the computer" to solve the given problem. The algorithm of the solution is, in this sense, the result of a "teaching plan", from which it follows that the programmer immediately performs some sort of teaching activity as well, while he or she is simultaneously – and consequently – expected to take into account the capacities and potentialities of the "student". It may also be interesting to highlight the analogy between the two main phases of the process:

general preparation – algorithm-construction independent of programming lang.

adjustment to a given class – adaptation to a concrete progr. language

Not only does the simile expressly show but it also makes the separation and sequence of the two phases indisputable. (True though it is that "adjustment" for the programmer is more or less continuous, whereas for the teacher it means rather the consideration of changing conditions). At any rate, studying the connection between the programming activity and teaching, the principle of "docendo discimus" comes to one's mind, which Joseph Joubert (1754–1824), the French classicist thinker, who wished to revive the Roman genius, expresses this way: "Those who teach, learn doubly". It is true that we do not simply have to be familiar with what we endeavour to teach, but we have to be deeply and analytically familiar with it. Thus preparing to teach others, on every occasion we unavoidably teach ourselves as well, which means: we learn.

REFERENCES

Gentner, D. 1983. Structure-mapping: a theoretical framework for analogy. *Cognitive Science*, 7.

Gyaraki, F.F. 1976. Algoritmusok es algoritmikus eloirasok didaktikai felhasznalasanak es optimalizalasanak lehetosegei, *Kandidatusi ertekezes tezisei,Budapest.*

Knuth, D.E. 1997. The Art of Computer Programming. Addison-Wesley, Harlow, England.

Lipman, M. 1991. Thinking in education. Cambridge: Cambridge University Press.

Pentelenyi, P. 1998. Development of algorithmic thinking. Budapest: Ligatura Ltd.

Pentelenyi, P. 1997. Can Problem Solving be Taught? *Engineering Education '97, IGIP Symposium, Klagenfurt.*

Polya, G. 1990. How to solve it? Princeton University Press.

Toth, P. & Pentelenyi, P. 2002. New Equipment and Methods in Technical Teacher Training. *In Litvinenko, V. – Melezinek, A. Prichodko, V. Ingenieur des 21. Jahrhunderts, Referat des 31. Internationalen Symposiums. Sankt-Petersburg.*

Treffinger, D.J., Feldhussen, J.F. & Isaksen, S.G. 1990. Organization and structure of productive thinking. *Creative Learning Today.*

Vigh, A. 1932. Az iparoktatas tortenete, Budapest.

System-based Vision for Strategic and Creative Design, Bontempi (ed.)
© *2003 Swets & Zeitlinger, Lisse, ISBN 90 5809 599 1*

A methodology for teaching "Computer Aided Design and Drawing": a didactical experience

E. Pezzuti, A. Umbertini & P.P. Valentini
Dipartimento di Ingegneria Meccanica, Università di Roma Tor Vergata, Rome, Italy

ABSTRACT: In this paper the results related to the activation of a course on computer aided design, following the passage by old academic organization of Faculty of Engineering of the University of Roma Tor Vergata (quinquennial degree) to the new (3 years basic degree and two years specialistic degree) are presented. The course, belonging to the formative packages of mechanical engineering degree, has been activated on the explicit demand of students and by personal initiative of the Prof. Eugenio Pezzuti.

1   INTRODUCTION

The Faculty of Engineering of the University of Rome Tor Vergata has been the first, in nation-wide, to upgrade the courses to the directives related to the three years engineering curricula. In fact, three years ago, according to the new organization, all the three years of basic degree have been simultaneously implemented. The possibility of sustaining exams and theses according to the old procedures has been given to the students that decided of maintaining the old curricula. At the same time, the possibility to migrate to the new curricula, computing the cumulated credits by means of conversion tables, has been given. Inevitably the transient phase involved some inconveniences for the students and a hard didactic work for the teachers. But has been also the occasion for a deep reorganization of the formative *iter* of the different courses offered at our faculty. In this context a completely new course of CAD has been offered. It fills an old gap of the Mechanical Engineering degree in our university where such course was never offered. The new course has five credits value and is scheduled inside a formative package. The need of a such course, further by obvious considerations on the actual importance of the CAD in design and industry, is also justified by the request of the students who asked for a long time its activation. By these considerations, on personal initiative of the Prof. Eugenio Pezzuti, the course has been activated. After the first year of adjustment it has obtained a good success attending an actual frequency, but constantly growing, of about 35 students. The result is very satisfactory considered that it isn't a mandatory course. The only true problem concerns

with the inadequacy of structures, with reference to the computer centre. In fact the personal computers are obsolete and insufficient in number and performance. Moreover there isn't support staff or a system manager. This situation has been partially solved by undertaking of all, of the students, of the tutors, and of the professor. However, the situation can become critical if, as expected, the course will be attended by an increasing number of students. Unfortunately the current economic situation doesn't allow adjustments in short times about structures and staff, but there is a probability of changing in the medium period.

2   THE STRUCTURE, THE CONTENTS AND THE ORGANIZATION OF THE COURSE

As reported the course has a value of 5 credits, for a total of about 50 hours, equally divided into theoretical lessons and practices. The practices are carried out in the computer centre. The course belongs to a formative package for mechanical engineering degree and is not mandatory, but the student must apply explicitly to attend the course and to include into his curriculum. About the theoretical lessons, the aim is to give the students, compatibly with the short time available, both a view on the mathematical part behind a CAD software and a methodology for drawing with CAD. The goal is to give the students a universal methodology, not specialized on a specific CAD software, but easily employable with others CAD programs. To clarify this concept, by personal opinion of the instructor, the teaching of a CAD course can be limited to explain the basic (or advanced) functions of a specific software, but must give to the students the fully understanding of the used tool and the philosophy of methodology of a CAD system. In practice, if a student uses a curve spline, he has to know, besides the commands, also the mathematics, the possibilities and the limits of the curve. Only in this way he can take all the advantages of the potentialities of the function. By understanding the mathematical and physical meaning of the software, the student will be able to use a different CAD system with a modest effort. For the same reasons it is also important the philosophy of use of a CAD system. It has been chosen to teach directly use a three-dimensional system CAD, by adopting the Autodesk software Mechanical Desktop. After a few lessons dedicated to the use of the 2D systems (substantially AutoCAD 2D), to conform the middle level of knowledge of the class, a 3D parametric variational CAD has been directly explained. This choice is motivated by the following reasons:

- on the ground of the experiences made in the other Italian Universities, during the national conventions of the professors of the ADM (Associazione Disegno di Machine) group, it has emerged that the teaching of a CAD 3D system has a great didactic and formative value, also in consideration that the students are able to use by themselves, in most cases, the functions of a 2D CAD software
- currently, due to the decreasing of the costs of the hardware and the software, the 3D CAD systems are widely spread also in the small-medium industry. There is also a growing demand of expert people.

Explaining the use of the program, besides the use of the single commands and functions, the philosophy and methodology of using have been widely discussed. In fact, while a 2D CAD system is substantially a expert drawing machine, a 3D parametric variational CAD allows to built a solid model, but, starting from a basic sketch, requires the constrains, the parametric dimensions and all the working phases (like extrusion, revolution etc.), in a way similar to the real building of the object. Therefore, it is important that the students make themselves familiar with these not intuitive concepts, but necessary to obtain a correct solid model. Also in this case a simple example can explain better. Let us assume that an object is created through extrusion of a basic sketch and that the object has a symmetry related to the extrusion; if we want automatically obtain the dimensions of symmetry during the automatic of the model, we have to extrude the sketch profile in two symmetrical directions. Otherwise if we extrude the sketch in only one direction, we obtain the same model, but during the automatic dimensioning phase the dimension displayed are different. The theoretical lessons are completed by a part, quickly explained because of the short available time, about the mathematics implemented in the CAD

Table 1. Theoretical lessons.

Theoretical lessons
Elements of vectorial calculus
Parametric formulation of the functions
The curves of CAD systems (interpolating curves, spline, b-spline, nurbs etc.) and di Hermite's and Lagrange's methods
Surfaces of CAD systems (bilinear, Coons, Spline, B-spline, Nurbs etc.)
The basic workings, the lofting
B-Rep and CSG systems
Rendering and shading

Table 2. Practices.

Practices
2D functions and 3D basic functions
Settings of the software, snaps, layer etc.
Drafting functions, basic primitives, editing functions, hatch, blocks etc.
Dimensioning, printing layout etc.
3D superficies and Boolean functions
Extraction of attributes and proprieties
Basic about lights, textures, materials, rendering and shading
3D parametric variational CAD
The basic sketch, its parametric dimensioning, constrains etc.
Basic workings: extrusion, sweep, revolution, lofting, chamfer, holes, use of features and parametric tables, etc.
3D editing functions
3D constrains and complex assembly, use of X-ref
Parametric libraries of standard components
Printing layout, automatic dimensioning, sections, etc.
Advanced rendering and shading

software. In the table 1, 2 the arguments of the theoretical part and the practices are synthetically summarised.

The software adopted for the course is the program Mechanical Desktop. This program is dedicated to mechanical applications and it has a very didactic formulation, very useful considered the aim of the course. For example, it is always explicitly displayed the generation tree of the model. Thus, it is possible, in every moment, to come back, modify and automatically update the model. Moreover, the software has all the sophisticated functions and features typical of the best CAD programs. It is also interesting the possibility of complex model assembly, specially related to the organization of practices, later illustrated, by single parts generated in single files. The single files can be linked to the file of the assembly and the complex model is automatically updated when a single file is modified. Finally, the possibility to use libraries of standard components, like cams, shafts, bearings, screw etc. has a remarkable didactic value. Besides the software is

2150

Figure 1. Ducati Monster motorcycle, full 3D solid model developed during the course by a group of students.

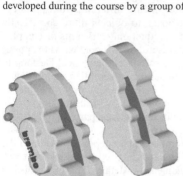

Figure 2. Details of the Ducati Monster motorcycle.

integrated by some calculations tools, like a 3D mesh generator and a linear fem module and modules to calculate the main mechanical components like screw, bearings, springs, shafts etc. Finally, but not less important, the economic aspect must be underlined: the AutoDesk, the producer of the software, has a educational policy that allows to the University, schools, professors and students to purchase the license of the software at discounted prices. The only true limit seems to be the requirement of using high

Figure 3. Bicycle Colnago, full 3D solid model developed during the course by a group of students.

Figure 4. Details of the back wheel of the bicycle.

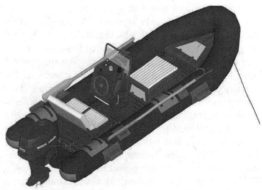

Figure 5. Raft MARSHALL, full 3D solid model developed during the course by a group of students.

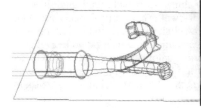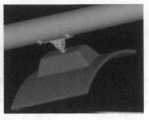

Figure 6. Details of the raft.

Figure 7. Details of the motor of the raft.

performance computers. In fact when the generated model is complex, the computing time is high. Regarding the didactic phase, the classes are delivered making use of a multimedial system, i.e. a laptop wired to a projector. This methodology, applied also in all the others courses taught by the first author, is very communicative and effective related to learning. It allows to visualise immediately the practical use and the effect of the explained concept, involves effectively and directly the students. After the theoretical classes, there is always a practice phase. In this last phase the students can apply and experiment the concept explained during the last lesson. About practices, the students are divided by groups; every group, at the end of the course, must prepare for the final examination a complex mechanical object using 3D solid modelling. The choice of the typology of the object is free for every group. This method has the double value to accustom the students to work in team, quality more and more appreciated in industry, and to stimulate the students to research the necessary data also by mean of instruments such as internet. The course is completed by some monothematic lessons about the possibility of interfacing a CAD system with other calculus and analysis programs. In detail examples of use of a 3D CAD model inside FEM or Multibody programs are illustrated in a synthetic way. As previously mentioned, for the final test every group must prepare a complex mechanical assembly. The object is freely chosen and this stimulates the imagination and the enthusiasm of the students. The typologies of mechanical objects chosen by the students are different, from motorcycles, bicycles or boats, to objects more specifically mechanics, such as reduction gears units or complex mechanisms. Very useful, in this context, is the possibility offered by the software Mechanical Desktop to build the full model by assembling single parts spreaded in single files. In this way every member of the group can autonomously work using single files of components, of its competence parts, and then they are assembled in a general file for the full model. The final project is completed by a report in which all the phases of modelling and the used functions and commands are explained in detail. This relationship allows the teacher an evaluation of the followed choices during development of the project, and also allows to the students to fix, in a clear way, the methodology of building a model and the techniques used in the realization of the project. In the following images there are some examples prepared during the course, by some different groups, as models and as details. All the models shown are assembled using 3D constrains for all the single parts. It is also possible to observe the good level achieved for complex solid modelling.

3 CONCLUSIONS

In this paper are discussed the issues related with the activation of a new course of computer aided design. The problems involved when changing the old academic organization of Faculty of Engineering of the University of Roma Tor Vergata (five years degree) to the new (3 years basic degree and two years specialistic degree) are presented. The results of applications of a methodology to teaching computer aided design, are shown. The method seems to have the possibility of stimulating effectively the learning capabilities of the students. Despite the length of the course is only 50 hours, the achieved level of the students, in the use of the software, is very high. This is witnessed also by the models shown in this paper. The success of the course is witnessed by the constant growing of the course participants, although it is not mandatory. The course is completely new and there is a constant upgrade of the CAD programs. For this reason, the method, that seems to be effective, is the same but the course is updated every year.

The models shown in this paper are developed by:

Ducati Monster: *Gattamelata Davide, Giordani Andrea*

Bycicle Colnago: *Andrea Domenico Pedretto, Giorgio Sola*

Raft Marshall: *Camplone Cristiano, Capolupo Domenico, Prosperoni Giancarlo, Sereni Alessandro*

System-based Vision for Strategic and Creative Design, Bontempi (ed.)
© 2003 Swets & Zeitlinger, Lisse, ISBN 90 5809 599 1

E-learning and university teaching: a virtual, interactive, distance learning course

C. Bucciarelli-Ducci & F. Fedele
Department of Cardiovascular and Respiratory Sciences, University of Rome "La Sapienza", Rome, Italy

R. Donati, R. Mezzanotte & D. Mattioli
Centro Studi Cuore, Rome, Italy

C. Grisoni
Teleskill Italia s.p.a., Italy

ABSTRACT: The Department of Cardiovascular and Respiratory Sciences, University of Rome "La Sapienza" has carried out an experimental project to compare e-learning with traditional teaching methods. University guaranteed the quality of the contents, Teleskill and Cisco System have provided the technology and network facilities. Thirty doctors participated to the virtual class, and thirty more doctors attended a traditional teaching course, based on the same topic. All the participants were asked to fill out an evaluation questionnaire both before and after the course (traditional or virtual). Moreover, they were asked to express their opinion on the new teaching method. The results have been very interesting, demonstrating a high level of satisfaction and interest toward new teaching technologies (82% of the participants), and an increase in knowledge comparable in the two groups (+22% in both). In the era of information technology, multimedia e-learning represents an available educational resource that demonstrated to be effective and interesting. We believe in a blended solution of virtual and traditional education.

1 INTRODUCTION

The increasing demands of continuous medical education and the credits doctors are continuously required to reach every year, justifies a scenario rich of scientific meetings, congresses and courses to which doctors have to attend. Moreover, the availability of new informational technologies and their application to educational issues represent an important challenge which the University cannot ignore. Medical institutions and health care providers have always sought innovative methods and instruments to improve the process of teaching and learning. The Department of Cardiovascular and Respiratory Sciences, University of Rome "La Sapienza" and the "Centro Studi Cuore" of Rome, are institutions involved in the teaching and the up-to-date process of cardiologists and cardiologists in training. They recently faced a new challenge: evaluate the effectiveness of a multimedia and interactive methodology of distance learning (e-learning) in terms of didactic impact, satisfaction and interest, by comparison

with the traditional methods of teaching. The virtual class, supported by a PC, was realized through the partnership of Teleskill Italia and the enabling technology of Cisco Systems for advanced multimedia support (Content Delivery Networking).

The aim of the research was to evaluate with a scientific method the didactic impact of a quality distance learning (advanced multimedia e-learning), through comparison with traditional classroom methods. In particular, we evaluated the parameters of "general acceptance", "increase in knowledge", and "overall teaching quality", with respect to topics and teachers.

2 METHODS

2.1 The course

The course *"Basic echocardiography: mitral and aortic valvular diseases"* was realized under the academic supervision of Professor Francesco Fedele and realized

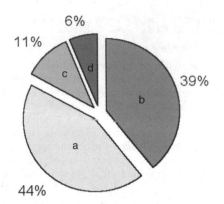

Figure 1. Satisfaction questionnaire. It shows the overall answer distribution. The answers were labelled according to the degree of acceptance: a = good; b = quite good; c = intermediate; d = negative.

by the collaboration of the University and the Centro Studi Cuore. The course focused on the application of a useful imaging tecnique, such as echocardiography, on the assessment of patients with mitral and aortic valvular disease. This topic represents a major chapter of cardiovascular diseases that cardiologists face every day in their clinical practice. The course emphasizes the physical principles of ultrasonography, focusing than on the qualitative and quantitative parameters of different pathologies. Echocardiography is highlighted as an irreplaceable tool in the clinical setting, directing diagnostic criteria and therapeutic options. The course was organized in five lessons, beginning with an introductive lecture on the imaging technologies available in the cardiovascular field. It was then articulted in two lessons on aortic valve diseases (insufficiency and stenosis) and two more lessons on mitral valve diseases (insufficiency and stenosis). The didactic plan included 120 minutes of audiovisual represented by the professor speaking and presenting slides (almost 300), with an integral transcription of the lessons flowing at the bottom of the screen. Every five minutes it was asked from the system to answer a question dealing with the concept just explained. This helped the student to focus on the concepts and to keep the concentration high. The use of all didactic materials was about 5 hours (300 minutes). Before entering the course, students were asked to fill a questionnarie of 50 self-evaluation, multiple choice questions. At the end of the course 50 more questions were asked to evaluate the increase in knowledge realized by the virtual, distance learning course. Moreover, we add 10 appraisal questions in order to investigate the interest and the general acceptance arised from such a course (the satisfaction was scaled from 1 to 4). Final evaluations verified the didactic

impact of the two different teaching methodologies, professors and the percentage of interest.

2.2 Participants

The participants were 30 doctors attending the first year fellowship in cardiology, I Postgraduate School of Cardiology, University of Rome "La Sapienza". They had limited or no experience neither with the innovative didactic method nor with echocardiography. The control group was represented by 30 more doctors participating to a traditional teaching course (face-to-face class), based on the same topic.

2.3 Technology involved

The technology involved was original and innovative.

The Teleskill Virtual Class is a collection of integrated technologies aimed to improve the educational experience of its users. The contents (video, HTML pages, pdf documents, activities, etc.) were delivered locally wherever a permanent Internet connection was available, regardless of geographical distances or bandwidth constraints. The course content had been actually published on a central server and then automatically replicated over the Internet into a local mirror server. The end-user, accessing the central web server, was automatically re-directed to the local storage from where the medical images and videos could be obtained with high-quality on a 10/100 mbps ethernet LAN. The solution also incorporates a secure end-user identification and tracking of the learning activities by ×509 digital certificates stored on Smartcard tokens.

The application platform utilized for the experiment was realized by Teleskill Italia, which also was responsible for the production of the multimedia contents. The system allowed students to learn in a completely natural way, by observing and listening to perfect replications of classroom lessons that are enriched with slides and reference materials, synchronized with the pace of the instructor. In addition, the system made available to the student a series of helpful tools such as a search engine of the spoken text, notes, e-books, and more references. Teleskill Virtual Class was initially designed to offer lessons over the Internet/Intranet, and today it is an innovative instrument that allows for the authentication of the user through a system of digital signatures, a statistical analysis of use, analysis of data and participation in virtual conferences. In this particular case the course was offered in a "wireless classroom," a classroom equipped with 5 portable computers with autonomous wireless access to the Internet. The course was produced using audiovisual materials, graphics and texts of the highest quality, in order to

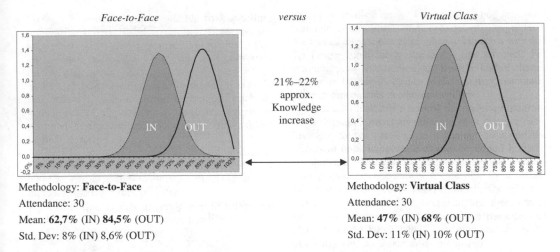

Face-to-Face	versus	Virtual Class

21%–22% approx. Knowledge increase

Methodology: **Face-to-Face**
Attendance: 30
Mean: **62,7%** (IN) **84,5%** (OUT)
Std. Dev: 8% (IN) 8,6% (OUT)

Methodology: **Virtual Class**
Attendance: 30
Mean: **47%** (IN) **68%** (OUT)
Std. Dev: 11% (IN) 10% (OUT)

Figure 2.   Learning effectiveness comparison in the traditional class (face-to-face) compared to the virtual class.

obtain the maximum educational efficiency and to guarantee a high level of satisfaction. However the high quality of the course conflicted with the limits imposed by the Internet connection made available by the University – a simple ADSL connection – decidedly insufficient with respect to the desired multimedia flow. For this reason it was adopted the Cisco Content Delivery Network (CDN) system. This allowed an easier and absolutely transparent transfer of all the course contents (audio, video, images, texts) to the students.

3   RESULTS

All the students accepted to answer the questionnaire and the analysis of the answers led to interesting results. The parameters that we chose to evaluate were:

– the percentage of general acceptance of the course;
– the percentage of correct answers pre-and post-test, and their distribution.

The virtual class received a high percentage of general acceptance and satisfaction among the students (82% of the participants) (see Figure 1), and they were very motivated to face this new teaching technology. Most of the students found the didactic method interesting and capable of stimulating their active involvement. Moreover, they considered it an effective way of learning, helpful to improve their study method. Although from the survey arose that many pupils considered the direct contact with the

professor very important, they advocated the virtual class as a partial supplier of the direct interaction with the teacher. The percentage of correct answers pre-test were experimentally distributed as a Gaussian curve with an average value of 67.7 and a standard deviation of 10%. Therefore, the Gaussian curve represented a good and easy way of presenting our results.

The answers pre- and post-test were graphed in two different curves, both for the virtual and traditional class, in order to compare the increase in knowledge in the two groups. The analysis of the didactic effectiveness of the face-to-face versus virtual class showed no significant differences in the knowledge increase: +22% in both groups (see Figure 2).

Furthermore, the system allowed a real time monitoring of the learning process, available to teachers and researchers on the web, at any time. It then allowed an immediate detection of knowledge "blind zones". In our opinion it could be possibly due to poor teaching quality and/or difficulties related to the subjects themselves. In order to understand the reason of a spread answer distribution, due to a high percentage of wrong answers, the questions asked were grouped in the following topics: methodology, physiopathology of the diseases, diagnosis and functional evaluation of the patient. We found out that the majority of the student considered methodology and functional evaluation more difficult topics, compared to physiopathology and diagnosis (see Figure 3). We then grouped the questions considering the teacher involved in the different parts of the course, and we observed no significant difference among the three teachers as far as efficacy of their teaching process.

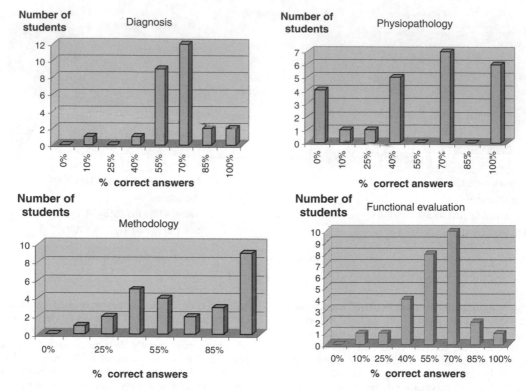

Figure 3. Learning effectiveness comparison with respect of topics.

4 CONCLUSIONS

The virtual class system showed to be an interesting and effective learning and teaching method. Both teachers and students were motivated to face a new experience related to educational issues.

The students participating to the project were interested and enthusiastic of the application of innovative technology to the teaching field. The learning process through a virtual class showed to be as effective as a traditional one.

We believe that this new, virtual, interactive and multimedia learning course could represent an interesting support to the traditional didactic methods for all the students that daily face time and distance constraints to reach university facilities and participate to educational activities.

Multimedia e-learning represents an immediately available educational resource, not to be considered alternative, but a complementary element to traditional teaching (blended solution), which better responds to the growing demands of continuous medical education.

Vocational guidance on the net: a pilot survey

P. Spagnoli & G. Tanucci
Dipartimento di Psicologia dei Processi di Sviluppo e Socializzazione Facoltà di Psicologia Università di Roma "La Sapienza", Rome, Italy

ABSTRACT: The basic assumption of this research is that a web site that supply vocational guidance services should be made following certain criteria. We propose an evaluation scale referred to the approach of usability concept used in Human-Computer Interaction theories based on to three constructs: navigability, information comprehensibility, graphic attraction. This study is a pilot survey meant to construct a web usability evaluation scale. Moreover it is meant to gather some data from different groups of students about web familiarity and the importance of the usability dimensions. The factorial analysis confirms the construct validity of the questionnaire, and the Cronbach's alpha test confirms the questionnaire reliability. The one way ANOVA results point out a non statistically difference between the groups of students evaluating the usability criteria.

1 SCENARY

Nowadays there is an incredible growth of vocational guidance web sites on the net, but not all of them are certainly made in consideration of the users needs.

Generally, web sites are not conceived according to specified criteria, but just following the competencies and the creativity of webmasters. Moreover, when the web site creation is based on an already existing project, usually it doesn't follow the user needs but just the "commercial needs".

This causes many problems for users who waste time trying to get the information they need from web sites (Nielsen 2000). They often give up searching not benefiting from the web service.

When projecting web sites, user needs and web site goals have to be kept in mind; particular attention is needed especially for those web sites which offer social services or information.

Usability theories are intended to solve this matter (Nielsen 2000; Visciola 2000).

The aim is to project a usability evaluation scale starting from the "vocational guidance service user" point of view.

2 METHOD

2.1 Participants

The sample included 62 participants. They are students at the Psychology Faculty of "La Sapienza" University aged between 20 and 31.

2.2 Apparatous

The usability construct used in this study is based on three dimensions:

- Navigability:
  as the quality level of a navigation system that helps to orient in the Web site and find the information.
- Information comprehensibility:
  as format and quality information and contents of Web sites.
- Graphic attraction
  as the web site pleasantness and graphic quality.

These three dimensions are chosen by studying in depth the usability theories and methods and considering the particular kind of Web sites as vocational guidance Web sites.

The goal is to explore the users importance perception of these three dimensions by evaluating the usability of a vocational guidance web site.

Important contributions derive from some Italian researchers in this field (Levialdi 2002; Ferlazzo Di Nocera & Renzi 2002).

The first questionnaire is a 36 items questionnaire where 18 items are evaluating items by a 1 to 5 Likert scale, and the others 18 are just "factual" items that indicate the web site feature:

- items from 1 to 20 are related to the navigability dimension
- items from 21 to 26 are related to information comprehensibility dimension

- items from 27 to 36 are related to graphic attraction dimension

2.3 Design and procedure

The main goals of this pilot survey are:

- Evaluating the importance for the users of the three usability dimensions.
- Projecting a first users evaluation scale of the three usability dimensions.
- Investigating on the hypothetical evaluation differences supplied by the four different sample groups which are related to the variable of familiarity with the World Wide Web.

In the design the participants have been sorted in four group according to their level of familiarity with the World Wide Web.

They have been asked to explore a vocational guidance web site in the computer laboratories of the Psycology Faculty's and then asked to fill in the questionnaire on the usability web site.

3 RESULTS

The analysis have been carried through the SPSS statistical program, version 8.0.

The results obtained by calculating the statistical means and the standard deviation of the three usability dimension, show that the most important dimension for the sample examined is Navigability, the second is Information Comprehensibility and the last is Graphic's attraction (See Table 1).

Even though the value of the three dimension means are fairly similar, the results point out how important is the efficiency of the navigation system to the sample.

From a graphic point of view the participants do not seem to be particularly interested.

The results of the factorial analysis, carried out by the questionnaire, point out a three dimensions structure confirming the three initial dimensions:

- Navigability
- Information comprehensibility
- Graphic attraction.

Three of the 18 evaluating items do not aggregate any dimension (item 3, 9 & 11 as in Table 2). Since words used in the items are not in common language, the problem of these three items may concern both the participants difficulty in understanding computer language and/or the bad items structure.

Moreover, two of the items (item 17 & item 20 as in the Table 2) previously referred to the Navigability dimensions seem to fit best with the Information Comprehensibility dimension (See Table 2).

Actually, the two items meaning expresses a kind of "predictability" dimension which could express the

Table 1. Average and standard deviations of the three usability's dimensions.

Dimension	Means	Std. dev
Navigability	3.80	0.66
Information comprehensibility	3.72	1.71
Graphic attraction	3.35	1.24

Table 2. Item's aggregation to the three factor.

Items	Navigability	Info comprehensibility	Graphic attraction
03**			
05	.58		
06	.87		
07	.82		
09**			
11**			
13	.85		
15	.83		
17*		.58	
20*		.80	
23		.81	
24		.55	
26		.55	
28			.83
30			.71
33			.86
32			.62
36			.42

* Items previously and hypothetically referred to navigability.
** Items that don't aggregate with any dimensions.

Information Comprehensibility dimension better than Navigability dimension.

Therefore, in this way the Information Comprehensibility Deviation Standard obtained in the previous analysis could be better explained by the hypothesis of a dimension with a double integrated meaning:

- Predictability
- Information comprehensibility

where Predictability is the measure of the easiness at identifying how the information search works.

In addition, the reliability analysis carried out on the remaining items (15, 5 referred to each dimension) points out the good consistence of the items (See Table 3) and shows a first perceptive evaluation scale of the importance of the usability dimensions.

Further studies will follow to construct a more exhaustive scale.

The third analysis goal is to find out hypothetical evaluation differences supplied by the four different sample groups. Those groups were sorted by the variable of familiarity with the World Wide Web.

Table 3. Reliability analysis results.

Dimensions	Items n°	Cronbach's alpha
Navigability	5	.75
Information comprehensibility	5	.88
Graphic attraction	5	.85

Table 4. ANOVA one way statistic analysis.

Dimensions	Fisher's F	Sig.
Navigability	0.86	0.46
Information comprehensibility	2.56	0.06
Graphic attraction	0.65	0.58

The results obtained by the one way ANOVA point out that there is not a statistical difference between the groups for the three usability dimensions with critical alpha 0.05 (See Table 4), therefore the familiarity with the World Wide Web doesn't seem to be influential to the importance evaluation of the three dimensions.

4 DISCUSSION

These results are very interesting to everyone who is approaching the web sites usability.

First at all the approach proposed in this study could be very useful to balance the dimension's weight in projecting different web sites. The specificity of the vocational guidance web sites lies on the information given to users. From the results showed in this study, this kind of web site needs particular attention to the information retrieval.

Moreover the approach is completely user centered as it takes into consideration the user perception of the dimension's importance.

This study is a useful contribute from the psychological point of view in projecting web site, derivable from cognitive studies and integrable with others disciplines. The belief is that only an interdisciplinary exchange of contributions could be very efficient in projecting web sites.

REFERENCES

Anderson, J.R. 1993. Psicologia Cognitiva e sue implicazioni. Bologna: Zanichelli

Badre, A.N. 1993. Nota didattica al seminario: *Methodological Issue for Interface design: a User-centered Approach.* Dipartimento di scienze dell'Informazione, Università degli studi di Roma La Sapienza

Calvo, M., Roncaglia, G., Ciotti, F. & Zela, M.A. 2000. Internet 2000: manuale per l'uso alla rete. Bari: La Terza

Card, S.K., Moran, T.P. & Newell, A. 1983. The psycology of Human-Computer Interaction. Hillsdale, NJ: Erlbaum

Di Fabio, A. 1998. Psicologia dell'Orientamento: problemi, metodi e strumenti. Firenze: Giunti

Di Nocera, F., Ferlazzo, F. & Renzi, P. 2002. Us.E.1.0 Note storiche ed istruzioni (pers.comm.)

Ercolani, A.P., Areni, A. & Mannetti, L. 1991. La ricerca in Psicologia. Roma: La Nuova Italia Scientifica

Duncan, J. & Humprheys, G.W. 1992. Beyond the search of surface: visual search and attentional engagement. Journal of experimental Psycology: Human Perception and Performance, 18: 578–588

Hutchins, E.L., Bain, J.D. & Pike, R. 1989. Different ways to cue a coeherent memory system: A theory for episodic, semantic, and procedural tasks. Psychological Review, 96: 208–233

Kieras, D.E. & Polson, P.G. 1985. An approach to the formal analysis of user complexity. International Journal of Man_Machine Studies, 22: 365–394.

Negroponte, N. 1995. Essere Digitali. Milano: Sperling e Kupfer

Nielsen, J. 2000. Web Usability. Milano: Apogeo

Norman, D.A. & Draper, S.W. 1986. User-centered system desygn: new prospectives on human-computer interaction. Hillsdale: Erlbaum

Penna, P. & Pessa, R. 1996. Interfacce Uomo-Macchina. Roma: Di Renzo

Spool, J.M. et al. 1999. Web Site Usability. San Francisco: Morgan Kaufmann Publ.

Sklar, J. 2000. Principi di web design. Milano: Apogeo

Visciola, M. 2000. Usabilità dei siti web. Milano: Apogeo

System-based Vision for Strategic and Creative Design, Bontempi (ed.)
© *2003 Swets & Zeitlinger, Lisse, ISBN 90 5809 599 1*

On some new emerging professions in the higher technical education

N.P. Belfiore
Department of Mechanics and Aeronautics, Rome, Italy

M. Di Benedetto & A. Matrisciano
CampusOne Management, Facoltà di Ingegneria, Rome, Italy

C. Moscogiuri
DIDAGROUP S.P.A. Rome, Italy

C. Mezzetti & M. Recchioni
Università degli Studi de L'Aquila, Facoltà di Scienze della formazione, L'Aquila, Italy

ABSTRACT: The need of new qualified and differentiated employment is herein analyzed, with special reference to the contest of the Italian Universities. Such requirement derive from the strong demand of innovation which the educational systems, at the highest levels, are called to deal with. The Social contest is firstly analyzed, together with its relationship with the recent evolution of the Enterprise's policy. Then, the effects of the new Italian regulations are studied. A special part is also dedicated to the fundamental roles that IT and e-learning are paying in the innovation of the University curricula. Finally, the management of the various activities is investigated, as suggested by the Italian CRUI. The performed analysis will show that there are many emerging professions which could significantly improve the quality of education in our technical Universities.

1 INTRODUCTION

In the last years, changes in Society, Economy, and in the Cultural Processes, have induced the experimentation of new and innovative approaches, such as, for instance, FaD and e-learning. On the other hand, new epistemological models have been investigated, on which the practical didactical activities could be based in traditional education.

The new didactical methods and the new efficient didactical strategies can significantly help in the development of higher technical curricula. However, these new opportunities, known in theory, require a serious application in practice. In fact, the theoretical concepts are not always simple to be interpreted and even understood. Furthermore, not always, the teachers are able, alone, to handle the process of curricula planning and their management. For all these reasons, the Authors of this papers have decided to begin a compared analysis of all, and only those, new competencies and capability which could be conveniently applied in managing the didactical activities of the Universities, with particular reference to the technical curricula in Italian Institutions.

The process of correctly and efficiently handling a higher technical course at the University is here regarded as a crucial task for the success of students' career and for the benefit of the whole Society (specially speaking with reference to the public Universities). This important work can be performed only if the staff is qualified for each assignment. Hence, in this paper, a study on the new emerging professions in education is presented. The existent new job assignments are reviewed, in general, and some other, simply proposed. Based on the acquired experience (Belfiore 2002), the Authors of this paper will focus their attention on the academic environment and will propose new types of professions.

2 EVOLUTION IN INDUSTRY, SOCIETY, AND INSTITUTIONS: THE NEED OF NEW QUALIFIED PROFESSIONS

The era of *globalization* has produced significant new varieties of effects in Society, Culture, Economy, and Politics. A progressive interdependency has grown among Countries, which has involved also the work market and the world of Education, with special effects on the Technical Education. The need of new professional specialization has now become critical with respect to the educational systems.

According to Parsons (Parsons 1976) *the more differentiated the Social structure is, the greater is the needs of qualified employment at the higher levels of the organization's structure.* Now, since our *western* Societies are affected by constant technological changes, which are stimulated by the developments in the research fields, our Educational Systems, at the highest levels, must promote a strategy of permanent education as well as of innovation's research and adaptation. In this way the products of an Educational System, in terms of innovations in research and culture's promotion, may play a fundamental role in the development of the whole Society.

2.1 Changes in industry

The importance of the role of management has recently grown in Industry, in such a way that Industrial Managers form a particular rank in Society. There are two main kinds of executives: the *leadership oriented* manager, who infuse courage and motivation to the employees, and *participation oriented* manager, who tries to stimulate the subordinates and considers important listening to them.

Nowadays, *globalization* in Economy forces the managers to strengthen their relationships with the various components of the Society. On the other hand, they have to deal with the market instabilities and the renewed welfare contributions, while the requirements of the Quality and the Environment constantly grow.

It is therefore clear how the new demands, at the medium and high levels of management and technical stuff, are enlarging the range of possible specialized tasks. According to Janossy (Janossy 1969) *the distance that nowadays exists between the individual's and the Society's (as a whole) knowledge is greater than the one existing at the beginning of the industrialization.* Such distance is expected to grow further as the work' subdivision grows. In fact, the professional interdependencies among individuals and among groups are increasing in such a way that today knowledge is necessarily *integrated* and *interdisciplinary*. Fundamental keys in knowledge are *specialization* and *depth*, not *quantity*.

2.2 Individual knowledge and job market' requests

As known, adaptability, flexibility, and dexterity have been typical demands in the Taylor-Ford era. Although these qualities are considered valuable still now, new capabilities are increasingly appreciated, such as

general culture, broadmindedness, rationality and communicativeness. The interpretative capabilities are now more appreciated than the manual ones, as well as the logical and abstract representations are now preferred rather than the trial and error accumulations (Acconero, 1994). In few words, the individual knowledge' changes, along generations following one another, is not *quantitative*, but *qualitative* (Janossy 1969).

2.3 Mutation processes in organized systems

Complex systems are very easily subject to changes, dues to various factors. The general approach may also change, as a consequence of the environment's inputs. However, it must be recalled that new solutions and new processes require not only to be accepted, but also to be supported, developed, and tested. For this reason, it seems that changes in organized systems, such as Institutions, Enterprises, and Organizations is not only a mere consequence of a crisis. It is a quite voluntary process, which involves Society, since individuals, as member of the community, change. Mutations involve relationships at each level, in the Organization and in Society.

Mutation must be, therefore, regarded as a transformation of the way we act and operate in the various contests, rather than an ineluctable step of the human history (Crozier & Friedberg 1977).

2.4 The importance of the knowledge in the mutation process

Finally, it is believed to be noteworthy to notice that a preliminary analysis of the contest over which the mutation is effective, must be always conducted. In particular, a system should know, somehow, its inner capabilities and resources in order to take the best choices and overcome the critical mutation periods (Crozier & Friedberg 1977).

3 NEW PROFESSIONALISMS IN THE TECHNICAL UNIVERSITIES

Mutations in Society and progress in technology have induced a great variety of changes in the higher technical education. In Italy, in particular, the process of development has been considerably strengthen by some recent national regulations which have involved the entire structure of the national public Universities. Such mutation is believed to be one of the most important reasons of the growing need of differentiation of professionalism. But there is another more general cause. Notwithstanding the temporary crisis of the so called New Economy, the acceleration of the Economy digitalization induces strong demands of changes in higher educational curricula, regarding both the methods and the contents. Teachers are called to improve their methods, and, in some cases, new approaches and new professions could be conveniently employed, even in teaching, with special recall to the adoption and diffusion of new learning media.

3.1 The new italian regulations for the universities and its effects on the needs of new professions

Since November 1999, the Italian Universities have radically changed their didactical structure. Regulation "*Decreto Ministeriale n. 509 del 3 novembre 1999*", among many other important changes, has introduced, for the students, the fundamental activity of a stage in an active Firms. Such activity, is no more regarded as a simply optional, but achieve the status of an actual part or the program, whom a certain number of credits correspond to. The sense of this innovation consists obviously in an attempt to make closer the fresh graduates knowledge to the ones expected by the market and the enterprises. However, the spirit of the ordinance, which must have a general approach to the problem, not always can be so easily accomplished. In fact, in many cities where a State University has a seat, there are no enough Structures or Organizations capable of receiving all the students. In these cases, the Programs can be modified in such a way to substitute the stage with a course having an equivalent number of credits, provided that that course is taught by a teacher who works in an external Structure (i.e. not in the University). This has, actually, introduced a new type of teacher, sometime called the "*docente laico*" (*lay professor*), who has the important task of giving students a first contact with the market world. Lay professors are typically selected inside the external structure top management and are called to use the *experiential* (Kolb 1974) and *imitative* (Fontana 1994) models of learning, in contrast to the academic teachers who usually prefer more conceptual approaches. The introduction and the permanent appointments of the *lay professors* are yielding some interesting collateral effects, such as the diffusion of more practical teaching methods, even among the academic teachers, who indulge more frequently on direct experiences in the transfer of the knowledge, specially in the description of the technical applications.

3.2 Renewed educational technologies and its effects on the needs of new professions

Among the effects of the above mentioned new regulation over the Italian Universities there is their increasing approach to the North American, and in general, Saxon models. Therefore, it can be useful to give a glance at those systems to inquire about a possible evolution of the Italian scenario. The results of this analysis will be described in the following part of this paragraph.

A quite recent instructional technology, called "learning objects", (LTSC 2000) has shown very useful characteristics in applications such as reusability, generativity, adaptability, and scalability (Hodgins 2000), (Urdan & Weggen 2000), and (Gibbons et al., 2000). Object-orientation leads to reusable modules, which can be adopted (Dahl & Nygaard 1966) in multiple contexts. Although the *learning objects* are commonly understood to be digital entities deliverable over the Internet, the idea suggests improvements in the didactical planning of an academic program, which can also incorporate small pieces of e-modules.

To facilitate the adoption of the learning objects approach, the Learning Technology Standards Committee (LTSC) of the Institute of Electrical and Electronics Engineers (IEEE) has developed, since 1996, instructional technology standards (LTSC 2000) which will be reasonably adopted by universities, corporations, and other organizations in order to achieve the interoperability of their instructional technologies, specifically their learning objects. According to these standards, a *learning objects is defined as any entity, digital or non-digital, which can be used, re-used or referenced during technology supported learning*. Examples of technology-supported learning include computer-based training systems, interactive learning environments, intelligent computer-aided instruction systems, distance learning systems, and collaborative learning environments (Wiley 2002). Examples of Learning Objects include multimedia content, instructional content, learning objectives, instructional software and software tools, and persons, organizations, or events referenced during technology supported learning (LOM 2000).

Learning objects and their behavior have been often explained through the LEGO metaphor, in the sense that small pieces of instruction can be assembled to create one larger instructional structure like the LEGO bricks can be stacked together to form scale models of houses, cars, or whatever. Furthermore, the object can be reused in other educational programs, like the elementary pieces can be reused to compose different scale models.

The metaphor, is however considered as very restrictive since it may mislead to wrong conclusions. In (Wiley 2002) for example, is addressed the importance of not extending some properties of the LEGO blocks to the learning object. In fact, the following properties cannot be correctly referred to the learning object: a block is combinable with any other block; blocks can be assembled in any manner you choose; blocks are fun and simple that even children can put them together.

Now, the interesting thing is that there are certain distance courses which give students credits that can be used in their educational contests. The educational credits, therefore, become a general accepted educational currency that provides a general accepted measure of learning outcomes achievable in notional hours at a given level and by any learning methodology.

Since the typical learning objects deal with basic topics such as Mathematics, Physics, Chemistry, they show a possible evolution of the didactical strategy of modularity and distance learning in planning Engineering Programs, which could be more flexible and efficient. This suggests also the introduction of the on-line teacher, which is called to have qualities which differ from the classical *talk and chalk* professor who teaches in a learning *place*. Teaching is rather hold in a learning space, made of virtual electronic shape (Recchioni 2001).

As a matter of facts, the on line teacher must have high capabilities in project management, since the aspect related to the course planning is here more relevant than in the traditional courses. A high competence in IT is also required, as well as in multimedia communications.

The most important concepts of adult education are also required, what is now improperly called *Andragogy*. We address the opportunity to change the term *Andragogy* (which seems referring to the male gender) with that we think to be the more appropriate, namely *"Antropogogy"*, since the Greek term ανϑρωποζ refers the *human kind*.

Finally, on line teachers should have a more strategic approach to the educational event, more conscious of the more general Company policy and more capable of identifying the individual and distinctive abilities (Costa & Rullani 1999).

4 RELATIONSHIP BETWEEN THE NEW I.T.'S AND THE E-LEARNING: INTRODUCTION OF NEW PROFESSIONS

Research and experimentation of new learning methods is of fundamental importance in our Era characterized by the enlargement of the class of people that, by virtue or by force, is interested to the acquisition of new knowledge and capabilities. This educational demand must be considered as permanent, as permanent is the nature of the education, nowadays. In fact, knowledge tends to become quickly obsolete while methods tends to remain the same. For these reasons, methods renewal is of crucial importance, at any level of education.

The general situation is rather critical. In fact, the above mentioned market globalization put enterprises under pressure, inducing them to a competitive attitudes which can interfere negatively with the valorization of their inner resources. On the other hand, University in involved in a process of renewal of

its didactical model, also oriented to communal and cooperative approaches rather than *one-to-many* strategies.

Distance FaD and electronic learning seems to fill the gap between the environment needs and the educational institutions. The main reason of this success is that the integrated procedures based on IT have a great upgrading capacity, as well as a more general attitude of stimulating the individual characteristics.

It is clear, for example, how a network based system offers resources that can be reached in any moment of the day (or night) and from any point of the globe! The learning opportunities grow, since that are more based on proper characteristics and needs. There are less constraints to the development of personalized educational programs. Furthermore, the information flows and the interactive techniques seem more appropriate to the mental processes, which are more easy to be guided by analogy and experience, rather than by logic and rationality.

Once the new ITs are becoming part of the educational programs, they can offer and suggest a new planning in terms of contents, methodologies, and instruments. However, the actual application of these new ideas, in spite of the appearances, is rather complex and cumbersome. In fact, renewals are always bringing the risk of the unknown. Such a huge development in didactical methods should be promoted and supported under the light of new specific researches whose results should be able to overcome the lack of experience. Unfortunately, these investigations have not been sufficiently developed, and is it may be due that, generally, innovation in methods is more difficult than in technology. This is also confirmed by the relative absence of this kind of academic debates in Universities. The supremacy of the methods should be extended, in education, over the simple technical means.

The need of new methods has suggested the idea of *blended* (or *integrated*) *learning*, which is characterized by the simultaneous presence of traditional and distance activities. *Constructivism* has ideologically supported the development of the systems of distance education which promote the collaborative and cooperative activities. According to Woolfolk (1993), *students actively construct their own knowledge; the mind of the student mediates input from the outside world to determine what the student will learn. Learning is active mental work, not passive reception of teaching. In this work other people play an important role by providing support, challenging thinking, and serving as coaches or models, but the student is the key to learning.* Blended learning seems well suiting this beliefs, since there are more occasions of individual based learning experiences.

It is, therefore, clear that teachers should pay attention not only to the technical media, but, mainly, to the didactical targets, to the analysis of the learners' groups and of the social and cultural contests in which the educational event was born. For example, the analysis of the cognitive style in that contest could be certainly useful.

From what above reported, it appears that new planning criteria, new roles and abilities, both in teachers and in management, new investigations on the operational aspects of the systems are urgently needed in higher education. In particular, planning an e-learning activity requires strategic plans, management system renewals, and structure upgrades. It is not sufficient to buy the hardware without developing proper using and maintaining capabilities. The first step should be the definition the real targets and the service characteristics, being aware that e-learning systems have a great impact over many variables in the organization and over the business.

The development of an e-learning object offers new chances in the teaching methods. The simple translation of the crude traditional material must be considered as a very restrictive exploit. In fact, it would be ignored many interesting features such as: the knowledge system management; the improvement in the capacity of analysis and testing of both the students' performances and potentialities; the creation of virtual meetings and communities; the identification of new instruments to information flows monitoring and control. Efficiency in planning an e-learning object is strictly related to the capacity of including in the system every different option and resource dedicated to learning: self-learning, virtual communities, virtual classes, practice, and tests. Media and activities have to be chosen by optimizing time and by promoting learning. Lessons should be arranged in didactical units not too much long, organized in such a way to promote the active interaction and the monitoring of the attention and comprehension levels. It is, therefore, believed that it is necessary to plan the following activities:

- analysis of the participants' characteristics and consequent planning the activities' time and media;
- analysis and valuation of the length of the planned activities;
- selection of the methodology;
- planning of the synchronous activities;
- planning of the asynchronous activities;
- planning of the monitoring and testing methods.

Once the didactical program and the activities are defined, the logical structure of the communications between teachers and students become of extreme importance. The adopted scheme will be expected to grant the correct management of the information flows among those who participate through the Internet.

On line teachers are, therefore, called to achieve new competencies in order to be able to plan the whole distance educational project and to manage the learning process. In other words, the traditional teaching capabilities must be supported by the new ones derived by the science of e-learning.

Finally, the whole process must be regarded as a sequence of important phases, namely, the target definition, project planning, selection of the didactical methods (both traditional and virtual), tutoring, and distance study.

5 MANAGING THE HIGHEST LEVEL OF TECHNICAL EDUCATION

As the experience matured during the last decade in the planning Undergraduate and Graduate Programs has shown, the assignment of a new profession called *Didactical Manager* (MD) seems, year by year, to have achieved an irreplaceable role that cannot be disregarded from the system.

The introduction of MDs, as well as many other fundamental innovations, much dues to the Italian CRUI (Conference of the Chancellors of the Italian Universities) which has edited a great amount of literature of valuable reference for the Italian Undergraduate and Graduate programs (CRUI 2000).

Among the various tasks that MDs have to deal with, there are some relatively new in the Italian Universities. Firstly, the function of *didactical coordination* among the different courses. Then, as of a *connection* between the students and the professors, as well as among the different professors. Furthermore, as of a possible reference point of novel activities related to the *development of new didactical methods*, to the *orientation service*, and to the system *monitoring and testing*, with particular reference to Quality and Customer Satisfaction Analysis. At last, but not least, as a source of new proposals, concerning activities based on the new Information Technologies and the relationships with Society (Unions, Environment, Secondary Schools, etc.).

All these tasks could be conveniently split into a variety of professions, which could help much the entire structure. In fact, the didactical management should be entrusted neither to a single person nor to an isolated office. It should be better considered as a *function*, which is necessary for the success of the educational system. Such a *function* should be intended as *a set of activities conducted by well co-ordinated operators*, who belong to many different divisions.

Other than didactics, there are important functions such as, for example, the economic administration and the *funds hounding*, which the structure has to deal with. Actually, the representatives of the teaching stuff who have the direct responsibility of the two above mentioned activities are not always adequately supported, while the demand of an upgraded and dynamic managing style is increasing. This leads to the conclusion that the technical supporting stuff should be strengthen in order to be more specialized and qualified, which can be obtained by developing a new strategy of job assignments.

From what has been said until now in this paragraph, it should be clear the importance of personnel's (or *personnel candidates'*) education and upgrading, with special reference to the Culture of Work. This concept has been, also, pointed out by the latest report of the Italian CENSIS (CENSIS 2001) on the status of the Nation.

Finally, it is worth commenting about the ancient concept that the Universities offer (theoretically) the highest level of education that one person can achieve. At least, as a free citizen. If this is true, as it probably is, then, those who teaches in the University are expected to posses the best qualification as possible over the subject to be taught. As a standard, this qualification is obtained at the cost of years and years of hard work, passed both studying and working in labs, and this capability is not replaceable. On the other hand, it is not said that a good researcher is also a good communicator. Hence, it may happen that at the highest levels (such as for example at the PhD Programs) professors pay a low (or even null) attention to the research of the most efficient teaching method. This drawback is, however, not much troublesome because at that level there are, mainly, *adult* students who are able enough to afford some efforts in order to learn from the best researchers. For example, in the North American Universities graduate students seek for qualified and referenced (better if *famous*) scientists, rather than for good communicators. This considerations put under the spotlight a twofold aspect of the instructional process in the University: professors should be good in teaching as well as in achieving results that are interesting for the scientific community. Such withstanding task is not easy. Sometime, it is not easy to try to talk to professors about new didactical methods, either. Certainly, professors *should* pay attention to the new didactical methods. At least, they should *care* about them. A Nobel Prize in Physics, Richard Feynman, for example, wrote about the argument something interesting that we report (Feynman 1985) as a stimulating conclusion of this paragraph: *"If you're teaching a class, you can think about the elementary things that you know very well. These things are kind of fun and delightful. It doesn't do any harm to think them over again. ... The questions of the students are often source of new research. They often ask profound questions that I've thought about at times and then given up on, so to speak, for a while. It wouldn't do any harm to think about them again and see if I can go any further now. ... I would never accept any position in which somebody has invented a happy situation for me where I don't have to teach."*

6 CONCLUSIONS

The Authors of this paper have attempted to give a contribution to the development of new didactical strategies in the planning of curricula at the University level of education.

Hybrid approaches, namely, those based on the blending of traditional and distance methods seem very promising, as well as the less traumatic. In fact, while the typical needs of the highest level of scientific and technical instruction are saved, the benefits of the new methodologies, such as those referring to the spheres of distance learning and of Information Technology, could be saved and made available to the system, with great advantage for students, teachers, and the whole Society.

The renewal process will be possible, provided that the theory came into practice correctly, and this means that new qualifications, such as those mentioned in the paper, namely MDs, *lay professors*, IT and e-learning experts, customer satisfaction experts, orientation and counseling operators, will be of fundamental importance for the educational systems.

REFERENCES

Aris Accornero. 1994. In Italian, *The World of Production*, Ed. Il Mulino, Bologna.

Belfiore, N.P., Di Benedetto, M., Matrisciano, A., Moscogiuri, C. 2002. The Development of an Integrated Approach to the Planning of the Higher Technical Education: past experience and proposals, *3rd Int. Conf. Inf. Technology Based Higher Education and Training*, Budapest, July 4–6.

CENSIS, 2001, in Italian, *35th Report on the status of Society in the Nation*, Roma: FRANCOANGELI, pp. 174–8.

Costa, G. & Rullani, E. 1999. *Il maestro e la rete, Formazione continua e reti multimediali*, Etas.

Crozier, M. & Friedberg, E. 1977. *L'Acteur et le systeme. Les contraintes de l'action collective*, Editions du Seuil, Paris.

CRUI , 2000, *Guida per Manager Didattici* , Roma.

Dahl, O.J. & Nygaard, K. 1966. SIMULA – An ALGOL based simulation language. *Communications of the ACM, 9* (9), pp. 671–8.

Feynman, R.P. 1985. *Surely You're Joking Mr. Feynman*, Norton & Company, New York.

Fontana, F. 1994. In Italian: *Lo sviluppo del personale*, Giappichelli.

Gibbons, A.S., Nelson, J. & Richards, R. 2000. The nature and origin of instructional objects. In D.A. Wiley (Ed.), *The instructional use of learning objects*. Bloomington, IN: Ass. for Educational Communications and Technology.

Kolb, D.A. 1974. On Management and Learning Process, in Kolb D.A. et al, *Organizational Psychology*. A book of Reading, Prentice Hall, New York.

Janossy, F. 1969. *Das Ende der Wirtschaftswunder*, Roma, It. Ed. Editori Riuniti, Roma, 1974.

Hodgins, Wayne 2000. *Into the future* [On-line]. Available: http://www.learnativity.com/download/MP7.PDF

Kapp, A. 1833. *Platon's Erziehungslehre, als Pädagogik für die Einzelnen und als Staatspädagogik*. Minden-Leipzig.

LTSC, IEEE, Learning Technology Standards Committee, (2000), *Learning technology standards committee website* [On-line]. Available: http://ltsc.ieee.org/

Merton, R.K. et al. 1959. *Sociology Today*, New York.

Parsons, T. 1976. *Explorations in general theory in social science. Essays in honor of Talcott Parsons.* Ed. J.J. Loubser [et al.]. N.Y./London: Free Press/Collier Macmillan.

Recchioni, M. 2001. *Formazione e nuove tecnologie*, Carocci., Roma.

Urdan, T.A. & Weggen, C.C. 2000. *Corporate e-learning: Exploring a new frontier* [On-line]. Available: http://wrhambrecht.com/research/coverage/elearning/ir/ir_explore.pdf

Wiley, D.A. 2000. *Connecting learning objects to instructional design theory: A definition, a metaphor, and a taxonomy*, In: D.A. Wiley (Ed.), Instructional use of learning objects, IN: Association for Educational Communication and Technology, Bloomington.

Woolfolk, A.E. 1993. *Educational Psychology*, Allyn & Bacon, Boston, pp. 485.

Increasing competitiveness by means of value analysis in higher education

R. Meszlényi, M. Gyulaffy & F. Nádasdi
College of Dunaújváros, Dunaújváros, Hungary
(Licensing and Administration Bureau of the Ministry of Economic Affairs)

The only eternal thing is change: only transformation stays constant. (Krug)

ABSTRACT: Competitiveness is influenced by financial conditions and the mobility of workforce in the global competition of globalization. To achieve these goals, we need a new strategy in higher education as well. The Bologna-process is trying to determine this new strategy in higher education and make a contribution by the grade-based conversion of higher education, creation of the students' mobility and the comparison of degrees. This new kind of strategy is to be introduced also in Hungary, which includes new management challenges as well. Besides the penetrability of the institutes penetrability inside the institutes must be fulfilled.

It is very important to ensure the penetrability between the flexible forms and other forms of institutions. It is essential to produce new syllabus which shows a modular structure. Value analysis provides a great assistance for establishing these criteria, adaptation to the new conditions in education and the comparison of various institutions and degrees.

1 INTRODUCTION

People must keep pace with changes, so that they could keep their jobs, to secure their positions in society and avoid being excluded, that is why they have to develop their knowledge and skills continuously. We are currently witnessing a process during which modern societies are gradually transformed into knowledge-based societies, and "among the numerous challenges of the future, education seems to be the most indispensable trump in our hand, with the help of which we can drive mankind towards peace, freedom and social equality". In this correlation, education itself, as one of the dimensions of social development, is also subject to transformation processes. Education, therefore, should enable those who participate in it to grow up to the challenges of future.

Under the pressure of market-oriented operation, the corporate expectations connected with labor have also changed. A mass need is felt for such middle-cadres who can play a key role in spheres of activity linking the lines of skilled worker and middle manager, both in the large enterprise and the small and medium-size enterprise sector, in the course of modernization and development of production, servicing and economic processes.

Within the correlation between the changing economic environment, the labor market tensions, as well

as unemployment, the most important question is that what should be the level and the specialization of the training which could offer the biggest chances of getting a job to those in need. On the other hand, in the dimension of the training institutions, schools and different firms, the question to be answered is that to what direction should they move, which specializations should they develop and which ones should they reduce in training. What is certain about labor market training is that it should be practice-oriented and it should follow flexibly the needs of the circle of users.

One of the basic conditions of a well-functioning market economy is the training of an appropriate level and system. Such flexible training forms should be favored which make it possible to the individuals to gain skills that suit their abilities and needs, and which offer them the chance of getting a job.

Within the framework of these training schemes, the acquisition of techniques that enhance the development of skills helping to search and find a job, as well as communication and enterprising skills is also possible, besides the acquisition of special knowledge conforming to the needs of labor market.

The pre-requisite of a functioning information society is a functioning cooperation society, where learning to learn, as well as individual and group competencies become more and more important, together with such abilities as learning skills, systematic thinking, as well as gaining and processing information.

In this situation, such human features necessary for the ability for future as mobility, risk-undertaking and learning skills, creativity, skills of achievement and activity, etc. are not developing themselves. This needs the support assured by life-long learning.

2 JOINING THE EU

2.1 Why is mobility a keyword?

We can divide the world's markets into three groups: world market, regions' markets and national markets, which are all classified by huge differences. The upper level is a capital market where there is a competition to get the unbound capital, the medium level is the commodity market with all the services and at the lower level we can find the national markets as the labor market. The different nations try to attract capital by improving educational systems, elaborating financial policies and establishing the right social-economic conditions.

The effective use of high technologies resulted in forming great economic regions and produced a global system of production and allocation. The world in globalization needs a special well-trained workforce which can adapt to the changing circumstances and which is mobile in aspects of geography and the various professions.

2.2 Why is student mobility very important?

Where we have to take action are the areas of the labor market and the structural and internal conversion of higher education. As the labor market assigns tasks to the various nations, this could be dealt with by the educational systems of the member states. A mobile workforce can be created by achieving student mobility. This process requires a new approach even from the higher education of the EU. A new feature of the modernization called Bologna-process is that joint objectives and strategies have been set for the first time by the representatives of universities, students and the member states. The new strategy for higher education of the EU focuses on the grade-based conversion of mass education, harmonization of the syllabus, creation of student mobility, establishing the system of degrees that are acknowledged mutually and a more effective education of the knowledge in respect of the EU. Sorbonne Declaration set the goals, the Bologna Declaration set the strategy and the Conference in Prague discussed the strategic actions.

Other subjects should be taught and they should be taught in a different way in the Hungarian higher education. It is essential for Hungary, which is trying to join the EU, to continue the reform of higher education

to increase competitiveness. Any failure would threaten the competitiveness of the domestic higher education: we have to face the possibility that we can not keep our students and may not attract other students from the EU countries or from all over the world. The situation can, of course, turn much gloomier: if the higher education does not contribute to training a qualified workforce of mobile students, the global competitiveness of Hungary will be diminished in the competition for unbound capital and developed technologies. Apart from these facts, the EU also demands the modernization of our higher education system since any failure to it could result in a decrease of the global competitiveness of the EU.

This process of modernization requires a new strategy from the domestic higher education, which also means new management challenges to face. I think that the institutional management structure of today is outdated and not capable of executing operations. In this conservative sphere of higher education it is really difficult to manage any changes.

What the new higher education management should be like? There is apparently a simple answer to this question: it should be like the management of the universities in the developed countries. The problem is not that easy to solve as management styles in the western countries differ very much one from the other in respect to their national characteristics, cultural traditions, economic development. Modernization can not be achieved by pure copying – a thorough analysis, considering and adaptation are needed with special attention to the different features of national cultures, traditions of the higher education, successes and failures.

3 MODERNIZATION AND MODULARIZATION

3.1 The objective of modularization

The social and economic changes as well as the international effects of the 90s conferred a new impetus to the process of the renewal and reform of Hungarian higher education. The so-called credit regulation also stimulates the introduction of the modular system.

The main objective of modularization can be found in the flexibility of students' options, in the compliance with market needs, in the improvement of cost efficiency and in the qualitative reform of education.

This flexibility makes it possible for the training to cope more easily with the many-sidedness of training objectives.

3.2 Necessity of the introduction of a modular training system

Due to the economic growth, the importance of the quality and level of education raised.

The training system and the teachers are compelled to continuously upgrading and improvement by factors as:

- the necessity to hold on in the sharp competition in the training (knowledge) market;
- the necessity of keeping and possible strengthening of the existing position in competition;
- the necessity of catching up with the educational level of the EU;
- the requirement of compatibility and continuous keeping of the level;
- as well as the recommendation of the EC from 1997, which states that "the scientific quality of research and the quality of the teaching staff are no longer guarantees for the quality of education, since it is equally important to assess the quality of the organization of programs, of teaching methods, as well as of the management, structure and communication of the institution".

All this went together with the apparition of a higher level quality assurance process, and the elaboration of the quality assurance system on institutional level.

The quality management and quality assurance standards are applied to a greater and greater extent on different levels of training. These standards compel the training suppliers to systematize and transparently restructure the training programs, which may lead to the development of a wide and comprehensive modular system.

3.3 Credit system

Credit system does not mean assigning the subjects on a faculty to different credit equalities. This would only ensure the students to go forward with his studies in a slower or faster way during a certain period. It is much more important that they can gather credits in a different faculty, at other institutions and may be at an institution abroad (see Peregrination) The Bologna Declaration goes even further: it suggest to involve the studies of long-life learning in the credit accumulation. This dimension requires once more the joint and harmonized syllabus development of colleges and universities and some measures to take for modernization: the first grade in both kinds of institutions must be – in mass education – credit compatible. It is necessary to establish a network of cooperation on credits with foreign institutions which train economic specialists. Cooperation and an extending of credit inclusion to post-secondary education (this is already legally required) and to long-life learning should also be considered. The greatest deficiency of the currently introduced credit systems is that they are valid only inside a certain institution: the students cannot take use of them abroad, only at other faculties within the same institution.

3.4 Modular training and quality

Quality is defined by Grosby as "compliance with the customer's needs". According to Juran's definition, "quality is nothing but usability", that is "we have to satisfy the customer's needs". A. R. Tenner and I. J. de Toro defines quality as the "satisfaction of the customer's needs". It can be, therefore, established that the customer-oriented approach necessarily came into the limelight. Accordingly, "the customer is the king again", stated Tenner and de Toro.

The "customer-centered" attitude is not far from the "student-centered" approach. The marketing view based upon customers' needs and satisfaction proved to be useful also in higher education.

With its flexibility, modularization (modular system) serves marketing needs either. Badley and Marshall (1995) draw a parallel between the two. Both have an effect on consumers, both supply "well-packed" products and services, in the appropriate time and place, in the quantity and quality conforming with the expectations.

Higher education institutions have to cope flexibly with the market challenges occurring in economy. Usually those institution, doing so-called short-cycle training, react most frequently to this challenge, which are able to accomplish a practice-oriented training.

The society's capacity of reaction to these changes and challenges depends on the capacity of reaction of higher education. If the capacity of reaction of higher education does not increase, the gap between higher education and society grows larger and larger. If we want to increase our capacity of reaction, we must be open to the future, to the environment and its changes. We have to be marketing-oriented, which adds the following to the traditional institutional culture. There is a need for:

- an environment-monitoring system, which would signal the new opportunities and threats on time;
- the continuous analysis of needs, so that we could react to them with innovative programs;
- the measurement of customers' satisfaction and the taking into consideration of their signals.

Basically interested in the training program as a process of satisfaction of a social need are:

- as consumers:
  - the students (customers)
  - the graduate (qualified) students
  - the employers who employ qualified students
  - chambers gathering the employers
  - civil trade organizations
- as producers:
  - the founders, maintainers and supporters of the training institution
  - the managers of the training institution

- the study material developers
- the teachers, tutors, mentors, instructors
- the technical assistance staff
- employers of the training institution
– as society:
  - the authorities controlling the training
  - the regional community
  - the interested social organizations.

4 SUPPORTING THE CREDIT-BASED MODULAR TRAINING WITH VALUE METHODOLOGY

In the humanities-based sphere (education, health care, etc.) in general, and in higher education especially, it is important that the quality assurance system of the institution should be elaborated basically according to the customer and function-oriented approach of TQM, in compliance with the spirit of value analysis. Therefore, value analysis is one of the most suitable methods for the comparison of different training systems.

The function is also characterized by a certain cost. The value of a particular function is determined by the fact on which level and at what cost we have developed it. Function analysis is a good method to model the future operation of a new system and in addition to this, the modernization of higher education and the introduction of the market approach requires some ascertainment and possible satisfying of the students' demands. Next we are going to demonstrate the most important steps and results of the ascertainment.

4.1 Comparison of the traditional and modular training systems with the help of value analysis

In the course of the research, we were trying to find answers to the following questions:

1. The achievement of what functions is expected in the course of the training process by teachers and students?
2. To what extent are these functions achieved in the traditional and in the modular training process?

(a) Hypothesis of the investigation
According to our hypothesis:
– The modular education system performs the functions expected by the students and the instructors on a higher level than the traditional system.

(b) Preparation of the investigation
Preparation and training of the team. The team was made up of students and teachers – by volunteering. The students got acquainted with the methodology of value analysis covered by the subject called "Value Analysis".

The subject is based upon the thematic of the Society of American Value Engineers International – SAVE International. The students doing valuable work can get the lowest level of SAVE International's international qualifications, that is the AVS – Associate Value Specialist.

(c) Results of the investigation
The questions raised during the investigation are answered as follows:

1. What functions do the teachers and students expect to be achieved during the training process?
   The results of the research show that the teachers and students expect the accomplishment of the following functions during the training period:
   – provision of permeability, flexibility and good information flow;
   – to detain the necessary staff and material conditions;
   – to set realistic requirements;
   – to detain an elaborated, well worked-out study material system;
   – to convey a high-level knowledge.
2. To what extent are these functions achieved in the traditional and in the modular training process?

In the course of function analysis, it was justified that the modular training achieves the functions expected by teachers and students on a higher scale.

5 SUMMARY

In the process of establishment of the knowledge-based society, education must undergo essential changes. Megatrends that determine the ability for the future give new answers to the questions concerning what, when and how needs to be learnt. Pre-requisites of the ability for the future, that is such human capacities as mobility, ability for risk undertaking and studying, creativity, disposing and productive capacity, etc. do not appear accidentally. To evolve, they need the support of life-long-learning. The acquisition of competencies and the life-long-learning places adult education in a new dimension, linking it closer and closer to higher education, which, in turn, also undergoes a wide process of transformation. One of the basic conditions of a well operating market economy is the training with an appropriate level and structure. Such flexible training forms should be preferred which make it possible for the individual to acquire a qualification that suits his/her abilities and needs, offering chances for employment.

The investigation of higher education with value methodology has contributed to explore new contexts which support our admission to the EU and a new form of education which could be more efficient as it is today.

REFERENCES

Jacques Delors: Our future looks both promising and worrying (New pedagogic review 1995).

Falusné Szikra Katalin: Devaluation of knowledge (KJK Bp. 1990).

Dr. Gyaraki F. Frigyes: Precedents and current position of modular training in higher education (Education Research Institute Bp. 1984).

Dr. Gyaraki F. Frigyes: Model of modular training (Higher education review 1987).

David Watson: Going Modular – Information Services (CNAA Dianion Paper 1989).

Dr. Kadocsa László: Modularization and credit system in the process of renewing higher education (ELTE NTT Phd PROGRAM 1999).

Meszlényi Rózsa: Efficiency and quality of higher education in respect of demands of the real sphere (Production technology 1999).

Meszlényi Rózsa: Comparison of various training systems with value methodology (Comments 2001. XXI).

Miles L.D.: Value analysis (KJK 1975).

Dr. Körmendi – Dr. Nádasdi: Theory and application in practice of value methodology (infoPROD 1996).

Dr. Papp Ottó: "Institution-based and multifunctional application of European quality assurance systems in higher education" Conference of Quality Trainers 2000. Ybl Miklós Technical Faculty of Szent István University, Bp. 2000.

Barakonyi Károly: Management of the strategic challenges of Bologna-process (BME IMVT Conference, Balatonfüred, August 23–24, 2002.

33. Special session on 'Information and decision systems in construction project management'

Evaluation of risks using 3D Risk Matrix

D. Antoniadis & A. Thorpe
Department of Civil & Building Engineering, Loughborough University, UK

ABSTRACT: Concerns and criticism are raised from Risk Management (RM) workshop facilitators and participants regarding the limitations of the current methodologies as well as the subjective nature of the outcomes. These concerns impede the acceptance of RM workshops and the actual process.

This paper introduces the concept of 3D Risk Matrix, risk prisms and pyramids and the methodology for analysing risks according to the position of their 3D 'coordinates'.

Comparison of results between the common 2D analysis and the proposed 3D concept will prove how the latter improves the decision making process, as well as supporting RM facilitators and participants to consider simultaneously a combination of factors, for example likelihood, impact on time and cost and thus achieve acceptance of the outcome (reducing subjective decision making).

The proposed methodology will also improve visualisation, communication and understanding of the RM workshop inputs as well as outputs.

1 INTRODUCTION

There are a number of issues, in the current Risk Management (RM) methodologies, which give rise to concerns amongst the practitioners and academics and render the process doubtful to most of the users (Hillson 1998, del Cano & Pilar de la Cruz 1998, Harmsworth & Chaffey 1997, Greene, Thorpe & Austin 2002, Hopkinson 2001). One concern is that current RM methodologies allow only for partial visualisation/comprehension of the assessed output so audiences, participants and recipients of the outputs many times find it hard to agree to the process, understand what is it they are contributing to and accept the result.

For example, it is extremely common to analyse risks against only two variables – Likelihood and Impact and obtain the product of the two variables in order to classify them as very high, or very low. Also in most cases Impact is considered as Impact in Cost, or as Impact in Time, but never combined.

The fact that the two major variables of Cost and Time cannot be combined coupled with the fact that only integer values can be used to represent the magnitude of the risk is inadequate.

In order to eliminate issues raised and improve the RM process during, as well as after, the RM workshops the 3-Dimensional (3D) Risk Cube/Matrix has been developed and presented in this paper.

It is proposed that individual risks are placed in a 3D Cube according to the magnitude and position of their coordinates. Further more it is proposed that the level of each risk is identified/prioritised using the risk pyramids (hereafter referred to as 'pyramids') that are formed within the smaller cubes/prisms.

The new methodology 3D Risk Matrix will:

a. Enable RM workshop facilitators to improve outcome,
b. Enable the expansion and consideration of risks in a 3D combination, e.g. Likelihood, Impact-Cost and Impact-Time,
c. Enable RM workshop participants to visualise and understand:
    i. How they contribute to the process,
    ii. How they can position and clearly indicate the risks,
    iii. How their knowledge of the risks analysed contribute to the outcome,
d. Allow the RM workshop facilitator to focus on specific areas – 'the pyramids' – within the Risk cube,
e. Improve qualitatively and quantitatively the process of identifying and analysing major risks,
f. Speed up the process by identifying immediately areas where discussion on risks should focus.

This paper puts into perspective, and for further discussion, the concept of 3D Risk Matrix. The three axes considered, in this case, are Likelihood, Impact-Cost and Impact-Time and risks are positioned in the three-dimensional cube in order of magnitude, minimum (1), medium (2), or maximum (3).

2 ANALYSIS

In two-dimensional risk analysis, and usually during RM workshops, risks are given an indicative level for Likelihood, Impact-Cost and/or Impact-Time using either classifications like Minimum/Low, Medium/Most Likely, Maximum/High, or numerical values 1, 2 and 3 so that a product can be created. The higher the product, or combination of letters the more attention is given to the risk and thus mitigating actions are drafted (CIRIA, 1996).

However, in the real world risks are more complex and a two-dimensional risk analysis can only be used as a stepping stone.

In this paper the case of 3D Risk Matrix methodology and the risk pyramids will be introduced, compared and discussed.

The three basic axes to be used are:

– Likelihood (L) in the 'x' direction
– Impact-Cost (C) in the 'y' direction
– Impact-Time (T) in the 'z' direction

The numerical values – 1, 2, 3 – will be used to indicate severity of the risk, as well as to create cross products so that levels can be easily identified.

For example a risk with:

Likelihood = 3	Will have a product of:
Impact Cost = 2	$3 \times 2 \times 3 = 18$
Impact Time = 3	

(Combinations of the products can be seen in Tables 2a–c.)

In practice, in the two-dimensional Risk Matrices, the highest products are selected for analysis and in most cases are given a coloured annotation for ease of identification.

However, when attempting to do this in the 3D risk matrix one cannot follow the same principal because of the wider spread of numbers/products. Also, since there is a 3D space each risk could be 'floating' within this space and therefore it is not appropriate to simply select a number of products.

Figure 1 below gives a 3D presentation of the cube that is formed by 27 combinations of risk and some of the coordinates which are the source of the various products (the coordinates of each point form the product of each risk).

Also the main cube is separated into smaller cubes which are defined by the individual coordinates. An expanded view can also be seen in Figure 1.

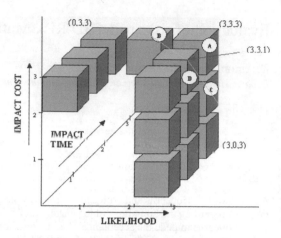

Figure 1. 3D presentation of a cube that is formed by 27 combinations of risk.

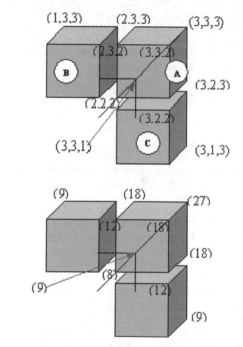

Figure 2. Expanded view of the "high level cubes".

For a example a risk with a Likelihood of 3, Impact Cost of 3 and Impact Time of 1 will have coordinates (3,3,1) and as can be seen above is located in one of the edges of cube 'D'.

Figure 2 below gives an expanded view of the 'high level' cubes that are formed from these coordinates.

As expected all high products are located around the four extreme cubes (A, B, C, D, see fig.1), with products 27, 18, 12 and 8 all situated on cube A and product 9 on the extreme edges of cubes B, C, and D.

Obviously any RM workshop will concentrate its efforts in establishing mitigating actions for coordinates/products that are at the higher end of the cube. That is where products of 'L', 'C' and 'T' are high, for example, (3,3,3), (3,3,2), etc.

One could stop the workshop at this point and start examining the above products. However, this still involves analysis of ten coordinates/products and it could lead to extended discussions about the exact level of a risk. Especially when it cannot be decided if a risk should be a 2 or a 3, with many participants wishing they could have sub-divisions, decimal numbers (for example 2.5, or 2.75), etc.

The 3D Risk Matrix allows for consideration of risks in the space formed between the coordinates, which are actually in the shape of pyramids. These pyramids are formed from the edges of cubes (coordinates) and the inner surfaces between the edges, thus allowing for the use of decimal numbers for the magnitude of Likelihood, Impact-Cost, or Impact-Time.

Points (2,3,3), (3,3,2), (3,3,3) and (3,2,3), see figure 3 below, form the top risk pyramid and contain the following products (18, 18, 27, 18). Therefore any risk that lies within this pyramid should be identified as top risk.

With a thorough analysis of cube 'A' one can easily identify 6 risk pyramids, which according to their position indicate the initial level of priority that can be given to the enclosed risks.

These pyramids are:

1. (3,3,2), (2,3,3), (3,3,3), (3,2,3) – (18,18,27,18)
2. (3,3,2), (2,3,3), (2,2,3), (3,2,3) – (18,18,12,18)
3. (2,3,2), (2,3,3), (3,3,2), (2,2,3) – (12,18,18,12)
4. (3,2,2), (3,3,2), (2,2,3), (3,2,3) – (12,18,12,18)
5. (2,3,2), (2,2,3), (3,3,2), (3,2,2) – (12,12,18,12)
6. (2,3,2), (2,2,3), (2,2,3) (3,2,2) – (8,12,12,12)

In addition to the above risk pyramids there are also those that are formed between cube 'A' and the products/coordinates of cubes B, C, D, namely (1,3,3), (3,1,3) and (3,3,1).

Therefore three more pyramids should be included in the analysis, which in order of magnitude of 'Likelihood' and 'Impact-Cost' are:

7. (3,3,1), (2,3,2), (3,3,2), (3,2,2) – (9,12,18,12)
8. (3,2,2), (2,2,3), (3,2,3), (3,1,3) – (12,12,18,9)
9. (1,3,3,3), (2,3,2), (2,3,3), (2,2,3) – (9,12,18,12)

Table 1 below gives a tabular representation of the risk pyramids from the highest to the lowest level according to their position within cube 'A', or between cube A and the other cubes (B, C, D), or to be more precise the position of the pyramid relevant to edge (3,3,3).

As mentioned earlier, in the case of 3D Risk Matrix a single product cannot accurately indicate the level/magnitude of risk unless it can be clearly, and with no

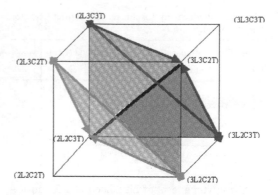

Figure 3. The risk pyramids.

Table 1. Tabular representation of the risk pyramids.

Highest risk pyramid	18,18,27,18 – 3L3C2T, 2L3C3T, 3L3C3T, 3L2C3T
2nd Highest risk pyramid	18,18,12,18 – 3L3C2T, 2L3C3T, 2L2C3T, 3L2C3T
3rd Highest risk pyramid$_{(1)}$	12,18,18,12 – 2L3C2T, 2L3C3T, 3L3C2T, 2L2C3T
3rd Highest risk pyramid$_{(2)}$	12,18,12,18 – 3L2C2T, 3L3C2T, 2L2C3T, 3L2C3T
4th Highest risk pyramid	12,12,18,12 – 2L3C2T, 2L2C3T, 3L3C2T, 3L2C2T
5th Highest risk pyramid$_{(1)}$	9,12,18,12 – 3L3C1T, 2L3C2T, 3L3C2T, 3L2C2T
5th Highest risk pyramid$_{(2)}$	12,12,18,9 – 3L2C2T, 2L2C3T, 3L2C3T, 3L1C3T
5th Highest risk pyramid$_{(3)}$	9,12,18,12 – 1L3C3T, 2L3C2T, 2L3C3T, 3L2C3T
6th Highest risk pyramid	8,12,12,12 – 2L2C2T, 2L3C2T, 2L2C3T, 3L2C2T

doubt, positioned in one of the pyramid edges. Especially when the three variables cannot be precisely represented with/by integers.

Positioning risks within the above pyramids enables facilitators as well as participants of RM workshops to create products using decimal numbers, especially for severity levels between 2 and 3 (perhaps as an output from statistical studies), and then identify in which pyramids these products belong, thus identify the level of risk.

For example:

1. A risk with a level of Likelihood of 2.5, Impact-Cost of 2.8 and Impact-Time of 1.8 (product: 12.6) falls within the 5th Highest Risk Pyramid$_{(1)}$ (by checking the pyramids in figure 4 and the above classification)

2.. A risk with a level of Likelihood of 3, Impact-Cost of 2.5 and Impact-Time of 3 (product: 22.5), falls within the Highest Risk Pyramid

Table 2a. Products of Likelihood and Impact-Cost for Impact-Time equal to 1.

Impact-Cost (C)	Impact-Time (T)	Likelihood (L)		
		1	2	3
		Products of (L × C × T)		
1	1	1	2	3
2		2	4	6
3		3	6	9

Table 2b. Products of Likelihood and Impact-Cost for Impact-Time equal to 2.

Impact-Cost (C)	Impact-Time (T)	Likelihood (L)		
		1	2	3
		Products of (L × C × T)		
1	2	2	4	6
2		4	8	12
3		6	12	18

3. A risk with a level of Likelihood of 2.5, Impact-Cost of 2.8 and Impact-Time of 2.5 (product: 17.5), falls within the 3rd Highest Risk Pyramid$_{(1)}$
4. A risk with a level of Likelihood of 2.5, Impact-Cost of 3 and Impact-Time of 2.5 (product: 18.75), falls on the edge between the Highest Risk Pyramid and the 2nd Highest Pyramid
5. A risk with a level of Likelihood of 2.2, Impact-Cost of 2.5 and Impact-Time of 3 (product: 16.5), falls within the 2nd Highest Risk Pyramid

As can be seen from the above examples, in particular examples 3 and 5, the magnitude of the product does not necessarily indicate the level of risk, but it is the coordinates that locate the level of risk.

Users can locate the pyramid in which their risks belong and thus take appropriate level of mitigating action(s) giving more, or less emphasis on what is required to be done.

Table 2c. Products of Likelihood and Impact-Cost for Impact-Time equal to 3.

Impact-Cost (C)	Impact-Time (T)	Likelihood (L)		
		1	2	3
		Products of (L × C × T)		
1	3	3	6	9
2		6	12	18
3		9	18	27

Level 1 Risk
Level 2 Risk
Level 3 Risk
Level 4 Risk   See note
Level 5 Risk
Level 6 Risk

Note: Level 4 Risk does not exist because all the coordinates that form the level 4 pyramid have been included in the highest-level pyramids.

3   ALTERNATIVE VIEW

The results from the above analysis can also be interpreted and presented in a two-dimensional way, which, will have to be stressed, does not give the same level of accuracy and can not accommodate the use of decimal variables for Likelihood, Impact-Cost and Impact-Time.

The interpretation is based on identifying the 2D levels of risk according to 'in which pyramid the coordinates belong'.

For example a risk with coordinates (2, 3, 2), see table 2b, is a level 3 risk because the highest risk level pyramid in which this coordinate appears for the first time is pyramid '3rd Highest Risk Pyramid$_{(1)}$', whereas risk (2, 2, 3) is a level 2 risk because this coordinate appears for the first time on risk pyramid '2nd Highest Risk Pyramid'.

Tables 2a–c below represent the level of risks as these have been interpreted using the above approach and a coloured code for each level.

If the above tabular presentation is to be used for classification of risks, it will still require for the user(s) to consider the levels of the three variables (Likelihood, Impact-Cost and Impact-Time). However, using the above tables does not allow for the use of decimal numbers between 2 and 3. Also in some cases, as can be seen in the examples below, it produces ambiguous results.

The examples below give an indication of how these tabular reports can be utilised.

1. A risk with a Likelihood 2, Impact-Cost 2 and Impact-Time 3 – product: 12, falls on Level 2 Risk, because its components can be located on table 2c.
2. A risk with a Likelihood 3, Impact-Cost 2 and Impact-Time 2 – product: 12, falls on Level 3 Risk, because its components can be located on table 2b.
3. A risk with a Likelihood 3, Impact-Cost 2 and Impact-Time 3 – product: 18, falls on Level 1 Risk, because its components can be located on table 2c.
4. A risk with a Likelihood 3, Impact-Cost 3 and Impact-Time 2 – product: 18, falls on Level 1 Risk, because its components can be located on table 2b.

Table 3. A comparison between 3D matrix and the alternative view.

Example	Risk coordinates	Alternative view risk level	3D matrix risk level
1	(2, 2, 3)	2	2
			3 Pyramid$_{(1)}$
			4
			5 Pyramid$_{(1)}$
			6
2	(3, 2, 2)	3	3 Pyramid$_{(2)}$
			4
			5 Pyramid$_{(1)}$
			5 Pyramid$_{(2)}$
			6
3	(3, 2, 3)	1	1
			2
			3 Pyramid$_{(2)}$
			5 Pyramid$_{(1)}$
4	(3, 3, 2)	1	1
			2
			3 Pyramid$_{(1)}$
			3 Pyramid$_{(2)}$
			4
			5 Pyramid$_{(1)}$

4 COMPARISON OF RESULTS

The table below presents a comparison between 3D Matrix and the Alternative view using the examples from the alternative view.

In comparing the results between the 3D Matrix and the Alternative view it can be seen, see table 3 above, that, as expected, the results agree on the highst-level risks due to the way the two-dimensional table has been constructed.

However, risks could have a variance of as much as 3, or 4 levels, with examples 3 and 4 having the biggest deviation.

Practically, and for a moderate level of Impact-Cost, this means that extensive effort, or large amount of funds can be locked in contingencies for risks that could be as low as level 6, see examples 1 and 2.

Also one could not avoid considering the accuracy/confidence level of a severity level 3 of Likelihood, Impact-Cost or Time, when looking at the breadth of risk levels in examples 3 and 4, especially when these involve the 'locking of funds' in contingencies.

5 ENABLING TOOLS AND FUTURE WORK

Previously it was made clear that the 3D Risk Matrix methodology, even though it allows for more flexibility when identifying the levels of the three variables

Figure 4. Pictorial example of risk prism/pyramid and vectors.

(Likelihood, Impact-Cost and Impact-Time), requires care and attention and obviously it is not a 'run of the mill' process.

Identifying 'in which pyramid the risk belongs' might not be possible with existing software, however, current packages can support practitioners in establishing better input data, or eliminating uncertainties, for the three variables.

For example using given a range/bandwidth of possible expectancies current risk software can be used to produce a statistical outcome of the most probable result. For example given a range of 2.5, 2.6 and 2.8 for Likelihood, and the appropriate distribution, the software can produce an indicative result that can be used on the 3D Risk Matrix.

Perhaps a future development/collaboration between CAD and Risk Analysis software houses could produce a combined product that will have a 3D model of the Risk Matrix and which will be used to pin point the risks within the pre-established risk pyramids and thus provide a visual display of the result.

With regard to the methodology presented in this paper a future development could be the use of vector mathematics in order to utilise magnitude of risk/opportunity at the point of application.

For example a risk that is positioned in the 2nd risk pyramid could be affected by other risks/opportunities, e.g. inflation, cost of money, etc. (sequential risks?). These other risks could be incorporated/accounted for as vectors applied at that point with a certain magnitude and direction. This pictorially can be seen in figure 4.

The introduction of Risk in the 3D space creates a number of opportunities especially in opening up the field to advanced mathematics and geometry.

6 CONCLUSION

The 3D Risk Matrix methodology considers risks as points in a 3 dimensional space.

The power of combining simultaneously three risk variables e.g. Likelihood, Impact-Cost and Impact-Time, as in the case presented, enables both RM workshop facilitators and participants to make informed and acceptable contributions about the risks to be considered.

Positioning individual risks within risk pyramids improves classification and allows for better decisions on the level of risks, especially when funds have to be locked on projects as contingency. Furthermore the 3D Risk Matrix methodology gives a pictorial view of the positioning of the risk(s), which enables improved communication to all audiences.

Visuallsing where the risk is, or can be, placed improves the levels of acceptance by the participants and improves the facilitator's level of comfort about the outcome.

The use of risk pyramids combined with the use of decimal levels of severity of risk enables the utilisation of detailed statistical data, as input to the process, thus improving utilisation of sources/databanks. Also the use of decimal levels of risk severity enables participants to be more accurate and comfortable with identifying the position of risks.

As it has been clearly indicated in the comparison carried out the 3D Risk Matrix methodology is thorough and enables the RM practitioners and/or workshops to arrive to an appropriate level of decision for mitigating actions against the pertinent level of risks.

Obviously the 3D Risk Matrix methodology is more cumbersome and one should be careful when identifying the appropriate risk pyramid. However, with the current advancements in software it should be reasonable to expect the development of a tool that will allow easy identification and placement of risks within the risk pyramids during RM workshops.

REFERENCES

Hillson, D. 1998. Project Risk Management: Future Development *International Journal of Project & Business Risk Management* Volume 2 issue 2, 181–196

del Cano, A. & Pilar de la Cruz M. 1998. The past, present and future of Project Risk Management. *The International Journal of Project & Business Risk Management* Volume 2 issue 4, 361–388

Harmsworth, B. & Chaffey, N. 1997. Practical Project Risk Management: Lessons from Yin and Yang. *The International Journal of Project & Business Risk Management* Volume 1 issue 2, 121–130

Greene, A., Thorpe, T. & Austin, S. 2002. As likely as not it could happen: Linguistic interpretations of risk. *ARCOM 8th Annual Conference* 2002 September 2–4, 637–646

Hopkinson, M. 2001. Schedule Risk Analysis: Critical issues for Planners and Managers. *4th European Project Management Conference* 2001

CIRIA Special publication 125 1996. Control of Risk. A Guide to the Systematic Management of Risk from Construction

System-based Vision for Strategic and Creative Design, Bontempi (ed.)
© *2003 Swets & Zeitlinger, Lisse, ISBN 90 5809 599 1*

Modelling change processes within construction projects

I.A. Motawa, C.J. Anumba & A. El-Hamalawi
Department of Civil & Building Engineering, Loughborough University, UK

P.W.H. Chung & M.L. Yeoh
Department of Computer Science, Loughborough University, UK

ABSTRACT: Changes constitute a major cause of delay and disturbance and it is widely accepted by both owners and constructors that change effects are difficult to quantify and frequently lead to disputes. Modelling change processes within construction projects is therefore essential for the project team to implement changes efficiently. Change identification in addition to predicting its impacts on project parameters can help in minimizing the disruptive effects of changes. Knowledge required for various construction disciplines to predict and simulate the effect of changes on projects may be imprecise, which is the main challenge for modelling. Existing systems for managing changes in multiview models are still inadequate in many respects. This paper presents a change process model that helps in managing changes through improved modelling of change prediction and evaluation. The model is developed to represent the key decisions required to implement changes and to make these decisions active in assisting the coordination process. The model can be used as an intelligent decision support system in predicting change scenarios on projects and also for evaluating change cases depending on the available information at the early stages of construction projects.

1 INTRODUCTION

Changes after a construction bid has been accepted are common and likely. Inconsistent management of the change process can result in long delays and overestimated costs. Changes are caused by different sources such as: design errors, a change in the functional requirement of the project, and unforeseen conditions. They always result in several consequences such as: breaking of project momentum, increased overhead and equipment costs, scheduling conflicts, rework, and decreased labour efficiency. Some of these consequences can be relatively easy to measure, while others are more difficult to quantify.

Change-control is considered an integral part of project management knowledge. Therefore, a change process model is required to help the management of change. The first part of this paper presents a model developed to represent the key stages of change implementation.

Research on modelling the change process in construction tended to focus on the identification of factors affecting the success of a change process, and recently introduced guidance for best practice in change management. However, appropriate strategies for managing change can be improved when a change can be predicted and thoroughly evaluated. Predicting changes can be based on modelling cause-and-effect relationships (dependencies) in projects. The evaluation of construction change often includes estimation of the possibility of implementing a change option. Therefore, the second part of this paper discusses a methodology for predicting change events and implementation based on the available information at the early stages of projects.

2 MANAGING CHANGE IN CONSTRUCTION – REVIEW

Change management is related to all project internal and external factors that influence project changes. It seeks to forecast possible changes; identify changes that have already occurred; plan preventive impacts; and coordinate changes across the entire project (Voropajev 1998).

Several generic models for change management have been developed. CII (1994) established a concept for project change management. Change is considered a modification to an agreement between project participants. The CII report defined the project elements that are subject to change and that will affect the change management process as: project scope, project organisation, work execution methods, control methods, contracts and risk allocation. The interaction of these elements becomes significantly more complex as the project proceeds.

Recommended practices for managing change efficiently were organised for each project phase of the project life cycle, CII conference 1996. A prototype change management system was proposed and included a set of 27 project change management best practices. Cox et al. (1999) specified a generic procedure for issuing a change order request after contract award. Stocks & Singh (1999) developed a method called "Functional Analysis Concept Design (FACD)" that aims to reduce the overall rate of construction change orders.

Karim & Adeli (1999) presented an object-oriented (OO) information model for construction scheduling, cost optimisation, and change order management. The model can be used to approve change order requests and to resolve change order conflicts. A best practice guide has been published by CIRIA (2001) to present best practice recommendations for the effective management of change on projects. The guide proposed three change management processes for: changes during design development, post-fixity changes that are urgent, and post-fixity changes that will be implemented during the remainder of the project process. A toolkit was developed that contains pro-forms, flow-charts and schedules for use in the implementation of an effective change management system.

Ibbs et al. (2001) introduced a project change management system that is founded on five principles: (1) promote a balanced change culture; (2) recognise change; (3) evaluate change; (4) implement change; and (5) continuously improve from lessons learned. The system composed of a level-one or "macro" flow chart showing the five change management principles necessary to manage change and a series of level-two flow charts that show the specific activities involved with each of the level-one functional activities.

Research has also been undertaken on evaluating the change effects on certain project elements. These studies dealt mainly with a single factor or a single project element such as: construction change order impacts on labour productivity at the craft level (Hester et al. 1991), effect of the size of change and its impact time on a project (Ibbs 1997), a linear regression model that predicts the impact of change orders on labour productivity (Hanna et al. 1999), the risk of changes to safety regulations and its effect on a project (Williams 2000).

Several researchers have investigated managing design changes; most of these dealt with developing integrated systems to represent design information, recording design rationale, facilitating design co-ordination and changes, and notifying users of file changes. Examples include models developed by Ahmed et al. (1992), Peltonen et al. (1993), Spooner & Hardwick (1993), Ganeshan et al. (1994), Krishnamurthy & Law (1995), Mokhtar et al. (1998), and Hegazy et al. (2001).

From the forgoing discussion of the literature, it can be seen that researchers have mainly focused on the identification of the change process and best practice recommendations for managing change during the project life cycle. While these recommendations are beneficial, they are not sufficient to manage the complex process of change, particularly when there are different change causes and consequences. Much of the discussion is presented in categorical ways with little attention being paid to modelling the dependent data. Therefore, there is a clear need in the construction industry for research work to focus on modelling this dependency, especially for multi-disciplinary causes and effects.

Cause-and-effect relationships (dependencies within the change process) are important for understanding how changes occur and how the change causes influence the effects. These dependencies are the bases for change evaluation within the overall change process. Although the causes of change are factual, they may not be readily identifiable due to a lack of information and they may also be interdependent. The likelihood and impact of change should be shown in the proposed modelling. The research reported in this paper has three aims which are: to model the change process so as to predict change scenarios on projects depending on the available information at the early stages of projects, to model the dependencies within the change process, and to model change options evaluation. This paper focuses on the first two aims.

3 CHANGE PROCESS MODEL FOR CONSTRUCTION

The model developed is based on the change process models reviewed in the literature and the process models of two case studies undertaken during this research. The proposed model, shown in Figure 1, is a generic change process model that can be applied to any change case, namely design changes or post-fixity changes during the construction phase. The following section gives a summary description of these generic change process stages.

3.1 Start up

This contains the proactive requirements that are essential for effective change management. These requirements enable the project team to respond readily to change, to manage change effectively, and to facilitate

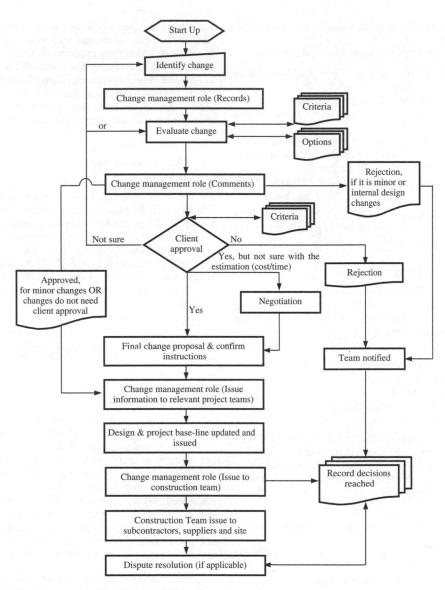

Figure 1. Generic change process model.

contingency plans for any anticipated change. Among these requirements are:

1. Project baseline and detailed cost and time plan. The project programme should be developed to manipulate change. A programme that satisfies this requirement should allow a late start for tasks subjected to change, identify the latest time for decisions to be taken and identify the time when the project information provided is complete.
2. Knowledge base that includes criteria for deciding on change and evaluation in terms of the key project objectives (cost, time, quality and value issues).

3. Integrated system for design management where management of the interfaces between designers and work packages are essential. It is also required for design rationale records for any change occurrence.
4. 3-D modelling that assists fast and more detailed assessment of the impact of proposed construction changes.
5. Procurement routes should consider change. The likelihood of changes becomes a criterion for selecting the procurement route.
6. Value management and VE systems.
7. IT communication facilities.

8. Dispute resolution mechanism for any change that occurs.
9. Risk analysis/management system that indicates:

- Various risks that may occur at different times within the project life cycle.
- The possibility of change occurrence that can be reasonably foreseen should be estimated (including its timing).
- Risk analysis at an early stage will enable appropriate procedures to be established and appropriate contingencies to be prepared.
- Scenario planning will help represent these changes.

The dependency model described later in this paper will help in modelling this last item of the proactive requirements.

3.2 Identify change

Full change identification will help in evaluating the change in the next step of the process and also during implementation. Change identification should cover change causes, change types and change effects. Change causes may be classified as:

- Client instruction/request for a proposed change;
- Design proposal change;
- Contractor change;
- Site conditions change; and/or
- Combination of these.

The change types will affect the degree of change effects and the evaluation criteria of change. Types of change may be minor/major, required/elective, or pre-/post-fixity. Various criteria can be used to identify the change type such as:

- Change that needs rework;
- Roughly costing of change works with respect to the project cost;
- Duration required to redirect the project work due to change; and
- Size of disruption to the workflow.

Change effects can be classified as:

- Direct effects (additional time and cost, change project information and outputs);
- Indirect effects (increased co-ordination failures and errors, increased waste in the process from abortive work and out-of-sequence working, reduction in productivity, uncertainty and consequently lower morale);
- Ripple effects. The cumulative effects of change on tasks located on successive orders and also on supply chain members. This will include number of resources working on time of change combined with the program float. Tasks with many logical links should receive special consideration.

The effects of change become greater with more live tasks at the time of change and with less available scope for reworking the programme. It is not easy to estimate effects accurately. The direct effects are easier to measure than the indirect effects. Even for direct effects, the ripple effect may not allow accurate assessment of impact. Indirect effects may be more onerous to the project than the original plan CIRIA (2001).

3.3 Change management role (Records)

It was concluded from the case studies undertaken that the role of a Change Manager is important to ensure effective change process management. The change management role can be executed by certain individuals that take the risks necessary to implement changes or can be executed by a member of the project team (e.g. project manager, architect). With advanced IT technologies, some roles of this manager can be carried out electronically such as information recording and propagating to the project team.

The change management role at this stage records information so far in the process and distributes them to the relevant project team (e.g. design team, construction team, cost team, programme input).

3.4 Evaluate change

Analysis of change options is required for decision-making – whether to go ahead with any of the change options or to undertake further investigations. Criteria are required to carry out this analysis which should include tangible and intangible criteria. It should cover time-cost and cost-benefit analyses. The evaluation stage needs experts' opinions. Evaluation steps include options evaluation, implications assessment and optimum selection of change options. Different models and decision support systems can be used to help decision-makers select an optimum solution. For example:

- Financial models that incorporate financial parameters to control the operation at any one time;
- Linear models where decision criteria are subjectively weighted and rated by a decision-maker and combined into a single measure;
- Linear models incorporating multiple ratings that add the corresponding probabilities for the multiple ratings of a given criterion and measures the imprecision and uncertainty associated with the process.
- Multi-attribute utility models that develop a method to combine qualitative and quantitative decision criteria that are aggregated to arrive at an expected utility where risk, uncertainty, and the decision-maker's preferences are modelled and considered.

- Fuzzy set models that are suitable for modelling qualitative criteria by determining the degree of membership in a set via membership functions that are elicited from decision-makers and combined into an aggregate measure.
- Statistical models to evaluate quantitatively criteria relevant in decision-making techniques such as least squares regression of logistic regression where a dependent variable and an independent variable exist.
- Artificial Intelligent systems which combine qualitative and quantitative criteria in the form of heuristic rules. These models enable learning mechanisms for future projects.
- Hybrid models which integrate any of the above models and decision support systems.

3.5 *Client approval*

Client approval is an important step in the process while different outputs are expected, as shown in Figure 1. The client needs to review change against the project baseline using tangible and intangible criteria. In many cases, clients need to use decision-making techniques for evaluation and comparison in order to decide on a change option.

The rest of the process stages, in Figure 1, involve integration between documentation and communication facilities. When dispute resolution is applicable, it requires investigation of direct and indirect causes of change, and analyses of the effect of multiple changes.

The model described above shows relationships between project characteristics inherent in the "start up" stage, causes of change in "identify change", and change effects in both "evaluate change" and "client approval". Analysis of such relationships is highly required to predict change occurrence or to diagnose change cases that have already occurred. The following section introduces such cause-and-effect relationships, or in other terms, dependency within the change process.

4 DEPENDENCY MODELLING

The main purpose of dependency modelling is to predict change events at the early stages of projects and therefore enable the taking of appropriate actions to minimize their disruptive effects. This is for proactive changes. On the other hand, in case of reactive changes, the dependency modelling is required for diagnostic purposes. To reduce the disruptive effects of change, it is important to identify what project characteristics lead to change causes and what these causes are, and then to understand how these causes are related to effects. How does one factor relate to another? What are the internal mechanisms by which a particular factor causes a

change in another factor? For example, how can poor communications lead to higher chance of change? How does an affected factor cause change in such a way that the former input factor ultimately gets affected? For example, poor communications leads to higher chance of change, but higher chance of change may eventually force improvements in communications.

Figure 2 illustrates a typical dependency diagram while Figure 3 gives an example of a change case. Multiple causes can lead to a change case and the effects of this change case cannot be added linearly. As shown in Figure 4 (the intersection areas), different causes of change may be responsible for a certain effect or a set of effects. In other words, the effect of a change cause C1 and C2, occurring together, may in general result in more than the sum of the effect of change cause C1 occurring on its own and the effect of change cause C2 occurring on its own.

An analysis is required to find out the possibility of occurrence of each cause of change with respect to each project characteristic. Multiple causes of change is a complex issue to determine the corresponding impacts

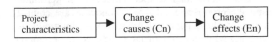

Figure 2. A typical dependency diagram.

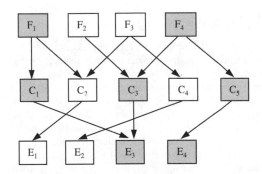

A change event example, which is based on F_1 and F_4 of project characteristics that lead to C_1, C_3 and C_5 of change causes, which in turn result in E_3 and E_4 of change effects

Figure 3. A dependency diagram for a change case.

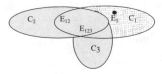

Figure 4. Multiple change causes.

2189

on projects. Lists of F_n, C_n, and E_n, were identified through case studies, details can be found elsewhere: Motawa et al. (2003). This helps project teams checking their project against these lists. A model based on a fuzzy logic technique is proposed to predict the overall effects of a change case by considering multiple causes of change led by certain project characteristics.

There are no predefined dependencies between each set of variables with another. Every project has its own case but the project team can define dependency routes for their projects. However, defining dependencies is not enough for predicting changes or effects. The likelihood of occurrence of change causes should be estimated. Change effects from multiple causes should also be simulated to reflect the uncertainty around the project towards change occurrence. The model output will estimate the amount of "a certain effect" due to "certain causes of change" under "specific project characteristics". This output can be useful in alerting the most effective project characteristics on change to take corrective actions and to minimise certain effects of change.

The generic change process model presented earlier and the proposed model of dependency are complementary. The dependency model is used for predicting changes while the generic model is used to monitor the process of implementing these change.

5 CONCLUSION

This paper has proposed a change process model that is intended to enable project teams to manage change effectively. It was concluded that modelling factors that affect decisions to change for various construction disciplines is deficient due to lack of the required linking knowledge. Therefore, the model addressed the cause-and-effect relationships (dependencies within the change process). As prediction of change scenarios on projects depends on the available information at the early stages of projects, the model is able to simulate anticipated changes and its effects considering the uncertainty around the project information at the early stages of projects.

REFERENCES

Ahmed, S., Sriram, D. & Logcher, R., 1992. Transaction-management issues in collaborative engineering. *J. Computing in Civil Engrg.* ASCE, 6(1), pp. 85–105.

Construction Industry Research and Information Association (CIRIA), 2001. Managing project change – A best practice guide.

CII conference, 1996. Project change management – implementation feedback report.

Cox, I.D., Morris, J.P., Rogerson, J.H. & Jared, G.E., 1999. A quantitative study of post contract award design changes in construction. *J. Constr. Mgmt. And Economics*, 17, pp. 427–439.

Ganeshan, R., Garrett, J. & Finger, S., 1994. A framework for representing design intent. *Design Studies*, 15(1), pp. 59–84.

Hanna, A.S., Russell, J.S. Vandenberg, P.J., 1999. The impact of change orders on mechanical construction labour efficiency. *J. Constr. Mgmt. and Economics*, 17, pp. 721–730.

Hegazy, T., Zaneldin, E. & Grierson, D., 2001. Improving design coordination for building projects – I: Information model. *J. Constr. Engrg. and Mgmt.* ASCE, 127(4), pp. 322–329.

Hester, W.T., Kuprenas, J.A. & Chang, T.C., 1991. Construction changes and change orders: their magnitude and impact. *Construction Industry Institute (CII), Source document 66, CII, Austin, Tex.*

Ibbs, C.W., 1997. Quantitative impacts of project change: Size issues. *J. Constr. Engrg. and Mgmt.* ASCE, 123(3), pp. 308–311.

Ibbs, C.W., Wong, C.K., & Kwak, Y.H., 2001. Project change management system. *J. Mgmt. in Engrg.* ASCE, 17(3), pp. 159–165.

Karim, A. & Adeli, H., 1999. CONSCOM: An OO construction scheduling and change management system. *J. Constr. Engrg. and Mgmt.* ASCE, 125(5), pp. 368–376.

Krishnamurthy, K. & Law, K., 1995. A data management model for design change control. *Concurrent engineering: Res. and applications,* 3(4), pp. 329–343.

Mokhtar, A., Bedard, C., & Fazio, P., 1998. Information model for managing design changes in a collaborative environment. *J. Computing in Civil Engrg.* ASCE, 12(2), pp. 82–92.

Motawa, I.A., Anumba, C.J., El-Hamalawi, A., Chung, P. & Yeoh, M., 2003. An innovative approach to the assessment of change implementation in construction projects. *Proc. of CICE conference, UK.*

Peltonen, H., Mannisto, T., Alho, K. & Sulonen, R., 1993. An engineering document management system. *Proc. ASME Winter Annu. Meeting.*

Project change management Research team, 1994. Project change management, Special publication 43-1. *Construction Industry Institute (CII), The university of Texas at Austin, US.*

Spooner, D. & Hardwick, M., 1993. Using persistent object technology to support concurrent engineering systems. *Concurrent engineering: Methodology and applications, Elsevier Science*, Amsterdam, pp. 205–234.

Stocks, S.N. & Singh A., 1999. Studies on the impact of functional analysis concept design on reduction in change orders. *J. Constr. Mgmt. and Economics*, 17, pp. 251–267.

Voropajev, V., 1998. Change management – A key integrative function of PM in transition economies. *I. J. of Project Mgmt*, 16(1), pp. 15–19.

Williams, T.M., 2000. Safety regulation changes during projects: the use of system dynamics to quantify the effects of change. *I. J. of Project Mgmt*, 18(1), pp. 23–31.

System-based Vision for Strategic and Creative Design, Bontempi (ed.)
© 2003 Swets & Zeitlinger, Lisse, ISBN 90 5809 599 1

Improving performance through integrated project team constituency

D.R. Moore
The Scott Sutherland School, Robert Gordon University, Garthdee Road, Aberdeen, UK

A.R.J. Dainty & M. Cheng
Department of Civil and Building Engineering, Loughborough University, Leicestershire, UK

ABSTRACT: Increasingly, project teams are becoming differentiated in terms of the extent of functional specialisms represented within them. This is particularly so in the context of large complex projects of long duration. The literature places much emphasis on the issue of teambuilding, and highlights considerations of team member type with the suggestion being that, along with their primary functional specialism, each member will bring with them their personal values, beliefs and behaviours. Advice is then proffered concerning how the differentiated types can be integrated so as to form an effective project team. However, it is arguable that "effective" is insufficient in the modern context of rapidly changing competitive environments and that project managers need now to consider how their teams may be moved towards a "superior" level of performance.

The paper places the findings of research on the differentiation between average and superior-performing managers carried out by the authors in the context of the management of projects characterised as having high levels of human difficulty (sensitive projects). A particular example relating to the need to achieve process integration (as identified by reports such as *Rethinking Construction*) is discussed as representing a sensitive process. The success of such projects is argued to be dependant upon the behaviours of project allies. In addition to the traditionally perceived "personal" characteristics, the research identifies a number of behaviours that underpin the performance of an individual. These behaviours are shown to have a significant impact on the quality of an individual's performance in the context of management and are therefore of relevance to the Project Manager's decisions concerning team constituency.

1 THE UK CASE

Within the UK construction industry there has been much emphasis placed, over the last decade or so, on the need for the industry to "rethink" itself. Sir John Egan's (1998) report, Rethinking Construction, in particular initiated a period of reflection by some of the industry leaders on how to go about their business in a more efficient manner. Concepts of business re-engineering and lean/agile construction appear to have featured prominently in this process. The concept of lean construction, for example has been vigorously adopted by the Construction Best Practice programme (CBP), and their website (cbpp.org.uk) contains examples of relevant case studies and workshops, all within the UK context. Whilst Egan did suggest lean construction as one possibility for achieving improved project performance, the report paid particular emphasis on the issue of project

process integration. This was defined in terms of several characteristics:

– It utilises the full construction team, bringing the skills of all the participants to bear on delivering value to the client.
– It is a process that is explicit and transparent, and therefore easily understood by the participants and their clients.
– It leads to increased efficiency of project delivery due to elimination of the constraints imposed through the traditional largely separated processes of planning, designing and constructing construction products.
– It avoids a contractual and confrontational culture.
– It avoids sequential working with regard to the input of designers, constructors and key suppliers.
– It facilitates the effective use of the skills and knowledge of suppliers and constructors in the design and planning of the projects.

- It does not assume that clients benefit from choosing a new team of designers, constructors and suppliers competitively for each project, and therefore does not inhibit learning, innovation and the development of skilled and experienced teams over a succession of contracts.
- It allows teams of designers, constructors and suppliers to work together over a number of projects, thereby continuously developing and innovating the product and the supply chain (Egan 1998).

For any project manager who is responsible for a construction industry product, Egan's emphasis on process integration is of importance in that integration without implementation is meaningless. While there may well be considerable emphasis placed on improving team performance through the implementation of integrated processes, there appears to be two areas within the objective that have not been sufficiently considered.

Firstly, Egan sets a goal of promoting learning and development of team members within an environment lacking the traditional boundaries between the functional specialisms of the traditional (UK) procurement process. There is no explicit recognition, however, of the problems that can be encountered in moving the structure (the formalised relationships between the specialisms) from a rigid, transactional hierarchy to one where boundaries are removed and the structure becomes more flexible. In essence, what is being attempted is a move from a transactional, or Newtonian, structure to a transformational, or Einsteinian, structure as discussed by Banner & Gagne (1995). This is particularly problematic with regard to the culture (values and beliefs) of those within the organisation structure. Individuals who have been inculcated with transactionalist values and beliefs may not immediately embrace the opportunities presented by a newly transformationalist structure. There is therefore a need to consider the behaviours that are desired of those individuals who are responsible for "transforming" the organisation culture within the context of the new environment represented by process integration.

Secondly, process integration itself should be regarded as representing a project (moving from a differentiated process to an integrated one), within which there is an element of what D'Herbemont and Cesar (1998) refer to as human difficulty. It is therefore argued that those who are seeking to implement process integration should be prepared to regard their activities as essentially being concerned with the management of a sensitive project. Performance levels in this context will therefore be dependant upon the emergence of project allies. This again emphasises the issue of behaviours by individuals. This is especially so when the objective is to produce an integrated project team (IPT) that actually functions in an integrated manner. It

is therefore relevant to consider the issue of behaviours in more detail prior to discussing the matter of performance in the context of energy release within a project.

2 BEHAVIOURS AND COMPETENCES

Much confusion surrounds the use of the terms competence, competency and competencies. Moore et al. (2002) suggested definitions that are adopted within this paper:

- competence – an area of work
- competency – the behaviour(s) supporting an area of work
- competencies – the attributes underpinning a behaviour.

Behaviours in general can be considered as being of two types: transactional and transformational.

2.1 Transactional versus transformational

Transformational organisations are typified by their lack of hierarchy and the resultant lack of positional power and authority available to players within them. Rather, the emphasis moves from positional power and authority to what is referred to as "sapiential" power and authority: authority resulting from knowledge and/or maturity (Banner & Gagne 1995). Maturity is defined in terms of "life experience" that causes an individual to be seen as having integrity and therefore trusted by those around them, rather than any reflection of their functional or formal knowledge.

Within transformational organisations the authority is fluid: the organisation chooses the most appropriate authority figure for any given set of circum- stances. In order to operate in this manner organisations require individuals who possess maturity and, unlike those within transactional organisations, are not driven by the question of "what is in this for me?" However, maturity is arguably a scarce resource. On the basis that it is difficult to identify a single organisation that could justifiably claim to be a fully transformational one (the closest seem to be chaordic organisations and Senge et al. (1999) argue that only one of these exists) the majority of organisations currently exhibit an essentially transactional culture. Thus it seems inevitable that the majority of players available to organisations will have "learned" to exhibit transactional behaviours. Wenger (1998) for example, notes that an interaction providing tension between experience and competence presents a rich basis for learning. However, if this tension is lost (such as by prolonged repetition of experience; a typical characteristic of transactional organisations) then the rate at which new competences are developed slows; the individual learns little that is new. Egan's championing of process integration results, at least in part, from a belief that this will act as a spur to learning on the part of those involved. This learning will not be achieved if process integration is approached in a transactional manner.

In the context of the work by Hutchins (2001), essentially transactional organisations place little or no pressure on the ruling paradigm. Individuals who have undergone learning within such a consistent paradigm are therefore argued by the authors as having a maturity deficiency with regard to the requirements of a fully transformational organisation (with its constant pressure on paradigms). In the event that an organisation decides it needs to adopt a transformational culture, it seems probable that it will be faced with a shortage of people possessing the required balance between functional knowledge and maturity. Such a situation will negatively impact on the achievement of the required culture change. The extent of that impact will be related to the degree of culture change required.

Transactional and transformational organisations represent opposite ends of a continuum, between which lie organisations referred to by Moore (2002) as being transforming (moving from transactional to transformational). This continuum is reflected in a further continuum representing the scope for change of an organisation. The work of Nadler & Tushman (1989) on what they refer to as organisational frame bending is of relevance to this second continuum. This has informed the suggestion that slow and small (incremental) changes lie at one end, while rapid and large (clean slate or neutron bomb) changes lie at the other, with the respective impacts on the organisation being at either operational or strategic level (Jasper Associates 1998). Within these extremes lie degrees of transforming organisations, with one frequently cited example that is toward the transformational end of the continuum being Gore Associates. This organisation is typified by a lack of traditional, transactional structure in terms of a hierarchy. Within such an organisation the behaviours of individuals are of significance to its success. However, this then raises the need to assess behaviours. Such assessment can be considered in the context of competences.

2.2 Competences and assessment

Over time the term "competence" has tended to become a synonym for "performance" in some areas. This is particularly evident in the UK with its emphasis on achieving a "competent performance" within the MCI (Management Charter Initiative) and NCVQ (National Council for Vocational Qualifications) approaches. These effectively set minimum levels of technical competence. In comparison, the American approach is to stress that competence is an "underlying characteristic". Such a characteristic may lead to effective or superior performance, but the two are not seen as being synonymous. The existence of this distinction appears to have contributed to the dominant approach to performance management in the USA (at least within private sector production industries) being centered on person-oriented job analyses as discussed by Cheng et al. (2003). These are typified by what are referred to as "behavioural event interviews", which are intended to identify characteristics that differentiate between superior and average performers. Through focussing in this manner on characteristics, competencies become identified in terms of skills (one of Egan's goals), personal characteristics or behaviours.

The dominant approach in the UK has, in contrast, been to use an analysis technique referred to as "functional analysis" to identify the necessary roles, tasks and duties required in order to carry out a specific occupation to an adequate standard. There is therefore little or no consideration of the characteristics exhibited by "successful" practitioners of a given occupation (Cheng et al. 2003). This situation arises, at least in part, because of methodological differences (Iles 1993). Methodologies within the UK, for example, typically involve attempts to firstly identify the key purpose or function of an occupation prior to the subdivision or desegregation of the purpose. The subdivision process is intended to establish those outcomes to be achieved in order to complete the key purpose, as referred to by Holmes and Joyce (1993), with the result that an extensive list of "elements of competence", grouped under major functional or key role areas, is produced. It then seems inevitable, within a transactional approach to performance management, that performance criteria are developed for each of these elements so to indicate appropriate (essentially adequate) competence levels. An example of this can be seen in the description of competence for professional, managerial and technical roles in the built environment as identified by the UK's construction industry standing conference (CISC). For one key purpose, 6 key roles, 26 units and 102 elements are identified, arguably from the perspective of the purpose and tasks of management rather than on the underlying generic behaviours or macro-competencies that the McBer (US) model identifies. The possibility of such a situation has led to researchers such as Mangham (1990) criticising the micro-competencies approach to performance management as trying to create identikit managers. Furthermore, transactional approaches such as the MCI framework invariably emphasise outcomes deduced from the performance of managers operating at what is regarded as an acceptable or adequate level of performance, thereby limiting the value of such approaches to the improvement of performance. By considering the behaviours exhibited by managers performing at a superior level, the management of the process integration "performance" will itself be enhanced.

3 BEHAVIOURS DIFFERENTIATING BETWEEN AVERAGE AND SUPERIOR MANAGEMENT PERFORMERS

Recognition of superior performance is problematic given that the abilities of an individual may not always appear to significantly influence the nature of the final outcome. Take, for example, a project manager who gels an ineffective team to ensure that it achieves at least a base level of performance; is he or she necessarily less effective than a project manager who merely empowers an already effective team to achieve superior performance? One way of overcoming this problem is to avoid focusing on the out-turn performance of the situation being managed, and concentrate instead on the *behaviours* of the manager and the impact that these have on the situation being managed. The transforming process has been argued as being essentially concerned with behaviours, therefore an emphasis on behaviours related to the learning performance of the organisation is posited as being appropriate.

It can be hypothesised that an average performer is "deficient" with regard to specific key behaviours within a given activity, whereas a superior performer is "endowed" in this regard. The questions then arise of what these behaviours may be, and how can their identification be used in the management of subsequent performance.

Performance management aims to generate better results from individuals, teams and the organisation as a whole. Essentially, it is about planning goals, targets and standards, continually monitoring progress towards achieving them and providing support where necessary. However, another way to use performance management is in recruitment and selection. Recruitment is the process by which managers attempt to locate people to fill identified positions. This is often a problematic process in construction projects, even in times of high unemployment, since both applicant and advertiser will have specific needs and expectations that will have to be matched. Any mismatch provides a potential basis for future conflict, on the basis of a conflict of interest (Katz & Kahn 1978), as both parties seek to achieve a performance that is "managed" so as to realise specific but different goals. Furthermore, conflict can arise as different individuals or groups within an organisation exert pressure through the expectation of differing outcomes and behaviours from an individual. This form of conflict is referred to as role conflict (Argyle 1994) and suggests expectations with regard to the level at which performance is deemed to be superior may vary between individuals and groups carrying out the assessment.

An important initial step when researching in this research is to identify the criteria or measures that define superior or effective performance in the job to be studied. Moore et al. (2003) reported research methodology and findings that illustrate this problem in the context of construction project management.

3.1 *Research methodology*

Three focus groups, comprising HRM practitioners, project construction managers and engineers, were conducted to identify performance effectiveness criteria. From the focus groups, 48 criteria for performance excellence were generated and used to develop questionnaires. Subsequently, questionnaires were completed by focus group members to rate how important each criteria was for the achievement of performance excellence. The results of factor analysis illustrated that nine factors were seen as being most important. The nine factors that were extracted from the completed questionnaires were as follows: team building, leadership, decision-making, trust, honesty and integrity, self-motivation and external relations.

Performance effectiveness criteria were used to identify a group of superior managers to form the basis of the main data collection phase and a comparison group of average performers. A panel of HRM specialists, construction managers, project managers and site managers from each participating company identified a total of 24 superior performers and 16 average performers in collaboration with the research team. A total of 40 "behavioural event interviews" were conducted, in which superior and average performers were interviewed and asked to describe the most critical situations they had encountered on their jobs. They were asked to consider factors such as what the situation or task involved, what events led up to it, who was involved? The interviewees were also asked what they thought, felt, or wanted to do in that situation? Finally, they were asked to describe what they actually did and what they thought the outcome was.

The 40 completed Behavioural Event Interviews were analysed to identify the personality, behaviours and skills that distinguish superior performers within the participating organisations. Any motives, thoughts, or behaviours that matched those provided by the McBer Competency Dictionary were coded. This coding process provided quantitative data that was used to test the findings for statistical significance. The resulting competency model was validated by a second criterion sample to see if the model based on the first criterion sample predicted the performance of the second criterion sample. Firstly, HRM specialists and senior line managers selected a second criterion sample of superior and average performers from each company. Then, managers and other knowledgeable observers were asked to rate members of that sample using a behavioural codebook developed from the interview data, and which describes those behaviours that predict superior job performance. The resulting scores were placed in a competency model intended to assess the probability of that person being a superior performer.

3.2 Findings

The ANOVA statistical technique was conducted to identify the competencies which distinguish superior managers from those performing at an average level. It was found that 12 competencies distinguish superior from average performers. They are: achievement orientation, initiative, information seeking, focus on client's needs, impact & influence, directiveness, teamwork & cooperation, team leadership, analytical thinking, conceptual thinking, self-control (composure) and flexibility. Logistic regression analysis (forward stepwise) was employed to distinguish which are the most predictive competencies among the 12 identified to generate a parsimonious model that can predict superior performers. Two behaviours were identified from the statistics as being the most significant in the context of a parsimonious model; self-control and team leadership. Self-control is highlighted by the model and contributes 92.5% accuracy. The next significant competency identified was team leadership which, when in combination with self-control, increased the overall accuracy of the prediction model to 95%.

As to the "predictive" behaviours themselves, it would seem obvious that self-control (composure) is an essential competency for the demanding, challenging and stressful construction environment. The ability to keep emotions under control and to restrain negative actions when tempted enables a person to maintain performance under stress and to respond constructively to the problem in hand. Also, construction is a project-based environment, it is all about teamwork, therefore, team leadership being one of the most predictive competencies may be expected.

One example of possible implications of this research is for the selection process, in that it may be appropriate to change the emphasis away from an individual's technical knowledge. The emphasis may be more appropriately placed towards their possession of particular behaviours, thus opening up the possibility of recruiting from outside the pool of "qualified" construction human resources. This idea has already been mooted as a possible solution to the current paucity of undergraduates studying construction courses in the UK.

A further possibility is that individuals possessing these behaviours may be appropriate to the management of sensitive projects, such as those related to process integration.

4 SENSITIVE PROJECTS

D'Herbemont & Cesar (1998) suggested that individual players could significantly affect project outcomes through their adoption of antagonistic and synergistic attitudes to the project. This is aprticularly relevant in cases of projects where there is a high level of human, rather than technical, difficulty. Projects characterised by a high level of technical difficulty can be managed satisfactorily on the basis of functional knowledge only and are not classed as sensitive. This suggestion should be considered in the context of players' expending energy in pursuit of project objectives.

4.1 Energy release

Around 40 to 80% of project players are said to expend little energy in this way. Such individuals can be considered to be broadly antagonistic to the project, although it should be noted that many players exhibit both antagonistic and synergistic attitudes towards a project.

Organisations can become "trapped" in patterns of behaviour that are not conducive to long-term survival, and these patterns are enforced through rules, procedures, regulations, policies, and so on. However, an organisation needs the right kind of behaviours amongst its people to adopt ways of working that are not reliant upon reinforcement by rules and procedures. In this context, it is important to recognise that the field of play is composed of individuals not groups: it is the individuals who will choose how involved they wish to be in the project. They will also decide how successful they wish the project to be. Players can be categorised under traditional headings such as the project team and institutional players such as planning and regulatory authorities: bodies who place obstacles in the way of the project. A non-traditional perspective regards players as being anyone who may benefit or suffer from the project (similar to the concept of stakeholders), but sensitive projects need only consider those who are able to have a direct influence on the project. The task then is to mobilise individual players in support of the project. It is in this context that "superior" management performers may be a particularly significant project resource in that their behaviours are well suited towards implementing techniques such as "launching the first circle"

4.2 Launching the first circle

This technique is part of a strategy referred to by D'Herbemont & Cesar (1998) as the herdsman strategy. This operates on the basis of realising that there is no such thing as "collective energy" in a sensitive project. There is only the sum of individuals" energy, and this may be spread across a number of projects, as players may operate on a number of projects at the same time, or may be focused against the project itself. The herdsman strategy has several key actions:

- identification of potential allies
- mobilisation of such allies
- provide allies with the means to persuade others
- show that you have confidence in your allies' ability to find the right way to persuade others
- move on to taking care of the stubborn players!

Launching the first circle comprises 5 steps:

- identify the waverers and golden triangles
- meet each of them (private meetings) and ask them to express their feelings about the project
- synthesis the comments and build a lateral project adapted to the needs of those interviewed
- invite allies from within the interviewees to form a group (more than 4 but less than 21) to listen to the synthesis of the meetings/interviews. This should act as a revelation that gets the dynamic in motion
- maintain the dynamic represented by the network of project allies through implementation of the lateral project.

Having completed the five steps, the technique is applied again with a new group, thereby launching the second circle which increases the network of allies represented by the first circle. A factor in the success of this technique is argued to be that of those who are responsible for its implementation exhibiting a superior level of management performance.

5 CONCLUSIONS

There is no intention to suggest that a team should be composed entirely of superior performers: an individual's role as a functional specialist outside of the context of manager must not be ignored. However, the situation where an IPT is constituted entirely of average performers is unlikely to produce the required synergy. A proportion of superior performers within the IPT is suggested as being a catalyst in bringing about a synergistic performance by the team as a whole. It is this synergy which arguably provides the team integration necessary to deliver the seamless project delivery mechanisms demanded by the proponents of the industry improvement agenda.

REFERENCES

Argyle, M. 1994. *The Psychology of Interpersonal Behaviour.* 5th Ed. London: Penguin Books.
Banner, D.K. & Gagne, T.E. 1995. *Designing Effective Organisations,* Thousand Oaks, California: Sage Publications Inc.
Cheng, M-I., Dainty, A.R.J. & Moore, D.R. 2003. The Differing Faces of Managerial Competency in British and American Production Industries. *The Journal of Management Development,* Bradford: MCB University Press. Paper in press. Publication expected in vol. 22.
D'Herbemont, O. & Cesar, B. 1998. *Managing Sensitive Projects: A Lateral Approach.* London: Macmillan Press Ltd.
Egan, J. 1998. *Rethinking Construction.* London: HMSO.
Holmes, L. & Joyce, P. 1993. Rescuing the useful concept of managerial competence: from outcomes back to process, *Personnel Review,* 22(6), 37–52.
Hutchins, T. 2001, *Unconstrained Organisations.* London: Thomas Telford Ltd.
Iles, A. 1993. Achieving strategic coherence in HRD through competence-based management and organisation development. *Personnel Review,* 22(6), 63–80.
Jasper Associates 1998. Organisational Change and Transformation. Capacity, Capability and Survival. http://www.aja4hr.com/services/organisational_capability_assessment.shtml
Katz, D. & Kahn, R. 1978. *The Social Psychology of Organisations.* 2nd ed., New York: John Wiley.
Mangham, I. 1990. Managing as a performing art, *British Journal of Management,* 1(2), 105–115.
Moore, D.R. 2002. Project Management: Designing effective organisational structures in construction. Oxford: Blackwell Publishing.
Moore, D.R., Cheng, M-I. & Dainty A.R.J. 2003. What Makes A Superior Management Performer: The Identification Of Key Behaviours In Superior Construction Managers. CIOB Innovation Awards competition, Premier Award winner. www.ciob.org.uk
Nadler, D.A. & Tushman, M.L. 1989. Organizational Frame Bending: principles for managing re-orientation. *The Academy of Management Executive,* 3(3) 194–204.
Senge, P., Kleiner, A., Roberts, C., Ross, R., Roth, G. & Smith, B. 1999. *The Dance of Change.* London: Nicholas Brealey Publishing.
Wenger, E. 1998. *Communities of Practice. Learning Meaning, and Identity.* Cambridge: Cambridge University Press.

An integrated information management system for the construction industry

S.P. Sakellaropoulos & A.P. Chassiakos
Department of Civil Engineering, University of Patras, Greece

ABSTRACT: The construction industry highly depends on effective information communication. Managing this information may be difficult due to the large amount of information and the fragmented nature of the construction industry, though critical for project success. Current research efforts have focused mainly on automating communication between computer tools, in order to reduce information management needs. This paper presents several design issues of an integrated information management system that can facilitate the communication among project participants. Construction information is stored in an appropriately designed database while information is accessed via the Internet. This is facilitated by intranet–extranet application with the desirable access provision and the employment of communication security tools. Information is transferred through web interfaces or by downloading/uploading software files. The proposed system can improve the information management and communication process leading to time and cost savings as well as to higher construction quality.

1 INTRODUCTION

Construction is one of the most information-dependent industries, mainly due to its extended fragmentation. Construction projects involve a large number of human resources with various specializations. In addition, projects are often complex and unique. Thus, the amount of information generated and exchanged during the construction process is enormous even for small-sized projects. The distance between the construction company headquarters and construction sites augments the communication problem. Thamhain & Wileman (1986) stated that communicating effectively among task groups is the third most important factor for the success of a project.

In order to surmount this communication deficiency, a new information vehicle is required. The Internet and the WWW seem to be attractive for transferring information rapidly and economically all around the world. Thus, the Internet provides a suitable platform on which organized attempts to manage construction information could be fruitful. This paper presents such an attempt. A central repository database, which contains construction information, is accessed by project participants through Internet. The objective of the research is not to provide an off-the-self commercial product but, rather, to experiment on the development and efficiency of such a system.

In the following section, previous research efforts related to information management systems for the construction industry are presented. An integrated information management system is then proposed. This section describes the basic conceptual framework and the system architecture, the development process, and some implementation issues. Finally, the expected benefits of the system are analyzed and the main conclusions of the research are reported.

2 BACKGROUND

Several research efforts can be found in the literature regarding the use of computers or information technology (IT) for improving the efficiency of the construction process. This is expected to be achieved by providing concise information and enhancing communication among project participants as well as by making more effective use of computer tools.

A number of terms can be found in the literature that reflect these objectives. The computer-integrated construction (CIC) or the construction information technology (CIT) focus more on the organized use of computer applications and the automated communication between them, considering the construction process as a whole. Allowing the computer tools to share data helps towards this direction. Since application tools used in the construction industry are numerous, the only feasible solution can be provided through a common data language, interpretable by all applications. The Internet-based construction project management (ICPM) focuses more on the use of the Internet and/or the WWW as a means for transferring information rapidly and inexpensively, facilitating thus communication among project participants. Some past research efforts on these issues are briefly presented below.

Opfer (1997) provides a review concerning intranet applications in the construction industry. In particular, this research deals mainly with specific applications and discusses issues concerning the practicality of the technology, cost savings, hardware and software used, and information security topics.

Garcia et al. (1998) present a combination of database and Internet technologies to support an improved and timely communication of project information between the construction site and company headquarters. In this application, project information is limited to those required for activity time and cost control, and for material needs and inventory.

Rezgui & Cooper (1998) give a comprehensive overview of a project named European Esprit Condor that provides a migration path from document-based to model-based approaches for information representation and structuring.

Shahid & Froese (1998) describe the need of project documentation and project information flow within the construction industry. The study mapped various types of project information against the documents that typically provide this information and the construction management functions that provide and access the information. This analysis led to the design of a user interface, which provides a conceptual backbone for the development of an information management system.

Tam (1999) examines the potential of IT in improving coordination between construction project participants and proposes a framework which comprises of six major components: data exchange through Internet, Internet chat enhanced with on-screen images/drawings, live video-cam for site-based data capture, link and search engine, email information exchange, and auxiliary services.

Dawood et al. (2002) present an IT-based tool for site document management as a first phase of a larger project. This tool uses the Internet to provide an automated integrated environment for communication, retrieval, storage, and distribution of project documents among the project team. The system involves resource procurement, site planning, progress reporting and contractor account, cost control, and measurement and valuation.

Mokhtar (2002) proposes an architecture for an information model that aims to use the Internet technology to overcome the problem of incompatibility errors in design information. The approach to develop the model is not to automate currently used coordination techniques, but rather to re-engineer these techniques using the capabilities of computers and the Internet over those of humans. The model also contains a component that enables the generation of customizable and multimedia construction documents. The focus of this work is mainly on drawings.

Sriprasert & Dawood (2002) propose a vision for the next generation of construction planning and control as multi-constraint, visual, and lean-based system. An implementation of this vision has resulted in a prototype web-based system. An elaboration on the system framework and an underpinning methodology to integrate information and constraint management with 4D planning and control system is the focus of the research.

Concluding this section, one can say that computer-integrated construction (CIC) aims at reducing information management needs by improving cooperation-communication of computer tools. It seems, however, that this direction is a long-term approach and benefits cannot be expected shortly. On the other hand, the application of Internet-based construction project management (ICPM) techniques can have a more rapid effect in improving the efficiency of the construction process. It is the authors' belief, however, that ICPM is still in its infancy. Existing research efforts on these issues have advanced to some point either by proposing conceptual frameworks (e.g. Tam 1999) or by developing partial components of such a system (e.g. Garcia 1998). However, there is no single information management system that addresses all needs and can be practically implemented. Research should advance further before such a system is developed. The complexity of the problem and the number of issues that need to be addressed provide a wide area of future research.

3 INTERNATIONAL STANDARDS

The lack of data standardization has been a major obstacle for computer-integrated construction research projects. The recognition of this problem has led to the ongoing development of such standards. The prominent efforts for the standardization of product models are the STEP (Standard for the Exchange of Product model data) standard by the International Organization for Standards (ISO) and the Industry Foundation Classes (IFCs) by the International Alliance for Interoperability (IAI).

The ISO-STEP 10303 is a standard for the human and computer interpretable representation and exchange of product data, with the objective of providing a neutral mechanism to describe the product data throughout the life cycle of the product, independently of any particular system or software. This standard refers to all industry-wide sectors that deal with a product rather than focusing on specific industries, e.g., the architecture, engineering, and construction industry (AEC).

The IAI is an industry-based consortium, which develops similar standards in the form of IFCs for exchanging data between computer systems within the AEC industry. The intention of the IFCs is to define how real things, such as doors, walls, etc. and abstract concepts, such as spaces, processes, etc. should be represented electronically.

Although these standards are very promising in facilitating the construction process, this can only be expected in the long run. More effort must be placed on them before they acquire practical applicability.

4 PROPOSED SYSTEM

The research presented in this paper is an effort within the areas of ICPM towards the development of a system that would take into consideration all information generated and exchanged during the construction process in an integrated and standardized manner. The objective of this work is to fill the gap between previous works that either simply set the conceptual framework or partly develop such a system for particular pieces of information only. The existence of a fully developed system increases its potential to be adopted by the construction industry.

4.1 System concepts/requirements

The proposed system is being built on four major concepts: integration, flexibility, generality, and simplicity.

- Integration. The construction process is seen as a whole. Thus, the information produced and dispersed during this process must circulate in a unified, standardized and integrated manner in order that the system achieves its goals. This requires substantial effort as, in practice, the information includes various farraginous formats: paper-based documents or drawings, software files, telephone calls, real-time/place conversations, and web interfaces.
- Flexibility. One characteristic of the construction industry is the uniqueness of its products. Thus, the information management system should be supple enough to adapt to specific project particularities. No limitations should be arising due to such particularities. The extensibility of the system, when required, is also implied into the flexibility.
- Generality. The concept of generality refers to the wide applicability of the system throughout the construction industry. It is different from the flexibility term since it defines the property that the system can be used as it is in various applications. For this purpose, the construction process and the corresponding information must be well identified before the necessary information is gathered.

- Simplicity. This property can be seen as a requirement for the system that improves its effectiveness and its application potential.

4.2 System development

An information management system for the construction industry delivers information to project participants. Hence, the development of the system should initially identify what kind of information is currently or must be in the future transferred and who are the sender and the recipient of it. However, the amount of construction information for even a small-sized project is quite large and this makes arduous the task of identification. The large number of project participants and the variety of their specialties and experience further complicate the problem.

Thus, the development of the system consists of two types of analyses, information analysis and participant analysis. It must be clear that such a system cannot substitute nor integrate all existing ways of information transfer. For example, even if a formal telephone conversation could be recorded and inserted in the system for future reference, a simple informal conversation between colleagues would be less valuable and could easily lead to information overflow. What information must or must not be inserted in the system is a critical issue. In addition, the information storage and transfer format needs to be defined.

There are two possible ways in which information can be transferred into a computer-based information management system: through a software file or through a web interface. The first consists of simply downloading (receiving information) or uploading (sending information) files from or to a web page respectively. These files may have been produced by various software tools (e.g. word-processing, CAD, project management programs, etc.). This method requires from everyone using the system to have the corresponding software packages installed in his/her personal computer in order to retrieve and modify the information.

The second way of electronic information transfer is through web interfaces (having .htm extension). Usually, web pages provide posted information but there are also cases where information is submitted in specially designed forms and sent to a particular recipient (e.g. submitting on-line applications). Web interfaces, although ideal for simply viewing or printing information, are not always suitable when increased interactivity is required. For example, a web interface depicting a drawing can be sufficient for someone to view it, or even to zoom in/out, without the need to have installed the corresponding CAD software package. On the contrary, an architect wishing to make some drawing corrections cannot do so simply by viewing it, but only by downloading the drawing to his/her PC, modifying it, and then uploading the new version.

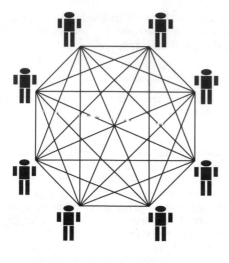

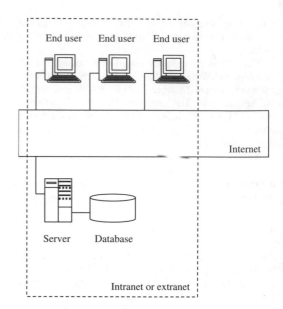

Figure 2.   The system architecture.

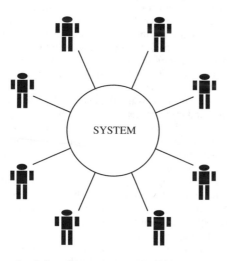

Figure 1.   Information management with current practice and with the proposed system.

It is apparent, therefore, that both ways should be combined for setting up the system, in order to take advantage of their effectiveness and reduce the effect of their limitations.

Concerning the participant analysis, the typical company structure needs to be coded, in terms of the personnel employed or involved in a project. Other external project participants (e.g. suppliers) should also be identified. Following, the information needs of each individual or group of participants should be identified along with the information chain (i.e., who is the sender and who is the recipient of each piece of information). Figure 1 shows the way that information is transferred among project participants with current

practice and with the proposed system. The lines in this figure depict (one way or bi-directional) information transfer. With current practice, this transfer commonly takes place directly between participant pairs. As the number of participants increases, the information communication process becomes complicated and difficult to organize and control. With the proposed system, however, information transfers between the system database and each project participant, facilitating thus the organization and control objectives.

4.3   System description

The proposed system consists of a repository database, a server, and end user (remote) terminals (Figure 2). The database is organized to accommodate storage of three types of information, project information, company information and general information. Project-specific information is the most important and massive part of data, and can include drawings, specifications, bills of quantities, etc. Company-specific information includes data concerning the company staff, headquarters operation and costs, etc. General information that may be stored in the database includes structural codes, material properties, etc.

End users access this information through Internet by simply using a web browser. The electronic data exchange is allowed among the project participants through a set of appropriately designed web pages, whose data are updated by project members as authorized. Each user is granted access to specific pieces of information and can perform certain actions. In general, users are allowed to store (or input), view or retrieve, search, delete, move, update or modify information. More advanced transactions are even allowed, such as locking or unlocking information (by the key holder), cross-referencing it, information update notifying, etc.

The system includes an intranet and an extranet. The intranet is used for information sharing among construction company members. The company members can be typically grouped in several functional departments dealing with the top management, the project management, the project planning and engineering, the administration/finance, the marketing, the maintenance, the quality management, and the research and development. The extranet allows outside participants to have access to particular pieces of project information. Outside participants can be the project owner, supervising authorities, external sources such as contractors, consultants, financial institutions, suppliers, and others.

Apart from the database, other tools are also developed, such as request for information (RFI) module and filters/viewers while security, scalability, and user friendliness issues are addressed. Other tools that can also be implemented in the system are electronic bulletin board, live site video cameras, and teleconference or virtual meeting rooms. The system could also be used for advertising/marketing purposes, as well as for e-commerce.

4.4 Implementation issues

The system implementation requirements are simple: PCs connected to the Internet, web browsers, and a server. This is an important characteristic of the system, as it is generally desirable for a computer-based system to be built up on simple and commonly used hardware and software. A system administrator may be necessary, in addition, to ensure that the system performs steadily and reliably. This administrator could also be responsible for setting up new services and handling any difficulty or particularity.

The Hyper-Text Mark-up Language (HTML) is used for the system interface while the eXtensible Mark-up Language (XML) seems to be more appropriate for setting up the system (managing the information) against the former because it provides a format for describing structured documents.

Security issues must also be addressed for the system. Both intranet and extranet are protected by firewalls against unauthorized use or hacking. A firewall is a computer tool, usually software, which isolates another computer from the Internet in order to prevent unauthorized access. Other security tools or methods such as encryption, authentication, digital signature, and message integrity can be used. In addition, the database and the server must be physically secured and protected. Extremely sensitive project information could be kept off the system.

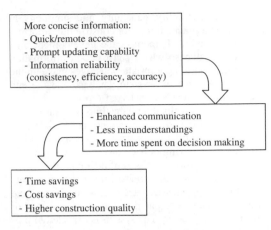

Figure 3.   Expected benefits of the system.

5   EXPECTED BENEFITS AND DRAWBACKS

The implementation of the proposed system in the construction process will produce a number of benefits, as seen in Figure 3. The system can provide more concise information to project participants compared to ordinary practice nowadays. As a result, communication among project participants will improve, misunderstanding will be reduced or eliminated, and more time will be allowed for decision-making. Ultimately, this will lead to time and cost savings, as well as to higher construction quality.

On the other hand, some problems may arise. In particular, despite all protection policies adopted, security remains an issue. The problem derives from the fact that the system facilitates information flows. A leakage of the system can result in large amounts of information being hacked in a short time.

Information owning must also be addressed and clearly defined, otherwise conflicts on distributing authorization may appear. As construction industry is rigid and strictly traditional, the acceptance of such a system may prove difficult. In addition, the necessary cost for setting up and managing the system as well as for training the personnel in its use may hinder the acceptance of the system in practice.

6   CONCLUSIONS

Construction is one of the most information-dependent industries, mainly due to its extended fragmentation. As construction projects become more complex and companies extend their works internationally, the need for efficient information management seems imperative.

In this paper, several design issues of an integrated information management system are presented. The system aims to conform to the basic principles of integration, flexibility, generality, and simplicity in order to be a practical tool for construction information management. It requires the development of a structured database where construction information is stored. A set of appropriate web pages make up the interface between the system database and the end users. Information is transferred through web interfaces or by downloading/uploading software files. Communication is served by an Intranet for information sharing within the construction company and an extranet for allowing outside projects participants to have access to specific pieces of information.

With such a system, project information is widely available (within the desirable access constraints) and easily obtainable compared to the traditional way of information exchange. The system is expected to provide more concise information to all project participants. This will lead to enhanced communication among them and less misunderstandings. In that way, time and cost savings as well as higher construction quality are expected.

REFERENCES

Dawood, N., Akinsola, A. & Hobbs, B. 2002. Development of automated communication of system for managing site information using Internet technology. *Automation in Construction* 11(5): 557–572.

Garcia, C., Garcia, G., Sarria, F. & Echeverry, D. 1998. Internet-based solutions to the fragmentation of the construction process. *Proceedings of the Congress on Computing in Civil Engineering, Reston, VA, USA, 1998*: 573–576.

Mokhtar, A.H.M. 2002. Coordinating and customizing design information through the Internet. *Engineering Construction and Architectural Management* 9(3): 222–231.

Opfer, N.D. 1997. Intranet–Internet applications for the construction industry. *ASC Proceedings of the 33rd Annual Conference, University of Washington – Seattle, Washington, 2–5 April 1997*: 211–220.

Rezgui, Y. & Cooper, G. 1998. A proposed open infrastructure for construction project document sharing. *Itcon (http://www.itcon.org/)* 3: 11–24.

Shahid, S. & Froese, T. 1998. Project management information control systems. *Canadian Journal of Civil Engineering* 25(4): 735–754.

Sriprasert, E. & Dawood, N. 2002. Next generation of construction planning and control system: the LEWIS approach. *Proceedings IGLC-10, Gramado, Brazil, 6–8 August 2002*.

Tam, C.M. 1999. Use of the Internet to enhance construction communication: Total Information Transfer System. *International Journal of Project Management* 17(2): 107–111.

Thamhain, H.J. & Wilemon, D.L. 1986. Criteria for controlling projects according to plan. *Project Management Journal* XVII(2): 75–81.

Public–private partnerships for infrastructure development in Greece

F. Striagka
Aristotle University of Thessaloniki, Thessaloniki, Greece

J.P. Pantouvakis
National Technical University of Athens, Athens, Greece

ABSTRACT: This paper sets out to add to the growing European literature and empirical database on public–private partnership by reporting the findings of several Greek pilot projects selected from across the sectors of the Greek l economy. Following a review and critique of what has been achieved in the field in Greece during the last years, it includes a methodology of pilot projects' identification and their development by taking into account characteristics and special features of the Greek market by applying the experience of the UK PFI. Also, it presents, the critical legal issues and the EU funding issues considering the aspects and requirements of the specific PFI projects in order to be feasible and viable in the Greek market environment. It concludes by examining how the PPP alliance in Greek practice may change the main organizational, social and cultural features that encourage or inhibit collaboration transition to a global environment.

Public–private partnerships (PPPs), worldwide represent a unique and flexible solution to implement infrastructure projects. PPPs are not only an alternative to Government budgetary constraints: they are also the means to improve the quality and delivery of public services and can also achieve social and environmental objectives. The intervention by the private sector can reduce inefficiency and respond effectively to user demands. Given their lack of budgetary resources and the enormous needs in the transition economies, PPPs are a strategic necessity rather than a policy option.

Ever since the mid-1990s public–private partnerships (more complex in appraisal and implementation than alternative procurement methods like BOOTs or BOTs) have been poised to take Europe and the rest of the world by storm. The formalisation in the shape of the private finance initiative, and its subsequent success in the UK has certainly sparked enthusiasm for public–private partnerships all over the world and, hence more and more countries are using the private sector to provide public goods and services. The perfect answer to decades of infrastructural underinvestment, PPPs were heralded as the way forward by businesses and the business press.

As PPPs can release states' resources which can be used for other purposes, following the success of their implementation in the UK, the rest of Europe has began to take up the idea and more and more countries, like Ireland, Portugal, Italy and the Netherlands are using the private sector to provide public goods and services (United Nations 2002). The Greek Government has also become increasingly as an advocate of PPPs, especially during the last two years. This was due to the fact that the implementation of PPP projects consist a pre-requisite in achieving funding by the 3rd Community Support Framework.

Currently in Europe, the Republic of Ireland has the most developed PPP programme (United Nations 2002). The involvement of private business in improving the country's infrastructure has been seen by a number of analysts as a key feature in the Irish economic miracle of the 1990s and this has been due to what has been referred to as the almost 'aggressive' nature of the Irish state in approaching the British expertise for getting PPP programmes off ground. This use of British know-how does not stop at the use of British consultancy; Ireland, the Netherlands, Italy, and Greece, have all set up central PPP units upon the Treasury Taskforce model.

1 CONCESSION CONTRACTING IN GREECE

In general, to achieve meaningful growth, developing countries like Greece, have to promote infrastructure development, which has a positive 'knock on' effect in catalyzing continuous economic development, apart from meeting basic needs. However, in proceeding towards this goal, these sort of countries face various constraints, among which, lack of advanced technology and inadequate public financial resources are two major drawbacks. To overcome, or alleviate these constraints developing countries are encouraging local and foreign private sector involvement in the provision of infrastructure project or services. Global trends of privatization and reduced governmental roles extend to countries like Greece.

Bearing in mind that the economic health of a country can to some extent be measured by the level of investment in its infrastructure, in the early 1990s, the Greek government has launched concession contracting in form of DBFO or BOT by promoting the implementation of four large scale complex infrastructure projects:

- the New Athens International Airport El. Venizelos (agreement signed in 1995) (Grant: 6.5% + 11% EU Cohesion Fund, State guarantees loans by EIB)
- Rio-Antirio Bridge (agreement signed in 1996) (Grant: 40%, standby loan by government-available during operation in case of lower than expected traffic)
- Attiki Odos (agreement signed in 1996) (Grant: 33%, State guarantees loans to EIB and other banks, from the completion of construction works onwards)
- Thessaloniki Metro (agreement signed in 1999) (Subsidy: 20% + ticket subsidies of 215% for number of passengers less than 30 million per year)

Only in the case of Attiki Odos the concessionaire comprises almost exclusively by Greek Contractors. The involvement of domestic contractors in Rio-Antirio project rises to 47% of the total concession company. However, in the case of the new Athens International Airport, local contractors participate only as sub-contractors. Finally, concerning the project of Thessaloniki Metro, which has not yet initiated due to a dispute between the two major concessionaire candidates, there is a possibility that sub-contracting is going to be assigned to Greek construction firms. Simultaneously, two parking garages in the city centre of Athens were also constructed under concession contracts as well as a new cemetery in Pireaus.

This positive improvement was the result of the macro-economic environment improvement, the country's accession in European Monetary Unit, and the EC policy objective to promote public–private partnership.

Advantages to be found in DBFO or BOT schemes include innovation and increased efficiency in services, possible reductions in unit cost of public services, technology transfer, development of local capital markets and transfer of risks to the private sector. However, BOT schemes are not a panacea. Owing to the many inherent uncertainties and risks projects of this kind cannot be successfully implemented unless the host government gives necessary support, ensures the right political and commercial environment and provides minimal guarantees to maintain a balanced risk-return structure (Ozdogan 1998). If one or more risks is not properly addressed, they could lead to under-achievement of the objectives, or even total failure of the projects.

Hence, experience in countries using the BOT procurement method demonstrates that the BOT formula becomes workable only if a strong legal basis and regulative framework exist plus adequate risk allocation between host government and project company (Ozdoganm et al. 2000). The reasons behind the lag in realization of BOT projects in Greece are similar to those presented in other developing countries as well and can be listed as unwillingness of the Government to provide guarantees against country risks, lack of adequate legislation, inexperience of the Government in packaging BOT projects, ineffective tendering and award mechanism, and a high level of bureaucracy resulting in delays.

Although the enthusiasm about concession schemes continues, the practical manifestation of a European PPP market appear to have stalled and the reality is much less dynamic. The British experience has been studied and used to inform each country's approach as well as Greece approach. However, this does not mean that home-grown knowledge should not be used. Greece has its own needs and economic culture, and so must tailor its PPP schemes to its own requirements. The issue of culture should not be overlooked, and, indeed, can sometimes present a barrier to establishing a sophisticated public–private partnership program.

Taking this into account, Greece is a country where the public and private sector have had little experience of working together (Simantira 2000). Thus, a change in culture should be made by both in order to form an effective partnership. The public sector has to accept that commercial business is essentially profit-driven and will want significant returns from the operation of infrastructure, whilst on the other hand, the private sector has to acknowledge that many features of operating public infrastructure are not profitable. In the UK this barrier was quickly overcome, but it has not been so straightforward for other European states like Greece. The problem of cultural readjustment has further implications as it might be a reason why foreign contractors may be reluctant to involve themselves in projects where PPPs are only just starting to take root due to the fact that the cost in getting through the learning curve is very high.

Since the Greek government and the private sector are both on learning curves, there are no records of successfully applied risk allocation principles. As a result, lengthy negotiations are the norm rather than the exception in the promotion procedure of BOT projects. Even if the various government departments do become more efficient in their procurement function, contractors and concessionaires are concerned about the high cost of preparing bids as well as the usual lengthy negotiations involved in the process (Simantira 2000). Such costs are normally not reimbursable and therefore the actual capital required, in case of successful bidding, must include the amount of large preparation costs on top of the equity stake. However, these costs are expected to be decreased as the firm becomes more experienced in bidding for particular types of projects. As there is an increased need for working capital under concession contracting unlike the conventional contracting business activities, there is a possibility that contractors may see their equity stake in concession projects as a means of counterbalancing the effects risen by intense fluctuations in construction demand. Some, on the other hand, may consider the opportunity costs of tying up capital as equity. Consequently, they may wish to sell their equity stake at some point after the initialization of operation, although there is a risk involved: though it is reasonable to expect that there will be a market for selling such equity, such markets are not necessarily established (Ive & Edkins 1998)

Once the learning curve has been overcome through, the potential for profit making emerges. The Europeanisation of PPP models is evidence of this, for not only are British firms seeking to involve themselves in the provision of public services abroad, but also world-wide firms are starting to get in on the act. Apart from the financial implications quantitative risk analysis brings, qualitative risk analysis on issues like organizational culture and structure may play an extremely important role concerning risk measurement and management (Kumaraswamy & Zhang 2001). Experience from the past has shown that the Greek construction companies are not compatible with the functions of facilities' operations and of service provision. The Greek construction sector has always been and still remains extremely fragmented. Contracting firms are too numerous and too small even for national standards. The underlying reason for the fragmentation is the family-oriented nature of the firms in accordance with a plethora of engineering professionals, who all want to exercise their enterprising skills. This kind of family-type organizational structure used to be quite common throughout the country's business environment and still poses major obstacles regarding rapid economic development of the country's economy. However, in recent years, due to the relative scarcity of work compared to the amount of contractors, which in turn cultivated intense animosity especially among larger contractors leading to destructive competition, contractors themselves started to structure permanent coalitions with each other.

This change was encouraged by the Government by making all the necessary legislative adjustments to support this effort. Also, the emerging markets of Balkans accelerated the process of Joint Ventures formation due to strategic purposes. Concession activity requires longer term involvement and commitment in a particular venture as well as revision acquirement of specialised resource utilisation.

Another additional fact responsible for the slow implementation of concession schemes in Greece compared with the UK approach is that laws tend to be far more prescriptive, precisely governing the relationship between the public and private sectors. A characteristic issue that must be mentioned is that concession contracts in Greece are granted by means of statute (Hatzopoulou-Tzika 1994). The first serious complication is obvious: any alterations to the initial agreement can only be made through further legislation ratified by a presidential decree. Another possible complication (which is not the case when concessions are granted under a private contract) concerns the third party action. If a contract is granted by means of statute, third parties may potentially appeal to courts 'for an order requiring the government to comply with the provisions of statute or to perform certain public duties' (Walker & Smith 1996). Third party intervention has occurred in Attiki Odos (appeal of Athens American College against the State Council concerning environmental issues) and in the project of Thessaloniki Metro when a disqualified bidder appealed against the State Council and later the case was proceeded to the European Court of Justice. Those interventions, of course, did not cancel the contracts, but obviously it seriously delayed their progress.

Another major issue for the Greek contractors is whether their firms' attributes are suited for concession type projects in addition to the kind of the role they would pursue in the process taking into account the various parameters that determine the Greek market. The key issues against which the sector's strengths and weaknesses are to be evaluated can be separated into the following categories:

1. Requirements for successful bidding
2. Requirements for successful completion of concession projects
3. Managing the effects of concession projects' undertaken on the construction firms business operation.

Adapting the six Critical Success Factors (Tiong et al. 1992), the feel in the Greek Contractors' market is that although they consider the concession market as an opportunity there is an entry barrier due to their small size and traditional organizational structure in accordance with their relatively low experience in concession schemes. Also, taking into consideration the requirements and demands of a concession contract regarding the multi-disciplinary know-how and the aspect of finance, it is crucial for them to form a strong team of stakeholders in order to come up with innovative and competitive proposals. However, their local knowledge of administrative and economic technicalities helps them in having a more realistic approach in evaluating risk and negotiating the risk allocation between them and the State (Simantira 2000).

Experience from Attiki Odos and Rio-Antirio Bridge has demonstrated that the culture of inefficiency that characterised the Greek contractors during the past years can be easily abandoned when the proper incentives are present. In the past, procurement procedures used to allow for extreme discounts (often higher than 60–70%) and the contractors faced with fierce competition and they systematically took advantage of the legislative inadequacies to ensure work at any cost. With the introduction of concession scheme and with the obligation generated from EU legislation Greek contractors have started creating an efficient operational management and to change their culture.

During the mid-1990s, the sector's poor performance led institutional investors to move on to shares' disposal as they did not encounter any trust to the market. The situation has gradually started to change, mainly due to sector's prospect of growth resulting from the new stream of public works of the 3rd CSF and of Olympic Games. Construction companies seem to enjoy more favourable attitude on behalf of the investors although the whole construction industry is currently at a transitional stage in terms of restructuring the contractors' organizations. However, apart from difficulties in equity provision, contractors face the problem of limited bank credibility due to their lack of experience in concession schemes and to their transitional stage of existence.

In order to create a new area of business opportunity, taking into consideration the general macroeconomic situation and the needs of the country, the Greek authorities and the European Commission have agreed within the Community Support Framework 1994–1999 (CSF II) and CSF III to maximize private sector partnerships in the development of transport related to the infrastructure, improving the competitive position of Greece in the global market environment.

In order to increase the flow of the private sector funds to road transport investments as was described above, a field that is known better, the Greek authorities are already in a procedure, which will lead to the concession of rights regarding these investments, in free market conditions. The procedure is aiming at establishing a proper institutional environment for the satisfactory and unobstructed operation of the private sector, subject to the proper supervision of the public authorities, and ensuring that these investments will generate the respective socio-economic development and benefits.

The Greek State without obstructing the on-going development of public works already under way, will choose private sector sponsors who are willing to commit the necessary funds and other resources to the construction of motorway sections not yet constructed and to the operation of sections that are already completed and will be included in the concession scheme. Such an arrangement will facilitate repayment of their investment through tolls and, where necessary, performance-related payment from the State.

The use of private funds contributes to the attainment of the target of completion of the trans-European road corridors in Greece, with utmost priority assigned to the road axes of the PATHE, Egnatia Odos, Korinthos – Tripoli–Kalamata and Ionia Odos.

The Ministries of National Economy (MNEC) and Environment, Physical Planning and Public Works (MEPPPW) had the opportunity to formulate a preliminary view of the concession schemes expected to be tendered in the near future. In October 7th 2000 the MEPPPW published in the Official Journal of the European Communities (OJEC) the preliminary announcements for the concession schemes tendering.

The Mixed Steering Committee and TENs Hellas have proceeded in providing necessary information to market participants about the schemes in order all interested parties to be aware of the projects in sufficient detail and to enable them to formulate their business and financial plans taking into account the opportunity of participating in one or more of the concession schemes. This information included data related to the maturity of projects, the technical difficulties involved for their implementation, the anticipated traffic volumes and potential uncertainty and risks involved, as well as the geographic location, progress of works, status of associated motorway sections, technical description of the sections to be included in the concession schemes etc.

Also, potential concessionaires have been informed about the State's intentions concerning legislative amendments, the risk allocation among the parties that would be involved in the various concessions, the tendering process and the main terms of the concession contract.

TENs Hellas (TENs Hellas & Mixed Steering Committee 2000) has examined more than 45 possible concession combinations in terms of construction cost and tolls combination. The responsible Authorities have given additional information and instructions mainly on the alternative scenarios concerning the amount of toll fare and the demanded financial public support. The adoption of a strategy concerning the selection of the toll fare method was an extremely sensitive issue with political and macroeconomic implications. Hence, TENs Hellas has implemented a series of alternative applications of the financial model with toll fares which result to the maximum return of income and with stable prices per km. The results of this financial model for different combinations and for the number of scenarios have been studied in relation to the priorities set by the CSF Operational Programmes. Also, it was taken into account the description of the geometrical characteristics, the current situation of different sections, construction cost estimate, timeschedules for the completion of the required expropriation and land acquisition, the design stage and possible problems that can arise as well as traffic loading data and elements in relation to the operation.

Thus, the responsible Ministries have concluded to the following six concession combinations:

1. Maliakos–Klidi.
2. Athens–Korinthos–Patra, which includes Patra–Pirgos–Tsakona as well as the infrastructure works for Korinthos–Patra railway.
3. Korinthos–Tripoli–Kalamata including the branch Lefktro–Sparti.
4. Antirrio–Ioannina (excluding construction works of Agrinio & Arta–Phillipiada by-passes) and Athens–Maliakos (which does not include any construction works) and the Schimatari–Chalkida branch of PATHE.
5. Central Greece motorway (based on a combination of real and shadow tolls or availability payments).
6. North extension of the eastern branch of the Imittos Western Peripheral Avenue, the Vouliagmenis Avenue–Spata airport I/C on the Athens Ring Road and the south section of the Imittos Western Peripheral Avenue.

Some of the key objectives of the Greek State with regard to road transport infrastructure development according to TENs Hellas and to the Mixed Steering Committee responsible for the concept of concession contracting in Greece are the following:

– The expedient completion of the main motorway axes of Greece.
– The reduction or, in certain cases, the elimination of the State funding for the construction, operation and maintenance of the axes.
– The maximization of the private sector participation, whose funding capability would be used to design, construct, maintain and operate the motorways in a manner that would ensure sufficient safety levels, minimum environmental implication and would offer high level of service to the users, without affecting the financial stability of the projects.

The projects' tendering is going to be executed via the negotiated procedure, in order the State to have the maximum flexibility concerning the best value for money.

All the concession schemes to be tendered during this phase will most probably launched simultaneously. The pre-qualification stage for the tendering of the projects is designed in a simple way in order to cover the most essential issues as well as the legal requirements and at the same time to minimize the cost for the interested parties to participate.

Following the pre-qualification stage, the State has reviewed a number of alternatives regarding the tendering process for the award of concessions.

According to the maturity of the technical studies of the various sections to be constructed, a phase for the compilation of studies will commence following the pre-qualification stage. More specifically, if pre-studies have not been prepared by the State, pre-qualified groups will be asked to assist in the preparation/execution of these studies. The pre-qualified groups will base their pre-studies on reconnaissance studies made available to them, as well as on instructions received by the State. The scope of this phase is to reduce the period between the invitation to tender, since the latter has to follow the approval of the environmental terms, a pre-requisite for the preparation of the environmental study and pre-study. Furthermore, tenders, through the preparation of pre-studies, will have the chance to develop and propose solutions that would minimize the whole-life cost of the projects and to facilitate the submission of their technical bid, which will follow. The tendering groups will not be evaluated based on the pre-studies that the pre-qualified groups will submit.

After the pre-studies submission, the Greek state's technical advisor will prepare the final pre-study and the environmental impact assessment in order to obtain official approval. Following the environmental terms approval or after the pre-qualification stage, the pre-qualified groups will be invited to submit their technical and financial bids based on the outcome of compiled pre-studies by the State's technical advisor. The evaluation of technical and financial bid will follow in a single stage in order to simplify the process and minimize the duration of the tendering process.

For those schemes where a reconnaissance study is also required, there will be an additional stage involving a similar process described above.

Alternatively, two more processes are under discussion and evaluation.

1. Submission of technical offer followed by the evaluation of technical solutions. It is intended that the evaluation of the technical offers will be performed on a pass/fail basis. The financial bid followed will enable the Greek state to make an evaluation of a number of bid options in order to identify and choose the most advantageous one.

2. Submission of technical offer and financial bid followed by evaluation of technical and financial bids and selection of candidates by best and final offer process (BAFO).

The Greek State has not decided yet on the precise awarding process. However, it is estimated that the whole procedure according to press articles is going to start after the Olympic Games. Main reasons for this delay is the Construction firms' overload due to simultaneous execution of CSF III projects and lack of financial credibility due to their στενότητα.

2 CONCLUSION

As to why more and more countries are looking to public–private partnerships, the answers are manifold. Most obvious is the economic argument. Governments want to reduce public spending. However, despite the fiscal restrictions of the Greek state for the past decades, the use of private finance for the provision of public goods and services has been relatively low. Many factors described in this paper have contributed to this hesitation on behalf of the Greek contractors to participate in the implementation of concession projects. It is this increased uncertainty regarding the size, characteristics and the content of the market that obstructs the rapid development of the sector in this field.

Change of culture for all the involved parties and particularly for the private and public sector need to accelerate. Contractors must restructure their organizations and their scope of works otherwise they may find themselves increasingly excluded from the construction demand arising from concession schemes and perhaps may limit themselves to subcontracting activities. On the other hand, the public sector must take all the necessary actions in order to provide proper institutional environment for the satisfactory and unobstructed operation of the private sector in the PPPs implementation.

REFERENCES

Hatzopoulou-Tzika, A. 1994. *Public Works Construction: National and E.U. legislation*, Athens: Papasotiriou.

HM Treasury Taskforce 2000. *Implementing Baker: Developing the Bridge between Public Sector Science and the Market*, London: Partnerships UK.

Ive, G. & Edkins, A. 1998. Constructors' Key Guide to PFI. *Construction Industry Council*, London: Thomas Telford.

Kumaraswamy, M.M. & Zhang, X.Q. 2001. Governmental role in BOT-led infrastructure development. *International Journal of Project Management*, Volume 19: 195–205.

Merna, A. & Smith, N.J. (2nd ed.) 1996. *Projects procured by privately financed concession contracts*. Hong Kong: Asia Law and Pracice.

Ozdoganm, I.D. & Birgonul, M.T. 2000. A decision support framework for project sponsors in the planning stage of build–operate–transfer (BOT) projects. *Journal of Construction Engineering and Management*, Volume 18: 343–353.

Simantira, V. 2000. The Market for Concession Contracting in Greece as a business opportunity for major domestic contractors, MSc Thesis, London: UCL.

TENs Hellas & Mixed Steering Committee 2000. Motorway Concession Contracts in Greece. Conference Proceedings, November 2000, Athens: Ministry of National Economy.

Tiffin, M. & Hall, P. 1998. PFI – the last chance saloon? In Civil Engineering (ed.); *Proc Instn. Civ. Engrs, London 12–18 February 1998*, London: Civil Engineering.

Tiong, R.L.K., Yeo, K.T. & McCarthy, S.C. 1992. Critical Success Factors in Winning BOT Contracts. *Journal of Construction Engineering and Management*, Volume 118 (Number 2): 217–229.

United Nations 2002. A Review of Public–Private Partnesrships for Infrastructure Development in Europe. United Nations, Economic and Social Council, Economic Commission for Europe, Committee for Trade, Industry and Enterprise development, Working Party on International Legal and Commercial Practice, New York: United Nations.

Walker, C. & Smith, A.J. 1996. Privatized Infrastructure: the BOT Approach. London: Thomas Telford.

A Microsoft Project add-in for time-cost trade off analysis

K.P. Anagnostopoulos
Dpt. of Civil Engineering, Democritus University of Thrace, Xanthi, Greece

L. Kotsikas
Civil Engineer, Democritus University of Thrace, Xanthi, Greece

A. Roumboutsos
Dpt. of Shipping, Trade and Commerce, University of the Aegean, Chios, Greece

ABSTRACT: A new software application program is presented, which permits to solve time-cost trade off problems of a project scheduled in Microsoft Project, in the case that discrete time-cost combinations are allowed on the project activities. Developed entirely in Visual Basic, the program works as a Microsoft Project add-in, in which it creates an integrated user interface (views and tables, a command menu, toolbars, a shortcut menu, controls etc). The program is based on two algorithms for solving the problems: (i) A genetic algorithm, that uses a one-point crossover based on the critical activities and a rank probability distribution to determine the parents for crossover; (ii) Two heuristics, that construct feasible solutions by compressing and relaxing the duration of critical and noncritical activities. In order to improve the time efficiency of the algorithm, the heuristics contain a subroutine that recalculates the critical paths by restricting the calculations to a partial network.

1 INTRODUCTION

The importance of the time-cost trade off problem was recognized over forty years ago, since the initial development of project scheduling techniques (Fulkerson 1961, Kelly 1961). Given that the duration of a specific activity is reduced but its direct cost is increased, as additional resources are required for its execution, decisions must be made on the execution of activities in order to balance the project total duration and the project total cost. This problem, known in the literature as the "time-cost trade off" problem, has become a central issue in construction planning and control (Liberatore et al. 2001, Shtub et al. 1994).

Whilst time-cost trade off problems have been extensively studied in the case of continuous time-cost relationships, little has been achieved in the more realistic case of the discrete time-cost trade off problem. The discrete time-cost trade off problem is a hard combinatorial problem, in the sense that exact procedures can only resolve problems of small dimensions (De et al. 1995, Demeulemeester et al. 1996). Consequently, efforts have been focused on the development of heuristic and metaheuristic algorithms (Anagnostopoulos & Kotsikas 2001, De et al. 1995, Feng et al. 1997, Li & Love 1997).

This paper presents a Microsoft Project add-in, named CRASH ADD-IN, which enables the solution of time-cost trade off problems for a project scheduled in Microsoft Project, in the case that discrete time-cost combinations are allowed in the project activities. The add-in is based on a hybrid genetic algorithm and heuristics to resolve the time-cost trade off problems. This user-friendly program enhances Microsoft Project with one of the most important aspects of construction planning and can be used both as a support for educational purposes and as a management tool for medium size real life projects.

2 THE TRADE OFF PROBLEMS AND THE ALGORITHMS

The CRASH ADD-IN resolves the three trade off problems described below. Assuming that $G = (N, A)$ is the activity-on-arc network of a project, where N is the set of n nodes, representing events and A the set of m arcs, representing activities. An activity may be executed by various discrete time-cost combinations. The direct cost is a non-increasing function of time. Let K_{ij} denote the set of all feasible time-cost combinations (d_{ij}^k, c_{ij}^k) for the activity (ij) with start event i and

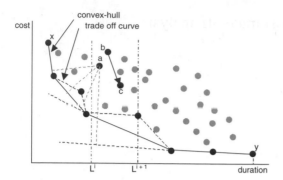

Figure 1. Trade off curve and convex-hull.

terminal event j, where d_{ij}^k is the duration and c_{ij}^k is the (direct) cost of the activity (ij) when it is performed with the combination k. Assuming that time-cost combinations for each set K_{ij} are ranked in increase order of magnitude as for the duration d_{ij}^k, i.e. if k and l are combinations such that $k < l$ than $d_{ij}^k < c_{ij}^k$ and $c_{ij}^k > c_{ij}^l$.

A network schedule (or solution) s is the network G containing an allowable duration in each of its activities. The network schedule has a total duration d^s equal to the longest path from node 1 to n, and total cost:

$$c^s = \Sigma c_{ij}^k \text{ for each activity } (ij) \text{ performed with the combination } k \text{ from } K_{ij}.$$

The maximum and minimum duration schedules are the two extreme situations. The network s^{max}, with the maximum selected duration d_{ij}^k for each activity, has the minimum total direct cost c^{min} and the longest total project duration d^{max} (point y, Fig. 1). In contrast, the network s^{min}, with the minimum duration d_{ij}^k selected for each activity, has the maximum direct cost c^{max} and the minimum total project duration d^{min} (point x, Fig. 1).

The CRASH ADD-IN provides a solution for the following three trade off problems (time given in dates):

- For a given total duration D of the project, find a solution s such that $d^s \leq D$ and c^s being minimal. The problem is solved with the heuristic H1.
- For a given budget B, find a solution s such that $c^s \leq B$ and d^s being minimal. The problem is solved with the heuristic H2.
- The generation of the project trade off curve, i.e., the efficient time-cost curve showing the relationship between project duration and cost over the feasible solutions. The trade off curve is the so called pareto front (or the non-dominated set), i.e., the set of solutions such that no other feasible solutions exist that have better objective values in both time and cost than the solutions in the nondominated set (Feng et al. 1997). Once the trade off curve has been established, the convex hull may be determined,

i.e., the subset of the trade off curve which forms the convex lower limit of all feasible solutions (Fig. 1). The trade off curve is generated by the hybrid genetic algorithm HG. It should be noted that the trade off curve solves the first two problems as well as the problem of minimising the project total cost addressing both the direct and indirect project cost.

2.1 The heuristics H1 and H2

Starting from the network s^{min}, the heuristic H1 generates successive feasible solutions $s^{current}$ such that $d^s \leq D$, where $D < d^{max}$. A new solution s^{new} is defined in the neighborhood of the current solution as follows: At least one activity must be relaxed to reduce the overall direct cost. If $d_{ij}^{current}$ is the current duration and d_{ij}^{new} is the relaxed duration of the activity (ij), then $\Delta d = d_{ij}^{new} - d_{ij}^{current}$ is the increase in the duration of activity (ij).

- If (ij) is a critical activity, then total project duration is increased by Δd, if $d_{ij}^{current} + \Delta d \leq D$.
- If (ij) is not critical, then: (a) if the total slack of activity (ij) is greater or equal to Δd, the duration of activity (ij) may be increased and the solution remains feasible. (b) If the total slack of activity (ij) is less than Δd, the duration of activity (ij) may be increased only if the solution remains feasible.

The process is repeated until no other activity may be relaxed without exceeding the project target duration. When the activity that provides the greatest improvement in cost is found, the network is recalculated. Suppose that activity (ij) is performed using the next greater duration d_{ij}. Calculations can be drastically reduced based on the following observations:

- In calculating the required time to complete the project, it is sufficient to calculate the early times $E(p)$ of the nodes $p > i$. The early time of nodes p, $p \leq i$, do not change when the duration of activity (ij) changes. Consequently, on average, only half of nodes early times have to be recalculated.
- The late times are calculated following the calculation of the early times. In the case, the late time $L(p)$ for each node $p > i$ is increased by the increase in the overall project duration. The late time of the remaining nodes is calculated as usual.

The heuristic H2, which minimizes project duration for a given project budget B, is similar to the heuristic H1. Starting with the network solution s^{max}, the heuristic H2 generates feasible solutions s such that $c^s \leq B$ by crashing the duration of critical and non critical activities.

2.2 The genetic algorithm YG

Genetic algorithms, developed initially by J.H. Holland, mimic the mechanisms of natural selection on computer.

They have been applied with satisfactory results in various combinatorial optimization problems and in engineering applications. A main characteristic of genetic algorithms is their problem independence, i.e., they only need a coding and minimum information taken from the structure of the particular problem. Hybridization attempts to enhance performance by imbedding into a genetic algorithm, for example, a problem specific heuristic (Aarts & Lenstra 1997). This approach is applied in the following algorithm YG, which generates the trade off curve.

Algorithm YG

Step 1. Generate an initial population P_0 with p solutions. Set *generation* = 0.

Step 2. Use heuristic H1 to substitute the set of p solutions in the population with p local optima. Determine the trade off curve and the convex hull of the population P_0.

Step 3. Increase the population by adding q offspring solutions (the population size becomes $p + q$).

Step 4. Use heuristic H1 to substitute the set of q offspring solutions by q local optima. Determine the trade off curve and the convex hull of the current population.

Step 5. Set *generation* = *generation* + 1. Create the new population $P_{generation}$ by selecting p solutions from the current population.

Step 6. Repeat steps 3 to 5 until *generation* $\leqslant$ *maxgenerations*.

An individual or solution of the initial population is generated by selecting randomly a time-cost combination $(d_{ij}^k, c_{ij}^k) \in K_{ij}$ for each activity (i, j). An individual corresponds to a network schedule and is represented by an m-dimensional table, where the index of the combination k is placed in the cell representing activity (i, j). The initial population, formed by p-2 randomly selected individuals and the two extreme network schedules s^{max} and s^{min}, evolves until the predefined number of generations *maxgenerations* is achieved.

In steps 2 to 4 of the algorithm, each individual is improved locally by using the heuristic H1. The segment (d^{min}, d^{max}) is divided into L equal segments. Let $S(L^i, L^{i+1})$ be the set of offspring solutions s^r with total duration d^r in the segment $[L^i, L^{i+1}]$: $L^i \leqslant d^r < L^{i+1}$ (Fig. 1). The heuristic improves the offspring solution s^r by defining a new total project duration L^{i+1} and generating a new solution s^{new} such that $d^{new} \leqslant L^{i+1}$ and the cost c^{new} being minimum (point b, Fig. 1).

Following this, the population is increased by adding q offspring solutions. The fitness value is equal to $d_i = \min(d_{ij}, \text{for all } j)$, where d_i = minimum distance between the solution i and each segment j of the convex hull (point a, Fig. 1) (Feng et al. 1997). The individuals are, then, ranked in descending order according to their minimum distance d_i and their

potential parents are selected by applying the rank probability distribution

$$p_r = \frac{2r}{r_{max}(r_{max}+1)}$$

where r is the individual order and r_{max} is the order of the best individual. In this selection mechanism the probability of selecting the best individual is double than in the case of the median, where the selection probability is $1/r_{max}$.

Once the two parents are defined, one is randomly selected. The crossover point is defined for the selected parent for the location where half of the critical activities are located. The first block of chromosomes of the one parent and the second of the other parent are used to form the offsprings.

After applying the heuristic H1 for the substitution of the q offspring solutions by q local optima, the population is reduced to its initial size by selecting n solutions from the current population. The selection is elitist in the sense that the new population contains all the members of the current trade off curve.

3 THE CRASH ADD-IN

The CRASH ADD-IN coded in Visual Basic 6.0 as a .dll file, operates only through MS Project 2000. The program is automatically installed on the hard disc and activated by the MS Project command *COM Add-in*. A new *view* (*Crash View*) and a new table (*Crash Table*) are created to insert data, while a toolbar as well as a command and macro menu are provided (Fig. 2). There are two groups of columns in the *Crash Table*. The first group is used to present the basic characteristics of each project activity (number, name, activity normal duration and cost). The second group consists of five pairs of columns. Each pair corresponds to an allowable time-cost combination, which may be feasible for the activity.

1. The generation of the trade off curve is realized through the YG algorithm. The curve is constructed using the *Trade Off Curve* command from the toolbar or the *Crash menu*. The program runs on parameter values of the YG inserted by the user or program default values. The trade off curve point values may be saved in a .crp type text file by a name and at a location prescribed by the user. Once the process is completed, the program creates a MS Excel 2000 book, where the results, are presented in a diagram and table format (Fig. 3).

2. The crash cost for a given project completion date is calculated through the command *Set Finish Date*. The dialog box *Set Finish Date* (Fig. 4) informs the user of the early completion date, the late completion

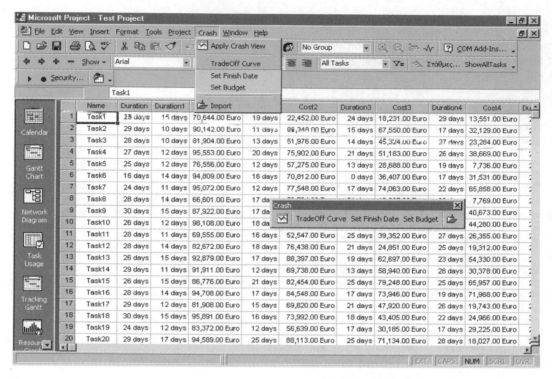

Figure 2.   The CRASH ADD-IN work environment.

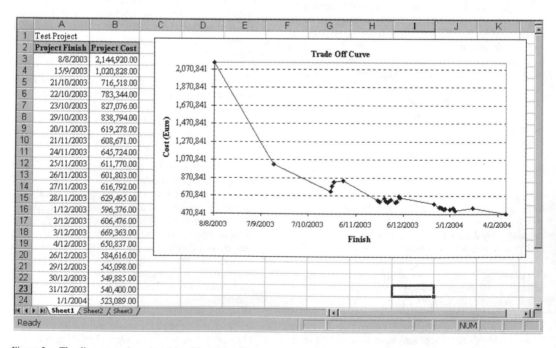

Figure 3.   The diagram and results table of the trade off curve.

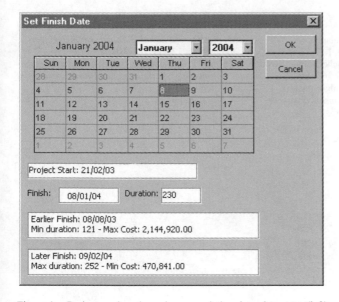

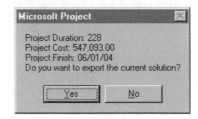

Figure 4. Project crash under a given completion date: data entry (left) and results (right).

date and the respective direct project costs, based on the given time-cost combinations of the activities. The user may either select the finishing date from the respective calendar, which is incorporated in the dialog box, or enter the date into the *Finish* text box in the form dd/mm/yyyy. The final solution is given directly and may be saved (Fig. 4, right).

3. The crash completion time for a given budget is calculated through the command *Set Budget*. The dialog box *Set Budget* informs the user of the least direct cost, the maximum direct cost and the respective completion dates based on the given time-cost combinations of the activities. Crashing is only possible if the set budget lies between the minimum and maximum cost. The final solution is given directly and may be saved.

4. As noted above the add-in provides the possibility of saving the solutions. Any solution saved as a *.crp* file may be applied to the project through the command *Import*. The program checks if the selected file corresponds to the current one with the particular time-cost combinations. In the list box of the dialog box lists all the saved solutions (completion date and respective direct cost).

4 CONCLUSIONS

In this paper a new software application program was presented, which permits to solve time-cost trade off problems of a project scheduled in Microsoft Project,

in the case that discrete time-cost combinations are allowed on the project activities. The add-in uses two heuristics and a hybrid genetic algorithm to confront these inherently difficult to solve problems. The heuristics provide solutions to crash cost problems for constrained completion dates and crash time problems for constrained completion costs. The hybrid algorithm combines a genetic algorithm used to evolve the population of solutions and a heuristic to optimize locally these solutions. This integrated (automatic installation, import and export of solutions etc.) and user-friendly add-in addresses one of the most important aspects of construction planning and control, and may be directly used for educational purposes.

REFERENCES

Aarts, E. & Lenstra, J.K. (eds.) 1997. *Local search in combinatorial optimization*. New York: J. Wiley.

Anagnostopoulos, K. & Kotsikas, L. 2001. Time-cost trade off in CPM networks by simulated annealing, *Operational Research, An International Journal* 1(3): 315–329.

De, P., Dunne, E.J. & Wells, C.E. 1995. The discrete time-cost trade off problem revisited. *European Journal of Operational Research* 64: 225–238.

Demeulemeester, E.L., Herroelen, W.S. & Elmaghraby, S.E. 1996. Optimal procedures for the discrete time/cost trade off problem in projects networks. *European Journal of Operational Research* 88: 50–68.

Feng, C.-W., Liu, L. & Burns, S.A. 1997. Using genetic algorithms to solve construction time-cost trade off problems. *Journal of Computing in Civil Engineering* 11(3): 184–189.

Fulkerson, D.R. 1961. A Network Flow Computation for Project Cost Curve. *Management Science* 7(3), 167–178.

Kelly, J.E. 1961. Critical path planning and scheduling: Mathematical basis. *Operations Research* 9(3): 167–179.

Li, H. & Love, P. 1997. Using improved genetic algorithms to facilitate time-cost optimization. *Journal of Construction Engineering and Management* 123(3): 234–237.

Liberatore, M.J., Pollack-Johnson, B. & Smith, C.A. 2001. Project management in construction: Software use and research directions. *Journal of Construction Engineering and Management* 127(2): 101–107.

Shtub, A., Bard, J.F. & Globerson, S. 1994. *Project Management: Engineering, Technology and Implementation.* London: Prentice Hall.

System-based Vision for Strategic and Creative Design, Bontempi (ed.)
© *2003 Swets & Zeitlinger, Lisse, ISBN 90 5809 599 1*

Perspectives' investigation of a systematic implementation of PPP/PFI model in Greece

F. Striagka
Aristotle University of Thessaloniki, Thessaloniki, Greece

J.P. Pantouvakis
National Technical University of Athens, Athens, Greece

ABSTRACT: This paper addresses the location of possible economic sectors of Greek economy where the PPP model can be implemented. It presents public sector investments, the implementation of which can be undertaken by suitably selected private sector bodies. As PPPs can embrace a wide range of structures and concepts, the research examines public works which are under construction or operation. The methodology process follows a certain structure, starting from a detailed analysis of programmed public investments and expenses, and going extensively through their prioritization taking into account predetermined criteria like level of investments, degree of investment repetition and attractiveness by the private sector. Collaboration processes are examined from the viewpoints of both the involved parties. The analysis of the data and the paper's conclusions stress some of the practical problems, limitations and paradoxes of partnering in Greek practice which prohibits the acceleration of the systematic implementation of PPP projects in Greece.

1 INTRODUCTION

Two arguments have traditionally been put forward for the initiative. The first refers to the PFI as an improved form of public procurement, which under the right circumstances could yield efficiency savings and value-for-money. The second major argument offered for the PFI is that it helps government overcome a perceived fiscal dilemma. In securing new financing for public investment, the PFI appears to allow government to reconcile the desire for higher capital spending with the commitment to maintaining a tight fiscal stance.

There are four possible reasons why PPPs (and, by future implication, PFI) have been introduced by the Greek government with enthusiasm. First, is the perceived political imperative of developing a new relationship between the public and private sectors, in an attempt to show that the traditional centre-left had dropped its historical ambivalence to the private sector and the profit motive. Second, it is hoped that the PPPs will improve the quality of public service, by increasing its efficiency through the introduction of managerial change and expertise drawn from the private sector. Third, as it was above mentioned, it is believed that the use of PPPs will help the government overcome a perceived fiscal dilemma. Last, but not least, the adaptation of the use of PPPs as an alternative public procurement method was set as a pre-requisite by the European Commission in order the relevant funding from the Third Community Support Framework for the period 2000–2006 to be awarded to Greece.

In general, it is rather difficult to give a clear and concise definition of the Greek PFI notion as it is currently at its embryonic stage, and this lack of clarity makes its evaluation not only complex but somehow impossible. However, in order to explore the suitability of the PPPs implementation in Greece the paper is divided into two substantive sections followed by a reflective conclusion. The first section will provide an overview of the emergence and development of PFI in Greece from its inception in 1998 to the present day to give a feel for the many parties involved and steps that have been made over the years followed in relation to the UK approach. The second section describes a methodology that may be followed in order to investigate the availability of specific sectors of the Greek Economy in terms of how actual PFI decisions should be made, paying particular attention to value-for – money and risk allocation issues. Simultaneously, it reviews the progress of the PFI implementation in Greece to date. This concluding section sets out a number of implications and questions on PFI in Greece as well as the greater flexibility needed to get it up and running.

2 THE EMERGENCE AND DEVELOPMENT OF THE PUBLIC – PRIVATE PARTNERSHIP IN GREECE

2.1 The precursor to PPPs (PFI)

The roots of PFI approach in Greece lies in the implementation of BOT procurement method in the early 1990s. Concession schemes in form of BOT (Build-Operate-Transfer) type have therefore provided an increasingly popular vehicle in facing the infrastructure bottleneck experienced in the country, since last decade, due to the increased investment undertaken by the Government with the support of the European Union. The government's major objective in the implementation of the BOT model has been the realization of urgent infrastructure projects, mainly of those satisfying the requirements of the TransEuropean Transport Networks, with the minimum possible financial burden and without affecting the limited borrowing capacity and raising income tax rates.

However, the philosophy and origins of concession schemes in Greece was not unknown. It can be traced back in ancient years, between the 4th and 5th century B.C., when Chaerephanes the Euboean signed a contract of service with the city of Eretrians regarding the draining and exploitation of the Ptechae Lake (Dystos Lake, Southern Euboea, 336-323 BCE). The agreement main clauses included the project funding by the contractor, Chaerephanes, who in return was granted the right to exploit the dried fields for 10 years, and the privilege of custom free import of all materials necessary to construction.

Nowadays, in the public sector field, traditional procurement methods are slowly being replaced by concession contracting. Within that aspect, in the mid 1990s, four major BOT projects were initiated in Greece, the New Athens International Airport, the Rio – Antirio Bridge, the Peripheral Athens Ring Road and Thessaloniki's Metro.

2.2 Early currently developments in PPPs (PFI)

With the initialization of the Third Community Support Framework (CSF) of period 2000–2006, the Greek government actively encourages the private sector to take the lead in joint ventures with the public sector. Thus, the public sector would have greater opportunity to use leasing where it involved significant transfer of risk to the private sector and offered good value for money (Private Finance Panel 1995, paragraph 2.4, p.7)

During the first semester of 1998, the Greek government, via the Ministry of National Economy, has initiated the procedure of an international competition for the selection of a Financial Consultant with international experience aiming at supporting the State in launching and shaping the structure and nature of PFI in Greece. TENs Hellas, a consortium of banks, technical and legal companies led by Bank of America and ETEBA and supported by Planet, KPMG and Marsh, was selected and appointed as the only Financial Consultant of the Greek state. TENs Hellas; role is to perform feasibility studies and cost estimations in the planning stage of a project before its tendering and it has been working on that issue since May 1999 with the support of GIBB Ltd, Dromos Consulting, Linklaters and the law firms Bernitsas, Kornilakis, Liakopoulos and associates (TENs Hellas & Mixed Steering Committee 2000).

The Ministry of National Economy moved on to the formation of a Private Finance Taskforce which was consisted of the Mixed Steering Committee for Public Works and TENs Hellas in order to control and organize the structural arrangements of the PFI procedure. Actually, the intension was a mixture of centralization the standardizing control of practice by the Ministry of National Economy and the Ministry of Public Works and the decentralization through encouraging Government Departments to build up their own expertise in PFI schemes by implementation. It was proposed by the international consultants that the Taskforce would have similar organizational structure to the British HM Treasury Taskforce. This means that the Taskforce will consist of separate "projects team" and "policy team". The projects team, a small group of leading PFI experts recruited from the private sector, would provide support for individual Departments and agencies on significant transactions ensuring high quality proposals, avoiding unnecessary bidding costs and signing off the commercial viability of significant projects before the procurement process commences. The policy team, consisting of high caliber public state staff, would have an on-going responsibility for the rules governing PFI and other public/private partnerships. To prevent repetition, overlap and inconsistency, the policy team in conjunction with the project team will continually develop and publish "standardized" models for key stages of the PFI procurement process. However, this sort of model has not worked yet due to the dismissed Mixed Steering Committee and to lack of regulatory context and legal framework concerning the PPP/PFI schemes.

3 CENTRALIZATION

One of the main working objects and issues of the Taskforce, that must be solved soon in order the PFI process to initiate in Greece, on one hand, is the determination of a legal framework and the development of standard procurement and contracting arrangements as well as education packages. On the other, it is presumably the resolution of how to account for PFI in order to have the best value-for-money. It is extremely vital the edition of a guidance which would act as a blueprint for the future development of PFI ensuring that future PFI contracts across different public services will be able to follow a consistent approach by incorporating standard conditions into the contracts. Finally, the developments in PFI in Greece, not only should be marked by standardization and centralization (Private Finance Treasury Taskforce, 1999) but also by an interest in value-for-money and particularly the role of the Public Sector Comparator (PSC) in making this judgment. As PFI is at its extremely early era in Greece, the PSC is seen to be the key to obtaining value-for-money.

Another important factor in the context of standardization will be the formation of a partnering body to the Taskforce like the British National Audit Office, which will undertake a series of examinations and value-for-money studies and drivers on PFI schemes under realization or on those having already been executed. The studies will concern the determination of terms during procurement and especially they will be looking for in judging whether the project is a viable deal as well as a prescription to those who are pursuing PFI contracts. More specifically, the body's investigation process should fall at least into four areas regarding "setting clear objectives", "proper procurement processes", "getting the best available deal" and "ensuring the deal makes sense" (National Audit Office 1999).

Obviously, the reports produced by this body would provide the opportunity to consider the cost savings for those PFI projects that the body has investigated bearing in mind that the cost savings will be the difference between the PFI cost and that of the Public Sector Comparator (Treasury Taskforce 1999). It should be reminded that the PSC is a fixture on the PFI landscape. A PSC provides a one – off snapshot of the value for money of a PFI project. It gives the answer to the question, given what we know now, is the PFI option likely to deliver best value for money? But in the absence of subsequent monitoring of contractor performance and contract payments against the assumptions made at the time the PSC was compiled, a firm view cannot be taken about whether value for money is being delivered in practice. To assure continuing value for money, public sector project managers will need to make full use of the provisions that now commonly feature in PFI contracts for periodic benchmarking of project costs. The purpose of this exercise is to provide a benchmark of the social, economic, and political objectives of the project against which agreement can be reached on affordability and the tenders evaluated. It will be essential in ensuring

that comparisons between the PFI model and the traditional model may be made on a quantifiable basis. The ongoing use of PSC's will require periodic review to ensure its continuing relevance and application as a benchmark. In principle, the best alternative available to the public sector should be used as the comparator. A point would be reached where in certain sectors it will be more valuable to compare a proposed PFI deal with the terms of previous PFI deals or an alternative benchmark based upon relevant private sector data and practice to delivering the service, rather than to a PSC (Anderssen & Enterprise LSE, 2000).

Also, the reports would illustrate various tensions between specifying services tightly enough to ensure that the bids received are comparable and yet allow scope for innovation or they can examine issues on competition as this is clearly seen as an important element in PFI. By nature of PFI schemes, there is a tension between the need for competition and the need for co-operation. In essence, PFI is seeking to build relationships based on competition, to avoid the alleged "comfortable" relationships fostered by a bureaucratic approach. Hence, reports may show the need that even in tightly specified contracts there should be a flexibility in order to maintain contractual relationships in a spirit of partnership.

4 THE CURRENT GREEK APPROACH

After a number of analyses conducted, needs of the transport sector in the context of the strategic development Plan for Transport Infrastructure – Greece 2010 were mostly determined. Subsequent Development plans have contributed in identifying the priority axes which are associated with the development of the country's basic road network, that affects the development of the country's major urban and economic centers, the country's regional development, and at the same time, conduce at the development of transport corridors facilitating access to/from the European Union and South – East Europe (TENs Hellas & Mixed Steering Committee 2000).

Six concession schemes have so far been identified by the Ministry of National Economy (MNEC) and the Ministry of Environment, Physical Planning & Public Works (MEPPPW) and formed the first group of schemes which have been announced and tendered according to a process which involves a pre-qualification stage and these are:

1. Maliakos – Klidi.
2. Athens – Korinthos – Patra,
3. Korinthos – Tripoli – Kalamata
4. Antirrio – Ioannina and Athens – Maliakos
5. Central Greece motorway
6. Imittos Western Peripheral Avenue

5 METHODOLOGY FOR UNDERTAKING A PILOT PROJECT

After the success of the 2nd Community Support Framework (CSF), the European Commission has granted approximately 28 billion Euro to the Greek State in order to support a great number of projects and activities in all sectors of the Greek economy for the period 2000–2006, which cover both the national needs, as well as the country's regional needs.

The national needs are described in Sectorial Operational Programmes under the responsibility of the relevant Ministries and include actions and activities which concern a specific field. In the meantime, the regional needs are described in Operational Programmes of the Regional authorities and include specific regional priorities and activities in different fields. Hence, the whole country is divided into 13 regions and 12 sectors, which are presented below:

Regional Operational Programs of:

Attiki	Epirus
Peloponnesos	Ionian Islands
West Greece	Sterea Ellada
Crete	East Macedonia
Central Macedonia	West Macedonia
North Aegean	South Aegean
Thessaly	

Sectorial Operational Programs

- Road Axes, Ports, Urban Development (MEPPW)
- Rail, Airports and Urban Transport (Ministry of Transport)
- Culture (Ministry of Culture)
- Information Society (Ministry of national Economy)
- Environment (MEPPW)
- Competitiveness (Ministry of Development)
- Health
- Education and initial professional education (Ministry of Education)
- Employment (Ministry of Employment)
- National Technical Assistance
- Fishery (Ministry of Agriculture)
- Agriculture Development (Ministry of Agriculture)

The European structural funds as well as the Cohesion Fund support and encourage the private sector participation by the implementation of PPP/PFI schemes. This would allow the member states to increase the available funding resources and to successfully enjoy the returns of this constructive and fruitful partnership between the private and public sector.

Each of the operational Programmes includes analytical information concerning the development needs of the 13 regions emphasizing on their special geographic characteristics, demographic trends, education status, economy, resources and employment, and it also presents a description of benefits and activities realized during the 2nd CSF. After a detailed analysis of the Programmes and the discussions followed with the responsible authorities and governmental staff, it was concluded that the prioritisation list reflects the list of the sectorial operational programmes. Thus, projects concerning environment, education, transport, sewage management, sanitation, health and culture are amongst the top priorities.

The research was based upon nine case studies of medium-to-large scale construction projects which must be undertaken by experienced contractors across a range of sectors within all industries. Case selection was based upon the following criteria:

1. Variation in project type – the projects' technical description and the sort of service that is supposed to deliver
2. Variation in project size –
3. Project's completion time schedule
4. Project maturity in terms of design and land expropriation
5. Amount of skills required for the construction, operation and maintenance of the project
6. Option of private finance and the project's prospect for profit-making
7. Partnerships – variety of involved parties
8. Legal issues – Estimation of required changes in the legal framework in order the concessionaire to function according to the Greek and EU legislation
9. Technical information – the project technical viability and assessment of environmental impact from its implementation
10. Environment – Progress of the required permits and licences in order the construction to proceed.
11. Financial feasibility – Indicative estimation concerning the income creation
12. Political Cost – reaction from the population

At this stage, it should be highlighted that the Operational Programmes included a plethora of axes and activities with no differentiation in size and scope of works. Also, it was noted that there was lack of evidence and data concerning the project's prospect for profit-making and assumptions regarding the operational cost. Finally, during the research, most of the operational programs were under continuous improvement and in a constant change.

Summarizing, the stages of the research included:

- Assessment of whether project is suitable for private finance
- Market consultation/testing
- Tender process

After a brief familiarisation and information review process, it was examined the universe of potential projects to select 2 or 3 for further scrutiny bearing in mind the Strategic Plan of the country for 2000–2010 and specifically the overall project objectives such as:

1. Usage requirements
2. The potential tariff which may be paid
3. The expected period for implementation

A key tool in the projects' scrutinisation was estimations which provided initial indications of the viability of the project, either in its own right or given the expected level of the Government/EU financial support.

Another key development in this phase was a draft allocation matrix, which defined the parties who would take various risks. There is a tendency in the public sector to try and transfer all project risks to the private sector, it has proven most economical to allocate risks to the party best able to manage it. If the private sector is forced to accept risks it can not manage, it is most likely to increase its price to protect its returns. The project should therefore reflect, a balanced risk allocation profile. This was a part of the project feasibility.

Also, it was taken into account the legislative framework concerning the PFI implementation mainly regarding taxation issues. Assumptions of changes were made before those projects go out to tender. A further action to undertake was a review of the relevant legal issues relating to the provision of EU funding alongside private sector funding.

6 THE USE OF PUBLIC SECTOR COMPARATOR

This phase was critical in confirming if the project remains consistent with the regulatory statutory and political requirements of the Government, whilst at the same time being commercially and financially supportable. Followed the case selection, a comparative analysis with the method of the Public Sector Comparator (PSC) was used to evaluate the projects procured either traditionally or by PFI scheme. The PSC model applies finance and operational assumptions in order to analyze the Net Present Value in relation to open financial flow and therefore evaluates by comparison of both systems the best option (Treasury Taskforce 1999). It must be emphasized that for the PPP/PFI scenario, the state provides annual payments in a form of rent or availability payment to the concessionaire as a return of service provision to the final users with the pre-requisite of satisfactory performance level. It is acknowledged that the amount of this payment equals to the amount needed by the concessionaire to reach the break-even point.

The assumptions made in the application of the PSC model relatively to performance levels were as follows and are presented in Table 1.

1. Operational performance 10%

Table 1. General and indicative funding assumptions for the research.

	Traditional Meth		PPP/PFI	
	Cost (%)	Capital (%)	Cost (%)	Capital (%)
Stakes	0		15–20	10
Private funding	0		9,0	40
Public funding	6,25	25	6,25	10
EU funding	6,25	75	6,25	40
EIB funding				

2. Increased income between 10%–20% (it differs depending on project's type)
3. Increase of 10% in basic capital expenses (PSC) in order the development cost and the cost of consultancy fees to be covered.
4. Overrun by 20%–30% of capital expenses
5. Nominal values
6. The public sector's lending cost is estimated to be 6,25% with a period of 20 years
7. The private sector's lending cost is estimated to be 9% with a period of 15 years
8. The cost of the private capital was estimated to be 17%.

It must be noted that all the assumptions made above were a result of interviews with banking financial advisers and with contractors rather experienced in undertaking PPP schemes. It is emphasized that the capital for PPP/PFI is indicative and depends on the projects characteristics and the prospect for profit-making. If there is a potential in profit-making then it is quite obvious that there would be greater private sector support. On the other hand if a project has a minimum prospect for profit - making, it is possible that the state may increase its capital participation. Support from the EIB has not been taken into account.

The model was quite simple due to the fact that there was a lack of financial information. It should be underlined that the procedure of the PSC model implementation was based on factors which can be quantified and to be expressed in financial terms. Other factors, like the quality of services, the leasing period, the politics are qualitative parameters which cannot be identified numerically. This note is extremely important concerning the PFI funding, bearing in mind that, in Greece, a great part of financial support comes from the EU in a form of zero cost funding. Although the EU resources are of zero cost, it is estimated that this form of funding has an opportunity cost. Therefore, the EU fund includes a cost of 6,25% (equals to the lending cost of the public sector). That would mean that there is a great possibility that the public sector funding would provide a lower NPV comparing to a PPP/PFI support. Of course, this does not mean that the use of private sector support is a non-valuable selection. On the contrary, the use of private finance means that the EU resources which were about to be used, they could be released and lead to social benefits.

Risk transfer can also be quantified using sensitivity analysis method in relation to the cost overrun (Anderssen & LSE Enterprise 2000). The value of the PSC is function of the assumptions quality. The model was based upon foundation assumptions relatively to the PPP/PFI schemes advantages. Although, the market experience was used to make all these assumptions, they were limited concerning their accuracy and uncertainty linked with the project type.

The conduction of the case study has shown that the PPP/PFI schemes have a greater funding cost as they require private capital and loans which have increased cost in relation to the lending cost. On the other hand, the public sector investments support capital expenses with huge low cost amounts or in a form of free funding (the EU structural funds) and/or public sector funding. If we attempt to compare the two procurement methods based only upon the financial cost, it is revealed that public sector support will be the best option. However, if evaluation is conducted in terms of total and combined benefits which occur from the PPP/PFI structure during the projects life cycle, then the PPP/PFI scheme has a competitive advantage. (Anderssen & LSE Enterprise, 2000).

7 CONCLUSIONS

To conclude we highlight a number of key issues and questions about the implementation of PFI scheme in Greece that remain to be addressed. In drawing these conclusions we are alert to the need to look to developments of PFI abroad. We are aware in this respect that the UK has been a key player in the policy transfer of PFI and as different nation with a different culture, evidence from the UK PFI experience will emerge.

A crucial question is how and who regulates the application of PFI schemes in Greece. In this paper, it has been emphasized the lack of legal framework and regulatory context which will form the base of the PFI initialization. It is well known that a particularly cooperative public-private partnership (PPP) is a precondition for successful project procurement.

Specifically, this lack of adequate legislation is due to disagreements raised between the Ministry of National Economy and the Ministry of Environment,

Physical Planning and Public Works. According to press publication, it is estimated that most of the six concession schemes will start at the end of 2004. The main reasons behind this delay are the difficulties in ensuring financial support as banks are in a period financial tightness and the simultaneous execution of the Olympic Games projects and the infrastructure projects funded by the 3rd CSF.

Also, in this paper, it was demonstrated the vital need of standardization and centralization of PFI especially now, at its early era. The Greek Taskforce should proceed sooner or later in the publish of guidelines in order clarification to be given in the interpretation and application of the PFI schemes in Greece, if, of course, the government still encourages the involvement of the private sector in infrastructure projects. Also, the Taskforce should define PFI in terms of value-for-money and risk allocation derived from the characteristics of the Greek market and its cultural characteristics. There is a need to explore the role, implications and effects on all stakeholders. In this connection it is important to be analyzed not only the public sector partners in this process but also the private sector companies and bankers involved in the consortiums making bids.

REFERENCES

Andersen, A. & LSE Enterprise 2000. Value for Money Drivers in the Private Finance Initiative. A report by Arthur Andersen and Enterprise LSE, Commissioned by the Treasury Taskforce. *Guides to PFI-Series One, January 2000,* London: Treasury Taskforce

National Audit Office 1999. *Examining the value for money of deals under the Private Finance Initiative, July 1999,* London: HM Treasury

Private Finance Treasury Taskforce 1999. How to account for PFI transactions. *Technical Note No 1 (Revised), June 1999,* London: HM Treasury

Private Finance Treasury Taskforce 1999. *Standardization of PFI contracts,* London: HM Treasury.

Private Finance Treasury Taskforce 1998. *Step by Step Guide to the PFI Procurement Process,* London: HM Treasury

TENs Hellas & Mixed Steering Committee 2000. Motorway Concession Contracts in Greece. Conference Proceedings, November 2000, Athens: Ministry of National Economy

Treasury Taskforce 1999. How to construct a public sector comparator. *Technical Note No 5, October 1999,* London: HM Treasury

System-based Vision for Strategic and Creative Design, Bontempi (ed.)
© 2003 Swets & Zeitlinger, Lisse, ISBN 90 5809 599 1

A computer-based feedback model for design/build organizations

G. Ozkaptan Alptekin
Department of Architecture, Beykent University, Istanbul, Turkey

A. Kanoglu
Department of Architecture, Istanbul Technical University, Istanbul, Turkey

ABSTRACT: Building production process is a multi-disciplinary and multi-phased activity. Various factors increased the complexity of building production process and consequently the number of participants taking place in the process. Co-ordination and integration of the production process became more important in the fragmented structure of construction industry. Obviously the lack of communication between designers and construction team causes quality problems. Insufficient communication between design and construction phases causes design failures regenerated in various projects. Design failures can be determined in both construction and occupation stages. Feedback of systematically recorded information of design failures, which were identified in construction stage would help to create and maintain the firm's memory and can be used as a medium to increase the quality of the design process as well as the design itself. In this paper, the conceptual and practical parts of a computer-based model developed for design/build organizations that aims to organize design failure information identified in construction stage are presented.

1 INTRODUCTION

1.1 *Background of the problem*

The fragmented structure of building production process resulting from multi-phased and multi-disciplinary nature of construction industry has brought out co-ordination and integration problems. Especially projects, which are delivered by fast-track approach, faced with integration problems because of information loss and quality problems.

Tan & Lu (1995) pointed out the most critical stage as planning and design stage throughout the building production process. Although the cost of this stage accounts relatively low (about 3–10 percent of the project cost), various specifications, layouts, schematics and procedures generated in this stage form the basis of the actual construction. The cost, scheduling, and performance problems can invariably be traced back to the problem of the quality of design, such as error, incompleteness, and lack of constructability.

Bubshait et al. (1999) reported a study by the Building Research Establishment (BRE) in which the causes of failure were analyzed to indicate whether they were due to faulty design, poor execution, the use of poor materials or unexpected user requirements. The percentages of failure, with some overlap between these categories, were found to be 58%, 35%, 12%, and 1% respectively. Faulty design was taken to include all cases where the failure could be attributed to not following the established design criteria.

Burati et al. (1992) identified causes of quality deviations in design and construction by analyzing nine fast-track industrial construction projects. The data were collected after the construction phase of the projects and identified the direct costs associated with rework (including redesign), repair, and replacement. It was concluded that deviations on the projects accounted for an average of 12.4% of the total project costs. Design deviations average 78% of the total number of deviations, 79% of the total deviation costs and 9.5% of the project cost. In this study, deviation data collected was classified as design, construction, fabrication, transportation and operability. Each of these areas was further subdivided by type of deviation. Design deviations categorized as design change/improvement, design change/construction, design change/field, design change/owner, design change/process, design change/fabrication, design change/unknown, design error, and design omission.

From design office's point of view, a design office is considered to undertake more than one complex project simultaneously and they have to work with different organizations to carry out these projects, which are in their various phases. In this circumstance, the intensity and complexity of the information flow is obviously high and the support of a computer-based information system is vital.

It is possible to increase the quality in building production process by the existing means such as choosing of appropriate organizational patterns and using of information technology, but these tools are not being used properly.

1.2 Related works

The idea of integrating building production process and improving design quality is not new. Various reference models for solving integration and co-ordination problems can be found in literature and also various computer models were generated. Conceptual models by Baldwin et al. (1999), Tan & Lu (1995), Matta et al. (1998), Chan & Chan (1999) were proposed to serve as interfaces of information management, design management and quality management.

Some of the application models which were developed by Kanoglu (2001), to serve integration of design and construction management; by Mokhtar et al. (1998), Hegazy et al. (2001) and Zaneldin et al. (2001), to serve design change management; by Miyatake & Kangari (1993) to serve computer integrated construction can be found in the literature.

These models are generated to provide integration of design and construction, to organize information flow between design and construction stages, to manage design changes, and to improve the quality of design. However, it has not been possible to locate an information system model to record and feedback design failures that can be identified in construction stage to improve the quality of design and design process.

1.3 Problem statement

The variety of building types, innovation of technical subsystems (security systems, intelligent buildings, etc.), improvement of technology and development of numerous materials causes complexity in building production process. Besides these developments, the number of organizations taking place in this process increase and the relationships among these organizations expose a fragmented structure.

Time becomes more critical in building production process and many efforts are seen to enhance projects complete rapidly. Fast-track approach to shorten the project delivery process causes an increase in change orders. Lack of communication between the parties arises integration problems among the designs of subsystems in buildings. These problems result as design failures.

Because of insufficient information flow among design and construction organizations, information of design failures do not turn back to design organization. That is why similar design failures continue to repeat, and deviations of cost, time and quality continue to occur.

In this context, one of the most important factors for design quality is identification of design failures that comes out in construction and occupation stages to avoid repeating these failures by the help of a feedback mechanism. However feedback from construction organization to design organization cannot be achieved in most cases due to the fragmented structure stated above. The dimensions and the tools of solution can be classified as:

- Defragmentation at *physical* dimension by new organizational approaches such as design/build organizations, strategic approaches such as joint-ventures, partnering, etc.
- Defragmentation at *conceptual* dimension by new philosophies, such as total quality management, just-in-time, supply-chain management, lean production, etc.
- Defragmentation at *virtual* dimension by information systems, using the tools such as integration, unification, standardization, interoperability, customizability, etc.

1.4 Aim of the study

In this study the content of design failure is taken as any design change that causes rework and results in additional cost and time consumption in a project. These failures may be identified in both construction and occupation stages (Fig. 1).

Ransom (1987) pointed out maintenance organizations as the repository of failures and their cause, and such knowledge that is fed back to design office would be of great benefit. However, impacting factors for identification of design failures in occupation stage have different characteristics relating to marketing,

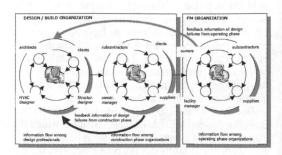

Figure 1. Information flow among the phases of building production process.

customer requirements and customer satisfaction, etc., and these may appear in long terms. Since feedback of information from occupation stage may differ related to these factors occupation stage was handled in a separate study.

In this study a conceptual model is proposed aiming to realize information feedback from construction to design organization. The conceptual model is then converted to a practical model using relational database architecture. In the model, design failures identified in construction stage are recorded systematically. This information is accessible by design office since an integrated information system is proposed to use by both parties. Thus, design failures are categorized and kept to establish firm's memory to prevent repetition of similar design failures.

The reliability and functionality of the proposed model depends on mutual confidence. Both design and construction organizations should be able to track other parties' records mutually and this may be possible with an appropriate project procurement system. Design/build system seems to be the most appropriate model for this purpose, so the model is proposed for this sort of organizations.

2 A COMPUTER-BASED FEEDBACK MODEL FOR DESIGN/BUILD ORGANIZATIONS

2.1 Background of the study

After investigating the current practice in detail by analyzing the information systems of some leading design offices and construction firms in Turkey it is determined that information systems even in large-scale firms do not incorporate any feedback mechanism that organizes the historical data transmitted between the architectural office and construction firm. Since the problem in current practice is the lack of a systematical approach and support of an integrated information system, the proposed model should include these tools.

2.2 The model and its components

The model was developed to function as part of an integrated information system of design/build organizations designed and developed in a research project conducted by Kanoglu (1999 and 2001).

As stated above, a certain part of the design failures can be determined during the construction phase although some other part can only be determined at operating phase. The corrective actions for the first part take place within the process of change management. Change order procedures should record all kind of modifications related to the scope of the projects. Changes may occur due to various reasons such as client's demand, external factors, etc., as well as design

failures. The model allows the construction team to indicate the reason of the change order (Fig. 2).

In a design/build organization, the construction team can use the model to record and report the changes of current project whereas the design group utilizes these outputs composed by using the data of past projects in the same category for avoiding the same sort of mistakes in the projects that will be undertaken in the future.

In a change order process more than one item related to each other can be recorded. The detail information should include the responsible participant who caused the failure, the building subsystem and the production items that are affected by the change as well as the quantity, cost and duration effects of the change (Fig. 3).

The model allows to filter the data of the past projects by project, change order ID, building type (category), responsibility, subsystem (heating, lighting, structure, etc.), sub-subsystem (radiators, electrical appliances, fittings, etc.), and production items (takeoff items). Analyzes regarding to cost and duration effects can be made by using various combinations of these parameters (Fig. 4).

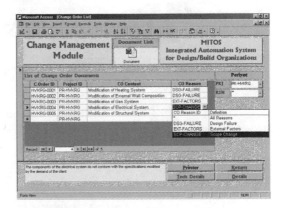

Figure 2. Change order information.

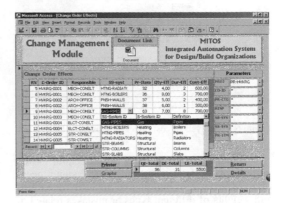

Figure 3. Change order detail information.

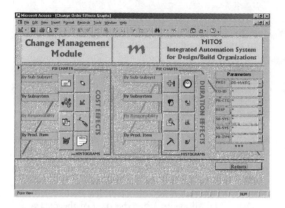

Figure 4.   Analysis by cost and duration effects.

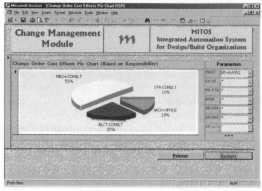

Figure 7.   Analysis of cost effects by responsibility (%).

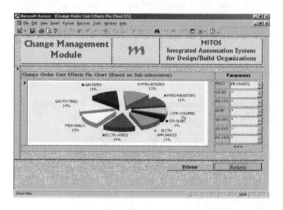

Figure 5.   Analysis of cost effects by sub-subsystem (%).

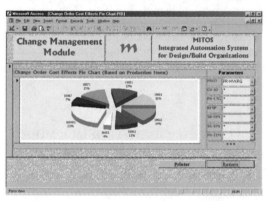

Figure 8.   Analysis of cost effects by production items (%).

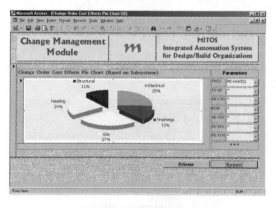

Figure 6.   Analysis of cost effect by subsystem (%).

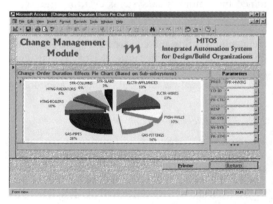

Figure 9.   Analysis of duration effects by sub-subsystems (%).

The outputs of the analyzes can be helpful for the participants of the design team in the design process of a certain type of building by indicating the ratios and weights of different types of mistakes related to subsystems (Figs. 5–12).

The graphical outputs are provided by the model to display the weights (%) in pie charts and quantities in histograms. The effects of the parameters (i.e., reason, responsibility, related subsystem, etc.) can be obtained in both tabular and graphical formats.

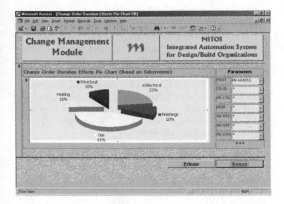

Figure 10. Analysis of duration effects by subsystems (%).

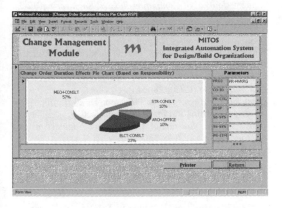

Figure 11. Analysis of duration effects by responsibility (%).

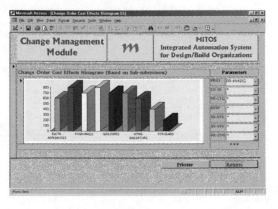

Figure 12. Analysis of cost effects by sub-subsystems ($).

These outputs can only be obtained from an integrated information system of both design offices and construction firms combined as a single entity that is referred to as design/build.

3 CONCLUSION

Design/build contracts combine design and construction under a single entity. Design/build is sometimes conducted by a company that has design and construction capabilities under the same roof or by a joint venture between a design firm and a construction company. This approach has gained the attention of the industry in the last decade and is expected to become the dominant project delivery system in the near future. This growth is justified by the reported benefits of the system, including faster delivery and lower cost. It is now time to have a management information system in place in design/build organizations that could facilitate the management of this delivery system. Such a system would facilitate the interactions of the members of the design/build team, but it would be particularly useful if the team members belong to different organizational cultures.

This paper describes efforts that are aimed at developing an integrated computerized automation environment that is expected to increase the efficiency and productivity of design/build project delivery organizations as well as design firms and construction companies that make it a point to effectively cooperate within the framework of a partnering arrangement or in traditional design-bid-build projects.

Although it has the stated advantages the design/build procurement route does not always achieve its claimed advantages. One aspect of this is the emphasis in the paper placed on integration, a claimed advantage of design/build. The suggestion of an automated system for design/build integration is worthy of examination. Such a support provided by an integrated information system will provide various advantages.

The paper addresses a problem that can be solved by an integrated information system developed for design/build organizations. The next step of the study will be the implementation of the model in selected design/build organizations in Turkey. This part of the research project is planned to last at least for two years.

Further analysis including the comparative evaluation of the design failures with their associated reasons and responsible parties are thought to be facilitating in the solution of various problems such as liquidated damages from claim management point of view in project level and to increase productivity from resource management point of view in corporate level as well. After collecting sufficient data the second part of the study is planned to start for testing the validity of the model and to determine the correlations among the parameters. According to the hypotheses some additional outputs may be added to the model.

ACKNOWLEDGEMENTS

The design/build model that was integrated with the model proposed in this study was developed in a previous research project supported by TUBITAK – The Technical and Scientific Research Council of Turkey.

REFERENCES

Baldwin, A.N., Austin, S.A., Hassan, T.M. & Thorpe, A. 1999. Modeling information flow during the conceptual and schematic stages of building design, *Construction Management and Economics*, 17: 155–167.

Burati, J.L., Farrington, J.J. & Ledbetter, W.B. 1992. Causes of quality deviations in design and construction, *Journal of Construction Engineering and Management*, 118(1): 34–49.

Bubshait, A.A, Farooq, G., Jannadi, M.O. & Assaf, S.A. 1999. Quality practices in design organizations, *Construction Management and Economics*, 17: 799–809.

Chan, E.H.W. & Chan, A.T.S. 1999. Imposing ISO 9000 quality assurance system on statutory agents in Hong Kong, *Journal of Construction Engineering and Management*, 125(4): 285–291.

Hegazy, T., Zaneldin, E. & Grierson, D. 2001. Improving design coordination for building projects I: Information model, *Journal Of Construction Engineering and Management*, 127(4): 322–329.

Kanoglu, A. 1999. *A Site Level Computer-based information system design for the construction companies*, Project Code INTAG-912: Research project supported by TUBITAK – The Scientific and Technical Research Council of Turkey, Istanbul.

Kanoglu, A. 2001. MITOS: Multi-phased integrated automation system for building production process, Information and communication technology in the practice of building and civil engineering, *Proceedings of 2nd worldwide ECCE symposium organized by VTT and RIL, Helsinki, Finland, 6–8 June*, 183 188

Matta, K., Chen, H. & Tama, J. 1998. The information requirements of total quality management, *Total Quality Management*, 9(6): 445–461.

Miyatake, Y. & Kangari, R. 1993. Experiencing computer integrated construction, *Journal of Construction Engineering and Management,* 119(2).

Mokhtar, A., Bedard, C. & Fazio, P. 1998. Information model for managing design changes in a collaborative environment, *Journal of Computing In Civil Engineering*, 12(2): 82–92.

Ransom, W.H. 1987. Building failures diagnosis and avoidance. London: E&FN Spon.

Tan, R.R. & Lu, Y.G. 1995. On the quality of construction Engineering design projects: Criteria and impacting factors, *International Journal of Quality & Reliability Management*, 12(5): 18–37.

Zaneldin, E., Hegazy, T. & Grierson, D. 2001. Improving design coordination for building projects II: A collaborative system, *Journal of Construction Engineering and Management*, 127(4): 330–336.

System-based Vision for Strategic and Creative Design, Bontempi (ed.)
© 2003 Swets & Zeitlinger, Lisse, ISBN 90 5809 599 1

A model for forecasting public risks created by large urban infrastructure construction projects

W. Korenromp, S. Al-Jibouri & J. van den Adel
Department of Civil Engineering, University of Twente

ABSTRACT: Large infrastructure construction projects in urban areas have considerable impacts on normal living as well as on their approximate vicinities and the environment. The nature of the project's surroundings can play a significant role in the decision-making process regarding the choice of construction methods to be used. The project surroundings and neighbourhood however can also be very diverse and complex and the impacts of construction on them are difficult to predict.

This paper describes a model, which has been developed to forecast the level and impact of various logistical, nuisances and societal risks that can be produced or caused by construction projects in urban environments. In this way it can be used to support decisions to choose alternatives and take measures that will avoid or minimise such risks. The model employs a combination of both quantitative and qualitative approaches for analysing these risks depending on the nature and amount of information available.

Experimentations with the developed model have shown that it is easy to use and that it produces useful analysis for predicting the level of communal risks associated with various construction alternatives.

1 INTRODUCTION

The role of risk management is becoming increasingly important within the current practice of the construction industry. More than ever, projects are required to be finished in a shorter time span and within stricter budgetary and quality requirements. The project profit margin is becoming smaller and the chance, of incurring a loss is getting larger, (Jaafari 2001). It is therefore of utmost importance that project risks are early identified, analysed and that measures are taken to limit their influences on achieving project objectives.

In general, there are many sources of risks for and from a construction project, (Avarot 1995, Lowe & Withworth 1996). One of these sources is the project physical surroundings. This is especially true in the case of large urban construction projects (Jaafari 2001, Miller & Lessard 2001, Avarot 1995). The project surroundings will be affected during the construction of the project in the form of possible damages caused by the project to the environment or hindrances to normal life within these surroundings. As results of such conditions and circumstances, the project organisation will also bear the burden of taking various measures to reduce these possible nuisances. Both the

project and its surroundings will thus be affected because of the nature of the environment within which these projects are being carried out (Feddema & de Bruijne 1994).

For estimating the risks resulting from carrying such project in urban areas, it is very important to identify all possible damages and disturbance that can be caused to the surroundings and by the construction of such projects. The construction project is considered here to be a source of such societal risks.

Despite the importance given to environment and the surroundings in urban areas, the nuisance, hindrances to normal living and logistics caused by construction are rarely taken into considerations when selecting alternatives for construction methods and control measures. Indeed even when such alternatives are considered this is often done in a very general manner and on a qualitative basis (Wit de 2001).

This paper describes a model which id developed for the purpose of forecasting social and logistical hindrances and will be sources of risk for the project organisation. It is an attempt to provide a structure for combining the information needed for estimating the extent of these disturbances.

2 PROBLEM BACKGROUND

In practice, the failure to analyse the risks resulting from construction projects on the surroundings and the hindrances they cause in urban areas is due to a number of reasons. One reason is related to the fact that known risk analysis techniques are not suitable for the analysis of societal risks that originate from construction projects. It is also rather complex to carry out quantitative analysis of this problem using these techniques due to the non availability of systematic methods and knowledge in this area.

Most risk analysis techniques are normally used to analyse direct risks that can affect construction projects and their consequences in terms of costs and durations. In the case of surrounding risks, there is usually another level of risk in an intermediate phase, called here as hindrance, which is not considered normally in risk analysis (DelftCluster; Expertpeiling 2002). In many cases such risks are neglected despite the fact that these can have serious consequences on the quality of living and the environment around the project. The construction project is considered here as an important source of such risks. In current practice however only the effects of conditions and surroundings around the projects on the project itself are analysed.

It is also important to mention that current risk analysis techniques don't offer a systematic manner for build up of information and experience that can be used for the analysis of future projects. This is rather unfortunate since good risk analysis requires good knowledge and/or quantitative data, which can reduce the time and effort to carry out such analysis. Instead and in order to make risk analysis more cost effective, such analysis is often carried out on the basis of personal judgment and/or qualitatively. As such it is often subjective and less controllable and in many cases lacks the scientific credibility. It can also be easily misinterpreted or manipulated in order for example to make political cases for approvals or disapprovals of major projects.

Although in general quantifying the analysis provides a better option, it is however not always possible or desirable to do so. A good quantitative analysis often can not be carried out without a qualitative analysis and thorough risk identification, see (Jaafari 2001). Also a qualitative risk analysis is desirable if it is based on project experience and knowledge since this can provide a good basis for quantitative analysis.

Furthermore many practitioners find the risk analysis methods to be complex. Some are used for specific objectives. Some are for example are used for identification, whilst others are used for quantification. There is therefore a need to have an integrated model, which can be used for all the phases of the process. Risk analysis is not a one off activity but rather a continuous process that takes the dynamic design process, changes and events etc. into consideration. It is however very important to implement risk analysis at early stages of the project so that consequences of choices can be identified early in the project (Chapman 2001). It is usually more expensive to take measures to correct an undesirable situation in a late stage of the project than in to identify such undesirable situation in the planning phase (Vermande & Spalburg 1998). In the early stages of the project however there is little information available for detailed planning. Ironically, risk analysis on the basis of such very limited information is required to be carried out in order to identify the risks associated with various construction choices. Also the initial planning process is an iterative process whereby choices are adjusted or enhanced which in turn create new risks. In order to effectively implement such process the model used should evolve with the initial planning process. All methods related to identification and quantitative risk analysis should therefore be brought under one integrated system.

3 BASIC CONCEPTS

In this work a model has been developed in order to deal with the problem described in the previous section. The basic principle is that an integrated and easy to used risk management tool is required with which quantitative risk analysis can be carried out within a dynamic design process. This can be made possible through the integration of project information and the use of a method that can be applied for all the phases of the process. Also both qualitative and quantitative analyses can be used. Furthermore attention is given in this model for the analysis of hindrances caused by construction projects to the surroundings around it.

The model is developed to deal with the surroundings and environment hindrances and in particular disturbances to societal and logistical processes. Unlike technical problems, social problems are complex and described as early as 1973 by Rittel & Webber as being "wicked". Also individual actions are difficult to forecast though with enough information it is possible to estimate them.

Societal hindrance from a construction project can for example be related to noise, visual, safety, and dust generated by the construction work. Logistical hindrances on the other hand are related to traffic jams caused by transport of construction materials and road detours.

Hindrances caused by construction can be considered as a two-dimensional concept, namely time and intensity.

The extent of the total hindrances will be judged on the basis of creating a hindrance profile during the construction project. The extent of hindrances will vary at different stages of the project depending on the type of work being carried out on the construction site at any specific stage.

4 MODEL STRUCTURE

Hindrance the issue of societal and logistical hindrances is a complex risk analysis subject and one of the objectives of this work is to provide a structure for this problem so that analysis can be carried out relatively easier (Lowe & Whitworth 1996).

There are many mechanisms with which the relationships between the sources of risks and damages caused by these can be represented. This work is focused on the mechanism related to societal and logistical hindrances caused by construction works in urban area. Three important examples of societal and logistical hindrances are nuisance to residents around the project site, hindrances to businesses near the project and the creation of traffic jams in the surroundings of the construction site. The damage caused by these hindrances will in turn lead to delay and financial risks to project organisation and/or society see Figure 1.

The mechanism can be described as a sub model which represents the relationships between two aspects (through various types of hindrances). Through this model an average estimate of the extent of a specific hindrance can be given. However this can be higher and lower depending on the project and the conditions under which the project is constructed. Such estimates for the extent of disturbance caused by a construction project can be made using Bayesian Belief Network (BBN), see [Jensen & Andersen 1990], whereby the relations between the factors affecting such estimates are represented in the form of networks. A commercial software program called Hugin, which is based on (BBN), is used to represent these networks. The analysis provided by these networks is semi quantitative.

Quantifying of results is not always possible and the program in cases also uses qualitative analysis where there is no qualitative information available.

Because hindrance is a 2-dimensional concept, two BBN's must be constructed in order to estimate the total amount of hindrance, one for time-forecast and one for intensity-forecast. With this information a profile can be formed.

Due to the uncertainties of forecasts of social response, the forecast for damage is divided into 2 parts, one determines the production of a hindrance due to construction activities and the second estimate is for the damage to the surroundings. The reason for this is that unlike the social response, the production can be accurately measured.

5 MODEL INTEGRATION

Risks to form a construction project are results of damage and hindrance to surroundings caused by construction activities. These activities are subject to changes from the original plans and hence risk analysis should also be changed accordingly. New information and solution will offer new possible risks. The construction activities are split according to Work Breakdown Structure (WBS) into more detailed activities. At each level of the WBS can be determined what the extent of the hindrance is. In this manner a hindrance profile can be determined. In a similar manner this can be done for the perception of hindrance. The projects can be classified as Infrastructural or building which can be verther split into housing, commercial buildings etc. As the planning becomes more detailed so will the forecasts of the hindrance become more accurate as can be seen in Figure 2.

There are as many modules that can be formed as the number of the various detail levels related to

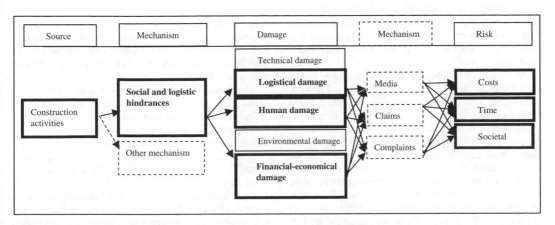

Figure 1.   Mechanisms showing the relationships between sources and hindrances.

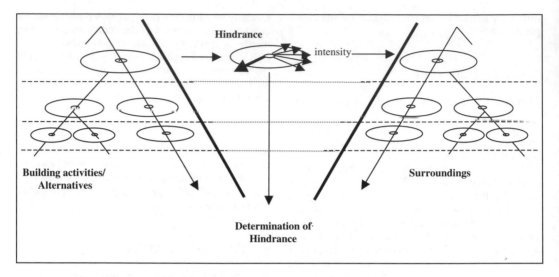

Figure 2. Different level of details for hindrance from construction activities.

construction activities, hindrance sorts and the type of projects and locations. The main model will combine and control all these modules. The necessary information required for each of these modules will be retrieved from databases.

In early stages of the planning phase, hindrance profiles can be determined on the basis of the general plan. The inaccuracies of forecasts on this basis will be still high. However an average hindrance profile can be used with possible expected variations. As this is done, the factors responsible for great variations can be examined using BBN/sub model in order to perform sensitivity analysis and to see what the consequences of these are. Attention should then be given to areas where is the variations and intensity are very high.

It is also possible here to integrate control measures in the network and investigate what the cost consequences of using a control measure are. There will be many possible control measures to take in order to reduce the production, perception of hindrance and to limit the effect of risks.

As a result of the above, a hindrance profile can be determined as can be seen in the example in Figure 3. This profile can be used to estimate the extent of the hindrance and the possible variations because of the incompleteness of the information. If the variation of the estimate is too large then it would be then possible to look in more detail in this specific area to investigate the factors and the possible control measures affecting this.

The level of detailed used will be dependent on the level of accuracy required and the phase of the plan. In early phases little information is available. The total hindrance will represent the total area under the hindrance profile.

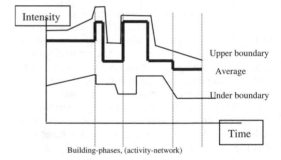

Figure 3. A noise hindrance profile.

6 CONCLUSIONS

It can be concluded that it is possible to build a model to forecast social en logistical hindrance caused by large urban construction projects. Such a model can also be used to estimate the risk to both the project and/or society.

The research has also shown that the model give reasonable and realistic forecasts of the production and perception of hindrance in the form of a profile.

The model can be used as a decision support tool during complex and dynamic design process.

Information can be integrated in the model so that effective risk analysis can be made with reasonably low effort.

The model is a first step in the development of a comprehensive risk management tool whereby all the information and construction methods are integrated within one system.

REFERENCES

Chapman, R.J. 2001. The controlling influences on effective risk identification and assessment for construction design management, *International Journal of Project Management*, vol.19, p.147–160.

DelftCluster project 04.03.01, 2001. *Integrale risicobeheersing-Delft beheerst ondergronds; Expertpeiling.*

Feddema, A. & Bruijne, de, V. 1994. *Boren of open bouwmethode? Een eerlijke vergelijking*, afstudeerverslag Bedrijfskunde hogeschool Alkmaar.

Jafaari, A. 2001. Management of Risks, uncertainties and opportunities on projects; time for a fundamental shift, *International Journal of Project Management*, vol.19, p.89–101.

Lowe, J. & Whitworth, T. (1996). Risk management and major construction projects, *The Organization and Management of Construction: shaping theory and practice*, vol.2.

Miller, R. & Lessard, D. 2001. Understanding and managing risks in large engineering projects, *International Journal of Project Management*, vol.19, p.437–443.

Rittel, H. & Webber, M. 1973. Dilemmas in a General Theory of Planning, *Policy Science*, vol.4, p.155–169.

Vermande, H.M. & Spalburg, M.G. 1998. *Risicomanagement, een verkenning* Stichting Bouwresearch (SBR), Rotterdam

Wit, de, M.S. 2001. *Integrale risicobeheersing-Delft beheerst ondergronds; Risman voorbij.*

A risk management model for construction projects

S. Smit, S. Al-Jibouri & J. van den Adel
Department of Civil Engineering, University of Twente, Enschede, The Netherlands

ABSTRACT: Risks are identifiably inherent in construction work and most risk management in construction is currently project based. The traditional approach to managing risk from projects has utilised the valuable assets of manager's experience and judgement. Interestingly, the term "risk management" has only been in common use for the last two decades. This is because the term refers to treating the risk management of risk as an explicit task. Traditionally, the management of risk, using experience and judgement has not been separated from other functions of managers and so not warranted a title. The paper describes a formal project risk model that employs risk management processes and technique for identifying, analysing and controlling risks. In addition the model also relies on making a risk management plan at the start of the project, which is tied to the project plan to determine the timing and frequency of risk assessment. The model is designed to be an integral function of the project organisation whereby responsibilities for implementing the risk management model are clearly identified. The model has been tested on cases of real projects and its practicality and usefulness are assessed.

1 INTRODUCTION

The construction industry in many countries is undergoing radical changes whereby many construction companies are entering into new types of contracts which allocate risks to them which have previously been held by others. This has resulted in considerable risks that have to be addressed and managed (Vermande 1998). An important source of risk is the changing contracting approaches being adopted. Traditionally a construction company is only involved during the construction phase. However in the new contracting approaches such as design & construct, the responsibility for design also lies within the construction company. Other sources of risk are related to the fact that increasingly more and more emphasis is given to the involvement of the customer in the process and the need to reduce project cost and time.

It is important to control these risks during the construction process since they may have a negative impact on costs and planning. With a formal risk management process, risks can be identified, analysed and controlled.

Construction Companies are aware of the need for risk management. In fact this is currently applied in many construction projects. However the problem with risk management is that in many cases risk management is seen as an extra project activity instead of being an integral part of the total project control. Amidst daily construction work, this "extra" activity is often neglected, because other activities have more priorities. Risk management is therefore applied as an ad hoc or as a corrective approach rather than being applied as an effective preventive tool.

This paper describes a model which is developed for a structured risk management process to be used by construction Companies during construction projects. The model is a link between risk management and the project management process which provides a framework for the implementation of risk management within a project organisation.

The development of the model is carried out on the basis of literature studies and interviews with experts from major construction companies in the Netherlands.

The developed model has been tested on cases from two real construction projects. The case studies used represent two different contract forms. The first case is of a concrete bridge with a traditional contracting approach. The cost of the project is about 40 million Euros. The second case is one of the six contracts from the Dutch high-speed rail track which is based on a design & construct type of contract. The cost estimate of this contract is 300 million Euros.

The case studies are used in order to:

- Asses the practical usefulness of the developed model
- Give recommendations for improving the present risk management process in construction projects.

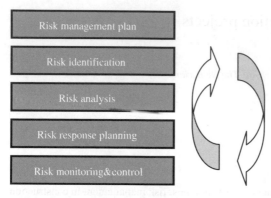

Figure 1. Risk management model developed by the Risk Management Specific Interest Group (2000).

Step 1: Risk management plan
The risk management plan describes the risk management activities during the project. The risk management activities are planned, budgeted and needed resources are reserved.
Step 2: Risk identification
Risk identification involves determining which risks will have an effect on the project and documenting their characteristics.
Step 3: Risk analysis
This step involves estimating the risk probability and the impact of risk on the project objectives. In risk analysis either quantitative or qualitative approaches can be used.
Step 4: Risk response planning
Risk response planning is the process of developing and determining actions to enhance opportunities and to reduce threats to the project objectives.
Step 5: Risk monitoring and controlling
This phase starts with the implementation of the risk response plan. Risk monitoring and control is the process of keeping track of the identified risks, monitoring residual risks and identifying new risks. During the project risks monitoring and controlling is an ongoing process. The effectiveness of the risk response actions must be monitored and evaluated.

Figure 2. Description of the steps in the risk management process.

2 BACKGROUND

2.1 *Basis of research*

In literature, there are quiet few different definitions of the term "risk". In general however risk is taken to mean the outcome of an uncertain event. Traditionally risks are characterised as threats or negative outcomes. However the outcome of an uncertain event can also be positive and has a beneficial effect on project objectives (Vermande 1998), in this situation the risk is called a project opportunity.

Risk management is a tool to manage the risks, threats and opportunities, during a construction project. The Risk Management Specific Interest Group (RM SIG) defines risk management as a systematic process of identifying, analysing and responding to potential project risk. It includes maximising the probability and impact of positive events and minimising the probability and consequences of events adverse to project objectives. So risk management is not only used to avoid or reduce threats to project objectives, but also to maximise project opportunities.

Current research shows that many models have been developed in the area of risk management. The basis of the majority of these risks management models however comprises three main steps or phases: *identifying*, *analysing* and *controlling*. Recent developments introducing more segmented risk management processes. This ranges from four phases (Smith 1999) to five (Vermande 1998; RM SIG, 2000) and even more. The phases of the different models are comparable. All models are extracted from the basic model, which contains the three steps. The developed model in this work is based on the risk management process model defined by the Risk Management Specific Interest Group (2000). Figure 1 shows the risk management process developed by the Risk Management Specific Interest Group.

2.2 *Risk management process*

The risk management process of the Risk Management Specific Interest Group distinguishes five steps. The different steps are described in Figure 2.

3 DEVELOPED MODEL

3.1 *Structure of the model*

The developed model is based on the risk management model of the Risk Management Specific Interest Group (2000) shown in Figure 1. Added to this model are two aspects:

- The link between risk management and the project management process;
- The integration of risk management procedure into the project organization.

In the following two sections, these two aspects of the model their relations with the risk management procedure are described.

3.2 *Risk and project management processes*

The link between risk management and project management are described in this section. Groote (2000) concludes that project management is primarily based on three important aspects:

- Project Phasing
- Decision making
- Monitoring & controlling

Project phasing means to divide a project development in several manageable stages. Examples of stages in construction projects are design stage, construction stage, maintenance stage etc. Every stage is a separated project part and ends with a concrete decision. At the end of each stage the results will be tested to make sure that they are in accordance with the project's requirements.

The end of each stage represents a *decision* moment whereby decisions can be made and documented before proceeding to the next stage. The documents produced form a formal end of the preceding phase. It is important that the risks are clearly assessed at the end of each stage and decisions based on reliable risk analysis are made accordingly.

The third aspect of project management is *monitoring & controlling*. Monitoring & controlling is the process representing all activities carried out to ensure that the project objectives are achieved. Groote (2000) distinguishes the five main project aspects that required to be monitored and controlled as being:

- Time *(When?)*
- Money *(How profitable?)*
- Quality *(How well?)*
- Information *(Based on what?)*
- Organisation *(For Whom?/by Whom?)*

During and at the end of every stage in the project, these five aspects are monitored and impacts of possible risks on them are analysed.

When occur, risks usually have impacts on one or more of these project aspects. For example the impact of risks can result in extra costs or exceeding the project planned schedule. Risk management should be seen as a tool for managing risks and achieving the project objectives. It is important that risk management is linked with other project management processes. This is because risk management process is related with other project management processes, such as cost control, planning and quality control.

3.3 Risk and project organisation

This section describes how to implement the developed risk management model within a project organisation. In general a project organisation of a construction project is similar to the organizational structure shown in Figure 3.

In order to integrate the risk management process into a project organisation effectively, the following conditions are required, see (Robillard 2001):

- A risk management function is established
- Risk management is implemented trough existing decision-making and reporting structures.

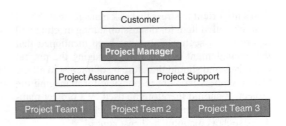

Figure 3.   Project organization of a construction project.

Bakker (2001) distinguishes four types of risk management functions:

- Risk auditor
- Risk manager
- Risk consultant
- Risk coach

For a *risk auditor*, the role is based on controlling of risks in adverse. The risk auditor has a consulting function to propose risk-mitigated action, but without any authority to execute these actions. This risk management function is mostly used during cases of projects with major costs or planning overrun.

The *risk manager* role is based on having responsibility for the total risk management process. In this function, knowledge and skills of risk management are directly used in daily practice.

The *risk consultant* has a supportive role. The responsibility of the risk consultant is to consult in matters related to risk management but without the responsibility for their implementations.

The *risk coach* role is comparable with the role of a risk consultant, but in this case the person is also responsible for the coaching of team members, like the project manager and team managers. This function is likely to be useful for organisations that want risk management process to be structurally integrated with other functions.

Another important point that is that risk management needs to be integrated within existing organizational decision-making procedures and reporting structures. In a project organisation the following reporting structures can be identified:

- *Project manager and Customer meeting*: In this meeting the project manager reports to the customer about progress of the project.
- *Team managers and project manager meeting*: Team managers report to the project manager about the progress of their project part and discus the total project.
- *Team meeting*: Team manager en team members report on daily work related to their parts of the project.

It's important to stress that risk management should be a scheduled item for discussion during meeting and reporting. In section 3.2 it has been mentioned that risk management is a tool for managing the project objectives and should therefore be of interest to all the team members who are responsible for managing the project. It should therefore be used by a project manager and team managers as a routinely as other procedures such as planning and cost control.

3.4 *The developed risk management model*

The five steps of the developed risk management model shown in Figure 1 can be described shortly as follows:

The *first step* required by the model is to develop a risk management plan. This plan describes the risk management activities during the project process. The risk management plan is the first important step required for a successful risk management process. During a risk management meeting *(with project manager, team manager's en risk manager)* the risk management approach will be discussed and determined. The risk management plan provides responsibilities of project members and the link of risk management procedures with project objectives. The risk management plan is tied with the project plan.

The *second step* of the risk management process is risk identification. It is recommended to start as early as possible with risk identification. If risks are early foreseen, they can be managed effectively. Risk identification is a cyclic process during the project whereby new risk will appear and other risks will disappear. At the project start it is recommended to carry out this procedure carefully in order to identify the most important risks with the highest impact on project objectives. It's not necessary to make a list of hundreds of all risk items to be envisaged. Examples of tools that can be used to identify risks are checklists, brainstorming, interviewing and SWOT analysis. Project members identify risks in their specific project part. The risk managers' role is to be a facilitator during the risk identification. The risk manager keeps an updated risk dossier with all project risks.

The *third step* of the risk management process is risk analysis. In this step identified risk will be analysed and evaluated. In risk analysis, two types of approaches can be distinguished, quantitative risk analysis and qualitative risk analysis. The quantitative risk analysis assess the probability of a risk and the impact in terms of exceeding costs and durations. Examples of tools used in quantitative risk analysis are Probability/impact matrices and Monte Carlo simulation. Qualitative risk analysis provides detailed information about risk sources and the impact of risks. Examples of tools that can be used in qualitative analysis are Event trees and Failure Mode and Effect Analysis. It's important to communicate the results of the risk analysis to all parties involved in order to determine their impacts on project objectives and to formulate appropriate response actions.

The *fourth step* of the risk management process is to develop a risk response plan. If from the previous step the analysis has shown that certain risks are unacceptable, response actions must be developed. Several response strategies can be identified:

- Avoid
- Mitigate
- Transfer
- Accept

In cases when the impact of a risk has a beneficial effect on project objectives other response strategies are developed in order to exploit such opportunities. No project manager wants to avoid a project opportunity unless its benefit is markedly low. If a risk has a beneficial effect the following response strategies can be used.

- Exploit
- Enhance
- Share
- Ignore

In this step a risk response plan is developed. The plan will contain all response actions and the responsible project individual or team for implementing the response action. It's important to show the effect of implementing the response actions on daily project processes such as quality plans or work plans.

The *fifth step* of the risk management process is risk monitoring and controlling. This step starts with the implementation of the risk response actions. During the project process new risk can arise and old risk can disappear. Risks are dynamic, so it's important to update the risk dossier as the project progresses. It's also important to monitor the effectiveness of the implemented response actions. Risks should be actively and continually managed and their impacts on project objectives are analysed

The project manager is responsible for managing the project and the project risks. The team managers are responsible for managing the progress and risks in their parts of the project. The risk manager responsibility would be to consult the risk management process and to coach team managers and project manager to use risk management as a tool to manage risks that may have impact on attaining the project objectives.

4 CASE STUDIES

The developed risk management model is applied on two construction projects. The first case represents a concrete bridge project with a traditional contract form. The costs of the project are about 40 million Euros. The second case is related to one of the six contracts from the Dutch high-speed rail track undertaken as a design and construct contract. The cost estimates of this contract are 300 million Euros. In these two case studies, comparisons are made between current risk management process applied by the construction company with that proposed by the developed model in order to test the applicability and usefulness of the developed model in relation to existing practice. The comparisons are based on three aspects:

- The steps used in the risk management process
- The implementation of risk management within the project organisation
- The integration of risk management and other project management activities.

The *first aspect* of comparison has shown that in general, the steps of the risk management procedure used by the construction company were similar to those suggested by the model. The only significant difference in approaching the risk management process between current practice and the proposed model was that in current practice only negative impacts of risks are identified and that no attention is given for identifying project opportunities. The proposed model encourages the identifications of both risks and opportunities. It is not only important to identify and control project risks but also to optimise project opportunities and strengths. A useful tool that can be used for this purpose is SWOT (Strength, Weakness, Opportunity, and Treat) analysis.

The *second aspect* of comparison is the integration of risk management procedure within the project organisation. It has been found that the role of the risk manager within a construction company is merely supportive. The risk manager identifies the risk, keeps an updated risk record, analyse risks and reports the results to the project manager and team managers. In the developed model the role of risk manager is not only supportive, but also coaching. In this role the risk manager coaches the project manager and team managers to use risk management in managing the project processes. The risk manager teaches the project manager and team managers how to use risk management process as tool to control their costs, planning and quality.

The *third aspect* of comparison was related to the link between risk management and project management processses. It has been found for example that, unlike the proposed model, the project risk management process employed by the company is not integrated within the project organisation. This has created a number of problems for the project as the responsibilities for risk management were not always clearly defined. The developed model suggests the integration of the risk management process within the project organisation in order to link this with other project processes such as cost control, planning and quality control. The proposed model also suggests that the risk management plan should be tied to the project stages and programmes. It's important to update the risk lists during each stage and at important milestones in the programme.

To be effective, risk management process and responsibilities are required to be integrated within the project organisational structure. Risk identification and assessment have to discussed and reported during regular project meetings. Also when decisions are made it is important that risks associated with such decisions are properly assessed and measures are taken to avoid or reduce effect of these the risks on the project objectives.

5 CONCLUSION

In this work a formal risk management model has been developed based on literature studies and interviews with experts in construction. This work has shown that many of the current risk management practices lack the structure required to carry out risk assessment efficiently. It has also shown that the majority of the risk management processes used in practice don't use a risk management plan that is tied to the project plan. Furthermore the risk management procedure is not structured and it is often carried out as ad hoc activities and not seen as an integral part of project management.

The applications of the developed model in two case studies have shown that this model is very beneficial and provides a formal risk management framework that is much systematic and therefore advantageous to the those used in current practice. The role of risk manager used in this model is to coach the project members, especially the project manager and team managers to use risk management as tool to control the project objectives.

REFERENCES

Bakker, K. 2001. Rollen in risicomanagement. deB Project & Risk.
Groote, G., Hugenholtz-Sasse, C., Slikker, P., e.a. 2000. Projecten leiden. Utrecht: Het spectrum.
Hillson, D. 2001. Extending the risk process to manage opportunities. 4th project management conference PMI Europe 2001. London.
Hillson, D. 1999. Developing effective risk response. Proceedings of the 30th annual project management institute. Philadelphia.

Hullet, D.T. 2000. Project risk management. Risk Management Specific Interest Group.

Robillard, L. 2001. Integrated risk management framework. Treasury Board of Canada.

Smit, S. 2003. Een model voor het risicomanagementproces van bouwbedrijven tijdens de uitvoering van civieltech- nische bouwprojecten. Faculteit Civiele Techniek. Universiteit Twente. Enschede.

Vermande, H.M., Spalburg, M.G. 1998. Risicomanagement in de bouw: een verkenning. SBR rapport 448.

ZBC Consultant. www.zbc.nl.

System-based Vision for Strategic and Creative Design, Bontempi (ed.)
© *2003 Swets & Zeitlinger, Lisse, ISBN 90 5809 599 1*

Multicriteria decision support systems in engineering planning and management

O.G. Manoliadis
Techn. Educ. Inst. of Western Macedonia, Kozani, Greece

J.P. Pantouvakis
National Technical University of Athens, Athens, Greece

ABSTRACT: The planning and management of engineering systems requires the integration of often very large volumes of disparate information from numerous sources; the coupling of this information with efficient tools for assessment and evaluation allow broad, interactive participation in the planning, assessment, and decision making process; and effective methods of communicating results and findings to a broad audience. Information technology, and in particular, the integration of data base management systems, and multi-criteria optimization models, provide some of the tools for effective decision support. The objective of the computer based decision support system proposed in this research is to improve planning and management through decision making processes by providing useful and scientifically sound information to the actors involved in these processes, including public officials, planners and scientists, and the general public. The Multicriteria Decision Support System uses the Compromise programming as Decision support tool that involves both rather descriptive information systems as well as more formal normative, prescriptive optimization approaches. The case of building life cycle is used as an example. A simplified example is used revolving around a choice between alternatives. The alternatives are analyzed and ultimately ranked according to a number of economic and environmental criteria by which they can be compared; these criteria are checked against the objectives and constraints (our expectations), involving possible trade-offs between conflicting objectives.

1   INTRODUCTION

Engineering planning and management problems are complex and multi-disciplinary in nature. They involve the need to forecast future states of complex systems often undergoing structural change, subject to sometimes erratic human intervention. This in turn requires the integration of quantitative science and engineering components with socio-political, regulatory, and economic considerations. Finally, this information has to be directly useful for decision making processes involving a broad range of actors. It seems obvious that no single method can address all these requirements credibly and satisfactorily.

However, methods which are based on modern information technology, and which are also embedded in the necessary institutional structures, offer at least some of the necessary ingredients of effective information and decision support systems. The integration of techniques such as data base management, multi-criteria decision support systems simulation models, expert systems seem to have the necessary

power and flexibility to support planning and management in engineering applications.

2   METHODOLOGY

2.1   *Decision support framework*

The ultimate objective of the computer based decision support system for engineering planning and management is to improve decision making processes by providing useful and scientifically sound information to the actors involved in these processes, including public officials, planners and scientists, and the general public.

Decision support is a very broad concept, and involves both rather descriptive information systems as well as more formal normative, prescriptive optimization approaches. Any decision problem can be understood as revolving around a choice between alternatives. These alternatives are analyzed and ultimately ranked according to a number of criteria by which they can be compared; these criteria are checked against the objectives and constraints (our expectations), involving possible trade-offs between conflicting objectives. An alternative that meets the constraints and scores highest on the objectives is then chosen. If no such alternative exists in the choice set, the constraints have to be relaxed, criteria have to be deleted (or possibly added), and the trade-offs redefined.

However, the key to an optimal choice is in having a set of options to choose from that does indeed contain an optimal solution. Thus, the generation or design of alternatives is a most important, if not the most important step. The selection process is then based on a comparative analysis of the ranking and elimination of (infeasible) alternatives from this set.

Modeling for decision support, or model based decision support systems for engineering problems have been discussed and advocated for a considerable time (Loucks et al., 1985). Success stories of actual use in the public debate and policy making process are somewhat rarer, in particular at the societal rather than commercial end of the spectrum of possible applications.

2.2 Data base model

The proposed system utilizes the usefulness of organized data collections, and various forms of data base management software is quite generally recognized. And modelers and certainly model users are quite aware of the fact that input data preparation is often the main effort in applied modeling. So the integration of the incorporated to the decision support system data bases and models allow users to automatically retrieve and load input data for complex engineering models.

2.3 Multicriteria system

The Multicriteria system can be used to assign a value to an output variable given a set of input variables by using rules and logical inference rather than numerical algorithms.

Generally, in the context of models, multicriteria systems are often used to help configure models (implementing an experienced modelers know-how to support the less experienced user) and estimate parameters. A number of these "intelligent front end systems" or model advisors have been developed in the environmental domain (Fedra, 1993). A rule-based approach can also be a substitute for a numerical model, in particular, if the processes described are not only in the engineering, but in socio-economic domain. An example could be environmental impact assessment based on a checklist of problems, which can be understood as a diagnostic or classification task. A qualitative label is assigned to potential problems,

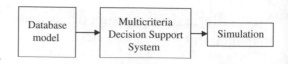

Figure 1.   Decision Support System flow chart.

based on the available data on environment and planned action, and a set of generic rules assessing and grading the likely consequences. Recent examples of such systems are given in Hushon (1990) and Wright et al. (1993) (Fedra, 1993). An example of such a rule-based system for impact assessment is described below. The role of the multicriteria system can be seen in Figure 1.

The flexibility to use, alternatively or conjunctively, both symbolic and numerical methods in one and the same application allows the system to be responsive to the information at hand, and the user's requirements and constraints. This combination and possible substitution of methods of analysis, and the integration of data bases allows to efficiently exploit whatever information, data and expertise is available in a given problem situation.

The approach used here is based on a model of human problem solving that recursively refines and redefines a problem as more information becomes available or certain alternatives are excluded at a screening level. Learning, ie., adaptive response to the problem situation and the information available, and the ability to modify function and behavior as more information becomes available, is a characteristic of intelligent systems.

The Multicriteria Decision analysis is based on the Composite programming. Composite programming was introduced by Zeleny (1973) and uses a single level analysis and a non-normalised distance. The first step behind the conceptual basis (Manoliadis, 2001) was to identify the criteria that are susceptible to change due to systems operation, and then categorize these criteria under main headings. The physical factors that were known to be responsible for changing the criteria conditions were then identified. After these basic indices have been identified, the process of building the main headings from its elementary components by defining the structure of the composite programming model begins. The process can be characterised as grouping the basic indicators into clusters based on either similar characteristics or the desire to contrast different features through trade-off analysis (for example economic vs environmental). The initial set of indicators that result from grouping basic indicators are called first level indicators. After the first level indicators have been specified, the grouping process continues until the highest level indicators have been specified.

The weight assigned to the particular management performance index reflects the relative preference of that element compared to other elements within the same group. Composite programming uses a double weighting scheme to reflect the importance of the maximal deviation between the indices used. Therefore the decision maker must specify a value for the balancing factors.

The ideal and worst values for each of the indicators are final components needed to describe the composite programming model structure.

After the structure of the management system has been formulated, an impact relationship, $R(x)$, for each of the basic indicators, using the ideal $R(i)$, and worst $R(w)$ values, of each criterium and for each management alternative, x, can be calculated:

$$R(x) = \frac{R(i) - R}{R(i) - R(w)} \qquad (1)$$

The impact relationship is a measure of the basic index value for each of the management options and serves as an input to the management system model. Values for basic indices are obtained from the simulation modelling of different alternatives.

Evaluation of the various management options is accomplished by calculating the composite distance within each group of indicators, beginning at the basic level and then progressing to successively higher levels.

The preferred management option is then identified from the candidates by locating the system nearest the ideal point in terms of the composite distance.

2.4 Simulation

The final step of the integration is the coupling of simulation of the preferred option: simulation usually requires an (often gross) simplification of the problem representation to become tractable. Simulation models, are capable of representing almost arbitrary levels of detail and complexity, are rarely capable of solving inverse problems, i.e., determining the necessary set of inputs or controls to reach a desired outcome.

One can, however, combine the approaches in that a simplified model (e.g., steady state and spatially aggregated) is used as the basis for optimization; the result is then used as the basis for a more detailed, e.g., dynamic and spatially distributed simulation model, that also keeps track of the criteria, objectives and constraints used for the optimization, but with a higher degree of temporal resolution, and possibly a more refined process description. If, in the simulation run, constraints are violated or objectives not met, the corresponding values can be tightened or relaxed in order to obtain a new solution which again is subjected to more detailed examination.

3 APPLICATION-RESULTS

An example of this methodology for complex analysis of a building life cycle is presented in the following paragraphs In the literature, numerous studies have been carried out with the aim of integrating the various project life-cycle phases through IT solutions (Bjork, 1989; Froese & Paulson, 1994; Wix, 1997). Due to the wide range of activities and professional views in the construction industry, many of the developed data and process models have either been developed for a specific stage or a few stages of the project life cycle. Consequently, many "isolated" decision models have been produced to tackle different aspects of the project life cycle. This isolation has reinforced current practices, failing to address fully the question of integration and related issues such as shared information, common processes, etc. (Yamazaki, 1995; Shaked & Warszawski, 1995). To provide formal yet practical decision support requires in this research a new approach, that supports a more open and participatory decision making process. A new paradigm of man-machine systems is needed where the emphasis is no longer on finding an optimal solution to a well defined problem, but rather to support the various phases of the problem definition and solving process by using a Multicriteria Decision Support System based on Composite Programming. For this purpose, a data base simulation model based on the BDA (Building Decision Analysis) was used (Fig. 2).

BDA is a computer program that supports the concurrent, integrated use of multiple simulation tools and databases, through a single, object-based representation of building components and systems. BDA acts as a data manager, allowing building designers to benefit from the capabilities of visualization tools throughout the building design process.

The BDA is implemented as a Windows™-based application for personal computers. In addition to the

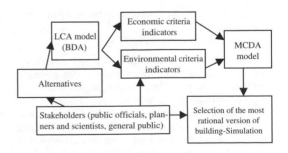

Figure 2. Simulation program flow chart.

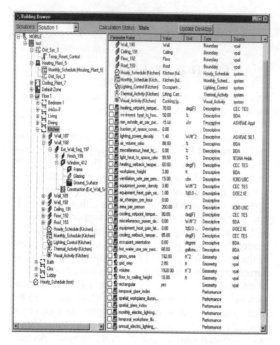

Figure 3. Schematic Graphic Editor for the application (Papamichael 1997).

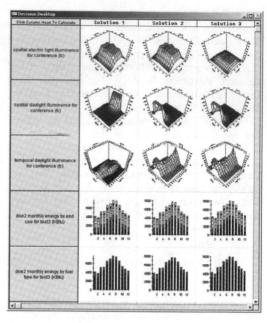

Figure 4. DBA output (Papamichael 1997).

The Schematic Graphic Editor, (Fig. 3) the current version of the BDA is linked <u>ATHENA</u> (lifecycle cost of materials). ATHENA v2.0 Environmental Impact Estimator uses that data in a building systems context to capture the full environmental story for a conceptual building design

A practical realization of the model BDA for this example was being developed step by step as follows:

- a comprehensive quantitative and conceptual description of the life cycle of a building (i.e office)
- its stages
- interested parties and environment expressed in terms of cost and related criteria (economic, environmental)
- development of the alternatives based upon feasible solutions
- Management alternatives.

A discrete set of three solutions based on lighting in an office building used for conference (Papamichael, 1997) form the possible alternatives. In general the range of options can vary from a small number to a large number of alternatives. Each solution has a number of alternative control technologies. Each option is associated with costs, and for a given overall budget (Fig. 4) and the proposed multicriterion decision analysis model finds the most effective (in terms

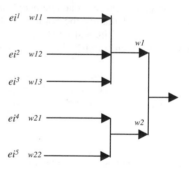

Figure 5. Schematic of indicators.

of cost and the environmental impact) investment strategy. That can be further analyzed by a discrete multi-criteria optimization tool by varying the budget, or the time horizon and discount rate for the cost estimates results in a large number of scenarios (Zhao et al., 1985).

(1) Environmental Indices and Composite Programming Structure

The performance indices designated for use in this case study were selected to provide values for basic indicators in the composite programming algorithm. Basic indicators were synthesised as to provide representative measures of performance with respect to economic and environmental criteria. The composite

Table 1. Calculation of composite distance.

Indicators	Weight	Sol. 1	Sol. 2	Sol. 3
Environmental	$I1 = 0.6$	$C1 = 0,35$	$C2 = 0,32$	$C3 = 0,39$
Economical	$I2 = 0.4$	$N1 = 0,67$	$N2 = 0,59$	$N3 = 0,57$
Composite distance		$S1 = I1*C1 + I2*N1$	$S1 = I1*C2 + I2*N2$	$S1 = I1*C3 + I2*N3$
		$= 0,48$	$= 0,39$	$= 0,46$

structure is formulated so that indices lead to trade-off between these criteria. Formulation of the case study is presented for illustrative purposes only. It should be noted that composite programming is a general purpose multicriterion decision making algorithm and it could be implemented to any type of performance measure. The case study five indices, (Fig. 3) namely spatial electric light illuminance for conference, spatial daylight illuminance for conference, temporal daylight illuminance, doe2 monthly energy by end use, monthly energy by fuel type.

These indices were aggregated through two higher level to obtain the final trade-off economic and environmental system performance. An illustration of the composite programming tree structure is presented in Figure 5.

Balancing factors, as for instance w_{12}, w_{22} can be used at any level of the multicriterion decision making.

(2) Analysis of Management Options

Selection of a preferred option is completed by determining which system is robust with respect to changes in preferences toward different criteria that measure system performance. These preferences may be expressed by various scenarios. The following scenarios were investigated individually within the context of the multicriterion decision making.

1) Express a strong concern for concern environmental impacts.
2) Stress the importance of minimum cost Preference weights that emphasise each of the overall objectives are given in Table 1.

Balancing factors are assumed equal to the overall index for each of the three scenarios

3 RESULTS

The calculation of the composite distance of each solution is presented in Table 1.

Examining the results in terms of the first management option "Express a strong concern for concern environmental impacts" *alternative* II is ranked first.

In terms of the performance of the system in cost alternative III ranked first.

The score of each alternative computed as the relative distance of the ideal solution (0.00,0.00,0.00),

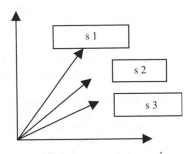

Figure 6. Composite distances.

assuming equal weight assigned to each index is as follows:

I(0,42), II(0.39), III(0,46)

Therefore the solution is that ranks better to all management options (or the better solution) is alternative number II since it has the closest distance from the ideal (Fig. 6).

Simulation of the preferred alternative as well as comparison of the results is then achieved by using the model DBA. An "optimal" emission scenario can then be used again at the level of the simulation model BDA – ATHENA2.0, and tested with a broad range of individual short-term scenarios (rather than the frequency data used for the long-term model) to test the abatement strategy under specific, including worst case assumptions

4 DISCUSSION

Engineering planning and management is an inherently and increasingly complex task. To provide formal yet practical decision support a new approach is introduced based upon composite programming, that supports a more open and participatory decision making process. A new paradigm of man-machine systems is needed where the emphasis is no longer on finding an optimal solution to a well defined problem, but rather to support the various phases of the problem definition and solving process.

Problem owners and various actors in the decision making process in building design have a central role;

supporting their respective tasks requires man-machine interfaces that are easy to use and easy to understand. An effective decision support system is present in order to first of all provide a common, shared information basis, framework and language for dialogue and negotiation. An information system that can cater to all these needs must be based on more than good science and solid engineering.

The dialogue between the actors in the decision making process is extended to a dialogue with the DSS, and BDA which plays the role of a technical expert and bookkeeper rather than an arbiter.

This requires that information provided is adequate for and acceptable to a broad range of users involved in the respective assessment and decision making processes.

The need for better tools to handle ever more critical environmental and engineering planning and management problems is obvious, and the rapidly developing field of information technology can provide the necessary machinery. The biggest challenge, however, seems to be the integration of new information technologies and more or less mature formal methods of analysis into institutional structures and engineering processes, that is, putting these tools to work in practice.

REFERENCES

Bjork B.C. 1989. *Basic structure of a proposed building product model, Computer Aided Design*, Vol. 21, No. 2, March, pp 71–78.

Fedra K. 1993. *Expert Systems in Water Resources Simulation and Optimization*. In: Stochastic Hydrology and its Use in Water Resources Systems Simulation and Optimization, J.B. Marco et.al., editors. pp 397–412 Kluwer Academic Publishers, The Netherlands.

Froese T. & Paulson B.C. 1994. *OPIS: an object model-based project information system*, Microcomputers in Civil Engineering, 9(1), pp 13–28.

Hushon 1990. *Expert Systems for Environmental Applications*, ACS Symposium Series 431. American Chemical Society, Washington DC.

Loucks D.P., Kindler J. & Fedra K. 1985. *Interactive Water Resources Modeling and Model Use: An Overview*. Water Resources Research 21(2), pp 95–102.

Manoliadis O., Baronos A., Tsolas I. & Vatalis K. 2001. *Environmental Impact Assessment of Irrigation Systems Using Sustainability Indices*, 9th Intrenational Conference of Environmental Science and Technology Syros Greece Vol. C, pp 291–298.

Papamichael K. 1999. *Application of information technologies in building design decisions*, Building Research & Information, Vol. 27, No. 1 pp 20–34.

Papamichael K., John La Porta & Hannah Chauvet 1997. *Decision Making through Use of Interoperable Simulation Software* Proceedings of the Building Simulation '97 Fifth International IBPSA Conference, Vol. II, September 8–10, Prague, Czech Republic.

Shaked O. & Warszawski A. 1995. *Knowledge – based system for construction planning of high – rise buildings*, Journal of Construction Engineering and Management, Vol.121, No. 2, June.

Yamazaki Y. 1995. *An integrated construction planning system using object-oriented product and process modelling*, Construction Management and Economic, Vol. 13, No. 5, September, pp 427–434.

Wix J. 1997. *Information models and modelling: standards, needs, problems and solutions.*, The International Journal of Construction Information Technology, pp 27–38.

Zhao Ch., L. Winkelbauer & K. Fedra 1985. *Advanced Decision-oriented Software for the Management of Hazardous Substances*. Part VI. The Interactive Decision-Support Module. CP-85-50, International Institute for Applied Systems Analysis, A-2361 Laxenburg, Austria.

Integrated project management information systems in construction: a case study implementation

P. Stephenson & I.C. Scrimshaw

School of Environment and Development, Sheffield Hallam University, Sheffield, UK

ABSTRACT: The generation and communication of information during a project life cycle is extensive owing to the complexity of design and construction processes, coupled with the parties involved at the various stages of development. Information, in its various forms, needs to be communicated to ensure project participants receive the required information to carry out the functions and processes required. In current project systems, information technology plays a major role in facilitating these tasks, but its employment and utilisation varies considerably depending upon the stakeholders involved, and the demands and requirements of a project. Integration of information is one means of providing the opportunity for a seamless flow of information to the parties concerned in a timely and consistent manner to facilitate effective communications.

Integrated Project Management Information Systems (IPMIS) utilise the benefits of information technology to provide an accurate and effective tool for the control and management of complex modern construction projects. This paper investigates the utilisation of IPMIS in the UK construction industry. A case study is introduced based on a major civil engineering project involving the construction of highways and bridges to determine the usefulness of an IPMIS. Information requirements and flows are determined, together with the specific information technology needs of the project. A conceptual model of an IPMIS is developed and implemented utilising appropriate software. Specific problems associated with the implementation are highlighted, together with identified benefits. Key lessons learnt from the systems implementation are also described.

1 INTRODUCTION

Over recent years there has been extensive growth in the specialist field of project management. Increasing numbers of organisations are now seeing the benefits of managing projects more effectively (PMI 1998; Zipf 1998). The management and control of project cost and programme, is an essential to facilitate successful project outcomes. However, as the constraints of project complexity, time and cost have become more important, so more sophisticated management systems have needed to be developed (Bjork 1999).

Effective control of the information flow during the construction process is critical to the success of the project (Dawood et al. 2002). The control of information flow has traditionally been achieved through the use of paper-based systems of files and registers to administer drawings and specifications, correspondence and progress. These systems, often developed in an *ad hoc* manner over time, have invariably differed between organisations, owing to the entry, processing and flow of information. Moreover, computers are increasingly becoming a central component of such

systems (Broyd 2000; Shahid & Froese 1998), which form the essential tools for the provision of accurate timely information to assist in the decision making process.

2 DATA MANAGEMENT IN ENGINEERING

The availability of cost effective computing resources and the utilisation of information technology for the transfer of data provides considerable opportunity for modern businesses to more effectively and efficiently control their projects (Froese et al. 1997). Engineering projects are particularly data rich in the form of text documents, drawings and CAD models; much of the data invariably changing and evolving throughout the life of a project.

However, there are still many parts of the construction industry which relies on hard printed copies of documents, although more recently, a greater proportion of construction professionals have begun to create and exchange data using information systems (Cowperthwaite et al. 1999). Data management system currently being used tend to have been adopted and evolved from other industries. Such systems include electronic documents interchange (Rezgui & Cooper 1998; Cowperthwaite et al. 1999; Debney 1999), electronic drawing management systems (Port 1994) and project portfolio management systems.

3 THE NEED FOR IPMIS

With the increasing use of information technology, more and more organisations are seeking to transfer both technical and commercial data electronically (Remenyi & Sherwood-Smith 1998; Gardner 1999). If this is to be achieved on a widespread basis, there must be an acceptance of new technology by the relevant parties involved (Doherty 1997; Howard et al. 1998; Rivard 2000).

Many organisations are now using *ad hoc* systems between their existing software programs, but these systems are sometimes difficult to use and much time is lost through incompatibilities in the different types of software. More recently there has been a trend with software developers to evolve more integrated management systems, with the aim being to bring all of a project's operations into a common shared environment (Russell & Froese 1997). As a result, project staff have a wider view of a project's financial information, scheduling and overall progression (Jaafari & Manivong 1998). However, for multi-disciplinary projects these packages are still unable to offer all of the operations required for complex projects.

4 CASE STUDY

The case study in this research was based on a UK motorway widening project, including strenghthening of an existing bridge structure. On award of the contract, the client (Highways Agency) in conjunction with its professional advisors (consulting engineers) decided to enter into an informal partnering arrangement with the contractor and its specialist subcontractor (bridgeworks). It was hoped a closer working relationship and information sharing would remove the adversarial and confrontational relationships between the parties.

A review of existing systems was carried out to determine the requirements and aspirations of the users and to discover the information flows (see Fig. 1).

Prior to the introduction of the partnering agreement, all flows of information followed the formal route, and this led to considerable delays (in the order of weeks) in the transmission of documents. The form of contract used was the ICE 5th edition which required that all contractual correspondence would be in writing with facsimile and e-mail communication being not acceptable.

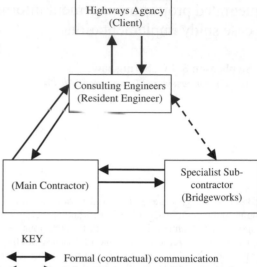

KEY

◀━━━━▶ Formal (contractual) communication
◀ ─ ─ ▶ Informal (non-contractual) communication

Figure 1. Information flows between users.

The introduction of partnering allowed informal communication to take place, with the proviso that formal communication continued to be used as required to meet contractual requirements. The introduction of the integrated project management information system was facilitated by the partnering agreement. Each of the parties had previously operated their own project management information system, but these were essentially paper-based and were individual to the organisation. While any of the parties could have introduced and operated their own integrated project management information system, utilising the benefits of IT, its effectiveness would have been diminished due to the existing contractual relationships.

5 SPECIAL INFORMATION TECHNOLOGY NEEDS

5.1 *Information requirements*

The information to be transferred between the members of the team included letters, reports, drawings for approval and requests for information, written instructions, construction programmes and progress updates. The scale of the project resulted in a considerable volume of correspondence passing between the individual members of the team with consequent transmission delays. To reduce the need for much of the paper-based communication, the team members required a robust secure storage environment for information and data relating to both technical and financial matters. The individual team members could then access and submit data relevant to their function. A central data storage and retrieval (data warehouse) facility with on-demand access was required. Commercial considerations of confidentiality of financial and personal information required the use of system security access rights to deny access to certain directories for unauthorised users. The most significant need was to reduce the volume of paper-based documents to be transferred.

5.2 Administration applications

The requirement for general administration applications on the case study was considerable. Several hundred personnel were working on the site, and the processing and efficient management of data was essential to the maintenance of good labour relations and efficiency. This information needed to be linked into the IPMS to allow calculations related to productivity and cost-recording. There was also a need for this information to be linked to the sub-contractor's head office and the joint commercial team for preparation of valuations and applications for payment on a monthly basis from the client. These systems also included procurement, ordering of material and accounts.

5.3 Project management applications

The project planning and control system required for this project was required to deal with the high level of complexity of the project. Large volumes of data had to be imported and exported throughout the life of the project to allow the monitoring of progress and costs. The planning software required for this project, therefore required the consolidation of individual work-package programmes into the overall master programme. There was also the need for a lower level planning software which allowed section managers to plan and monitor their element of the project. Additionally, there was the need for the management of contract drawings and approvals.

5.4 Technical applications

A requirement of the project was appropriate design software packages for the structural design of the contract works in both 2D and 3D. The data produced from this software also needed to be linked to software to produce detailed fabrication drawings.

5.5 Works applications

The works applications that were required on the project included a material and production control system. These needed to be interfaced with the detailed fabrication drawing system and also to the project management applications systems. The labour hours and production records were required to interface with the materials control system, the accounts system and also the project management system.

5.6 Financial applications

All of the information and application systems listed above were required to be interfaced with the project cost control system. Such data would be consolidated and monitored by the project commercial manager and his team of quantity surveyors. Information gained in this manner would be utilised in the preparation of applications for payment of the works completed.

6 CONCEPTUAL MODEL OF THE IDEAL INTEGRATED SYSTEM

The identification of the project information technology requirements and specific problems concerning the bridge project, along with the knowledge of the need for information flow between identified parties, allowed the determination of a conceptual ideal integrated project management systems for the project. The model needed to connect individual applications and thus conceptually allow the sharing of information between relevant departments (see Fig. 2). Once this conceptual model had been determined, efforts were made to identify potential software that could be used to meet the tasks required.

7 IMPLEMENTATION OF IPMS

Once the key software had been identified, it was decided to implement an integrated project management system. The speed of introduction of the system was of major importance to increase productivity and to recover the existing delays. The introduction process was to be phased to minimise disruption to the existing systems, while construction works continued on the project. Due to the requirement of speed, it was agreed that currently available proven technology and software would be utilised.

The IT strategy was developed to improve workflow by introducing IT solutions. These solutions were put in place to assist with the flow of information and to enable contract teams and departments to concentrate on the critical areas of their work. It was essential to develop buy-in and acceptance from the project teams.

It was also recognised that the success of the system introduction would be dependent upon the enthusiastic adoption and utilisation by members of the project team.

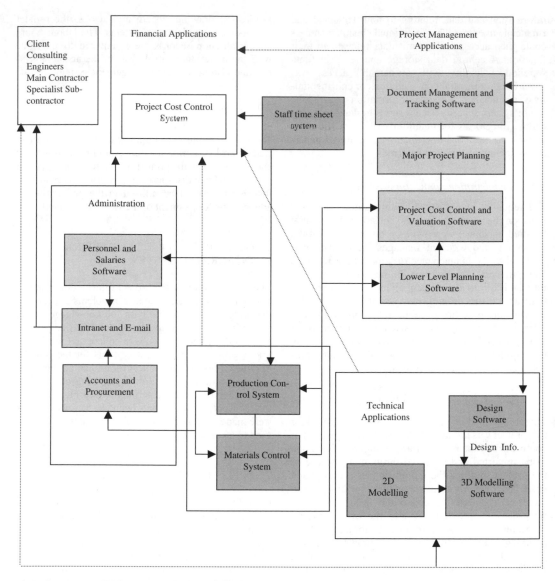

Figure 2. A conceptual model of an Integrated Project Management System.

The software required for this project was broken into five main areas, including, administration, works, technical, project management and financial applications. The selected software to be utilised on the project is shown in Table 1.

7.1 Implementation issues

The integrated project management system described was implemented in stages over a period of three months. Those elements of the system that were already commercially available were introduced first as stand-alone elements. Other elements were developed during this period. The speed of introduction was imperative for improving the performance of the contract, in order for allowed benefits to be realised more quickly.

Training of staff was introduced at the time of implementation. There was general reluctance in the acceptance of the system from a number of employees, and this was mainly due to unfamiliarity of techniques used, computer hardware, and a fear of staff reduction once the system had been implemented. On-going assistance was available throughout the

Table 1. Software used for the IPMIS.

	Software used	Description
Project Management Software		
Project cost control and valuation software	Primavers Expedition	Commercially available project valuation software control software, interfaced to the other systems
Lower level planning software	Suretrack Project Manager	Project planning software package produced by Primavera for the non-planning specialist, whilst offering full integration with P3, used for workfront planning
Document management and tracking software	CYCO	Commercially available electronic document management software
Major project planning	Primavera Project Planner P3	Project planning software for complex projects but requiring specialist operation, interfaced to the other systems
Administration Software		
Personnel and salaries software		Bespoke software developed by specialist sub-contractor to meet administrative requirements
Accounts and procurement		Existing software systems of specialist sub-contractor
Intranet and e-mail	MS Exchange	Intranet to provide project-wide communications
Works Applications		
Production control system	GODATA	A bespoke system developed to meet the unique requirements of the project
Materials control system	GODATA	As above
Technical Applications		
Design software	CSC Design Suite	Commercially available structural design software
3D modelling software	XSTEEL	Commercially available structural steelwork fabrication software
2D modelling	ME10	Commercially available structural design software
Financial Applications		
Project cost control system	MS Access	A database system developed to meet the requirements of the joint commercial team, interfacing into the other applications

introduction of the system in the form of a help desk. Additionally, there was a member of the IT implementation team available on site throughout the implementation period.

7.2 Benefits of implementing the system

The benefits of the implementation were identified as:

- Productivity improvements achieved by providing users of business management, project management and financial management systems with improved access to supporting documents.
- Saving in production of paperwork and transmission time.
- Reduction in the overall number of staff required to administer the contract.

- Improved efficiency of works carried out.
- Faster delivery of documents to users' desks.
- Improved awareness and availability of document data.
- Availability of information to the project team.

7.3 Key lessons learnt during the implementation

Several lessons were learnt during the implementation of the system. These included:

- Objectives and goals of the system should be clearly defined.
- System implementation should be planned to ensure economy of effort and integration with other products.

- System implemented should be attractive to users and easy to use.
- End users need to be consulted early in the project; and information of potential problems needs to be passed to users as early as possible.

It is important that users do not have an over-inflated expectation of the use and performance of the system.

8 CONCLUSION

Technology currently exists to facilitate the requirements of the construction industry. The continuing development of commercial IT provides applications which lead the development of the construction industry, (a "technology push" viewpoint) rather than the construction industry providing a demand for new applications which are then fulfilled by the IT industry. There is a requirement, therefore, for users to remain in touch with IT developments and for organisations to maintain a proactive culture in response to change.

The level of utilisation and effectiveness of an IPMIS is heavily dependent upon the attitudes of the users. At the commencement of the study, many members of the project team had either limited or no IT skills whatsoever. Therefore training in basic skills (word processing, spreadsheets, e-mail) was a precursor to the operation of the system. Their training could be considered as a business benefit, but due to the transient nature of construction project teams, senior managers viewed the training as a project cost to be borne at site level rather than staff development to be funded by their company training budgets. Resistance to the use of the system was minimised by a positive management insistence on its adoption and utilisation.

The introduction of an IPMIS is motivated by the cost and time savings which are gained by its use. However, failure to address the human element of the system will invariably affect the overall effectiveness of the system. Construction organisations traditionally work on short-term time horizons for investment decisions which results in only larger projects having the opportunity to utilise IPMIS, and where the costs of setting up a system can be justified on that project alone. Smaller projects where an IPMIS could provide even greater benefits are unable to justify the costs involved and continue to operate in a piecemeal manner.

The implementation of the IPMIS was carried out to utilise the benefits of IT in order to provide an accurate and effective tool for the control and management of a complex project. However, the case study revealed that many problems concerning the usage of these systems. While the software and hardware is now available to allow technological progression within the construction sector, the acceptance and utilisation of these systems is, in many cases, of greater importance than the technological issues.

REFERENCES

Bjork, B.C. 1999. Information Technology in Construction: domain definition and research issues, *International Journal of Computer Integrated Design and Construction*, 1(1), 3–16.

Broyd, T. 2000. The Impact of IT on Design and Construction, *Journal of the Institution of Civil Engineers*, 138 (2), 87–96.

Cowperthwaite, S. Raven, G. & Richards, M. 1999. Integrating Culture and Technology, *The Structural Engineer*, 77 (1), 16–20.

Dawood, N. Akinsola, A. & Hobbs, B. 2002. Development of automated communication of system for managing site information using internet technology, *Automation in Construction*, 557–572.

Doherty, J.M. 1997. A Survey of Computer use in the New Zealand Building and Construction Industry, *Electronic Journal of Information Technology in Construction*, Vol. 2, 1–13.

Debney, P.M. 1999. CAD today is only just the beginning, *The Structural Engineer*, 77 (3), 16–20.

Froese, T. Rankin, J. & Yu, K. 1997. Project Management Application Models and Computer-Assited Construction Planning in Total Project Systems, *International Journal of Construction Information Technnology*, 5(1), 39–62.

Gardner, P.J. 1999. IT in Engineering Consultancies, *The Structural Engineer*, 77 (3), 13–15.

Howard, R. Kiviniemi, A. & Samuelson, O. 1998. Surveys in the Construction Industry and Experience of the IT Barometer in Scandinavia, *Electronic Journal of Information Technology in Construction*, Vol. 3, 45–56.

Jaafari, A. & Manivong, K. 1998. Towards a smart project management information system, *International Journal of Project Management*, 16(4), 249–265.

Port, S.R. 1994. Data Management in Engineering, *Proceedings of the Institution of Civil Engineers*, 92, 133–137.

PMI (Project Management Institute) 1998. *Value of Project Management*, Project Management Institute.

Remenyi, D. & Sherwood-Smith, M. 1998. Business benefits from information systems through an active benefits realisation programme, *International Journal of Project Management*, 16(2), 81–98.

Rezgui, Y. & Cooper, G. 1998. A proposed open infrastructure for construction project document sharing, *Electronic Journal of Information Technology in Construction*, Vol. 3, 11–24.

Rivard, H. 2000. A survey on the impact of information technology on the Canadian Architecture, Engineering and Construction Industry, *Electronic Journal of Information Technology in Construction*, Vol. 5, 37–56.

Russell, A. & Froese, T. 1997. Challenges and a Vision for Computer-Integrated Management Systems for Medium-Sized Contractors, *Canadian Journal of Civil Engineering*, 24(2), 180–190.

Shahid, S. & Froese, T. 1998. Project Management Information Control Systems, *Canadian Journal of Civil Engineering*, 25(4), 735–754.

Zipf, P.J. 1998. An integrated project management system, *ASCE Journal of Management in Engineering*, 14(3), 38–41.

System-based Vision for Strategic and Creative Design, Bontempi (ed.)
© *2003 Swets & Zeitlinger, Lisse, ISBN 90 5809 599 1*

Multi-project scheduling in a construction environment with resource constraints

P. Stephenson & Y. Ying
School of Environment and Development, Sheffield Hallam University, UK

ABSTRACT: The simultaneous management of projects by organisations is a common occurrence within the construction sector. Many contracting organisations now work in an environment carrying out work for a variety of clients under different procurement methods. This often results in complex situations associated with multi-project scheduling where projects have to compete for, and share, common resources.

The limitation of resources also exacerbates the problem and makes the scheduling process much more difficult. The traditional project management tools such as, bar charts, CPM, and PERT have their own limitations in multi-project situations; and while many heuristic rules have been developed to address scheduling problems, it is often difficult to select the most appropriate rule(s) to suit the continually changing conditions in multi-project environments.

This paper addresses the issues surrounding multi-project scheduling. This includes traditional and heuristic approaches for scheduling projects covering both time and resource implications. Different approaches are covered in dealing with these situations, for the control of project duration. Details are also provided of an industrial survey concerned with multi-project scheduling which identifies current practice within the UK construction industry. The implementation of strategies in multi-project planning are also covered by way of stand-alone, integrated and combined systems, and a combined implementation model is developed for the multi-project planning process.

1 INTRODUCTION

The management of multiple projects constructed simultaneously with an organisation is now becoming common place in construction. However, the context of the multi-project tends to be one of instability for project participants. Change in multi-project planning and scheduling is inevitable and often frequent, therefore the traditional scheduling approaches are no longer feasible in a dynamic, multi-project setting.

On the other hand, the potential for improvement in the management of simultaneous multiple projects is significant. It is estimated that up to 90%, by value, of all projects are carried out in a multi-project context (Turner 1993). The influence of even a small improvement in their management on the project-management field could therefore have great impact on both time and cost.

2 APPROACHES TO PROJECT SCHEDULING

Ever since the development of critical path methods in the 1950s problems have been recognised concerning resource-constrained multi-project scheduling. Its importance has increased in recent years with the increasing cost of both direct labour resources and indirect costs which tend to increase the cost of project delay, often disproportionately (Kurtulus & Davis 1982).

The three popular project scheduling techniques include Bar chart, Critical Path Method (CPM) and Project Evaluation and Review Technique (PERT) which have been the focus of researchers over several years. This includes resource constraints with CPM (Badiru 1993; Gemmill & Tsai 1997), and also with PERT (Badiru 1993; Gordon & Tulip 1997). Traditionally, the resource-constrained project-scheduling problem is approached by using visual inspection, and arranging resource utilisation sequence by visually inspecting the project network to provide limited resources to the activities in the critical path (Badiru 1993; Coman & Ronen 1995; Demeulemesster & Herroelen 1994). Additionally, heuristic methods can be employed for large and complex projects, where the near optimal project schedule can be addressed using criteria concerning time and sequence of activities (Badiru 1993; Demeulemeester & Herroelen 1994; Farid & Manoharan 1996).

To date, many heuristic scheduling rules have been applied to the single-project scheduling problem and have been shown to be effective providing near optimal solutions (Davis & Patterson 1975). Some studies (Patterson 1976; Kurtulus & Narula 1985; Allam 1988) have discussed the effects of problem characteristics on the performance of the heuristic procedures. The results of these studies revealed that careful choice of the most appropriate scheduling rule is extremely important since the scheduling factors may have considerable effects on the performance of the scheduling rule. For example, the minimum slack first (MINSLK) (Davis & Patterson 1975), the minimum late start time first (MINLST) (Moder & Phillips 1964), the activity-time (ACTIM) and activity-resource (ACTRES) criteria (Bedworth 1973), the resource over time (ROT) criterion (Elsayed 1982) and the activity resource controls (ACROS) criterion (Elsayed & Nasr 1986) fall into this type of heuristics. Additionally, weighted combination search methods have been employed (Whitehouse & Brown 1979), including the weighted ROT-ACTIM criterion, the weighted ROT-ACTRES criterion (Elsayed 1982) and the weighed ACTIM – ACROS (TIMROS) criterion (Elsayed & Nasr 1986).

Due to the variety of network structures and resources, no single heuristic method can always produce the best project schedule for all problems. Boctor (1990) presented a three-heuristic approach which increased the possibility of obtaining the best or even the optimal solution. Other previous studies in this area include those of Morse & Whitehouse (1988), Boctor (1993), Moselhi & Lorterapon (1993) and Bell & Han (1991).

To solve the multi-project scheduling problem, researchers such as Patterson (1976) have used the single-project approach. Kurtulus & Davis (1982) presented the multi-project approach which dealt with each project as an independent project, indicating that the two approaches might produce different project schedules even while using the same heuristic scheduling rules. Chiu & Tsai (1993) further investigated the effectiveness of the two approaches using four heuristic rules under two different objective functions (i.e. the minimum total project delay and the minimum total project duration). Conversely, the multi-project approach was found to be superior to the single-project approach under the criterion of the minimum total project delay. Therefore, it is particularly important to analyse the network structures and objective functions and to choose an appropriate heuristic in order to solve the multi-project problem effectively.

The theory of constraint (TOC) proposed by Goldratt (1997) makes emphasis on the systematic management of project by identifying the uncertain factors hindering project implementation, and suggests the global deployment resources. Thus, the TOC approach focuses on the constraint that blocks the achievement of goal of the project (Rand 2000; Wei, Liu & Tsai 2002). Traditionally time estimates for individual activities contain some provision for contingencies. Critical chain scheduling aggregates these provisions into a project buffer, and as a result of aggregation, the total contingency reserve and project duration can be reduced (Steyn 2002).

In order to provide a deeper understanding of multi-project management, some important areas must be considered such as, capacity, conflict, commitment and context (Reiss 1996). Resource conflicts, in particular, have to be a major consideration when planning a series of projects, since many projects vie for the same resources (Harrison 1985). Conflicts arise because more than one project manager has planned more than one project containing more than one task to occur simultaneously. The result of these tasks is that the resources available may not always satisfy the required demand. To alleviate such situations, Reiss (1996) proposes seven stages in managing a portfolio of projects. These include planning, transmission, consolidation, evaluation, experimentation and decision making, dissemination and the measurement of achievement.

3 INDUSTRY SURVEY

The sample for the survey on multi-project scheduling was based on large organisations operating in the UK construction industry. The top 150 companies (ranked by turnover and employees) were selected for the sample questionnaire survey. Specific questions were grouped into sections comprising general information, multi-project scheduling, resources constraints, scheduling approaches and the use of software systems for multi-project management.

Over 65% of responses were from project managers and construction managers, and all respondents in the survey indicated that their organisations were in the process of planning multi-projects. The most important factor considered in multi-project scheduling was identified as resources followed by project duration and cost estimation. The prioritisation of projects in a multi-project environment was also found to be based on profitability followed by urgency and project size. Furthermore, while resource issues were identified as problematic, over 90% of companies never refused to tender for work. In dealing with resource issues, delaying start times, obtaining additional resources and the transfer of resources from existing projects were seen as actions to deal with acquiring new projects. In scheduling projects, over 60% used CPM, closely followed by bar charts with only 5% of organisations using PERT.

Table 1. Use of heuristics in multi-project scheduling.

Choices	Responses	%
SOF – the shortest operation first	2	7
MINSLK – the minimum slack first	5	18
SASP – the shortest activity from the shortest project	2	7
LALP – the longest activity from the longest project	4	14
MOF – the maximum operation first	1	4
MAXSLK – the maximum slack first	3	11
MINTWK – the minimum total work content	0	0
MAXTWK – the maximum total work content	0	0
MINLFT – the minimum latest finish time first	7	25
SRD – the smallest resource demand first	2	7
GRD – the greatest resource demand first	2	7
GENRES – the maximum GENRES value first	0	0
ROT1 – the maximum ROT-ACTIM value first	0	0
ROT2 – the maximum ROT-ACTRES value first	0	0
TIMROS – the maximum TIMROS value first	0	0
TOTAL	28	100

With regard to respondents following heuristics to multi-project scheduling there were relatively few responses. The acceptability of heuristics varied from 0–25%. Most organisations applying traditional methods in project scheduling, indicating they would need time to examine the developed heuristics before acceptance. Table 1 shows the response to the choices presented.

With regard to software utilisation, three application packages featured strongly; these being Microsoft Project (45%), P3 or SureTrak (25%) and Open Plan (15%) with other software representing 10% or less.

Implementing strategies of multi-project planning process utilising software tools would appear to fall into three categories. Firstly, a stand-alone systems where planners work with their own copy of a popular PC based project planning systems to plan their individual projects. Secondly, an integrated system where the organisation normally purchases a site licence of a programme planning system, and planners have access to the tool through a local terminal. And thirdly, where other organisations have created combined systems based on the previous two described. Each project team uses a simple stand-alone PC planning system and the project office use a much more powerful system to integrate the many individual project plans on a computer network. The individual plans are created and kept up to date using the popular single-project based tools and the files are transferred to a consolidation system. Consolidation can be achieved by a tool specifically designed for the purpose or by the use of a database driven project planning tool. The project office team manipulates the data within the consolidation system and can report on conflicts across the projects.

4 A DELEGATION MODEL

There are three main strategies aimed at following the current multi-project management model employed by organisations, and these tend to follow the availability of software tools. The methods employed by these organisations to plan and monitor their workload do not represent a model of an organisation's management. Many software tools are predominantly based on a single project philosophy, and such models are therefore structurally flawed and often inappropriate. Hence, an alternative approach which follows an organisation's structure is likely to be more appropriate and applicable.

Most organisations run their project workload by delegation where senior staff identify projects and delegate them to project managers. These project managers take responsibility for the projects and may plan their own workload in appropriate detail. Furthermore, project managers normally require the efforts of resources within the organisation and these may be obtained through a subcontract matrix, a secondment matrix or the resource pool.

A multi-user tool may be installed over a computer network and each user is given a work plan within which an individual's or a team's workload can be planned. Connections between work plans are created by the act of delegating work, and this permits each user to plan at a level that is appropriate to their needs. The act of delegation would establish a link between the two work plans which would carry updated information automatically. In a sub-contract matrix organisation work would eventually be delegated to resource managers or departmental managers who would balance the work load from the many project and sub-project managers.

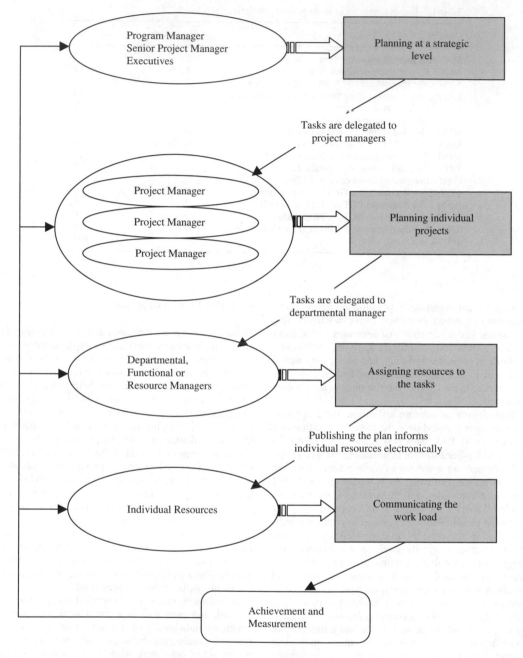

Figure 1. An organisational delegation model for multi-project planning.

In a secondment matrix, resources would be loaned by the resource managers or departmental managers to the many project and sub-project managers on a full time or part time basis. Individual team members would have their own personal work plan, and once individual resources have been assigned to work, a link between the departmental plan and the personal work plan would be established. Resources would receive work instructions through the system and would report actual achievement.

The achievement measurement data would be transmitted through the assignment link and data describing updated plans would be transmitted up the delegation links. Each user on the system would see a view of the total workload which would be appropriate to their specific need and would be able to investigate in greater detail by inspection of lower level plans. An organisational delegation model for multi-project planning is shown in Figure 1.

5 CONCLUSION

The management of multiple simultaneous projects by an organisation is a common occurrence within the construction sector. The context of the multi-project is one of instability and constant change for the participants. However, today's concept and perceptions in project management theories are not sufficient to deal with multi-project situations, because they are based on different assumptions. The traditional scheduling approaches have their significant drawbacks in dealing with multi-project scheduling.

Many heuristic scheduling rules have been designed for the single-project scheduling problem which have been used in multi-project environments. However, the performance of heuristic scheduling rules is significantly affected by network structures, performance criteria and the degree of resource availability. None of the heuristic rules can always produce the best solution for all problems, and it is necessary to develop a system and choose adaptive and stable heuristic rules in order to ensure that they can be effectively applied to practical problems.

The utilisation of project management software within multi-project planning process tends to be based on stand-alone systems, integrated systems and combination systems, which all attempt to make this process work efficiently. One additional approach was also identified as a delegation model to provide better co-ordination within an organisation's structure in order to facilitate improvements to the multi-project planning process.

REFERENCES

Allam, S.I.G. 1988. Multi-project scheduling: a new categorization for heuristic scheduling rules in construction scheduling problems, *Construction Management and Economics*, 6(2), 93–115.

Badiru, A.B. 1993. Activity resource assignments using critical resource diagramming, *Project Management Journal*, 24(3), 15–21.

Bedworth, D.D. 1973. *Industrial Systems: Planning, Analysis and Control*, Ronald, New York.

Boctor, F.F. 1990. Some efficient multi-heuristic procedures for resource-constrained project scheduling, *European Journal of Operational Research*, 49(1), 3–13.

Chiu, H.N. & Tsai, D.M. 1993. A comparison of single-project and multi-project approaches in resource-constrained multi-project scheduling problems, *Journal of the Chinese Institute of Industrial Engineers*, 10, 171–179.

Coman, A. & Ronen, B. 1995. Information technology in operations management: a theory of constraints approach, *International Journal of Production Research*, 33(5), 1403–1415.

Davis, E.W. & Patterson, J.H. 1975. A comparison of heuristic and optimum solutions in resource-constrained project scheduling, *Management Science*, 21(12), 944–955.

Demeulemeester, E. & Herroelen, W. 1994. Solving in the multiple resource constrained, single project scheduling problem, *European Journal of Operational Research*, 76, 218–228.

Elsayed, E.A. 1982. Algorithms for project scheduling with resource-constrained, *International Journal of Production Research*, 20(1), 95–103.

Elsayed, E.A. & Nasr, N.Z. 1986. Heuristics for resource-constrained scheduling, *International Journal of Production Research*, 24(2), 299–310.

Farid, F. & Manoharan, S. 1996. Comparative analysis of resource allocation capabilities of project management software packages, *Project Management Journal*, 27(2), 35–44.

Gemmill, D.D. & Tsai, Y.W. 1997. Using a simulated annealing algorithm to schedule activities of resource constrained projects, *Project Management Journal*, 28(4), 8–20.

Goldratt, E.M. 1997. *Critical Chain*, The North River Press, Great Barrington, MA.

Harrison, F.L. 1995. *Advanced Project Management – a structured approach*, Gower.

Kurtulus, I.S. & Davis, E.W. 1982. Multi-project scheduling: categorization of heuristic rules performance, *Management Science*, 28(2), 161–172.

Kurtulus, I.S. & Narula, S.C. 1985. Multi-project scheduling: analysis of project performance, *IIE Transactions*, 17(1), 58–66.

Moder, J.J. & Phillips, C.R. 1964. *Project Management with CPM and PERT*, Reinhold, New York.

Morse, L.C. & Whitehouse, G.E. 1988. A study of combining heuristics for scheduling projects with limited multiple resources, *Computers and Industrial Engineering*, 15(1–4), 153–161.

Patterson, J.H. 1976. Project scheduling: the effects of problem structure on heuristic performance, *Naval Research Logistics Quarterly*, 23, 95–123.

Rand, G.K. 2000. Critical chain: the theory of constraints applied to project management, *International Journal of Project Management*, 18(3), 173–177.

Reiss, G, 1996. *Programme Management Demystified*, E & FN Spon, London.

Steyn, H. 2002. Project management applications of the theory of constraints beyond the critical chain scheduling, *International Journal of Project Management*, 20(1), 75–80.

Turner, J.R. 1993. *The Handbook of Project-Based Management*, McGraw Hill, London.

Wei, C.C. Liu, P.H. & Tsai, Y.C. 2002. Resource-constrained project management using enhanced theory of constraint, *International Journal of Project Management*, 20(7), 561–567.

System-based Vision for Strategic and Creative Design, Bontempi (ed.)
© *2003 Swets & Zeitlinger, Lisse, ISBN 90 5809 599 1*

Workflow technology and knowledge management in construction process

Marco Masera & Andrea Stracuzzi
Department of Civil Engineering, University of Pisa, Italy

Saverio Mecca
Department of Architectural Technology and Industrial Design, University of Florence, Italy

ABSTRACT: In the construction context the production process involves a large number of operators with heterogeneous roles and diffused knowledge. Knowledge flows in communication process is a key issue for analysing the knowledge management in organisational systems goal oriented and for managing human resources.

The paper is focussed on knowledge capturing through an analysis model of the communication processes supported by a Workflow technique of knowledge representation. A Workflow technique allows an analysis of the production process specifying, for each role: actions, communication and knowledge. The approach aims at developing the analysis of the decisional process, process of extraction and creation of knowledge emphasizing a self-elicitated knowledge representation. This study is devoted to the analysis of Workflow technology in the field of the construction process control.

1 KNOWLEDGE FLOW IN CONSTRUCTION MANAGEMENT

The construction industry is a turbulent environment in which processes are continuously transformed for adapting their relation to changeable conditions. The project re-engineering and the project reorganization imply a continuous fragmentation and destruction of the social and technical processes of knowledge production and reproduction. Corporate constitute an ever more strategic level for organizing and producing knowledge. In the perspective of many studies (Nonaka 1996) Information and knowledge management represent basic sources for developing corporate able to sustain the actual challange.

The relation between knowledge and communication assumes a strategic wage for developing IT and organizations because all technicians involved in the construction process explicit their cognitive actions through communication. Organization works as a self-activating net in which people orient the action by mean of language. Nonaka and Takeuchi (1996) conceptualized a constitutive character of organization operating by mean of projects. The elicitation of knowledge plays a key role in construction projects.

Following the observation of Egbu and Botteril (2002) the empiric evidences suggest Information Technology is overwhelming applied in explicit knowledge transfer. Face-to-face interaction and oral communication using common language are often a crucial source of tacit knowledge. The common, informal, not IT supported language match a kind of knowledge very important in organizational matter and for creating the context of more formalized and objective knowledge (such as texts, manuals, contracts and so on).

This kind of scenarios offer interesting application fields for emphasizing workflow approaches in knowledge transformation processes and particularly in construction management. The shift of process workflow analysis on knowledge transformation started just from studies on communication processes such in Winograd and Flores (1987). The workflow approach offers an interesting conceptual and theoretic frame for analyzing the communication process in the organizational context. The operability of the model can be extended to the field of diffuse knowledge management through the representation of the operative interfaces between the system operators. To sump up, through the workflow models it is possible

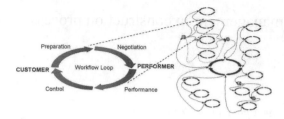

Figure 1. Workflow loop in the context of the global work-flow process.

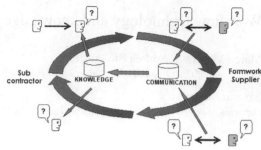

Figure 2. Communication and knowledge in workflow loop.

to describe the process cognitive maps associating them with people, times and project activities.

2 WORKFLOW APPROACH IN CONSTRUCTION MANAGEMENT

The basic figures of the model are a generic *Customer* to which corresponds a generic *Performer*. The two actors elaborate some performance applications and undertake to perform them through the production of systematic networks of linguistic act.

A process of communication is articulated into four fundamental phases:

– *Preparation*: constituted by the networks of linguistic acts that are used to formulate a request;
– *Negotiation*: constituted by the activities of negotiation to integrate the knowledge and to reciprocally direct the action of the two partners;
– *Performance*: constituted by the activities planned during the preparation and negotiation phases;
– *Control*: constituted by the activities that are necessary to check and to decide the acceptance of the work performed.

A programming or planning process can be mapped into workflow loops that can be structured either hierarchically or in a distributed way, either in sequence or in parallel.

3 KNOWLEDGE FLOW IN DECISION PROCESS

The observation on field of a paradigmatic relation between sub-contractor for concrete work and formwork performer offers a interesting case of a typical relation scheme in knowledge workflow. During the **preparation** phase the decisional process evolves through a number of states achievable by mean communicating and searching for structured knowledge. In the next phase the **negotiation** starts to define the operative aspects of work. During negotiation the partners exchange a relevant mass of information

supported by fax, mail, e-mail, oral language, technical specification and so on. This sort of communication is crucial for explicating all knowledge requested for assuming the task and for exploring details and solutions. This kind of experience is relevant as a cognitive performance for structuring the organizational relation. During the **performance** phase, the communication process improves articulating the management process for proactive prevention of failures, breakdown or issues. The language used in the operative context is more informal and solution-oriented; the intensity of knowledge exchange is very high and is characterized by personal experience, rapidity of decision and scarcely oriented to a systematic reification of results.

The **control** phase is relevant in the communication process for developing the quality management procedures and non-conformity treatment. The control phase tends to emphasize the formal communication in relation to control level or the non-conformity type.

4 REQUIREMENT OF A KNOWLEDGE FLOW MANAGEMENT SYSTEM

From the on field analysis the principle requirements that inform the social system of knowledge management are differentiated in the communication and in the knowledge. For the communication domain the requirement are the following:

– *Traceability of the cognitive processes*: the traceability is a key criterion to manage the processes quality; the tracing of the processes from the resolution to the action constitutes a meaningful aim in the field of the evaluation of the planning decisions quality;
– *Facility and rapidity of the creation of prototypes and implementation of the communication cycles*: the episodic nature of the building projects requires to particularly focus on the facility of use and on the reduced times of plan setting up;

– *Use of the communication processes for the elaboration of the plan documents*: the Information Technology applications make it possible to transform the communication flows into document elements that can efficiently be used.

In the knowledge domain the requirement are the following:

– *Reification of knowledge*: elicitated knowledge belongs to the natural context and doesn't require a generalization process; fuzzy knowledge, a-systematic knowledge, experience add value for developing projects;
– *Redundancy and completeness of the knowledge*: the redundancy caused by the dialogue processes of knowledge construction and the formalisation of the communication process development make it possible to transform the commitments of the partners into concrete actions, less subject to the uncertainty compared with the systems with a lack of knowledge;
– *Reusability of the processes*: the reusability of the processes is an aim that should be carefully valued in order to facilitate the planning of new cognitive processes;
– *Transferability of the knowledge through case-based learning*: the corporate knowledge management can more and more be identified with the application of the skills directly to the production processes according to models that don't require any formalisation of the organisational structure.

5 ORGANIZATION DESIGN BY MEAN WORKFLOW ANALYSIS

An organization can be described paradigmatically through a knowledge flow represented by the communication processes in a IT system. This assumption is a reduction of the organizational complexity and requires to be manage carefully. The hypothesis widely accepted in knowledge management studies emphasizes the control of knowledge production by mean of the information technology. Therefore the reification of knowledge and the participation to a project mediated by an IT system can not easily achieved in a high competitive environment. Fundamental objective is to support the decision of human to share knowledge in a cooperative way of operating. Anyway the knowledge management has to achieve the conditions for practically orienting the partners of a project in sharing their knowledge.

Web application are currently used through a client-server protocol that constitutes a market standard for many software companies and a mass of current products can be interfaced in a coherent approach to an organizational support for a project. Anyway the application of workflow management on the web requires the satisfaction of some basic requirements:

– to consent a distributing organization of project in space with production unit positioned in different places and with different tasks;
– to use the system easily and transparency of the technology;
– open source of programs and customizability to specific context.

Specific objectives in construction sector are to be oriented for a network integration of producers, suppliers and designers and managers only episodically assembled and distributed in many places. Web connection can break down the separation between off site and on site as a functional, temporal and managerial separation.

6 THE SUPPORT OF KNOWLEDGE FLOW MANAGEMENT

To support the decision of people to share explicit knowledge is probably a decisive challenge in developing knowledge management IT systems. IT systems cannot "create" new knowledge and "elicitation" is necessary a self consciousness of shared knowledge. A Cycle of knowledge production can be articulated in the following phase:

– *The tacit knowledge is shared*: partners influence each other in assuming decision for action. Workgroups build knowledge for problem solving and share experience and discuss opinion. The tacit knowledge is so elicitated;
– *Concepts creation*: the easier media for supporting knowledge is used (photos, texts, videos);
– *Concepts validation*: knowledge is managed for relational use in project domain by modifying the preparation procedure and control procedure. The learning space is enlarged, redundant concepts are eliminated and transferability and reusability criterion are verify. This stage is assumable as a real reification of knowledge;
– *Knowledge diffusion*: at the decisional level knowledge is diffused and modality of diffusion are decided.

This kind of process produce knowledge basically strong hierarchical controlled or the criterion of reification and diffusion can be share when the IT system permits the control by the producer. In episodic and atomized organizational structures as well as in construction sector , the participation to the project management is sustainable through an individual control of shared knowledge. Otherwise partners are not driven on to improve their participation in process development.

7 PROTOCOL ANALYSIS OF THE KNOWLEDGE SYSTEM

An experimentation on field is dealt in the domain of concrete works on site for describing the base of knowledge and the principle protocol for communicating knowledge. The domain requires a typical structure of knowledge subdivided in more level. At the first level using web forms is it possible to pilot the search in the knowledge base across a gradual level of formalized knowledge.

At a second level mapping the partners conversation in the process is it possible to build the diagrams of knowledge workflow. Conversation protocols help subjects in explicating the state of the knowledge production and permit to trace the conversation aiming at verifying the consistency and the robustness of the technical specification. This "learning by conversation" helps designers and mangers in clarifying better their respective actions and to mutual recognize the activity of partners. Decision can be also more contextual to the stream of decision making in the project.

An effective and efficient use of the formwork technological systems requires a progressive effort of focusing on the commitments and specification of the quality plan. In the operational context, the following operators have to collaborate in the planning:

- the performer of formwork system holds the specialist knowledge concerning the equipment;
- the sub-contractor of concrete works is entrusted with the realisation of works in reinforced concrete and he has, with his own specialist skills, to respond to the technical-organisational requirements for the integration of the basic processes for the production of works in reinforced concrete;
- the main contractor has, with his managerial skills, to manage the fundamental parameters of the process and to guarantee the construction specifications required.

The conditions for the production of consistent knowledge flows in particular between these construction partners regarding the correct integration of the processes are therefore determined. Through a workflow approach, we can represent these knowledge flows through repeated cycles of communication in which the main knowledge flows are specified.

By mean of borrowing concepts and elicitation methodology of Winograd's and Flores approach a communications protocol, between customer and performer, can be described and used in the performance phase (see table 1).

It is observable like the protocols that concern to customer and performer are generally different. Particularly the customer can interrupt one-sidedly a communication process, while the performer doesn't

Table 1. Proposed protocol in performance phase for customer and performer.

Performer Protocol	
Request of information	It is a protocol used by the performer for starting a communication process toward the customer.
Observation	It is a generic protocol used by the performer for responding to the customer.
Report progress	It is a protocol used by the performer for communicating the work in process.
Offer without change	The performer responds to a "request of offer" of the customer after a "request of offer" from the customer; in this case there is not any costs increase.
Offer with change	The performer responds to a "request of offer" of the customer after a "request of offer" from the customer; in this case there is a costs increase.
Final report	It is a protocol used for completing work.
Customer Protocol	
Question	It is a protocol used by the customer for starting a communication process toward the performer.
Observation	It is a generic protocol used by the customer for responding to the performer.
Request of offer	It is a protocol used by the customer for asking a request of offer to the performer.
Accept	It is a protocol used by the customer for accepting the performer offer. The protocol follows an offer with/without change.
Agree	It is a protocol used by the customer for accepting a generic observation of the performer.
Cancel	It is a protocol used by the customer for breaking off a communication process.

dispose of this protocol. An excessive number of protocols can be cause of errors in the operators and attention must be set in their definition.

So that a communication process correctly proceeds is necessary that the operators are stimulated to the communication. The Workflow experiences in the sector of management of the document processes suggest the use of "temporal ties" for the time control of the communication process.

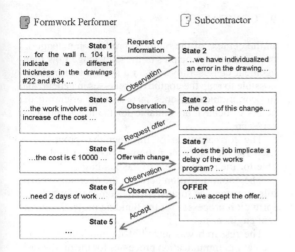

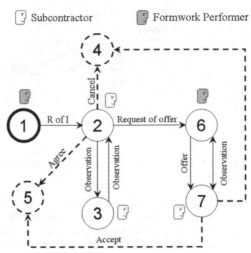

Figure 3. Example of a schematic communication flow, between formwork performer and subcontractor.

Figure 4. Scheme of communication flow with the use of protocols.

Nevertheless the management of temporal ties in turbulent environments must be faced with an approach that allows autonomy and responsibility operators. For instance the operators, responsibly, manages the own alerts time; or it is possible that customer and performer, together, decide the temporal ties of the communication process.

The communication proceeds through the succession of conversation states that are modelled by the linguistic acts of one of the partners. A process of communication states typically takes on a cyclical evolution on that can be turned into schemes (Fig. 4)

The protocols support communication by mean of mapping the communicative states of the process. The example of Figure 3 shows the elicitation technique for analysing a state of the communication process (Fig. 4).

The state 1 is the initial state of the process that is modified through a formwork provider request from addressed to the subcontractor. The issue opens a state of unresolved situation regarding a generic problem.

In the state 2 the sub contractor answers with an observation based on the analysis of the issue received. He can answer through a communication break or by mean of accepting the performer's affirmation and moving toward the closing states 4 and 5. If a change of the work needs, then he uses the protocol "request of offer" linked to state 6.

The state 3 is similar to the subcontractor 2nd state and it is determined by the answer given to the formwork performer in consequence of the first question.

The states 6 and 7 are similar to the states 2 and 3 but the discussion is turned on the specific operative and economic of the changed work.

8 TOWARD WEB APPLICATION OF FLOW MANAGEMENT

Web applications help us in building a knowledge flow system for managing a process. In the preparation and control phases, in which the knowledge flows are consistent, the use of an implemented model through web forms permits us to experiment a prototype for verifying the requirement system satisfaction.

On a first level the designer is supported by an explanatory documentation; schematic knowledge is gradually verified for deciding adequately the characteristics of the needed formwork system, through correlation between input data, and useful knowledge for correct design select; for instance if casting velocity is required, the correlation is with the necessary information for the choice (knowledge): force on the wood panel, stress on the formwork looms, and so on.

A further level of knowledge graduation is represented by the training course, turned for student users, on a specific formwork system. In this way, two levels for use of application are possible: an expository level, for student and generic user, and an expert level for searching technical knowledge. Nevertheless, the learning must be considered like a collective activity in which the student and designer roles are not predetermined, but subordinate of individual experiences and particular knowledge context.

This type of structure proves to be of particular interest for the overlapping of the representation fields of processes that present a stratified and organised knowledge.

A Workflow system works effectively needs that the involved operators, when they activate a

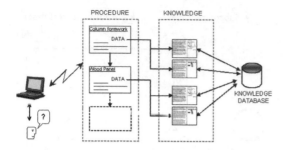

Figure 5. Structure of web application in preparation and control phases.

communication flow, follow an explicit procedures. It is necessary, preliminarily to use, to establish the communication forms; it is important, in fact, to use homogeneous communication forms, in electronic format, in way to facilitate scheduling, filing, search. The use of technologies based on e-mail communication results simple, low cost, it requires a reduced operator training, it allows the transfer of any digital data. Nevertheless, for a full scheduling needs also to manage the e-mail attachments with an only digital format; a possible solution is the use of PDF format files that allows to transfer texts, drawings, images, etc., in compressed size and it is broadly supported in internet communication.

The value added by a workflow system is represented by a system that is based on a communication protocol open to the accumulation of knowledge and flexible in front of the operators requirements of adaptation and self-regulation.

The traceability of the communication process through the use of communication protocols allows a formalisation level for ordering and tracing the communicative process.

The experimentation applied on the case study demonstrated an increased attention and self-consciousness in order the following issues:

– *Process analysis*. For instance, it is possible to locate the process phase that has required the greatest number of "informative requests" and to understand if this is due to design problem. It is possible to establish the times required for completing a communication cycle and to deduce if it has been source of delay in the work plane.
– *Operators* training for improvement of the decision making process. The comprehension of communicative flow through a learning process that we can call "learning by communication" allows to increase knowledge and experience of the operators.

Each of these choices has a different economic impact on the organisation: it requires great or less economic resources and involves human resources with different levels of participation.

9 CONCLUSION

In the context of the building production managerial processes, the principles of the knowledge flow management can usefully sustain a system that makes it possible to map the cognitive processes in general and the communicative ones in particular, orienting the actors to behaviours suitable for the times and ways foreseen.

The formalization of the cognitive processes is used for the experimentation of planning tools suitable for the management of the building quality.

The communication between the operators constitutes a reference to analyse the company knowledge formation processes for the building projects concurrent engineering.

The research was aimed at the modelling of elementary communication processes for the predisposition of quality control tools in the realisation of works in reinforced concrete and for the improvement of the consistence of the plans through the control of the commitments of the partners involved in the process.

The project knowledge flow can be modelled by analysing the production processes from which the unchanging elements but also the changing ones like the episodic nature of the organisational structure and the specific project conditions, can be individualised. It is above all the uncertainty of the project that originates communication flows that through the workflow methods can be captured, mapped, traced and made available for the case-based learning and then reused.

REFERENCES

Al-Ghassani A., Kamara J., Anumba C., Carrello P., 2002. *A Tool for developing Knowledge Management strategies*, ITcon Vol. 7 2002. pp. 69–82, at http://www.itcon.org/2002/8.

Al-Jibouri S.H., Mawdesley M.J., 2002. A knowledge based system for linking information to support decision making in construction, ITcon Vol. 7 2002. pp. 83–100, at http://www.itcon.org/2002/6.

Becker J., Muehln M., 2002. *Workflow application architectures: classification and characteristics of workflow-based information system*, in WorkFlow Handbook 2002 edited by Layna Fisher, pp. 39–49, Future Strategies Inc., Book division, Lighthouse Point, Florida.

Egbu C., Bates M., Botterill K., 2001. *A conceptual framework for studying Knowledge Management in project-based environments*. Proceedings of the International Postgraduate Research Conference in the Built and Human Environments, 15–16th March 2001, University of Salford (UK).

Egbu C., Botterill K., 2002. *Information Technologies for Knowledge Management*: their usage and effectiveness, ITcon Vol. 7 2002. pp. 125–135, at http://www.itcon.org/2002/8.

Hamzah A.R., Berawi M.A., 2001. *Developing a Knowledge Management system for Construction Contract Management,* Proceedings of the 14th International Conference on Applications of Prolog, 20–22 October 2001, University of Tokyo, Tokyo, Japan, pp. 358–378.

Lienhard H., 2002. *Workflow as a Web Application*, in WorkFlow Handbook 2002 edited by Layna Fisher, pp. 65–80, Future Strategies Inc., Book division, Lighthouse Point, Florida.

Maturana H.R., Varela F.J., 1980. Autopoiesis and cognition. The realisation of living, Dordrecht.

Nonaka I., 1996. *The knowledge – creating company*, in How Organization Learn, edited by Ken Starkey, International Thomson Business Press, London, 1996, pp. 18–31.

Nonaka I., Takeuchi H., 1997. *The knowledge – creating company,* Guerini Editore, Milano.

Winograd T., Flores F., 1987. *Understanding computers and cognition*, Addison-Wesley.

34. Special session on "Innovative materials and innovative use of materials in structures"

Flexural strengthening of RC members by means of CFRP laminates: general issues and improvements in the technique

A. Nurchi
University of Cagliari, Department of Structural Engineering, Cagliari, Italy

S. Matthys
Ghent University, Department of Structural Engineering, Magnel Laboratory for Concrete Research, Ghent, Belgium

ABSTRACT: In the last decade Fibre Reinforced Polymers (FRP) have become a promising category of materials in Civil Engineering. Their use as externally bonded reinforcement (FRP EBR) for strengthening of structural bearing members is already a well documented technique in several countries. Often it appears in the design that FRP EBR can only be used at a fairly limited stress, due to e.g. debonding failure modes. Newly developed FRP materials offer new possibilities to improve the effectiveness of the strengthening. In this paper the use of special multi-directional laminates with extra bolted anchorage is discussed, based on a test programme recently carried out at the Magnel Laboratory for Concrete Research. The test results show how this technique can prevent or postpone some debonding failure modes, thus leading to a higher strengthening factor. Also, the deformability at ultimate can be significantly improved (so-called "pseudo-ductile behaviour").

1 INTRODUCTION

The use of Fibre Reinforced Polymers (FRP) as structural reinforcement in concrete construction has gained considerable interest. In the shape of bars, rods, tendons, they can be used as an alternative high strength non-corroding material for reinforcing or prestressing concrete elements. However the most successful application up to now has been in strengthening or repairing existing structures. For this purpose the application of FRP laminates, strips and fabrics has many advantages with respect to traditional techniques such as steel plate bonding. These advantages basically relate to the ease of application, low disrupt costs, as well as the advanced material. Given the various research programmes and applications, the use of externally bonded FRP reinforcement (FRP EBR) for strengthening in flexure, shear and for confinement has become a well documented and standard technique in several countries.

The structural behaviour of concrete elements strengthened with FRP EBR has been extensively investigated and reported by several authors (e.g. Matthys 2000), and design recommendations for FRP EBR have been recently published by *fib* (International Federation for Structural Concrete) (*fib* 2001).

In the following sections some general aspects on the structural behaviour and design of FRP EBR for strengthening of reinforced concrete members are presented, focusing on the ultimate (limit) state in case of flexural strengthening. From this discussion it will appear that in certain circumstances the efficiency of the flexural strengthening is limited, so that improvements are of interest. Following, the use of special multi-directional laminates is addressed, presenting a test programme recently conduced at the Magnel Laboratory for Concrete Research.

2 FRP EBR STRENGTHENING TECHNIQUE

Different FRP EBR systems are available on the market. Usually they are classified in "wet lay-up" systems and "prefab" systems (Fig. 1). The first system consists of dry fibre sheets, fabrics or tows, which are applied to the surface of the concrete member by means of a proper resin. The latter has both the function of impregnating the fibres (thus realizing the FRP composite) and gluing the composite to the concrete. In the "prefab" systems pre-manufactured elements such as straight strips (or laminates), jackets or angles are applied to the concrete elements by means of an adhesive.

Figure 1. Example of FRP EBR systems: "prefab" (left) and "wet lay-up" (right).

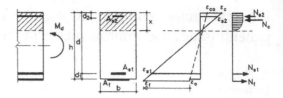

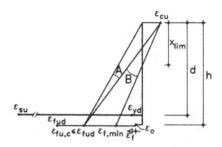

Figure 2. ULS analysis for a strengthened rectangular section.

The basic strengthening technique, which is generally used in practice, consists in the manual application of one of the described systems, with the fibres parallel to the direction of the principal tensile stresses. Typically, epoxy is used as structural adhesive.

Figure 3. Possible strain distributions at ULS in the critical section of a strengthened flexural member.

3 DESIGN OF FLEXURAL STRENGTHENING

3.1 Preliminary analysis of the initial member

When dealing with the design of the strengthening of a reinforced concrete member, the first step is always the analysis of the "initial condition" of the element before strengthening. This preliminary study can give helpful information for the choice of the most appropriate strengthening technique and material. The load level during strengthening may also be decided.

A proper analysis of the initial state of the member should give insight in causes of possible defects (such as reinforcing steel corrosion) and may indicate the amount of concrete repair prior to strengthening.

Furthermore, in case of flexural strengthening of a beam, also the shear strength should be verified. In fact the increase of service load on the beam will result not only in higher bending moments, but also in increased shear forces. The shear resistance of the beam must be higher than the acting shear force on the strengthened element. If this is not the case, the shear resistance will determine the maximum strengthening factor, unless shear strengthening is provided as well.

3.2 Analysis based on full composite action

At a first stage the flexural strengthening may be designed based on the evaluation of the resisting bending moment of the composite cross section. This calculation is performed assuming full composite action between the concrete and the FRP EBR, which means that, in the strengthened element, the increase of strain in the FRP equals the one in the concrete at the same level (Fig. 2). This is the same approach normally used in the analysis of reinforced concrete sections.

With this hypothesis, failure of the structural member will happen when (Fig. 3):

- the FRP reaches failure in tension (zone A), or
- the concrete reaches crushing in compression (zone B).

For both conditions the internal steel reinforcement may be in the elastic field or already at yielding (which is normally regarded as a more favourable situation). The failure mode depends on the ratios of both steel and FRP reinforcement.

This analysis is used to derive the minimum amount of FRP reinforcement needed to achieve the required resisting design moment.

It has to be noticed that by adding FRP reinforcement the "available" tensile force in the cross section is increased. This only corresponds to an increase in the resisting moment if sufficient capacity of the concrete in compression is available (equilibrium between tensile and compressive force). Hence, the maximum strengthening is limited by the compressive capacity of the concrete. When this is reached, failure due to concrete crushing will occur. This is typically the case for over-reinforced members (with high steel reinforcement ratios). In this situation the effect of adding FRP reinforcement becomes less pronounced and relatively high amounts of additional reinforcement are needed, unless special provisions are taken.

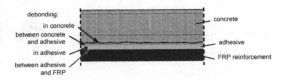

Figure 4. Possible debonding interfaces (*fib* 2001).

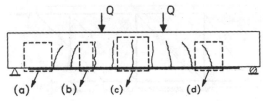

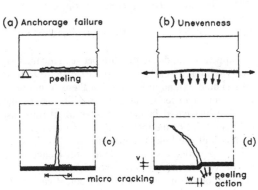

Figure 5. Debonding (peeling) mechanisms.

3.3 Debonding mechanisms

3.3.1 General

Assuming that the amount of FRP EBR has been determined based on the "full composite action" condition, further analysis with respect to the bond interface is needed. Indeed, the hypothesis of full composite action until failure may not be correct. There are a number of bond failure mechanisms which may be activated, leading to the loss of composite action between FRP reinforcement and concrete. These debonding failure modes result in an additional limit to the load bearing capacity of the strengthened element. As debonding may happen when the stresses (or strains) in the FRP are still relatively low, the reinforcing material is not exploited to a large extent.

In principle bond failure may take place at different interfaces, in particular (Fig. 4):

- in the concrete,
- in the adhesive (cohesion failure),
- at the interface concrete/adhesive or adhesive FRP (adhesion failure),
- in the FRP (interlaminar shear failure).

Given the high shear strength of the resins used as matrices and adhesives, as well as their good adhesion properties, bond failure will normally occur in the concrete substrate, either close to the surface or along a weak layer (such as along the internal reinforcement). Only in case of high temperatures, extremely high concrete strength or bad surface preparation bond failure may happen in the adhesive or its interface.

3.3.2 Peeling

Debonding can be due to different reasons and initiate at different locations, as shown in Figure 5. Starting at the free end, the tensile force in the FRP has to be built up. Hence, relatively high bond stresses arise in the anchorage zone. If a critical value (depending on concrete and FRP properties and on the available anchorage length) is exceeded, "peeling" failure at the anchorage will occur (Fig. 5a). Peeling forces may also arise as a consequence of extensive local unevenness of the concrete surface, as shown in Figure 5b.

Other critical points for bond are cracks in the concrete. With this respect the role of flexural and shear cracks is significantly different. When a flexural crack opens (Fig. 5c), localised bond stress concentrations arise near the crack. If they overcome

a certain value, micro cracks will appear in the concrete close to the surface (Fig. 5c). This will yield local debonding, which only tends to propagate when reaching an advanced state. This local debonding is to be avoided at service conditions.

Cracks in regions with high acting shear forces and bending moments are much more dangerous. In this case (Fig. 5d) both horizontal and vertical displacement components are obtained. Due to the vertical displacement a direct peeling action is obtained, which will result in progressive debonding. Furthermore, in such regions relatively high bond stresses are present, due to the variation of the force in the EBR. Hence, debonding starting at a shear crack may easily propagate along the beam. Generally, in regions with high acting shear forces but almost no bending moments (less steep cracks), the vertical crack displacement is arrested in a better way by the internal shear reinforcement.

3.3.3 Concrete rip-off

A special type of anchorage failure mechanism is the so-called concrete rip-off. This can occur in case the EBR is curtailed at a certain distance from the support and a shear crack is typically formed at the end of the EBR. At a first stage this crack propagates at 45°. If no stirrups are available, shear failure is obtained. The presence of stirrups can bridge the EBR-end shear crack, in which case the crack may propagate along the internal reinforcement, resulting in the detachment of the concrete cover. The concrete rip-off mechanism can be explained based on the truss model, as shown in Figure 6.

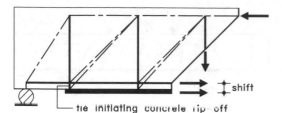

Figure 6.   Truss model and concrete rip-off mechanism.

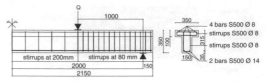

Figure 7.   Dimensions and reinforcement of test specimens.

It appears that, due to the shift between the internal (steel) reinforcement and the (FRP) EBR, the shear links should extend to the level of the external reinforcement in order to have the equilibrium of the forces in the truss. In fact the internal stirrups just reach the level of the internal reinforcement, and the extension of the link is represented only by the concrete cover which is subjected to tension. Hence, once the shear crack starting at the end of the EBR reaches the level of the internal reinforcement, it can easily develop along the latter, thus leading to the concrete rip-off failure mode and the complete loss of the external reinforcement.

4   IMPROVEMENTS IN THE STRENGTHENING TECHNIQUE

From the discussion it appears that the effectiveness of the basic technique may be limited under given circumstances. Improvements are needed in order to achieve higher strengthening factors (better use of the material), or to avoid debonding failure modes.

The fundamental hypothesis of the EBR technique is to realize the composite action between FRP and concrete. Hence, it is of primary interest to improve the bond behaviour and the anchorage capacity of FRP strips, in order to prevent (or at least postpone) debonding.

Improved anchorage capacity is also beneficial in case of prestressed FRP EBR or in the following situation, considering the case of limited concrete compressive capacity. In this case the basic technique is less effective, as it has been addressed above. If a high strengthening factor is needed, a very effective solution is to increase the distance between the external reinforcement and the compressed zone (increased lever arm between tension and compression force). In this way the resisting moment can be increased while keeping the stresses in the materials relatively low. The increased lever arm can be achieved by using a (lightweight) spacer between the FRP and the concrete. In this case a strong end anchorage of the FRP and spacer to the beam is needed.

Different systems to improve the anchorage capacity of EBR can be thought of. A simple method is the use of extra mechanical fixings by means of bolts. This technique has been first studied for strengthening by means of steel plate bonding. The bolts can provide anchorage to the plate when debonding occurs. Furthermore, if the bolts are prestressed, also the bond performance at low slip values will improve, due to the compressive stresses at the interface (Pichler & Wicke 1994).

However, the feasibility of this solution is not always obvious in case of FRP. In fact the unidirectional laminates generally used can not be bolted, since they are very weak in transferring forces to the bolt anchorage. Newly developed multi-directional FRP laminates offer a good solution in this case. Additional fibres in other directions than the longitudinal one (often ±45°) provide sufficient stress transfer capacity and allow for the use of bolt anchorages.

To prove the effectiveness of the use of bolts and hence the advantages of the new type of laminate, a test programme was set up at the Magnel Laboratory for Concrete Research.

5   EXPERIMENTAL RESEARCH ON BOLTED MULTI-DIRECTIONAL FRP LAMINATES

The feasibility of bolted anchorages for the new multi-directional CFRP laminates has been demonstrated by preliminary tests (Matthys et al. 2003). Following, a test programme was conducted to investigate the role of these anchorages in the structural behaviour of FRP strengthened RC members. For this purpose, a series of four-point bending tests have been executed on RC T-beams with a span of 4 m. The main aspects of this test programme are shown in the following. A more detailed discussion is presented in Nurchi et al. (2003).

The dimensions and details of the reinforcement of the T-beams and the test set-up are given in Figure 7. A total of 5 beams have been tested, including one reference (unstrengthened) beam, and 4 beams strengthened with multi-directional CFRP laminate, with or without the use of additional mechanical fixings by means of bolts (see Table 1).

Flexural strengthening was provided to Beams 2 to 5, by gluing one layer of multi-directional PC CarboComp Plus, with a length of 3.66 m. The strips are about 1.8 mm thick and 100 mm wide, with fibres

Table 1.	Configuration of the beams.

Beam	Description
1	Unstrengthened
2	CFRP
3	CFRP + 2 × 2 bolts
4	CFRP + 2 × 6 bolts
5	CFRP + 2 × 6 bolts (stronger anchorage)

Table 2. Concrete mechanical properties.

Beam	Age of test [days]	f_c [MPa]	$f_{c.cube}$ [MPa]	$f_{ct.bend}$ [MPa]	$f_{ct.split}$ [MPa]	E [MPa]
1	34	35.9	52.3	4.47	4.59	38850
2	63	38.7	52.1	4.65	3.94	45970
3	94	46.4	59.9	5.76	3.91	43110
4	141	44.4	58.7	5.40	4.18	38530
5	204	46.7	59.9	6.59	4.35	39010

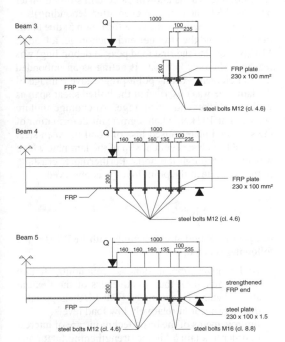

Table 3. Mean tensile properties of the reinforcement.

Material	Ø or t_n [mm]	f_y [MPa]	f_u [MPa]	ε_u [‰]	E [MPa]
Steel S500	14	590	680	100	210000
PC Carbo-Comp Plus	1.00	–	2700[*]	15.3	190000[*]

[*] Based on nominal thickness t_n.

Table 4. Main test results.

Beam	Age of test [days]	Q_y [kN]	Q_{max} [kN]	Q_{max}/Q_{ref}	At Q_{max}	At ultimate
1	34	53	68	1.00	YS/CC	
2	63	72	85	1.25	YS/CR	
3	94	75	92	1.35	YS/PV	AF
4	141	79	102	1.50	YS/PV	AF
5	204	75	118	1.74	YS/PV	AF/CC

YS: yielding of steel, CC: concrete crushing, CR: concrete rip-off, PV: peeling due to vertical crack displacement, AF: anchorage failure.

Figure 8. Layout of the bolts.

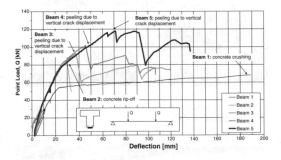

Figure 9. Load vs. deflection curves (Beams 1 to 5).

over the full length in the longitudinal as well as in the plus and minus 45° directions.

Different layouts of bolts were used in Beams 3 to 5, as shown in Figure 8. The bolts were installed with epoxy adhesive, with a bond length of 180 mm above the internal reinforcement. They were pretensioned with a force of approximately 10 kN. In Beam 5 steel end plates and bolts with increased diameter and strength were used to realise the end anchorage. These bolts were pretensioned with a force of approximately 20 kN. The configuration and prestressing force of the 4 bolts along the shear span was the same as for Beam 4.

The mechanical properties of the concrete and the reinforcing materials, determined by testing, are reported in Tables 2 and 3.

An overview of the test results in terms of yielding load, maximum load, strengthening ratio and failure aspect of the tested beams is given in Table 4.

The load vs. midspan deflection curves are given in Figure 9.

In the unstrengthened beam yielding of the internal steel was reached at 53 kN. After that, large plastic deformation developed until concrete crushing occurred at 68 kN.

The strengthened beams, due to the presence of the external reinforcement, are significantly stiffer than the reference beam after cracking and yielding of the internal steel occurs at a much higher (about 40%) load level. Even after yielding of the steel reinforcement the load considerably increases. The maximum attainable load is limited, in case of this test programme, by debonding of the laminate.

In Beam 2 the concrete rip-off mechanism was activated (Fig. 10 left), and failure occurred at 85 kN (corresponding to a strenghtening factor of 1.25). As the laminate is lost the load bearing capacity of the beam drops considerably.

The test of Beam 3 proved the effectiveness of the end bolts in preventing the concrete rip-off mechanism. Figure 10 (right) clearly shows that when bolts are provided, the crack starting at the end of the laminate is not propagating along the reinforcement. The load could be increased in this case up to 92 kN (strengthening factor 1.35), when debonding occurred away from the anchorage zone due to vertical crack displacement (Fig. 11 left).

After debonding, the bolted beam did not fail immediately. The bolts provided anchorage of the laminate, which could further act as an unbonded tension member. This is also confirmed by strain gauge measurements, showing an almost constant strain distribution along the whole length of the laminate between the end anchorages after debonding.

Since bond is missing and the bolted anchorage deforms (bolts are bent, slip of the laminate at the bolts), the stiffness of the beam is low, resulting in large deflections. At about 80 kN the anchorage failed

Figure 10. Concrete rip-off in Beam 2 (left); Bolts prevent rip-off in Beam 3 (right).

Figure 11. Peeling due to vertical crack displacement (left) and anchorage failure (right) in Beam 3.

(Fig. 11 right) and the laminate was lost. At this point the beam tends to behave like the unstrengthened one.

In Beams 4 and 5 peeling due to vertical crack displacement occurred as in the previous case. However, it was postponed and its propagation was much more gradual. The maximum load attained for Beams 4 and 5 was respectively 102 kN (strengthening factor 1.50) and 118 kN (strengthening factor 1.74). Furthermore, higher residual strength and stiffness after debonding were obtained (bolts allowing a certain stress transfer between FRP and concrete even after debonding).

This was especially the case for Beam 5: due to the inner bolts and the stronger end anchorage, a load of 118 kN (equal to the first load peak at debonding) was reached when the laminate is acting as an unbonded tension element. At this point progressive damage of the laminate was observed at the bolted shear span as well as at the end anchorage. Anchorage failure occurred at 107 kN, when significant deformation of the steel end plate caused loss of bond between steel and CFRP, with subsequent slip of the laminate which was pulled through the bolts. Furthermore, crushing of the concrete at the point loads was observed.

6 CONCLUSIONS

In case of flexural strengthening with FRP EBR, the following is noted:

– Under given circumstances debonding failure modes may limit the effectiveness of the flexural strengthening by means of FRP EBR. Especially, this may occur at relatively low load levels.
– Improvements of the basic technique are of interest in order to attain a higher strengthening factor and to obtain a more favourable behaviour at ultimate.

Tests on RC beams strengthened with multidirectional CFRP strips with extra mechanical fixings have demonstrated that:

– By means of bolts at the laminate ends, anchorage failure such as concrete rip-off can be prevented.
– The use of bolts over the shear span significantly postponed debonding of the laminate due to vertical crack displacement.
– Due to the bolt anchorage, after debonding the laminate acts as an external tension member. This results in increased deflections at ultimate and less brittle failure modes.
– If the end anchorage is designed sufficiently strong, the bearing capacity of the strengthened beam after debonding (external reinforcement acting as an unbonded tension member) equals at least the initial bearing capacity of the beam with bonded external reinforcement. In this way a pseudo-ductile behaviour is obtained with large deformation capacity of the beam at ultimate load.

REFERENCES

fib Task Group 9.3, 2001. *Externally bonded FRP reinforcement for RC structures*. Lausanne: Fédération Internationale du Béton.

Matthys, S. 2000. *Structural behaviour and design of concrete members strengthened with externally bonded FRP reinforcement*, Doctoral thesis. Ghent: Ghent University.

Matthys, S. Nurchi, A. & De Vuyst, T. 2003. Anchorage capacity of multi-directional CFRP strips with additional mechanical fixings. In *Proc. "Composites in Constructions" International Conference (CCC). Rende, September 16–19 2003*. (in press).

Nurchi, A. Matthys, S. Taerwe, L. Scarpa, M. & Janssens, J. 2003. Tests on T-beams strengthened in flexure with a glued and bolted CFRP laminate. In *Proc. 6th International Conference on Fibre Reinforced Polymer Reinforcement for Concrete Structures (FRPRCS-6). Singapore, July 8–10 2003*. (in press).

Pichler, D. & Wicke, M. 1994. Verstärken von Betonbauteilen durch angeklebte Stahllamellen mit angepresster Endverankerung. *Beton- und Stahlbetonbau* 89(10): 261–264.

System-based Vision for Strategic and Creative Design, Bontempi (ed.)
© 2003 Swets & Zeitlinger, Lisse, ISBN 90 5809 599 1

Experimental and numerical study of buckling glass beams with 2 m span

J. Belis, R. Van Impe, M. De Beule, G. Lagae & W. Vanlaere
Laboratory for Research on Structural Models, Ghent University, Ghent, Belgium

ABSTRACT: Glass is earning its place in the slow-changing realm of structural building materials. Glass beams can be a good solution for the structural design of transparent roofs or floors in buildings. Strength and stiffness of this material are characteristics which are dealt with in literature. Glass beams, however, usually have a slender cross section, which makes them vulnerable to the instability phenomenon of lateral torsional buckling. Numerical and experimental tests on small glass beams at the Laboratory for Research on Structural Models have been reported by the authors on previous occasions. This paper deals with experiments on larger beams. Several experimental tests are carried out on single pane glass beams with a span of little more than 2 m, providing data on critical lateral torsional buckling load and coinciding displacements of the beam's geometry. Results are compared to FEM simulations and to theoretical values, derived from classical structural analysis theory.

1 INTRODUCTION

Although difficult to prove scientifically, it is common sense that working in natural daylight has a positive influence on the psychological feeling of wellness of employees (Norris 1997).

Transparent layers in buildings do not only give way to an improvement of spatial qualities and spatial perception of nowadays architecture, they also permit natural light to penetrate deeper than ever in the building's core.

It is obvious that these effects will only be fortified when the glass supporting structure itself will also be designed with transparent materials.

Glass can do the job. Several constructions in Europe and even in an earthquake-sensitive area like Japan are realised with load-bearing glass beams or columns. The expenses for such structural glass applications are high because of exaggerated design safety factors and obliged full-scale laboratory tests: structural codes for this kind of applications are missing.

Mainly for this reason, several scientific institutes are doing fundamental research in this field of structural engineering. The Laboratory for Research on Structural Models (Laboratorium voor Modelonderzoek) of the Ghent University in Belgium is currently working on the subject of glass beams.

Most of the time, real applications require the lamination of several glass panes to form one beam for safety reasons. Before the complicated structural behaviour of such composites can be understood in depth, it is an absolute requirement to examine the structural behaviour of single pane glass beams.

Generally, glass beams have a cross section which is very slender. This makes most glass beams sensitive to lateral torsional buckling. In the following paper, results of research on the buckling behaviour of single-pane glass beams is presented.

2 GEOMETRY

The geometry of the numerical models has been determined by the geometry of the glass test specimens that were available for experimental laboratory tests in order to obtain comparable results. The test beams that are referred to consist of only one panel of fully tempered glass. All specimens have the same dimensions, i.e. a height of 400 mm, a thickness of 10 mm, and a total length of 2200 mm. Taking into account the boundary conditions of the test set up (cf. §5.1), the actual span of the beams is 2100 mm, which will consequently be used in the following numerical and theoretical calculations.

3 NUMERICAL STUDY

3.1 *Material behaviour*

Glass is a brittle material, of which the mechanical behaviour depends on temperature. Under normal thermal serviceability circumstances, i.e. room temperature, crack propagation and fracture of the single pane glass beam are not preceded by plastic deformation. The material behaviour before fracture is linear elastic.

In this study, the material is considered to be ideally elastic.

3.2 General

Numerical simulations have been performed with the commercially available finite elements method software Abaqus (HKS 2002). Because of the relatively small thickness of the beams and the homogeneity of the beam's material, shell elements are chosen to build the model with. A fine mesh of 110 by 30 elements, based on a mesh convergence study as described in the references (De Beule, in press), is used.

3.2.1 Simplifications
3.2.1.1 Geometry

Due to the shell elements chosen, the cross-section of a single glass pane is modelled as a slender rectangle. This is an approximate simplification of the real geometry, since sharp edges of industrially produced glass panes are usually chamfered by a polishing treatment (Fig. 1).

The chamfer radius is about 2 mm. Since the removed glass area is negligible compared to the total cross-section, this geometrical simplification is accepted.

3.2.1.2 Residual stresses

The panes under consideration are made of fully tempered glass: by means of a very well controlled thermal treatment, tensile stresses are introduced in the internal part of the glass pane, while the outer surfaces are subjected to compressive stresses.

The overall bending strength of the glass is improved considerably this way. However, these residual stresses are not considered in this study. This simplification is justified by a study of Van Impe (Van Impe, in prep.), where it is shown that the Wagner effect of longitudinal compressive residual stresses has little effect on the torsional and buckling behaviour of tempered glass panes.

3.2.2 Analysis

The effect of lateral torsional buckling is examined with an Abaqus buckling analysis, which permits to calculate the solution of eigenvalue problems. This is acceptable since we want information on the critical buckling load and the according eigenmodes.

3.3 Imperfections

When experimental tests are done, one has to be aware of the effect of set-up imperfections. Especially of interest here is the effect of imperfections of the loading position, since those could be expected in the laboratory.

Eccentricity of the loading point compared to the centre of the beam at mid-span is examined below in longitudinal and transversal directions. Loading eccentricities referring to the height are not considered, since glass beams in real applications are practically always loaded on the upper rim. This is actually a more unfavourable situation than when the load is applied in the shear centre of the beam.

3.3.1 Longitudinal eccentricity

A series of simulations is executed in order to visualize the effect of loading positions varying in the longitudinal direction along the upper rim. In Figure 2, the critical (buckling) load is plotted against the longitudinal position x, normalised to the length of the beam L.

Because of the symmetry in the set-up, values of x/L are only shown up to x/L = 0.5.

Buckling loads are increasing rapidly when the load is positioned more towards the supports. In Figure 2, discrete points are interconnected for reasons of clarity. In the direct vicinity and above the supports (x/L is close to zero), failure will probably occur due to local stress concentrations instead of buckling, but those loading situations are beyond the scope of this contribution.

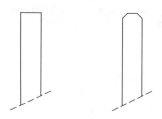

Figure 1. Assumed glass edge geometry and real edge geometry.

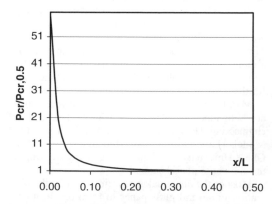

Figure 2. Critical (buckling) load against the longitudinal position x, normalised to the length of the beam L.

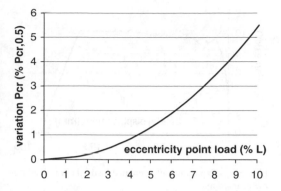

Figure 3. Detailed plot of effect of longitudinal loading point eccentricity.

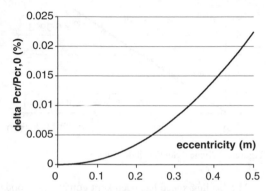

Figure 4. Effect of transversal eccentricity on buckling load, represented by an exponential trend line.

Set-up errors will be in the order of magnitude of mm, so a detailed plot is made of the area close to mid-span. The line connecting the calculated points in Figure 3 shows that a load eccentricity of 8.5 cm (approximately 4 percent of the length) causes an increase of less than 1 percent of the buckling load. Apparently, small deviations in the longitudinal direction do not have a significant effect on the buckling load.

3.3.2 Transversal eccentricity

In analogy to paragraph 3.3.1, eccentricity of the loading position in the thickness direction is studied. Since two-dimensional shell elements do not have multiple nodes in the directions normal to their surface, an additional two-node beam element is added.

This extra element is modelled as a rigid body, connected to the upper node at mid-span of the beam. The force is applied on the free end of the rigid body, of which the length is varied throughout the analysis from zero to 0.5 m. Results are shown in Figure 4.

In a buckling analysis, the effect of load eccentricity in thickness direction seems to be very limited for a perfect elastic and straight beam which is focussed on here.

Table 1. Values of the buckling load P_{cr} after the first iterations.

Preliminary analysis	First iteration	Second iteration
$P_{cr,0} = 9846.500$	$P_{cr,1} = 9846.526$	$P_{cr,2} = 9846.526$

All loads are given in N.

3.4 Pre-loading

The buckling loads obtained so far, are ideal elastic buckling loads, obtained from perfectly straight models. In reality however, a beam will deform slightly because of load factors that are still lower than the critical buckling load factor.

This initial imperfection can be taken count of by executing a preliminary static analysis due to a pre-loading (below the critical load P_{cr}) before the actual buckling analysis. When this iterative process is repeated until convergence of pre-loading and buckling load is found, the real critical buckling load is reached. Generally, such convergence is found already after very little iteration.

Results of such operations are shown in Table 1.

4 THEORETICAL CALCULATIONS

4.1 Lateral torsional buckling

In classical structural analysis theory, several authors have proposed expressions to determine the critical lateral torsional buckling load of prismatic beams (Timoshenko 1991, Trahair 1993, Young 1989). Comparison of results, obtained with different expressions, reveals that the overall buckling load deviation for I-beams is quite small (Pattheeuws 2001).

We arbitrarily chose the expression of Eurocode 3 (NBN-ENV 1993) as a reference for numerical and experimental results.

The warping stiffness I_w for a rectangular section of these dimensions is negligible and supposed to be equal to zero. After simplification, the following expression is obtained:

$$M_{cr} = C_1 \frac{\pi^2 EI_z}{L^2} \left[\left(\frac{L^2 GI_t}{\pi^2 EI_z} + \left(C_2 z_g \right)^2 \right)^{0.5} - C_2 z_g \right] \quad (1)$$

4.2 Application

For a point load at mid-span, the maximum bending moment is given by:

$$M_{cr} = \frac{P_{cr} L}{4}. \quad (2)$$

In this expression, I_t is the torsion constant, I_z is the second moment of area about the minor axis, $L = 2100$ mm is the length of the beam between points which have lateral restraints and $z_g = 200$ mm is the distance between the point of load application and the shear centre.

The values of C_1 and C_2 are given in Eurocode 3 for various load cases. For a point load P at mid span, values are respectively $C_1 = 1.365$ and $C_2 = 0.553$. As a result, we find $P_{cr} = 10,168.30$ N.

5 EXPERIMENTAL STUDY

5.1 Test set-up

The laboratory boundary conditions are created such that the test specimens would have the same degrees of freedom as those that were applied in the numerical and theoretical calculations. All beams are simply supported at both ends by so-called fork bearings. The centre of the end supports is shifted away from the ends of the beam for 50 mm inward, to avoid exaggerated stress concentrations at the corners of the pane and to permit a stable fitting of the beam between fork bearings. The only restricted rotational degree of freedom is the rotation about the longitudinal axis at the supports.

Since the structural performance of glass is dependent on the quality of its surfaces, surface damage because of direct contact between glass and steel should be avoided. In the set-up, rubber strips with small thickness are applied as contact interfaces. In real glass constructions, it is general practice to prevent direct contact between glass and other hard materials with a similar technique.

A special mechanical device is built in to assure the verticality of the applied load, even if large lateral displacements (i.e. order of magnitude of 100 mm) occur during the buckling process.

Loads are applied at mid-span of the beam. An eccentricity in the thickness direction of half the glass pane thickness (5 mm) has been imposed on purpose, in an attempt to predict and control the side to which buckling would occur. As noticed previously, this eccentricity has no further meaningful effect on the magnitude of the buckling load. Apparently, the applied eccentricity seemed not to be sufficient to manipulate the bifurcation process and neither to predict the critical elastic buckling load.

5.2 Experimental results

Several series of buckling tests have been performed. Some glass panes have been gradually loaded with 75 percent of the buckling load, smoothly unloaded afterwards, and loaded again. Ten such loading cycles have been repeated prior to destructive buckling tests.

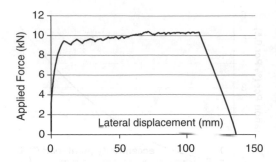

Figure 5. Experimentally obtained load – lateral displacement function.

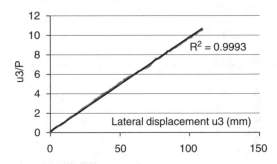

Figure 6. Typical Southwell diagram.

The loading speed has been kept constant as good as possible during all experiments.

5.2.1 Load-displacement

A typical load – lateral displacement function is shown in Figure 5.

In a first stage, the relationship is almost perfectly linear. In a second stage, a rapidly increasing lateral displacement of the upper rim can clearly be distinguished. However, it is not possible to deduct a reliable value for P_{cr} directly from this Figure.

5.2.2 Southwell diagram

In order to find the critical buckling load, the so-called Southwell method can be used (Southwell 1932). This method results in a typical Southwell diagram, as shown in Figure 6. In the Figure, the best fitting linear function is drawn based on linear regression. The inverse of the slope coefficient of this line gives the value of P_{cr}.

Some test results are shown in Table 2, together with the deviation from the theoretical critical buckling value.

5.2.3 Pathology

Fully tempered glass has a high internal energy level, held in equilibrium by internal residual stresses. When stresses reach a critical value, a crash that disturbs the

Table 2. Comparison of theoretical and experimental buckling load results.

Theoretical value	Southwell result	Difference
10,168.3	10,207.9	39.6

All buckling loads are given in N.

Table 3. Comparison of the buckling loads obtained from theoretical, numerical and experimental work.

$P_{theoretical}$	$P_{numerical}$	$P_{experimental}$	$\Delta_{numerical}$	$\Delta_{experimental}$
10,168.3	9846.5	10,207.9	3.16	0.39

Critical loads are given in N; Δ in % of the theoretical buckling load.

Figure 7. Typical fracture pattern of single-pane tempered glass beam after total collapse.

internal equilibrium is initiated. This causes an immediate energy release, which explains why crack initiation is followed by quasi-instantaneous and total crack propagation, together with an explosion-like noise.

A typical post buckling fracture pattern of a test beam is shown in Figure 7. From the crack pattern, the authors deduct that cracks initiated at the outer side of the upper rim at mid-span. Failure was due to excessive tensile bending stresses. The thick line in the middle of the picture is a steel device used to introduce the point load in the system.

6 COMPARISON OF RESULTS

Table 3 compares theoretical, numerical and experimental values of the buckling load.

7 CONCLUSIONS

7.1 Theory against numerical analysis

The accordance between theoretical and numerical values of the critical elastic lateral torsional buckling load is in the order of magnitude of 3%. The numerically obtained buckling load is slightly conservative compared with the theoretical value. In the authors' opinion this is a reasonably good result, which means that the used numerical modelling method is considered reliable within the context of the assumptions as stated above.

7.2 Theory against experiments

Compared to theoretical values, the experimental test set-up of the Laboratory for Research on Structural Models gives very good to excellent results for the experimental determination of the lateral torsional buckling load of single-pane tempered glass beams. There is almost no distinction between theoretical and experimental values.

7.3 Load imperfections

The buckling load of single-pane glass beams seems to be rather insensitive to small imperfections of the loading position in the longitudinal direction as well as in the transversal direction.

7.4 Residual stresses

Residual stresses are not taken into account in this study. Slight improvements on the realism of the numerical determination of the lateral torsional buckling load are expected when the residual stresses would be implemented in the model.

7.5 Post-critical behaviour

Experimental test results seem to indicate that the post-buckling behaviour could be slightly stable. However, the post-critical behaviour was not examined in this paper, and needs further research before proper conclusions can be drawn.

ACKNOWLEDGEMENTS

The authors wish to acknowledge the Glaverbel Company in Belgium, which has been willing to provide for the glass beams, necessary for this research.

REFERENCES

Belis, J. 2002. Buckling of Glass Beams, *Proc. Third FTW PhD Symp., Ghent, December 11, 2002.*

De Beule, M., Belis, J., Van Impe, R., Lagae, G. & Vanlaere, W. 2003. Torsional behaviour of laminated glass beams with a varying interlayer stiffness. *Abstract accepted for presentation at ISEC 02, Rome, 23–26 September 2003.* Rotterdam: Balkema.

HKS 2002. *Abaqus User's Manuals version 6.3*, USA: Hibbitt, Karlsson & Sorensen, Inc.

Norris, D. 1997. Daylight and productivity: is there a causal link?, *Proc. Glass Processing Days, Tampere, 13–15 September 1997.* Tampere: Tamglass Ltd. Oy.

NBN-ENV 1993-1-1, Eurocode 3 1992. *Design of steel structures – Part 1 – 1: General rules and rules for buildings*, Brussels: Belgian Institute for Normalisation (BIN) vzw.

Pattheeuws, S. 2001. *Calculation of lateral torsional buckling loads with finite elements method, Graduate thesis* (in Dutch), Ghent: RUG.

Southwell, R. V. 1932. On the Analysis of Experimental Observations in Problems of Elastic Stability, *Proc. of the Royal Society*, Vol. 135A, London: 601.

Timoshenko, S. & Gere, J. 1991. *The Theory of Elastic Stability, 2nd edition*, New York/Toronto/London: McGraw-Hill Book company Inc.

Trahair, N. 1993. *Flexural-Torsional Buckling of Structures*, London: E&FN Spon.

Van Impe, R., Belis, J., Buffel, P., De Beule M., Lagae, G. & Vanlaere W. 2003. The effect of residual stresses on the torsional and buckling behaviour of glass beams (in Dutch), *Abstract accepted for presentation at 6e Nationaal Congres over Theoretische & Toegepaste Mechanica, Ghent, 26–27 May 2003.*

Young, W. 1989. *Roark's Formulas for Stress an Strain, 6th edition*, New York/Toronto/London: McGraw-Hill Book company Inc.

Torsional behaviour of laminated glass beams with a varying interlayer stiffness

M. De Beule, J. Belis, R. Van Impe, W. Vanlaere & G. Lagae
Laboratory for Research on Structural Models, Ghent University, Ghent, Belgium

ABSTRACT: As glass is a brittle material, it seems at first sight neither safe nor suitable to fulfil a primary role in the structural system of buildings. The composition of beams of several glass panes with a soft interlayer can reduce this problem of safety, related to a high probability of brittle failure, to acceptable proportions. Failure of one panel still results in an important residual strength of the total laminate, and glass pieces are not scattered everywhere. Moreover, inner panes are relatively well protected against impacts. Numerical FEM simulations are performed on laminated glass beams with different interlayer stiffness to examine the effect on the torsional behaviour. Results are compared to a single pane reference beam. To conclude, the relevance of industrial efforts for developing stiffer interlayers is put into the right perspective for this kind of applications.

1 INTRODUCTION

To provide an acceptable safety level, beams are normally composed of several glass plates with strong, transparent adhesive layers.

Increasing the thickness of the beam increases its load bearing capacity accordingly. After collapse of one or more of the composed panes, the remaining are still able to ensure a certain residual strength. Usually, outer plates are not counted on in strength calculations for safety reasons. They are thinner and only used for protection of the inner plates against surface damage, which is critical for the overall strength of glass. Although no practical residual strength remains after collapse of all composed panes, the broken pieces remain fixed to the adhesive layer and are prevented from falling down. Well-designed laminated beams have an acceptable safety level, though they still partly or completely behave in a brittle manner.

It is estimated that about ninety percent of all laminated glass beams have polyvinyl butyral (PVB) interlayers; others are connected by means of e.g. other polymers or epoxy resins.

The interlayer stiffness affects the mechanical characteristics of the laminate. The effect of an increased stiffness is a current research topic in the field of architectural glazing, where the resistance against dynamic impacts (e.g. wind, hurricane, earthquakes, ...) is investigated. In the study of the behaviour of (laminated) glass panes as structural building elements, the torsional behaviour plays an important role. The effect of an increase of the interlayer stiffness is investigated

here numerically, using an Abaqus FEM analysis. The results are compared to a single pane reference beam with a length of 2100 mm and a height of 400 mm. These dimensions match with an experimental model used at the Laboratory for Research on Structural models at Ghent University.

2 FINITE ELEMENT MODEL

2.1 *Geometry and material properties*

In this paper the authors are not interested in pure torsional problems, but in lateral torsional buckling, in which torsion is combined with in plane bending and bending out of plane. The parameter chosen to evaluate the structural behaviour in buckling is the elastic torsional buckling load P_{cr}.

The numerical results are put to the test using the expression of Eurocode 3 (NBN-ENV 1993) for the determination of the critical lateral torsional buckling load of prismatic beams. This formula assumes a perfect linear beam geometry, no residual stresses, no eccentric forces and perfect linear elastic material properties. Due to the considered dimensions of the rectangular cross-section, the warping stiffness I_w is negligible and supposed equal to zero. After simplification the following expression is found:

$$M_{cr} = C_1 \frac{\pi^2 EI_z}{L^2} \left[\left(\frac{L^2 GI_t}{\pi^2 EI_z} + \left(C_2 z_g\right)^2 \right)^{0.5} - C_2 z_g \right] \quad (1)$$

P_{cr} for a load at mid-span can be found using the equation for the maximum bending moment

$$M_{cr} = \frac{P_{cr} L}{4}$$ (2)

In the Expression 1 I_t is the torsion constant, I_z is the second moment of area about the minor axis, $L = 2100\,mm$ is the length of the beam between points which have lateral restraints and $z_g = 200\,mm$ is the distance between the point of load application and the shear centre. The values of C_1 and C_2 are given in the Eurocode 3 for various load cases. For a point load at mid-span the values are $C_1 = 1,365$ and $C_2 = 0,553$.

The simply supported beam has a laminated rectangular cross-section with 3 glass panes of 5 mm thickness divided by 2 layers of 0,38 mm of thickness (cf. Beam 2 on Figure 3). Young's modulus of the glass panes is $70.000\,N/mm^2$ and Poisson's ratio amounts 0,23. The elastic parameters of the interlayer are variable. The point of application of the concentrated load is the centre point on the upper rim of the central glass pane.

2.2 Element type

Previous investigation by Pattheeuws (Pattheeuws 2001) at Ghent University showed that the general-purpose shell element S4(R) is an accurate and powerful tool for elastic buckling load calculations of simply supported beams. The use of these shell elements has advantages as well as disadvantages. The main advantages are the limited calculation time and the possibility to create a "sandwich" shell (modelling of laminae with different mechanical characteristics).

As the response of the shell is assumed linear elastic and its behaviour is not dependent on temperature changes, the "shell general section option" is chosen to define the laminae. The section response can be specified by associating the section with several different material definitions. For a composite shell the equivalent section properties can then be determined automatically by Abaqus. A first disadvantage in this type of modelling is the fact that there is only one element (with the equivalent section properties) in the thickness direction of the shell. This means that the borders of the laminated pane remain straight when it is bending. A second is that local instabilities can arise if the interlayer modulus of Young gets very low (i.e. $400\,N/mm^2$ or lower). These disadvantages seem to be of lower importance for this investigation, as the modulus of Young for adhesive interlayers is in the order of $700\,N/mm^2$.

2.3 Mesh

There is a direct relationship between the number of elements at the one hand and at the other hand the

calculation time and the accuracy of the calculations. The higher the number of elements, the higher the calculation time and the higher the accuracy. To perform a realistic, accurate and workable study an optimum number of elements is determined.

The number of elements (l) is related to the number of nodes in the longitudinal direction (m) and the number of nodes in the height direction (n) by the following equation:

$$l = (m-1)(n-1)$$ (3)

The determination of the optimum is done by calculating several configurations with varying m and n. These calculations demonstrate that the number of nodes in the length direction is much more important than the number of nodes in the height direction. When a minimum elastic buckling load needs to be determined, it is much more effective to augment m (for a constant n) than to increase n (for a constant m) as is shown in Figure 1.

For the same number of elements and for almost the same computational time, the highest m results in the lowest P_{cr}. Further increase of m would not decrease $P_{cr,min}$ much, but would nevertheless increase the calculation time.

An overall optimum (for this geometry) is found in a mesh with m = 110 and n = 30. This mesh is used for all further calculations.

2.4 Analysis type

In order to obtain the correct elastic buckling load, there are two possible simulation methods: an iterative eigenvalue determination method ($P_{cr,iter}$) and a static "modified Riks method". The first method is described by Belis (Belis, in press). The second method is very useful if one is interested in post-buckling behaviour (which is not the case here). As this study searches only the maximum in the load-displacement diagram, the first method is used here. Because the parametric study involves a lot of numerical simulations, a method is

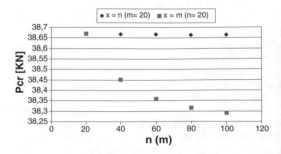

Figure 1. Effect of the number of nodes on the elastic buckling load P_{cr}.

searched to find a relationship between P_{cr} and $P_{cr,iter}$. In some characteristic points $P_{cr,iter}$ is calculated and these calculations prove that $P_{cr,iter}$ is about 0,2% higher than P_{cr}. Since this difference is most likely the same for all simulations, the iterative calculations do not have to be examined and the effect of the interlayer stiffness can be examined using the (non iterative) elastic torsional buckling load P_{cr}.

2.5 Validity of the FEM

The authors would like to emphasize that all calculations are made in the assumption of linear elastic material behaviour. The effect of the different layers is investigated using S4 shell elements and using the "shell general section option". The elastic buckling load (used to investigate the effect of the interlayer stiffness) is simulated by a non-iterative eigenvalue calculation.

3 VARYING INTERLAYER STIFFNESS

3.1 Parameters

Executing the parametric study to vary the interlayer stiffness, there are two main parameters, which can be modified: the modulus of Young (E) and Poisson's ratio (v). A few hundred simulations are done varying v from 0,02 to 0,48 and E from 600 N/mm^2 to 70.000 N/mm^2. The authors have chosen these boundaries for the following reasons. According to Verhegghe (Verhegghe 2001) Poisson's ratio can never exceed 0,5 for isotropic simple (non composite) materials, because otherwise one would be able to create free energy. The following formula proves that the modulus of compression (K) has to be strictly positive for the compression of a structure:

$$\frac{\Delta V}{V} = \frac{\sigma_m}{K} = \frac{\sigma_m \, 3(1-2v)}{E} \qquad (4)$$

In Equation 4 σ_m represents the mean stress (a negative value for compression) and $\Delta V/V$ the volumetric change.

The modulus of young of PVB lies in the neighbourhood of 700 N/mm^2. Since the effect of increasing the interlayer stiffness is the topic of this investigation, the value of 600 N/mm^2 is chosen as the lower limit. The considered upper limit for the modulus of Young is the one from glass. There are translucent materials that are stiffer, but the authors are in the opinion that these materials are at this time not suitable for use in structural glass beams for several reasons, amongst which costs, serviceability temperatures, adhesion problems and others.

3.2 Results

In order to obtain a clear graphic, not all calculated values are shown in Figure 2.

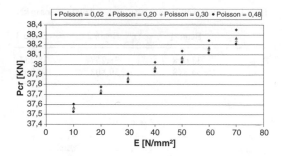

Figure 2. Effect of the variation of E and v on the elastic buckling load P_{cr}.

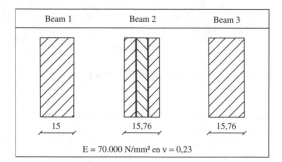

Figure 3. Single pane beam verification.

As clearly shown in this figure, the modulus of Young and Poisson's ratio have opposite effects. An augmentation of Young's modulus causes an increase in P_{cr} (for v = constant value) and an augmentation in Poisson's ratio causes a decrease in the elastic buckling load (for E = constant value).

The highest resistance against torsional buckling can then be found for E = 70.000 N/mm^2 and v = 0,02, i.e. P_{cr} = 38,352 kN. The minimum value (not shown in Figure 2) of P_{cr} is 35,850 kN for E = 600 N/mm^2 and v = 0,48. This means a decrease of $\pm 7\%$ of the torsional resistance against the obtained maximum value.

3.3 Verification

As neither experimental data, nor literature concerning this topic is available at the moment, a few logical verifications are carried out to gain confidence in these strictly numerical results.

3.3.1 Single pane beam
Three characteristic configurations, shown in Figure 3, are studied. The first and the third beam are modelled as a single glass pane; the second one is modelled with laminates (all layers with the mechanical characteristics of glass). The obtained numerical FEM and

Table 1. Single pane beam verification. All dimensions in kN.

	Beam 1	Beam 2	Beam 3
$P_{cr,FEM}$	33,018	38,257	38,257
$P_{cr,EC3}$	34,318	39,803	39,803

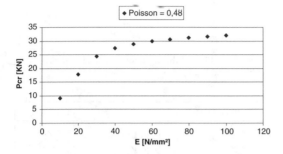

Figure 4. Composite pane beam verification.

the analytical values, using the Equation 1, are compared in Table 1.

The numerical data show that there is no difference in modelling a single pane and a composite pane with the same mechanical characteristics for each layer. The numerical critical torsional buckling load is about 4% lower than the "safe" values calculated with the theoretical Formula 1.

3.3.2 Composite pane beam

If theoretically the interlayer stiffness is zero, which would mean that the three glass panes act separately, the Formula 1 gives the following (absolute minimum) critical buckling load: $P_{cr} = 9,277$. This configuration cannot be simulated exactly using shell elements (cf. §2.2). The only possibility to model this problem is to reduce the interlayer stiffness to a very low value. Figure 4 shows the relation between P_{cr} and E (for $\nu = 0,48$).

The figure shows that the critical buckling load decreases very quickly for small values of E. Attention should be paid to local instabilities that arise in the numerical simulation. These instabilities are important for low values of E and become smaller for higher values of E. They completely disappear if E exceeds $400\,N/mm^2$. Since the global study involves only values of E beyond $600\,N/mm^2$, the local instabilities do not play an important role. If one is interested in these low values of the modulus of Young, this problem could possibly better be investigated using solid elements.

4 CONCLUSIONS

The numerical study shows that theoretically an improvement of $\pm 7\%$ of the critical elastic torsional buckling load can be obtained by changing the interlayer stiffness. An important disadvantage is a decrease of the impact resistance when the interlayers get stiffer. So one has to bear in mind multiple factors (for example depending on the application field) before manufacturing these stiffer interlayers.

Further investigation is necessary to examine the non-linear material behaviour of laminated glass beams and the importance of imperfections. A "solid" finite element model can probably provide more clarity in the behaviour of soft interlayers (E $< 400\,N/mm^2$).

REFERENCES

Belis, J., Van Impe R., De Beule M., Lagae G. & Vanlaere W. 2003. Experimental and numerical study of buckling glass beams with 2 m span. *Abstract accepted for presentation at ISEC 02, Rome, 23–26 September 2003*. Rotterdam: Balkema.

Pattheeuws, S. 2001. *Calculation of lateral torsional buckling loads with finite element method, Graduate thesis* (in Dutch), Ghent: RUG.

Verhegghe, B. 2001. Elasticiteit en sterkteleer.

Confronting non-linear and evolutional numerical analyses of horizontally loaded piles embedded in soils with experimental results and the concept of safety

R. Van Impe, J. Belis, P. Buffel, M. De Beule, G. Lagae & W. Vanlaere
Laboratory for Research on Structural Models, Department of Structural Engineering, Ghent University, Ghent, Belgium

ABSTRACT: The paper shows the computational steps to calculate the non-linear horizontal response of piles that are partly embedded in soils, mooring dolphins are a well-known practical example, where effects of the loading history, e.g. loading followed by partial unloading and reloading, are accounted for. The behaviour of the structure is studied in a realistic manner, which accounts for the logical succession of loading states and their interaction for determining the value of the soil pressures and internal forces distribution – e.g. bending moments, shear forces – along the member. That is, a true evolutional approach is conducted in the numerical analysis. In order to check the validity and practical value of the computer program a comparison with – admittedly scarce – available test data has been carried out. It is shown that the computer model gives satisfactory results and is able to predict the response of the structure, even for cyclic loading histories. Finally an attempt is made to address the concept of safety in the context of the design philosophy that is typical for a limit states approach.

1 INTRODUCTION

Several methods of analysis of laterally loaded piles partly embedded in soils are available to-date. They can be classified into two main categories. The first category deals with so-called plastic analyses of the ultimate limit state in keeping with the design philosophy of a partly deterministic and partly probabilistic approach. Because the analysis is concerned with ultimate limit state design, results for horizontal displacements of the pile section are meaningless.

One the other hand, the elastic-plastic methods of analyses try to predict the response of the pile, i.e. deflections and rotations of the pile sections together with the distribution of internal forces and lateral pressures along the embedded portion when subjected to some loading history in the serviceability limit state.

In the latter category it is imperative to model the behaviour of the soil-structure interaction as accurately as possible. To this end, hyperbolic functions are introduced to describe the elastic-plastic nature of horizontal pressures exerted by the soil on a retaining structure as a function of the relation between the horizontal pressures, exerted by the soil and the horizontal displacements of the retaining structure.

The method exposed in (Van Impe 2002) makes it possible to carry out incremental and iterative analyses and also makes the theory readily applicable for practical purposes using the matrix displacement method.

2 EQUILIBRIUM EQUATION AND THE DISPLACEMENT METHOD FOR BEAM MEMBERS

2.1 *General method*

It has been shown in (Van Impe 2002 and Van Impe et al. 2003) that equilibrium of a short slice of the pile, with the symbols used there, can be written as

$$EI \frac{d^4 w}{dx^4} + kbw = bq(x) \qquad (1)$$

The net pressure on the slice q(x) results from earth pressures at the left and right-hand sides of the contact area between pile wall and soil, from water pressures or from another external, but a priori quantifiable source. Figure 1 shows the generalized displacements for a particular beam slice, i.e. lateral displacements w and rotations α, together with the adopted sign convention for external and internal forces, i.e. lateral loads, external couples (if any) and bending moments and shear forces respectively.

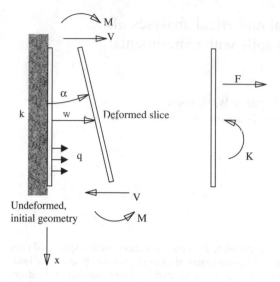

Figure 1. Notations and sign conventions.

According to the philosophy of the matrix displacement (or finite element) method the differential Equation 1 is equivalent with the matrix formulation

$$[S]\,\{u\} = \{F\}, \tag{2}$$

in which [S] denotes the stiffness matrix, {u} the vector of lateral displacements w and rotations α of the upper and lower node of the member and {F} the load vector.

$$[S] = \frac{2EI\lambda}{\sinh^2\beta - \sin^2\beta}.$$

$$
\begin{bmatrix}
2\lambda^2(\sin\beta\cos\beta + \cosh\beta\sinh\beta) & & & \\
\lambda(\sinh^2\beta + \sin^2\beta) & \sinh\beta\cosh\beta - \sin\beta\cos\beta & & \\
-2\lambda^2(\cosh\beta\cdot\sin\beta + \sinh\beta\cos\beta) & -2\lambda\sinh\beta\sin\beta & 2\lambda^2(\cosh\beta\sin\beta + \sin\beta\cos\beta) & \\
2\lambda^2\sinh\beta\sin\beta & \cosh\beta\cdot\sin\beta - \sinh\beta\cos\beta & -\lambda(\sin^2\beta + \sinh^2\beta) & \sinh\beta\cosh\beta - \sin\beta\cos\beta
\end{bmatrix}
$$

$$k > 0 \; ; \; \lambda^4 = \frac{kb}{4EI} \; ; \; \beta = \lambda\ell$$

$$\tag{3}$$

$$[S] = \frac{EI\lambda\sqrt{2}}{1 - \cosh\beta\cos\beta}.$$

$$
\begin{bmatrix}
2\lambda^2(\sinh\beta\sin\beta + \cosh\beta\sinh\beta) & & & \\
\lambda\sqrt{2}(\sinh\beta + \sin\beta) & \cosh\beta\sin\beta - \sinh\beta\cos\beta & & \\
-2\lambda^2(\sin\beta + \sin\beta) & -\lambda\sqrt{2}(\cosh\beta - \cos\beta) & 2\lambda^2(\cos\beta\sin\beta + \cosh\beta\cdot\sin\beta) & \\
\lambda\sqrt{2}(\cosh\beta - \cos\beta) & \sinh\beta - \sin\beta & -\lambda\sqrt{2}\sinh\beta\sin\beta & \sinh\beta\cosh\beta - \sin\beta\cos\beta
\end{bmatrix}
$$

$$k < 0 \; ; \; \lambda^4 = \frac{kb}{4\,EI} \; ; \; \beta = \lambda\sqrt{2}\cdot\ell$$

$$\tag{4}$$

Equations 3–4 above give the explicit formulation of the (symmetric) stiffness matrix for positive and negative values of the stiffness parameter k respectively. Depending on the adapted iteration scheme during a particular load step or construction stage, the stiffness matrix matches a tangent, secant or initial stiffness formulation. [S] and {F} are, generally spoken, functions of w and are updated continuously during the incremental, iteration approach.

It is worthwhile to insist on the fact that this equation is apparently similar to the one that governs equilibrium of a beam on elastic foundation according to Winkler's theory. In reality however, there are substantial differences.

• The equation is a non-linear differential equation because the right-hand side and the second term of the left-hand side are functions of w, which is a consequence of the non-linear nature of the soil-structure interaction and by this
  (a) Requires an iterative solution strategy
  (b) Requires the stiffness parameter k to be formulated as a function of w and consequently of x. Besides that, k may become negative during the iteration cycles, exception made for the case where iterations are carried out with constant stiffness.

• In the "finite element" displacement formulation used here, it is assumed that k does not vary along an elementary beam slice. This, of course, has at least some influence on the obtained numerical results. Because k is a function of w and because w varies with the depth of envisaged beam slice under the surface of the soil, the authors have chosen to select the value of k corresponding with the middle of the slice. Selecting a sufficiently dense mesh of beam slices minimizes the error caused by this choice.

2.2 Course of the computations

At the start of the analysis, all lateral displacements of the pile or wall are zero. Subsequently, the pile is subjected to "loads" and the programme computes the bending moments and shear forces in the member and its corresponding deflections and rotations, together with the lateral soil pressures. The word "loads" should be interpreted in a very broad sense: it may be an excavation at e.g. the left-hand side of the pile, the lowering of the level of the water table, the application of temporary struts or anchors, which may be prestressed or not, the removal of temporary struts, the application of external loads and so on....

In order to evaluate the effect of the present construction stage on the internal forces distribution in the wall/pile, the following method is adopted:

• The stiffness matrix of the elementary slices into which the pile has been subdivided, is computed,

and the equilibrium equations for the whole structure are assembled.
- At the same time, the stiffness of ground anchors or frame struts is taken into account.
- Eventually, the set of (linearized) equilibrium equations [S] {u} = {F} is solved for the (incremental) displacements {u}.

Because the solution technique involves a linearization process, the found solution {u} is evidently only an approximation of the "exact" generalized displacements. Therefore, a Newton-Raphson (or modified Newton-Raphson) iteration scheme is carried out in which the displacements, stiffness-matrices (if iterations are not carried out with constant stiffness) and load vectors are gradually adapted and/or updated. In this context, reference is made to the handling of non-linear, evolutional and elastic-plastic constitutive equations set forth in (Van Impe 2003).

The computation steps, outlined in the above, are repeated until convergence is achieved. Several convergence criteria are built into the computer code to check for convergence: there are convergence tests on the (variable) lateral stiffness parameter k of the elements, on the soil pressures at the left-hand side and at the right-hand side of the wall, on the stiffness of ground anchors that may show a non-linear behaviour, especially when they exert forces that are approaching their ultimate carrying capacity; on the displacement and load vector of the structure.

3 CONFRONTATION OF THE METHOD WITH EXPERIMENTAL EVIDENCE

3.1 General

In the past, a number of tests on laterally loaded piles have been carried out and well documented. Yet the available experimental data in this field is still rather scarce. The authors want to confront the validity of the proposed calculation method with these experiments.

3.2 The experimental findings of Michel Adam and Jacques Lejay: static loading

In total, thirty tests have been carried out in clay soils of Lannois at the site of Romainville in France, with seven different types of piles imbedded in the soil. The piles with a box section are composed of steel boards of type Larssen and they are subjected to different combinations of shear force and bending moments at ground level. The embedded part of the piles varies from 2.5 m to 8.3 m, which means that the experiments cover a rather large range of responses, going from the one that is typical for an "infinitely" long pile to the one that corresponds with an "infinitely stiff pile" over a vast number of intermediate ones, typical for relatively flexible piles.

Table 1. Pile sections.

Pile	A	B	H
Width b (cm)	43	43	95
Section A (cm^2)	888	1100	6390
Stiffness EI (kNm^2)	16000	26000	775000

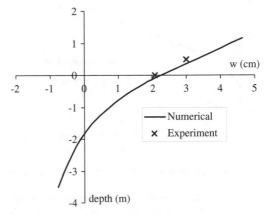

Figure 2. Deflections for pile A.

The soil, in which the tests have been carried out, consists of a very homogeneous clay layer with an angle of internal friction $\varphi = 22°$ and cohesion $C = 55 \text{ kN/m}^2$.

The test piles are equipped with inclinometers; displacement measuring instruments and pressure cells so that for each particular test results of rotations and displacements, of the bending moment and shear forces and of the net lateral pressures on the piles are available. In the present contribution, three test cases shown in figure are selected and a comparison is made with the numerical results from the authors' computer programme.

The cross-section of piles A and B are the same but the length of the embedded portion is different: d = 4.90 m for the latter, termed "infinitely long pile" and d = 3.50 m for the former, termed "flexible pile". On the other hand, the embedding for pile H is the same as for pile A, but its cross-section has a much larger bending stiffness so that it remains almost straight during the loading of the pile. Details of the cross-section of the members are given in Table 1.

3.2.1 Comparisons for pile A
The member, termed "flexible" by Adam et al., is subjected to a gradually increasing horizontal load F applied at 1.15 m above ground level. Figure 2 shows that the calculated deflections for the final value of

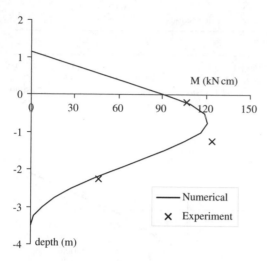

Figure 3.   Bending moment distribution for pile A.

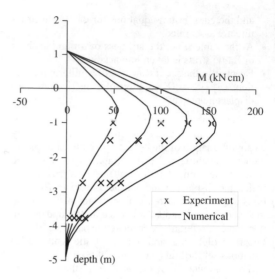

Figure 4.   Bending moment distribution for pile B.

F = 80 kN are in good agreement with the experimental values.

Adam et al. also registered the bending moments along the member and their results are compared with the numerically obtained values (Fig. 3). It is clear that the agreement may be qualified as excellent from a practical point view.

3.2.2   Comparisons for pile B

In this case, the embedded portion of the pile is substantially longer than for pile A whereas the overall cross-sectional dimensions are the same. Adam et al. speak of an "infinitely long pile".

Here too, the pile is subjected to some load history: 35 kN – 55 kN – 75 kN – 90 kN. Before the horizontal load test was carried out, it has been preceded by a vertical compression test during which the uppermost layer of the soil has been disturbed. As a matter of fact, Adam et al. speak of a disturbed depth of about 30 cm. In order to carry out the numerical simulations and to account for the lower shear strength properties of the disturbed layer, the latter has been partly discarded from the analysis. Figure 4 shows the bending moment distribution corresponding with each particular load step. When the numerical results are inspected and compared against the results from measurements, the agreement is quite satisfying.

3.2.3   Comparisons for pile H

This member has much larger cross-sectional properties than pile A or B and is termed "infinitely stiff". The latter is confirmed when the displacements of the pile are plotted (Fig. 5): it is seen that the centre-line of the member stays almost perfectly straight, which has also been observed during the experiments.

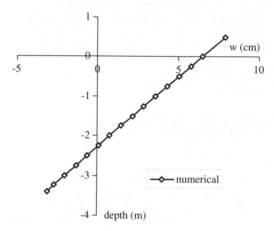

Figure 5.   Displacements for pile H.

The soil pressure distribution is shown in Figure 6. It is seen that the soil pressures on the left-hand side of the pile vanish over a considerable depth, measured from the ground level. Likewise, the pressures on the right-hand side of the pile become zero almost at the point where the lateral displacements become negative. Both observations are explained by the presence of a cohesive soil: it has been set out right from the beginning that the computer programme ignores any effect of adhesion.

Nothing is said about the experimental observations until now. The authors must admit that the agreement is not good at all. Moreover, they have no idea, at least for the moment, about the reason for the discrepancy. That

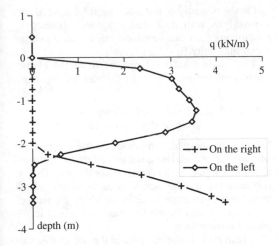

Figure 6. Soil pressure distribution.

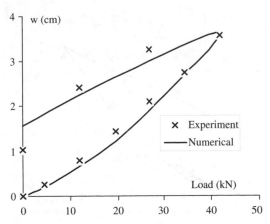

Figure 7. Load–displacement plot.

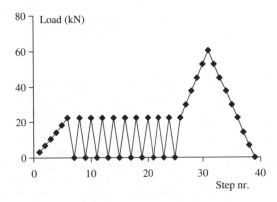

Figure 8. Load history.

the ratio of bending moment over the shear force at ground level is much smaller than the ratio observed for piles A and B certainly cannot be the cause. It is surmised that due to the overall dimensions of the cross-section of the pile that there may be a resisting couple at the lower end of the pile, which is not taken into account in the calculations: the dynamic boundary conditions in the computer model set both bending moment and shear force equal to zero at the lower end. When the experimentally observed drawings, given in (Adam et al. 1971), are examined, one sees that substantial bending in the member is present near to that edge, which as stated before, cannot be taken into account in the computer programme in its present version. Further research is needed to clarify the mentioned discrepancy.

3.3 Loading followed by unloading and reloading the experiments of Baguelin and Jezequel

One of the tested piles is a H-section with the following properties: $I = 13673\,cm^4$, $A = 97.3\,cm^2$, total length $= 8\,m$ with an embedded portion of length $6.10\,m$, frontal width $= 28\,cm$. It has been driven in a trench near the Arguenon River in France. Near the surface of the trench, a layer of clay is present, at depths larger than $3.5\,m$ the soil consists of fine sand. Details of the measuring equipment are given in (Baguelin et al. 1972).

In a first stage, the pile is subjected to a step-wise increasing lateral force, applied at a height of 45 cm above ground level, and is subsequently unloaded. Figure 7 gives the displacements as a function of the applied loads.

In a second stage, the pile is reloaded step-wise, incorporating 10 loading and unloading cycles

$0\,kN – 23\,kN – 0\,kN$, until a maximum horizontal load of $61\,kN$ is applied, and finally unloaded – again in small load steps. Figure 8 may visualize the rather complex loading history.

Baguelin et al. give experimental values of the lateral displacements starting from zero when the lateral load is also zero. The authors believe that this is not correct: during the previous test – one month earlier – there was a residual displacement when the load had been removed, as is seen in Figure 7.

Because Baguelin et al. did not make the proposed shift in their diagram, the authors of the present contribution decided to run the application from scratch, i.e. as if the previous test had not been carried out at all, in order to be able to make comparisons with the available data. The response of the pile to the loading history is shown in Figure 9 and it may be concluded that the agreement between numerical results and experimental evidence is quite satisfactory.

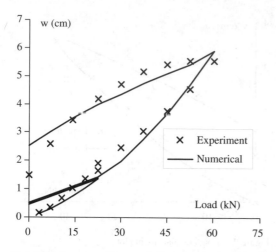

Figure 9. Response of the tested pile to complex load history.

4 ON THE CONCEPT OF SAFETY OF LATERALLY LOADED PILES

4.1 *General*

Some of the shortcomings of the plastic methods are given in Vandepitte (1988). The conclusion is that plastic methods for studying the ultimate limit state cannot be rationally justified, except when the retaining structure is statically determinate in the ultimate limit state and the horizontal displacements can become great enough to give rise to either an active or a passive state of deformation of the soil all along (or nearly all along) both faces of the wall. The authors believe that it more rational to apply plastic-plastic methods for analysing both the ultimate limit state and the serviceability state.

4.2 *Elastic-plastic methods and the ultimate limit state*

In keeping with a generally observable trend in the civil engineering profession the ultimate limit state is analysed with a design value φ_d of the angle of internal friction and, when occasion arises, a design value c_d of the cohesion. The design values φ_d and c_d are obtained through dividing the characteristic value of the corresponding shear strength of the soil by a material coefficient exceeding unity; $\tan \varphi_d = \tan \varphi_k/\gamma_\varphi$ and $c_d = c_k/\gamma_c$. φ_k and c_k are "characteristic" values of φ and c, respectively. Practically speaking, they are close to the lowest values obtained from a series of measurements pertaining to the soil layer in question. In view of the fact that the magnitude of c is more uncertain and less well known than that of φ it is not surprising to see in the Structural Eurocodes that $\gamma_c > \gamma_\varphi$.

On the other hand, the loads must be entered into the analysis with their design values, obtained by multiplying their characteristic value with a number larger than unity $F_d = \gamma_F \cdot F_k$.

The required safety is present when the elastic-plastic method converges for a particular building stage.

4.3 *Evolutional, plastic plastic method and the serviceability limit state*

Only the plastic-plastic methods enable a civil engineer to properly analyse a retaining structure or laterally loaded pile under service conditions. This is substantiated by the scarce measurements, which have been carried out in situ and which the authors are aware of, and by the test cases examined in the present contribution.

A realistic calculation implies the use of true values of the properties of the soil and of the retaining structure or pile as opposed to design values of the material characteristics.

For example φ_d and c_d are naturally smaller than the actual angle of internal friction and the actual cohesion respectively; they lead to an overestimated active pressure p_a and an underestimated pressure p_b, and are therefore unfit to be employed in a calculation relating to service conditions. The authors are of the opinion that it is not desirable to use even the characteristic values φ_k and c_k, since these are on the low side for the soil layer concerned, but that an average value of φ and of c for each statue is to be preferred. Neither should load factors be introduced into such a calculation. In order to get an idea of the overall safety – from a point of view of the shear resistance of the soil – one can proceed as follows:

- An examination of the deflections of the earth retaining structure or pile, especially the deflections for the embedded portion, makes it possible to indicate the zones where the soil is in state of compression and where it is in a state of expansion.
- For the compressed portion, one can draw a diagram where the soil pressures are set equal to the passive pressures.
- The ratio of the area measured between the reference line and the curve that gives the calculated pressures and a similar area measured between the same reference line and the passive pressures may be taken as the inverse of the margin of safety for the examined construction stage.

5 CONCLUSIONS

A method for analysing the behaviour horizontally loaded piles and earth-retaining structures in a realistic manner has been presented. The method accounts for the logical succession of loading states and their interaction for determining the value of the soil pressures and internal forces distribution – e.g. bending moments, shear forces – along the member or lining. That is, a true evolutional approach is conducted in the numerical analysis.

In order to check the validity and practical value of the computer program a comparison with – admittedly scarce – available test data has been carried out. It is shown that the computer model gives satisfactory results and is able to predict the response of the structure, even for cyclic loading histories.

Finally an attempt is made to address the concept of safety in the context of the design philosophy that is typical for a limit states approach.

REFERENCES

Adam, Michel & Lejay, Jacques 1971. Etude des pieux sollicités horizontalement. Détermination du module de réaction dans un sol donné. *Annales de l'Institut Technique et des Travaux Publics* 280: 125–146.

Baquelin, F. & Jezequel, J.F. 1972. Etude éxperimentale du comportement de pieux sollicités horizontalement. *Annales de l'Institut Technique et des Travaux Publics* 297: 153–204.

Vandepitte, Daniël 1988. Slurry walls – Evolution of methods of analysis. *Tunnels et ouvrages souterrains* 87: 139–151.

Van Impe, Rudy 2002. *Berekening van Constructies 5.* Ghent: Laboratory for research on structural models.

Van Impe, Rudy et al. 2003, Hyperbolic functions for modelling the elasto-plastic behaviour of soils in the evolutional, non-linear analysis of slurry walls and sheet-piling structures. *Proceedings of the ISEC-02 Conference,* Rome (to be published).

System-based Vision for Strategic and Creative Design, Bontempi (ed.)
© 2003 Swets & Zeitlinger, Lisse, ISBN 90 5809 599 1

Hyperbolic functions for modelling the elasto-plastic behaviour of soils in the evolutional, non-linear analysis of slurry walls and sheet-piling structures

R. Van Impe, J. Belis, M. De Beule, G. Lagae & W. Vanlaere
Laboratory for Research on Structural Models, Department of Structural Engineering,
Ghent University, Ghent, Belgium

ABSTRACT: The concept of modelling lateral soil pressures on slender slurry walls and sheet-piling earth retaining structures by means of hyperbolic functions is introduced in the paper. It is shown how this mathematical formulation is applied to perform a non-linear calculation of the structure in the serviceability state. A common approach – termed classical – in the study of retaining structures is that several construction stages are being analysed separately, i.e. as if they all would have arisen all of a sudden, disregarding their mutual interaction and impact on the response of the finished structure. The present approach brings a remedy for the obvious shortcomings of such calculation methods and shows how a true evolutional analysis strategy, where the logical succession of building stages is respected, can be conducted. Differences between the classical approach and the method outlined in this contribution are show by means of practical design examples.

1 CONSTITUTIVE EQUATIONS FOR SOILS EXERTING LATERAL PRESSURE

1.1 *Plastic methods and their shortcomings*

Plastic methods are used most often by civil engineers for design purposes in their everyday practice. They are based on the premise that the earth on both sides of the wall is in a limit state and that consequently, the lateral pressure it exerts on the wall at any point is either the active or the passive earth pressure. Wherever the direction of the lateral displacement of the lining at a certain level is such that it would recede from the soil, the pressure transmitted to the wall by the soil is assumed to be the active lateral earth pressure p_a. Wherever the wall compresses the soil laterally, the earth is supposed to be in a passive state of deformation and to develop the greatest possible counter-pressure, namely the passive pressure p_p. Yet, plastic methods present a number of obvious shortcomings (Vandepitte 1988).

The analysis is not based on the real behaviour of the soil. It is assumed off-hand that the lateral earth pressure at any point is either the active or the passive pressure, without making sure that the local horizontal displacement of the sheeting is sufficient to cause the pressure to decrease to the active one or to increase to the passive one.

The stiffness of the retaining structure and of the soil do not appear at all in the analysis. Whereas the assumptions made are credible for the ultimate state of failure by overturning of unbraced and unanchored and hence vertically cantilevering sheeting, the horizontal movement of the embedded part of a concrete wall with several lateral supports is not great enough to bring the soil into its passive state of deformation.

The extensional deformation of braces and tiebacks is neglected. It is true that the lateral supports are often concrete floors and that these are hardly compressible. The assumption of immobility is less realistic for ground anchors.

The plastic methods do not enable the designer to evaluate the horizontal displacements of the slurry wall.

1.2 *Method based on a non-linear constitutive law*

The basic premise is the existence of a definite relation between the horizontal movement of the slurry wall at any level and the pressure applied at the same point by the soil on the wall, on the understanding that the earth pressure can neither drop below the active pressure p_a nor rise above the passive pressure p_p.

In the elastic-plastic method expounded in (Vandepitte 1979) the hyperbolic functions below are supposed to reflect the behaviour of the soil (see Fig. 1).

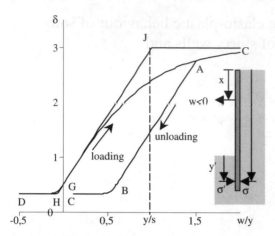

Figure 1. General notations and plot of the variation of the pressure coefficient with the relative deflection of the wall.

Depending on the sign of the wall movement, the lateral earth pressure coefficient δ on the right-hand side of the earth retaining wall is given by

$$\delta = \lambda_n + (\delta_p - \lambda_n)\tanh s\frac{w}{y} \qquad (1.a)$$

$$\delta = \lambda_n + (\lambda_n - \delta_a)\tanh\left(s\cdot\frac{\delta_p - \lambda_n}{\lambda_n - \delta_a}\cdot\frac{w}{y}\right) \qquad (2.a)$$

for $w > 0$ and $w < 0$ respectively.

Similar equations hold for the left-hand side:

$$\delta' = \lambda_n - (\lambda_n - \delta_a)\tanh\left(s\cdot\frac{\delta_p - \lambda_r}{\lambda_n - \delta_a}\cdot\frac{w}{y'}\right) \qquad (1.b)$$

$$\delta' = \lambda_n - (\delta'_p - \lambda_n)\tanh s\frac{w}{y} \qquad (2.b)$$

for $w > 0$ or $w < 0$ respectively.

When one replaces σ' and y' by σ and y, noting further that a negative w has the same effect on the soil to the left of the wall as a positive w has on the soil to the right of the wall, one realises that the Equations 1.b & 2.b are as a matter of fact the same as Equations 1.a & 2.a.

In the above: w is the lateral displacement of the wall; σ (σ') is the vertical intergranular pressure at the level defined by y (y'); λ_n is the neutral lateral earth pressure ($\lambda_n = 0,5$ or $\lambda_n = 1 - \sin\varphi$ is often used); φ is the angle of internal friction; y (y') is the depth under the surface; $\delta_p = \lambda_p\cos\psi$ with λ_p the coefficient of passive earth pressure; $\delta_a = \lambda_a\cos\psi$ with λ_a the coefficient of active earth pressure. It is well

known that the angle of wall friction ψ affects the active earth pressures only slightly but has a considerable influence on the passive earth pressures p_p. Here, its effect is taken into account by means of the following, partly pragmatic, procedure: taking a value of ψ that is a fraction of the internal angle of friction, e.g. $\psi = 2/3\varphi$ for rather rough contact surfaces between soil and wall or $\psi = 1/3\varphi$ for smooth surfaces; attributing the correct sign to it, having regard to the direction of the relative movement of the wall and the soil accompanying the deformation of both; then taking values of λ_a and especially of λ_p from tables as a function of φ and ψ, for example the tables given by (Caquot et al. 1973). Admittedly, this procedure requires earth pressure tables and seems more complicated than using the expressions, found in (Bureau Seco 1978): $\lambda_a = tg^2([\pi/4] - [\varphi/2])$ and $\lambda_p = n\cdot\tan^2([\pi/4] + [\varphi/2])$, with n being a factor reflecting the beneficial influence of wall friction: $n = 1$ for $\psi = 0°$, $n = 1.25$ for $\psi = 15°$, $n = 1.75$ for $\psi \geq 35°$.

The parameter s, appearing in Equations 1–2 is a stiffness parameter, usually taken equal to 100. By substituting a higher or a lower number than 100 for s while keeping λ_a, λ_p, λ_n and all other data being equal, one attributes more or less horizontal stiffness, respectively, to the soil. Decreasing the stiffness parameter may also be interpreted as allowing the soil to "creep".

For a given depth y, the Equations 1–2 between δ and w are represented graphically by curve DHGC in Figure 1. The ordinate of point G corresponds with a state of neutral pressure, i.e. $\delta = \lambda_n$. The factor s/y appearing in the argument of the hyperbolic tangents in Formulae 1–2 is the reciprocal of the abscissa of the intersection J of the tangent GJ at G and the horizontal having the limiting ordinate δ_p.

2 CLASSICAL ELASTO-PLASTIC METHOD

2.1 General outline

The method advocated by Vandepitte is based further on the differential equation:

$$EI\frac{d^4w}{dy^4} = b\cdot(p' - p - p'') \qquad (4)$$

where EI is the bending stiffness of the wall; p'' denotes the lateral pressure acting towards the left of the lining and possibly due to a local surcharge on the surface to the right of the retaining structure, or to a groundwater level which may be higher on the right-hand side than on the left-hand side; b is the width of the lining (e.g. 1 m for a continuous earth retaining structure or the actual frontal width of a pile) evaluated

2294

independently. In (Vandepitte 1979) the differential Equation 4 is transformed into a difference equation whose first member is linear in the wall displacement w at a few points in the vicinity of the level for which Equation 4 is written. The expressions 1–2 are then substituted for p and p'. Because those are not linear in the lateral displacements w the calculation has to be carried out iteratively. The final values of w are found by linearising the difference equation and by means of successive approximations, each one obtained as the solution of a set of linear algebraic equations.

2.2 Drawbacks of the classical method

The drawback of the elasto-plastic method, described above and of the computer programme implementing it lies in the fact that it is not amenable to an evolutional calculation; i.e. that it is assumed that the analysis relating to any given stage of excavation is carried out as if the corresponding state of structure materialized all of a sudden. That it is often the result of more than one construction stage, which *de facto* must have an influence on the considered state, and that lateral supports and/or ground anchors are not installed is not allowed for in the computation. Quite obviously, this must lead to (considerable) differences between the calculated internal forces – i.e. bending moments and shear forces in the lining – and loads in the struts and/or anchors and the actual values.

3 EVOLUTIONAL ANALYSIS

3.1 Description of the method

In the present paragraph it is shown how the defect of the "classical" approach can be removed by refashioning the method allowing the designer to analyse a sequence of construction stages. To that end the diagram in Figure is amplified by a curve – the crosses in Figure 1 – which represent the effect of a load reversal.

When for example, the soil has been compressed laterally at the left-hand side of the sheeting due to excavation on the left, until w is equal to the abscissa of point A in Figure 1 and when it is subsequently decompressed, e.g. as the result of installing and prestressing a ground anchor, the line ABC is adopted as the path followed by the load which represents the evolving state of deformation; path ABC consists of a straight line AB parallel to GJ (elastic recovery) and of the curve BC obtained by horizontal translation of GHD. The second part of this statement means that the soil loses its "memory" when it goes from a "passive" state ($\delta > \lambda_n$) to an "active" state ($\delta < \lambda_n$) during plastic unloading.

A recompression following a decompression that ended with w being equal to the abscissa of C, is

likewise assumed to be described by a straight line during elastic recovery and by a curve resulting from a horizontal translation of curve GH during plastic recompression. For an analytical treatment of all conceivable situations, the reader is referred to (Van Impe 1990 & 2002).

3.2 Differential equation of equilibrium and Newton-Raphson or modified Newton-Raphson iteration schemes

For reasons of simplicity, the pressures on the wall q" that are not due to soil action are left out of the following discussion. From §3.1 equilibrium is governed by the differential equation.

$$EI \frac{d^4 w}{dx^4} = b \cdot (\delta' \sigma' - \delta \sigma) \qquad (5)$$

It is also known that both σ' and σ are non-linear functions of the wall displacement w, which means that an iterative strategy is required to solve Equation 5 for w. Let w_i be an approximation for the actual wall displacement, found during iteration i. Different approaches to find a "better" approximation, w_{i+1}, are possible. These are briefly discussed below.

3.2.1 Tangential stiffness method
The relation, $\delta = \delta(w)$, is linearized a priori:

$$\delta(w) = \delta(w_i) + \frac{d\delta(w)}{dw}\bigg|_{w=w_i} (w_{i+1} - w_i) \qquad (6)$$

which is a Taylor series expansion of the first order, i.e. neglecting second and higher order terms. The direction coefficient $d\delta(w)/dw$ can be calculated with the known analytical expression of the soil pressure.

Substitution of Equation 6 into the differential equation yields

$$EI\frac{d^4 w_{i+1}}{dx^4} + b \cdot \left(\sigma \frac{d\delta(w)}{dw} - \sigma' \frac{d\delta'(w)}{dw} \right)\bigg|_{w=w_i} \cdot w_{i+1}$$

$$= b \left[\sigma' \left(\delta'(w_i) - \frac{d\delta'(w)}{dw}\bigg|_{w=w_i} \cdot w_i \right) \right.$$

$$\left. - \sigma \left(\delta(w_i) - \frac{d\delta(w)}{dw}\bigg|_{w=w_i} \cdot w_i \right) \right] \qquad (7.a)$$

or

$$EI\frac{d^4 w_{i+1}}{dx^4} + kb \cdot w_{i+1} = b \cdot p(w_i) \qquad (8)$$

In essence, Equation 8 does not differ from the differential equation for a beam on elastic foundation, with spring stiffness k given by

$$k = \sigma' \left.\frac{d\delta'(w)}{dw}\right|_{w=w_i} - \sigma \left.\frac{d\delta(w)}{dw}\right|_{w=w_i}, \qquad (9.a)$$

which can be solved readily (Van Impe 2002). With the "better" approximation w_{i+1} correspond new soil pressure coefficients δ and δ', and another linearization process with w_{i+1} taking the place of w_i can be carried out, yielding a differential equation in w_{i+2} etc.

3.2.2 Secant stiffness method

The terms of the right-hand side of Equation 5 are rewritten as

$$\sigma \delta(w_{i+1}) = \sigma \frac{\delta(w_{i+1})}{w_{i+1}} \cdot w_{i+1} \cong \sigma \frac{\delta(w_i)}{w_i} w_{i+1},$$

yielding the modified differential equation

$$EI\frac{d^4 w_{i+1}}{dx^4} + b\left(-\sigma'\frac{\delta'(w_i)}{w_i} + \sigma\frac{\delta(w_i)}{w_i}\right) \cdot w_{i+1} = 0$$
$$(7.b)$$

which is the differential equation for a beam on elastic foundation with spring stiffness

$$k = \sigma\frac{\delta(w_i)}{w_i} - \sigma'\frac{\delta(w_i)}{w_i'} \qquad (9.b)$$

Because the right-hand side of Equation 7.b vanishes when no other external pressures than those arising from soil action are present, it may be necessary to start the iteration cycle in an artificial way.

3.2.3 Initial stiffness method

As the name suggests, iterations for all load steps and/or construction stages are conducted with a constant stiffness equal to the initial stiffness corresponding with the inclination of the straight line GJ in Figure 1. Hence, the following approximations are used:

$$\delta(w_{i+1}) \cong \delta(w_i) + \frac{s}{y}(\delta_p - \lambda_n) \cdot (w_{i+1} - w_i)$$

and a similar expression for $\delta'(w_{i+1})$. Consequently, the differential equation is refashioned as follows:

$$EI\frac{d^4 w_{i+1}}{dx^4} + b\left[\sigma\frac{s}{y}(\delta_p - \lambda_n) + \sigma'\frac{s'}{y'}(\delta_p' - \lambda_n)\right]w_{i+1}$$

$$= b\left[\sigma'\left(\delta'(w_i) + s'\frac{\delta_p' - \lambda_n'}{y'}w_i\right)\right.$$
$$\left. - \sigma\left(\delta(w_i) - s\frac{\delta_p - \lambda_n}{y} \cdot w_i\right)\right] = 0 \qquad (7.c)$$

3.3 Selection of an algorithm

The initial stiffness iteration method is numerically the simplest but may converge slower. The stiffness parameter k is always positive, except when $\sigma = \sigma' = 0$. The latter situation arises when the upper portion of the retaining structure extends above the ground level, but here the equilibrium equation simplifies to that of a beam in bending which hardly provokes sleepless nights for structural engineers.

The tangent stiffness approach normally assures a rapid convergence. k may also occasionally be zero in this approach. Suppose for instance that w gets so large that $\delta = \delta_p$ and $\delta' = \delta_a'$ and correspondingly $d\delta/dw = d\delta'/dw = 0$. Here too, the seeming difficulty is removed by realizing that equilibrium is governed by the well known differential equation for a simple beam in bending.

Eventually, it is possible that the secant stiffness approach yields a stiffness parameter k that not only may be vanishing but may get negative as well. Even the latter possibility does not confront the analyzer with a numerically impossible-to-solve problem.

An ill-conditioned situation arises when w_i tends to vanish because the secant stiffness k then gets infinitely large. In that event, it is recommended that the iteration process is temporarily conducted with constant or tangent stiffness.

In their eager to devise a robust computer programme the authors in fact implemented all of the aforementioned solution strategies and made the programme smart enough to switch between schemes as it is confronted with unforeseen convergence problems.

4 EXTENDING THE THEORY FOR COHESIVE SOILS

The equation for the active and passive pressure, given in (Vandepitte 1979):

$$p_a = \lambda_a\sigma - 2c\sqrt{\lambda_a} = \lambda_a\sigma - 2ctg\left(\frac{\pi}{4} - \frac{\varphi}{2}\right) \qquad (10)$$

with $\lambda_a = tg^2([\pi/4] - [\varphi/2])$ and

$$p_p = \lambda_p\sigma + 2c\sqrt{\lambda_p} = \lambda_p\sigma + 2ctg\left(\frac{\pi}{4} + \frac{\varphi}{2}\right) \qquad (11)$$

with $\lambda_p = tg^2([\pi/4] + [\varphi/2])$ result from Coulomb's theory based on plane slide surfaces. When the angle of wall friction ψ differs from zero, the theory leads to results that are not theoretically exact but nevertheless sufficiently adequate as far as the active pressure is concerned.

In the case of passive earth pressure and also when $\psi \neq 0$, the assumption of plane slide surfaces often, but not always, leads to a considerable and dangerous overestimation of the passive earth pressure. Especially when ψ is not small, one can do better and apply Vandepitte's sound but rather laborious method to evaluate the passive pressure p_p.

Alternatively, one can borrow a more reliable (and generally lower) value of λ_p from the tables of Caquot et al. and refashion the expression of the passive earth pressure as follows:

$$p_p = \lambda_p\,\sigma + 2c\sqrt{\lambda_p} = \sigma\left(\lambda_p + \frac{2c\sqrt{\lambda_p}}{\sigma}\right) \qquad (12)$$

The elasto-plastic evolutional method exposed in the foregoing discussion can now be adapted quite straight-forwardly as follows: δ_a and δ_p are redefined as $\delta_a = (\lambda_a - [2c\sqrt{\lambda_a}]/\sigma)\cos\psi$ and $\delta_p = (\lambda_p + [2c\sqrt{\lambda_p}]/\sigma)\cos\psi$, respectively, and introduced in the elasto-plastic constitutive equations given by Equations 1–2. The hyperbolic functions appearing in these behavioural laws remain the same and nothing else has to be modified in the given evolutional, non-linear analysis sketched above. One has to remark, though, that δ_a may easily become negative if the cohesion is rather important and also when $\sigma = \gamma y$ is a small quantity, e.g. for soil layers near to the surface. It means that instead of exerting a pressure on the wall on the "active" side, the soil adheres to it and exerts a negative pressure. It is the authors' conviction that it does not seem wise to account on the adhesion and therefore it is suggested to use the modified expression for the coefficient of active earth pressure:

$$\delta_a = \left(\lambda_a - \frac{2c\sqrt{\lambda_a}}{\sigma}\right)\cos\psi \quad \text{but } \delta_a \geq 0 \qquad (13)$$

5 NUMERICAL EXAMPLE: COMPARISON OF THE CLASSICAL APPROACH WITH THE EVOLUTIONAL METHOD OF ANALYSIS

An earth retaining structure composed of steel sheeting is shown in Figure 2.

Details of the levels of the ground at both sides of the wall, the initial and final position of the ground

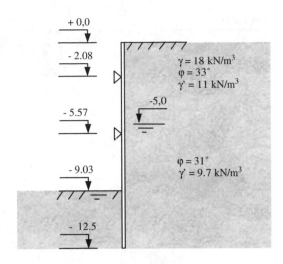

Figure 2.   Practical example of retaining structure.

water table, and the soil characteristics are given. They are borrowed from (Vandepitte 1979).

During the excavation of the site, two struts are installed, at the levels -2 m and -5.625 m respectively. The steel struts are 16 m long and spaced 4 m apart the upper ones are hollow circular members with an outer diameter of 406 mm and a wall thickness of 4.5 mm; the lower struts are hollow rectangular sections with a cross-sectional area of 124 cm^2.

For calculating the water pressures it is assumed that the lost head varies linearly along the shortest stream line that follows both sides of the wall. At the same time the influence of flowing ground water on the soil pressures is accounted for. A downward flow pressure behind the sheeting increases the apparent soil weight whereas an upward flow pressure decreases the apparent soil weight in front of the wall.

In the classical approach the bending moments and shear forces in the wall, the earth pressures and compressive forces in the steel struts are calculated for the final situation, i.e. as if the realization of the site would have materialized all of a sudden.

An evolutional approach is conducted as follows: it is assumed that the excavation is completed in five successive stages:

a) Removal of the soil on the left-hand side to level -3.0 m, followed by the emplacement of the first row of struts at level -2.00 m.
b) Lowering of the ground water table at the left-hand side to a level of -6.50 m.
c) Further excavation to level -6.50 m, followed by emplacement of the second row of steel struts at the level -5.625 m.
d) Lowering the ground-water table until the level of -9.00 m is reached.

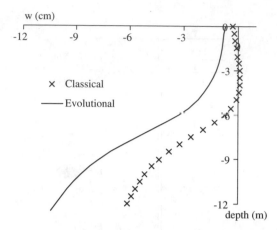

Figure 3. Displacements.

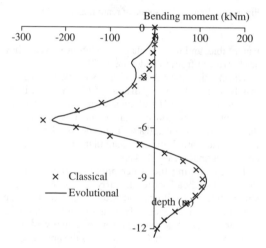

Figure 4. Bending moments.

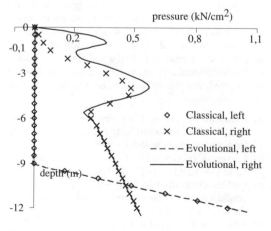

Figure 5. Soil pressures.

e) Final excavation to the level -9 m.

The displacements and bending moments in the sheeting and the soil pressures at the end of each construction stage are given in Figures 3–5 together with their corresponding values for the final state according to the classical approach. It may be clear that the classical approach assumes the struts to be fully effective right from the start of excavation, whereas in reality they are applied when the wall has already undergone a certain amount of displacement. This explains why the displacements of the earth retaining structure resulting from an evolutional analysis are substantially larger than those from the classical approach.

The forces in the struts at the end of stage (e) are 80 kN and 262 kN respectively. Here too, a remarkable difference is noted when compared with the classical analysis, especially for the upper struts: D. Vandepitte gives a value of 2.7 kN.

In the present example, the influence of accounting for a logical succession of building stages is less pronounced if one compares the bending moment distributions for the finished state, except for the upper portion of the wall where they are seriously underestimated by the classical approach (Fig. 4).

The evolutional method also clearly shows that the bending moment distribution in intermediate stages differs substantially from those in the final state.

The soil pressure on the right-hand side of the wall at levels beneath the lower struts is approximately equal to the active pressure. Above that level this is not the case anymore (Fig. 5). The reason is that during the construction stages that took place before the lower struts have been installed, the receding movements of the wall were larger than in the final state. This entailed a reversion of the sense of wall movement whereby the right-hand side of the wall recompressed the soil particles. Obviously the magnitude of wall movement towards the right was large enough to provoke a substantial increase the lateral soil pressure coefficient.

6 CONCLUSIONS

The main disadvantage of the classical elasto-plastic calculation method for studying slender earth retaining structures or laterally loaded piles, namely that the analyzed situation arises all at once and that a logical succession of construction stages, with mutual interaction effects, is neglected altogether, has been removed by adopting an evolutional method.

To this end, the elasto-plastic constitutive equations which make use of hyperbolic functions have been reworked to meet the requirements of an incremental and iterative solution technique. The reformulation of the behavioural equations, which in essence is relatively simple, makes it possible to account for the influence of a reversal of the direction of movement of different points of the retaining structure. The used mathematical models seem to be adequate to reflect the physical nature of the investigated problem.

Finally, a practical example has been documented to show the differences between the classical and evolutional elasto-plastic analyses.

REFERENCES

Bureau Seco 1978. *Les parois moulées dans le sol.* Brussels.
Caquot, A. et al. 1973. *Tables de butée et de poussée.* Paris, Brussels, Montréal: Gauthier-Villars.
Vandepitte, Daniël 1988. Slurry walls – Evolution of methods of analysis. *Tunnels et ouvrages souterrains* 87: 139–151.
Vandepitte, Daniël 1979. *Berekening van Constructies; Bouwkunde en civiele techniek, Boekdeel I.* Ghent: E. Story-Scientia.
Van Impe, Rudy 1990. De verplaatsingenmethode toegepast bij de evolutieve berekening van grondkerende wanden in de gebruikstoestand. Tijdschrift der openbare werken van België 2: 109–122.
Van Impe, Rudy 2002. *Berekening van Constructies 5.* Ghent: Laboratory for research on structural models.

Innovative aluminium structures

G.C. Giuliani
Professional Engineer Milano-Italy Edmonton-Canada

ABSTRACT: To overcome some constraints of the normal jointing systems of aluminium structures, innovative ideas have been applied resulting in a general new fastening method and in an improved structural efficiency.

The results of the non linear analysis of a 12.00 m span prestressed string beam as well as the ones recorded during the relevant test are illustrated.

1 IN GENERAL

Aluminium is a very interesting structural material which features a very favourable ratio between the yielding stress and the specific weight if compared to the similar value for standard rolled steel (mechanical efficiency around 1.0 E4 m. for standard aluminium alloys such as AlMgSi1 and 0.66 E4 m. for high tensile steel or 0.46 E4 m. for standard steel rolled sections).

The use of extrusion enables the engineer to search for the optimum cross section shape of the structural members raising the buckling load without increasing the weight and taking into account that the Young's modulus is around 1/3 of the steel one. The problem of joints is a very important one.

The use of welding reduces the strength of the sections of around 50% even with modern technologies like MIG system and all the structure has to be over designed to overcome this local weakness.

The use of bolting requires a lot of plates, holes, washers, nuts and bolts; it is possible to reduce the number of the aforesaid components, as it is normal in steel construction, by using high strength bolts to provide for friction connections but, in this case there is the danger of corrosion arising form use of materials having different electrochemical potentials; the consequently requested electrical isolation is not cheap neither are stainless bolts.

The use of bolted joints with the rational box section, which can be obtained by extrusion, requires spacers in order to avoid local transverse bending arising from the load applied by the bolt tightening.

To reduce the number of the plates requested in the joints it is convenient to provide for special member sections shaped with extensions which allow for direct reciprocal connection.

Innovative ideas for joints are then requested but, to achieve good results, the whole structure has to be designed in a shape consistent with the selected connections.

2 JOINT SYSTEM IMPROVEMENT

In general an axial force passing through a joint is carried by shear stresses arising in fillet welding or in bolts; axially working bolts are seldom used because of the secondary bending of the plates that are connected.

The use of prestressing is a solution (Fig. 1)

a) A strand pulled through the whole chords of the structure (truss or beam) is capable of tightening all joints requiring end anchors only, irrespective of the number of members to be connected along the span.

No additional pieces are requested for the joint stability; secondary bolts are useful for speeding up the assembly of the whole structure.

b) Even the joints which are subjected to compressive forces under superimposed loading can be designed in the same way without loss of efficiency; the

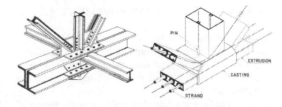

Figure 1. Schemes of a conventional bolted joint and of a prestressed one.

Figure 2.   Details of the strut to tie connection.

Figure 3.   Extensions of the sections for bolts placing and an "ad hoc" shaped connector.

compressive strain produces a shortening of the members and consequently a reduction of the postensioning force.

c) No machining of members is requested but only a saw cut of the ends; should the contact surfaces not be exactly aligned within prescribed tolerances of 1/10 mm, during postensioning a local plastic deformation arises without any structural damage.

d) The danger of intermetallic corrosion can be avoided by a cheap plastic sheathing applied around the strand in the mill; this kind of wrapped strand is of normal use in postensioning of reinforced concrete structures.

e) To avoid the use of plates and to provide at the same time a solution for spacers it is possible to apply extruded elements, featuring in their shape the proper number of holes, which can be saw cut and placed inside the box section members to be connected. (Fig. 2)

f) Where several members have to be connected at the same joint, a proper solution can be achieved by means of casting encasing all sections and to be subjected to few machining operations; the cost of the model for sand core forming is irrelevant if the number of similar castings is higher than 20.

At joints where the anchoring devices for postensioning are located, the use of alloy casting is again very useful.

3   MEMBER SYSTEM IMPROVEMENT

Because of the ratio between the Young's modulus and the yielding stress which, for aluminium alloys, is about 1/3 of the steel one, the problem of elastic buckling is very important in determining the optimum structural solution.

A rational approach to this problem is the design of the member sections in a box shape, easily obtained by the extrusion process.

For connection purposes it is convenient adding to the box section continuous flat extensions which can bear holes for the fastening bolts. (Fig. 3)

The members which house postensioning strands must have built in ducts in their cross section; these ducts can be properly placed so that they act as longitudinal stiffeners against buckling of webs and of flanges.

4   STRUCTURAL SYSTEM IMPROVEMENT

The use of postensioning as a jointing system yields an improvement of the structural system; the connecting strands, being continuous along a chord, constitute an additional axially resistant member resulting in a reduction of the overall weight.

A proper pattern of the postensioning system yields lifting forces at joints where deviations occur and, for this reason, a part of the load is directly supported by the strands. (Fig. 4)

No buckling of the members compressed by the postensioning can occur if their sections feature ducts containing the strands thus preventing lateral sway and allowing for their easy pulling through. Because of the prestressing, in case of alternate loads, the fatigue behaviour of the structure is improved being the maximum and minimum stresses always in the compression range.

5   STRUCTURAL ANALYSIS

Some particular features of the aforesaid jointing system need a specific approach to the structural analysis which has to be performed according to the ultimate load capacity method, because of the following reasons:

a) The postensioning force has to be high enough to avoid decompression of the joints under the effects of the service loads; design has to take into account the different coefficients of thermal expansion of aluminium and steel;

b) Each strand is continuous through the aluminium members and fixed at its ends only; for this reason its supplementary strain due to the deflection of the structure subjected to the superimposed loads

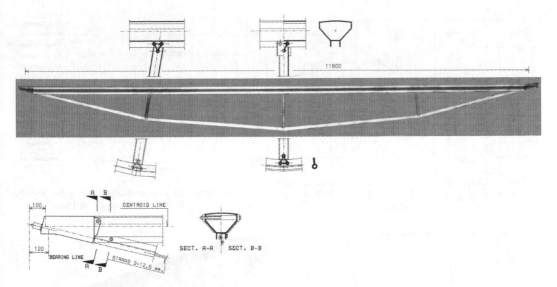

Figure 4. A prestressed string beam.

is constant over all the length resulting in a elongation which has a value lower than the sum of the elongations of the corresponding aluminium members, requiring again an ultimate load analysis.

c) Because of the aforesaid reasons the prestressing has to be designed for the ultimate force requested at the connection.

The compression force in the members due to the postensioning is balanced by the tension in the strands and the system is self equilibrated, therefore the use of postensioning has no effect on the buckling behaviour of the structure; the unintentional eccentricities of the applied prestressing forces, are upper bounded by the gap between the duct and strand diameters and result in low value second order moments.

Because of the requested ultimate load capacity method, it is worth to proceed with a non linear analysis taking into account the stress–strain relationship of the aluminium alloys which features a very good postyielding behaviour.

In order to properly understand the results, a structural analysis software based on finite three-dimensional elements (FEM) was used. Because of the important changes of the geometry occurring with the load steps and of the material constitutive laws extended over the proportionality limits, the non linear behaviour had to be investigated.

Indeed, not to neglect the stiffening effects caused by the structure deformation, which cannot be caught in a normal analysis where all equilibrium equations are always related to a non deformed configuration, and considering the important traction forces arising in the strand, a 2nd level analysis on wide displacements

was performed. This way through the loading procedure is monitored step by step, so to update the structure response to any geometry variations, and to write the equilibrium equation of finding the deformed configuration. The analysis continued up to the convergence criteria satisfaction, which was expressed in terms of nodal forces; the method used for the solution it is the iterative one of Newton-Raphson.

In Figure 5a the load–displacement relationship is reported, while in Figure 5b the variations of the strand tension during the prestressing and the loading are plotted; both the aforesaid actions are applied in steps in order to detect the non linear response of the structure.

The analysis clearly shows that the ultimate load carrying capacity of the structure is determined by the yielding of the elements which connect the struts to the lower prestressed string, while the tendon stress is over the proportionality range but does not break.

In the service condition the mid span upper joint of the structure is laterally restrained by a bracing system; the out of plane buckling of the truss depends on the stiffness of the horizontal supports and therefore a sensitivity analysis was implemented also in order to get information for the bracing design. The ultimate load versus the horizontal stiffness of the restraints is shown in Figure 6, while the horizontal displacements of the upper and the lower chords calculated for different values of the aforesaid stiffness and plotted against the ultimate load are shown in Figure 7.

The analysis was extended to the buckling modes of the top chord in order to eventually refine the relevant sectional design for further research and development of this kind of innovative structure; the results are illustrated in Figure 8.

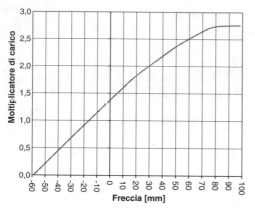

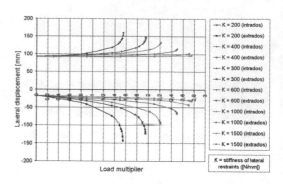

Figure 7. Horizontal displacement of the upper and the lower chord versus the horizontal support stiffness.

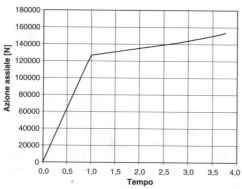

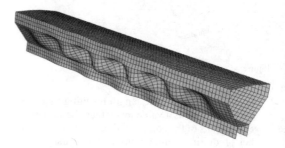

Figure 8a. Buckling mode for horizontal bending moment Mh = 108 kNm.

Figure 5 (a) Midspan deflection versus the applied load. (b) Axial force in the tendon versus the applied load (Pre-stressing = Tempo 0–1, load increasing = Tempo 1–4).

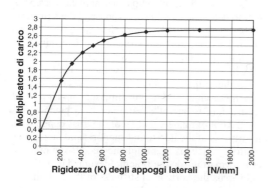

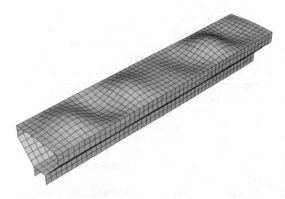

Figure 8b. Buckling mode for axial load − 407 kN + vertical bending moment Mv = 19,6 kNm.

Figure 6. Ultimate load versus the horizontal support stiffness.

6 TEST ON PROTOTYPE AND RESULT ANALYSIS

Loads tests have been carried out in order to verify the exactness of the hypothesis and of the design checks. As a result the structure, loaded for 2.28 times the limit service condition, enabled the validation of the project.

The equipment used for testing are those of Figure 9. During the test, the worse deformations recorded close to the span centre had a regular trend; the diagram of the truss middle displacement related to the subsequent loading and unloading cycles is shown in Figure 10.

Figure 9. Testing equipment for the prestressed string beam Figure 4.

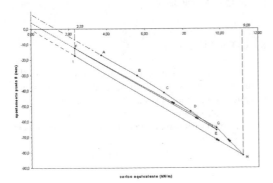

Figure 10. Load–deflection experimental diagram for the prestressed string beam of Figure 4.

When loaded with 10.27 kN/m a slight bearing softening in a few connections was observed, though the structure, in spite of its components subjected to stresses falling into the non linear branch of the constitutive law, did not show any sign of incipient collapse.

The yielding of the connectors was expected according to the analysis illustrated in the preceding paragraph.

7 EXAMPLES OF CONSTRUCTION

7.1 Beams

The aforesaid innovative ideas for the design of aluminium structures were applied to a wider main structure which is composed of main trusses bearing on columns and featuring the same construction principles used in the secondary trusses (Fig. 11).

7.2 Others structures

The principles of innovative design illustrated in the paragraphs 2-3-4 of this article have been recently used for the stair structure of the new Barcelona air control tower, in order to achieve either an attractive architectural aspect and a cost saving in comparison to a steel solution, taking into account the incidence of the maintenance against corrosion in a close to sea environment.

Figure 11. Rendering of a 50 m span prestressed trussed beam al 10.00 centers.

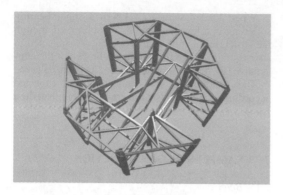

Figure 12. The structure inside the tower.

Figure 13. A joint detail.

The self supported structure has the total height of 43.38 m and an octagonal plan with a 2.70 m side; it is constituted by four trussed triangular columns which house piping and ducts, the two elevators and a double stair with joggling ramparts (Fig. 12).

All the sections and the connecting elements were specially designed in order to achieve either the structure optimization and a pleasant architectural aspect; the joints feature rust free bolts which hold special extrusions located inside the vertical stanchions and the diagonal members and, in some cases, cast elements (Fig. 13).

The alloy used for the extrusions is the 6082 grade with T6 tempering.

The structure was conceived and designed to effect, by means of special connections, the precise in space positioning of the precast concrete elements bearing the tower during their erection and to resist the relevant actions due to weight and wind. At the completion of each ring of the tower supporting elements, the whole stair case is permanently restrained by this part of the shaft and the actions and the displacements due to the wind are reduced thanks to the interaction of the two structures.

8 CONCLUSIONS

The proposed innovative ideas for the design of aluminium structures lead to important savings in material and labour, concepts can be extended with similar advantages to a wide range of different cases.

It is worth to outline that no new materials have been used but only a selection of them among the ones available for common constructions.

REFERENCES

G.C. Giuliani & G. Valentini, Trondheim 1988. Modern Design Principles, *Innovative Ideas for Aluminium Structures.*

G.C. Giuliani, ABCD N.2/2002. Strutture in Alluminio per l'industria delle costruzioni.

G.C. Giuliani, 06/2001. Lezioni ai corsi TAS, *Progettazione strutturale con l'Alluminio.*

G. Barbieri, ABCD N.1/2003. Analisi non lineare di capriata in alluminio con impiego di precompressione.

35. Invited session on "Recent advances in wind engineering"

System-based Vision for Strategic and Creative Design, Bontempi (ed.)
© 2003 Swets & Zeitlinger, Lisse, ISBN 90 5809 599 1

Some experimental results on wind fields in a built environment

C. Paulotto, M. Ciampoli & G. Augusti
Dipartimento di Ingegneria Strutturale e Geotecnica, Università di Roma "La Sapienza", Roma, Italia

ABSTRACT: The assessment of the response of structures to the action of strong winds is essential for safety and economic reasons. In designing structures in a built environment, account should be taken of both shield effect and amplification of wind velocity due to the presence of obstacles, as well as of the specific turbulence structure. A preliminary problem is the check of the adequacy of the simplified models of representation of the variation of the wind speed with height, and, in case of adoption of the logarithmic law, the characterization of the surface roughness by means of the aerodynamic roughness length, z_0. In this paper, the results of some tests on the influence of different arrays of buildings are presented: the tests have been carried out in the wind tunnel of CRIACIV in Prato, Italy. The roughness has been realized by gluing on a plywood baseboard wood cubes in a staggered pattern; varying the surface density of the cubes, three different roughnesses have been obtained. The influence on z_0 of the geometric aspect ratio of the obstacles has been studied.

1 INTRODUCTION

The collapse of a structure due to wind action, the dispersion of industrial releases, the windborn debris are important sources of risk in very densely populated areas, such as urban areas.

The comprehension of the specific features of wind in such environment, as well as a reliable assessment of the structural response to the action of strong winds, are essential for both safety and economic reasons, and for planning actions devoted to the mitigation of wind-related risk.

If present codes and regulations are considered, it can be noted that specific guidelines on the characterization of the wind mean velocity profile in urban areas are still poor: moreover, in some cases, it may be essential to take account of local effects, such as the shield effect, the amplification of the wind velocity due to the presence of obstacles, and the alteration of the turbulence structure.

In order to assess a more refined model of the wind mean velocity profile in urban areas, it is necessary to model the boundary layer. To this aim, the first problem to be solved is the characterization of the surface roughness.

Two alternative parameters are commonly used to describe the surface roughness, that is, the aerodynamic roughness length z_0 or the drag coefficient. The former parameter is usually preferred because, over homogeneous terrains, it depends only on the roughness, and

is also height-independent for a sizeable height interval (Wieringa 1993).

In this paper, the results of tests on the wind field over models of various arrays of buildings are presented and discussed, as well as the possibility of extending the usual representation of the wind mean profile to urban areas. The influence on z_0 of the geometrical aspect ratios of the obstacles has been studied. The roughness has been realized by gluing on a plywood baseboard wood cubes of the same size in a staggered pattern; varying the surface density of the cubes, three different roughnesses have been obtained.

2 CHARACTERIZATION OF THE SURFACE ROUGHNESS

The *boundary layer* over a rough surface can be ideally considered made by an *outer layer* and a *surface layer*. In turn, the surface layer can be subdivided into: an *inertial sublayer*, a *roughness sublayer* and a *canopy sublayer*.

The canopy sublayer can be defined as the region extending between the base surface and the mean height of the obstacles. The roughness sublayer is the region immediately above the canopy sublayer, within which the flow is strongly influenced by the individual roughness element, and therefore is not spatially homogeneous. [In technical literature, it can be found that the depth of the roughness sublayer varies between

1.5 and 5 times the height of the roughness elements.] The inertial sublayer is the uppermost part of the surface layer, where both the "velocity defect law" and the "law of the wall" are simultaneously valid (Raupach et al. 1991).

The vertical variation of the wind mean velocity U in the inertial sublayer can be described by the logarithmic law (Tennekes 1973):

$$\frac{U}{u_*} = \frac{1}{k}\ln\left(\frac{z}{z_0}\right) \qquad (1)$$

where: u_* is the friction velocity, which is related to the wind stress τ_0 on the ground, and can be taken as $u_* = (\tau_0/\rho)^{0.5}$ with ρ the air density; k is the von Karman constant; z_0 is the roughness length.

If the rough surface corresponds to a dense array of obstacles, in order to take into account the upward displacement of the flow determined by the surface protrusions, Equation 1 must be modified into:

$$\frac{U}{u_*} = \frac{1}{k}\ln\left(\frac{z-d}{z_0}\right) \qquad (2)$$

where d is the so-called *zero-plane displacement height* (Jackson 1981).

For fully developed flow, z_0 is a surface property, determined by the roughness geometry; z_0 can be rigorously defined, and Equations 1 or 2 applied, only when an *inertial sublayer* exists, in which the turbulence statistics are approximately constant with height.

The just described structure of the boundary layer applies to situations that are characterized by a very long and homogeneous fetch; but, in practice, roughness changes and limited fetch usually occur.

Due to roughness changes, an *internal boundary layer* develops over the new roughness, and grows in its height with downwind distance (Wieringa 1993, Cheng & Castro 2002b). Over this internal boundary layer, the flow field is characteristic of the upstream condition. The internal boundary layer is composed by: an *equilibrium layer* that exists near to the ground, where the wind profiles have completely adapted to the local surface conditions, and, a *transition zone*.

In the equilibrium layer (that can be approximately assumed to extend over the lowest 10% of the internal boundary layer), a new constant stress layer can be achieved, and the roughness characterized by the logarithmic profile only at sufficient distance from the discontinuity. The constant stress layer has first to encompass the roughness sublayer before any thickness of inertial sublayer can be developed.

Therefore, in practical cases like in urban areas, where roughness frequently changes, it is possible

that a significant inertial sublayer cannot develop at all. As a consequence, the question arises how Equations 1 or 2 could be applied, and, in this case, how a significant value of z_0 could be obtained from point measurements of wind velocity (Cheng & Castro 2002a).

3 EXPERIMENTAL INVESTIGATION

In order to investigate some features of the boundary layer, experiments were carried out.

3.1 Wind tunnel

Tests were conducted in the CRIACIV wind tunnel at Prato, Italy, which is a low speed open circuit tunnel, with a test section 1.6 m high, 2.4 m wide and 2.4 m long. They were carried out at a nominal free stream velocity U_∞ of 24 m/s (x, y, z) are the streamwise, lateral and vertical coordinates, where $z = 0$ is the top surface of the baseboard on which the roughness elements were glued. The origin of the coordinate system is in the centre of the test section. Mean and instantaneous velocities will be denoted as (U, V, W) and (u, v, w), respectively.

3.2 Roughness surfaces

The roughness elements were sharp-edged wood cubes with a side dimension of 90 mm. They were glued onto a plywood baseboard (4 m long and 2 m wide) in staggered patterns: each pattern was characterized by a constant value of the parameter λ, the ratio between the frontal area of the obstacles and the lot area of the obstacles.

Experiments were conducted on three values of λ (that is, $\lambda = 9\%$, 14% and 18%), corresponding to three different roughnesses, denoted as R1, R2 and R3, respectively. Measurements were taken along the three vertical lines P1, P4, P5 shown in Figure 1.

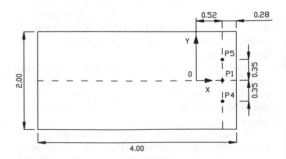

Figure 1. Plan locations of the vertical lines where measurements were taken (dimensions in m).

3.3 Instrumentation

Hot wire anemometer was the main instrumentation adopted in the experiments. The signals from the bridges of a constant temperature system (DANTEC 56C17) were filtered, and then digitized via a 16 bit A/D converter (IOTECH ADC488/8SA). A 45° x-wire probe (DANTEC P61) was used. A sampling frequency of 2 kHz and 65,536 samples were used at each measurement location.

4 RESULTS

4.1 Flow properties and depth of roughness sublayer

In what follows, the first results of the experiments are discussed: they are expressed in terms of velocity measurements and turbulence statistics, for the three different roughness values R1, R2, R3 (§3.2).

Looking at Figures 2 and 3, it can be noted that the mean velocity U (non-dimensionalised with respect to the free stream velocity U_∞) and the turbulence intensity (defined as the ratio between the standard deviation of the fluctuations of u and its mean value U) in the streamwise (x) direction do not change significantly with y over the whole investigated range of heights z for the three roughness values (R1, R2, R3).

On the contrary, Reynolds shear stresses τ_R, that are represented in Figure 4, exhibit some larger variations with y, especially at lower heights.

The roughness sublayer (Section 2) is defined as the region where the flow is strongly influenced by each individual roughness element.

An approximate evaluation of the depth of the roughness sublayer can be obtained by evaluating, for each roughness value (R1, R2, R3) and at different heights, the ratio between the standard deviation of the Reynolds shear stresses (each corresponding to the different locations P1, P4, P5) and the maximum absolute value of the same quantity.

It has been assumed to be within the roughness sublayer when this ratio is larger than 0.05. According to this definition, the corresponding depths of the roughness sublayer were found respectively equal to 0.27 m, 0.31 m and 0.31 m. According to a widely accepted definition, the boundary layer is set by the limit $U \leqslant 0.99 \, U_\infty$: in our experiments, the heights of the boundary layer for the three roughness values were found respectively equal to 0.29 m, 0.31 m and 0.33 m. Therefore, in the considered cases, practically almost the whole boundary layer appears to be formed by the roughness sublayer.

4.2 Assessment of the roughness length

As illustrated in Section 2, the assessment of the roughness length z_0 in the situations that were considered in

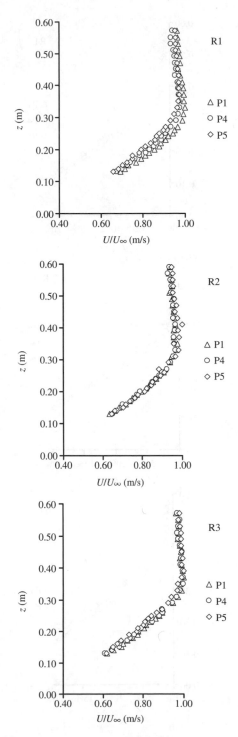

Figure 2. Mean velocity profiles along the verticals P1, P4 and P5 for the three investigated roughness values (R1, R2, R3).

2311

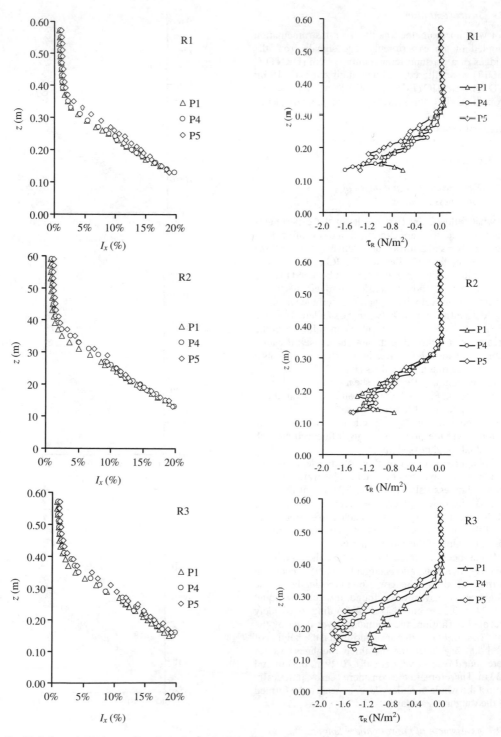

Figure 3. Turbulence intensity profiles along the verticals P1, P4, P5 for the three investigated roughness values (R1, R2, R3).

Figure 4. Reynolds shear stress profile along the verticals P1, P4, P5 for the three investigated roughness values (R1, R2, R3).

the experimental programme, is a controversial problem. Moreover, as the flow within the roughness sublayer is inherently three dimensional, taking measurements along only one vertical line cannot lead to a satisfactory description of the flow.

However, in order to save the format of Equations 1 and 2, in (Macdonald et al. 2000) and (Macdonald 2000) lateral averages of the mean velocity profiles have been considered.

Following this suggestion, the roughness length z_0 was estimated by means of a least square fitting of the lateral averages (at different heights) of the three vertical mean velocity profiles, assuming that the roughness sublayer and the inertial sublayer could be described together by a single logarithmic law.

The resulting values of z_0 were then compared with previous experimental results, and with the values obtained by empirical equations.

The considered experimental programmes used roughnesses similar to our experimental set.

In (Jia et al. 1998), a 6.1 m long fetch of wood cubes was used. The cubes had a side dimension of 0.09 m and were arranged in staggered patterns.

In Hall et al. (1996), the data were taken from wind-tunnel experiments over staggered arrays of 0.10 m cubes. The results were based on the mean velocity profiles averaged for five different lateral locations at a fetch of about 2.20 m.

The z_0 values obtained from our experiments are shown in Figure 5, together with the values obtained by Jia et al. (1998) and by Hall et al. (1996): they are in satisfactory agreement with Jia's values, and generally larger than Hall's values.

Note that Jia's data were obtained with a method similar to the method adopted in our calculations (but applied to two wind mean velocity profiles measured on the same vertical line and corresponding to two values of the nominal free stream velocity U_∞); on the contrary, Hall's data were obtained with a method proposed in (Petersen 1997), that is quite different from the method previously outlined.

To investigate the effect of these two calibration procedures, our data have been elaborated also with the alternative method adopted by Hall.

The values resulting from calibration by the two methods are compared in Figure 6. The differences are relatively small; therefore it is fair to assume that the differences between ours and Hall's values depend more on the different values of the height h of the wood cubes in relation to the depth of the boundary layer (Hall et al. 1996) than on the method of calibration of z_0.

From Figure 7, it can be noted that our experimental results are in very good agreement with the relation proposed by Macdonald et al. (1998):

$$\frac{z_0}{h} = \exp\left(-0.52\lambda^{-0.5}\right) \qquad (4)$$

while, Lettau equation:

$$\frac{z_0}{h} = 0.5\lambda \qquad (5)$$

does not fit at all with the test results.

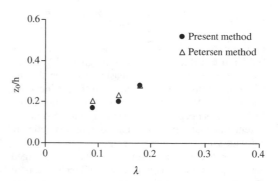

Figure 6. Value of the ratio z_0 to the roughness height, h, obtained by the method presented in this study and by Petersen method (Petersen 1997).

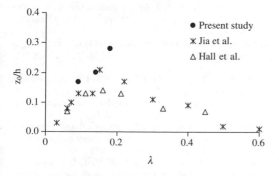

Figure 5. Variation of the ratio z_0 to the roughness height, h, as a function of the frontal area density.

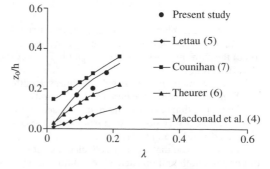

Figure 7. Variation of the ratio z_0 to the roughness height, h, as a function of the frontal area density, λ the different expressions quoted in the paper are considered.

Indeed in (Macdonald et al. 1998), an analytical derivation of Equation 5 is presented, and it is pointed out how Lettau equation is more representative of rounded obstacles then of rectangular prisms: in order to take into account the greater drag coefficient of the sharp edges roughness, Equation 4 must be used.

Figure 7 shows also the plots of relations (6) and (7), derived respectively by (Theurer 1993) and (Counihan 1971):

$$\frac{z_0}{h} = 1.6\lambda(1 - 1.67\lambda) \tag{6}$$

and

$$\frac{z_0}{h} = 8.2\frac{h}{x_F} + 0.65\lambda - 0.08 \tag{7}$$

where x_F is the fetch dimension. The agreement with test results is much worse than Equation 5.

5 CONCLUDING REMARKS

Preliminary results of wind-tunnel tests on the flow field above surfaces of different roughness have been presented and discussed. Measurements have been taken along three vertical lines at different locations normally to the flow direction, and practically no variation has been found in the mean velocity and turbulence intensity profiles; therefore, they practically coincide with their lateral averages.

More recently, the idea of a spatially averaged vertical profile has been developed (Cheng & Castro 2002a). Spatial averaging over (at least) one repeated unit of a uniform array will lead to a representative profile within the roughness sublayer, which could then be assumed as horizontally homogeneous on scales larger than the unit. It is not immediately obvious whether such a spatially averaged profile would be logarithmic.

It is therefore planned to conduct further experiments along lines at different locations also in the direction of the flow, in order to allow spatial (and not only lateral) averages.

The experimentally estimated roughness lengths have been plotted versus the obstacle density, and compared with previous experiments and proposed analytical relations: a very good agreement has been found with Macdonald et al. (1998).

It must be noted that all proposed relations between roughness length and obstacle aspect ratio fail when $\lambda > 20 \div 25\%$, due to the aerodynamic interaction between the obstacles and the development of a finite displacement height in the velocity profile.

Table 1. Values of frontal density for some European and American cities (Ratti et al. 2000).

	London	Toulouse	Berlin	Salt Lake	Los Angeles
λ	0.32	0.32	0.23	0.11	0.45

This limitation is a problem in many practical applications, and especially in urban situations, where λ commonly exceeds 20% (as it can be noted in Table 1).

The presented experiments constitute only a very first stage of an experimental program on the wind flow field over terrains representative of densely built situations, ultimately aimed at suggesting specific design rules for urban areas.

ACKNOWLEDGEMENTS

The experiments reported in this paper are part of a research in progress in the framework of the "Research Programme of National Interest" (PRIN) WINDER-FUL (2001–2003). The tests have been carried out in the wind tunnel of CRIACIV in Prato, Italy. Sincere thanks are due to Lorenzo Procino, for his invaluable help in performing the tests.

REFERENCES

Cheng, H. & Castro, I.P. 2002. Near wall flow over urban-like roughness. *Boundary-layer Meteorology* 104: 229–259.

Cheng, H. & Castro, I.P. 2002. Near wall flow development after a step change in surface roughness. *Boundary-layer Meteorology* 105: 411–432.

Counihan, J. 1971. Wind-tunnel determination of the roughness length as a function of the fetch and the roughness density of three-dimensional roughness elements. *Atmospheric Environment* 5: 637–642.

Hall, D.J., Macdonald, R., Walker, S. & Spanton, A.M. 1996. Measurements of dispersion within simulated urban arrays – A small scale wind tunnel study. BRE Client Report CR 178/96. Garston, Watford: Building Research Establishment.

Jackson, P.S. 1981. On the displacement height in the logarithmic velocity profile. *Journal of Fluid Mechanics* 111: 15–25.

Jia, Y., Sill, B.L. & Reinhold, T.A. 1998. Effects of surface roughness element spacing on boundary-layer velocity profile parameters. *Journal of Wind Engineering and Industrial Aerodynamics* 73: 215–230.

Lettau, H. 1969. Note on aerodynamic roughness parameter estimation on the basis of roughness element description. *Journal of Applied Meteorology* 8: 828–832.

Macdonald, R.W., Griffiths, R.F. & Hall, D.J. 1998. An improved method for the estimation of surface roughness of obstacle arrays. *Atmospheric Environment* 11: 1857–1864.

Macdonald, R.W. 2000. Modelling the mean velocity profile in the urban canopy layer. *Boundary-layer Meteorology* 97: 25–45.

Macdonald, R.W., Carter, S. & Slawson, P.R. 2000. Measurements of mean velocity and turbulence statistics in simple obstacle arrays at 1:200 scale. Thermal Fluid Report 2000–1. Canada: University of Waterloo.

Petersen, R.L. 1997. A wind-tunnel evaluation of methods for estimating surface roughness length at industrial facilities. *Atmospheric Environment* 1: 45–57.

Ratti, C., Di Sabatino, S., Britter, R., Brown, M., Caton, F. & Burian, S. 2000. Analysis of 3-D urban databases with respect to pollution dispersion for a number of European and American cities. http://www.dmu.dk/AtmosphericEnvironment/trapos/wg.htm.

Raupach, M.R., Antonia, R.A. & Rajagopalan, S. 1991. Rough-wall turbulent boundary layers. *Applied Mechanics Review* 44: 1–25.

Tennekes, H. 1973. The logarithmic wind profile. *Journal of the Atmospheric Sciences* 30: 234–238.

Wieringa, J. 1993. Representative roughness parameter for homogeneous terrain. Boundary Layer Meteorology 63: 323–363.

System-based Vision for Strategic and Creative Design, Bontempi (ed.)
© 2003 Swets & Zeitlinger, Lisse, ISBN 90 5809 599 1

Active control of a wind excited mast

M. Breccolotti, V. Gusella & A.L. Materazzi
Department of Civil and Environmental Engineering, Perugia, Italy

ABSTRACT: In this paper are reported the experimental tests conducted using the aluminum mast realized by the University of Perugia to asses the effects of active control strategies to reduce the vibrations of flexible structures subjected to wind loads. In the analyzed cases, the algorithm used to generate the control forces was a closed loop type in which the control force is proportional to the state space coordinates and to their derivatives (PD).

Two kinds of tests were realized with artificially generated external forces (free response analysis) and with wind-induced external forces. During the first tests the effect of the control was evaluated in term of structural damping of the first vibration mode. From the comparison between the tests results and some numerical simulations it was possible to underline the effect of imperfections of real systems in the performance of the active control device. In the second tests the efficiency of the active control device was determined by means of statistical analysis on the dynamic responses.

1 INTRODUCTION

Time dependent environmental forces such as that due to earthquake, wind, vehicular traffic etc. affect many structures (buildings, towers, bridges, antennas, etc.) both at the service limit state and at the ultimate limit state. Excessive vibrations can in fact reduce the comfort of inhabitants, can disturb the transmission of signals but can also cause stress peaks exceeding materials strength or can cause fatigue damages. Several techniques can be used to reduce the effects of these dynamic forces. Among them the active control technique uses an external power source to generate control forces that reduce the effect of the environmental actions according to different control strategies (Soong 1990, Housner et al. 1997). Within a multidiscipline project, a aluminum mast was realized by the University of Perugia to verify the applicability of the active control methodology to flexible structure subjected to wind loads (Fig. 1). Several experimental tests have been planned and carried out to identify the dynamic characteristics of the structure and to put in evidence the positive effect of the control device in reducing the vibration due to external time dependent forces. In a first phase the active control device has been used as a vybrodine with frequency of the applied load close to the first eigenfrequency of the mast. Later on a shield has been added on the top of the antenna to simulate the presence of transmission instruments and other tests have been realized to evaluate the effectiveness of the active control during windstorms.

Figure 1. The aluminum mast at the University of Perugia.

2 EXPERIMENTAL DEVICE

Two triangular cross section trusses compose the aluminum mast with an overall height equal to 10.2 m and a first eigenfrequency of 1.35 Hz experimentally determined (Gioffrè et al. 2000). The two elements can slip one respect the other varying the total height of the structure. The boundary condition of the cantilever beam can be easily changed adding two order of cables. On the top of the structure is located the active control system, an active mass damper (AMD) controlled by a dedicated PC (Fig. 2). It is made up of two rotating engines connected to two steel masses, whose weight is approximately 122 N, that can move along two perpendicular directions. An optical encoder

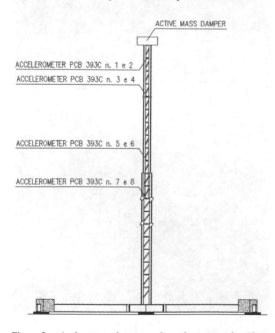

Figure 2. The AMD placed on the top of the structure.

Figure 3. Active mass damper and accelerometers location.

connected to the rotating engine allows us to know the relative displacement between the mass and the top of the structure. The acquisition system is composed of 8 accelerometers type PCB 393 C disposed on 4 levels (Fig. 3).

3 NUMERICAL MODEL

A finite element model has been realized to investigate the possible dynamic behavior of the structure in the controlled and un-controlled configuration. The structure has been modelled with 7 beams with stiffness equivalent to that of the triangular cross-section truss with 14 degree of freedom (d.o.f.) (un-controlled configuration) or 15 d.o.f. (controlled configuration).

The numerical simulation has been carried out using the first order theory in the state space. Figure 4 is depicted the structure with the labels of the nodes. The matrix equation of motion is

$$M\ddot{x}(t) + C\dot{x}(t) + Kx(t) = Du(t) + Ef(t) \qquad (1)$$

where M, C and K are, respectively, the mass, damping and stiffness matrices, x(t) is the displacement

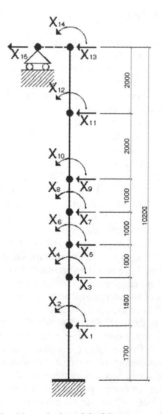

Figure 4. Numerical model of the structure.

2318

vector, f(t) is the vector representing the external load, u(t) is the control force vector and D and E are location matrices for the control forces and the external loads respectively.

In the state space this equation becomes

$$\dot{z}(t) = Az(t) + Bu(t) + Hf(t) \qquad z(0) = z_0 \qquad (2)$$

where

$$z(t) = \begin{bmatrix} x(t) \\ \dot{x}(t) \end{bmatrix} \qquad (3)$$

$$A = \begin{bmatrix} 0 & I \\ -M^{-1}K & -M^{-1}C \end{bmatrix} \qquad (4)$$

$$B = \begin{bmatrix} 0 \\ M^{-1}D \end{bmatrix} \qquad (5)$$

$$H = \begin{bmatrix} I \\ M^{-1}E \end{bmatrix} \qquad (6)$$

4 ADOPTED CONTROL ALGORITHM

A proportional and derivative algorithm (PD) has been used to command the AMD. The feedback parameters used in the closed loop control strategy are the relative displacement (proportional term) and the relative velocity (derivative term) between the top of the mast (d.o.f. no. 13) and the steel mass (d.o.f. no. 15). The control force has been calculated using Equation 7 in which k_p and k_d are control gains respectively for the proportional and the derivative terms

$$u(t) = k_p \cdot [x_{15}(t) - x_{13}(t)] + k_d \cdot [\dot{x}_{15}(t) - \dot{x}_{13}(t)] \qquad (7)$$

The control force can also be expressed in the following different way

$$u(t) = K_P x(t) + K_D \dot{x}(t) \qquad (8)$$

Upon substituting Equation 8 into Equation 1 is possible to see that the introduction of the feedback control force in the system practically consist in an alteration of the damping and stiffness matrices (Ficola et al. 1996a, b)

$$M\ddot{x}(t) + (C - DK_D)\dot{x}(t) + (K - DK_P)x(t) = Ef(t) \qquad (9)$$

5 FORCED VIBRATION TESTS

During the first phase of this research a set of forced vibration tests have been carried out. In these tests the control device was used initially to excite the structure applying a sinusoidal force on the top of the mast for a prescribed time interval (30 sec). This sinusoidal force was a 1,35 Hz frequency force, close to the first eigenfrequency. It is worth to note that to apply this force the software that command the AMD was written with a higher frequency to take into account the time delay caused by the limited computing velocity.

Passed this initial period the law that command the AMD changes and the device starts to work as an active control device in accord with Equation 7. Several attempts were made to find the couple of control gains k_p and k_d that maximize the effect of control, based on the evaluation of the damping of the first mode. It turned out that this parameter are $k_p = 16$ and $k_d = 24$. With these parameters the damping increased from 2.51% to 7.23%, calculated using the logarithmic decrement of acceleration, and the free oscillations of the mast were reduced faster as can be seen from the acceleration data acquired by accelerometer no. 1 (Fig. 5).

At the same time a numerical investigation was realized using the finite element model described in Figure 4. The results of this latter foresaw a much higher first mode damping (10,64%) obtained for the control gains $k_p = 32$ and $k_d = 48$. The difference between the theoretical behavior and the experimentally observed response can be explained if we consider same flaws, such time delays, doubtful boundary conditions, limited available energy etc., always present in real systems.

Moreover the effect of control is negligible for small vibration as can be seen in Figure 6 where is plotted the acceleration time history in a semi-logarithmic scale. In fact, after an initial phase where the oscillations were strongly and promptly reduced, the dynamic response

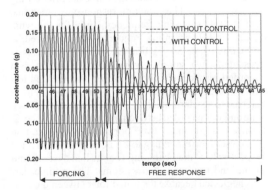

Figure 5. Free response of uncontrolled and controlled structure.

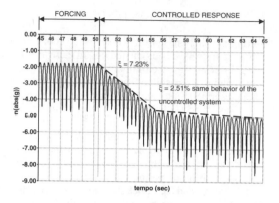

Figure 6. Logarithmic decrement of the controlled free response.

of the system turns to be the same of the un-controlled structure. This behavior is due to the friction present in the mechanism: for small vibration the relative displacement and the relative velocity between the AMD masses and the top of the structure are small and the corresponding control force is smaller than the friction.

In this way the AMD masses move together with the antenna and the effect of active control disappears.

6 WIND LOAD VIBRATION TESTS

Other tests were executed during windstorms to evaluate the reduction of oscillation using the active control. Each tests was one hour long, divided in six equally long intervals. The control is used during intervals no. 1, 3 and 5 while during intervals no. 2, 4 and 6 the response of the un-controlled structure is recorded. Since it wasn't possible to contemporarily check the controlled and un-controlled response a ultrasonic anemometer was installed close to the antenna. The data obtained from the anemometer were used to statistically compare the wind velocities time histories during the six intervals. The structural response was recorded with 100 Hz sampling frequency while for the wind velocity a 4 Hz sampling frequency was employed.

In this paper only a brief résumé of two tests is presented. Two samples from the acceleration data recorded respectively during the 1st and the 2nd test are depicted in Figures 7 and 8. Both of them is referred to accelerometer no. 1. The former is 1150 sec long and was recorded between intervals no. 1 and no. 2. From this sample, two portions with duration 327.68 sec (corresponding to 2^{15} points) were extracted respectively at the beginning (controlled system) and at the end (un-controlled system). The middle part was neglected because the transition between the controlled and the un-controlled system can affect the actual response of the mast.

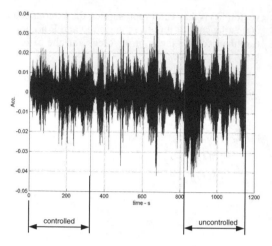

Figure 7. Test no. 1: acceleration time history.

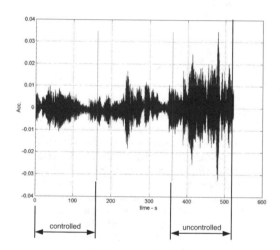

Figure 8. Test no. 2: acceleration time history.

Even at a first glance can be observed that the horizontal wind speed components u and v are really similar for the controlled and the uncontrolled phases (Fig. 9) while a relevant reduction of the maximum acceleration can be observed during the controlled phase (Fig. 7).

The effect of the control device is even more evident from the results of the second test. From this latter a 549 sec long sample was extracted between intervals no. 5 and no. 6. Two 163.84 sec long portion (corresponding to 2^{14} points) were obtained at the beginning (controlled system) and at the end (un-controlled system) of this sample (Fig. 8). The horizontal wind components velocities for these two portions are plotted in Figure 10.

In table no. 1 are summarized the results of the two tests in term of velocities and acceleration standard

2320

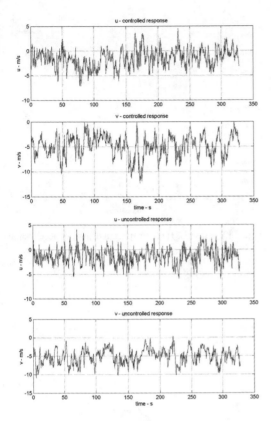

Figure 9.   Test no. 1: horizontal wind speed components.

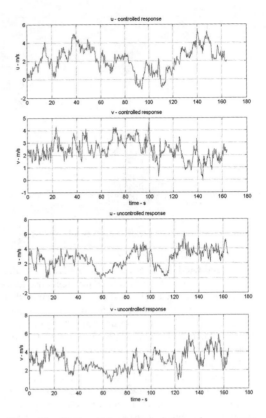

Figure 10.   Test no. 2: horizontal wind speed components.

deviation. The subscript c indicates the controlled phase while nc is for the un-controlled phase.

It can be noticed that in both tests the ratio between the standard deviation of the acceleration in the controlled phase and the one relative to the un-controlled phase is 0.55 for the 1st test and 0.38 for the 2nd test while the analogous ratio for the wind components velocities are generally bigger than 1.

In Figures 11 and 12 are depicted the acceleration probability densities (bars) of the analyzed time history portions. In each figure are represented the controlled (narrow bars) and un-controlled (wide bars) responses. The solid line describes a gaussian process with zero mean and standard deviation equal to that experimentally determined. It is clear that the presence of the control reduce the tails of the distributions, reducing also the stress peaks caused by the time varying forces.

7   CONCLUSIONS

In this paper the tests regarding the use of active control system in flexible structures subjected to wind loads have been presented. Two different type of experiment

Table 1.   Summary of the wind load vibration tests.

	$\sigma(u)_c/\sigma(u)_{nc}$	$\sigma(v)_c/\sigma(v)_{nc}$	$\sigma(acc)_c/\sigma(acc)_{nc}$
Test no. 1	1.16	1.09	0.55
Test no. 2	1.09	0.7626	0.38

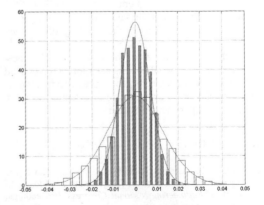

Figure 11.   Test no. 1: acceleration probability density.

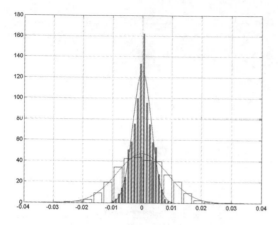

Figure 12. Test no. 2: acceleration probability density.

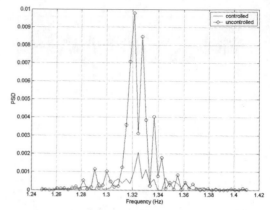

Figure 13. Test no. 1: power spectral density of acceleration.

were conducted to evaluate the benefits on the dynamic behavior of the structure: free response tests and wind-induced vibration tests. Both confirmed that the introduction of this type of control device ensures a significant increment of the structural damping with the consequent reduction of the wind-induced vibrations. This latter aspect causes positive effects for both ultimate (peak stresses reduction, fatigue) and service limit states (deflection and vibrations reduction).

During the free response tests a numerical analysis made possible to detect the effects of friction, imperfection, time delay and other aspects, always present in real systems, that can affect the control device performances.

For the wind-induced vibration tests the wind velocity and the dynamic response of the structure have been stored at the same time. The wind velocity served to compare from a statistical point of view the uncontrolled and the controlled structure behavior for comparable external loads.

The efficiency of the active control system was observed through a relevant reduction of the maximum of the floating part of the dynamic response. This behavior is a consequence of the reduction of the variation of the process, linked to the resonant part of the load spectrum close to the first eigenfrequency of the structure (Figs. 13, 14).

Further developments of this research will consist in the execution of other tests for more severe wind condition and the realization of finite elements model and numerical studies to compare the effective response with those experimentally observed.

REFERENCES

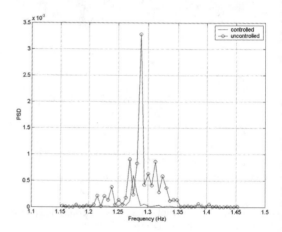

Figure 14. Test no. 2: power spectral density of acceleration.

Breccolotti, M. & Materazzi, A.L. 2001. Active control experiments on deformable structures, *Workshop on vibration problems on civil structures and mechanical construction*, Proceedings, Perugia, Italy (in Italian).

Ficola, A., La Cava, M. & Materazzi, A.L. 1996a. Use of mass damper for the active control of wind action sensitive structures. In *4th Wind Engineering National Congress*, Proceedings, Trieste, Italy (in Italian).

Ficola, A., La Cava, M. & Materazzi, A.L. 1996b. Advanced techniques for the active control of wind exposed flexible structures. In *3rd European Conference on Structural Dynamics*, Proceedings, Florence, Italy.

Gioffrè, M., Gusella, V. & Materazzi, A.L. 2000. Dynamic identification of the University of Perugia stayed mast. In *6th Wind Engineering National Congress*, Proceedings, Genova, Italy. (in Italian)

Housner, G.W. et al., 1997. Structural control: past, present, and future, *Journal of Engineering Mechanics*, Special Issue, Vol. 123, n. 9.

Soong, T.T. 1990. Active structural control: theory and practice. Harlow (UK), Longman Scient. & Tech.

Breccolotti, M., Gusella, V. & Materazzi, A.L. 2002. PD control experiment on a mast subjected to wind loads. In *7th Wind Engineering National Congress*, Proceedings, Milan, Italy (in Italian).

System-based Vision for Strategic and Creative Design, Bontempi (ed.)
© 2003 Swets & Zeitlinger, Lisse, ISBN 90 5809 599 1

Representation of the wind action on structures by Proper Orthogonal Decomposition

L. Carassale, G. Solari
Dept. of Structural and Geotechnical Engineering, University of Genova, Italy

ABSTRACT: The wind acting on the nodal points of a discretized structure is usually modeled as a Gaussian, stationery random vector whose components are correlated with each other according to characteristic patterns. Numerous techniques usually applied in structural dynamics require having the input constituted by a vector of independent random processes or even white noises. The application of such techniques in the wind-engineering context requires the introduction of suitable pre-filters to reproduce the exact correlation and harmonic content of the wind excitation. The present paper describes the development and the implementation of digital filters based on the Proper Orthogonal Decomposition of the wind turbulence. Two different approximated approaches are discussed and verified through numerical examples.

1 INTRODUCTION

Wind load on structures is traditionally idealized as a linear or nonlinear mapping of the incoming atmospheric turbulence, which is considered as the driving process for the dynamical system representing the wind force or the structural response.

Turbulence is represented as a multi-dimensional random process (dependent on time and space) characterized by typical correlation patterns (Solari & Piccardo 2001). In practise, it is often desirable to discretize the force and the structure and represent the turbulence on the nodes of the structure as a multivariate random process.

If the structure has linear behaviour and the fluid-structure interaction is linearized, the response can be obtained by the classic spectral analysis (Elishakoff 1983). If nonlinearities are taken into account, some alternative strategies including Monte Carlo simulation, Markov diffusion process approaches (Soong & Grigoriu 1993) and Volterra series approach (Schetzen 1980) have been applied.

The correlation between different components of the random vector representing the incoming turbulence constitutes a major difficulty in the application of the previous techniques. Some of them require to have the input constituted by a vector of white noises; in other cases, the input can be a vector of colored, but independent random processes; in some other cases there are no specific requirements that the input vector has to comply, but the condition of independent-component input strongly increases the computational efficiency.

The issue of having an independent-component input even in the cases in which the natural input has a correlation structure, likewise it is typical in wind engineering, has been tackled by introducing suitable pre-filters (Chen & Kareem 2001). Among the possible choices, some authors have recognized some advantages in adopting filters based on the Proper Orthogonal Decomposition (POD) (Benfratello & Muscolino 1999).

POD represents a multi-variate random process as a sum of components fully coherent in space called the modes of the process. The most attractive characteristic of POD is the ability of representing the target process by a small number of modes enabling a strong reduction of the space in which the process itself is defined (Carassale et al. 2001).

A pre-filter defined in this way can be easily implemented within all the techniques based on a frequency-domain representation of the system, while its time-domain application requires some further developments that complicates significantly the solution scheme (Di Paola & Gullo 2001, Tubino & Solari 2003).

The present paper describes the development and the implementation of POD-based digital filters for the representation of the driving excitation process of dynamic systems. Two forms of approximation based on the assumption of frequency-invariant POD eigenvectors are proposed. The effect of such approximations is investigated through two wind-engineering-related numerical examples.

2 CORRELATING AND COLORIZING FILTERS

Let $v(t)$ be an N-variate random process, function of time t, constituting the excitation for some dynamic system. Assuming $v(t)$ as zero-mean, weakly-stationary, and Gaussian, it can be, represented by the Fourier-Stieltjes integral (Priestley 1981):

$$v(t) = \int_{-\infty}^{\infty} e^{i\omega t} \, d\, Z_v(\omega) \tag{1}$$

i being imaginary unit; $Z_v(\omega)$ is an N-variate complex-valued random process (as function of the circular frequency ω) whose frequency increments $d\, Z_v(\omega) = Z_v(\omega + d\omega) - Z_v(\omega)$ are zero-mean and orthogonal in the following sense:

$$E\left[d\, Z_v(\omega) d\, Z_v^*(\omega') \right] = \begin{cases} S_{vv}(\omega) d\omega & \text{if } \omega = \omega' \\ 0 & \text{otherwise} \end{cases} \tag{2}$$

where $E[\bullet]$ is the expectation operator and the superscript * denotes the conjugate transpose; $S_{vv}(\omega)$ is the cross-power-spectral density matrix (cpsdm) of $v(t)$.

Let $C_{vv}(\tau) = E[v(t + \tau)v^{\mathrm{T}}(t)]$ be the covariance matrix of $v(t)$ at the time lag τ. Combining Eqs. (1) and (2), it results:

$$C_{vv}(\tau) = \int_{-\infty}^{\infty} e^{i\omega\tau} S_{vv}(\omega) d\omega \tag{3}$$

Many techniques for the analysis of randomly-excited dynamic systems require that the excitation is expressed in the form:

$$v(t) = \mathcal{R}[y(t)] \tag{4}$$

where $\mathcal{R}$ is a linear operator and $y(t)$ is an orthogonal-component process, i.e. in a vector of uncorrelated (independent because of the Gaussianity) processes. Such process can be represented as:

$$y(t) = \int_{-\infty}^{\infty} e^{i\omega t} d\, Z_y(\omega) \tag{5}$$

in which the increment process $d\, Z_y$ is orthogonal in the sense of Eq. (2) and:

$$E\left[d\, Z_y(\omega) d\, Z_y^*(\omega) \right] = \Gamma(\omega) d\omega \tag{6}$$

where $\Gamma(\omega)$ is a diagonal matrix; the elements of Γ represent the Power Spectral Density function (psdf) of the components of y. The operator $\mathcal{R}$ is called correlating filter since given an orthogonal-component process y it returns an output with the correlation structure specified by Eq. (3).

In order to represent rigorously the excitation $v(t)$, $y(t)$ must be, in general, at least an N-order vector. A suitable choice of the operator $\mathcal{R}$, however, usually leads to reasonable results even using an input vector of strongly reduced order.

In some circumstances the condition of orthogonal-component input is no longer sufficient and must be restricted with the condition of white-noise input; i.e. the excitation $v(t)$ must be expressed in the form:

$$v(t) = \mathcal{R}\left[\mathcal{F}\left[w(t) \right] \right] = \mathcal{H}\left[w(t) \right] \tag{7}$$

where $w(t)$ is a vector of Gaussian, independent white noises with unitary intensity, i.e.:

$$E\left[w(t + \tau)w(t) \right] = I\delta(\tau) \tag{8}$$

where I is the identity matrix. Coherently with the frequency-domain representation of v and y, w is expressed as:

$$w(t) = \int_{-\infty}^{\infty} e^{i\omega t} d\, B(\omega) \tag{9}$$

where $B(\omega)$ is a standard Wiener process. The operator $\mathcal{F}$ is called colorizing filter, since given a vector of independent white noises it returns a vector of independent processes whose psdf's are defined by Eq. (6).

The filters $\mathcal{R}$ and $\mathcal{F}$ must be determined on the base of the statistical properties of v. Among the infinite linear operators that satisfy the requirements for Eqs. (4) and (7), it is convenient to chose the ones that allows the best reduction of the input process order. Many authors have recognized the ability of POD in achieving this result.

Such technique is described in the next section as it is usually applied in wind engineering and fluid dynamics. Section 4 shows the implementation of POD-based linear filters for Eqs. (4) and (7).

3 PROPER ORTHOGONAL DECOMPOSITION

The Proper Orthogonal Decomposition (POD) has received uncountable applications in many fields of physics and engineering (for reference on structural dynamics applications see the review paper (Solari & Carassale 2001). In the context of stationery processes,

POD has been derived in two forms (Carassale et al. 2001): a weak form referred to as Covariance Proper Transformation (CPT) and a strong form called Spectral Proper Transformation (SPT).

3.1 Covariance Proper Transformation

Let $\lambda_1, \ldots, \lambda_N$ be the eigenvalues of $\boldsymbol{C}_{vv}(0)$, called covariance eigenvalues; $\boldsymbol{\phi}_1, \ldots, \boldsymbol{\phi}_N$ are the corresponding covariance eigenvectors. $\boldsymbol{C}_{vv}(0)$ is real, symmetric and positive definite, thus its eigenvalues are real and positive; its eigenvectors are real and can be orthonormalized to result:

$$\boldsymbol{\phi}_h^{\mathrm{T}}\boldsymbol{\phi}_k = \delta_{hk}; \qquad \boldsymbol{\phi}_h^{\mathrm{T}}\boldsymbol{C}_{vv}(0)\boldsymbol{\phi}_k = \delta_{hk}\lambda_k \qquad (10)$$

Thanks to these properties, $\boldsymbol{C}_{vv}(0)$ can be represented by the spectral decomposition (Loeve 1945):

$$\boldsymbol{C}_{vv}(0) = \sum_{k=1}^{N} \boldsymbol{\phi}_k \boldsymbol{\phi}_k^{\mathrm{T}}\lambda_k = \boldsymbol{\Phi}\boldsymbol{\Lambda}\boldsymbol{\Phi}^{\mathrm{T}} \qquad (11)$$

where $\boldsymbol{\Phi} = [\boldsymbol{\phi}_1 \ldots, \boldsymbol{\phi}_N]$ and $\boldsymbol{\Lambda} = \mathbf{diag}(\lambda_1 \ldots \lambda_N)$. On the base of such decomposition, any possible realization of the process v can be expressed as:

$$\boldsymbol{v}(t) = \sum_{k=1}^{N} \boldsymbol{\phi}_k x_k(t) = \boldsymbol{\Phi}\boldsymbol{x}(t) \qquad (12)$$

where $\boldsymbol{x}(t)$ is a zero-mean, weakly-stationary random vector whose components are referred to as covariance principal components of $\boldsymbol{v}(t)$; Eq. (12) is called Covariance Principal Transformation (CPT). Combining Eqs. (10) and (11), it yields:

$$\mathrm{E}\left[\boldsymbol{x}(t+\tau)\boldsymbol{x}^{\mathrm{T}}(t)\right] = \begin{cases} \boldsymbol{\Lambda} & \text{if } \tau=0 \\ \boldsymbol{\Phi}^{\mathrm{T}}\boldsymbol{C}_{vv}(\tau)\boldsymbol{\Phi} & \text{otherwise} \end{cases} \qquad (13)$$

The properties of $\boldsymbol{x}$ stated by Eq. (13) have two important consequences.

First, the covariance eigenvalues provides the variance of the covariance principal components. This implies that, if the eigenvalues are sorted in decreasing order, then the summations in Eqs. (11) and (12) can be truncated retaining only a limited number of terms, say $N_c < N$. Coherently, $\boldsymbol{\Lambda}$ is reduced to an N_c-order diagonal matrix containing the N_c largest eigenvalues and $\boldsymbol{\Phi}$ is reduced to an $N \times N_c$-order matrix obtained assembling the corresponding eigenvectors; $\boldsymbol{x}$ is an N_c-order vector containing the first N_c covariance principal components.

Second, the covariance principal components are uncorrelated with each other only for $\tau = 0$; therefore Eq. (12) meets only partially the requirements for

defining a correlating operator in the sense of Eqs. (4)–(6). In order to satisfy rigorously such specification, a more sophisticate is transformation required.

3.2 Spectral Proper Transformation (SPT)

Let $\gamma_1(\omega), \ldots, \gamma_N(\omega)$ be the eigenvalues of $\boldsymbol{S}_{vv}(\omega)$, called spectral eigenvalues; $\boldsymbol{\theta}_1(\omega), \ldots, \boldsymbol{\theta}_N(\omega)$ are the corresponding spectral eigenvectors. $\boldsymbol{S}_{vv}(\omega)$ is Hermitian and non-negative definite, thus its eigenvalues are real and non-negative; its eigenvectors, in general complex, can be orthonormalized to result:

$$\boldsymbol{\theta}_h^*(\omega)\boldsymbol{\theta}_k(\omega) = \delta_{hk}$$
$$\boldsymbol{\theta}_h^*(\omega)\boldsymbol{S}_{vv}(\omega)\boldsymbol{\theta}_k(\omega) = \delta_{hk}\gamma_k(\omega) \qquad (14)$$

Thanks to these properties, $\boldsymbol{S}_{vv}(\omega)$ can be factorized as:

$$\boldsymbol{S}_{vv}(\omega) = \sum_{k=1}^{N} \boldsymbol{\theta}_k(\omega)\boldsymbol{\theta}_k^*(\omega)\gamma_k(\omega) \qquad (15)$$

Introducing this decomposition into Eq. (2), it can be observed that the orthogonal increments $\mathrm{d}Z_v(\omega)$ must be given in the form:

$$\mathrm{d}\boldsymbol{Z}_v(\omega) = \sum_{k=1}^{N_s} \boldsymbol{\theta}_k(\omega)\mathrm{d}Z_{y_k}(\omega) = \boldsymbol{\Theta}(\omega)\mathrm{d}\boldsymbol{Z}_y(\omega) \qquad (16)$$

where $\boldsymbol{\Theta} = [\boldsymbol{\theta}_1 \ldots, \boldsymbol{\theta}_N]$ and $\mathrm{d}\boldsymbol{Z}_y(\omega)$ satisfies Eq. (6) with $\boldsymbol{\Gamma}(\omega) = \mathbf{diag}(\gamma_1 \ldots, \gamma_N)$. The components of the process y defined in this way through Eq. (5) are referred to as the spectral principal components of $v(t)$. The linear relationship between y and v, whose frequency-domain expression is given by Eq. (16), is referred to as Spectral Proper Transformation (SPT).

The spectral eigenvalues represent the psdf of spectral principal components, therefore, sorting the eigenvalues in decreasing order, the summations in Eqs. (15) and (16) can be truncated retaining only a limited number $N_s < N$ of terms. Coherently, $\boldsymbol{\Gamma}$ is reduced to an N_s-order diagonal matrix containing the N_s largest eigenvalues and $\boldsymbol{\Theta}$ is reduced to an $N \times N_s$-order matrix containing the corresponding eigenvectors.

The spectral principal components are uncorrelated for any time lag, therefore SPT meets entirely the requirements of the correlating filter defined by Eqs. (4)–(6). Combining Eq. (1) and (16), the process v is expressed as:

$$v(t) = \sum_{k=1}^{N_s} \boldsymbol{v}^{(k)}(t) = \sum_{k=1}^{N_s} \int_{-\infty}^{\infty} e^{i\omega t}\,\boldsymbol{\theta}_k(\omega)\mathrm{d}Z_{y_k}(\omega) \qquad (17)$$

where $\boldsymbol{v}^{(k)}$ ($k = 1, \ldots, N_s$) is a fully coherent vector referred to as the k-th modal contribution to v.

3.3 Algebraic SPT

SPT has been defined as the linear relationship between the processes v and y accordingly to Eqs. (17) and (5) or to the frequency-domain expression of Eq. (16). An explicit time-domain expression of SPT is in general not available, with the relevant exception of the particular case in which the spectral eigenvector matrix Θ does not depend on frequency. In this case SPT simply represents a memoryless transformation and Eq. (17) becomes:

$$v(t) = \sum_{k=1}^{N_s} \theta_k y_k(\omega) = \Theta y(t) \qquad (18)$$

It can be demonstrated that the spectral eigenvectors do not depend on frequency if and only if they are also covariance eigenvectors. This implies that, if all the significant spectral eigenvectors do not depend on frequency, then CPT coincides with SPT and the covariance principal components x_k coincide with the spectral principal components y_k.

The invariance of the spectral eigenvectors is highly desirable since Eq. (18) can be easily implemented in any time-domain numerical scheme. For this reason, many authors adopted expressions like Eq. (18) also when the condition of invariance was not met substituting the eigenvector matrix $\Theta(\omega)$ with a constant matrix $\bar{\Theta}$. Since it has been noted that whether the spectral eigenvectors are invariant then they coincide with the covariance eigenvectors, it seems natural to adopt the approximation:

$$\Theta(\omega) \simeq \bar{\Theta} = \Phi \qquad (19)$$

but sometimes also different choices have been judged as appropriate (Benfratello et al. 1999).

Often the property of SPT of defining independent principal components is determinant. This capability can be retrieved in this approximate context adopting two alternative approaches. Following the first approach, the process $v(t)$ represented, in general, by Eq. (17) is approximated by the relationship:

$$v(t) \simeq \bar{v}(t) = \Phi y(t) \qquad (20)$$

where the components $y_k(t)$ of $y(t)$ are statistically independent and are defined by the spectral eigenvalues through Eqs. (5) and (6). Within the second approach, $v(t)$ is approximated by:

$$v(t) \simeq \tilde{v}(t) = \Phi \tilde{x}(t) \qquad (21)$$

where the components $\tilde{x}_k$ of $\tilde{x}$ are a modification of the covariance principal components in which their cross-correlation is suppressed; they are defined as:

$$\tilde{x}_k(t) = \int_{-\infty}^{\infty} e^{i\omega t}\, dZ_{\tilde{x}_k}(\omega) \qquad (22)$$

where the increments are orthogonal in the sense of Eq. (2) and:

$$E\left[dZ_{\tilde{x}_h}(\omega)\, dZ_{\tilde{x}_k}^{*}(\omega)\right] = \delta_{hk}\phi_h^{T} S_{vv}(\omega)\phi_k\, d\omega \qquad (23)$$

The two approximations of the process $v(t)$ obtained trough Eqs. (20) and (21) are in general different: no theoretical reasons can be claimed for preferring an approach or the other.

4 IMPLEMENTATION OF POD-BASED CORRELATING AND COLORIZING FILTERS

It has been demonstrated that SPT satisfies the requirements of the correlating filter $\mathcal{R}$ defined by Eqs. (4)–(6). The frequency-domain relationship of Eq. (16) determines the transfer function $R(s)$ of the correlating filter $\mathcal{R}$ along the imaginary axis.

$$R(i\omega) = \Theta(\omega) \qquad (24)$$

Analogously, the transfer function of the colorizing filter $\mathcal{F}$ is a diagonal matrix and is determined along the imaginary axis by the relationship:

$$F(i\omega)F^{*}(i\omega) = \Gamma(\omega) \qquad (25)$$

The knowledge of the transfer function along the imaginary axis is sufficient for the implementation of frequency-domain-based applications likewise linear and nonlinear spectral analyses (Carassale & Kareem 2002) and digital simulation via harmonic superimposition (Di Paola & Gullo 2001).

The implementation of time-domain expressions for $\mathcal{R}$ and $\mathcal{F}$ requires, on the contrary, completing the definition of their transfer functions on the whole s-plane. This can be done assuming that the transfer function $H(s) = R(s)F(s)$ of the composite operator $\mathcal{H}$ (Eq. (7)) is expressed in the form:

$$h_k(s) = b_k(s)/a_k(s) \qquad (k = 1,...,N_s) \qquad (26)$$

where $h_k(s)$ is the k-th column of $H(s)$; a_k and b_k are, respectively, scalar and N-order-vector polynomial functions defined as:

$$a_k(s) = s^n + a_k^{(1)} s^{n-1} + ... + a_k^{(n)}$$
$$b_k(s) = b_k^{(1)} s^{n-1} + ... + b_k^{(n)} \qquad (27)$$

for some n representing the order of the filter. Eq. (26) represents the transfer function of a system that provides the k-th modal contribution $v^{(k)}$ to the excitation process v given a standard white noise, say the k-th component w_k of the vector $\boldsymbol{w}$. On the base of this transfer function, it is possible to build the following state-space model (Kailath 1980):

$$\begin{cases} \dot{z}_k = A_k z_k + c y_k \\ v^{(k)} = B_k z_k \end{cases} \qquad (28)$$

where z_k is an n-order state variable, $\boldsymbol{B}_k$ is the $N \times n$-order matrix obtained assembling by columns the vectors $\boldsymbol{b}_k^{(1)}, \ldots, \boldsymbol{b}_k^{(n)}$; A_k and c are, respectively, the $n \times n$-order matrix and the n-order vector defined as:

$$A_k = - \begin{bmatrix} a_k^{(1)} & a_k^{(2)} & \cdots & a_k^{(n)} \\ 1 & 0 & \cdots & 0 \\ & & \ddots & \\ 0 & & 1 & 0 \end{bmatrix} \qquad c = \begin{bmatrix} 1 \\ 0 \\ \vdots \\ 0 \end{bmatrix} \qquad (29)$$

The most difficult step of this technique consists in the evaluation of the polynomial coefficients $a_k^{(j)}$ and $b_k^{(j)}$ ($k = 1, \ldots, N_s; j = 1, \ldots, n$) that satisfy Eqs. (24) and (25) with the constrain of having all the roots of $a_k(s)$ (the poles of the filter) in the left-half s-plane. Procedures based on the minimization of the mean square error are proposed in Tubino & Solari (2003) and Carassale & Solari (2003).

The problem is much simpler if the spectral eigenvectors are frequency invariant. In this case the correlating filter $\mathscr{R}$ is memoryless and the transfer function of the colorizing filter $\mathscr{F}$, that is a diagonal matrix, can be assumed by choosing its diagonal elements in the form (Kailath 1980):

$$f_k(s) = \frac{\prod_{j=1}^{n-1}\left(s - \beta_k^{(j)}\right)}{\prod_{j=1}^{n}\left(s - \alpha_k^{(j)}\right)} \qquad (k = 1, \ldots, N_s) \qquad (30)$$

where $\alpha_k^{(j)}$ and $\beta_k^{(j)}$ represent, respectively, the poles and the zeros of the filter and can be estimated imposing the condition of Eq. (25). It can be demonstrated that if γ_k are strictly positive and limited functions, then the requirement of having the poles on the left-half s-plane can be automatically satisfied without imposing any constrain in the error minimization procedure (Kailath 1980).

5 NUMERICAL EXAMPLES – INFLUENCE OF THE INVARIANT-EIGENVECTOR ASSUMPTION

It has been noted that the assumption of invariant spectral eigenvectors produces great advantages in the implementation of the filter for representing the system excitation. In this section the consequences of this assumption are discussed through a couple of

wind-engineering-related applications. Applications regarding the representation of turbulence along bridges are presented in Tubino (2003).

The first application regards a 3-dof structure, referred to as Model 1, representing a simple model of cantilever bridge under construction in which the mass of the deck and its surface exposed to the wind in concentrated into $N = 3$ nodal points (Fig. 1). The second application regards a model of still chimney, referred to as Model 2, discretized by $N = 16$ nodal points (Fig. 2).

The excitation process $v(t)$ is constituted by the wind turbulence acting on the nodal points along the direction x. Such a process is defined through the turbulence model described in (Solari & Piccardo 2001). Details and numerical values adopted in the examples are provided in Carassale & Solari (2003).

Figures 3 and 4 show, respectively, the spectral relative error in the representation of the base bending moment M_b and torsional moment M_t in Model 1. The

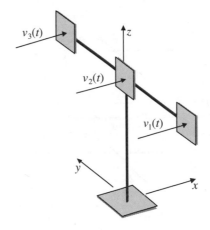

Figure 1. Model 1 – structure and turbulence components.

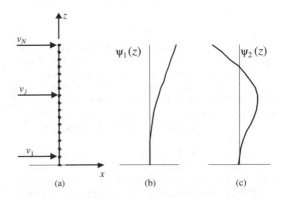

Figure 2. Model 2 (a), first vibration mode (b), second vibration mode (c).

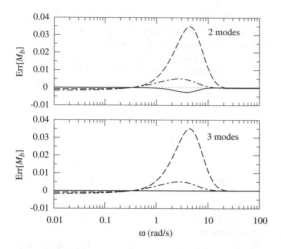

Figure 3. Model 1: spectral error in the representation of base bending moment M_b.

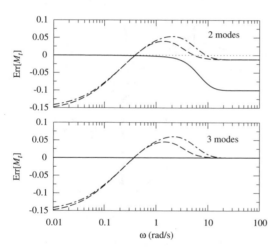

Figure 4. Model 1: Spectral error in the representation of base torsional moment M_t.

result of the rigorous application of the POD (solid line) is compared to the approximations based on Eq. (20) (dashed line) and Eq. (21) (dash-dot line). It should be noted that in the case of $N_s = N = 3$ the rigorous POD returns the exact spectrum of the force, while the approximations, even considering all the modes, lead to errors of 10–15% on M_t in the low-frequency range.

Figures 5 and 6 show, respectively, the spectral relative error in the representation of the first modal force p_1 and the second modal force p_2 in Model 2. The POD representation is performed retaining 1, 2 and 5 modes. The rigorous application of POD leads to a representation convergent to the exact value of the

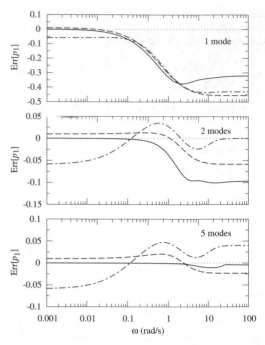

Figure 5. Model 2: spectral error in the representation of the first modal force p_1.

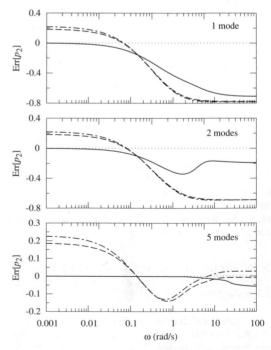

Figure 6. Model 2: spectral error in the representation of the second modal force p_2.

modal forces (with 5 modes the error is less than 5%), while the approximate filters converge to results affected by the 20% error in some frequency ranges.

6 CONCLUSIONS

POD-based filters are convenient tools for representing the external excitation of dynamical systems in wind engineering and other similar contexts. The ability of POD in reducing the order of the excitation has been demonstrated in the numerical examples. In the case of Model 2, the excitation order passed from $N = 16$ to $N_s = 5$ within a very reasonable approximation.

Invariant-eigenvector approximations have been found effective, leading, in most of the cases, to substantially correct numerical results. Good results seem to be achieved whether the energy content of the process is concentrated in a rather narrow spectral band like in the case of wind. The safe application in wind engineering of such approximation concepts requires, however, careful parametric studies and further investigation.

REFERENCES

Benfratello, S. & Muscolino, G. 1999. Filter approach to the stochastic analysis of MDOF wind-excited structures, *Prob. Engrg. Mech.*, **14**, 311–321.

Kailath, T. 1980. *Linear systems*. Prentice-Hall, Inc., Englewood Cliffs, N.J.

Carassale, L., Piccardo, G. & Solari, G. 2001. Double modal transformation and wind engineering applications, *J. Engrg. Mech. ASCE*, **127**(5), 432–439.

Carassale, L. & Solari, G. 2003. POD-based filters for the analysis of random vibration in structures. In preparation.

Chen, X. & Kareem, A. 2001. Aeroelastic analysis of bridges under multicorrelated winds: integrated state-space approach, *J. Engrg. Mech. ASCE*, **127**(11), 1124–1134.

Di Paola, M. & Gullo, I. 2001. Digital generation of multivariate wind processes, *Prob. Engrg. Mech.*, **16**, 1–10.

Loeve, M. 1955. *Probability theory.* Van Nostrand, N.Y.

Priestley, M.B. 1981. *Spectral analysis and time series.* Academic Press, London, UK.

Schetzen, M. 1980. *The Volterra and Weiner theories of nonlinear systems.* John Wiley & Sons, Inc., N.Y.

Solari, G. & Carassale, L. 2000. Modal transformation tools in structural dynamics and wind engineering, *Wind & Structures*, **3**(4), 221–241.

Solari, G. & Piccardo, G. 2001. Probabilistic 3-D turbulence modeling for gust buffeting of structures, *Prob. Engrg. Mech.*, **16**, 73–86.

Soong, T.T. & Grigoriu, M. 1993. *Random vibration of structural and mechanical systems.* Prentice-Hall, Inc., Englewood Cliffs, N.J.

Tubino, F. 2003. *Wind Actions on Long-Span Bridges*, Ph.D. Thesis, Dept. of Structural and Geotechnical Engineering, University of Genova, Italy.

Tubino, F. & Solari, G. 2003. Principal representation of turbulence fields: a double POD procedure, Proc. 11*th International Conference on Wind Engineering*, 11ICWE, June 2003, Texas, USA, In press.

A model for vortex-shedding induced oscillations of long-span bridges

P. D'Asdia & V. Sepe
Università degli Studi di Chieti – Pescara "G. D'Annunzio" – Dipartimento di Progettazione,
Riabilitazione e Controllo delle Strutture Architettoniche, Pescara, Italy

L. Caracoglia
University of Illinois – Department of Civil and Environmental Engineering, Urbana, Illinois, USA

S. Noè
Università degli Studi di Trieste – Dipartimento di Ingegneria Civile, Trieste, Italy

ABSTRACT: The paper describes the extension to bridge decks of a model for the vortex-shedding induced loads, recently proposed by some of the writers, and already successfully applied to chimneys.

The model, which operates in the time domain, schematically reproduces the load effect on the structure, due to the vortex-shedding, as an alternating force per unit length, with variable direction, frequency and phase, as a function of the relative wind velocity.

The lock-in phenomenon is simulated by imposing the instantaneous correspondence of the excitation frequency with the oscillation frequency of the structure in a pre-selected range. Outside this interval, the frequency of the pulsating force is calculated through the Strouhal's Law. The limited number of aerodynamic parameters that are required, can be obtained from the usual section model tests.

The results of a numerical investigation are compared with experimental results from section-model tests and full-scale measurements reported in the scientific literature.

1   INTRODUCTION

Recent studies have highlighted the relevance of vortex shedding from the bridge deck for wind speed in the lock-in interval, during or after the erection of the structure (Larsen & Jacobsen 1992; Lee et al. 1997; Scanlan 1998; Kawatani et al. 1999; Frandsen 2001). An example from the past is also represented by the Menai Strait Suspension Bridge in England (1826), corresponding to the first recorded event of vortex-induced vibration observed during the construction stages of the structure (Buonopane & Billington 1993).

In 1997, the Storebaelt suspension bridge in Denmark experienced low-frequency vertical oscillation during the final stage of deck erection for wind speeds between 4 and 12 m/s. On-site monitoring of the vibration and wind speed measurements indicated that the cause of the vibration was periodic vortex shedding; specific countermeasures were adopted, including the design of appropriate guide vanes (Larsen et al. 2000).

Vortex induced steady-state vibrations have been also observed on other categories of bridges, characterized by extreme slenderness of the deck. An interesting example, recently documented, is concerned with the multi-span cable-stayed bridge of the Trans-Tokyo Bay Highway Crossing. The two longest spans of the bridge, completed in 1997, experienced significant vibrations due to vortex shedding under prevailing winds orthogonal to bridge axis, with wind velocity around 16–17 m/s, resulting in maximum amplitudes exceeding 50 cm (Fujino & Yoshida 2002). Another example can be derived from Ozkan et al. (2001), related to the full-scale long-term monitoring of the Fred Hartman Bridge, a twin-deck cable-stayed bridge crossing the Houston Ship Channel between Baytown and La Porte, Texas. One of the recent and unexpected recorded events is what is believed to be co-existing vortex-induced vibration of the deck at a frequency corresponding to lower vertical modes and buffeting effects at higher modes (Ozkan et al. 2001). Another example includes the Rio-Niteroi bridge, spanning across the Guanabara bay in Rio de Janeiro (Battista & Pfeil 2000).

The key aspect of all these examples is the fact that, as a consequence of the extremely low values of the inherent mechanical damping, significant vertical oscillation of the deck can be noticed at relatively low wind speed (15–25 m/s), considerably lower than the flutter instability threshold, and likely to occur very often in the lifetime of the structure. Consequences associated with this kind of cyclically wind-induced oscillations might potentially include fatigue problems on the structural elements (e.g. hangers, in particular at the anchorages) and the reduction of travel safety and/or comfort levels for both road and railroad users.

This paper describes the extension to bridge decks of a model for vortex-shedding induced loads, recently proposed by some of the writers, and already successfully applied to chimneys (D'Asdia & Noè 1998), as it results from comparison with wind tunnel tests performed by the writers and others (Noè et al. 1998).

The model, described in Sect. 3 below, operates in the time domain and requires a limited number of aerodynamic parameters, which can be obtained from the same section models, commonly used for bridge aerodynamic tests.

The proposed model, applied to the current design of the suspension bridge over the Messina Strait (Caracoglia et al. 2000), has shown vertical deck acceleration values very close to those obtained by Diana et al. (1995) from full-bridge aeroelastic model tests, in particular for wind speeds at deck level in the range of 10–20 m/s.

In Sect. 4, the results of a numerical investigation performed with the proposed model are compared with experimental results from section-model tests and full-scale measurements reported in the scientific literature.

2 THE FINITE ELEMENT CODE

The model for vortex-shedding, described in Sect. 3, has been implemented in an *ad hoc* nonlinear F.E.M. code (*Tenso*), already used by the writers in other studies (e.g. Caracoglia et al. 2000). The program allows for finite-element analysis in the time domain of suspension bridges and other large structures, simulated by one-dimensional elements, taking into account the geometrical nonlinearities of the elements. The other aerodynamic forces acting on deck, towers and cables are evaluated through the static coefficients; the effective angle of attack is evaluated at each time instant by combining the effects of the wind speed, bridge deformation and motion, in accordance with the quasi-stationary method.

Any configuration of the wind field (mean value, correlation functions and turbulence) can be introduced to evaluate the wind speed on each element (deck, hangers, cables, towers, etc.).

The dynamical integration is performed through a total Lagrangian approach by means of the New-mark step-by-step method; mechanical damping is described through the Rayleigh method, in terms of mass and stiffness matrices.

3 THE PROPOSED MODEL

The across-wind force due to vortex shedding from a fixed cylindrical body can be described as a sinusoidal force per unit length, the frequency f of which is proportional to the cross-flow speed V_w. The Strouhal number $St = fH/V_w$ is a constant depending on the shape of the cross-section. The deck across-wind dimension is denoted by H. The total force value on a finite length of the cylinder, depends on its longitudinal correlation and it is therefore influenced by the turbulence of the oncoming wind. The same applies to an elastically suspended cylinder if the shedding frequencies are well separated from the eigenfrequencies and then the oscillation amplitude due to the across-wind force is sufficiently small.

When the shedding frequency is close to one of the system eigenfrequencies, and if the structural damping is sufficiently small to allow large-amplitude oscillation (i.e. at least a few percent of the section transverse dimension), an aeroelastic interaction between the vortex-induced force and the structural response can be observed, leading to the synchronization of the shedding to the oscillation frequency of the system (*lock-in*). The magnitude of the across-wind force and the range of the lock-in wind velocity can be related to some characteristic factors, depending on the structural response (oscillation amplitude and frequency) and, therefore, on the mechanical and aerodynamic behaviour of the deck.

The *lock-in* phenomenon typically shows a hysteretic and self-limiting behaviour, i.e., the oscillation amplitude never exceeds values of the order of the transverse dimension of the section, also for very low structural damping. A comprehensive fluid-dynamic model is not yet available in the literature for an application to a wide-range of cases, although many models have been presented but only tested on a restricted number of bluff sections (Simiu & Scanlan 1996; Scanlan 1998; Goswami 1991; Goswami et al. 1992; Hansen 1998; Ruscheweyh 1994). However, all these models require a preliminary evaluation of some aspects of the response itself, e.g., the limiting amplitude of oscillation. Their ability to forecast the maximum amplitude of the response is therefore subjected to the correct assessment of the parameters and the adequate description of the structural behaviour in the entire range of wind velocities. As a consequence, the evaluation of the structural response obtained from different codes, based on different theoretical models, may be very different, in particular for weakly damped systems.

The numerical simulations described in the next sections of this paper have been obtained through a model for the vortex-shedding induced loads, originally developed by D'Asdia & Noè (1998) for circular chimneys and subsequently extended to other categories of structures (Caracoglia 2000; Caracoglia et al. 2000).

Implemented within the structural code described in Sect. 2, the proposed model is able to provide a reasonably correct response of the structure under vortex shedding, including the steady-state oscillation amplitude (limit cycle). This procedure only requires standard aerodynamic parameters of the deck, usually available from wind-tunnel tests (Strouhal number St, drag coefficient C_D and vortex-shedding coefficient C_{L-she}, cf. Eq. 2–8 below), and an approximate initial estimate of the parameters Ω_L and Ω_U, defining the lock-in frequency interval.

As shown in Sect. 4 below, due to the intrinsic stability of the proposed model of the load, also an approximate assessment of the above mentioned parameters (e.g. based upon results from the literature on similar sections) is capable of capturing the system response with the same order of accuracy.

The model is based on the hypothesis that the response of an elastically suspended body, caused by vortex shedding, can be treated as a nonlinear oscillation phenomenon. A detailed description can be found in D'Asdia & Noè (1998) and in Noè et al. (1998).

The model operates in the time domain. Its formulation schematically reproduces the load effect on the structure, due to the variation of the surface pressure field induced by the detachment of the vortices, through an alternating force per unit length, with variable direction, frequency and phase, as a function of the relative wind velocity. The direction of the force is always perpendicular to the relative velocity.

The lock-in phenomenon is simulated by imposing an instantaneous correspondence of the excitation frequency with the oscillation frequency of the structure in a pre-selected range of frequencies. Outside this interval, the frequency of the pulsating force is calculated, according to Strouhal's Law, as a function of the relative wind velocity. As the frequency varies, the continuity of the exciting force is ensured by a suitable modification of its phase.

The forces acting on the cross section are represented in Figure 1.

The term Δt represents the integration time-step; the total force per unit-length due to wind action at time t is:

$$\vec{F}_{w(t)} = \vec{F}_{she(t)} + \vec{F}_{drag(t)} \qquad (1)$$

The component due to vortex shedding at time t is the force $\vec{F}_{she(t)}$, assumed as perpendicular to the relative velocity $\vec{V}_{w,rel(t-\Delta t)}$ at time $t - \Delta t$. The modulus of $\vec{F}_{she(t)}$ is evaluated at time t as a function of the

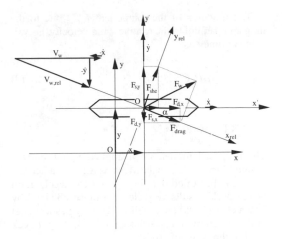

Figure 1. Cross section.

values at time $t - \Delta t$, according to the following expressions:

$$F_{she(t)} = \frac{1}{2}\rho H\, C_{L-she}\, V^2_{w,rel(t)} \sin\left(\omega_{she(t)}t + \varphi_{she(t)}\right) \qquad (2)$$

$$V_{w,rel(t)} = \sqrt{V^2_{w(t-\Delta t)} + \dot{y}^2_{(t-\Delta t)}}\;; \qquad (3)$$

$$\alpha_{(t)} = \tan^{-1}\left(\dot{y}_{(t-\Delta t)}\,/\,V_{w(t-\Delta t)}\right); \qquad (4)$$

$$\omega_{St(t)} = 2\pi St\frac{V_{w,rel(t)}}{H}\;;\quad \widetilde{\omega}_{(t)} = \frac{\pi}{\tau_{(y=\bar{y})_n} - \tau_{(y=\bar{y})_{n-1}}}\;; \; (5,6)$$

$$\begin{cases} \omega_{she(t)} = \omega_{St(t)} \; if \; \dfrac{\omega_{St(t)}}{\widetilde{\omega}_{(t)}} < \Omega_L \; or \; \dfrac{\omega_{St(t)}}{\widetilde{\omega}_{(t)}} > \Omega_U \\[4mm] \omega_{she(t)} = \omega_{(t)} \quad if \;\; \Omega_L \leq \dfrac{\omega_{St(t)}}{\widetilde{\omega}_{(t)}} \leq \Omega_U \end{cases} \qquad (7)$$

$$\varphi_{she(t)} = \left[\omega_{she(t-\Delta t)} - \omega_{she(t)}\right]t + \varphi_{she(t-\Delta t)} \qquad (8)$$

where V_w denotes the wind speed, y denotes the across-wind displacement, ρ the air density, and the dot symbol denotes the time-derivative; $\widetilde{\omega}_{(t)}$ is the estimated value of the instantaneous circular frequency of the across-wind oscillation due to vortex shedding; $\bar{y}$ is the mean across-wind displacement of the section, while $\tau(y-\bar{y})_n$, $\tau(y-\bar{y})_{n-1}$ represent the time instants corresponding to the last two zero-crossings of the function $y - \bar{y}$.

2333

The modulus of the component $\vec{F}_{drag}$ due to the drag and parallel to the relative wind velocity, is evaluated at time t as:

$$F_{drag(t)} = \frac{1}{2}\rho H\, C_D\, V^2_{w,rel(t)} \qquad (9)$$

4 AN EXAMPLE

The proposed model, when applied to the current design of the bridge on the Messina Strait (Caracoglia et al. 2000), turned out, for the vertical acceleration of the deck, results very close to those obtained by Diana et al. (1995) on the full-bridge aeroelastic model of the structure, in particular for wind speeds at deck level in the range of 10–20 m/s.

In this paper, the model is applied to the Great Belt Bridge (main span L = 1624 m, Fig. 2). In particular, comparisons are shown with tests on a 1:80 section model reported in Larsen (1993) and the full scale measurements performed by Larsen et al. (2000).

The tests on the 1:80 section model, reproducing the vertical frequency of the 8th natural mode (0.39 Hz), were carried out in the wind tunnel of the Danish Maritime Institute (Larsen 1993). For a logarithmic

damping $\delta = 0.01$ and for a wind speed V_w between 12 and 16 m/s, a maximum rms value of across-wind displacement $\eta_{max,rms} = 0.063\,H$ was recorded, corresponding to $\eta_{max} = 0.09\,H$, assuming a lock-in peak factor of 1.41, where H denotes the transverse dimension of the deck. For the same damping, for an estimated Strouhal number $St = 0.12$ and assuming, according to previous studies (D'Asdia & Noè 1998), the values $\Omega_L = 1.0$ and $\Omega_U = 1.1$, the proposed model is capable of reproducing the experimentally derived displacement (broken line in Fig. 3) when a lift coefficient $C_{L-she} = 0.28$ is considered, compatible with values currently reported in the technical literature. In Figure 3 V/V_{cr} denotes the ratio between the actual value of the wind speed and the value for which the Strouhal's shedding frequency coincides with the structural eigenfrequency.

Simulations based upon the value $C_{L-she} = 0.40$ (closer to values suggested by the Eurocode 1, part 2.4, for similar sections), are also reported in Figure 3; in the same figure the results obtained for different damping coefficients are plotted, for the sake of comparison.

In consideration of the uncertainties related to the assessment of the aerodynamic parameters St, C_{L-she}, C_D, not reported in the quoted paper (Larsen 1993), the results obtained can be considered as quite satisfactory, in terms of both the maximum value of displacements and the overall shape of the lock-in curve.

The full-scale measurements reported in Larsen et al. (2000), turned out maximum values of the across-wind displacements between 0.05 and 0.07 H for wind speed between 5 and 10 m/s, with longitudinal shapes corresponding to the 3rd, 5th and 6th vertical natural mode (natural frequencies 0.13, 0.21 and 0.24 Hz).

Figure 4 shows the comparison between the maximum value of the measured displacement (broken

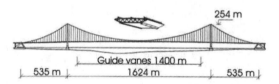

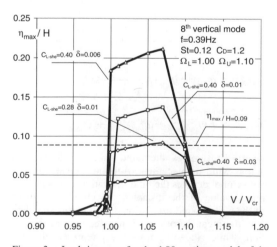

Figure 2. Great Belt Bridge (from Larsen et al. 2000).

Figure 3. Lock-in curve for the 1:80 section model of the Great Belt Bridge.

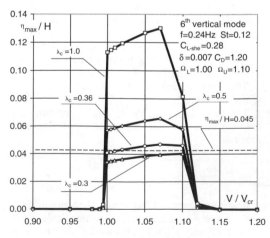

Figure 4. Lock-in curve for the Great Belt Bridge.

line) in case of lock-in with the 6th mode and the curve, obtained for the same mode by the proposed model; a lift coefficient $C_{L-she} = 0.28$, corresponding to the value identified from the section model analysis (cf. Fig. 3), and a logarithmic damping $\delta = 0.007$ have been adopted. The curves refer to different values of the non-dimensional correlation length λ_c, defined as:

$$\lambda_C = \frac{l_c}{l} \qquad (10)$$

where l is the distance between two consecutive nodes of the modal shape and l_c is the length of application of a perfectly correlated load ($\lambda_c \leq 1$).

Good correspondence between the numerical results and the experimental values is achieved for λ_c around 1/3.

In consideration of the fact that such a correlation length is reasonable for this category of structures, the reliability of the proposed model is confirmed.

5 CONCLUDING REMARKS

The paper describes a time-domain model for the vortex-shedding induced loads, extending the model by D'Asdia & Noè (1998), successfully applied to chimneys.

The model schematically reproduces the load effect on the structure, as an alternating force per unit length, with variable direction, frequency and phase, as a function of both relative and absolute wind velocity.

The limited number of aerodynamic parameters needed in the proposed model can be obtained from the same section model commonly used for bridge aerodynamic tests.

The results of a numerical investigation are compared with experimental data reported in the scientific literature, obtained from wind tunnel section-model tests and full-scale measurements of the Great Belt Bridge.

Despite the uncertainties on some aerodynamic and mechanical parameters (approximately estimated from literature data for similar sections), the numerical results are quite satisfactory, confirming the reliability of the model.

ACKNOWLEDGEMENTS

This work has been carried out in the framework of the National Research Program "Wind and INfrastructures: Dominating Eolian Risk For Utilities and Lifelines" (WINDERFUL), co-financed by the Italian "Ministry for Education, University and Research" (MIUR 2001–2002).

REFERENCES

Battista, R.C. & Pfeil, M.S. 2000. Reduction of vortex-induced oscillations of Rio-Niteroi Bridge by dynamic control devices, *Journal of Engineering and Industrial Aerodynamics*, 84, 273–288.

Buonopane, S.C. & Billington, D. 1993. Theory and history of suspension bridge design from 1823 to 1940, *Journal of Structural Engineering ASCE*, 119(3), 954–977.

Caracoglia, L. 2000. Wind-structure oscillations on long-span suspension bridges, *Ph.D. Dissertation*, University of Trieste, Italy.

Caracoglia, L., Noè, S. & Sepe, V. 2000. Aspetti non convenzionali della dinamica indotta dal vento nei ponti sospesi di grande luce, *Proceedings of the 6th Italian National Conference on Wind Engineering (IN-VENTO-2000)*, Genova, (in Italian).

D'Asdia, P. & Noè, S. 1998. Vortex induced vibration of reinforced concrete chimneys: in situ experimentation and numerical previsions, *Journal of Wind Engineering and Industrial Aerodynamics*, 74–76, 765–776.

Diana, G., Falco, M., Bruni, S., Cigada, A., Larose, G.L., Damsgaard, A. & Collina, A. 1995. Comparisons between wind tunnel tests on a full aeroelastic model of the proposed bridge over Stretto di Messina and numerical results, *Journal of Wind Engineering and Industrial Aerodynamics*, 54/55, 101–113.

Frandsen, J.B. 2001. Simultaneous pressures and accelerations measured full-scale on the Great Belt East suspension bridge. *Journal of Wind Engineering and Industrial Aerodynamics*, 89, 95–129.

Fujino, Y. & Yoshida, Y. 2002. Wind-induced vibration and control of trans-Tokyo bay crossing bridge. *Journal of Structural Engineering*, 128(8), 1012–1025.

Goswami, I. 1991. Vortex-induced vibrations of circular cylinders, *PhD Dissertation*, The Johns Hopkins University, Baltimore, MD (USA).

Goswami, I., Scanlan, R.H. & Jones N.P. 1992. Vortex shedding from circular cylinders: experimental data and a new model, *Journal of Wind Engineering and Industrial Aerodynamics*, 41–44, 763–774.

Hansen, S.O., 1998. Vortex-induced vibrations of line-like structures, *Cicind Report*, 15, n.1.

Kawatani, M., Toda, N., Sato, M. & Kobayashi, H. 1999. Vortex-induced torsional oscillations of bridge girders with basic sections in turbulent flows, *Journal of Wind Engineering and Industrial Aerodynamics*, 83, 327–336.

Larsen, A. 1993. Aerodynamic aspects of the final design of the 1624 m suspension bridge across the Great Belt, *Journal of Wind Engineering and Industrial Aerodynamics*, 48, 261–285.

Larsen, A., Esdahl, S., Andersen, J.E. & Vejrum, T. 2000. Storebalt suspension bridge – vortex shedding excitation and mitigation by guide vanes, *Journal of Wind Engineering and Industrial Aerodynamics*, 88, 283–296.

Larsen, A. & Jacobsen, S. 1992. Aerodynamic design of the Great Belt East Bridge. In Larsen, A. (ed), *Aerodynamics of Large Bridges*, Balkema, Rotterdam, 269–283.

Lee, S., Lee, J.S. & Kim, J.D. 1997. Prediction of vortex-induced wind loading on long-span bridges, *Journal of Wind Engineering and Industrial Aerodynamics*, 67–68, 267–278.

Noè, S., D'Asdia, P. & Fathi, S. 1998. Simulazione della risposta dinamica di strutture cilindriche elastiche soggette a distacco dei vortici. Studi su un modello numerico, *Proceedings of the 5th Italian Conference on Wind Engineering INVENTO-98*, Perugia, Italy, 307–324 (in Italian).

Ozkan, E., Main, J.A. & Jones, N.P. 2001. Full-scale measurements on the Fred Hartman Bridge, *Proceedings of the 5th Asia-Pacific Conference on Wind Engineering*, Kyoto, Japan.

Ruscheweyh, H. 1994. Vortex excited vibrations, in Sockel, H. (ed), *Wind excited vibrations of structures*, Springer-Verlag, New York, 51–87.

Scanlan, R.H. 1998. Bridge flutter derivatives at vortex lock-in, *ASCE Journal of Structural Engineering*, 124(4), 450–458.

Simiu, E. & Scanlan, R.H. 1996. *Wind effects on structures* (*3rd Edition*), John Wiley and Sons Inc., New York, London.

Stochastic characterization of wind effects from experimental data

M. Gioffrè

Department of Civil and Environmental Engineering, University of Perugia, Perugia, Italy

ABSTRACT: Experimental records of wind pressure fluctuations are used to estimate the stochastic features of internal forces in the main resisting system of a typical low-rise gable roof building. The load bearing structure is made of steel truss frames connected by bracing systems. Tributary areas on the building envelope and influence coefficients are used to calibrate appropriate weights to sum the simultaneous time histories of pressures and calculate internal forces at any cross-section of any member of the wind load resisting system. Statistical analysis is finally performed on the obtained response time series and their main probabilistic features are estimated. The spatial variation and the non-Gaussian properties of the wind effect are carefully highlighted since they might significantly affect the predicted response peak values.

1 INTRODUCTION

In the last decades significant improvements have been made in the field of wind engineering, nevertheless windstorms still cause severe damage especially to low-rise buildings. One of the main difficulties is the accurate modelling of the wind pressure fields impinging on the building surfaces given the complexity of the flow-structure interactions on bluff bodies. Wind tunnel studies have shown that these interactions are responsible for the non-Gaussian features of the pressure fluctuations that are then reflected in higher peak values then those predicted by the standard gust factor approach.

The combined use of experimental measurements and stochastic simulation represents a reliable methodology to model the wind pressure fields on bluff bodies and to develop effective economical designs to withstand the effect of extreme wind events. For this reason the scientific community is moving toward the concept of Database-Assisted-Design (DAD) as an alternative to the current standard provisions. Simiu started the first research project aimed to this goal with specific reference to low-rise buildings using wind pressure time histories for estimating internal forces time series in the main resisting frames (Whalen et al. 1998) and their maximum expected values (Gioffrè et al. 2000, Sadek & Simiu 2002). Experimental measurements and stochastic simulation were also used to investigate the local effects on cladding of prismatic tall buildings (Gioffrè et al. 2001a, Gioffrè et al. 2001b) and to evaluate the effects of the spatial probabilistic characteristics of pressure fields on the aggregate uplift acting on roof panels of low-rise gable roof buildings

representative of typical homes (Cope et al. 2002). Other studies investigated the peak structural response associated to wind pressure distributions on low-rise buildings using the quasi-static method (Tamura et al. 2001) and the static approach adopted in the current standard provisions (Gioffrè et al. 2002).

This study seeks the spatial stochastic characteristics of wind effects in the main resisting structure of low-rise gable roof buildings made of steel truss frames connected by a bracing system. The wind pressure load is obtained using the measurements conducted at the boundary layer wind tunnel of the University of Western Ontario (UWO) (Lin & Surry 1997). The internal forces time histories are computed using a modified version of the numerical procedure developed at N.I.S.T. (National Institute of Standards and Technology) for the pilot project WiLDE-LRS (Wind Load Design Environment for Low Rise Structures) (Whalen et al. 2000) that uses full records of pressures measured at hundreds of taps over the building envelope. Finally, the obtained internal forces time series are characterized by statistical analysis in order to give effective descriptions of the structural response. In particular, the main focus is on the non-Gaussian features that would affect the appropriate peak values to be used for reliable design.

2 WIND LOAD MODEL

The wind load model used in this study is calibrated from experimental data recorded at the wind tunnel facility of UWO and consisting of pressure coefficient time histories on gable roof building models.

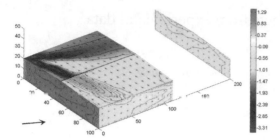

Figure 1. Maps of the pressure fluctuations skewness coefficient γ_3 for $\beta = 45°$.

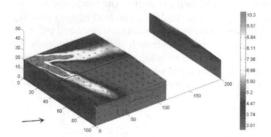

Figure 2. Maps of the pressure fluctuations kurtosis coefficient γ_3 for $\beta = 45°$.

The tested buildings had rectangular plan (61 m by 30.5 m), roof pitch of 1 on 24, and two different eave heights: 6.1 m and 9.75 m. The length scale used were 1:200 and 1:100 while the velocity scale was fixed to 1:2.5 giving the corresponding time scales of 1:80 and 1:40. The boundary layer wind tunnel simulated open country and suburban terrain using appropriate roughness elements. Combining the features described above, a total of six model configurations were used for data recording.

The pressure coefficient fluctuations were simultaneously recorded at 500 tap locations for a time lag of 60 seconds and sampling frequency 400 Hz. The incoming wind had different directions varying from 0° to 180° with 5° steps. The complete description of the experimental setup can be found in (Lin & Surry 1997).

The main statistical properties of the wind pressure load on the building surfaces were reported in (Gioffrè et al. 2002) using a set of interactive numerical procedures giving plots of pressure coefficient fluctuations, histograms, auto- and cross-correlation functions, auto- and cross-spectral functions at each desired tap or couple of taps, for each available configuration. Furthermore, the first four statistical moments associated to the pressure fluctuations at each tap location were described using 3D maps that are very effective to localize areas where the wind pressure deviate from the Gaussian model (skewness coefficient $\gamma_{3G} = 0$, kurtosis coefficient $\gamma_{4G} = 3$).

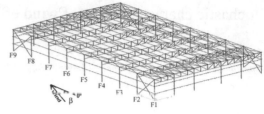

Figure 3. View of the main wind load resisting system.

The sets of data used in this study were recorded on the 1:200 building model, with 6.1 m eave height, open country, and wind angles $\beta = 0°$ (wind perpendicular to the gable end) and $\beta = 45°$ (cornering wind) shown in Figure 3. Figures 1 and 2 show an example of strong deviation from the Gaussian behavior in the separated flow regions associated to cornering wind (e.g. the suction delta-wings on the windward side of the roof).

3 RESPONSE TIME HISTORIES

The pressure coefficient time histories described in the previous section are used in the basic DAD framework to obtain the response time series at any cross-section of any member of the wind load resisting system of low-rise gable roof buildings.

Calculation are carried out using a modified version of the numerical procedure WiLDE-LRS developed at N.I.S.T. (Whalen et al. 2000). This application program is a pilot project aimed to overcome the oversimplification inherent in conventional standards that reduces the highly complex nature of wind-structure interactions into simplified tables and plots both for the lack of comprehensive information and the need of simplified design procedures. The growing availability of wind tunnel facilities, wind tunnel data, and new communication technologies validate the effort in this direction that is promising to give more risk-consistent, safer, and economical designs.

The WiLDE-LRS procedure is based on four main stages. In the first stage the influence coefficients for the building structural system constituted by rigid portal frames hinged at the column bases are calculated from the geometric frame properties. In the second stage the wind pressure time series are computed from the aerodynamic databases containing pressure coefficients as functions of wind directions and from information on extreme wind speeds. The wind load transferred to the frames by the girts and purlins is determined in the third stage. Finally, in the fourth stage the internal forces time histories are computed by adding up the effects of the wind loads transmitted by the girts and purlins at each time instant weighting with the respective influence coefficients.

The basic assumptions that are used in the program at the present stage are the linear elastic behavior of the structure, and that effects arising from cross-bracing between the frames and wind-induced dynamic effects are negligible. This last assumption is warranted for low-rise buildings with natural frequencies of vibration not less than about 2 Hz. Work is in progress on the extension of WiLDE-LRS to other types of structural systems with linear and nonlinear behavior, and that experience wind-induced dynamic effects.

A slightly different structural system made of nine steel truss frames fixed at the column bases and connected by bracing systems is used in this study (Fig. 3). The new influence coefficients are computed using the finite element approach and are used as input in the WiLDE-LRS numerical procedure to obtain the response time histories in each truss element of the nine frames. Figure 4 shows a detailed view of the i-th truss frame constituting the structure with the element numbering.

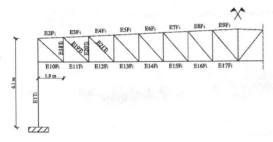

Figure 4. Detail of the i-th truss frame with the element numbering.

4 RESULTS

The structural system investigated in this study is mainly constituted by trusses that transfer the wind load to the columns. Each element in the truss is double hinged, the self weight is neglected and the load is transferred by the girts and purlins on the hinged joints. It follows that each element in the truss is only affected by the axial force. As anticipated earlier, two cases are studied in this work: wind perpendicular to the gable end ($\beta = 0°$), and wind blowing with $\beta = 45°$ (cornering wind). For each case the obtained time histories are used to estimate the response probabilistic features by statistical analysis. Histograms of the axial force fluctuations are plotted together with the associated Gaussian distribution calibrated from the mean and the standard deviation (continuous line). Furthermore, the skewness and kurtosis coefficients are estimated to quantify the degree of deviation from the Gaussian model.

4.1 Wind perpendicular to the gable end

Figure 5 shows the axial force time series at six elements of the most upwind frame (frame F1). It is sudden evident that the force fluctuations are asymmetric about positive mean values and have variability increasing from the eave (E2) to the ridge (E9). The negative pressures arising from flow separation at the gable end induce positive axial forces in the truss upper elements while the lower elements experience compression. This force inversion has to be carefully considered in order to avoid instability phenomena.

The probabilistic features of the axial forces of Figure 5 are reported in Figure 6. The non-Gaussian properties that characterize the wind pressure fluctuations in the separated flow area at the gable end are still reflected in the response time series that experience

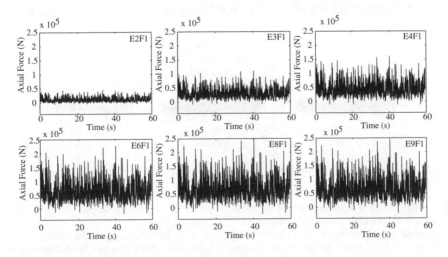

Figure 5. Axial force time histories at six elements of frame F1 for $\beta = 0°$.

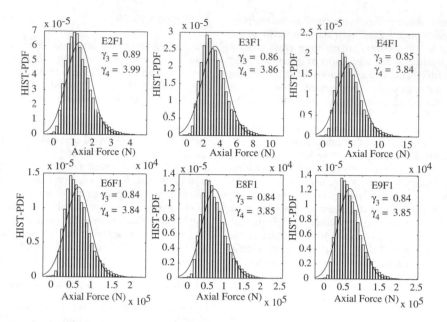

Figure 6. Axial force histograms, skewness and kurtosis coefficients estimated at six elements of frame F1 for $\beta = 0°$.

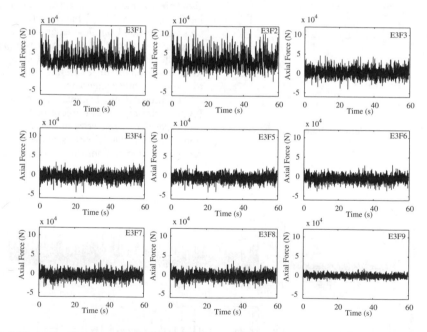

Figure 7. Axial force time histories at element E3 in the nine frames for $\beta = 0°$.

high values of the skewness coefficients. This result confirm that it is not possible to claim the central limit theorem when adding up strongly non-Gaussian and correlated pressure fluctuations. All the elements considered in frame F1 have similar stochastic features given the symmetry of the structure about the wind flow direction.

In order to seek into the spatial variation of the wind load effects the axial force fluctuations at element E3 in the nine frames are also reported in Figure 7.

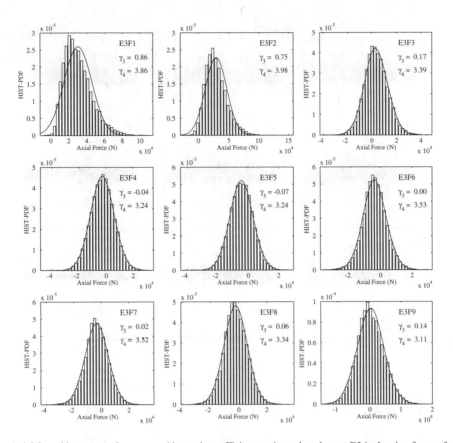

Figure 8. Axial force histograms, skewness and kurtosis coefficients estimated at element E3 in the nine frames for $\beta = 0°$.

Positive mean values of the forces are experienced in the first three frames while the remaining six frames are characterized by negative mean values. This trend reflect the flow reattachment on the roof surface. Also the probabilistic features are influenced by the pressure fluctuation properties highlighting the close connection between wind load and wind load effects in the structural system. Moving from the windward frame the axial forces lose the non-Gaussian properties to become quite Gaussian at the fourth frame.

4.2 Cornering wind

The results for cornering wind ($\beta = 45°$) are reported in Figures 9–12. If one moves from frame F1 to frame F9, element E3 has always positive mean values of the axial forces and asymmetric fluctuations (Fig. 9). The probabilistic features closely reflect the pressure patterns on the roof surface shown in Figures 1 and 2 where the suction delta-wings on the windward side of the roof are highlighted by the skewness and kurtosis coefficients maps. The element on the upwind frame is strongly influenced by the flow separation at

the roof corner and the axial force manifest the higher deviation from the Gaussian model. In the second frame (F2) both the skewness and kurtosis coefficients in element E3 decrease significantly and rise up again in the following frames. The trend in the last frames is decreasing with the distance from the corner.

The axial force time series at six elements of the most upwind frame (frame F1) are reported in Figure 11. Again the force mean values are positive and the variability of the fluctuations increase from eave to ridge. Similar trend is observed for the non-Gaussian features that manifest the higher values of skewness and kurtosis coefficients near the corner where the complex flow-structure interactions significantly influence the pressure fluctuations.

All the results presented in this section demonstrate the variability in space and time of wind load effects and confirm the need to update conventional codes and standards using the growing availability of wind tunnel facilities, shared databases, powerful computational systems, and fast internet connections. This update would lead to improve the risk-consistency, safety, and economy of the structural design. The

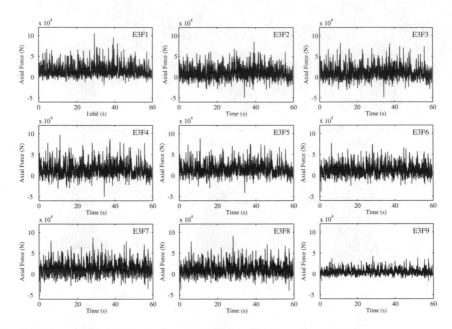

Figure 9.   Axial force time histories at element E3 in the nine frames for $\beta = 45°$.

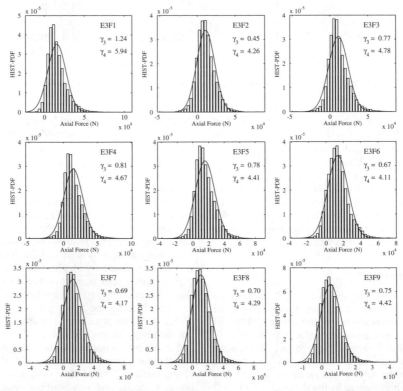

Figure 10.   Axial force histograms, skewness and kurtosis coefficients estimated at element E3 in the nine frames for $\beta = 45°$.

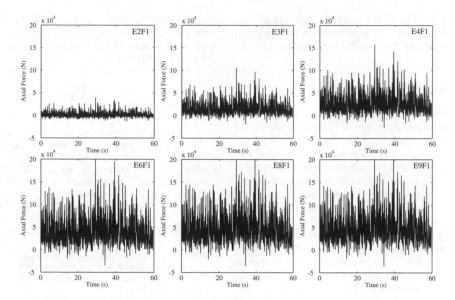

Figure 11. Axial force time histories at six elements of frame F1 for $\beta = 45°$.

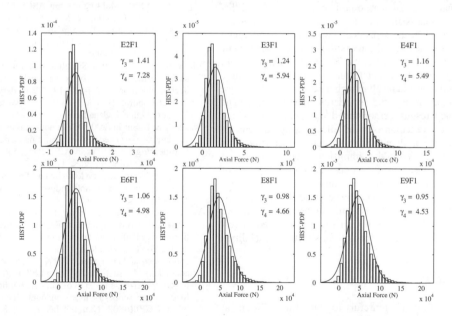

Figure 12. Axial force histograms, skewness and kurtosis coefficients estimated at six elements of frame F1 for $\beta = 45°$.

limited information capacity of the current standard provisions can be easily enlarged using the DAD approach. The internal forces time series can be used both for design purposes and for reliability analysis based on the probability information associated to the structural response. In particular, peaks of internal forces with specified mean recurrence intervals can be accurately estimated using basic theory of stochastic processes that account both for Gaussian and non-Gaussian behavior of the time histories of interest (Gioffrè et al. 2000, Sadek & Simiu 2002). Furthermore, new modules for the WiLDE-LRS software are currently under development to add simulation capabilities. It will be possible to calibrate suitable stochastic models on the available data and to generate artificial response time histories to be used to enlarge the information needed for reliability analysis.

5 CONCLUSIONS

The Database-Assisted-Design (DAD) approach was used to investigate the response stochastic features of a typical low-rise gable roof building affected by wind load. Calculation were performed with a modified version of the WiLDE-LRS software developed at NIST using sets of pressure coefficients recorded at the UWO wind tunnel facility. The influence coefficients of the truss frames constituting the structural system were evaluated by the finite element method. The internal forces time histories at any cross-section of any member of the wind load resisting system were obtained summing up the pressure fluctuations on the building envelope weighting with the tributary areas and the influence coefficients. Dynamic effects were assumed to be negligible and the structure had linear elastic behavior. Finally, statistical analysis was used to estimate the internal forces stochastic properties. It was found that the non-Gaussian features associated to the pressure fluctuation in the separated flow regions significantly influence the internal forces time histories, which show similar stochastic patterns. There are members with strongly non-Gaussian features of the response time series that would result in higher peak values than those predicted using the standards prescriptions. The obtained results demonstrated the complexity of the accurate evaluation of wind effects, which depend on several factors like the flow-structure interactions that are not included in the simplified descriptions of wind loads proposed by the codes. The combined use of experimental measurements and simple numerical procedures within a DAD context represent a valid alternative to the current standard provisions to improve significantly the description of wind-induced effects and obtain useful information for structural reliability.

ACKNOWLEDGMENTS

This author is grateful to Dr. Emil Simiu at N.I.S.T., USA, for providing the wind tunnel data and to Andrea Grazini at the Department if Civil and Environmental Engineering, Perugia, Italy, for his help with the numerical analysis. This work was partially supported by the M.I.U.R. research project (Cofin 2001: WINDERFUL).

REFERENCES

Cope, A.D., K.R. Gurley, M. Gioffrè, & T.A. Reinhold (2002). Low-rise gable roof wind loads: Characterization and stochastic simulation. submitted for review.

Gioffrè, M., A. Grazini, & V. Gusella (2002). Reliability of low-rise buildings: Experimental wind load modelling vs. building codes. In *Reliability and Optimization of Structural Systems – Proceedings of the 10th IF IP WG 7.5 Working Conference on Reliability and Optimization of Structural Systems,* March 25–27, Osaka, Japan.

Gioffrè, M., M. Grigoriu, M. Kasperski, & E. Simiu (2000). Wind-induced peak bending moments in low-rise building frames. Journal of Engineering Mechanics, ASCE 126(8), 879–881.

Gioffrè, M., V. Gusella, & M. Grigoriu (2001a). Non-gaussian wind pressure on prismatic buildings I: Stochastic field. Journal of Structural Engineering, ASCE 127(9), 981–989.

Gioffrè, M., V. Gusella, & M. Grigoriu (2001b). Non-gaussian wind pressure on prismatic buildings II: Numerical simulation. Journal of Structural Engineering, ASCE 127(9), 990–995.

Lin, J. & D. Surry (1997). Simultaneous time series of pressures on the envelope of two large low-rise buildings. Technical Report BLWT-SS7-1997, Boundary-Layer Wind Tunnel Laboratory, The University of Western Ontario, London, Ontario, Canada.

Sadek, F. & E. Simiu (2002). Peak non-gaussian wind effects for database-assisted low-rise building design. Journal of Engineering Mechanics, ASCE 128(5), 530–539.

Tamura, Y., H. Kikuchi, & K. Hibi (2001). Extreme wind pressure distributions on low-rise building models. Journal of Wind Engineering and Industrial Aerodynamics 89, 1635–1646.

Whalen, T., E. Simiu, G. Harris, J. Lin, & D. Surry (1998). The use of aerodynamic databases for the effective estimation of wind effects in main wind-force resisting systems: application to low buildings. Journal of Wind Engineering and Industrial Aerodynamics 77 & 78, 685–693.

Whalen, T.M., V. Shah, & J.S. Yang (2000). A pilot project for computer-based design of low-rise buildings for wind loads – the wilde-lrs user's manual. Technical Report NIST GCR00-802, National Institute of Standards and Technology, Gaithersburg, MD.

Inclined cable aerodynamics – wind tunnel test and field observation

M. Matsumoto & T. Yagi
Kyoto University, Kyoto, Japan

ABSTRACT: This paper describes the inclined cable aerodynamics highlighted to rain-wind-induced vibration in comparison of their response characteristics observed for prototype stay cables of cable-stayed bridges, a large-scale cable model in the field and a rigid yawed cable model in wind tunnel. Then, the rain-wind-induced vibrations are classified into high speed restricted-response caused by peculiar low frequency vortices and galloping instability.

1 INTRODUCTION

The inclined stay cables of cable-stayed bridges often show various wind-induced vibrations, such as buffeting, Karman vortex-induced vibration, rain-wind-induced vibration and galloping instability for single cable and additionally wake galloping and wake induced flutter for twin or multi-cables. In particular, rain-wind-induced vibration has recently become much concerned to bridge engineers, because it frequently occurs for many cable-stayed bridges in the world in rainy and windy days and shows rather large amplitude. Wind velocity is mostly not so high, such as 10 m/s to 15 m/s. Furthermore, in the observed wind-induced vibrations in rainy days, some particular vibrations with remarkably large amplitude like galloping were included, therefore, there might be two different types of vibrations in rain-wind-induced vibration, such as velocity-restricted response at high reduced velocity (shortly, it is called high speed restricted-response, hereafter) and galloping. The fundamental characteristics of rain-wind-induced vibration have been gradually verified but, as a matter of fact, many subjects remain unsolved yet, such as the mechanism of high speed restricted-response of inclined cables, the similarity between the aerodynamic responses observed for the rigid section model with/without the upper rivulet on cable surface in wind tunnel and for the prototype cables in natural wind, the effect of the fixed/moving upper water rivulet on the cable aerodynamics, the applicability of quasi-steady theory on rain-wind-induced vibration and so on. In this paper, these unsolved subjects are discussed by comparison of the aerodynamic characteristics of the inclined cable in wind tunnel test with the measured response characteristics of the prototype cables and large-scale cable model in the field.

2 RAIN-WIND-INDUCED VIBRATION OF PROTOTYPE STAY CABLES

Figure 1 and Figure 2 show Vr (Reduced velocity: V/fD, V: wind velocity, f: vibration frequency, D: cable diameter)-A (vibration amplitude) diagrams of cable rain-wind induced vibration, which have been observed for several cable-stayed bridges in Japan (Matsumoto et al. 1989) and for Fred-Hartman Bridge in US (Main et al. 2001), respectively. The response mostly appeared at the certain restricted reduced velocity range of $Vr = 20$ to 40, and it seems to be the velocity-restricted type like as vortex-induced vibration. Furthermore, at those reduced velocity range,

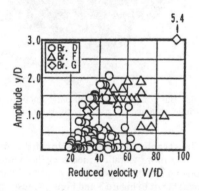

Figure 1. Reduced velocity – amplitude (Vr-A) diagram of prototype stay cables in Japan.

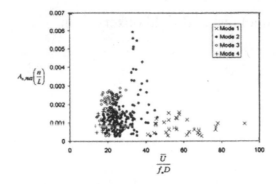

Figure 2. Reduced velocity – amplitude (Vr-A) diagram of stay cables of Fred-Hartman bridge.

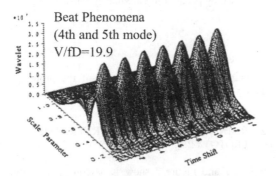

Figure 3. Beat response of cable of Meiko-West Bridge, obtained by wavelet analysis (4th and 5th mode).

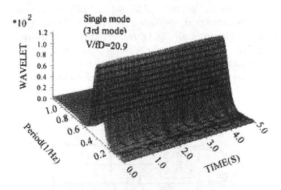

Figure 4. Harmonic vibration with single mode of Meiko-West Bridge, obtained by wavelet analysis (3rd mode).

the inclined cable shows the beat-vibration composed with two or more vibration modes or the typical single mode vibration. Those typical response characteristics are shown in Figure 3 and Figure 4, respectively, which have been observed for Meiko-West Bridge in Japan.

3 WIND-INDUCED VIBRATION OF A LARGE-SCALE CABLE MODEL IN THE FIELD

3.1 Field test apparatus and cable models

The large-scale cable model test has been carried out in natural wind to observe and investigate more precisely the inclined cable aerodynamics, which is expected to have advantages of response measurement without any disturbance by traffics and in similar condition with the one of prototype cables. The Shionomisaki Wind Effect Laboratory of the Disaster Prevention Research Institute, Kyoto University at Kushimoto, far from Kyoto City by about 300 km south and the south edge of Kii Peninsular, was selected as the observation site, because of its windy area and the good arrangement of wind velocity measurement facilities. The stiff pylon with height of 24 m was constructed there and a large scale elastic cable model with length of 30 m was set up with the vertical angle of about 45°. A cable model was set in E-W direction. The field measurement started since September, 2000 up to today. For first one year, a cable model, which was aluminum pipe lapped by polyethylene membrane with 110 mm diameter was used, but its structural damping could be large because of the friction between two materials. The vibration of this model was fundamentally a pipe bending vibration. For the second one year, the first cable model was replaced by the another cable model, which consisted from about 58 cable segments with 500 mm length and 160 mm diameter, attached to the steel wire with 30 m length and 20 mm diameter with the expect of much more sensitive system to vibration. Each segment consisted in aluminum pipe lapped by polyethylene membrane and about 10 mm gaps in between near-by segments were smoothly modified by the particular tape with the similar water repellency property with polyethylene material not to prevent formation of the upper water rivulet as much as possible. For the last 6 months, the aluminum pipe painted by the particular material with similar water-repellency with 30 m length and 150 mm diameter has been used. These three models are called as M1, M2 and M3, respectively. Figure 5 shows the picture of the pylon and the cable model (M1) at the site. The response was measured by 2-axis accelerometer installed on the cable model surface at the point from the down end by 2 m approximately. Wind velocity was also measured by near-by 3-components ultra-sonic anemometer at 10 m height from ground. The acceleration data of cable and the wind velocity data were stored in a digital recorder.

3.2 Response characteristics

Unfortunately, the intensity of turbulence at the test field is unexpectedly large, that is 20% to 30%, therefore, a typical rain-wind-induced vibration has seldom

Figure 5.　Cable setup.

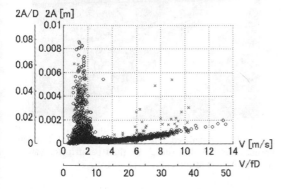

Figure 6.　V-A diagram (in-plane 2nd mode), where o: without rainfall, x: with rainfall (model M1).

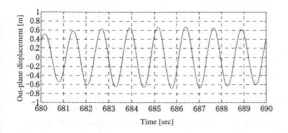

Figure 7.　Response of out-plane (1st mode, model M2).

been captured. However, as shown in Figure 6 (model M1), rain-wind-induced vibration clearly appeared at the characteristic reduced velocity at around 20, 30 or 40 through Wavelet analysis, by comparison of

responses under condition of the similar wind velocity and the similar wind direction but under different condition of with/without rain. On the other hand, model M2 showed the out-plane vibration with rather large amplitude during Typhoon Pabuk (T0111) passing through near-by the site, in 2001. The maximum instantaneous and 20 minutes average wind velocities were almost 13.2 m/s and 35.3 m/s, respectively, and the wind direction was ESE-SE, which means wind blew with 36° angle approximately and from the cable down-end to up-end. The rain volume was 37 mm/h at that time. This wind direction to inclined cable can't generate rain-wind-induced vibration at all because of no formation of upper water rivulet on the cable surface. The maximum double amplitude was about 1.36 m, as shown in Figure 7 and it is much larger than the evaluated one from buffeting analysis. Therefore, galloping instability was thought to appear.

4　AERODYNAMICS OF INCLINED CABLES IN WIND TUNNEL

4.1　Typical response characteristics – high speed restricted-response and galloping

The typical cross-flow response characteristics of inclined/yawed cable observed by wind tunnel tests are classified into Karman vortex-induced vibration, high speed restricted-response, which shows the velocity-restricted vibration at higher reduced velocity than the one related to Karman vortex, and galloping instability. These characteristics are rather sensitively influenced by the yawing angle (β) of wind to the cable axis, the existence of fixed upper rivulet and its position on the cable surface, Scruton number ($Sc = 2m \delta/\rho D^2$, m: mass per unit length, δ: logarithmic damping decrement, ρ: air-density, D: cable diameter), flow turbulent level, Reynolds number and the cable end condition. In particular, the rigid yawed cable model with $\beta = 45°$ shows galloping instability even though at the without rivulet state, as shown in Figure 8. Furthermore, it should be noted that the latent high speed restricted response can be observed at approximately $Vr = 40$ and/or 80 and so on by the typical aerodynamic damping features in V-A-δ diagram as shown in Figure 9 ($\beta = 45°$, without rivulet, $Sc = 1.14$, in smooth flow). The high speed restricted-response is observed under the condition of $\beta = 45°$, $\theta = 66°$ (θ: rivulet position measured from front stagnation point), $Sc = 1.10$ and in smooth flow (see Figure 10).

4.2　Exciting factors

Based upon a series of wind tunnel tests, galloping instability can be generated by the "axial flow" in near wake of yawed/inclined cable and the existence of the

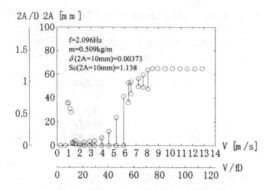

Figure 8. V-A diagram (in smooth flow, $\beta = 45°$, without water rivulet).

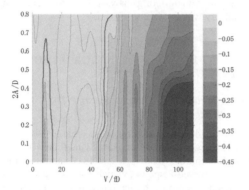

Figure 9. V-A-δ diagram (in smooth flow, $\beta = 45°$, without water rivulet).

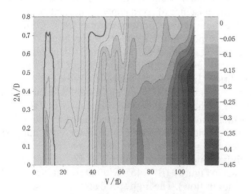

Figure 10. V-A-δ diagram (in smooth flow, $\beta = 45°$, $\theta = 66°$)

"upper rivulet", individually. Recently, it is reported that the unsteady movement synchronized to the cable motion must be substantial for the appearance of galloping or high speed restricted-response (ex. Xu & Wang 2003). However, it should be noted that galloping

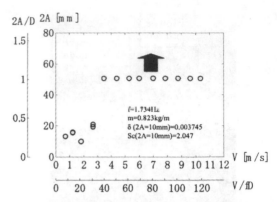

Figure 11. V-A diagram (in smooth flow, $\beta = 0°, \theta = 54°$).

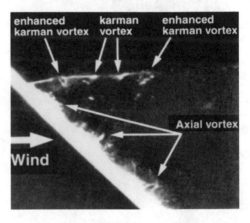

Figure 12. Visualized intermittently amplified Karman vortices and "axial vortices" by fluid paraffin for yawed ($\beta = 45°$) cable in smooth flow ($V = 1.0$ m/s).

instability could be observed under the without rivulet state and the fixed rivulet state. The role of the "axial flow" on galloping in-stability can be confirmed by the galloping appearance of non-yawed cable ($\beta = 0°$) by the artificial axial flow (Matsumoto et al. 1990), playing a similar effect like installation of splitter plate in a wake. On the other hand, the existence of upper rivulet at the particular position, despite fixed one, also can generate galloping instability, which is indirectly confirmed by V-A test for the non-yawed cable ($\beta = 0°$) with $\theta = 54°$ in smooth flow as shown in Figure 11. As far as high speed velocity-restricted response, the particular "low frequency fluctuating component" appears in the lift force, the pressure on cable surface and the velocity in a wake, in particular, at the up-stream end of yawed cable ($\beta = 45°$) and at the particular reduced velocity range, such as 20, 40 and so on (Matsumoto et al. 1999, 2001ab). This particular low frequency feature is surely caused by the low frequency flow fluctuation around cable, and this flow field can be

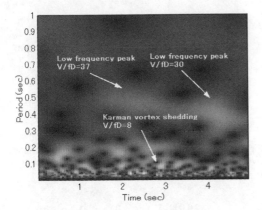

Figure 13. Wavelet analysis of fluctuating wind velocity in the wake of stationary cable model (without rivulet, $\beta = 45°$, $V = 4$ m/s, in smooth flow).

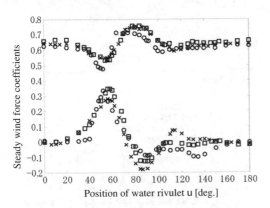

Figure 14. Steady wind force coefficients of cable model ($\beta = 45°$, $V = 8.0$ m/s, o: in smooth flow, □: in turbulent flow ($Iu = 5.7\%$), x: in turbulent flow ($Iu = 10.7\%$).

related to the particular vortex shedding with much longer period than the one of Karman vortex. The flow field was visualized as shown in Figure 12 (Matsumoto et al. 1998), in which Karman vortex is intermittently amplified. The low frequency fluctuation of the lift force or the velocity in a wake shows extremely unsteady characteristic as shown in the results based on wavelet analysis (see Figure 13). Therefore, these peculiar vortices must be related to Karman vortex shedding, more or less. The more detail discussion is shown in 4.4.

4.3 Role of upper rivulet and axial flow on the response and the steady force (lift and drag) and applicability of quasi-steady theory

The wind-induced vibration of yawed/inclined cable must be caused by "axial flow" in a wake and "formation of upper rivulet", as described above. The

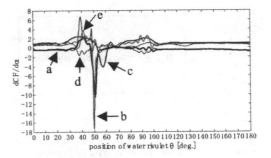

Figure 15. Slope of lateral force coefficient C_F against position of water rivulet (in smooth flow, $V = 8.0$ m/s), (a): $\beta = 0°$, with PSP, (b): $\beta = 0°$, with fixed PSP at $\theta = 180°$, (c): $\beta = 0°$, without PSP, (d): $\beta = 0°$, with PSP and fixed water rivulet, (e): $\beta = 45°$, without PSP.

steady lift and drag force coefficients (C_L and C_D) are characterized by the position of rivulet, defined by θ, for yawed ($\beta = 45°$) cable as shown in Figure 14. For non-yawed ($\beta = 0°$) cable as shown in Figure 15 (c), the lateral force coefficient ($C_F = C_L\cos a + C_D\sin a$, a: relative angle of attack) shows the negative slope between $\theta = 53°$ and $\theta = 61°$, where galloping instability might appear based upon quasi-steady theory. On the other hand, for yawed ($\beta = 45°$) cable, the question of how to take into account the effect of axial flow in a wake arises for application of quasi-steady theory. Because the cable motion can change the initial rivulet position (θ) by the motion-induced relative angle of attack α, but the position of axial flow should remain in a wake center during cable motion. Therefore, it is impossible to measure the steady forces (C_L and C_D) needed for the quasi-steady theory. For example, the yawed ($\beta = 45°$) cable without rivulet or with rivulet at $\theta = 0°$ or $\theta = 180°$ shows galloping instability, but the lateral force coefficient (C_F) at those situations without rivulet state, or with rivulet at $\theta = 0°$ and $\theta = 180°$, based upon the direct measurement of C_L and C_D, does not show negative slope. In this study, a 30% perforated splitter plate (PSP) was introduced in a wake of non-yawed cable to simulate the axial flow in a wake of yawed/inclined cable. Thus, it might be possible to measure the steady forces needed for application of quasi-steady theory. The lateral force coefficients (C_F) of non-yawed cable without rivulet and with rivulet at $\theta = 0°$ and $\theta = 180°$ show negative slope, corresponding well the galloping appearance of yawed ($\beta = 45°$) cable without rivulet and with rivulet at $\theta = 0°$ and $\theta = 180°$, shown in Figure 15(a). However, the V-A characteristics were not explained by use of quasi-steady non-linear analysis. The main reason might be significant effect of the non-stationary low frequency vortex flow on the quasi-steady flow. Further research is, of course, needed, but there are unsolved subjects in application of quasi-steady

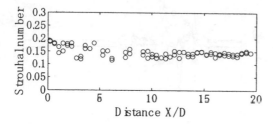

Figure 16. Strouhal number measured from unsteady pressure at $\theta = 135°$ along the cable axis (in smooth flow, $\beta = 45°$, $V = 4.0\,\text{m/s}$).

theory to explain the complicated wind-induced behavior of inclined cables.

4.4 Strouhal number and high speed restricted-response

A rigid yawed cable model similarly shows high speed restricted-response at the particular reduced velocity of 20, 40, 80 and so on with the prototype cable in natural wind. However, the question of where is corresponding part to the up-stream end of a rigid cable model in wind tunnel for a prototype cable remains, because the exciting cause of high speed restricted-response is almost produced there, as described above. Because prototype cable does neither have, of course, the boundary wall nor a window at the upstream cable-end, like wind tunnel test. On the other hand, axial velocity intensity changes along a cable axis, in particular, at near the up-stream end. This axial velocity variation produces the variation of Strouhal number at near upstream cable end as shown in Figure 16. Strouhal number variation along cable axis could generate a peculiar vortex with significantly longer period than that of Karman vortex. Then what about a prototype cable case? It is quite natural to image that the position of upper water rivulet is not constant along cable axis, but changes along cable axis, because of a prototype cable length, usually more than 100 m associated with rain-wind-induced vibration. Besides, it has been confirmed by wind tunnel test that Strouhal number of yawed cable ($\beta = 45°$) changes with the rivulet position as shown in Figure 17. Therefore, the low frequency fluctuating flow or vortex could also be produced by the variation of Strouhal number along cable axis, in similar with a rigid cable model case in wind tunnel. In order to prove this hypothesis, the fixed rivulet position was artificially changed for a non-yawed cable ($\beta = 0°$) in smooth flow by $\theta = 65°$, 55° and 65° as shown in Figure 18. The non-yawed cable with constant rivulet position of $\theta = 55°$ shows galloping instability, but doesn't at $\theta = 65°$. The Strouhal numbers of the cases at constant position of $\theta = 55°$ and $\theta = 65°$ are 0.18 and 0.16, respectively.

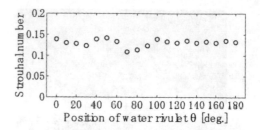

Figure 17. Strouhal number against position of water rivulet measured from unsteady lift force (in smooth flow, $\beta = 45°$, $V = 8.0\,\text{m/s}$).

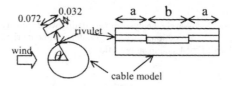

Figure 18. Position of artificial water rivulet.

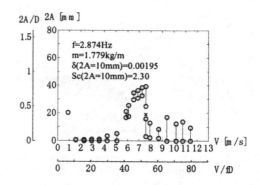

Figure 19. V-A diagram ($\theta = 65°\text{-}55°\text{-}65°$, 249 mm-167 mm-249 mm).

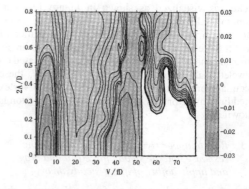

Figure 20. V-A-δ diagram ($\theta = 65°\text{-}55°\text{-}65°$, 249 mm-167 mm-249 mm, white part corresponds to indeterministic zone due to unstable response).

2350

This combination of two rivulet positions can generate low frequency velocity fluctuation and high speed restricted-response at around $Vr = 40$ as shown in Figure 19. Also, V-A-δ diagram is shown in Figure 20. This high speed restricted-response must be related to a peculiar vortex with longer period, thus it could be called as high speed vortex-induced vibration.

5 CONCLUSION

This paper describes the inclined cable aerodynamics highlighted to rain-wind-induced vibration in comparison of their response characteristics observed for prototype stay cables of cable-stayed bridges, a large-scale cable model in the field and a rigid yawed cable model in wind tunnel. The main conclusions are as follows: (1) rain-wind-induced vibrations are classified into high speed restricted-response caused by peculiar low frequency vortices and galloping instability, (2) majority of observed rain-wind-induced vibrations of prototype stay cables shows the former response type, however, a few violent vibrations seem to show the later type, (3) the wind-induced vibrations observed for a large-scale cable model in the field are mainly buffeting, because of high turbulent level of wind at that site, but the responses at the particular reduced velocity such as 20–40 at rainy days are significantly amplified, it mean rain-win-induced vibration might occur even in such high turbulent flow of 20–30% turbulent intensity, (4) Galloping instability might be observed for a large-scale cable model under precipitation and strong wind due to Typhoon, which may be the galloping instability type without rain, which is observed for a cable model in wind tunnel, taking into account of the relative wind direction to inclined cable, (5) It was verified by wind tunnel test that the aero-steady forces, such as lift and drag force, of yawed/inclined cables are characterized by both axial flow in near wake and the rivulet on cable surface, (6) the aero-steady forces for application to quasi-steady theory to analyze its aerodynamic response of yawed/inclined cable is discussed by use of a peculiar perforated splitter plate in a wake to simulate the axial flow position during cross flow vibration and (7) the wind velocity-restricted re-sponse at high reduced velocity range, around 20, 40 and so on, was discussed from the point of the non-uniform distribution of Strouhal number along cable axis, in consequence, it is implied that the unsteady and complicated influence of Karman vortices to this vibration. Therefore, it can be called as "high speed vortex-induced vibration".

Finally, the authors would like to acknowledge Professor H.Shirato, Mr. S.Sakai, Mr. J.Ohya, Mr. T.Okada and Mr. T.Oishi of Kyoto University for their contributions to site measurement and a series of wind tunnel tests. Some parts of this study have been carried out under the support of Grant-in-Aid for Scientific Research (A) No. 12305030 from Japan Society for the Promotion of Science.

REFERENCES

Matsumoto, M., Yokoyama, K., Miyata, T., Fujino, Y. & Yamaguchi, H. 1989. Wind-induced cable vibration of cable-stayed bridges. *Proceedings of Japan-Canada Joint Work-shop on Bridge Aerodynamics*, Ottawa, Canada: 101–110.

Matsumoto, M., Shiraishi, N., Kitagawa, M., Knisely, C.W., Shirato, H., Kim Y. & Tsujii, M. 1990. Aerodynamic behavior of inclined circular cylinders – cable aerodynamics. In M. Ito, M. Matsumoto & N. Shiraishi (eds), *Bluff Body Aerodynamics and Its Applications*: 63–72. Elsevier.

Matsumoto, M. 1998. Observed behavior of prototype cable vibration and its generation mechanism. In Larsen, Larose & Liversey (eds), *Bridge Aerodynamics*: 189–211. Rotterdam: Balkema.

Matsumoto, M., Yagi, T. & Tsushima, D. 1999. Inclined cables aerodynamics – velocity rsetricted response at high reduced velocity. *Proceedings of the Third International Symposium on Cable Dynamics*, Trondheim, Norway: 91–96.

Matsumoto, M., Yagi, T., Shigemura, Y. & Tsushima, D. 2001a. Vortex-induced cable vibration of cable-stayed bridges at high reduced wind velocity. *Journal of Wind Engineering and Industrial Aerodynamics* 89: 633–647.

Matsumoto, M., Yagi, T., Goto, M. & Seiichiro, S. 2001b. Cable aerodynamic vibration at high reduced velocity. *Proceedings of the Third International Symposium on Cable Dynamics*, Montréal, Canada: 43–50.

Main, J.A., Jones, N.P. & Yamaguchi, H. 2001. Characterization of Rain-Wind Induced Stay-Cable Vibration from Full-Scale Measurement. *Proceedings of the Forth International Symposium on Cable Dynamics*, Montréal, Canada: 235–242.

Xu, Y.L. & Wang, L.Y. 2003. Analytical study of wind-rain-induced cable vibration: SDOF model. *Journal of Wind Engineering and Industrial Aerodynamics* 91: 27–40.

System-based Vision for Strategic and Creative Design, Bontempi (ed.)
© *2003 Swets & Zeitlinger, Lisse, ISBN 90 5809 599 1*

Database-assisted design, structural reliability, and safety margins for building codes

S.M.C. Diniz
Federal University of Minas Gerais, Belo Horizonte, Brazil

E. Simiu
National Institute of Standards and Technology, Gaithersburg, MD, USA

ABSTRACT: We present a database-assisted design (DAD)-compatible probabilistic procedure for estimating (a) wind effects for any return period and (b) wind load factors. Unlike earlier reliability estimates, we account for the randomness of peak effects and for knowledge uncertainties. We found that, in contrast to load factor estimates based on the Extreme Value Type I distribution of the wind speeds, estimates based on the Type III distribution of the largest values yield load factor estimates consistent with accepted standard values. The ratio between 3-s and 1-hr wind speeds, which comes into play when converting meteorological data to wind tunnel reference wind speeds, can affect the estimates significantly. An increase of the extreme wind data set from, say, 30 to 50 years, has a marginal effect on the estimates. Research on the effect of terrain roughness uncertainties for buildings in built-up terrain may require an expansion of the aerodynamic databases currently available. Consideration of uncertainties in isolation, as commonly performed by some wind tunnel operators, especially for peak effects estimation, can be misleading; for structural applications, uncertainties need to be considered collectively.

1 INTRODUCTION

For codification and design purposes we are interested in the estimation of peak wind load effects F'_{pk} (τ, N) (i.e. wind-induced bending moments, axial forces, and shear forces in main wind-load resisting systems such as frames) corresponding to a storm of duration τ (typically taken to be one hour) and a mean recurrence interval of the wind speed, N. The peak wind load effects, in turn, depend upon the peak load effect coefficient C_{pk} induced by a unit wind speed at some reference height, and the wind speed V upwind of the structure at that reference height. In engineering practice and standards various levels of refinement have been used in the estimation of the wind load effects $F_{pk}(\tau, N)$. These levels differ in the way in which the peak load effect coefficients C_{pk} and the wind speeds V are defined as functions of the uncertainties affecting them.

The pressure diagrams used in the ASCE 7 Standard (ASCE, 1998) were developed from wind tunnel measurements for a small number of building shapes. The time series of these pressures, weighted by the respective tributary areas pertaining to a frame (viewed as typical) and by the respective influence coefficients

for the knee of that frame, were summed up on line to yield the peak bending moment at the knee. A similar procedure was used to obtain peak uplift force. The pressure diagrams were developed "by eye" so that they produce in the frame a bending moment at the knee and an uplift force roughly equal to their counterparts obtained on line. The wind speed corre-sponding to a given mean recurrence interval was obtained in the ASCE 7 Standard by assuming, for non-hurricane winds, the validity of the Extreme Value (EV) Type I distribution, and disregard significant knowledge uncertainties.

A more elaborate approach uses the database-assisted design (DAD) procedure (Whalen et al. 1998). This eliminates large sources of errors affecting the ASCE 7 Standard and therefore results in far more accurate, differentiated, and risk-consistent standard provisions.

We report in this paper progress in our development of a reliability-based design procedure (DA-RBD) that extends DAD capabilities and incorporates, within a probabilistic framework, the effects of uncertainties in the definition of wind effects. The procedure accounts for all the relevant random uncertainties (including those associated with the extreme wind speeds and the peak wind effect coefficients C_{pk}), and knowledge uncertainties (climatological, including those associated with sampling errors in the estimation of extreme wind speeds and aerodynamic peaks, and those of a micrometeorological type, e.g., associated with terrain roughness lengths). We use and complement within a DAD framework approaches and results developed in (Diniz et al. 2002).

Once the probabilistic descriptions of the wind load effects corresponding to specified mean recurrence intervals have been obtained, wind load factors that account for all the relevant uncertainties can be estimated. We present numerical examples which show that knowledge uncertainties neglected in the derivation of load factors in ASCE 7 Standard have a considerable effect on the estimation of wind effects and wind load factors. We also summarize results of sensitivity analyses of the wind load factors to the assumed statistics of the basic random variables.

2 WIND LOAD EFFECT MODEL

For a storm of duration τ corresponding to a mean recurrence interval N the wind load effect can be written as

$$F_{pk}(\tau, N) = 0.5 \, \rho \, C_{pk}(\tau) \, V^2(\tau, H_{aero}, z_0, N) \quad (1)$$

where ρ is the air density, $C_{pk}(\tau)$ is the peak wind effect coefficient for a wind effect time history of duration τ, and $V(\tau, H_{aero}, z_0, N)$ is the wind speed with a mean recurrence interval N, averaged over time τ, at elevation H_{aero} in terrain with roughness length z_0. The peak factor $C_{pk}(\tau)$ and the wind speed $V(\tau, H_{aero}, z_0, N)$ are random variables, which are themselves functions of several other random variables. Moreover, as mentioned earlier, both the peak factor and the wind speed are affected by knowledge uncertainties and sampling errors that must be included in the probabilistic modelling of wind load effects. If the probabilistic descriptions of all the variables involved, as well as their functional relationships, are known, Equation 1 may serve as the basis of a Monte Carlo simulation procedure for the generation of samples of the wind effect $F_{pk}(\tau, N)$. The requisite statistical information for the computation of load factors may then be obtained from the generated samples. The probabilistic descriptions of the wind speeds and the peak coefficients are presented in the following sections.

3 KNOWLEDGE UNCERTAINTY MODELS

3.1 Wind speeds

To describe the wind speeds $V(\tau, H_{aero}, z_0, N)$ we first need to estimate, by accounting for sampling errors,

the speed $v(T, H_{met}, z_{01}, N)$ with a mean recurrence interval N, averaged over time T ($T = 3$ s), at standard elevation H_{met} in open terrain with roughness length z_{01}. To obtain wind speeds V from the wind speeds v, micrometeorological information, with its attendant uncertainties, must be included in the procedure. The following expression is used to relate the wind speeds V and v:

$$V = u_H \, t(\tau, T) \, q \, v \left(\frac{s \, z_0}{u \, z_{01}} \right)^{d\delta} \frac{\ln\left(\dfrac{H_{aero}}{s \, z_0}\right)}{\ln\left(\dfrac{H_{met}}{u \, z_{01}}\right)} \quad (2)$$

where the random variables d, s, and u represent the uncertainties with respect to the actual values of the empirical constant $\delta = 0.0706$ and of the roughness lenghts z_0 and z_{01}, respectively; q denotes wind speed observation errors; $t(\tau, T)$ is the wind speed conversion factor from averaging time T to τ); u_H reflects the uncertainty with respect to the applicability of this model to hurricane wind speeds (for extratropical storm winds u_H is, by definition, unity).

3.2 Peak coefficients

For low-rise structures, the peak coefficients are in general non-Gaussian. Their probability density function is obtained by using the entire information inherent in the time series of the wind effect. Since the length of the record is finite, the distribution of the peaks will be affected by sampling errors. For 1-hr long records, typical sampling errors in the estimation of the peaks are about 5%. For shorter records that study found that for 30-min or 20-min records the sampling errors are about 1.5 times or twice as large, respectively Sadek & Simiu (2002). This uncertainty can be accounted for by multiplying the peak with the factor e, a normal random variable with unit mean and standard deviation suggested, on the basis of numerical simulations, in Sadek & Simiu (2002).

3.3 Aerodynamic effects

Knowledge uncertainties in the estimation of the aerodynamic effects are modeled as random variables a, b, and c, where a reflects aerodynamic errors inherent in the wind tunnel being used; b reflects aerodynamic errors due to the use of wind tunnel rather than full-scale measurements; c reflects errors in the transformation of aerodynamic effects into stresses or other structural effects. For Monte Carlo simulation purposes, in Equation 1 we therefore substitute for C_{pk} the product $abce \, C_{pk}$.

3.4 *Wind load effects*

Probabilistic descriptions of wind load effects are thus obtained by simulation using the following relationship:

$$F_{pk}(\tau, N) = 0.5 \, \rho \, a \, b \, c \, e \, C_{pk}(\tau) \, V^2(\tau, H_{aero}, z_0, N) \tag{3}$$

where a, b, c, and e were defined earlier and the wind speed V is given by Equation 2. A proposed probabilistic description of all the random variables is summarized in Table 1, which displays their assumed statistics (mean, standard deviation, coefficient of variation (c.o.v.), and type of distribution). The Monte Carlo simulation procedure was implemented in the MATLAB-based software WiLEP (Wind Load Effects, Probabilistic description). WiLEP generates samples of F_{pk} and other related variables, and is a component of a broader NIST project aimed at developing procedures, including user-friendly software, for the reliability assessment of low-rise structures subjected to wind loads estimated principally by using the DAD approach.

In the next sections our approach is illustrated by numerical examples on the probabilistic description of wind speeds and of wind load effects, including the effects of knowledge uncertainties. Wind speeds are assumed to have Extreme Vale distribution. We examine the cases of EV Type I and EV Type III distributions of the largest values. Expressions for these distributions and for the sampling errors in the estimation of the extreme wind speeds are given, e.g., in (Minciarelli et al. 2001), which is closely followed throughout this paper.

4 NUMERICAL EXAMPLE

The results presented in this section are dependent on the assumed statistics of the basic variables related to the wind speeds, some of which must be based on judgment. Table 1 summarizes the probabilistic descriptions of random variables assumed in the calculations for this section.

We calculated wind load effects at the knee of a frame of a low-rise steel building in open terrain with 6.1 m eave height, described in (Whalen et al. 1998), and assumed to be located near Helena, Montana. The aerodynamic pressures were measured at UWO at about 500 pressure taps (Lin & Surry 1997). The time histories of the internal forces were computed by using the software described in (Whalen et al. 2000). It was assumed that the length of the climatological data set is 50 years, and that the mean and standard deviation of the wind speeds at 10 m elevation in open terrain are 20 m/s and 2.10 m/s, respectively. The simulated bending moments were derived for a 1-hr storm duration by using a 1-hr (prototype) wind tunnel record length. The results of the calculations are shown in

Table 1. Assumed statistics of random variables reflecting uncertainties in estimation of wind effects.

Basic var.	Mean	SD	COV	Distr.
a	1.0	0.05	0.05	Norm. (N)
b	1.0	0.05	0.05	Norm.
c	1.0	0.025	0.05	Norm.
d	1.0	0.10	0.10	Trunc. N
e	1.0	*	*	Norm.
C_{pk}	**	**	**	Type I
s	1.0	0.30	0.30	Trunc. N
t	***	–	0.05	Norm.
u	1.0	0.30	0.30	Trunc. N
u_H†	1.0	0.05	0.05	Norm.

* Sampling errors in estimation of $C_{pk}(\tau)$ depend on τ and the length of the wind tunnel record. For $\tau = 1$ hr, and record lengths 20, 30, and 60 min, the corresponding COV's may be taken as 0.09, 0.07, and 0.05, respectively (Sadek & Simiu 2002).
** Values obtained in WiLDE-LRS02.
*** Depends on averaging times T and τ e.g. for $\tau = 1$ hr and $T = 1$ min, mean(t) = 0.8065 (Diniz et al. 2002).
† u_H is unity for non-hurricane winds.

Figs 1 and 2 for $N = 50$ yrs and 500 yrs, under the assumptions that the extreme wind speeds have EV Type III and EV Type I distributions. Note the significant effects of accounting for uncertainties (i.e. the large dispersion of the histogram reflecting them), on the one hand, and of the assumed distributions of the extreme wind speeds, on the other.

Assumptions on uncertainty distributions may be modified as judged necessary by standard-writing bodies or other interested parties.

5 WIND LOAD FACTORS

In the ASCE 7 Standard the following expression is used for the wind load factors:

$$LF_{(wind)} = \left(\frac{500 - yr \, wind \, speed}{50 - yr \, wind \, speed} \right)^2 \tag{4}$$

where the 500-yr and 50-yr wind speeds are point estimates, and uncertainties in the peak load effect coefficients are neglected, as are knowledge uncertainties and sampling errors associated to the estimation of wind speeds. We computed load factors that account for the uncertainties in the climatological, micrometeorological, and aerodynamic parameters, using the expression

$$LF_{(wind)} = F_{pk}(\tau, N_2)_{p2} / F_{pk}(\tau, N_1)_{p1} \tag{5}$$

where $F_{pk}(\tau, N_2)_{p2}$ is the $p2$-percentage point of the peak internal forces corresponding to mean recurrence interval N_2; and $F_{pk}(\tau, N_1)_{p1}$ is the $p1$-percentage

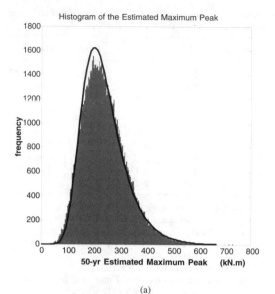

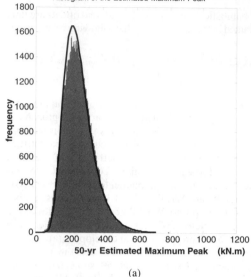

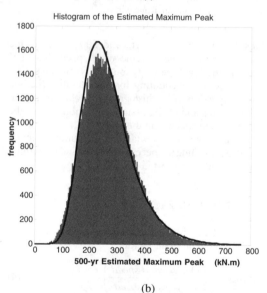

(b)

Figure 1. Histograms of: (a) the 50-yr, and (b) 500-yr estimated maximum moment peaks based on an EV Type III distribution ($c = -0.2$).

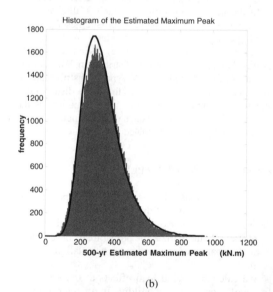

(b)

Figure 2. Histograms of: (a) the 50-yr, and (b) 500-yr estimated maximum moment peaks based on an EV Type I distribution ($c = 0$).

point of the peak internal forces corresponding to mean recurrence interval N_1. For example, if $N_2 = 500$, $N_1 = 50$, $p2 = 0.90$ and $p1 = 0.50$, as was proposed by Ellingwood et al. (1980), the load factor is computed as follows:

$$LF_{(wind)} = F_{pk}(\tau, 500)_{90\%} / F_{pk}(\tau, 50)_{50\%} \qquad (6)$$

where 50% and 90% denote the 50- and 90-percentage points, respectively. Note that the ASCE 7 Standard uses wind load factors that are dependent upon the parameters of a "generic" (typical) building used for the estimation of what are referred to in the Commentary to the Standard as "pseudo" pressure coefficients listed in the Standard (see, e.g., ASCE 7 Standard-95,

p. 162; the same approach is implicit in later versions). The load factors derived in this paper also depend upon building parameters, except that those parameters pertain not to a "generic" building, but rather to the actual building being analyzed.

Wind load factors computed by Equation 6 are summarized below for the moments at the knee of a frame in the building considered in Section 4. It was assumed c.o.v.$(t) = 0.075$, s.d.$(s) = $ s.d.$(u) = 0.3$ (open terrain), with the other variabilities taken from Table 1. The wind load factors were found to have the values 1.79, 1.64, and 1.54, under the assumptions that the extreme wind speeds have an EV Type I distribution, an EV Type III distribution with tail length parameter $c = -0.1$, and an EV Type III distribution with $c = -0.2$, respectively. It was found that the parameters that had a significant influence on the results were c.o.v.(t) and the tail length parameter of the EV distribution of the extreme wind speed (recall that, in the limit a Type III distribution of the largest values converges to the EV Type I distribution as c becomes vanishingly small). It is noteworthy that the length of the wind tunnel record (between 20 min and 60 min), the size of the extreme wind speeds sample (between 30 and 50), and the standard deviation of the error in the estimation of the open terrain roughness lengths (between 0.1 and 0.3) had no significant effect on the results.

Our calculations also confirmed that estimated load factors are higher for hurricane than for non-hurricane regions. For example, for $n = 50$, $T_1 = 60$ min, $\tau = 60$ min, $sd_s = 0.3$, c.o.v.$(t) = 0.05$, and under the reasonable assumption $c = -0.2$ (see Heckert et al. 1998), the computed load factors are 1.49 and 2.00 for the non-hurricane and hurricane regions, respectively. The ASCE 7 Standard accounts for such differences by keeping the nominal load constant throughout the United States and adjusting hurricane wind speeds upward in its wind speed map. Note that in Equation 2 we also include an additional uncertainty, embodied in the variable u_H, that accounts for the modelling errors inherent in the simulation of hurricane wind speeds.

6 SUMMARY AND CONCLUSIONS

1. The type of Extreme Value distribution that best fits extreme wind speeds has a significant effect on the magnitude of estimated load factors. For non-hurricane regions, estimates based on the assumption that the distribution is of the EV Type III distribution yield load factors that are consistent with values specified by building codes and judged by broad consensus to be appropriate for design purposes. This assumption is supported by results of modern extreme value estimation methods that were not used for structural engineering applications until the 1990's.

Ellingwod et al. (1980) remarked that reliability with respect to wind loads appears to be relatively low when compared to that for gravity loads, according to the simplified methods for calculating safety indexes then in use. (These methods assumed the validity of the EV Type I distribution as a model of the extreme wind speeds.) Ellingwood et al. nevertheless decided, in our opinion judiciously, not to propose an increase in the wind load factors specified by the predecessor of the ASCE 7 Standard. In contrast with reliability calculations performed on or before 1980, our calculations are consistent with wind load factor values specified in the ASCE 7 Standard, while accounting in what we believe is a far more realistic manner for all uncertainties affecting the estimation of wind effects.

Our calculations also confirmed that estimated load factors are larger for hurricane-prone than for non-hurricane regions.

2. The variability in the conversion ratio t that accounts for wind speed averaging times, e.g., the conversion from a 3-s to 1-hr averaging time, can have an important effect on the estimated wind load factor. This suggests that additional research is necessary on the statistics of ratio t, and that to the extent that they are available, sustained wind speed data, which have lower variability than peak gusts, should be made use of as much as possible.

3. An increase in the size of the extreme wind data set from, say, 30–50, has a relatively marginal effect on the estimated values of the load factors.

4. For buildings in open terrain load factors are typically affected relatively weakly by even large uncertainties with respect to the roughness lengths of the open terrain over which the wind speed data are measured at weather station sites, and of the terrain upwind of the building. Additional research is needed, however, to cover buildings in built-up terrain. Such research may require an expansion of the aerodynamic databases currently available.

5. The relatively weak effect of uncertainties associated with typical lengths of wind tunnel records, terrain roughness lengths for buildings in open terrain, and with speed sample sizes, is due to the fact that, in the overall reliability scheme underlying our estimates, these uncertainties are drowned by other, more dominant uncertainties. For example, the variability of the peaks of a given record may be overwhelmed by the variabilities of the ratio of 3-s to 1-hr speeds and of the extreme wind speeds. This observation is obvious for structural reliability practitioners. Nevertheless, it has not always been taken into account in wind engineering practice, where criteria based on very high quantiles of the peak load effects distribution have been used, without regard for the interaction among the variabilities of the relevant factors affecting wind load effects.

ACKNOWLEDGMENTS

We thank M. Grigoriu for helpful and stimulating interactions, and Fabio Minciarelli for the work that allowed the development of the software used in this paper. S.M.C. Diniz worked on this paper during her tenure as a Guest Researcher at the National Institute of Standards and Technology.

REFERENCES

American Society of Civil Engineers, 1998, ASCE Standard ASCE 7–98, *Minimum Design Loads for Buildings and Other Structures*, American Society of Civil Engineers, Reston, Virginia.

Whalen, T., Simiu, E., Harris, G., Lin, L. & Surry, D. 1998, The Use of Aerodynamic Databases for the Effective Estimation of Wind Effects in Main Wind-Force Resisting Systems: Application to Low Buildings, *Journal of Wind Eng. Ind. Aerodyn.*, 77–78, pp. 685–693.

Diniz, S.M.C., Sadek, S. & Simiu 2002, Wind Speed Estimation Uncertainties: Effects of Climatological and Micrometeorological Parameters, *4th Computational Stochastic Mechanics Conference*, Corfu, Greece, June 9–12, 2002 (in press).

Sadek, F. & Simiu, E. 2002, Peak Non-Gaussian Wind Effects for Database-Assisted Low-Rise Building Design, *Journal of Engineering Mechanics*, 128, pp. 530–539.

Minciarelli, F., Gioffrè, M., Grigoriu, M. & Simiu, E. 2001, Estimates of Extreme Wind Effects and Wind Load Factors: Influence of Knowledge Uncertainties, *Probabilistic Engineering Mechanics*, 16, pp. 331–340.

Lin, J. & Surry, D. 1997, *Simultaneous Time Series of Pressures on the Envelope of Two Large Low-Rise Buildings*, BLWT-SS7-1997, Boundary Layer Wind Tunnel Laboratory, University of Western Ontario, London, Ontario, Canada.

Whalen, T.M., Shah, V. & Yang, J.-S. 2000, *A Pilot Project for Computer-Based Design of Low-Rise Buildings for Wind Loads – The WiLDE-LRS User's Manual*, NIST/ TTU/Purdue University Contractor Report, National Institute of Standards and Technology, Gaithersburg, Maryland.

Ellingwood, B.R. et al., Development of Probability-based Load Criteria for American National Standard A58, NBS Special Publication 577, National Bureau of Standards, Washington, DC, June 1980.

Heckert, N.A., Simiu, E. & Whalen, T. 1998, Estimates of Hurricane Wind Speeds by "Peaks Over Threshold Method", *Journal of Structural Engineering*, 124, pp. 445–449.

System-based Vision for Strategic and Creative Design, Bontempi (ed.)
© 2003 Swets & Zeitlinger, Lisse, ISBN 90 5809 599 1

Wind action and structural design: a history

P. Spinelli

Civil Engineering Department, University of Florence, Florence, Italy

ABSTRACT: In the following a short, non strictly scientific, history of the relation of wind and structures is reported. Wind and structures, that in ancient centuries have different field and meanings, in recent times have become closer and closer. Wind is now one of major design loads for tall buildings, for long bridges, for large roofs and for energy wind structures installations. The history of the study of wind and its effects on structures is here delineated underlining the significant steps of development.

1 INTRODUCTION

History of wind is usually not associated with structures. Also in common language, the conception of wind is hardly associated to catastrophic events, ruins or structural problems. On the contrary, and especially in climatic conditions that do not favor the tornadoes and hurricanes formation, like Italian ones, it is very often associated to the bioclimatic: psychological and physical reactions of personal comfort or discomfort, irritation or pleasantness, cold sensitiveness or motion difficulties.

In case, people more frequently think to minor damages to objects, roofs, windows, trees, etc.

A very impressive date can be observed looking to some 2002 statistics (Munich Re 2003). In those the wind storm and the floods are at the top of the year's list of natural catastrophes with just under 500 events (previous years: 450) and also dominate the insurance claims burdens (99% of the insured natural catastrophe loses; previous year: 92%).

Also in consideration of the entity of the wind damages, wind storms and hurricanes are comparable with every other causes. The consideration that this natural

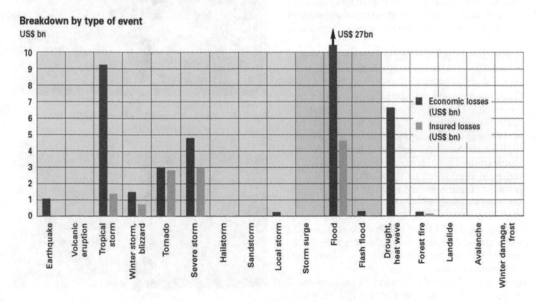

Figure 1. Economic and insured losses in the year 2002 reported by Munich Re Insurance (Munich Re 2003).

event is felt like less catastrophic than other natural events it is essentially due to the fact that the average damage caused by the wind for each event is relatively small. Therefore the wind creates damages that are diffused on a wide area and are much more frequent than other catastrophic events like for instance earthquakes.

Nevertheless, thinking to negative effects of the wind on buildings, rarely it's thought as important constructions in engineering. At most it is considered some modest structures like road signs, or to very light structures (like the straw house in the Three little Pigs fable, that could be swept away by the breath of the wicked wolf); but in archives it is also possible to find remarkable documents of eclatant wind damages (see Figs 2–4).

2 THE WIND IN THE ANCIENT WAY OF THINKING

Actually, since ancient time, the most important structures had very often problems in bearing their own weight instead of facing the strength of the wind. Therefore the problems to solve were linked to gravity laws and not to natural events. Structural disasters were linked to very rarely wind damages, but they were nearly always connected to wrong materials, foundations failures, fire or earthquakes. In ancient centenary, this was very true as the weight of natural (stone, masonry) and the low height of the building let wind effect not effective at all in constituting any statical problem to the construction itself. Lateral force due to wind are in fact up to an order of magnitude less than the vertical force of self weight or lateral force due to arch trusses. And again it is well known that stresses of these major causes results in stresses of an order of magnitude less than the material (stone) compressive strength.

So The Babele Tower (see Fig. 5) (a ziggurat construction built up in 562 a.C., that in the Christian tradition was built up by Noe's descendants "to reach the sky") did not have any problems due to wind action.

The Busink Construction, that was reconstructed thanks to the indications of Erodoto, who saw its ruin in 460 a.C., was in plan a square building with its sides of 91.60 mt and 80 mt tall and stood inside Babilonia city between the Tigri river and the Eufrate river.

And even the Pharos of Alexandria (see Fig. 6) (huge square building covered by white marble began in 305 a.C. and finished in 280 a.C. thanks to Alessandro Magno's lieutenant-general, Sostrato, located in Pharos Peninsula), was 120–140 mt tall. It collapsed because of its burden and degradation in 1326, spared by Giulio Cesar in 47 a.C. and by Arabs in 646 d.C. (1600 years after its construction).

However the studies on the wind, in meteorological sense, began with the ancient Greeks. The Andronico Tower or the Tower of the winds Fig. 7, octagonal

Figure 2.   Hurricane Iris, Belize October 2001.

Figure 3.   This is a wind turbine in northern Germany that was buckled by the storm Jeanett (2001).

Figure 4.   Roofs damaged by a storm wind.

Figure 5.   Representation of the Babele Tower (562 a.C.).

building on which are represented wind divinities) stands nowadays in Athens and represent through human figures different wind (for instance Zephiros, young male dressed up with a mantle of flowers – divinity of the western wind – and Boreas, old male dressed up with heavy clothes to warm himself – divinity of the northern wind).

Aristotle wrote in his famous work "Meteorologica" about the essence and the characteristics of winds, and his concept of wind starts with incorrect interpretation of air motion and humidity of the air. In particular he interpreted the wind like an exhalation coming from the earth and due to the heat of the sun. He says that we recognize two kinds of evaporation, one moist, the other dry. That in which moisture predominates is the source of rain, while the dry evaporation is the source and substance of all winds. Aristotle moreover, defines many types of wind depending on the zone and on direction and states on erroneous conclusion. He says that is "plainly false" that the wind is a "motion of the air" comparing the wind flow in more directions to the system of the rivers on the earth. If one asserts that all the winds are one wind, it is just like thinking that all rivers are one and the same river.

But already before Aristotle, Kahun, city of the ancient Egypt built in 2000 a.C., was constructed in order to locate the richest quarters on the best side of the wind, the coolest, and the poorest quarters on the worst, the warmest (see Fig. 9).

Later, Vitruvio, in the first century a.C., indicated the right places to build up the cities according to the wind position and to the bioclimatic rules. According to Vitruvio's indications the cities did not have to be

Figure 6.   Pharos of Alexandria (305 a.C.).

Figure 7.   Andronico Tower or Tower of the Winds.

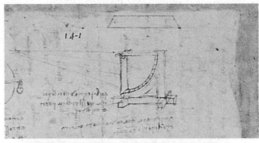

Figure 8.   "Leonardo da Vinci" pendulum-anemometer his drawing.

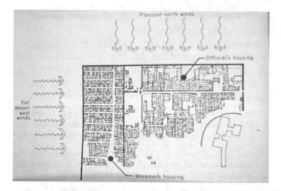

Figure 9. Plan of Kahun, city of the ancient Egypt built in 2000 a.C.

Figure 10. Old windmill with horizontal axes in Pakistan.

constructed against the dominant wind, because "if the roads run against the dominant winds, the winds, that come from the open country, are canalized in narrow places and sweep away the cities".

And during the South America colonization the new Spanish cities are built paying much attention to the specific bioclimatic rules. For instance, the streets in Buenos Aires are perpendicular to one another but oblique to the dominant winds.

The most important studies on meteorology took place from the Middle-Age on, from Leonardo da Vinci, who created the first pendulum-anemometer (see Fig. 8), and studied first hydrodynamics and vortical motion of fluids in his work "Del Moto e misura dell'acqua".

3 WIND AND "WIND ENERGY" STRUCTURES

In speaking of effect of wind on structures let us start with a not conventional concept of structures that are structures for machines that utilized the energy of wind. Probably this is the first connection between wind and structures. It is to underline the fact that in most famous XIX century treaties of architecture books wind are neither mentioned. In Valadier (1832) only one chapter of a total of 79 is entitled "Wind Force" and applies to wind force on windmills reporting some studies of Smeaton on optional blade inclination.

The horizontal forces due to wind action are on the other hand well known in the past due to the sailing technical ability. From ancient times the effectiveness of wind to give translation force to a ship is well known and have conducted the ship travels that develops human civilization. Anyway forces due to wind were known in a sort of dynamic sense, and in fact the utilization of wind energy by means of windmills had been going on for about 4000 years. This development started with windmills with vertical axes and only later in the Middle Age wind

Figure 11. Post-mills: the rotor mechanism moves on a circular track to face the mill into the wind (England 1803).

wheels changed to a horizontal axis to achieve higher efficiency.

The first vertical axes windmills (called also "post-mill", Fig. 11) were a rudimental construction in which the whole body of the mill was moved, by the miller, around a large vertical post when the wind direction changed.

The studies about the relation between the wind velocity and the pressure on the surface were developed firstly by the British engineer John Smeaton. He proposed a constant of proportionality 0.005 (with units in lb/ft^2) between the resistance force and the square of velocity of the object (Smeaton coefficient).

4 THE WIND, THE BIG DISASTERS AND THE DEVELOPMENT OF THE LARGE BRIDGE CONSTRUCTION

Rarely, we find any trace, in ancient chronicle, of negative effects of the wind on constructions.

Perhaps one trace is that one we find in a English news (Chronical Anglie) which regarding to a wind storm on 15th January 1362 (known like St. Mary's wind) reports: "... a wild wind blew with such a strength that destroyed tall houses and buildings, towers, bell-towers, trees and other resistant and durable things ...".

In Italy, Poleni recalls the whirl that caused on 17th August 1756 the ruin of the "Palazzo della Ragione" (the Justice Palace in that time) with a wooden covering composed of arches, with dimension in plan of 35 for 85 mt (see Fig. 12). Poleni dismayed tells that "... a bulk of large solidity and magnificiently built like that, and during many times adorned, was the object of our right obligingness and of the admiration of visitors ... in that baneful day the impact of the whirl collided violently with the magnificent vault of Hall, so that, in spite of all chains and steel spikes, the wind detached it from the walls, and a large part of it ruined on the floor and the other fell down on the arches of the Northern Lodge ... among so many ruins (nevertheless) no person was injured ...".

A witness of the rests of that terrible ruin can be found in a Fossati's drawing (see Fig. 12).

It is not mere chance that the worst damages can be remembered only in very light structures and coverings.

But with inventions of the modern science in construction materials, lighter and more resistant than ever, the action of wind becomes more and more important for building the right structure.

But as often happens in engineering, one has unfortunately the knowledge of a structural problem only after big disasters, in structural engineering this happens with bridges disasters.

A very impressive sequence of structural collapses, happened directly or indirectly because of wind, took place in the last century. In particular, during the first part of the last century, many bridges collapse took place in Europe. Von Karman lists 7 bridge collapses in 35 years (from 1818 to 1853), that happened almost in England. The bridges typology regards almost ever suspension or cable stayed bridges, in which the slenderness of the suspended deck contrasts with heavy piles of stones.

The skeptic-like attitude of engineers with regard to suspension Bridges is reported in the same architectural treatise book of Valadier (1832). He cites a

Figure 12. "Il Paazzo della Ragione di Padova" after the wind storm.

Figure 13. Drawing of the Brighton Chain Pier Bridge (Britan) reported by Colonel Reid (1836).

famous bridge engineer I. Cordier who states that suspension bridge is not at all advisable either for durability or for reliability in comparison with wood or stone bridges. He says that also economically is not sustainable, and criticizes very much the type of construction because of its vulnerability. He reminds that vibration caused by a group of cows in United states, or by three person in Britain or a "wind hit" was sufficient to destroy some of the first suspension bridges.

The most famous suspension bridge in that time in Europe is the Menai Bridge in Wales, made by Thomas Telford, with a span 177 mt long and still perfectly maintained nowadays. The collapse of the Brighton Chain Pier in Brighton (England) happened the 30th November 1836 and is documented by painting and drawing of the age. The Brighton Chain Pier was a suspension bridge with span length of 78 m that, and as the Colonel Reid reported, started to rock in considerable way during a wind storm (Fig. 13). In Britain, in the second half of XIX century, the construction of two strategic bridge was undertaken: the first one was the bridges across the Tay river close to Dundee, in Scotland, and the second one was the bridge on Forth river, close to Edinburgh. Such bridges had to connect the land divided by the firths of two rivers, (the Forth close to Edinburgh and the Tay close to Dundee).

The bridge on the Tay opened in 1877 and was at that time the bridge with longest development of the world. It was composed by 84 spans of steel truss beams 60 m long. The bridge was designed and directed during the construction by Sir Thomas Bouch, one of most eminent engineer of the time, to whom the design of another important suspension bridge across the Forth river was entrusted. The night of 29th December 1879 during a wind storm the Tay Bridge collapsed while a train with six carriages was crossing it, causing 75 victims. The following investigation established that the designer hadn't held on account in any way the action of the wind, that was thought not important for civil engineering structures.

The management of the project of the Forth Bridge, close to Edinburgh, was withdrawn to the designer and it came assigned to engineers Baker and Fowler. The first one was a young researcher who was considered (in the technical environment) how one "builds bridges only on the paper". The other one was a known and famous engineer, since he designed the London Underground Tunnels and many other hydraulics work in Egypt.

Baker, conscious of the importance of the action of the wind, made some experiments, measuring the action of the wind on panels of several dimensions and he did a important discovery: the wind pressure decreases to increasing of the exposed surface. In fact, the effect of the non contemporaneity of the wind action did not render the maximum peaks contemporaneous on the all surface.

However the precaution that was suited at the moment carried Baker not to hold account of this effect that would have diminished a lot the design load; he designed the structures with a lateral wind pressure very high, about the double of the pressure taken into account in the present British Standard Code. This fact clearly appears by impressiveness of the bridge that is characterized by the particularly static behaviour of cantilever connected by Gerber beams (see Figs 14 and 15).

However for more than a century, because of the disaster happened, in Europe the diffusion of the suspension and cable stayed bridges was practically given up. On the contrary in the United States, since the first bridge across Niagara Waterfalls in the US and specially after the invention of steel cables by Roebling, the suspension bridge technique spread and developed rapidly. The Niagara bridge was a structure in steel cables and wood, and designed by the Roebling who

Figure 14.   Firth Bridge (1889).

Figure 15.   Model utilized by Baker to explain the static behaviour of his bridge (1889).

designed even the Brooklyn Bridge in New York (1883 span length 480 m). Nevertheless also in US had been some sensational collapse due to the wind pressure. Between the most clamorous collapses we remember the Niagara Waterfalls Footbridge and the bridge of Wheeling on the Ohio river which was completed in 1849 and ruined under a wind storm, six years later in 1854.

The length of the span of 305 m made the Wheeling Bridge the longest in the world in that time (see Fig. 15). Of this collapse it is preserved as a track in the story published on the local newspaper of Wheeling, and after few days republished even on the New York Times. From the original article we can read that "... with anxiety for some instant we observed it to move itself like a ship in a storm: at a certain point it rose almost up to the height of the pier ... at the end there seemed to happen a determinant torsion along the whole span and roughly half of the deck was turned upside down ...".

The scene described is very similar to that we have already seen drawn for the Brighton bridge and which will be repeated, almost a century later, for Tacoma Narrows Bridge. The collapse of the Tacoma Narrows is may be the most famous ruin of the story. The bridge had a span 853 m long and a very high deck slenderness (D/L = 1/350).

The bridge immediately after his construction became famous for his extreme flexibility and actually it attracted the local motorbikes who crossed the bridge while it was rocking, like in a sort of rodeo. But there wasn't any reason to be worried about its stability, so that the local bank invented the slogan "... as safe as the Tacoma Bridge".

The cause of the collapse was determined by a new phenomenon not known now till that time in civil engineering. This phenomenon, called flutter or aeroelastic instability, was instead well known in aeronautical engineering, and it was responsible of the wing oscillation in the first biplane. Since then there has been a considerable development of the aeroelasticity applied to the civil engineering. From an engineering point of view the preventive experimentation of long span bridge by a section-model in boundary layer wind tunnel (BLWT) represents one of the first steps in the structural bridge design process, and the aerodynamic requirement modify the section of the suspension bridge.

The beginning of this studies signs the birth of bridges with the classic aerodynamic section shape.

The first bridge with the aerodynamic deck section has been the bridge on the Severn river in England (1966, span length 990 m) and since the construction of this structure, many other ones have taken the aerodynamic shape. Among the latest we remember the bridge across the Humber in England (1980, span length 1410 m), and the bridge on the Bosforo strait (1973, span length 1074 m).

And also the recently longest bridge take into account these aerodynamic aspects and so they are the result of long and complex experimentation in boundary layer wind tunnel. Of these last ones, it must be remembered the Great Belt Bridge (see Fig. 17), in Denmark and finally the bridge on the Akashi Kaikyo built in Japan (with reaches span of 1990 m) (see Fig. 18) and the design of Messina Bridge possessing a sort of wind transparent section (see Fig. 19).

Figure 17.   The Great Belt Bridge (Denmark).

Figure 16.   Bridge of Wheeling on Ohio river, collapsed in 1854.

Figure 18.   Akashi kaikyo Bridge (Japan).

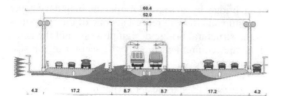

Figure 19.   Section of Messina Bridge.

5   THE WIND AND THE DEVELOPMENT OF TALL BUILDINGS

Besides the structures which extend horizontally, like bridges, also tall buildings, which extend vertically, have the wind as dominant and decisive load.

Also for tall buildings the most impressive increase in height, with respect to the past, was favoured by a technological jump, represented by the advent of "steel". This jump happened in 1889 with the erection of the Eiffel Tower (see Fig. 20) and marked all at once the jump from 162 m of the obelisk in Washington to 312 m, which was the absolute record of height for tall buildings (the base width was 125 m). Eiffel was very criticized both for the aesthetics and functionality of the tower, which besides appearing clumsy to Parisians eyes lacking utility and meaning. Eiffel tried to utilize to use it in every way, for a series of scientific experiments, also linked to the study of structural vibrations. His registrations of the top displacements during many wind storms show an elliptic movement with maximum displacements of about 14 cm (see Fig. 21). With reference to tests on existing structures, another interesting structure whose wind induced vibrations were measured experimentally is the Arnolfo Tower of Palazzo Vecchio in Florence.

Using an horizon (that is a basin) of mercury and a device similar to a seismograph, in 1903 Padre Alfani registered the vibrations of the Arnolfo Tower and presented the results in 1909 in a Communication to the Conference of Italian Engineers and Architects (see Fig. 22).

But after some years of French record, this record passed to the United States with the famous Empire State Building in New York (completed in 1931) which increased to 380 m the height record of the skyscraper typology.

The vibrations at the top of the skyscraper, together with the wind pressures on its lateral facades, were the subject of analytical studies and experimental campaigns. A full scale experimental campaign was set up, in a way very impressive for that time, as many experimental devices were used: 38 fixed manometers, 20 cameras to take simultaneously photos of lectures of manometers, and anemometers, extensimeters, etc.

Figure 20.   Eiffel Tower, chronological phase of construction.

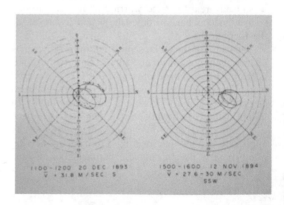

Figure 21.   Studies on the structural dynamic of the Eiffel Tower.

This very large amount of data was not appropriately analyzed because at that time there were neither calculation tools nor appropriate knowledge about the dynamics of structures and response of structures to wind action, which were developed only some decades later. Moreover full-scale data were very different from those coming from wind tunnel tests performed some

Figure 22. Proceedings of the Conference of Italian Engineers and Architects.

Figure 23. Sears Tower Chicago (US).

years before, as these tests have been performed in an aeronautical tunnel, which is not appropriate to the aim.

Wind tunnel tests have been developed since the first tests on the blades of windmills. In the following years many other wind tunnel tests were performed, also on models of buildings. Only in 1930 Bailey performed in England full-scale tests on buildings, of which previously wind tunnel tests have been done. Results were different in an impressive way. Bailey guessed that so different results depended essentially on the fact that in wind tunnels, used up to that time, the flow was uniform, while in nature, due to obstacles encountered by the wind and to the friction with the earth surface, both the flow vorticity and the variation of the mean speed with the height are different. Therefore, for the first time, he built at the National Physical Laboratory in Teddington a very long tunnel and put a lot of obstacles on the floor, repeating many times the tests. Obtained results were absolutely more convincing. Since that time the model techniques and the measurements in the wind tunnel underwent a prodigious acceleration.

Results are surely adherent to reality; and the boundary layer wind tunnel (BLWT) confirms to be a prodigious support device for the structural design. However the dynamic response of structures was not fully investigated yet. This behaviour leads the structure to oscillate in the wind with frequencies equal to the natural frequencies, so that the vibration amplitudes of motion become an important design restraint. The comfort of the living people of the last floor represents very often the most severe restraint in the design of tall buildings. And to calculate exactly these vibrations it was required to wait for the complete development of Davenport's theory (in 1963), which represents the final ring of the "wind chain" for a serious and definitive engineering provision of the behaviour of structures under wind action.

BLWT tests combined with the statistical theory of wind turbulence by Davenport together with the powerful support represented by the design of numerical simulations on computers, led finally to an accurate design process of civil engineering structures, confirmed impressively from full-scale measurements under wind storms.

In the last years the height records followed one another: Twin towers of the World Trade Center in New York, 412 m tall, built in 1972 and 1973, after accurate studies in the BLWT, and dramatically collapsed in 2001; the Sears Tower in Chicago (see Fig. 23), (443 m tall, completed in 1974) which is still the highest

Figure 24.   Petronas Tower Kuala Lumpur (Malaysia).

Figure 25.   Olympic Stadium Rome (Italy).

building for offices in the world; the Petronas Tower in Kuala Lumpur Malaysia (see Fig. 24) (452 m completed in 1998); and the CN Tower in Toronto, the highest TV antenna in the world (the top reaches 553 m, completed in 1976).

Also the large roofs of our time are structures extremely sensitive to wind action, which determines shape and dimensions of the structure. Just to remind some among the most impressive, we can mention the roof of the Olympic Stadium in Rome (see Fig. 25) (tensile structure with an elliptic plan with diameters of 308 m and 237 m), subjected to an intensive experimental campaign in the BLWT combined with numerical simulations of the dynamic behaviour performed using software codes for the structural analysis; and it must be also mentioned the large pneumatic structure of the Pontiac Stadium (Michigan) with plan dimensions of 230 m × 180 m, and the Big Egg in Tokio (about 160 m of side in plan), where the light roofing-structure is supported by an inner air overpressure; so when the weight is definitely defeated, the wind action becomes the decisive design action.

6   CONCLUSIONS

At the beginning of the third millennium, and with the giant's steps made especially in the last two decades, the Engineer can say to have succeeded in dominating efficiently the wind action, which is one of the most powerful environmental actions. If this knowledge could not make real for the Engineer the irreverent dream of Noè's descendants to "reach the sky", the hope is that he will succeed in avoiding those disasters and those damages registered in the history of Civil Engineering, and in designing with safety all the large and small structures, which are required for the man life.

REFERENCES

Munich, Re 2003. *Topics Annual Review Natural Catastrophes 2002*, Munich Re Group.
Valadier, G. 1832. *L'Architettura Pratica*, Rome.
Koerte, A. 1992. *Firth of Forth Firth of Tay*, Birkhauser Verlag, Basel.
Aynsley, R.M., Melbourne, W. & Vickery, B.J. 1997. "*Architectural Aerodynamics*", Applied Science Publishers Ltd.
Vickery, B.J. 1990. Istituto di Tecnica delle Costruzioni, Università di Bologna, Prof. B.J. Vickery, Course on "*Effetti del vento sulle costruzioni*", Novembre 1990.
Spinelli, P. 1990. L'Ingegneria del Vento in Italia: Alcune Riflessioni Introductive, *Proceedings of 1° Convegno Nazionale di Ingegneria del Vento ANIV*, 28–30 October, Florence 1990.
Solari, G. 1990. L'Evoluzione Storica e Scientifica dell'-Ingegneria del Vento, *Proceedings of 1° Convegno Nazionale di Ingegneria del Vento ANIV*, 28–30 October, Florence 1990.

Performance of buildings in extreme winds

K.C. Mehta
Wind Science and Engineering Research Center, Texas Tech University, Lubbock, Texas

ABSTRACT: Researchers at Texas Tech University have documented damage caused by extreme wind events for more than three decades. Their observations have identified storm-related, environment-related, and building-related variables that affect damage to buildings. This paper gives examples of damage observations in multi-story, large metal, small commercial, and timber buildings and delineates some general lessons for structural engineers to glean from the damage documentation experiences of Texas Tech personnel to reduce damage to an acceptable level.

1   INTRODUCTION

Extreme winds are found in storms such as hurricanes, tornadoes, and downbursts. Hurricanes (also called typhoons and cyclones in different parts of the world) are tropical storms that form over warm ocean waters and affect coastal areas when they make landfall. The storms are rotating vortices with the diameter of the eye ranging from 10 to 40 km and damaging winds extending up to 160 km from the center of the storm. Hurricane storms are large enough where buildings experience straightline winds, though the direction of winds can change as the storm passes over a building.

Tornadoes and downbursts are found embedded in thunderstorms, which can occur in the coastal areas as well as in the interior of landmasses. Tornadic storms are also vortices, though they are much smaller in size compared to hurricanes. Tornadoes have a short life, but can have wind speeds as high as 90 m/s. In the United States, an average of 1000 tornadoes occur annually. The average width of damaging winds is about 100 m (with maximum of 1500 m), and the average length is one kilometer, though damage paths as long as several hundred kilometers have been recorded. The size of a severe tornado is large enough to impact most buildings with winds coming in a straight line and changing directions as the storm passes over a building. Downbursts are rapidly descending parcels of air which fan out when they hit the ground. In general, a building experiences winds coming in a straight line whether the storm is a hurricane, a tornado, or a downburst. Our damage investigation experiences have shown that damaging mechanisms of buildings are similar, irrespective of the storm type.

At Texas Tech University, we started documenting windstorm-induced damage when a severe tornado devastated almost half the city of Lubbock (population 150,000 on May 11, 1970). Because Lubbock is the hometown of Texas Tech, we documented in-depth building and structure damage. Since then, Texas Tech personnel have documented building damage in more than 120 windstorm events. Data of these building damage documentation efforts are archived and kept on file for future use. These damage experiences have shown several common themes. The paper delineates lessons learned from these experiences.

2   VARIABLES AFFECTING DAMAGE

The extent of damage that a building experiences during an extreme wind event depends on several variables (Minor et al. 1983). Variables can be separated into three groups: storm-related, environment-related, and building-dependent variables.

2.1   *Storm-dependent variables*

Storm-dependent variables affecting the extent of damage are governed by meteorological parameters of the storm and location of the building with respect to the storm. The most important parameter is the magnitude of wind speed. If the wind speed exceeds the design basis wind speed, there is the likelihood of at least some damage. If the wind speed exceeds the limit state load, a total collapse of a building is possible. Also, the location of a building with respect to the path of the storm affects the extent of damage. If the center of a hurricane or tornado passes over a building, that building will experience not only close to maximum wind speed of the storm, but also winds from different directions. If the building is located on a side of the storm path, it will experience winds only from one direction; the magnitude of wind speed will depend on the distance of the building from the center of the storm. Tornadic storms have a high gradient in wind speed from the center of the storm; the extent of damage reduction is noticeable as the distance from the center of the storm increases.

2.2 Environment-related variables

The surrounding terrain affects the extent of damage in two ways: (1) friction of terrain affecting wind speed and (2) potential of windborne debris. These two effects are counteracting. Built-up terrain provides higher friction of ground, which in turn reduces wind speed. It also, in some cases, provides shelter to the building, thus reducing the extent of damage. However, built-up terrain has a larger amount of windborne debris, which increases the extent of damage. Windborne debris has been found to be a significant damaging mechanism in windstorms.

2.3 Building-dependent variables

Building-dependent variables include engineering attention to design, redundancy in structural systems, connections, materials, and quality of construction. Since there are a large number of structural systems which can use steel, concrete, masonry, or timber or a combination of these materials, it is difficult to separate buildings by structural systems or materials. However, our experiences of damage documentation in the field have suggested a delineation of buildings related to a combination of engineering attention, redundancy in structural systems, and construction materials. Thus, damage observations can be made for groups of buildings representing multistory buildings, large metal buildings, small commercial buildings, and timber buildings.

3 DAMAGE OBSERVATIONS

3.1 Multistory buildings

Buildings with more than four stories have specific structural systems to resist gravity and lateral loads. Buildings with four to sixty stories in height that have steel or concrete framing systems have been exposed to extreme winds. Visible images of the buildings convey catastrophic damage (Figs 1, 2). However, neither of these buildings sustained any structural damage. The damages are strictly to the exterior façade and the interior contents.

Figure 1. A ten-story building damaged by windborne debris in a tornado.

Figure 2. A five-story building damaged in a hurricane.

The building shown in Figure 1 was impacted by a tornado in Fort Worth, Texas, in March 2000. The tornado was not a severe one. However, upstream of this building was a workshop with several buildings. The tornadic storm destroyed the workshop and picked up a large amount of debris from it. The windborne debris impacted the façade of the building shown in Figure 1 and broke virtually every glass window. The wind entered the building through the broken windows and devastated interior walls, ceilings, and office contents. Structural engineers can take satisfaction in a good design of the building, though the public blames engineers for the amount of damage. In the future, structural engineers will have to learn to design for windborne debris and mitigate breach of façade to reduce damage to the interior.

The apartment/condominium building in Figure 2 was in the path of Hurricane Andrew that made landfall in South Florida in 1992. The devastating damage was to the exterior façade, which in turn resulted in damage to the interior of the building. A similar building in the same area (experiencing similar wind speed) experienced very little damage. The difference between the two buildings was the façade. Exterior wall panels, non-structural elements, had poor connections for the building shown in Figure 2. The panels were pulled out as units. Loss of the exterior panels put the occupants in a dangerous situation for bodily harm and generated catastrophic loss to the interior. In windstorms, breaches in exterior walls have catastrophic effects.

3.2 Large metal buildings

Industrial plants, commercial warehouses, and aircraft hangars use large metal buildings. Steel framing permits large spans, providing clear working space without interference from columns. When the steel framing is custom-designed specifically for that building, the building has suffered little damage. One component of these buildings that has shown weakness is overhead doors. Commercial overhead doors are flexible and often come off their guides under wind pressures. Figure 3 shows failed overhead doors. This

failure causes additional internal pressures and leads to additional damage.

Large metal buildings are often pre-engineered. Through sophisticated structural analysis and design, the pre-engineered buildings are made highly efficient in the use of material. In metal buildings, saving in steel material has a significant impact on the economy of the building. This structural efficiency has detrimental effects on the structural stability of the building. Lateral bracing and redundancy tend to be overlooked. The damaged metal building in Figure 4 illustrates the result of stability. The roof purlins buckled when wind pressures on them caused a combination of uplift load on the roof and axial load coming from the windward wall.

Figure 5 shows a complete collapse of a metal building. Efficiency of design leaves this type of building without any redundancy. If one of the components fails because of a small design flaw or because of workmanship, the entire building tends to collapse in hurricane winds, which may last from 30 minutes to several hours. Wind is not forgiving; it will find a weak link in the system and fail it. If there is a lack of redundancy in the structural system, the entire building can collapse.

Figure 3. Loss of overhead doors contributed to uplift of roof panels.

Figure 4. Instability of roof purlins caused damage to windward wall.

Figure 5. Collapse of pre-engineered metal building.

Figure 6. Collapse of loadbearing wall causes destruction of the building.

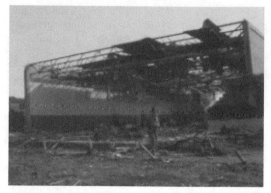

Figure 7. Structural system survived, but building is severely damaged and contents are lost.

Unfortunately, structural engineering research has not developed an algorithm that guides the designer how much quantitative redundancy is needed. As yet, research has not produced scenarios of progressive collapse of a building in the high winds of hurricanes.

3.3 Small commercial buildings

Strip malls and retail stores are part of small commercial buildings. Structural systems for these buildings can vary from beam and columns to loadbearing walls and roof joists. The construction materials can be steel, concrete, masonry, or wood or a combination of these materials in the same building. These buildings do not always receive custom engineering attention of analysis and design. It is common to use prescriptive requirements to size a member. For example, a loadbearing masonry wall may be specified based on height to thickness ratio rather than on a calculation of stresses.

These types of buildings sustain varying extents of damage. Figure 6 shows the collapsed wall and roof in an extreme wind event. The loadbearing wall collapsed under lateral wind pressure; the collapse of the wall is catastrophic since it was unreinforced. The roof, which was supported on the loadbearing walls, collapsed when the walls failed. Failure of one component resulted in total collapse of the building.

The basic structural system of steel columns, steel beams, and open web joists remained standing in the building shown in Figure 7. The exterior wall collapsed, and the roof panels were lifted, thus exposing all of the interior to wind and rain. Even though the structural system remained intact, the loss in this building was very high. Exterior walls and roof panels are very important in mitigating damage in windstorms.

3.4 Timber buildings

In the United States, single-family and two-story apartment buildings are generally constructed of timber material. Most of these buildings are constructed using prescriptive requirements of codes, without

Figure 8. Failure of connection uplifts entire roof structure.

Figure 9. Uplift of roof and collapse of exterior walls destroys the building.

calculations of loads and resistance. Damage to buildings is related to load path, connections, workmanship, and material quality.

Connections are the most important items in timber buildings. Failure of connections between roof and walls uplifts the entire roof structure, as shown in Figure 8. Loss of the roof can cause collapse of exterior walls and virtual destruction of a building (Fig. 9).

Figure 10.   Debris of pieces of timber from failed buildings.

Over a period of time, timber can deteriorate if it is not properly maintained. Extensive damage, at times, can be a result of deteriorated material or poor workmanship or a combination of these factors.

4   WINDBORNE DEBRIS

Buildings can sustain a large amount of damage in windstorms due to windborne debris, as indicated in Figure 1. In a built-up area, windborne debris is always generated in windstorms. The windborne debris can be as small as roof gravel and increase in size up to pieces of timber. Figure 10 shows debris of pieces of timber generated in a residential area where timber residences failed. The windborne debris can fail window glass and penetrate exterior walls. Thus, the debris damaging mechanism causes a breach in the exterior, resulting in additional damage due to internal pressures. Research continues to examine the size, shape, and weight of windborne debris and its impact speed. In addition, window glass and exterior wall and roof surfaces have been developed which can resist the debris and yet to be economical.

5   LESSONS FOR STRUCTURAL ENGINEERS

There are several lessons that structural engineers can learn from observing wind-induced damage. Some of the significant lessons are:

- Extent of damage depends on many variables.
- Multistory building structural systems sustain little damage, but breaches in exterior walls cause a large amount of damage to the interior.
- Large metal buildings can collapse if there is a lack of bracing and redundancy.
- Loadbearing wall systems suffer catastrophic collapse if walls are not able to resist out-of-plane bending forces.
- Failure of connections in timber buildings leads to severe damage.
- Windborne debris is inherently prevalent in built-up areas during a windstorm. Structural engineers will need to respond to this damaging mechanism.

ACKNOWLEDGEMENT

This paper is based on windstorm damage documentation conducted by many faculty members and students of the Wind Science and Engineering Research Center at Texas Tech University. The transcript was prepared by Mrs. Lynnetta Hibdon, technical writer of the Center. Contributions of these individuals is acknowledged.

REFERENCE

Minor, J.E., McDonald, J.R. & Mehta, K.C. 1983. *The tornado: An engineering-oriented perspective*. National Oceanic and Aeronautic Administration (NOAA) Technical Memorandum NWS SR-147. p. 196.

Flutter optimization of bridge decks: experimental and analytical procedures

G. Bartoli & M. Righi

Dipartimento di Ingegneria Civile, Università degli Studi di Firenze, Firenze, Italy

ABSTRACT: In the paper, some preliminary results of an ongoing research on the bridge deck's performance under wind are reported. The research aims to determine an approximate way to predict the critical flutter wind speed of a given deck by analyzing an "aerodynamic performance index" which allows to estimate the stability behavior in comparison with the one of an "equivalent" flat plate. Both the analytical aspects of the problem as well as some of the results obtained from wind tunnel experiments are reported.

1 INTRODUCTION

As it is well known, bridge decks could exhibit an unstable behaviour under wind loading due to coupling of natural frequencies (coupled flutter) or simply along one degree of freedom (torsional or single-mode flutter) due to the interaction between the motion of the bridge and the incoming flow. According to classical Scanlan's approach (Scanlan & Tomko 1971), the critical wind velocity value could be estimated in an analytical way by using the so called "flutter derivatives"; these quantities can be estimated either by measuring time-decaying oscillations of the bridge decks after given initial conditions (and their simultaneous identification is often a hard task, see e.g. Scanlan et al. 1994), or by forcing an harmonic motion on the bridge and measuring the force necessary to maintain the imposed motion (as reported e.g. in Jensen & Hoeffer 1998). Once flutter derivatives have been experimentally evaluated for different reduced velocity values, the flutter condition can be analytically estimated and an optimisation can be achieved by trying to increase the velocity at which flutter can occur. This approach is obviously time-consuming, especially when, even in an early stage of the design, a high number of different configurations have to be tested in order to achieve a satisfactory value for the critical wind velocity.

For this reason, a different approach with respect to the two "classical" ones has been adopted in the present study; bridge stability boundaries has been directly detected by wind tunnel tests, and the critical wind speed has been compared with the one of a flat plate possessing the same dynamic characteristics (in the following referred to as "equivalent" flat plate). In this way, the aerodynamic performance of the bridge deck can be expressed by means of an "aerodynamic performance coefficient" β, which represents the ratio between the critical flutter wind speed of the investigated section with respect to the one of the equivalent flat plate.

The adopted procedure will be sketched in the following paragraphs, together with a description of the main results obtained during the wind tunnel experimental tests.

2 FLAT-PLATE THEORY

Term "flat-plate" (FP in the following) refers to a plate with a negligible thickness and a width (chord length) denoted by B, invested by the wind along this latter dimension; the other dimension (transversal with respect to the wind direction) is supposed to be large enough to be considered as infinite. Under the previous hypotheses, the flow around the body can be treated as a bi-dimensional one.

FP is prone to coupled flutter: the value of the critical velocity can easily be obtained using the circulatory complex function C(K) obtained by Theodorsen (1934); the function is defined as

$$C(K) = F(K) + \iota \cdot G(K) \tag{1}$$

where ι denotes complex unity while F(K) and G(K) are the two real functions

$$F(K) = \frac{J_1(k)[J_1(k) + J_0(k)] + Y_1(k)[Y_1(k) - J_0(k)]}{[J_1(k) + J_0(k)]^2 + [Y_1(k) - J_0(k)]^2}$$

$$(2a)$$

$$G(K) = -\frac{J_1(k) \cdot J_0(k) + Y_1(k) \cdot Y_0(k)}{[J_1(k) + J_0(k)]^2 + [Y_1(k) - J_0(k)]^2}$$

$$(2b)$$

In the previous formulas, $J_i(k)$ and $Y_i(k)$ respectively represent first and second kind Bessel's functions. The two parameters K and k inserted into the equations are values for the reduced frequency evaluated with respect to the chord length (B) and the half chord length (b), that is

$$K = \frac{\omega \cdot B}{U} = \frac{2 \cdot \pi \cdot f \cdot B}{U}$$

$$(3a)$$

$$k = \frac{\omega \cdot b}{U} = \frac{K}{2}$$

$$(3b)$$

Starting from C(K) values, following expressions can be used for estimating flutter derivatives, as reported in several references (i.e. see Dyrbye & Ole Hansen 1997):

$$H_1^*(K) = -\frac{\pi \cdot F(k)}{k}$$

$$(4a)$$

$$H_2^*(K) = \frac{\pi}{4k}\left[1 + F(k) + \frac{2G(k)}{k}\right]$$

$$(4b)$$

$$H_3^*(K) = \frac{\pi}{2k^2}\left[F(k) - \frac{kG(k)}{2}\right]$$

$$(4c)$$

$$H_4^*(K) = \frac{\pi}{2}\left[1 + \frac{2G(k)}{k}\right]$$

$$(4d)$$

$$A_1^*(K) = -\frac{\pi \cdot F(k)}{4 \cdot k}$$

$$(4e)$$

$$A_2^*(K) = -\frac{\pi}{16k}\left[1 - F(k) - \frac{2G(k)}{k}\right]$$

$$(4f)$$

$$A_3^*(K) = \frac{\pi}{8k^2}\left[F(k) - \frac{kG(k)}{2}\right]$$

$$(4g)$$

$$A_4^*(K) = \frac{\pi}{4}\frac{G(k)}{k}$$

$$(4h)$$

The critical wind velocity can then be obtained by means of the solution of the classical flutter equations:

$$M\left[1 + 2\xi_h\omega_h + \omega_h^2\right] = \frac{1}{2}\rho U^2 B$$
$$\left[KH_1^* \frac{\dot{h}}{U} + KH_2^* \frac{B\dot{\alpha}}{U} + K^2 H_3^* \alpha + K^2 H_4^* \frac{h}{B}\right]$$

$$(5a)$$

$$I\left[1 + 2\xi_\alpha\omega_\alpha + \omega_\alpha^2\right] = \frac{1}{2}\rho U^2 B^2$$
$$\left[KA_1^* \frac{\dot{h}}{U} + KA_2^* \frac{B\dot{\alpha}}{U} + K^2 A_3^* \alpha + K^2 A_4^* \frac{h}{B}\right]$$

$$(5b)$$

where h and α represent the two degrees of freedom of the dynamic system (vertical deflection and angular rotation); m and I the mass and moment of inertia per unit of length; ξ_h and ξ_α the damping ratios; ω_h and ω_α the angular frequencies of the system in "still air"; ρ and U the density of mass and the mean speed of the fluid. Argument K of the flutter derivatives has been omitted for the sake of brevity. Critical condition (as a function of the wind speed U) can be achieved by imposing a non-decaying harmonic motion of the system, then solving the eigenvalue problem that turns out to be a complex one.

The critical wind speed (U_{fp}) can also be estimated using (or by the use of) some relationships which have been proposed by several authors. All the approximated expressions can be achieved starting from the definition of the divergence wind speed, that is the wind speed that can cause a divergent motion of the FP because of the loosing of the torsional stiffness due to the wind action. It can be easily shown that the divergence velocity is equal to

$$U_d = \sqrt{\frac{2 \cdot K_t}{\pi \cdot \rho}}$$

where K_t represent the torsional stiffness expressed by

$$K_t = I \cdot \omega_\alpha^2$$

The critical flutter wind speed is normally defined as a fraction of the divergence speed, by assuming

$$U_{fp} \propto U_d \sqrt{1 - \left(\frac{\omega_h}{\omega_\alpha}\right)^2} = U_d \sqrt{1 - \frac{1}{\gamma_\omega^2}}$$

where γ_ω denotes the frequency ratio, which is the ratio between the torsional and the vertical natural frequencies.

Table 1. Approximated expressions for the critical wind speed of a flat plate.

Author	Reported expression
Selberg (1961)	$U_{fp} = 0.44 \cdot \omega_t \cdot B \cdot \sqrt{\left[1 - \dfrac{1}{\gamma_\omega^2}\right] \dfrac{\sqrt{\upsilon}}{\mu}}$
Selberg (in Gimsing 1997)	$U_{fp} = 0.52 \cdot U_d \cdot \sqrt{\left[1 - \dfrac{1}{\gamma_\omega^2}\right]} \cdot B \cdot \sqrt{\dfrac{m}{I}}$
Rochard (1963)	$U_{fp} = 0.443 \cdot \omega_t \cdot B \cdot \sqrt{\left[1 - \dfrac{1}{\gamma_\omega^2}\right]} \cdot \dfrac{1}{\mu} \cdot \dfrac{2 \cdot \upsilon}{(1 + \upsilon)}$

The approximation holds true for values greater than 1, and various formulas has been proposed for the proportional term, as reported in following Table 1.

In the table two non dimensional quantities have been introduced:

$$\mu = \frac{2 \cdot \pi \cdot \rho \cdot B^2}{m} \qquad (6a)$$

$$\nu = \frac{2 \cdot I}{B^2 \cdot m} \qquad (6b)$$

The first one represent the lightness ratio, that is the ratio between the air "moved" by the flat plate and its mass; the lower is μ and the lighter is the system. Normally μ is between the range [0.07–0.20].

The second one gives information about the mass distribution: a high value of ν means that masses are eccentric with respect to the center of gravity of the system.

The critical wind speed is directly related to the frequency ratio γ_ω: more stable conditions can be obtained by increasing this ratio. A good stability (that is high values for the critical wind speed) is obtained for γ_ω greater than two (modern suspended bridges are normally characterized by ratio in the range [1.40–2.00]).

3 THE AERODYNAMIC PERFORMANCE INDEX

The flat-plate theory represents an important tool for a better comprehension of flutter mechanism.

Unfortunately, the same analytical approach cannot be used for bridge deck's sections, because, up to now, no analytical expressions have been found for the circulatory function C(K) apart from the FP case.

Nevertheless, from a design point of view, it will be very useful to have a tool able to give a first approximation of the critical flutter wind speed, as a function of the geometry of the section only. This can be obtained by introducing the "aerodynamic performance index", β: the coefficient can be estimated by wind tunnel tests, these allowing to evaluate the critical flutter wind speed of the given section (in the following denoted by U_c). Then the critical flutter wind speed for the equivalent FP (that is a flat plate with the same dynamic properties and the same width) can be estimated, either by the classical approach or by using one of the approximate relationships. The aerodynamic performance index can then be defined as

$$\beta = \frac{U_c}{U_{fp}}$$

being U_{fp} the theoretical critical flutter wind speed of the equivalent FP.

The coefficient β is an index of the aerodynamic stability of the bridge deck's section, and its value is normally within the range from 0.4 and 0.8 for bluff bodies and from 0.8 and 1.1 for streamlined sections, (see Table 2); larger values of β (up to 1.25 and higher) can be reached for streamlined slotted profiles (that is aerodynamic profiles with some openings on the deck).

Main advantages of this approach can be resumed as follow:

- critical wind speed for the equivalent FP can be determined analytically, being its flutter derivatives expressed in an analytical form;
- the aerodynamic performance index β depends only on the geometric characteristics of the bridge deck's section, while it is independent from other parameters characterizing its dynamic response. In a first design stage is then possible to estimate the critical wind speed of a given section without making provision for expensive wind tunnel tests, then looking for some refinements leading to a better performance of the bridge. Only in a second design phase, the real behavior of the bridge can be evaluated by

Table 2. Aerodynamic performance index (β) values for some typical bridge deck's cross sections.

Deck section	β = U_c/U_{fp}	
	Gimsing (1997)	Dyrbye (1991)
	0.43	0.4–0.6
	0.62	–
	0.91	0.8–0.9
	0.77	0.6–0.8

some specific wind tunnel tests, where the influence of the actual geometry can be better estimated.

4 EXPERIMENTAL SET-UP

Wind tunnel tests were performed at the CRIACIV-DIC Boundary Layer Wind Tunnel (BLWT) in Prato. The tunnel is of the open type and the target wind speed (in the range 0–35 m/s) is obtained by means of both the regulation of the pitch of the 10 blades constituting the fan and the rotating speed. The wind tunnel cross-section is a rectangular one with sides of 2.2 m by 1.6 m, slightly divergent from the inlet to the working section whose dimension are 2.4 m by 1.6 m. The global length of the wind tunnel (from inlet to the end of diffusers) is about 24 m.

Tests have been performed on section models whose profiles are box section with two slenderness ratio B/D = 12.5 and B/D = 5, they are 920 mm long (L), 375 mm (or 200) wide (B) and 30 mm (or 40) thick (D). The maximum allowable L is fixed by the CRIACIV BLWT working cross and by the end-plates dimension; in order to consider the model as a section model (two-dimensional) the ratio L/B should be at least 2.5 (usually it is about 3) (Larsen 1992). The models were fabricated from a plywood deck with transversal stiffening wooden girders, a special kind of wood was used to make models as light as possible; rotation axis was made of stainless steel 16 mm diameter and 1.5 mm thickness. The model was supported by an elastic system made up of eight helical springs mounted vertically and two aluminum made arms whose task was to transfer forces from the aeroelastic model to the endplates. The mounting positions of springs were adjusted so that the

elastic centre and the centre of gravity of the cross-section coincided. Also the ratio between the natural frequencies of the heaving and torsional modes was controlled by the mounting position of springs. Model characteristics are reported in following Table 3.

4.1 Characteristics of the flow

Flow within the end-plates was characterized in smooth flow regime with respect to the reference wind speed U_{ref} taken from a Pitot-Prandtl tube positioned above the end-plates; actual fluctuating wind speed profile has been measured via X-probe anemometers. A ratio $U_{deck}/U_{ref} = 1.19$ was obtained, being U_{deck} the mean wind speed at deck level.

One example of the obtained profiles is reported in Figure 1.

5 RESULTS

Initial analyses have been performed in still air to find out all dynamic and damping properties of a given configuration, via a free decay test and using the linear least-square analysis method.

Figures 2 and 3 sketch the time evolution of free-decaying motion for both the modes. Correspondent PSDFs are reported in Figure 4.

Analysis with flow includes tests made with the model left in forced free vibration under incoming smooth flow in order to obtain direct measurement of flutter speed; this one can be seen both as the speed where the system torsional and heaving frequencies are coupled and the speed that induces in the system unbearable deck displacements. Calculated data from records are: Maximum, RMS and Standard deviation of Torsional, Heaving and Rocking mode; Frequency

Table 3. Model configurations.

Description	Symbol	Unit	Configuration R12.5 (B/D = 12.5)	Configuration R5 (B/D = 5)
Geometry				
Height	D	[m]	0.030	0.040
Width	B	[m]	0.375	0.200
Geometry ratio	B/D	[–]	12.500	5.000
Length	L	[m]	0.920	0.920
Dynamics				
Mass per m	M/m	[kg/m]	4.024	3.157
Inertia	I	[kgm^2]	0.033	0.009
Mass ratio	I/M	[m^2]	0.009	0.003
f_h	f_h	[Hz]	3.40	3.90
f_t	f_t	[Hz]	4.63	7.70
Frequency ratio	γ_ω	[–]	1.36	1.97
Heaving stiffness	K_h	[N/m]	1690	1744
Torsional stiffness	K_t	[Nm]	27.72	21.07
Damping				
Heaving	ξ_h	[%]	0.24	0.16
Torsional	ξ_t	[%]	0.49	0.55
Critical speed				
Divergence velocity	U_d	[m/s]	14.171	23.161
Selberg (1)	U_{fp}	[m/s]	10.490	20.650
Selberg (2)	U_{fp}	[m/s]	9.999	19.683
Rochard	U_{fp}	[m/s]	10.259	20.498
Theodorsen	U_{fp}	[m/s]	8.980	20.140
U_d/U_{fp}		[%]	139	114

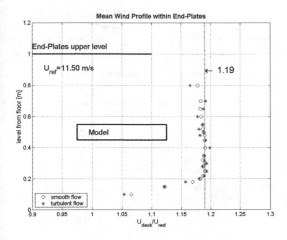

Figure 1. Mean wind velocity within the end-plates.

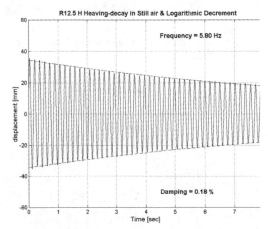

Figure 2. Logarithmic damping of heaving mode in still air (model R12.5).

of vibration of Torsional, Heaving and Rocking mode; Instantaneous Centre of Rotation (CR) of the model.

Figure 5 depicts the evolution of the actual frequency of oscillation as a function of the wind speed, while in Figure 6 the evolution of torsional displacement is reported. In both figures, several different tests conditions are reported, all being characterized by the same geometry of the deck section but different dynamic properties.

Displacement of any chosen point of the deck (that is leading or trailing edge) can be seen as sum of a vertical displacements of the deck as rigid body and a rotation. If these two components are normalized with respect to bridge deck width B, a graph (reported in Figure 7) can be drawn, emphasizing the main mode of vibration.

Samples of vibration patterns are given in Figures 8 and 9. Shaded lines select the area within the deck moves; these lines converge to the mean value of the instantaneous centre of rotation at a given U_{red} (reduced velocity, $U_{red} = U/fB = 2\pi/K$), reported as ordinate in the graphic.

In order to find out the β factor, calculation of critical wind speed U_c must be made. Assessing the exact wind speed corresponding to U_c it is not an easy task as it depends on what phenomena one it is interested

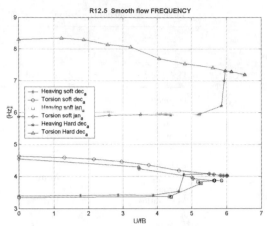

Figure 5.   Mode coupling (model R12.5).

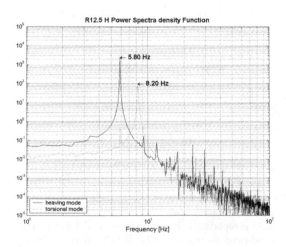

Figure 3.   Logarithmic damping of torsional mode in still air (model R12.5).

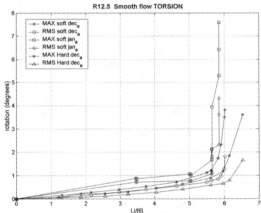

Figure 6.   Torsional displacement (model R12.5).

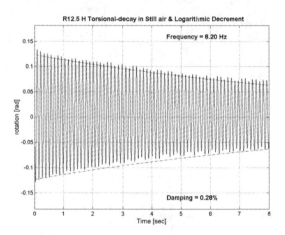

Figure 4.   PSDF of heaving and torsional mode in free decay in still air (model R12.5).

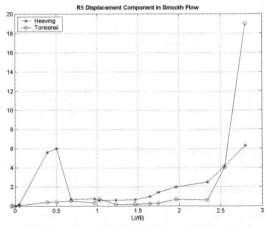

Figure 7.   Normalized displacement components (model R5).

2380

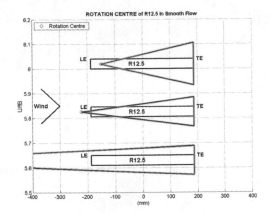

Figure 8.   RC position in smooth flow (model R12.5).

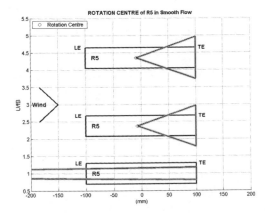

Figure 9.   RC position in smooth flow (model R5).

on. For designing purposes some bridge designer might prefer monitoring the oscillation amplitude (Bartoli & Righi 2001), whereas according to the Scanlan eigenvalues analysis *coupling* has occurred in the range of incipient instability. In the present study four *failure criterion* to define the occurrence of U_c are established: 1. *Coupling* of torsional and heaving frequencies (direct measure); 2. *Divergence* on deck displacements in free vibration under flow (direct measure); 3. Torsional *RMS* of $0.5°$ are exceeded in free vibration under flow (direct measure); 4. *Eigenvalues analysis* according to the Scanlan's approach (using flutter derivatives from experimental data).

6   CONCLUDING REMARKS

Results confirm that excellent consistence from all configurations has been attained in terms of maximum and rms displacements, in terms of frequency coupling

Table 4.   Aerodynamic performance index β.

			Configuration	
	Symbol	Unit	R12.5	R5
Coupling				
Critical speed	U_c	[m/s]	8.40	6.90
U_c/U_{fp}	β	[–]	0.83	0.34
Divergence				
Critical speed	U_c	[m/s]	9.20	7.60
U_c/U_{fp}	β	[–]	0.90	0.37
Rotation RMS = 0.5°				
Critical speed	U_c	[m/s]	7.00	5.90
U_c/U_{fp}	β	[–]	0.69	0.29
Eigenvalues analysis				
Critical speed	U_c	[m/s]	8.25	6.20
U_c/U_{fp}	β	[–]	0.80	0.30

speed and in terms of RC pattern at different wind speeds. It follows that the model dynamic properties do not effect critically BLWT tests, so that the results can be used for any other section model of the same geometry but different stiffness and masses.

Results then confirm that the proposed approach by the aerodynamic performance index can be an effective way to get a reliable estimation of the critical flutter wind speed.

REFERENCES

Bartoli, G. & Righi, M. 2001. Wind tunnel experiments on the aerodynamic performances of the cross section of a long span cable stayed bridge. In *Proc. of 3rd EACWE 2001*, TU/e Eindhoven, The Netherlands.

Dyrbye, C. & Ole Hansen, S. 1997. *Wind Loads on Structures*. New York: John Wiley & Sons.

Gimsing, N. 1997. *Cable Supported Bridges: Concept and Design* (Second Edition). New York: John Wiley & Sons.

Jensen, A.G. & Hoeffer, R. 1998. Flat plate flutter derivatives – An alternative formulation. *J. Wind. Engrg. Ind. Aerodyn.* (74–76): 859–869.

Larsen, A. 1992. *Aerodynamics of Large Bridges*. Rotterdam: Balkema.

Rochard, Y. 1963. Instabilité des ponts suspendus dans le vent-Experiences sur modèle rèduit. *Nat. Phys. Lab., Paper 10*, Teddington, England.

Sarkar, P., Jones, N. & Scanlan, R.H. 1994. Identification of aeroelastic parameters of flexible bridges. *ASCE J. Engrg. Mech.* 120(8): 1718–1743.

Scanlan, R.H. & Tomko, J.J. 1971. Airfoil and bridge deck flutter derivatives. *ASCE J. Mech. Div.* (97): 1717–1737.

Selberg, A. 1961. Oscillation and aerodynamic stability of suspension bridges. *Acta Polythechnica Scandinavica, Civ. Eng. and Build. Constr. Series No. 13*, Oslo, Norway.

Theodorsen, T. 1934. General theory of aerodynamic instability and the mechanism of flutter. *NACA Report no. 496*, Washington DC.

System-based Vision for Strategic and Creative Design, Bontempi (ed.)
© *2003 Swets & Zeitlinger, Lisse, ISBN 90 5809 599 1*

A wind tunnel database for wind resistant design of tall buildings

C.-M. Cheng & C.-T. Liu
Department of Civil Engineering, Tamkang University, Taipei, Taiwan

Po-C. Lu
Department of Water Resource and Environmental Engineering, Tamkang University, Taipei, Taiwan

ABSTRACT: Tall building models with various geometry shapes were tested in wind tunnel for their wind loads. The tested models can be categorized into two sets of wind tunnel studies. The first set is to study the wind load acting on buildings with different cross-sectional shapes. The second set of study emphasized on the effects of minor variations on building shape. The wind loads of building models were measured by high frequency force balance in the turbulent boundary layer flows. With sufficient wind tunnel data, then collaborating with proper structural dynamics procedure, a wind tunnel databased wind resistant design guide for tall buildings can then be built.

1 INTRODUCTION

Wind effects on high-rise buildings include wind load on structural system, cladding pressure, habitants' serviceability and pedestrian level wind environment. Wind load acting on structural system, which could be essential for certain high-rise buildings, should be evaluated base on accurate buildings' response estimations. For most of the tall buildings, the design wind loads are determined by elaborate physical modelling via wind tunnel experiment. Prior to wind tunnel experiment, the building geometry and structural system are decided, in other words, the two most important factors that affecting buildings' wind loads are set and, in most case, will be costly to change. At the present, wind code is used to provide the preliminary design wind load. However, wind code is constructed based on the wind loads data of isolated square (or rectangular) shaped buildings. It could be very conservative for tall buildings other than rectangular shape or buildings with shielding effect. On the other hand, it could be underestimation for very tall buildings or buildings with flexible structural systems. If the preliminary design wind load can be obtained handily and with reasonable accuracy, then it could be used interactively with the building design process. So, the objective of this research project is to build a wind tunnel databased wind resistant design guide for tall buildings, which has an intermediate function between wind code and actual wind tunnel simulation.

The high frequency force balance technique developed by Tschanz (Tschanz 1982) was quickly adopted by most wind tunnel laboratories and became a standard measurement procedure around the world. More elaborated data analysis procedure has been proposed to improve the accuracy and reliability of this technique (Yip & Flay 1995). Due to the simplicity, this technique has been adopted by researchers to investigate the effects of buildings geometry on wind loads (Hayashida & Iwasa 1990). Many researchers used this technique to study the effectiveness of reducing wind loads by minor moderation on building shape (Miyashita et al 1993, Kawai 1998, Kikitsu & Okada 1999, Tamura & Miyagi 1999).

From building designing point of view, most tall buildings have little aeroelastic effect to concern. Therefore, collaborate high frequency force balance measurements with proper structural dynamics procedure, buildings' design wind loads: base shear, overturning moment, torque, as well as distribution of design wind loads along building height can be accurately deduced. With sufficient data from the wind tunnel measurement of (i) models in various flow fields, (ii) models with various geometry shapes, (iii) models under various wind attack angle, and carefully deduced correlations among those factors, a workable wind load design guide for tall buildings can be established. In this project, more than 30 isolated tall building models were tested in both open terrain and city environment flow fields ($\alpha = 0.15, 0.33$), to obtain wind force coefficients and reduced force spectra in the alongwind, acrosswind and tortional directions. All models were tested at wind attack angle increment of 22.5°, so that the wind direction effect can be considered in the future.

Table 1. List of building models tested in wind tunnel.

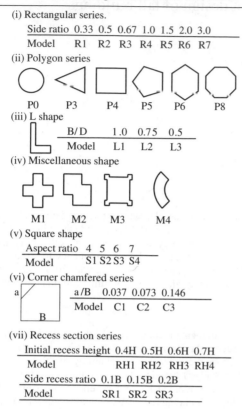

(i) Rectangular series.

Side ratio	0.33	0.5	0.67	1.0	1.5	2.0	3.0
Model	R1	R2	R3	R4	R5	R6	R7

(ii) Polygon series

P0	P3	P4	P5	P6	P8

(iii) L shape

B/D	1.0	0.75	0.5
Model	L1	L2	L3

(iv) Miscellaneous shape

M1	M2	M3	M4

(v) Square shape

Aspect ratio	4	5	6	7
Model	S1	S2	S3	S4

(vi) Corner chamfered series

a/B	0.037	0.073	0.146
Model	C1	C2	C3

(vii) Recess section series

Initial recess height	0.4H	0.5H	0.6H	0.7H
Model	RH1	RH2	RH3	RH4
Side recess ratio	0.1B	0.15B	0.2B	
Model	SR1	SR2	SR3	

2 EXPERIMENTAL SETUP

The building model tests were conducted in a boundary layer wind tunnel at Tamkang University. This wind tunnel has a 18.0 m × 2.0 m × 1.5 m test section. Two sets of atmospheric boundary layer flows, BL1 and BL2, were generated to represent flows over open and urban terrain, respectively. BL1, the open terrain flow field, has a $\alpha = 0.15$ mean velocity profile, with turbulent intensity varying from 20% near ground to 3% at gradient height. BL2, the urban terrain flow field, has a $\alpha = 0.32$ velocity gradient with turbulent intensity varying from 35% to 6%. The gradient height is 120 cm ± 10 cm for both flow fields. During model testing, velocity at model height, U_H, was taken as the normalization factor for the reduced velocity, $U_r = U_H/f_0 D$. Blockage ratio is less than 5%, therefore, its effect ignored. Reynolds number was kept greater than 4×10^4 which is higher than $R_{e,cr} \approx 2.0 \times 10^4$ required for Reynolds number similarity.

Total of 30-plus building shapes were studied in wind tunnel. It can be categorized into two sets of models. The first set of wind tunnel test is to study the wind load acting on buildings with different cross-sectional shape. Four types of cross-section were chosen: (i) rectangular with different cross sectional side ratio, (ii) polygons (circular, triangular and rectangular included), (iii) L shaped cross-sections, and (iv) some irregular shapes. During this part of investigation, building models have same aspect ratio, H/B = 7, and same volume as the square model. The second set of study, concentrating on buildings with basic square cross-section shape, consists of four geometric variations: (i) building aspect ratio, (ii) corner chamfered ratio (iii) recess sections with various initial recess height (iv) recess sections with various recess ratio. This set of models has same volume as a square model with aspect ratio of 5. The tentative scaling ratio of building models is 1:400. The wind loads of building models were measured by high frequency force balance. Wind direction effects were taken into consideration at 22.5° increment. The description of geometry and the nomenclatures of all models are listed in Table 1.

3 EXPERIMENTAL RESULTS

Although all force balance measurements are carried out at every 22.5° increment of wind attack angle, generally speaking, majority of the building tend to have the largest wind load when wind is normal to the building. Also considering the fact that most of the testing models have symmetric geometric shapes, exhibit small acrosswind force (mean) and torsional force (both mean and RMS) at zero wind attack angle. Therefore, only the mean and fluctuating alongwind force and the fluctuating component of the acrosswind force are reported and discussed in this article.

3.1 Turbulent boundary layer effects on buildings wind loads

The normalized force coefficients measured in both the two turbulent boundary layers are listed in Table 2 & 3. The data indicate that nearly all testing models have noticeable higher mean drag, lower RMS drag and lower RMS lift in the open terrain flow field, BL1, than in the urban terrain flow field, BL2. The flow field features of smaller velocity gradient and lower turbulence intensity in BL1 would induced effects on the wind loads coincides with the observed higher mean drag and lower RMS drag. Normally, bluff body tends to have stronger and better coherent vortex shedding process in a smooth flow field, consequently, inducing larger fluctuation lift force. Comparing the RMS lift coefficient and the lift force spectra, shown in Figure 1(b), the lift force spectra measured in open terrain

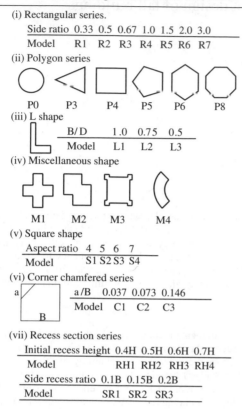

Table 2. Wind force coefficients of tall buildings with various cross sectional shapes.

Model	BL1 ($\alpha = 0.15$)			BL2 ($\alpha = 0.32$)		
	C_D	C_D'	C_L'	C_D	C_D'	C_L'
R1	1.79	0.141	0.099	1.77	0.339	0.114
R2	1.53	0.126	0.078	1.44	0.317	0.190
R3	1.38	0.136	0.152	1.20	0.264	0.219
R4	1.10	0.109	0.209	0.87	0.211	0.235
R5	0.79	0.090	0.223	0.64	0.155	0.233
R6	0.63	0.069	0.225	0.52	0.123	0.253
R7	0.49	0.059	0.168	0.35	0.085	0.296
P0	0.78	0.102	0.126	0.50	0.124	0.160
P3	1.67	0.131	0.098	1.77	0.322	0.239
P4	1.10	0.109	0.209	0.87	0.211	0.235
P5	0.59	0.046	0.099	0.60	0.110	0.196
P6	0.76	0.053	0.056	0.76	0.139	0.122
P8	0.83	0.078	0.074	0.74	0.167	0.129
L1	1.89	0.152	0.076	1.87	0.376	0.178
L2	1.74	0.144	0.116	1.67	0.337	0.263
L3	1.58	0.134	0.161	1.45	0.292	0.335
M1	1.21	0.091	0.065	1.15	0.225	0.146
M2	1.08	0.104	0.137	1.00	0.229	0.191
M3	1.22	0.096	0.178	1.08	0.241	0.242
M4	1.60	0.111	0.051	1.66	0.292	0.102

Table 3. Effects of shape modifications on wind loads.

Model	BL1 ($\alpha = 0.15$)			BL2 ($\alpha = 0.32$)		
	$\tilde{F}_D$	$\tilde{F}_D'$	$\tilde{F}_L'$	$\tilde{F}_D$	$\tilde{F}_D'$	$\tilde{F}_L'$
S1	0.88	0.881	0.867	0.86	0.954	0.893
S2	1.00	1.000	1.000	1.000	1.000	1.000
S3	1.15	1.018	1.138	1.178	1.035	1.159
S4	1.19	1.027	1.046	1.023	1.061	1.065
C1	0.88	1.138	1.102	0.723	0.701	0.764
C2	0.66	0.917	0.770	0.660	0.636	0.671
C3	0.71	0.752	0.628	0.755	0.740	0.715
RH1	1.13	1.018	0.755	1.149	0.909	0.768
RH2	1.03	0.917	0.796	1.138	0.905	0.732
RH3	0.99	0.853	0.770	1.096	0.952	0.764
RH4	0.95	0.862	0.796	1.060	0.918	0.764
SR1	0.99	0.853	0.770	1.096	0.952	0.764
SR2	1.07	0.939	0.730	1.029	0.887	0.759
SR3	0.98	0.881	0.730	1.067	0.871	0.714

Note: "~" denotes normalized wind loads w.r.t model S2.

flow field indeed show higher spectral peaks, however, the lift force spectra obtained in urban terrain exhibit broader spectral bandwidth, therefore, larger RMS lift coefficient. This spectral characteristic implies that, near the critical wind speed of vortex shedding resonance, tall buildings tend to have greater acrosswind dynamic response when located in an open terrain. At

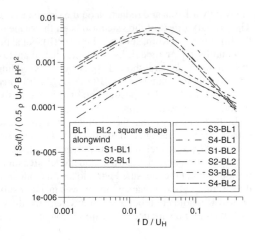

Figure 1(a). Alongwind force spectra of square shape models.

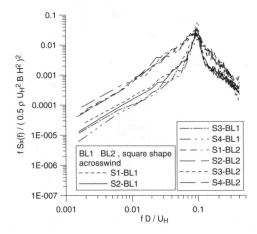

Figure 1(b). Acrosswind force spectra of square models.

lower wind speed, tall buildings will have relatively larger vortex induce vibration in an urban environment.

3.2 Effects of building cross sectional shape on wind loads

Four groups of building shape were studied in this project, namely, (i) rectangular shape (ii) polygons (iii) L shape (iv) miscellaneous. During this part of investigation, all models have same height and same cross section area. The measured wind loads were normalized by same building width and building height, so that the wind force coefficients and force spectra can be compared on the same basis. For the rectangular shaped buildings, the R series models, when the side ratio B/D increases, the alongwind force decreases and acrosswind force increases monotonically, which reflects the simple fact of the increase on the windward area and reduction on the side face area. The primary effect of the side ratio is on the shape of the lift force spectrum, shown in Figure 2(b). For short rectangular shape model, i.e. side ratio less than 1.0, the lift force spectra show distinct vortex shedding peak. For long rectangular, i.e. B/D > 1.0, the lift force spectra show broader spectral bandwidth due to the reattachment phenomenon.

In this investigating, the circular, triangular and square shapes, designated model P0, P3 and P4, respectively, were included in the polygon series along with pentagon, hexagon and octagon models. It should be noticed that the measured wind loads on model P0, the circular cylinder shaped model, was taken at subcritical Reynolds number without any remedy, therefore, it should be applied with care. Among the tested

models, the triangular shape, model P3, has the largest alongwind force (both mean and RMS) and the square shaped model has the largest fluctuating acrosswind force. The pentagon shape model subjected to the lowest wind loads. The mean and RMS drag coefficients generally reflect the magnitude of the drag force and the alongwind structural response. However, the RMS lift coefficient may not truly represent tall buildings' acrosswind response, which is significant to the design wind load. The lift force spectra measured in BL1, shown in Figure 3(b), indicate that the square shape mode, P4, has the vortex induced spectral peaks noticeably greater than the rest of the tested models. The circular and the triangular shape models have about the same spectral peaks one order of magnitude less than the square model. The lift force spectra of the pentagon and hexagon shape models become broadband nature, and do not exhibit spectral peaks. However,

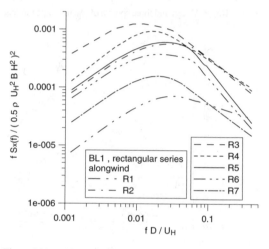

Figure 2(a). Alongwind force spectra of rectangular models.

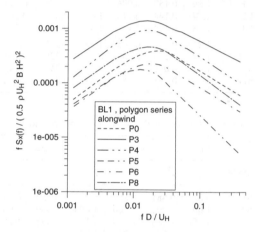

Figure 3(a). Alongwind force spectra of polygon models.

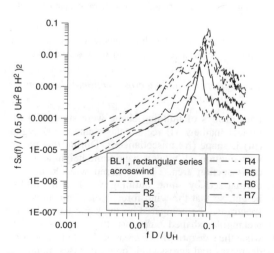

Figure 2(b). Acrosswind force spectra of rectangular models.

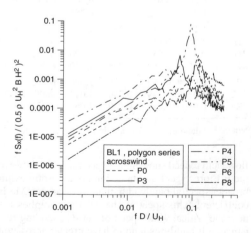

Figure 3(b). Acrosswind force spectra of polygon models.

the lift force spectrum of the octagon model regains the narrowband characteristics, and the spectral peak is only slightly less than the circular and triangular models.

The L shaped models, in general, have the largest wind loads both in alongwind and acrosswind directions among all model series. However, the lift force spectra of the three L shape models, shown in Figure 4(b), exhibit only weak vortex shedding peaks. In other word, in spite of the large RMS lift force coefficient, tall building of L shape has little potential of vortex shedding resonance.

Four other shaped models, model M1–M4, were tested in this project. The results, shown in Table 2 and Figure 5(a)–5(b) indicate that the arch shaped model has the largest alongwind force but the least acrosswind force among them. Model M2 and M3, which are alterations from square shape, have moderate force

coefficients but show distinct vortex induced peak in the lift force spectra.

3.3 Effects of cross sectional modifications on wind loads

The second phase of this wind tunnel testing program is to study the effects of minor cross sectional modifications on tall buildings' wind loads. Four geometry modification effects were studied, namely, effects of aspect ratio, effects of corner chamfering, and the effects of two type of recess sections. The square shape is the basic cross sectional geometry, and the wind loads of the $H/B = 5$ square model is the basis of this comparison. All wind loads taken from the geometry modification models were normalized by the corresponding wind load of model S2.

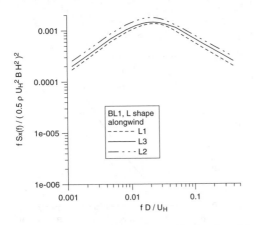

Figure 4(a). Alongwind force spectra of L shaped models.

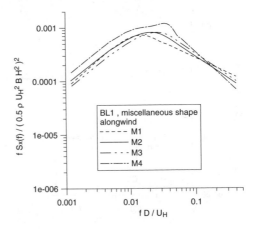

Figure 5(a). Alongwind force spectra of miscellaneous models.

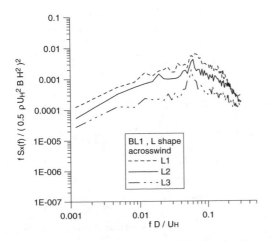

Figure 4(b). Acrosswind force spectra of L shaped models.

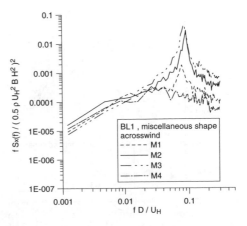

Figure 5(b). Acrosswind force spectra of miscellaneous models.

The normalized wind force w.r.t the wind loads of model S2 are listed in Table 3. Increasing the model aspect ratio from 4 to 7, which put building model in a smoother and higher wind speed zone, could increase mean alongwind force by 30%. The effect of aspect ratio on the RMS drag force is insignificant. Chamfering building's sharp edge corner can effectively reduce both building's mean and fluctuating drag force. Using recess section as building's geometry will introduce two effects on wind loads: (i) increase of building height will put building in higher wind speed and less turbulent zone; (ii) the discontinuity caused by recess tends to weaken the flow separation. As the result of these two effects, there is only minor variation on mean drag, and moderate (<15%) change on the fluctuating drag force.

Change of building aspect ratio and adopting recess section cast only minor influence on the nature of lift force; all lift force spectra show similar narrow-band characteristics as the square building. However, chamfering building's sharp edge corner makes noticeable impact on the across wind force spectra. The spectra, shown in Figure 6, becomes lower in peak and wider in bandwidth.

4 BUILDINGS DESIGN WIND LOAD

The wind force spectra obtained in this project were then applied to a prototype tall building for the estimation of design wind load. The prototype structure is assumed to be 200 m in height, H/B = 7, square shape building. The structural density is $200 \, kg/m^3$, damping ratio $\xi = 0.01$ and natural period of 5 seconds. Design wind speed is taken to be 42.5 m/s. Since the current Taiwan wind code is similar to the

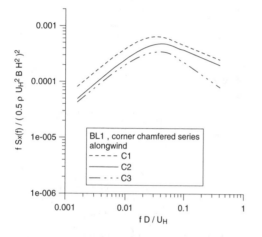

Figure 6(a). Alongwind force spectra of corner cut models.

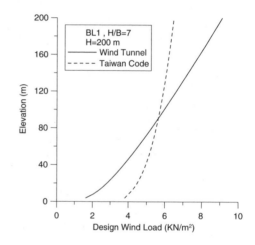

Figure 7(a). Tall building's design wind load.

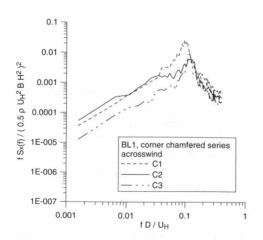

Figure 6(b). Acrosswind force spectra of corner cut models.

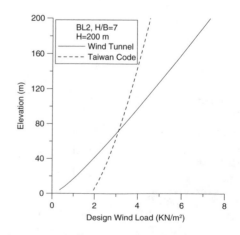

Figure 7(b). Tall building's design wind load.

ANSI/ASCE 7–88, only the wind load in the along-wind direction is taken into consideration in this study. During calculation of design wind load at building's various elevation, following assumptions were made: (i) velocity profile observes power law; (ii) only the linear fundamental mode was taken into consideration; (iii) gust response approach is used for structural maximum response (iv) the distribution of the resonant part along building height is proportional to the inertia force; (v) the distribution of the background part is proportional to the mean wind load. The results are shown in Figure 7. For both open terrain and urban terrain flow fields, the design wind load calculated based upon wind tunnel data is noticeably larger than the current Taiwan wind code in the upper half of the building height. In the lower half of the building height, although of less engineering significance, the design wind load based on wind tunnel data is smaller than the wind code. The reason of this discrepancy is partly due to that the wind code is actually based upon the so-called "point-like" structural concept; the vibration mode shape of tall building is not taken into consideration. For a more realistic distribution of design wind load, the "line-like" structure concept should be adopted in the next version's wind code.

5 CONCLUSIONS

This paper describes partial work of a project that intends to construct a wind load database that could be beneficial to tall building design. Up to this moment, wind loads of more than 30 building geometry shapes in two type of boundary layer flow fields were included in this database. These data can be divided into two categories: (1) core database which consists of 20 different cross sectional shapes; (2) auxiliary database which consists of flow field effects and minor geometry variation effects on building wind loads. These wind load data laid a foundation of tall building's wind resistant design. More wind tunnel testing results should be compiled into both categories

to enrich this wind load database. At the meantime, a parallel project on expert system has just begun. Combining the efforts of these two projects, a wind tunnel databased wind resistant design guide for tall buildings can then be built.

ACKNOWLEDGEMENT

The authors wish to express their appreciation to the National Science Council of the Republic of China for partial financial support under Contract No. NSC89-2211-E032-024.

REFERENCES

Hayashida, H. & Iwasa, Y. 1990. Aerodynamics shape effects of tall building for vortex induced vibration. *Journal of Wind Engineering and Industrial Aerodynamics* (33): 237–242.

Kawai, H. 1998. Effect of corner modifications on aeroelastic instabilities of tall buildings. *Journal of Wind Engineering and Industrial Aerodynamics* (74–76): 719–729.

Kikitsu, H. & Okada, H. 1999. Open passage design of tall buildings for reducing aerodynamics response. *Wind Engineering into 21st* Century, Larsen, Larose & Livesey.

Miyashita, K. et al. 1993. Wind induced response of high-rise buildings: effects of corner cuts or openings in square buildings. *Journal of Wind Engineering and Industrial Aerodynamics* (50): 319–328.

Tamura, T. & Miyagi, T. 1999. The effect of turbulence on aerodynamic forces on a square cylinder with various corner shapes. *Journal of Wind Engineering and Industrial Aerodynamics* (83): 135–145.

Yip, D.Y.N. & Flay, R.G.J. 1995. A new force balance data analysis method for wind response predictions of tall buildings. *Journal of Wind Engineering and Industrial Aerodynamics* (54/55): 457–471.

Tschanz, T. 1982. Measurement of total dynamic loads using elastic models with high natural frequencies. *Proceedings, International Workshop on Wind Tunnel Modeling Criteria and Techniques in Civil Engineering Applications*: 296–312. Maryland USA.

36. Special session on "Geomechanics aspects of slopes excavations and constructions"

System-based Vision for Strategic and Creative Design, Bontempi (ed.)
© *2003 Swets & Zeitlinger, Lisse, ISBN 90 5809 599 1*

Identification of soil parameters for finite element simulation of geotechnical structures: pressuremeter test and excavation problem

Y. Malécot, E. Flavigny & M. Boulon

University Joseph Fourier-Grenoble I, Laboratory "Sols, Solides, Structures", Grenoble, France

ABSTRACT: This paper is dedicated to the identification of constitutive parameters of the Mohr-Coulomb constitutive model from *in situ* measurements. A general definition of an objective function is proposed. A direct approach of inverse analysis is used to identify the shear modulus and the friction angle in four different situations. The first two examples deal with a "numerical" and with a real pressuremeter curve. A difficult convergence and a strong non unicity of solution is observed, which is classical in inverse analysis (ill posed problems). In a second stage, the horizontal displacements related to two excavation problems are used for identifying the two mechanical parameters. A clear minimum of the objective function is detected, giving a unique solution. The reasons of these differences are discussed and some ways of improving the interpretation of the pressuremeter test results are proposed.

1 INTRODUCTION

In most construction projects the geotechnical study is limited to the realization of some in situ tests which results don't allow the direct identification of the constitutive parameters of the soil layers. The use of the finite element method for the design of geotechnical structures is consequently strongly limited by the rough knowledge of the mechanical properties of the soil.

This paper presents preliminary results of a study having the aim of creating an inverse analysis tool permitting to identify a part of the soil parameters from *in situ* measurements.

In order to have a suitable identification method able to adapt itself to different kinds of geotechnical structures, we have chosen to use a direct approach to solve the inverse problem (Gioda & Maier 1980, Gens et al. 1988, Lecampion et al. 2002).

Trial values of the unknown parameters are used as input values in the finite element code "PLAXIS" to simulate the associated direct problem until the discrepancy between measurements and numerical results is minimized. It is well known that the solution of an inverse problem is not necessarily unique without a regularization by external constraints.

2 IDENTIFICATION METHOD

2.1 *The error function*

The discrepancy between the N measurements Ue_i and the associated numerical results Un_i is expressed

as a scalar error function F_{err} in the sense of the least square method:

$$F_{err} = \left(\frac{1}{N} \sum_{i=1}^{N} \frac{(Ue_i - Un_i)^2}{(1-\alpha)Ue_i^2 + \alpha \frac{1}{N}\sum_{j=1}^{N} Ue_j^2} \right)^{1/2} \quad (1)$$

Depending on the physical meaning of the experimental data, α is a coefficient which has to be chosen between zero and one. When α is equal to zero F_{err} refers to a rate of a relative error whereas when α is equal to unity F_{err} represents a rate of an absolute error. For the application cases presented in this paper, we have taken $\alpha = 0$ for the pressuremeter test and $\alpha = 1$ for the excavation problem (see section 3).

The error function $F_{err}(\mathbf{p})$ being defined as a scalar for each set of N_p unknown parameters, noted as a vector $\mathbf{p}$, the inverse problem is "solved" as a minimization problem in the N_p-dimension space restricted to authorized values of $\mathbf{p}$ between $\mathbf{p_{min}}$ and $\mathbf{p_{max}}$.

2.2 *Minimization algorithm*

In order to have a method able to adapt to a non-convex error function, we have chosen to perform this minimization by an iterative algorithm based on a gradient method.

Stage 1: Let us give an a priori $\mathbf{p_0}$ set of unknown parameters, $F_{err}(\mathbf{p_0})$ and its gradient $\nabla F_{err}(\mathbf{P_0})$ are evaluated by simulations of the direct associated problem.

Stage 2: The next set of unknown parameters $\mathbf{p}_1$ is then chosen such as:

$$\mathbf{p}_1 = \mathbf{p}_0 + x\mathbf{d}_0 \qquad (2)$$

where $\mathbf{d}_0$ is a vector indicating the going downward:

$$\mathbf{d}_0 = -\frac{F_{err}(\mathbf{p}_0)}{\left\|\nabla F_{err}(\mathbf{p}_0)\right\|^2}\nabla F_{err}(\mathbf{p}_0) \qquad (3)$$

and x is a non-dimensional scalar step which optimum value is determined by a quadratic estimation of F_{err} in the $\mathbf{d}_0$ direction.

Stage 3: The optimization program is iterating from *Stage* 2 to *Stage* 1 until F_{err} is less than the "measured" error or until the norm increment parameter vector, $\|\mathbf{p}_1 - \mathbf{p}_0\|$, is less than the desired precision.

3   APPLICATION CASES

The case studies presented in this paper deal with soils made of one or two layers of homogeneous sand modelled by a five-parameter Mohr-Coulomb model. The effect of water is limited to hydrostatic pressure.

To reduce the number of unknown parameters we assumed *a priori* values for parameters which influence is weak in their possible variation range or which value could be known from empirical relation. Hence, we assumed:

- cohesion $c \approx 0$
- Poisson's ratio $\nu = 0.25$
- dilatancy angle $\psi = \phi - 30°$

To take into account the increase of the shear modulus G with the depth z, we also assumed:

- $G(z) = G_{ref}\left(\dfrac{z}{z_{ref}}\right)^{0.4}$

Finally assuming a normally consolidated behaviour for the sand we used the Jaky relation to determine the initial stress field:

- coeffficient $K_0 = 1$-sin ϕ

All things considered, we have applied the identification method to determinate the shear modulus G_{ref} and the friction angle ϕ in two kinds of problems: The pressuremeter test and an excavation problem.

In both cases, to test the method, we have first used, as measurements, numerical results of a simplified problem before using true experimental data. It is clear that the identification method presented in this

paper has a chance to be pertinent only if the numerical model is able to reproduce the experimental data correctly. Assuming this important hypothesis, it seems, then, possible to use the numerical tool to explore what can be obtained from inverse analysis in more complex cases and what kind of measure is needed to identify a significant part of the soil constitutive parameters.

3.1   *Pressuremeter test*

3.1.1   *Numerical pressuremeter test*
We have simulated a pressuremeter test at 3 m depth. The finite element 2D axisymmetrical model is shown on Figure 1 (see details on Table 1).

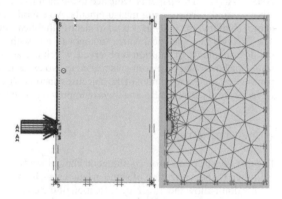

Figure 1.   Numerical pressuremeter test: 2D axisymmetrical model and associated mesh.

Table 1.   Characteristics of the numerical models

Numerical pressuremeter test
Problem size D = 2*3 m H = 5 m
Probe size d = 2*2.5 cm h = 40 cm
Axisymmetric, Elements type = 15 Nodes
219 Elements, 1932 Nodes, 2628 stress points

Pressuremeter test in calibration chamber
Problem size D = 2*0.6 m H = 1.5 m
Probe size d = 2*2.7 cm h = 16 cm
Axisymmetric, Elements type = 15 Nodes
287 Elements, 2521 Nodes, 3444 stress points

Numerical excavation problem
Problem size L = 50 m H = 25 m
Excavation size h = 6 m l = 2*10 m
Wall high h_w = 9 m
Plane strain, Elements type = 15 Nodes
246 Elements, 2185 Nodes, 2952 stress points

Sheet pile wall field test in Hochstetten
Problem size L = 70 m H =24 m
Excavation size h = 5.35 m l = 4 m
Wall high h_w = 6 m
Plane strain, Elements type = 15 Nodes
464 Elements, 3987 Nodes, 5568 stress points

The probe, is modelled as an homogeneous pressure P which is applied on the soil without unloading of the latter (molded or self-drilled pressuremeter). The volumetric strain of the probe $\Delta V/V$ is integrated from the nodes displacement. We have arbitrarily taken $G_{ref} = 30000\,\text{kPa}$ and $\phi = 35°$ to create numerically an "experimental" pressuremeter curve $P(\Delta V/V)$ (see Fig. 2). From this point G_{ref} and ϕ have been considered as unknown parameters.

We have first explored exhaustively the error function $F_{err}(G_{ref}, \phi)$ to have an idea of its surface topology. As shown in Figure 3, F_{err} is not convex and doesn't present a unique minimum. One can even notice that the F_{err} surface is crossed by a long flat valley of minimums (F_{err} less than 5%).

Even if it is clear that this result is hopeless for the inverse analysis of a pressuremeter curve, we have tried to apply the method presented in section 2.2. The way followed by the inverse analysis program is shown on Figure 4. It takes three calculation points to obtain an error function value F_{err} less than 5%. As attempted, in spite of this fast convergence, we can also see on Figure 4 that the results of the inverse analysis program are strongly dependent on the choice of the initial parameter vector $\mathbf{p_0}$. For the two different $\mathbf{p_0}$, namely $\mathbf{p_0} = (60000\,\text{kPa}, 39°)$ or $\mathbf{p_0} = (30000\,\text{kPa}, 45°)$, we obtain two different optimized parameter vectors $\mathbf{p_{min}}$ which minimized F_{err}, $\mathbf{p_{min}} = (50000\,\text{kPa}, 31°)$ or $\mathbf{p_{min}} = 5\,(15000\,\text{kPa}, 42°)$. The pressuremeter curves associated to these values are plotted on Figure 2. We observe that the curves are to close to be discriminated even if the parameters value are very different.

3.1.2 Pressuremeter test in calibration chamber

In spite of the previous results, we have simulated a molded pressuremeter test which has been performed by Mokrani in the calibration chamber of the

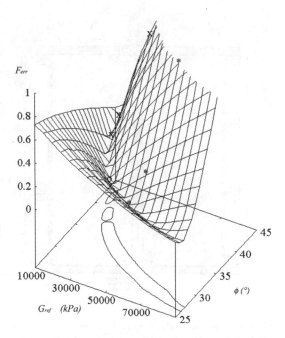

Figure 3. Numerical pressuremeter test: error function F_{err} versus G_{ref} and ϕ; (o) reference values $G_{ref} = 30000\,\text{kPa}$ $\phi = 35°$; way followed by the inverse analysis program (*) with $\mathbf{p_0} = (60000\,\text{kPa}, 39°)$; (x) with $\mathbf{p_0} = (30000\,\text{kPa}, 45°)$.

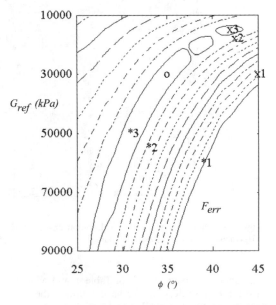

Figure 4. Numerical pressuremeter test: contour values of F_{err} versus G_{ref} and ϕ; (o) reference values $G_{ref} = 30000\,\text{kPa}$ $\phi = 35°$; way followed by the inverse analysis program (*) with $\mathbf{p_0} = (60000\,\text{kPa}, 39°)$; (x) with $\mathbf{p_0} = (30000\,\text{kPa}, 45°)$.

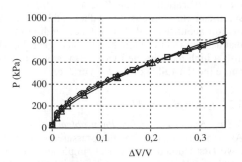

Figure 2. Numerical pressuremeter test – pressuremeter curve: pressure P versus the volumetric strain of the probe $\Delta V/V$; ($\square$) reference values: $G_{ref} = 30000\,\text{kPa}$, $\phi = 35°$; ($\diamond$) first optimised values: $G_{ref} = 50000\,\text{kPa}$, $\phi = 31°$; ($\triangle$) second optimised values: $G_{ref} = 15000\,\text{kPa}$, $\phi = 42°$.

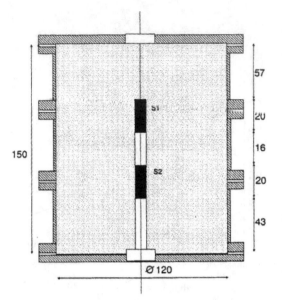

Figure 5. Pressuremeter test in calibration chamber: cross section of the experimental device.

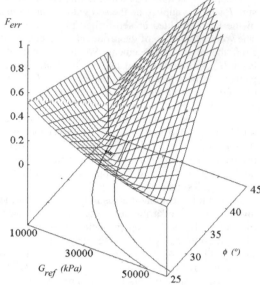

Figure 7. Pressuremeter test in calibration chamber: error function F_{err} versus G_{ref} and ϕ; (*) way followed by the inverse analysis program with $\mathbf{p_0} = (55000\,\text{kPa}, 40°)$.

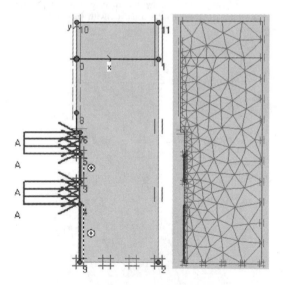

Figure 6. Pressuremeter test in calibration chamber: 2D axisymetrical model and associated mesh.

3S laboratory (see Figs 5, 6, Table 1 and Mokrani 1991 for more details). We have followed the same process as it is described in section 3.1.1.

The error function $F_{err}(G_{ref}, \phi)$ is plotted on Figure 7. It is worth noticing that its shape is very similar as the one which has been obtained with the

numerical pressuremeter test. We can particularly observe that the F_{err} surface is crossed by a long flat valley of minimums. However, we can also notice that, with true experimental data, the level of the valley is higher (F_{err} equal to 10%). This result is not surprising, since it is impossible to reproduce perfectly the experimental data with the numerical model both because of the experimental error and because of the simplified hypothesis of the numerical modelling.

As in the previous case, the inverse analysis process has a fast convergence which result depends on the initial parameter vector $\mathbf{p_0}$. Consequently, it makes the identification of both G_{ref} and ϕ from a pressuremeter curve impossible without additional information about the soil. For instance, adding the fact that usually a sand which the friction angle is high has also a high modulus and vice versa, would probably curve the flat valley of the F_{err} surface.

3.2 Excavation problem

3.2.1 Numerical excavation problem
As it has been done for the pressuremeter test, we have first studied a numerical simplified excavation problem. The finite element 2D model (plane strain) is shown on Figure 8 (see also numerical details on Table 1). The symmetric excavation, which is six meter depth and twenty meter broad, is supported by a sheet pile wall which head is stabilized by a strut.

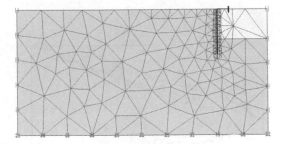

Figure 8. Numerical excavation problem: 2D model (plan strain) and associated mesh.

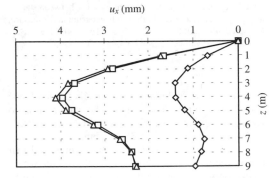

Figure 9. Numerical excavation problem: horizontal displacements of the sheet pile wall u_x versus the depth z; ($\square$) reference values: $G_{ref} = 22500\,\text{kPa}$, $\phi = 35°$; ($\lozenge$) a priori initial values; ($\triangle$) optimised values: $G_{ref} = 23000\,\text{kPa}$, $\phi = 34.5°$.

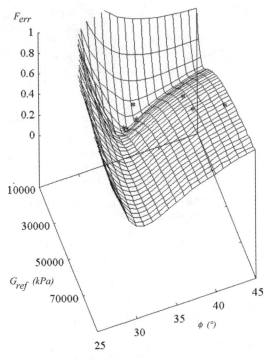

Figure 10. Numerical excavation problem: error function F_{err} versus G_{ref} and ϕ; (o)reference values: $G_{ref} = 22500\,\text{kPa}$, $\phi = 35°$; (*) way followed by the inverse analysis program with $\mathbf{p_0} = (60000\,\text{kPa}, 44°)$.

The horizontal displacements of the sheet pile wall u_x are obtained from the nodes displacement of the wall at each depth z.

As in section 3.1.1 we have arbitrarily taken $G_{ref} = 22500\,\text{kPa}$ and $\phi = 35°$ to create numerically an "experimental" wall displacements curve $u_x(z)$ (see Fig. 9) before exploring exhaustively the error function $F_{err}(G_{ref}, \phi)$. This function is plotted on Figure 10. We can observe that F_{err} is more convex than it was for the pressuremeter test and, more important, that it has a unique and well defined minimum.

The way followed by the inverse analysis program applied to this problem is shown on Figure 10. It takes around six calculation points to obtain an error function value F_{err} less than a few percents. As attempted, for two different initial parameter vectors $\mathbf{p_0}$ we obtain optimized parameters which are very close to the experimental ones independently on the choice of $\mathbf{p_0}$ (not represented on Fig. 10). Some wall displacement curves $u_x(z)$ associated to this problem are plotted on Figure 9.

3.2.2 Sheet pile wall field test in Hochstetten

To confirm the results obtained in the previous section, we have simulated the sheet pile wall field test in Hochstetten (Von Wolfffrsdorff 1994, Mestat & Arafati 1998). The excavation which is about five meter depth and four meter broad was achieved in six stages. Since an important cracking of the soil has been observed before the struts were placed in stage 3 (Von Wolffrsdorff 1994), we have used this stage as a reference one. So, we have chosen as experimental data the horizontal displacement of the wall $u_x(z)$ between the end of stage 3 (excavation up to 1.75 m depth, struts placed) and the end of stage 6 (excavation up to 5.35 m depth). This curve is plotted on Figure 13. The associated Plaxis model which has been used, is shown on Figure 11 (see Table 1). We have followed the same process as described in previous sections.

The error function $F_{err}(G_{ref}, \phi)$, plotted on Figure 12, shows as in the pressuremeter case that its shape is very similar as the one which has been obtained with the numerical excavation problem. We can particularly observe that the F_{err} surface has also a unique and well defined minimum. However, since

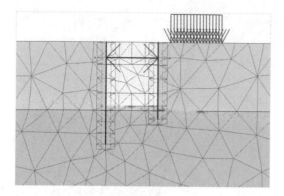

Figure 11.   Sheet pile wall field test in Hochstetten: central part of the 2D model (plan strain).

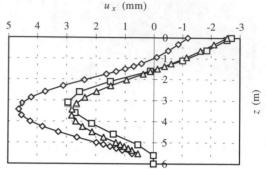

Figure 13.   Sheet pile wall field test in Hochstetten: horizontal displacements of the sheet pile wall u_x between stage 3 and stage 6 versus the depth z; ($\square$) experimental measure; ($\diamond$) *a priori* initial values: $G_{ref} = 16000\,$kPa, $\phi = 36°$; ($\triangle$) optimised values: $G_{ref} = 8000\,$kPa, $\phi = 45°$.

We can observe on Figure 12, that it takes around nine calculation points, depending on the choice of $\mathbf{p_0}$, to reach this optimum which is around $G_{ref} = 8000\,$kPa and $\phi = 45°$ with $z_{ref} = 5.35\,$m depth.

One can notice that ϕ is relatively high and that G_{ref} is relatively low comparing to the values which were proposed in Von Wolffrsdorff (1994) but Figure 13 shows that these values permit a good reproduction of the experimental data.

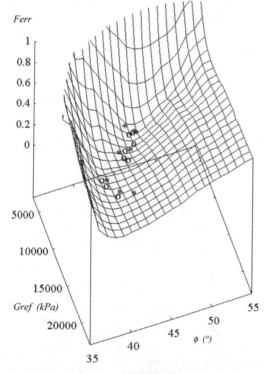

Figure 12.   Sheet pile wall field test in Hochstetten: error function F_{err} versus G_{ref} and ϕ; way followed by the inverse analysis program (o) with $\mathbf{p_0} = (16000\,\text{kPa}, 36°)$; (*) with $\mathbf{p_0} = (18000\,\text{kPa}, 42°)$.

we are not able to reproduce perfectly the experimental data with the numerical model, we can also observe that it is slightly flatter than it was in the numerical case.

As in the previous section, the inverse analysis program converges to optimized parameters which values are independent on the initial parameter vector $\mathbf{p_0}$.

4   DISCUSSION AND CONCLUSION

The present paper discussed the identification of some parameters of the Mohr-Coulomb model from particular *in situ* measurements. On one hand, it doesn't seem possible to simply identify both a friction angle and an elastic modulus from a pressuremeter curve. On the other hand, this identification is possible from the knowledge of the horizontal displacement of a sheet pile wall when the soil layering is simple. We can deduce that the pressuremeter curve results in a loss of local information and is not pertinent in terms of mechanical soil identification. Adding *a priori* information about the soil, using the unloading of the soil or changing the definition of the error function are some of the trails which have to be explored to obtain a unique solution.

In both studied case types, the error function obtained rather from numerical data or from true experimental data have the same shape. It seems, then, possible to use the numerical tool to explore what can be obtained from inverse analysis in more complex cases and what kind of measures is needed to identify a significant part of the soil constitutive parameters.

The simple minimization algorithm presented in this paper is efficient to minimize non convex two parameter function. However, it had to be tested for higher number of parameters, and for more realistic constitutive equations for soil.

REFERENCES

Gens, A. Ledesma, A. & Alonso, E.E. 1988. Back analysis using prior information – Application to the staged excavation of a cavern in rock. In Swoboda (ed.), *Numerical Methods in Geomechanics, Innsbruck* 1988. Rotterdam: Balkema.

Gioda, G. & Maier, G. 1980. Direct search solution of an inverse problem in elastoplasticity: Identification of cohesion, friction angle and in situ stress by pressure tunnel tests. *International Journal for Numerical Methods in Engineering.*

Lecampion, B. Constantinescu, A. & Nguyen Minh D. 2002. Parameter identification for lined tunnels in viscoplastic medium, *International Journal for Numerical and Analytical Methods in Geomechanics* 26: 1191–1211.

Mestat, P., Arafati, N. 1998. Modélisation par éléments finis du comportement du rideau de palplanches expérimental de Hochstetten. *Bull. des Laboratoire des Ponts et Chaussées.*

Mokrani, L. 1991. Simulation physique du comportement des pieux à grande profondeur en chambre de calibration. *Thèse de l'I.N.P.G.*

PLAXIS, Finite Element Code for Soil and Rock Analyses http://www.plaxis.nl/ie.html.

Von Wolffersdorff, Peter-Andreas 1994. Results of field test and evaluation of the predictions and subsequent calculations, *Workshop Sheet Pile Test Karlsruhe, Delft University, Holland, October* 1994.

Displacements of river bridge pier foundations due to geotechnical effect of floods

F. Federico
Department of Civil Engineering, University of Rome "Tor Vergata", Rome, Italy

G. Mastroianni
Civil Engineer, Rome, Italy

ABSTRACT: Safety of river bridge piers is analysed through the comparison of limit states of the soil (G)-foundation (F)-piers (P) system, potentially occuring during a flood. The displacements of a rigid, shallow F resting on compressible G, subjected to vertical, horizontal forces and destabilizing moments transmitted by P are first analysed through an original constitutive model (c.m.). The tilt and the stability of the G-F-P system are then investigated on the base of the c.m., taking into account the complex relationships among the tilt of P, the scour of G and the hydrodynamic forces (horizontal force on P and uplift on F). Results show that, depending on properties of G, geometry of P and F, water level during a flood and displacements history, the tilt of P may slowly increase towards a final, stable value, or quickly reach a critical value corresponding to the rotational instability. This one occurs before that the bearing capacity of the F-G sub-system is achieved.

1 INTRODUCTION

Exceptional alluvial events affected Italian territory in the last years; in particular, floods damaged highway and railway bridges. Most bridge failures was caused by scour of soils (G) and sediments surrounding piers (P) and supporting their foundations (F) (Ballio et al. 2000). Vulnerability analysis processed so far did not sufficiently take into account the main geotechnical factors governing the displacements evolution of the (G)-(F)-(P) system, caused either by the applied loads and bearing capacity reduction, both in turn coupled to floods (Federico et al. 2003).

Scour reduces the bearing capacity of the G-F geo-technical sub-system: plastic displacements may follow this reduction. Hydrodynamic pressures are applied on the front of P: a destabilizing moment is thus applied to the G-F sub-system. Moreover, it must be taken into account that the P are often tall structures: as a consequence, an additional destabilizing moment on G-F is associated to their tilt.

Thus, the safety of river bridge P must be evaluated through the analysis of the limit states achievable by the G-F-P system or its composing sub-systems, taking into account the bearing capacity reduction of G-F and the rotational instability of G-F-P, both related to the applied loads.

2 PROBLEM SETTING

2.1 Geometry

Reference is made to a rigid, leaning (tilt angle ϑ) P, with height h_P and rectangular cross section (b, width; l, length), under the action of a flood (Fig. 1). P is connected to a rigid, shallow F, rectangularly shaped (B, width; L, length) with thickness h_f, resting on cohesionless and compressible G; D is the depth of the F plane and $d_S(y)$ is the scour depth of G surrounding F depending on the elevation y of the river during the flood; h_W is the distance between the center of the F-G contact area (point O, Fig.1) and the center of gravity of the F-P sub-system; $h_T = h_p + h_f$ is the total height of the F-P sub-system.

2.2 Acting forces

The slenderness of tall river bridge P is the main responsible of their high sensitivity to perturbation factors (excavations, vibrations, earthquakes, water level variations, wind, …). The limited horizontal extension of F compared to the height and the not negligible weight of the P-F sub-system, generally cause *high contact pressures* on the F plane.

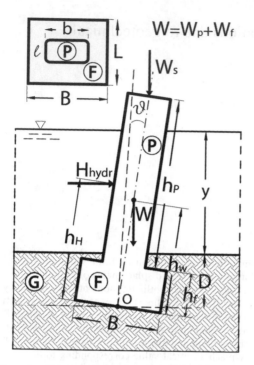

$W = W_p + W_f$

Figure 1. Forces acting upon a river bridge pier (P) and its foundation (F), resting on compressible soils (G). The reaction forces by G on F and the uplift force are not shown.

These, in turn, promote the yielding of G since the construction stage and, of primary importance, may cause an *initial* tilt angle.

Plastic settlements and rotations follow (Gudehus 1978), as it is confirmed by laboratory small scale tests on centrifuge models of tall structures resting on compressible clayey layers (Abghari et al. 1988).

The F of a tilted P is statically acted upon by eccentric vertical *loads* that steadily promote a *destabilizing moment* on F; this one may be strongly increased by horizontal force on P resulting from hydrodynamic pressures.

Conversely, the G-F geotechnical system may generate a *stabilizing moment* of limited extent, due to the small sizes of F (B, L, Fig. 1) as compared to the height h_P of P, depending on the cumulated displacements (in particular, rotations). So, especially if the P-F sub-system rests on on compressible G, depending on the elevation (y) of the fluvial current, the possible equilibrium tilt angle falls in a narrow range (Jappelli & Federico 1995).

The *lateral resistance* offered by the soil, that strongly depends upon cumulated settlements and rotations of the F, might contribute to the overall stability of the structure but, for river bridge P, due to scouring, it must be neglected.

The forces acting on the F-P sub-system are analytically expressed as follows:

– the total weight $W_T = W_S + W_p + W_f$ (W_S, applied, vertical forces; W_f, weight of F; W_P, weight of P) acting at a distance h_W from the F center (point O, Fig. 1);

– the uplift force $S(y)$ depending on the elevation y of the river: $S(y) = \gamma_w\,BLh_f + \gamma_w\,(D-h_f)bl$; $S(y)$ (not shown in Fig. 1) acts at a distance $h_S(y)$ from the point O (Fig. 1), approximately expressed as: $h_S(y) = (D + y)/2$, being γ_w the specific weight of the water;

– a horizontal force $H_{hydr}(y)$ applied at a distance h_H from O (the resultant of the horizontal actions on the P as wind or/and hydrodynamic pressures, seismic actions);

– the effective normal force and the shear force along the foundation plane (not shown in Fig. 1).

The increase of the river discharge causes the rise of the elevation y, as well as the increase of the following variables: current velocity, horizontal force H_{hydr} acting on the P, uplift force $S(y)$, scour depth $d_s(y)$; on the contrary, the bearing capacity of the G-F sub-system decreases.

2.3 *Scour depth and hydrodynamic forces*

The scour depth $d_s(y) \leq D$ has been estimated through the formula (Breusers et al. 1977):

$$d_s(y) = 2f_1f_2f_3l\,tanh\left(\frac{y}{l}\right) \qquad (1)$$

The coefficients f_2 and f_3 assume the values $f_2 = f_3 = 1$ for the considered case.

Being v_0 the river current velocity and v_e the critical velocity (scour beginning) defined as:

$$v_e = 0.85\sqrt{2gd_{50}\frac{\gamma_s - \gamma_w}{\gamma_w}} \qquad (2)$$

the coefficient $f_1 = 0$ if $v_0 \leq v_e$; otherwise ($v_0 > v_e$), it is assumed $f_1 = 1$. In the relationships (1, 2): l, length of the F measured along the direction orthogonal to the river flow; γ_s, specific weight of the G particles; d_{50} representative grain size of G; g, gravity acceleration.

The scour is thus neglected ($f_1 = 0$) if $0 \leq v_0/v_e \leq 1$.

The average velocity v of the river current is obtained under the following hypotheses: (a) steady state and uniform, slow flow of the river current ($i < i_c$), i being the slope of the river-bed and i_c the critical slope; (b) the river current is orthogonal to the short side of P; (c) the changes of water levels due to the reduction of the cross section of the flowstream, caused by the presence of the P, are neglected; (d) validity of the Gauckler – Strickler formula to describe the uniform motion of the current:

$$v = c_s R_H^{\frac{2}{3}} i^{\frac{1}{2}} \cong c_s y^{\frac{2}{3}} i^{\frac{1}{2}} \tag{3}$$

R_H, hydraulic radius (for wide channels, $R_H \approx y$). The elevation y_e of the river surface corresponding to the scour beginning is thus obtained:

$$y_e = \left(v_e / (c_s \sqrt{i}) \right)^{\frac{3}{2}} \tag{4}$$

Referring to a rectangular cross section of the river, with constant width $b_0 = 35\,\text{m}$, in absence of cobbles or bushes, the resistance coefficients c_s is approximately equal to 40 [$\text{m}^{1/3}\,\text{s}^{-1}$]. The discharge Q is thus expressed vs y as $Q = vb_0 y = c_s b_0 \sqrt{i} y^{5/3}$. The hydrodynamic force H_{hydr} acting on P is:

$$H_{hydr} = \frac{\gamma_w v^2 a}{g} = \frac{\gamma_w l y c_s^2 i y^{\frac{4}{3}}}{g} = c_1 y^{\frac{7}{3}} \tag{5}$$

where $a = ly$, surface of P exposed to the hydrodynamic pressures during floods and $c_1 = \gamma_w l c_s^2 i/g$.

3 G-F INTERACTION CONSTITUTIVE MODEL

3.1 Bearing capacity

The bearing capacity (q_{lim}) of the G-F sub-system, under vertical, centred loads, for drained conditions, assumes the following expression:

$$q_{lim}(y) = N_\gamma \zeta_\gamma \gamma' \frac{B}{2} + N_q(y) \zeta_q \gamma' [D - d_s(y)] \tag{6}$$

γ', submerged unit weight of G; N_q, N_γ, dimensionless factors, depending on the shear resistance angle φ' of the G surrounding F; ζ_q, ζ_γ shape factors (Vesic 1975);

$$d_q(y) = 1 + 2\frac{(D - d_s(y))}{B} tg\varphi' (1 - sen\,\varphi')^2 \tag{6'}$$

for ($D/B < 1$), correction (depth) factor.
The ultimate value of the vertical, centred load acting on the G-F system is $N_R(y) = q_{lim}(y)BL$.

Scouring of soils surrounding the P and F, due to the rise of the river water level as well as to the increase of its average velocity, induces a decrease of the G-F bearing capacity, mainly proportional to the decrease of the foundation depth ($D - d_s(y)$).

3.2 Constitutive model

The displacements of the G-F sub-system, subjected to variable (during floods) vertical force N, horizontal force H and overturning moment M have been globally modelled through an original rigid-plastic constitutive model (Mastroianni 2002).

The motion of the rigid, shallow F is identified through the rotational component ϑ, the vertical (settlement) and horizontal displacements v, u, respectively; v, u and ϑ are considered as generalized strains while N, H and M as generalized stresses (Nova & Montrasio 1991).

The collapse (f_R) and yielding surfaces (f), the plastic potential (g) and the hardening rule (χ) are written as a function of non-dimensional variables:

$$f_R(h, n, m,) = \pm \frac{m}{\rho B} \pm \frac{h}{\alpha} - n(1 - n) \tag{7}$$

$$f(h, n, m, n_0) = \pm \frac{m}{\rho B} \pm \frac{h}{\alpha} - n\left(1 - \frac{n}{n_0}\right) = 0 \tag{8}$$

$$g(h, n, m, n_g) = \pm \frac{m}{\rho \xi B} \pm \frac{h}{\alpha \lambda} - n\left(1 - \frac{n}{n_g}\right) = 0 \tag{9}$$

$$\chi(h, n, m) = \overline{K}(1 - n)\left(1 - \frac{m}{m_R}\right)\left(1 - \frac{h}{h_R}\right) \tag{10}$$

$$h = \frac{H}{N_R}; \ n = \frac{N}{N_R}; m = \frac{M}{N_R}; n_0 = \frac{N_0}{N_R}; n_g = \frac{N_G}{N_R} \tag{11}$$

$$\overline{K} = \frac{K}{N_R} \tag{12}$$

$$m_R(h, n) = \rho\left[n(1 - n) - \frac{h}{\alpha}\right] \tag{13}$$

$$h_R(n, m) = \alpha\left[n(1 - n) - \frac{m}{\rho}\right] \tag{14}$$

The yielding surface f is omothetic to the collapse surface f_R; N_0 and N_G respectively fix the position of the curves $f = 0$, $g = 0$ in the M, N plane ($H = 0$). This parameter doesn't play any practical role because,

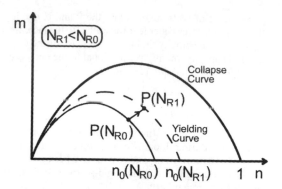

Figure 2. Scheme of yield loci and collapse curve ($H = 0$).

to determine the plastic strains, only the derivates of g occur. K depends upon G properties and F geometry.

A *non* associated flow rule is assumed. The *plastic generalized strain increment* may be expressed as (Chen 1975, Hill 1950):

$$du^p = \frac{1}{\chi(h,n,m)} \frac{\partial g}{\partial h}\left(\frac{\partial f}{\partial m}dm + \frac{\partial f}{\partial n}dn + \frac{\partial f}{\partial h}dh\right)$$

(15a)

$$dv^p = \frac{1}{\chi(h,n,m)} \frac{\partial g}{\partial n}\left(\frac{\partial f}{\partial m}dm + \frac{\partial f}{\partial n}dn + \frac{\partial f}{\partial h}dh\right)$$

(15b)

$$d\theta^p = \frac{1}{\chi(h,n,m)} \frac{\partial g}{\partial m}\left(\frac{\partial f}{\partial m}dm + \frac{\partial f}{\partial n}dn + \frac{\partial f}{\partial h}dh\right)$$

(15c)

As N_R becomes less than the minimum value achieved in the past ($N_{R1} < N_{R0}$), due to the scour of G (N assuming a constant value), the collapse surface (generalized stress M, N plane) contracts. Due to the increase of n_0 value (Equation 11), the expansion of the yielding curve in the dimensionless m, n stress plane ($n_0(N_{R1}) > n_0(N_{R0})$ in Fig. 2) and corresponding plastic displacements follow.

3.3 Estimate of parameters values

Parameters α, ρ, λ, ξ, K and N_R characterise the proposed model: α and ρ depend on the F depth D. According to the relationships proposed by Montrasio (2001) and taking into account the scour effect:

$$\alpha(y) = \alpha_0 + 0.72(D - d_s(y))$$

(16a)

$$\rho(y) = \rho_0 + 0.3(D - d_s(y))$$

(16b)

$\rho_0 = 0.5$ has been assumed on the base of experimental results (Musso & Ferlisi 1999); $\alpha_0 = 0.48$ is the

value proposed by Montrasio (2001). Parameters λ and ξ have been determined through a comparison between theoretical results and experimental data (Mastroianni 2002): $\lambda = 1.8$; $\xi = 5.5$.

The relationship proposed by Nova & Montrasio (1991) has been first considered for parameter K:

$$\frac{KB}{N_R} = 1 + 30D_R \Rightarrow \overline{K} = \frac{K}{N_R} = \frac{(1 + 30D_R)}{B}$$

(17)

D_R being the relative density of sandy G.

To better fit experimental results carried out on actual rectangularly shaped footings (Wolffersdorff 1991), a slightly modified K value has been assumed in computations (Mastroianni 2002).

4 STABILITY OF PIERS

For a general equilibrium configuration of the G-F-P system ($\vartheta = \vartheta_{eq}$), it is possible to define a safety factor referred to the stability of the tilt of P (Desideri & Viggiani 1994):

$$\eta_{\partial M} = \frac{\left(\dfrac{\partial M_{stab}}{\partial \vartheta}\right)_{\vartheta = \vartheta_{eq}}}{\left(\dfrac{\partial M_{dest}}{\partial \vartheta}\right)_{\vartheta = \vartheta_{eq}}}$$

(18)

Acting a generic perturbation (e.g., a horizontal force) with intensity (scalar value) S, and assuming that, for slender structures, the generalized stresses depend on the tilt ϑ too, H, M, N may be written as:

$$H = H(S,\theta);\ N = N(S,\theta);\ M = M(S,\theta)$$

(19 a,b,c)

The increase dM of the destabilizing moment due to an increase dS of the perturbation induces the evolution of the yielding surface, the reduction of χ, the increase $d\vartheta$ of the current tilt. A crucial importance assumes the *critical value* ϑ^* for which the G-F-P system achieves the instability condition, for assigned kind and intensity of the applied perturbation. This particular value may be determined (Como 1993; Federico 1996) by imposing

$$\frac{d\theta}{dS} \to \infty$$

(20)

Assuming small tilts ($\sin\theta \approx \theta$, $\cos\theta \approx 1$), the expressions (19 a, b, c) may be rewritten as:

$$N = N(S);\ H = H(S)$$

(21 a,b)

$$M = M(S,\vartheta) \cong H(S)h_H(S) + N(S)h_W(S)\theta$$

(21c)

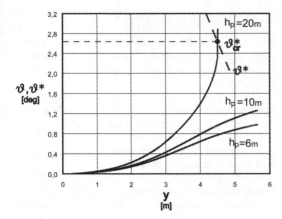

Figure 3. Current (ϑ) and critical (ϑ^*) tilts of the P vs water elevation y, for three values of the height h_p of the P ($D_R = 0.6$).

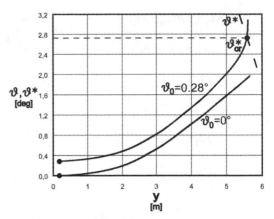

Figure 5. Current (ϑ) and critical (ϑ^*) tilts of the P vs water elevation y, for two values of the initial tilt ϑ_0 of the P ($h_p = 15$ m; $D_R = 0.6$).

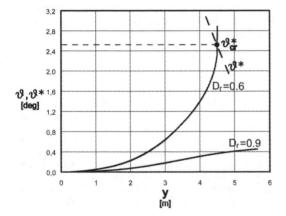

Figure 4. Current (ϑ) and critical (ϑ^*) tilts of the P vs water elevation y, for two values of the relative density D_R of G ($h_p = 20$ m).

Rewriting the hardening function, the infinitesimal increase of tilt and destabilizing moment, as a function of S, it is obtained (Mastroianni 2002):

$$\left(\frac{M_{cr}(S)}{M_R(S)}\right) = 1 - \frac{\dfrac{h_W N(S)}{\rho B}}{K\left(1 - \dfrac{N(S)}{N_R(S)}\right)\left(1 - \dfrac{H(S)}{H_R(S)}\right)} < 1 \qquad (22)$$

The Relationship 22 analytically expresses that the critical destabilizing moment ($M_{cr}(S)$) of the G-F-P global system, corresponding to the critical tilt ($\vartheta^*(S)$, see Figs 3, 4, 5) is always less than the ultimate value of the destabilizing moment ($M_R(S)$) acting only upon the G-F sub-system (the corresponding generalized

stress is represented by a point that lies on the collapse curve ($f_R = 0$, Equation 7).

Thus, the G-F-P system attains an instability condition (critical tilt) before that the G-F sub-system attains its limit state for bearing capacity under an eccentric and inclined load.

5 TILT OF PIERS DURING FLOODS

5.1 Setting and analysis of the problem

The tilt of P and the rotational instability of G-F-P are finally investigated. The analysis is carried out by assigning the maximum discharge Q (500 m³/s), obtaining y_{max} and imposing the generalized stress path corresponding to the progressive increase of y, first from zero to y_e, for which the scour begins, and then up to y_{max}.

The displacement evolution of P is modelled by a system of ordinary differential equations (15 a,b,c), taking into account the evolution of applied loads during a flood (21 a,b,c), scour of G, hydrodynamic pressures on P and uplift force on F.

The *geometry* of the system is characterized by the following values: (G): D equal to $d_s(y_{max})$ [m]; (F): $B = 8$ [m], $L = 5$ [m], $h_f = 1$ [m]; (P): $b = 7$ [m]; $l = 2.2$ [m]; $h_p = 6 \div 20$ [m]; initial tilt ϑ_0.

It results $N_R(y = 0) = 16278$ [t] and $(y = y_{max}) = 2653$ [t] (eq.6, 6′).

Under the hypotheses of small tilts, by assuming $h_H(y) = (D + y/2)$, the forces applied to the P-F subsystem and the hardening rule are written as follows:

$$N(y) = W_T - S(y); H(y) = H_{hydr}(y) \qquad (23\ a,b)$$

$$M(y,\theta) = F_1(y,\theta) + [F_2 - h_S(y)S(y)](\theta + \theta_0) \qquad (23c)$$

2405

$$F_I(y,\theta) = h_H(y)H_{hydr}(y) \qquad (24a)$$

$$F_2 = (W_p + W_f)h_W + W_S h_T \qquad (24b)$$

$$h(y) = \frac{H(y)}{N_R(y)}; \; n(y) = \frac{N(y)}{N_n(y)} \qquad (25\ a,b)$$

$$n_0(y) = \frac{N_0(y)}{N_R(y)}; \; n_g(y) = \frac{N_G(y)}{N_R(y)} \qquad (26\ a,b)$$

$$m(y,\theta) = \frac{M(y,\theta)}{N_R(y)} \qquad (27)$$

$$\chi(y,\theta) = \overline{K}(1 - n(y))\left(1 - \frac{m(y,\theta)}{m_R(y)}\right)\left(1 - \frac{h(y)}{h_R(y,\theta)}\right) \qquad (28)$$

The system of differential equations is numerically solved through the Runge-Kutta method. Particular importance assumes the tilt (ϑ) evolution.

5.2 *Results of computations*

The proposed model and results of numerical analyses put into evidence (Figs 3, 4 and 5) that:

- applied loads, displacements and the instability condition (critical tilt ϑ^*) of the G-F-P system evolve during floods in *a priori* unknown way, being coupled to the elevation y of the river current as well as to the current tilt (ϑ) of the P-F sub-system;
- the critical tilt ϑ^* (instability condition, eq. 20) of slender P may occur even for very small tilts; it significantly depends on the initial, plastic tilt of the F-P sub-system before (ϑ_0) the beginning of floods (Fig. 5); this means that the history of displacements of F and particularly past, progressively cumulated tilts, plays a fundamental role on the safety of P;
- critical tilt ϑ^* occurs before that the conventional bearing capacity of the F-G sub-system is achieved. Thus, the safety factor referred to the bearing capacity of the G-F sub-system (depending on the scour depth $d_s(y)$) does not represent the safety of the global G-F-P system (eq. 22);
- in proximity of ϑ^*, before that the instability condition is achieved, small increases of the water level (i.e. of hydrodinamic $H(y)$ and uplift forces $S(y)$) may determine appreciable increases of displacements (serviceability limit state), that could impair the functionality of superimposed structures (Figures 3, 4, 5);
- the instability condition is not attained by P of limited height (Figure 3) or if F rests on stiff soils (compare results shown in Figure 4).

6 CONCLUDING REMARKS

An original procedure is proposed to assess the vulnerability of bridge piers (P) in rivers, taking into account the phenomena governing fluvial dynamics during flood events.

The procedure requires: (i) an estimation of the maximum scour depth of the soil (G) surrounding both the P and F; (ii) an analysis of the displacements of the P-F-G system, in function of the physical and mechanical properties of the G, the shape of the F and P, as well as the destabilizing effects (horizontal force, scouring) induced by hydrodynamic forces. The coupling of both the hydraulic and geotechnical analyses enables to foresee displacements evolution, up to rotational instability, and to identify the most significant factors affecting P vulnerability.

Results of analyses show that the instability of P during floods may occur even for very small tilts, depending on the slenderness of P, on its initial tilt (before flood), on scour depth; this ultimate limit state occurs before that the conventional failure for bearing capacity of the F-G sub-system is achieved.

REFERENCES

Abghari A., Cheney J.A. & Kutter B.L. 1988. Leaning stability of tall towers. *Int. Conf. on Geotechnical Centrifuge Modelling*, J. Corté ed., 435–442, Balkema.

Ballio F., Bianchi A., Franzetti S., De Falco F. & Mancini M. 2000. Vulnerabilità idraulica di ponti fluviali. Dipartimento di Ingegneria Idraulica, Ambientale e del Rilevamento, *Politecnico di Milano, Ferrovie Dello Stato S.P.A.*.

Breusers H.N.C., Nicollet G. & Shen H.W. 1977. Local scour around cylindrical piers. *Journal of Hydraulic Research*, vol.15, no. 3, p. 211–252.

Chen W.F. 1975. Limit Analysis and Soil Plasticity. *Developments in Geotechnical Engineering*, 7, Elsevier.

Como M. 1993. Plastic and visco-plastic stability of leaning towers. *Convegno "Fisica Matematica e Ingegneria delle Strutture" in ricordo di Giulio Krall*, C.N.R., Elba.

Desideri A. & Viggiani C. 1994. On the stability of towers. *Symp. "Development in Geotechnical Engineering", Bangkok*, Balkema.

Federico F. 1996. Perturbation factors in the stability of leaning towers. *2nd Int. Conf. on "Multi-purpose High – Rise Towers and Tall Buildings"*, Singapore, July.

Federico F., Silvagni G. & Volpi F. 2003. Scour vulnerability of river bridge piers. *A.S.C.E., Journal of Geotechnical and Geoenvironmental Engineering*, in press.

Hill R. 1950. The Mathematical Theory of Plasticity. The *Oxford Engineering Science Series*, Clarendon Press.

Gudehus G. 1978. Engineering approximations for some stability problems in geomechanics, *"Advances in Analysis of Geotechnical Instabilities"*, Univ. of Waterloo, 1–24.

Jappelli R. & Federico F. 1995. Remarks on geotechnical safeguarding measures for leaning towers. *1st Int. Cong. on Science and Technology for the Safeguarding of Cultural Heritage in the Mediterranean Basin*, Catania-Siracusa.

Mastroianni G. 2002. Displacements of shallow foundations of river bridge piers. Thesis, *Department of Civil Engineering*, University of Rome "Tor Vergata".

Montrasio L. 2001. Cedimenti di Fondazioni su Sabbia – *Hevelius, Argomenti di Ingegneria Geotecnica*.

Musso A. & Ferlisi S. 2001. Displacements of a model on dense sand under vertical eccentric load. (*in press*).

Nova R. & Montrasio L. 1991. Settlements of shallow foundations on sand. *Géotechnique*, 41, n° 2, 243–256.

Vesic A.S. 1975. Bearing capacity of shallow foundations. In: Winterkorn, Fang, editors: *Foundation Engineering Handbook*. Van Nostrand Reinhold, pp. 121–147.

Wolffersdorff. 1991. Probebelastung zur Baugrundtagung 1990. *Versuchsergebnisse und Auswertung des Prognosewettbewerbes*.

A study of the microstructure to assess the reliability of laboratory compacted soils as reference material for earth constructions

C. Jommi
Politecnico di Milano, Milano, Italy

A. Sciotti
Università degli Studi di Roma "La Sapienza", Roma, Italy

ABSTRACT: Specifications for earth constructions are based on criteria derived from the hypothesis that the compacted soil in the field will behave as the reference material compacted in the laboratory through standard procedures. It is well known that different compaction procedures may create different microstructures for the same soil. As the microstructure plays an important role in the overall hydro-mechanical behaviour of compacted soils, it should be carefully considered when a reference material is adopted to characterise a given soil as construction material. An experimental investigation, aimed at devising and suggesting appropriate techniques to get a reliable reference material for earth constructions, was programmed. The investigation included a study of the microstructure of field-and laboratory-compacted clayey soil, performed by means of scanning electron microscope observations and mercury intrusion porosimetry. The main results of the investigation are summarised in this paper. The differences already observed at the microstructure level pose the question whether the soil compacted in the laboratory adopting the standard procedures should be effectively considered as a correct reference material.

1 INTRODUCTION

When soil is used as construction material, the specifications for placement techniques aim at assuring the stiffness, strength and permeability properties, which the soil should possess to guarantee the proper service conditions of the construction. Field-compaction specifications require the water content and the dry unit weight of the fill material to be a stated percentage of the optimum water content and maximum dry unit weight resulting from laboratory compaction tests on the borrow material. The design relies on the assumption that the overall hydro-mechanical behaviour will depend mostly on the compaction dry density and on the moulding water content, irrespective of compaction techniques.

During construction, field measurements of water content and volume unit weight are performed to verify the observance of design specifications. These controls substitute laboratory determination of mechanical and hydraulic properties on the field-compacted soil, which are expensive, time consuming and not feasible at the same rate of fill placement. Field tests of mechanical and hydraulic properties are rarely performed and

only for special structures. As a consequence, assuming a proper field compaction has been realised, the prediction of earth structure behaviour relies on mechanical and hydraulic properties determined on laboratory-compacted soil, considered the reference material.

Lambe (1958) first suggested a dependence of the overall behaviour of compacted soils on the microstructure created during compaction. He proposed different models for soils compacted dry and wet of optimum and succeeded in explaining the main features of the different compacted soils as a function of the different microstructure models.

Scanning electron microscope (SEM) observations and mercury intrusion porosimeter (MIP) data contributed to highlight the predominant influence of the moulding water content on the microstructure built up during compaction. The dry of optimum samples are characterised by aggregates of clay platelets several μm in size and two main levels of porosity can be distinguished: the largest pores between aggregates, having dimensions comparable with the size of the aggregates, and the small pores inside aggregates. The aggregates are hardly evident in samples compacted wet of optimum, which display a more massive structure without the large inter-aggregates voids (Sridharan et al. 1971; Ahmed et al. 1974; Delage et al. 1996).

It is a common belief that compaction at water contents equal or higher than optimum (OMC) tends to produce a homogeneous material, whose behaviour is similar irrespective of the compaction energy (e.g. Hilf 1991; Cabot & Le Bihan 1993). This assumption will be valid as long as the same microstructure is expected. Although direct comparisons between undisturbed samples retrieved from field compacted earth works and laboratory compacted samples are scarce, the soil compacted in the laboratory has proved to behave quite differently from the same material compacted in the field (Colombo 1965; Benson & Daniel 1990; Høeg et al. 2000). The observed discrepancies may be partly due to the role of microstructural features that compacted soils developed mainly during the construction stage, depending on the employed methods and procedures.

An experimental programme was thus initiated, with the aim of devising and suggesting appropriate techniques to get a reliable reference material for earth constructions. The investigation included a study of the microstructure of field-and laboratory-compacted clayey soil, performed by means of scanning electron microscope observations and mercury intrusion porosimetry. The main results of this part of the investigation are summarised in the following. The contribution of a proper description of the microstructure is expected to:

1. help explaining the differences in the overall hydro-mechanical behaviour between field-and laboratory-compacted soils already observed in preceding experimental investigations;
2. help choosing proper models for constitutive behaviour;
3. provide the necessary information to assign correct values to soil parameters when indirect methods are adopted for their estimation, taking into proper account the microstructure features which dominate the mechanical and hydraulic behaviour.

2 MATERIAL AND SAMPLES

Undisturbed samples were retrieved from a river embankment about seven months after the end of the construction. In the construction of river embankments a reduced dry density and a high moisture content allow the in-place fill to be deformable enough to adapt to the foundation soils settlements, hence avoiding cracking and fissuring. The borrow pit soil compacted in the laboratory with a somewhat reduced energy with respect to standard Proctor is then a better reference material for earth construction than the standard Proctor (Colombo 1965).

Table 1. Classification properties of the compacted soil.

Sample	Clay fraction (%)	Liquid limit (%)	Plasticity index (%)	Void ratio —	Water content (%)
Field-compacted	27	52.4	24.4	0.860	27.9
Laboratory compacted	30	59.0	29.5	0.825	26.5

The soil used for the construction is a medium plastic clayey silt with activity $A = 0.9$ and specific gravity $G_s = 2.71$. Its relevant classification properties are summarised in table 1. The reference samples were compacted in the laboratory to ½ standard Proctor energy, by reducing the number of blows per layer to half of the standard. Compaction in the laboratory followed the procedures usually adopted to guarantee a homogeneous reference material. The borrow soil was initially dried, passed to sieve n. 200, mixed to the target water content with tap water, cured and dynamically compacted.

Both SEM and MIP require the soil to be completely dehydrated prior to testing. All the samples were carefully fractured before drying in order to obtain undisturbed surfaces, and they were then air-dried at room temperature. The drying process is accompanied by a moderate reduction in volume. The sample compacted in the laboratory suffered a volume reduction on the order of 9%, while the undisturbed sample shrinkage was 7%. The difference in the volume reduction of the two samples is the first evidence of different microstructure features (Seed & Chan 1959).

3 MICROSTRUCTURE INVESTIGATION

3.1 Mercury intrusion porosimetry

Mercury intrusion porosimetry is based on the principle that a non-wetting fluid will not intrude the pore space of a porous medium unless a positive pressure is provided. The mercury pressure, p_m, required for intrusion depends on the surface tension of the fluid, γ_m, the contact angle between the fluid and the particles, θ_m, and the pore size.

Pores that are intruded under a given pressure are referred to as having an equivalent cylindrical diameter $D = -4\gamma_m \cos\theta_m/p_m$. In a standard porosimeter the voids of the soil sample are intruded by gradually increasing mercury pressure, while the cumulative volume of intruded mercury is recorded. By assimilating the soil pore system to a bundle of parallel capillary tubes of different diameters, an inferred pore size distribution is provided.

It is worth noting that MIP can detect pores of a given diameter only if they are correlated, and that the entire volume of a pore will be assigned to its entrance diameter, D_e, i.e. the diameter of its neck. Hence, MIP does not provide, in general, the real distribution of pore sizes, but it gives relevant information on the connection structure of the pore system. MIP data will be presented as a function of the pore entrance diameter, to permit a comparison with SEM observations.

Mercury intrusion porosimetry has been performed with a porosimeter allowing for a maximum injection pressure of 200 MPa. Pores of entrance diameter in the range 100 μm ÷ 0.007 μm can be detected.

Two specimens of each sample were tested, at least. The experimental data are completely repeatable for the samples compacted in the laboratory. For the undisturbed samples small differences have been observed in the cumulative volume of pores of equivalent diameter greater than 20 μm.

The intrusion curves of both the undisturbed sample and the laboratory-compacted sample, shown in Figure 1, present the typical shape expected for clayey soils compacted at the optimum water content. A relatively small volume is intruded before reaching a breakthrough point, representing the onset of the effective intrusion of the pore clusters. The breakthrough diameters are 1 μm and 0.80 μm for the undisturbed sample and for the laboratory-compacted sample, respectively. The first intrusion fills progressively all the interconnected pore space. Upon extrusion, performed by gradually decreasing the mercury pressure, only the mercury filling *nonconstricted* pores is recovered.

Most of the intruded mercury remains entrapped in pores whose mean diameter is larger than the exit diameter. Delage & Lefebvre (1984) proposed to use the extrusion data to distinguish the *free porosity*, recoverable upon extrusion, from the *entrapped* porosity, represented by the difference between the intruded and extruded mercury volume.

The free volume of the undisturbed sample determined on the basis of extrusion data is slightly higher than 20% of the total intruded volume. For the laboratory compacted sample it is around 19%. The free volumes correspond to pores with equivalent diameter smaller than 0.2 μm and 0.09 μm on the first intrusion curves of the field-compacted and the laboratory-compacted samples, respectively. These diameters coincide with the upper limiting value for reversible penetration.

The log-log representation of the cumulative intrusion curves, shown in Figure 2, may better highlight the nature of the pore system connection mechanisms. The range of free porosity is indicated by III. A transition zone, II, is observed corresponding to the pressure range in which most of the pores volume is intruded (62% for the field-compacted sample and 80% for the laboratory-compacted material). This zone is delimited by entrance diameters $0.2\,\mu m \leqslant D_e \leqslant 1\,\mu m$ for the undisturbed sample, and $0.09\,\mu m \leqslant D_e \leqslant 0.8\,\mu m$ for the sample compacted in the laboratory.

The penetration in the largest pores of the undisturbed sample occurs with two clearly different mechanisms, separated by a dotted vertical line in figure 2.

A small volume of pores of entrance diameter $D_e \geqslant 20\,\mu m$ is easily intruded, as shown by the steep first part of the cumulative intrusion curve (I a). A second intrusion mechanism characterises pores with $1\,\mu m < D_e < 20\,\mu m$. In this range (I b), the intrusion curve can be well approximated by a straight line.

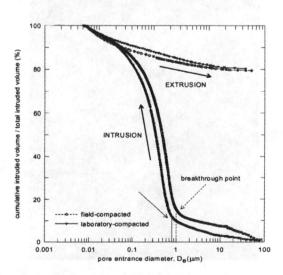

Figure 1.   Cumulative intrusion curves.

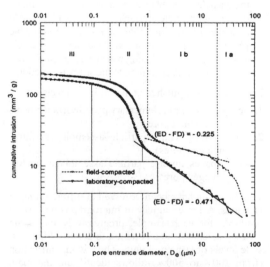

Figure 2.   Log-log representation of the cumulative intrusion curves of field-and laboratory-compacted samples.

2411

A straight line in the log-log representation of a cumulative intrusion curve denotes a range of entrance diameters of self-similar pores with a constant fractal dimension (Meng 1994).

The slope in the log-log plot equals the difference between the pore fractal dimension, FD, and their Euclidean dimension, ED. For the undisturbed sample, this difference $(FD - ED)$ is 0.225.

The sample compacted in the laboratory presents a significant difference from the undisturbed sample. The range of large pores is characterised by a single intrusion mechanism, identified by the straight line approximating all the range of pores with $D_e > 0.8 \, \mu m$. The first intrusion mechanism, observed for the undisturbed sample for $D_e > 20 \, \mu m$, is lacking. The slope of the straight line delimiting the range of self-similarity is $(FD - ED) = -0.471$.

3.2 Scanning electron microscope observations

Microstructure studies on SEM aim at describing spatial arrangements of particles and associated pore spaces. The description of particle arrangements includes type of particles, type, size and levels of particle aggregations, interactions between particles and between aggregations. To describe pore spaces means to identify their recurrent dimensions and distribution. In this study the microstructure description is mainly based on the characterisation scheme developed by Collins & McGown (1974).

Two main levels of aggregations may be identified:

1. the elementary particle arrangement, which is the basic form of particle interaction at the level of individual clay, silt or sand particles (elementary level); different sub-forms have been described;
2. the particle assemblage, which is the result of many elementary particle arrangements and is identifiable as a unit of particle organisation having definable physical boundaries (assemblage level). Particle assemblages can be distinguished in aggregates, connector, matrices, according to the specific function.

Porosity is distinguished at the different levels into:

1. pore spaces within elementary particle arrangements (intra-elemental pores);
2. pore spaces within particle assemblages (intra-assemblages pores);
3. pore spaces among particle assemblages (inter-assemblages pores).

The materials have been observed at different magnifications. A large number of micrographs have been produced for each sample, progressively increasing the magnifications in an attempt to detect all the possible levels of structural units. The low magnification (from 200× to 500×) allows identifying the main features of the fabric and the higher level of particle assemblages, defining their dimensions and shape, and the connections among them; high magnification (2000 ×) allows examining the elementary particle arrangements constituting the particle assemblages and details of clay particles interaction.

3.2.1 Laboratory-compacted sample

The microstructure of the sample compacted in laboratory is characterised at the higher level by aggregates, which consist of clay particles and silt grains. The aggregates are clearly distinguishable, even if the pore spaces between them are not large (Fig. 3).

In the elementary arrangements clay particles are not clearly identifiable and the whole structure seems to be veiled by a coating of clayey platelets (partially discernible structure) (Fig. 4). Interparticle contacts are mostly face-to-face or edge-to-face with a low angle of inclination.

Silt grains are evenly distributed and surrounded by clay particles, so that they don't stand out from the clay portion.

Porosity can be distinguished in:

1. inter-aggregates (inter-assemblages) porosity: pores between aggregates are largely anisometric with maximum widths up to ten of μm (Fig. 3);
2. intra-aggregates (intra-assemblages) porosity: pores between sets of elementary particle arrangements have dimensions extremely variable: isometric pores have a maximum size of $2 \div 6 \, \mu m$ (Fig. 4);
3. intra-elemental porosity: pores have dimensions lower than $0.5 \, \mu m$, which are hardly discernible by SEM.

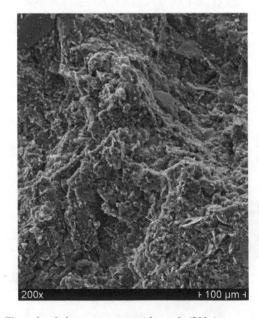

Figure 3.   Laboratory-compacted sample (200x).

Figure 4. Laboratory-compacted sample (2000×).

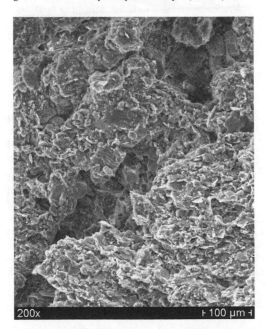

Figure 5. Field-compacted sample (200×).

3.2.2 *Field-compacted samples*

The higher level of particle organisation are aggregates whose dimensions are greater than 100 μm (the range is 100 ÷ 300 μm) (Fig. 5).

The aggregates are the main structural element, but in some cases a matrix, which seems to fill the larger

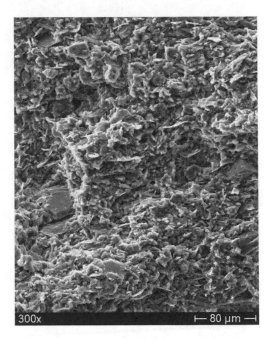

Figure 6. Field-compacted sample (300 ×).

voids between aggregates, is evident (Fig. 6). This matrix is made up of clay particles and silt grains: it differentiates from aggregates for a lower level of structural organisation and a different pore spaces distribution.

The aggregates are constituted by elementary particles arrangements of clay platelets and silt grains (Fig. 5).

At the elementary level clay particles are clearly discernible having well defined edges (Fig. 7); the arrangement is mostly random: inter-particle contacts are edge-to-face or edge-to-edge, giving rise to a sort of bookhouse structure (Fig. 7). The clarity of the particle edges is better than in the case of the laboratory-compacted samples, suggesting an increase in the particle spacing at the smaller pore sizes.

Silt grains are clothed with clay particles, they are clearly visible and appear evenly distributed; interactions occur through the clay coating or a clay matrix.

Porosity can be distinguished in:

1. inter-aggregates (inter-assemblages) pores: these are large and irregular voids, with size of tens of μm (30–80 μm) (Fig. 5);
2. intra-assemblages porosity:
   a) intra-aggregates pores between sets of elementary particle arrangements are largely isometric and regular in shape (Fig. 7); their dimensions are around 1.5 ÷ 5 μm;
   b) matrix porosity: pores have irregular shapes and a large variety of dimensions, in the range between 2 ÷ 10 μm (Fig. 6);

2000x ⊢ 10 µm ⊣

Figure 7. Field-compacted sample (2000×).

3. intra-elemental pores: they have dimensions below 1 µm, which are hardly discernible by SEM.

4 DISCUSSION AND CONCLUSIONS

The data presented highlight that the differences between the microstructure created during laboratory and field compaction are significant, at any level of particle organisation, although compaction was performed to similar dry density and moulding water content.

The soil compacted in the laboratory OMC shows the microstructural features expected for a similar preparation of cohesive soils, as described by previous studies (Cabot & Le Bihan 1993; Delage et al. 1996). The undisturbed sample globally appears to possess more structure levels than the sample prepared in the laboratory.

The different forms of aggregations found in the latter case may be due to the natural structural features of the pit soil, which are not eliminated by field compaction. On the contrary, the standard procedures for laboratory compaction tend to erase the original microstructure.

The observations confirm an early suggestion by Barden & Sides (1970), who stated that differences should exist between the microstructure of fine grained field-and laboratory-compacted soil. By comparing MIP data of samples compacted in the laboratory and in the field, Prapaharan et al. (1991) had concluded that the laboratory procedures are not capable of always reproducing the structure of field-compacted soil. Nevertheless, only the combined investigation through SEM and MIP, presented herein, may help understanding how the microstructural differences between the observed materials could affect the overall hydro-mechanical behaviour.

The different forms of aggregations are expected to play a significant role especially in unsaturated conditions, which characterise the state of a compacted earthwork during most of its service life. In unsaturated soils the microstructure effects are enhanced by the interaction of the soil pore system with the two permeating fluids.

The main relevant aspects expected on the basis of the observed microstructural features are summarised below:

1. the different intrusion mechanisms appearing from the MIP curves, confirming the different forms of particle organisation highlighted by SEM observation, will affect the hydraulic behaviour, i.e. permeability and retention properties;
2. the field-compacted soil will exhibit a stiffer response in the pre-failure range than the laboratory reference material. From the data presented here, this holds true, at least for the shrinkage following air-drying. The field-compacted material possesses a slightly higher volumetric stability than the same soil compacted in the laboratory;
3. significant differences in the deviatoric response of field-compacted and laboratory-compacted cohesive soils have already been observed (e.g. Colombo 1965; Ahmed et al. 1974; Høeg et al. 2000), both in the pre-failure range and at the peak. These features seem to be consistent with the different forms of aggregation observed for the two materials.

The observed microstructural differences suggest that the borrow pit soil compacted in the laboratory with standard procedures might not represent a reliable reference material for field-compacted earthworks. Laboratory compaction methods, better preserving the microstructure of the original soil, should be devised to overcome the difficulty in testing undisturbed samples of the material compacted in the field.

ACKNOWLEDGMENTS

The authors are indebted to Mr. Iscandri, Mr. Negrotti and Mr. Panzironi for their help with MIP and SEM instrumentation.

REFERENCES

Ahmed, S., Lovell, G.W. jr. & Diamond, S. 1974. Pore sizes and strength of compacted clay. *ASCE Journal of the Geotechnical Engineering Division*, 100(4): 407–425.

Barden, L. & Sides, G. 1970. Engineering behaviour and structure of compacted clay. *ASCE Journal of the Soil Mechanics and Foundation Division*, 96(4): 1171–1200.

Benson, C.H. & Daniel D.E. 1990. Influence of clods on hydraulic conductivity of compacted clay. *ASCE Journal of Geotechnical Engineering*, 116(8): 1231–1248.

Cabot, L. & Le Bihan J.-P. 1993. Quelques propriétés d'une argile sur la ligne optimale de compactage. *Canadian Geotechnical Journal* 30: 1033–1040.

Collins, K. & McGown, A. 1974. The form and the function of microfabric features in a variety of natural soils. *Géotechnique*, 24(2): 223–254.

Colombo, P. 1965. Materials for embankment construction. A study of the characteristics of the silty clay of river Po levees (in italian). *La Ricerca Scientifica*, Anno 35, Serie 2, 8(4): 13–16.

Delage, P., Audiguier, M., Cui, Y. & Howat, M. 1996. Microstructure of a compacted silt. *Canadian Geotechnical Journal*, 33, 150–158.

Delage, P. & Lefebvre, G. 1984. Study of the structure of a sensitive Champlain clay and of its evolution during consolidation. *Canadian Geotechnical Journal*, 21, 21–35.

Hilf, J.W. 1991. Compacted Fill. In H.-K. Fang (ed.), *Foundation Engineering Handbook*: 249–316. New York: Van Nostrand Reinhold.

Høeg, K., Dyvik, R. & Sandbækken, G. 2000. Strength of undisturbed versus reconstituted silt and silty sand specimens. *ASCE Journal of Geotechnical and Geoenvironmental Engineering*, 126(7): 606–617.

Lambe, T.W. 1958. The structure of compacted clay. *ASCE Journal of the Soil Mechanics and Foundation Division*, 84(2): 1–34.

Meng, B. 1994. Resolution-dependent characterization of interconnected pore systems: development and suitability of a new method. *Materials and Structures*, 27: 63–70.

Prapaharan, S., White, D.M. & Altschaeffl, A.G. 1991. Fabric of field-and laboratory-compacted clay. *ASCE Journal of Geotechnical Engineering*, 117(12): 1934–1940.

Seed, H.B. & Chan, C.K. 1959. Structure and strength characteristics of compacted clays. *ASCE Journal of the Soil Mechanics and Foundation Division*, 85(5): 87–128.

Sridharan, A., Altschaeffl, A.G. & Diamond, S. 1971. Pore size distribution studies. *ASCE Journal of the Soil Mechanics and Foundation Division*, 97(5): 771–787.

Stability of motion of detrital reservoir banks

A. Musso, P. Provenzano
University of Rome "Tor Vergata", Rome, Italy

A.P.S. Selvadurai
McGill University, Montreal, Canada

ABSTRACT: A stability analysis of motion of a detrital reservoir bank has been developed on the basis of surface displacement data, collected over a 25-year period. Since the reservoir bank has a characteristic saddle-shaped profile, due to the accumulation of debris on a concave bedrock, the slope has been modelled as a combination of a steep *feeder block* that is elastically connected to a *accumulation block* on a flatter profile. The blocks are at rest with frictional contact on planar surfaces, which are assumed to lie parallel to slope profile. The motion of the *two-block system* under stress increments induced by reservoir level variations at the toe region has been investigated and the stability-sliding conditions have been determined in case of a rate-effect-dependent frictional contact. The results are presented as time histories of displacements related to the history of the variations in reservoir level. The effects of monotonic as well as cycling rising of the water level have been investigated. Although highly idealized, the model provides a useful guide that accounts for some unexplained aspects of post-failure slope movements, well documented in technical literature.

1 INTRODUCTION

The banks of artificial reservoirs, often made up of debris, are frequently in condition of incipient motion, that results from a safety factor slightly larger than unity. In such reservoir banks, slope movements can be triggered, as a consequence of changing reservoir levels. Depending upon the state of the mobilized strength (either residual or intermediate between peak and residual), the motion has a different meaning within the frame of the evolution of displacements toward the collapse. When the governing strength is at the residual value, or the mobilized resistance is equal to the maximum of a stress–strain curve characterized by a brittleness index equal to zero, post-failure movements are said to occur. In these cases, a relationship between action and response could be established, leading to some form of empirical tool that could be used for a variety of purposes, including development of policies for land utilization, based on forecasting of slope movement. In technical literature, many studies have focused on this subject, however, only a few focus on an adequate period of monitoring of the kinematics of a slope that would permit the analysis of the time-history of the slope movement phenomenon and its relationship to external triggering actions.

The conventional approach of analysis for such landslides considers only limiting stability conditions and relatively little account is taken of the actual motions of the sliding earth mass, despite the fact that slow stable creep may be acceptable in many situations whereas accelerating motions are undesirable. Moreover, the influence of rapid shearing on the strength of shear zones must be considered in the study of stability of these movements. A generalized analysis of the stability of motions of an idealized accumulation landslide is proposed. The landslide is idealized in the conventional way (Davis et al. 1992) by assuming parallel planar ground and failure surfaces, and a rigid sliding mass. The analysis differs from the conventional limit state analysis in two aspects. First the inertia of the sliding mass is taken into account. Second the limiting shear stress on the failure surface is assumed to depend on the sliding velocity. It is proposed that a small amount of velocity dependence is the simplest mechanism that permits the analysis of the range of motions observed in actual landslides.

2 AN EXAMPLE OF DETRITAL RESERVOIR BANK

The Ragoleto Reservoir in Sicily was constructed in 1963 by damming the Drillo River. On the occasion of the first filling of the Ragoleto reservoir, a section of the right bank, involving a detrital volume of about

$6 \times 10^6 \, m^3$, began to experience movements. The slope was immediately put under observation, with the continuous and uninterrupted recording of displacements of points located on the ground surface as well as of reservoir water levels. Along with this continual observations, inclinometric measurements and records of ground water levels in piezometers have been carried out over shorter periods of time, thus allowing the sliding mechanism to be identified. Records of observations for nearly twenty five years are available.

A mantle of debris overlies the bedrock surface, dipping at an angle to the horizontal variable between a maximum of about 15° at top of slope to 4° ~ 6° at the toe. The typical vertical cross section is pictured in the geological profile of Figure 1. The sliding mass exhibits a steep feeder slope lying immediately above a flatter accumulation slope. In the entire area under observation, delimited by two rather deep gullies (V. Barone & V. Interlandi) particularly in the South-West zone, appreciable displacements have been measured. The movement direction, nearly the same for all observation points, is towards the reservoir (South) with a relevant component towards the West at the top of the slope and a smaller one in the eastern direction at the toe. The maximum values of velocity, recorded during the semester April–October 1973, are close to 10 cm/month.

It is interesting to note that the control points located outside of the area are stationary or rather displace with a velocity much smaller than that prevailing in the central region. The vertical components of displacement are much smaller than the corresponding horizontal ones. This behaviour is in good agreement with the relatively small inclination of the bedrock surface. Referring to the last reading (January, 1992) the maximum displacement recorded was at the location B (270 cm). There does not appear to be a correlation between slide behaviour and rainfall distribution, probably due to the high permeability of soils. On the other hand a correlation between rate of displacement and reservoir water level appears plausible (Fig. 2). For each water level fluctuation, the slide rate history is a function of the rate of water fluctuation in the reservoir: starting from the same water level, the lower the rate of fluctuation, the higher the rate of slide movement and the lower the time lag required to reach the peak value of the rate of slide movement. The correlation between rate of water fluctuation and kinematic behaviour of the reservoir bank can be assumed as a clue of the influence of fast shearing on constitutive relation for frictional slip. Particularly when the water table reaches the maximum level (remaining constant) stable sliding occurs till the bank stops. The proposed stability analysis will show that, according to a growing body of evidence (Davis 1992, Wedage et al. 1997, Chau 1999), stable sliding can only occur when the strength-velocity relationship is positive. A simplified modelling of the landslide motion is presented, with an explanation of the surging motion and the mean acceleration, which may occur during water level fluctuations.

3 MODEL FORMULATION

Figure 3 shows the accumulation profile that is modelled. Both the upper feeder slope and the lower accumulation slope rest on a continuous slip surface,

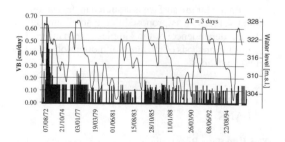

Figure 1. Typical vertical cross section of subsoil (Musso 1994).

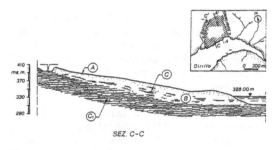

Figure 2. Histogram of displacements at the toe of detrital bank and correspondent reservoir water level values (Musso 1994).

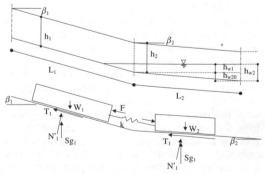

Figure 3. Accumulation slide profile and idealized model for accumulation slide.

which is assumed to be parallel to the upper surface. The piezometric surface is assumed to be horizontal, coinciding with the height of the reservoir water level, at a distance h_{w2} above the base of the slope. Horizontal dimensions L_i of both upper and lower slopes are sufficiently large, which permits the use of an infinite slope approximation and the omission of side and end effects. The response of the landslide can be ideally represented as a rigid block sliding on planar surfaces, which are parallel to the slope profile.

The block masses m_i are the masses of material contained within the two slides. We assume that they are invariant, even if the masses can be affected both by addition of material from uphill-slope and by transfer of material to the accumulation slope. The blocks move with velocities v_i. Interaction between the feeder and the accumulation slides is idealized by an elastic spring connecting the two blocks. For purposes of evaluating the forces on the upper block and parallel to the accumulation slope and for evaluating the forces on the lower block, the direction of the elastic force is assumed to be parallel to the feeder slope (Davis 1992). The model oversimplifies the physical problem, particularly with regard to the interaction between the two sliding masses. Nevertheless, the model has the virtue of simplicity and allows motions of the accumulation slide to affect the stability of the feeder slide. Moreover, as we will show below, the model is suited to represent all the important responses observed in the reservoir bank at Ragoleto. Motions of the two-blocks system are described by the three non-linear differential equations:

$$\begin{cases} m_1 \dfrac{dv_1}{dt} = W_1' \sin \beta_1 - T_{f1} - F \\ m_2 \dfrac{dv_2}{dt} = W_2' \sin \beta_2 - T_{f2} + F \\ \dfrac{dF}{dt} = k(v_1 - v_2) \end{cases} \qquad (1)$$

where

$$\begin{aligned} W_1' &= W_1 - S_{g1} \\ W_2' &= W_2 - S_{g2} \end{aligned} \qquad (2)$$

being

$$S_{g1} = 0$$

$$S_{g2} = \frac{1}{2} \cdot \gamma_w \cdot \frac{h_{w2}^2}{\tan \beta_2} \qquad \text{for } h_{w2} < h_{w20} \quad (3.a)$$

$$S_{g1} = \frac{1}{2} \cdot \gamma_w \cdot \frac{h_{w1}^2}{\tan \beta_1}$$

$$S_{g2} = \gamma_w \cdot \frac{h_{w1} + h_{w2}}{2} \cdot L_2 \cos \beta_2 \qquad \text{for } h_{w2} > h_{w20} \quad (3.b)$$

$W_i = m_i g$ are the block weights, g is the acceleration gravity, γ_w is the pore water unit weight, h_{w1} and h_{w2} are the water levels shown in figure 3, T_{fi} are the sliding resistances on failure surface, F is the elastic force, k is the spring constant and β_i are the slope angles. Motivated by the studies by Dieterich (1979), Rice & Ruina (1984) and from the experimental results of Skempton (1985) it is suggested to use a positive rate-dependent relationship for the friction resistances, by the usual Coulomb failure shear strength:

$$T_{fi} = N_i' \left[\tan \varphi' + A \log \left(1 + \frac{v_i}{v_r} \right) \right] \qquad \text{for } i=1, 2 \qquad (4)$$

where A is a constant, φ' is the stress friction angle and v_r is a constant reference velocity, while

$$\begin{aligned} N_1' &= W_1' \cdot \cos \beta_1 \\ N_2' &= W_2' \cdot \cos \beta_2 \end{aligned} \qquad (5)$$

The reactived landslide has no cohesion (well formed failure surface). Introducing the dimensionless ratios (see Appendix) the Equations 1 can be represented in the following non-dimensional forms:

$$\begin{cases} \dfrac{dV_1}{dT} + p \cdot \mathfrak{I} = \sin \beta_1 - \cos \beta_1 \cdot \left(\tan \varphi' + A \log_{10} \left(1 + \dfrac{V_1}{Vr} \right) \right) \\ \dfrac{dV_2}{dT} - \mathfrak{I} = \left[\sin \beta_2 - \cos \beta_2 \cdot \left(\tan \varphi' + A \log_{10} \left(1 + \dfrac{V_2}{Vr} \right) \right) \right] \cdot \\ \qquad \cdot \left[1 - \Gamma \cdot H_2^2 \cdot \dfrac{h_2}{L_2 \cdot \sin \beta_2} \right] \\ \dfrac{d\mathfrak{I}}{dT} = V_1 - V_2 \end{cases}$$

$$\text{for } H_2 \leq H_{20} \qquad (6.a)$$

$$\begin{cases} \dfrac{dV_1}{dT} + p \cdot \mathfrak{I} = \left[\sin \beta_1 - \cos \beta_1 \cdot \left(\tan \varphi' + A \log_{10} \left(\dfrac{V_1}{Vr} \right) \right) \right] \cdot \\ \qquad \cdot \left[1 - \Gamma \cdot H_1^2 \cdot r^2 \cdot \dfrac{h_1}{L_1 \cdot \sin \beta_1} \right] \\ \dfrac{dV_2}{dT} - \mathfrak{I} = \left[\sin \beta_2 - \cos \beta_2 \cdot \left(\tan \varphi' + A \log_{10} \left(\dfrac{V_2}{Vr} \right) \right) \right] \cdot \\ \qquad \cdot \left[1 - \Gamma \cdot (H_1 + H_2) \right] \\ \dfrac{d\mathfrak{I}}{dT} = V_1 - V_2 \end{cases}$$

$$\text{for } H_2 > H_{20} \qquad (6.b)$$

3.1 Static stability analysis

There are well-defined limits for static behaviour of the two-block model. Assuming static stability conditions at $t = 0$, when $h_{w2} = 0$, the shear strength forces are:

2419

$$T_{mi} = \frac{T_{fi}}{\eta_G} = \frac{T_{fi}}{\eta_G} = \frac{N_i' \cdot \tan\varphi'}{\eta_G} = \frac{W_i' \cdot \cos\beta_i \cdot \tan\varphi'}{\eta_G} \qquad (7)$$

where the global safety factor, in dimensionless form, is

$$\eta_G = \frac{\dfrac{m_1}{m_2} \cdot \cos\beta_1 \cdot \tan\varphi' + \cos\beta_2 \cdot \tan\varphi' - \xi_2 \cdot H_2^{\,2}}{\dfrac{m_1}{m_2} \cdot \sin\beta_1 + \sin\beta_2 - \Gamma \cdot \dfrac{h_2}{L_2} \cdot H_2^{\,2}}$$

for $H_2 \le H_{20}$ (8.a)

$$\eta_G = \frac{\dfrac{m_1}{m_2}(\cos\beta_1 \tan\varphi' - \xi_1 r^2 H_1^{\,2}) + \cos\beta_2 \tan\varphi' - \Gamma(H_1 + H_2)}{\dfrac{m_1}{m_2}(\sin\beta_1 - \Gamma r^2 \dfrac{h_1}{L_1} H_1^{\,2}) + \sin\beta_2 - \Gamma(H_1 + H_2)\sin\beta_2}$$

$\cdot \cos\beta_2 \tan\varphi'$
for $H_2 > H_{20}$

(8.b)

In a similar manner, the safety factors for each block can be defined as follows:

$$\eta_1 = \frac{T_{f1}}{T_{m1}} = \frac{\dfrac{m_1}{m_2} \cdot \cos\beta_1 \cdot \tan\varphi'}{\dfrac{m_1}{m_2} \cdot \sin\beta_1 - \Im}$$

$$\eta_2 = \frac{T_{f2}}{T_{m2}} = \frac{\cos\beta_2 \cdot \tan\varphi' - \xi_2 \cdot H_2^{\,2}}{(\sin\beta_2 - \Gamma \cdot \dfrac{h_2}{L_2} \cdot H_2^{\,2}) + \Im}$$

for $H_2 \le H_{20}$ (9.a)

$$\eta_1 = \frac{T_{f1}}{T_{m1}} = \frac{\dfrac{m_1}{m_2} \cdot (\cos\beta_1 \cdot \tan\varphi' - \xi_1 \cdot r^2 \cdot H_1^{\,2})}{\dfrac{m_1}{m_2} \cdot (\sin\beta_1 - \Gamma \cdot r^2 \cdot \dfrac{h_1}{L_1} \cdot H_1^{\,2}) - \Im}$$

$$\eta_2 = \frac{T_{f2}}{T_{m2}} = \frac{\cos\beta_2 \cdot \tan\varphi' - \Gamma \cdot (H_1 + H_2) \cdot \cos\beta_2 \cdot \tan\varphi'}{\sin\beta_2 - \Gamma \cdot (H_1 + H_2) \cdot \sin\beta_2 + \Im}$$
for $H_2 > H_{20}$
At $t=0$ is
(9.b)

$$\eta_{G0} = \frac{\dfrac{m_1}{m_2} \cdot \cos\beta_1 \cdot \tan\varphi' + \cos\beta_2 \cdot \tan\varphi'}{\dfrac{m_1}{m_2} \cdot \sin\beta_1 + \sin\beta_2}$$

(10)

The partial safety factors depend on the initial elastic force, which is derived by setting in the first or in the second equation of (1), the time-derivatives equal to zero. The initial elastic force, in dimensionless form, is:

$$\Im_0 = \frac{\cos\beta_2 \cdot \tan\varphi'}{\eta_{G0}} - \sin\beta_2 \qquad (11)$$

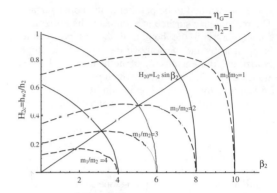

Figure 4. Critical values of water table for the static stability of the accumulation slope and of the overall two-block system, $\varphi = 12°, \beta_1 = 14°, h_1/L_1 = 1/10, h_2/L_2 = 1/5$.

Starting from the initial condition, if the water level H_2 increases, when $\eta_i = 1$ the i-th block will move and the elastic force will change. It can be verified that the feeder block does not slide before the accumulation block slides. Imposing $\eta_{10} = 1$, $\eta_{20} = 1$ or $\eta_{G0} = 1$, it possible to derive the dimensionless critical value of water table H_{2c} which corresponds, respectively, to either the sliding of the accumulation or the feeder, block-system.

Figure 4 shows the H_{2c} values, which corresponds to increasing values of accumulation slope inclination β_2 and of mass ratio m_1/m_2, both for $H_2 < H_{20}$ and for $H_2 > H_{20}$. Fixing in the system configuration β_2 and m_1/m_2, the graph shows that, as the water table rises, first the accumulation-block slides and later the entire system slides ($\eta_{G0} = 1$).

3.2 Reservoir level changes

The stability of motion of the accumulation sliding is analysed by numerically integrating Equation 6. Since the system of ordinary differential are non-linear and non-autonomous, their solution can be achieved if the appropriate initial conditions are specified. These can be described by fixing the hydraulic boundary conditions. In particular the aim of the simulation is to determine the response of a reservoir bank for a drawdown water table time-history. Starting from the initial condition $H_2(T = 0) = 0$ the water table increases linearly to the maximum value H_{2max}, according to $H_2 = C^*T$, where C^* is the dimensionless (raising) velocity. When the level reaches a fixed maximum value H_{2max} it remains constant for a $\Delta T = T_{H2max} - T_{H1max}$; afterwards it reduces to the initial value. In particular, if T_1 is the time related to the accumulation-block instability $[\eta_2(T_1) = 1]$ and H_{2c} is the corresponding water level, at first, when $H_2 < H_{20}$, we have to solve the Equation 6 for

$$H_1[T] = 0$$
$$H_2[T] = H_{2c} + C * \cdot (T - T_1) \qquad \forall \ T \geq T_1 \qquad (12)$$

and for

$$H_1[T] = C^* \cdot (T - T_2)$$
$$H_2[T] = H_{20} + C^* \cdot (T - T_2) \qquad \forall \ T \geq T_2 \qquad (13)$$

when $H_2 > H_{20}$, assuming T_2 the time corresponding to $H_2 = H_{20}$.

Afterwards, maintaining the water level constant, for $H_{2max} > H_{20}$ we have to solve the Equation 6 for

$$H_1[T] = H_{1max} = cost$$
$$H_2[T] = H_{2max} = cost \qquad \forall \ T_{1Hmax} \leq T \leq T_{2Hmax} \qquad (14)$$

When the water level comes down to the initial value, we have to solve the Equation 6 for

$$H_1[T] = H_{1max} - C^*(T - T_{2Hmax})$$
$$H_2[T] = H_{2max} - C^*(T - T_{2Hmax}) \qquad \forall \ T \geq T_{2Hmax} \qquad (15)$$

when $H_2 > H_{20}$, and imposing

$$H_1[T] = 0$$
$$H_2[T] = H_{20} - C^* \cdot (T - T_3) \qquad \forall \ T \geq T_3 \qquad (16)$$

when $H_2 < H_{20}$, being T_3 the time corresponding to $H_2 = H_{20}$.

4 STABILITY ANALYSIS OF THE RESERVOIR BANK

On the basis of surface displacements data a stability analysis of motion has been developed using the mechanical model described previously. The problem parameters are collected in Table 1.

The block system models a unit strip of the actual landslide. As compared with the site conditions, the geometry parameters are reduced by a factor $\lambda = 2$. In a λ-scaled model the self-weight stresses at homologous

Table 1. Reservoir bank parameters.

Geometric parameters

L_1	160 m	h_1	15 m	β_1	12°
L_2	80 m	h_2	15 m	β_2	6°

Physical-mechanical parameters

ϕ	12°		γ	17 KN/m³ γ_w 10 KN/m³
$K = E/L_2$	10⁴ [kN/m³]			A 0.005
$\alpha = h_{w2}/t$	1 [m/day]			

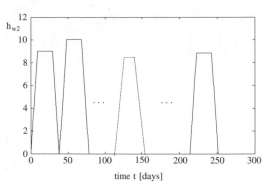

Figure 5. Water level time history.

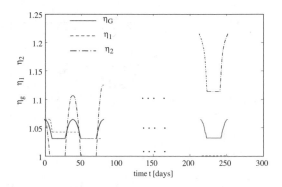

Figure 6. Safety factors of each block and of the system.

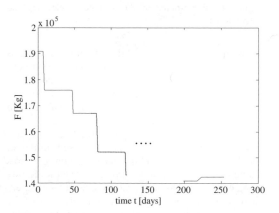

Figure 7. Elastic force time history.

points of model and site masses, and particularly along the corresponding failure surfaces developing on site and scaled block-system are thus the same. The mobilised strength has been derived by a back analysis of the slide in a limit equilibrium state, referred to the sections of Figure 1. The angle of residual shear

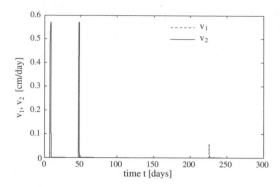

Figure 8. Slope velocity time history.

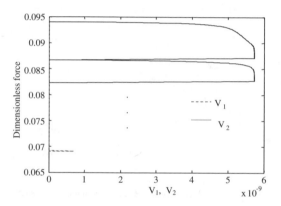

Figure 9. Phase-plane of dimensionless velocities and elastic force.

strength determined, by a linear failure criterion, is $\varphi' = 12°$. The elastic constant and the friction constant have been chosen, through a trial and error procedure. A K value of $10^4 [\text{kN/m}^3]$ and an A value of 0.005 give better agreement between calculated and observed velocities.

To illustrate the model response, the λ scaled watertable elevation time history has been used. In agreement with the actual condition the model response to a sequence of reservoir water level rise-water level lowering operations has been evaluated, in terms of safety factors, interaction elastic force and velocities of the blocks. For time $t < 0$ the slope is assumed to be stable ($\eta_{10} = \eta_{20} = \eta_{G0} = 1.06$); the initial elastic force is $F_0 = 1.91 \cdot 10^5 \text{kg}$ ($\Im_0 = 0.094$). At $t = 0$ the water table rises at velocity of 1m/day: the factor η_2 decreases, while the factor η_1 and the elastic force remain constant. At time $t_1 = 7.15$ days ($T_1 = 526.4$) η_2 reaches a value of unity and the accumulation block begins to move. At $t_2 = 8.36$ days the water table reaches h_{w0} and η_1 begins to decrease. In the first stages of water level rise, the maximum water level

reached on site at the elevation 326 m.s.l., corresponded to $h_{w2max} = 18 \text{m/}\lambda$ and $H_{max} = 0.6$, reached at $t_{1Hmax} = 9$ days. For $t_2 < t < t_{1Hmax}$ the motion of accumulation block is stable and reaches a stationary velocity of 0.58 cm/day. For $t_{1Hmax} < t < t_{2Hmax}$, when h_{w2max} remains constant, the velocity of accumulation block quickly decreases and the stable velocity is zero. The Phase Plane of Figure 9 is useful to understand the behavior of the solution and the stability condition. In the same period the elastic force reaches its minimum value $F = 1.76 \cdot 10^5 \text{kg}$ ($\Im = 0.087$). At $t > t_{2Hmax}$ the water level begins to fall: being the block at rest, the safety factors increase again. Reducing the elastic force, η_2 reaches a value greater than η_{20}, η_1 reaches a value smaller than η_{10}, while η_G resets to $\eta_{G0} = 1.06$.

At the end of this water level rise-water level lowering cycle, a second one starts again, reaching a value of $H_{max} = 0.67$ (corresponding to the site elevation of 328 m.s.l.). The velocity time history repeats itself: the accumulation block begins to move only when $H_2 = H_{2max}^{(1 \text{ cycle})} = 0.6$. The maximum velocity in the second cycle is no larger than the maximum value reached in the first cycle, while the elastic force reduces to $F = 1.67 \cdot 10^5 \text{kg}$ ($\Im = 0.082$). Repeating the water level rise-water level lowering cycles, the elastic force reduces to $F = 1.48 \cdot 10^5 \text{kg}$ ($\Im = 0.07$), corresponding to the limit equilibrium state of the feeder block ($\eta_1 = 1$). At this time if the water table rises again, when $H_2 > H_{20}$ the feeder block begins to move, while the accumulation block experiences no movement. When $H_2 = H_{2max}$ the motion is stable with stationary velocity equal to zero and the feeder block stops. Repeating the water level rise-water level lowering cycles the model is able to provide a prediction of the characteristic intermittent motion observed at the site.

5 SUMMARY AND CONCLUSIONS

A model capable of examining the stability of postfailure movement, occurring when the governing strength is at the residual value was investigated. An accumulation profile, derived from the accumulation of debris on a concave potential sliding surface has been considered. Frictional resistance is assumed to depend on the velocity of sliding. The case history of a reservoir bank provided an opportunity to examine rate effects of strength on the residual strength values. Introducing a simplified two-block system, the stability condition of actual motion under stress increments induced by reservoir level variations at the toe, has been investigated. The results are presented as time histories of displacements related to the history of reservoir level variation. The effects of monotonic as well as cycling rising of water level have been investigated. Although highly idealized, the model has the capability to take into account certain unexplained relevant features of post-failure slope movements. Among these, successive stick-slip movements due to safety factor gradually approaching unity, certain "strain-hardening" effect in the slope response, involving a progressive rise of the critical reservoir level responsible for the triggering of the slide, for different events of reservoir filling. The proposed model can be extended to an elastic continuum, even if the overall stability properties are not significantly different from the much simpler rigid-block analysis.

APPENDIX: DIMENSIONLESS RATIOS

$$H_1 = \frac{h_{w1}}{h_2} \qquad H_2 = \frac{h_{w2}}{h_2}$$

$$\Gamma = \frac{\gamma_w}{2 \cdot \gamma} \qquad p = \frac{m_2}{m_1} \qquad r = \frac{h_2}{h_1}$$

$$\xi_1 = \Gamma \cdot \frac{h_1}{L_1} \cdot \frac{tan \ \varphi'}{tan \ \beta_1} \qquad \xi_2 = \Gamma \cdot \frac{h_2}{L_2} \cdot \frac{tan \ \varphi'}{tan \ \beta_2}$$

$$\omega_2 = \sqrt{\frac{k}{m_2}} \qquad T = t \cdot \omega_2 \qquad \Im = \frac{F}{m_2 \cdot g}$$

$$X_1 = \frac{x_1 \cdot \omega_2^2}{g} \qquad X_2 = \frac{x_2 \cdot \omega_2^2}{g}$$

$$V_1 = \frac{v_1 \cdot \omega_2}{g} \qquad V_2 = \frac{v_2 \cdot \omega_2}{g} \qquad C* = \frac{C}{\omega_2}$$

where $C = tan \ \alpha = H_2/t$ [sec^{-1}] and α the water raising velocity.

REFERENCES

Chau, K.T. 1999. Onset of natural terrain landslides modelled by linear stability analysis of creeping slopes with a two-state variable friction law, *International Journal for Numerical and Analytical Methods in Geomechanics* 23: 1835–1855

Davis, R.O. 1992. Modelling stability and surging in accumulation slides. *Engineering Geology* 33: 1–9

Davis, R.O., Desai, C.S. & Smith, N.R. 1993. Stability of Motions of translation landslides. *J. of Geotechnical Engineering*: 119, 3: 420–432

Dieterich, J.H. 1979. Modeling of rock friction. Experiment results and constitutive equations. *J. of Geophysical research* 10: 2161–2168

Musso, A. 1994. Limitazione dell'uso dei pendii con ridotto margine di sicurezza. *Proc. Interventi di stabilizzazione dei pendii* (In Italian), Ed. A. Pellegrino, CISM

Restivo, E. 1999 Interpretazione di meccanismi di dissesto di sponde di serbatoi artificiali in condizioni di post-rottura. PhD Thesis (In Italian) University of Palermo. Italy

Rice, J.R. & Ruina, A.L. 1984. Stability of steady frictional slipping, *J. of Applied Mechanics* 50: 343–349

Skempton, A.W. 1985. Residual strength of clays in landslides, folded strata and the laboratory. *Geotechnique*: 35, 1: 3–18

Wedage, A.M.P., Morgenstern, N.R. & Chan D.H. 1997, Simulation of time-dependent movements in syncrude tailing dyke foundation. *Can. Geotech. J.* 35: 284–298

Hybrid structural steel beam – reinforced concrete column connection

J.C. Adajar
AJIISS, Dallas, Texas, USA

T. Kanakubo
University of Tsukuba, Ibaraki, Japan

M. Nonogami & N. Kayashima
Daisue Construction Corp., Osaka, Japan

Y. Sonobe
Ashikaga Institute of Technology, Japan

M. Fujisawa
Tsukuba College of Technology, Ibaraki, Japan

ABSTRACT: Nine specimens were tested to evaluate the behavior of a hybrid steel beam – reinforced concrete column connection under axial and antisymmetrical cyclic loads. The connection is made of welded plates forming an encasement that confines twelve reinforcing bars at the corners and four bars at the center core. Experimental and theoretical results on loads, displacements, stresses and strains are presented to describe the general structural performance of the connection.

1   INTRODUCTION

Newly developed hybrid frames made of reinforced concrete (RC) columns connected to steel beams have been used in Japan. These frames can provide practical and economical merits by combining longer steel beams with high compression resistant RC columns. To further develop such hybrid structures, it is necessary to determine the strength and ductility of the connection. This study focuses on beam-column connections, as shown in Figure 1, that consist of welded steel plates forming an encasement confining the concrete and the reinforcing bars. The connections are designed to effectively resist varying column axial loads especially in tension. Openings in the encasement are made to provide the necessary arrangements for reinforcing bars. A square tube at the center of the encasement serves as the inner core confinement and the outer plates act as outer confining reinforcements. The specimens are modeled after some beam-column connections of a nine-story building that can be subjected to antisymmetrical bending moments and varying axial loads.

1.1   Objective

The main objective of this research is to investigate the general structural performance of the hybrid connection and its design method.

2   EXPERIMENTAL INVESTIGATION

2.1   Specimens

The specimens are about 1/3 the actual size of the designed structural members at the second floor of a nine-storey building. Table 1 shows the properties of the specimens. The column cross section is 320 mm ×

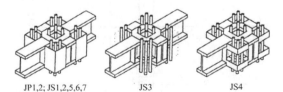

JP1,2; JS1,2,5,6,7   JS3   JS4

Figure 1.   Beam-column connections.

Table 1. Properties of specimens.

Specimen	JP1	JP2	JS1	JS2	
BxD(mm)	320x320				
main bars	12-D13		12-D16		column
core bars	4-D13		4-D16		
pg(%)	1.98		3.11		
stirrups	4-U5.1@50		4-U6.4@50		
pw(%)	0.50		0.75		
DxBxt1xt2	H-200x100x5.5x8		BH-200x100x6x25		beam
inner tube	□ 120x120x6		□ 125x125x6		connx
outer plates	3.2		2.3	4.5	

Specimen	JS3	JS4	JS5	JS6	JS7
BxD(mm)	320x320				
main bars	12-D16				
core bars	4-D16				
pg(%)	3.11				
stirrups	4-U6.4@50				
pw(%)	0.75				
DxBxt1xt2	BH-200x100x12x25				
inner tube	□ 125x125x9			□ 125x125x4.5	
outer plates	0.0	2.3	4.5	2.3	4.5

Table 2. Material properties.

Steel Plates				
plate properties	yield kgf/cm^2	ultimate kgf/cm^2	elongation %	part
SS400 t=2.3 mm	3610	4440	34	outer plate
SS400 t=4.5 mm	2900	3800	42	
SM490 t=6 mm	4030	5720	28	beam web
SM490 t=12 mm	3780	5550	28	
SM490 t=25 mm	3500	5400	48	flange
SM490 t=28 mm	3650	5440	31	diaphragm
SM490 t=9 mm	3640	5320	28	
SN490 t=4.5 mm	4060	5620	34	steel tube
SN490 t=6 mm	4170	5650	32	
Steel Bars				
designation	yield kgf/cm^2	ultimate kgf/cm^2	elongation %	part
D13	3680	5510	26	main/core
D16	4820	6880	19	bars
U6.4	13900	14900	9	stirrups
Concrete				
specimen	comp. strength kgf/cm^2	E kgf/cm^2 x10^{-5}	tensile strength kgf/cm^2	part
JP1	433	2.49		
JP2	456	2.50		
JS1	217	2.14	22.0	
JS2	234	2.35	20.6	
JS3	238	2.23	23.7	column
JS4	237	2.47	23.3	
JS5	190	2.12	21.1	
JS6	201	2.37	21.5	
JS7	211	2.27	21.1	

320 mm. Beam size is 200 mm × 100 mm. The column/ beam depth proportion Dc/Db is 1.6 and the width proportion Bc/Bb is 3.2.

2.2 Material properties

Table 2 shows the material test results. In flexural type specimens JP1 and JP2, D13(SD345) (diameter = 13 mm; yield strength = 345 Mpa) reinforcing bars and high strength ties U5.1(SBPD1275/1420) were used for RC columns. The concrete strength was designed to be 360 kgf/cm^2. H200 × 100 × 5.5 × 8 (SS400) was selected for the beams, In shear type specimens JS1–JS7, D16(SD390) steel bars and high strength stirrups U4.6(SBPD1275/1420) were utilized for columns. The design strength was 210 kgf/cm^2 for concrete. Steel beams were BH200 × 100 × 6 × 25 (SM490) and BH200 × 100 × 12 × 25 (SM490).

2.3 Loading method and sequence

As illustrated in Figure 2, an oil jack was used to apply the axial load to the column, and actuators at beam-ends provided the antisymmetrical bending moments. The loading cycle and sequence, as shown in Figure 3, started from a drift angle R = ±1/800 and ended at R = ±1/25. The drift angle was doubled after every two loading cycles. At R = ±1/25, the cycle was done only once.

2.4 Displacement and strain gauges

Figure 4 shows the setup to measure relative displacements and curvatures in beams and column. Gauges to record shear deformations within the beam-column connection are shown in Figure 5. Strain gauges placed

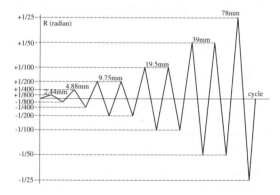

Figure 2. Loading cycle and sequence.

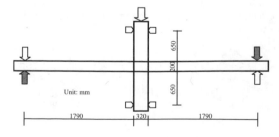

Figure 3. Loading method.

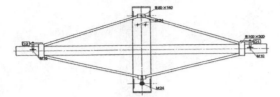

Figure 4.   Setup of displacement gauges.

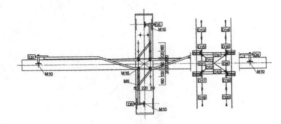

Figure 5.   Shear deformation gauges.

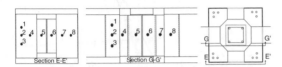

Figure 6.   Strain gauges on steel plates.

at preferred locations to monitor strains on steel plates
are shown in Figure 6.

3   TEST RESULTS

3.1   *Loads, displacements, cracking, and yielding*

Load-displacement diagrams are shown in Figure 7.
In bending failure type specimens JP1 (cross shape)
and JP2 (half cross shape), the maximum loads for
both specimens are 4.2 tonf during the positive cycle
and 4.4 tonf during the negative cycle. More visible
cracks occurred in the column of half-cross specimen
JP2 than in JP1. Cracks in JP1 are very minimal and
occurred only near the beam-column connection.
There was no indication of yielding in the steel reinforcements of the column. However, the web of the
beam was yielding. In a shear failure type specimen
JS1 that has steel encasement around the beam-column connection, cracks due to bending occurred in
the column during a drift R = 1/100 cycle. During
this cycle, the end of the steel beam and some parts
inside the beam-column connection were yielding.
Encasement plates yielded during the last cycle when
R = 1/25. Shear cracks on the column was seen during R = 1/25 cycle. A maximum load of 8.1 tonf was

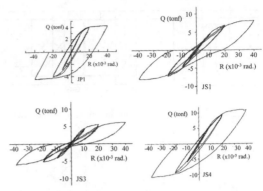

Figure 7.   Load – displacement diagrams.

reached during this cycle. The maximum load was
still less than the ultimate flexural strength of the column. In specimen JS3, where outer plate encasement
was not present, the flexural cracks in the column
occurred when R = 1/200. During this cycle, shear
and flexural cracks were noticed at the beam-column
connection. Yielding of steel plates inside the beam-
column connection and at beam-ends occurred during
R = 1/100. A maximum load of 6.2 tonf was attained
at R = 1/25. The ultimate flexural strength of specimen JS3 was not reached. In specimen JS4, flexural
cracks started to occur when R = 1/200. The beam
started to yield upon reaching R = 1/100. Plates
inside the beam-column connection began yielding
during R = 1/50. In this cycle, shear cracks appeared
in the area of beam-column connection. A maximum
load of 11.1 tonf was reached when R = +1/25. This
maximum load was almost equal to the beam ultimate
bending capacity.

3.2   *Strain distributions*

In flexural type specimens, the maximum strain in the
square tubing was 900μ for cross shape specimen JP1
and 400μ for half cross shape specimen JP2. The
outer plates of the beam-column connection indicated
a maximum strain of 500μ for JP1 and 300μ for JP2.
Figure 8 shows the strain values on the plates inside
the connection when R = 1/25. Strain readings for
shear type specimens JS1, JS3 and JS4 when R =
1/25 are plotted in Figure 9. The maximum load was
thought to be governed by the beam-column connection. This was true for specimens JS1 and JS3
where the inner plates and square tubing indicated large
strains. However, in JS4, the specimen that was thought
to fail in bending at beam-ends, relatively small strains
on the square tube were observed. Although strains on
the outer plates of JS1 were smaller than those on the
square tubing, yielding occurred on the outer plates.

2427

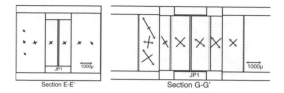

Figure 8. Typical strain indication in flexural type specimens.

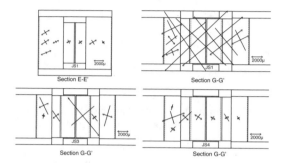

Figure 9. Typical strain indications in shear type specimens.

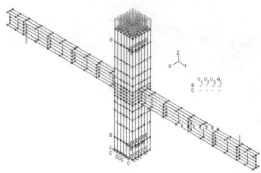

Figure 10. FEM model with load and boundary conditions.

4 ANALYSIS

4.1 Results of first analytical investigation

A structural model was created using ADINA finite element analysis program to simulate the behavior of a flexural type specimen JP1. The model indicating the load and boundary conditions is shown in Figure 10. Full fixity was assumed at the bottom end of the column. It was also assumed that there was no translation along y-axis at intermediate points near the top and bottom ends of the column where confining oil jacks were located. The model consisted of 2-node line elements for steel reinforcing bars, 4-node shell elements for steel plates, and 8-node solid elements for concrete for a total of 1340 nodes. Bilinear model with Von Mises yield condition was used for steel elements. Figure 11 shows that the experimental and theoretical load-displacement relations for the beam of specimen JP1 are in good agreement. Both analytical and experimental investigations indicated that when $R = 1/25$, the deflection of the beam is about 95% of the total deformation of the specimen as can be observed in Figure 12. During testing, cracks were observed on the surface of the column in the area of the beam-column connection but analytical results indicated that cracks formed beneath the surface as shown in Figure 13. Theoretical results for stresses on plates inside the connection were notably different from the experimental results. A response model for bond between concrete and steel plates was not provided. This may account for the difference between analytical and actual stress results

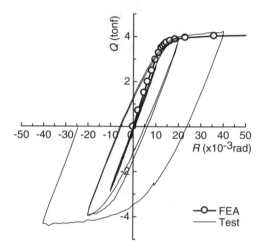

Figure 11. Analytical and experimental load-displacement relations for JP1.

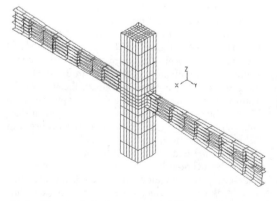

Figure 12. Analytical exaggerated deflection of JP1.

in the inner part of the connection. Because of this, a second analytical investigation was done to reexamine the stress distributions on steel plates.

4.2 Results of second analytical investigation

Finite element models for beams and beam-column connections without concrete and reinforcing bars were created as shown in Figure 14. Since it is the reinforced concrete column that supports and restrains the beam, the resistance of column was modeled as boundary pin supports along the outer and inner square perimeters at the top and bottom of the connection. The main objective was to investigate the stress proportion and flow from the beam to the plates in the

beam-column connection. Models that were identical in shapes and sizes to actual specimens were created using 4-node shell elements. Figures 15 and 16 show stress contours at maximum loads obtained from one-load-step calculations. It was observed that in all specimens, the stresses indicated by the strain gauges were closely approximated by analytical results. Slight differences in the stress results were found in the square tube and in the inner webs. The assumption for

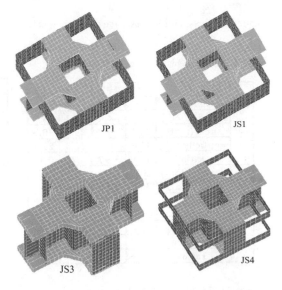

Figure 14.   Typical models for beam-column connections.

Figure 13.   Theoretical crack formations in JP1.

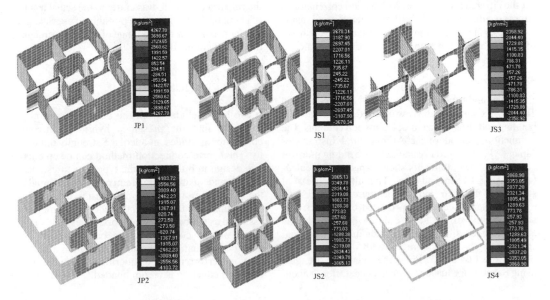

Figure 15.   Stress distributions in the connections of specimens JP1, JP2, JS1, JS2, JS3, and JS4.

2429

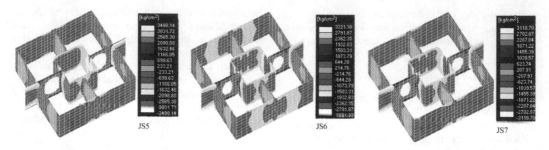

Figure 16.   Stress distributions in the connections of specimens JS5, JS6, and JS7.

Table 3.   Experimental and analytical strengths.

	Specimen	JP1	JP2	JS1	JS2
c o l u m n	axial(tonf)			67.0	
	flexural(tonf)	29.5	34.4	28.3	28.9
	shear(tonf)	22.2	23.0	29.3	30.1
	applied(tonf)	11.6	5.8	22.3	23.7
	ratio	1.91	3.97	0.79	0.82
b e a m	flexural(tonf)	3.5	3.5	10.1	
	shear(tonf)	30.6	30.6	20.9	
	max. load(tonf)	4.2	4.2	8.1	8.6
	ratio	1.2	1.2	0.8	0.85
c o n x	shear(tonf)	668	690	94.4	113.0
	applied(tonf)	65.0	32.1	124	132
	ratio	10.3	21.5	1.31	1.16

	Specimen	JS3	JS4	JS5	JS6	JS7
c o l u m n	axial(tonf)	67.0		57.0		
	flexural(tonf)	29.1	28.9	26.6	27.7	28
	shear(tonf)	30.2	30.1	29.1	29.8	28.9
	applied(tonf)	17.1	15.3	26.4	23.1	24.8
	ratio	0.59	0.53	0.99	0.83	0.89
b e a m	flexural(tonf)	11.4				
	shear(tonf)	39.3				
	max. load(tonf)	6.2	11.1	9.6	8.4	9.0
	ratio	0.54	0.97	0.84	0.74	0.79
c o n x	shear(tonf)	99.2	105	142	109	126
	applied(tonf)	94.9	84.9	147	129	138
	ratio	0.96	0.72	1.04	1.18	1.09

concrete as resisting boundary supports may account for the differences. However, with a good correlation in stress results, finite element models formed using exact sizes and shapes can be suggested to be the bases of actual structural design of this hybrid connection.

4.3   Strength

Empirical and experimental strengths are presented in Table 3. Existing equations were used to calculate the bending and shear strengths of beams and columns. The e function method of the Architectural Institute of Japan (AIJ) was used to calculate the bending moment capacity of the column. Its shear capacity was calculated using Urbon equation. The ultimate flexural strength of the beam and the shear strength were estimated using the equations provided by the AIJ. The shear strength of the beam-column connection was approximated using Sakaguchi equation. The actual shear strengths for specimens with outer confining plates were 1.04 to 1.31 times the calculated strengths using Sakaguchi equation for an average factor of about

1.16. For specimen JS3 without outer confining plates, the calculated strength was closer to the actual at a ratio of about 0.96. It is therefore possible to use Sakaguchi equation to determine the range of the shear strength of the hybrid connection.

5   CONCLUSIONS

It can be concluded from the observed general performance that the presented hybrid beam-column connection provides an adequate strength and ductility. The hybrid connection method can be an alternative design in building frame type structures. Design can be made using finite element analytical model of the steel beam and steel plates in the beam-column connection with the reinforced concrete column action assumed as the boundary supports. The shear strength range of the connection can be calculated using Sakaguchi equation. Further investigation of the connection performance considering the bond between steel and concrete is recommended.

Semi-rigid composite moment connections – experimental evaluation

A. Fernson, K.S.P. de Silva & Indubhushan Patnaikuni
School of Civil and Chemical Engineering, RMIT University, Melbourne, Australia

K. Thirugnanasundaralingam
Ainley Engineering Consultants Pty. Ltd., Melbourne, Australia

ABSTRACT: Steel beam-concrete slab composite floor systems are widely used in Australia due to inherent advantages such as construction speed, economy and better quality control. However, the current Australian Code AS 2327, (1996) for composite construction covers only the design of simply supported composite beams. Even though the British Code BS 5950, (1990) and Eurocode 4 (1985) provide design guidelines for continuous composite beams, the beams need to have continuity over the supports or full moment connectivity if the beams are discontinued at the supports. This paper investigates the effectiveness in achieving continuity of simply supported composite beams, with standard cleat plate end connections, by providing additional top reinforcement in the slab over the support. The test results show, with the addition of a few additional reinforcing bars over the simply supported ends, semi-rigid connectivity can be achieved and the beams tend to behave as a partially continuous member. This in effect improves the load carrying capacity and the stiffness of the floor system with minimum cost. This also enhances the performance of the floor under fire as the proposed method providing adequate end anchorage for the beams to act in catenary under large plastic deformations.

1   INTRODUCTION

Figure 1 shows a typical configuration of a composite floor system where secondary composite beams are supported by a primary non-composite steel beam. The secondary beams are bolted to a cleat plate welded to the web of the primary beam and the secondary beams are designed as simply supported composite beams spanning between primary beams. This is the most simple and commonly used connection type in Australia. A full range of connection types is given in Standardised Structural Connections (1985).

There is increasing interest world wide in understanding the behaviour of support connections in composite beams in general, but particularly if the concrete slab is continuous over the simply supported ends as given in Figure1.

The presence of slab over the supports, properly designed, can be utilised to convert simply supported ends of the steel beam to act as a semi-rigid moment connections providing partial-continuity. This would improve load carrying capacity, deflection and fire resistivity of the floor system quite effectively. This paper presents the results of an experimental investigation into such semi-rigid composite connections and

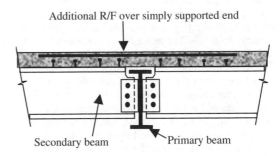

Additional R/F over simply supported end

Secondary beam   Primary beam

Figure 1.

their behaviour. Though a number of researchers have investigated the behaviour, the experimental data is limited and warrants further work to formulate a reliable design procedure for such joints. Leon, R.T. (1990) observed that provision of longitudinal reinforcement in the slab over the support significantly increases the strength and stiffness of a simple connection, without additional shear studs or bolts in the connection. This cost-effective approach provides a number of advantages.

They are:

i. Lower structural depths through reduced beam sizes.
ii. Improved vibration and deflection characteristics through higher stiffness.
iii. Reduced floor-to-floor and therefore the overall building height resulting cost savings in cladding, façade, services.
iv. Better crack control in the slab over the support.

2 EXPERIMENTAL PROGRAM

2.1 Test specimens

Two non-composite and four composite proto-type connections were tested to investigate the rotational characteristics and the moment capacities of simply supported end connections with additional top reinforcements over the supports as shown in Figure 1. The secondary steel beams are designed as composite members, which are simply supported off the primary steel beam. Primary steel member is non-composite. The added top reinforcements over the internal support are also shown. Specimens are designed to the Australian standards AS 2327 (1996). Table 1(a) summarises the specifications and the relevant standards adopted in the experimental program. Table 1(b) gives the member sizes and connection details.

2.2 Testing rig and instrumentation

Figure 2 shows the testing rig, the specimen and the instrumentation. The specimen was loaded up to failure with a hydraulic jack and measured using a 50 tonne capacity load cell. The underside of the transfer beam had housing for rollers welded to it at 2805 mm centers. The semi-circular housing was designed to facilitate the slip of the 50 mm diameter × 750 mm long rollers.

Instrumentation included six Clinometers, four transducers and a number of strain gauges. Six Clinometers were placed, two on the primary beam and two on each secondary beam. The vertical displacement was measured with two transducers located under the bottom flange of the secondary beams, 150 mm from each end of the specimen. Similarly, the horizontal slip was measured with transducers placed horizontal to the two vertical plates welded to the cast-in-plates located at each end of the slab. Strain gauges were used to measure the strains in the reinforcement and in the steel beam. The gauges were 5 mm linear and 10 mm rosette 120 W electrical resistance type strain gauges. The strains on the steel beam were measured with a Vishay Instruments Model P-350A Digital Strain Indicator, and the strains in the reinforcement were recorded with the Datataker.

All readings from load cell, strain gauges, six inclinometers and transducer were logged with a computer interface data logger. During the experiments, the crack width measurements were carried out with digital calipers. Local buckling of the web and web-side-plate were observed visually, and the strains were measured using strain gauges.

Similar experimental work has been carried out by Leon, R.T. (1990) and Xiao et al. (1994). The additional reinforcements provided over the negative moment regions, as given in Table 1(b), were comparable with the reinforcements provided by Xiao et al. (1991), in their experimental work. Acknowledging the fact that the material properties and the testing rig parameters could influence a meaningful comparison, measured data of this study is benchmarked against their work. The moment capacities, rotation capacities, yielding mechanisms of steel, and concrete cracking patterns of the tested samples were recorded. There are a number of methods available to calculate the theoretical moment capacity although the method proposed by Xiao et. al. (1991) is adopted for compatibility of results.

Table 1(a). Test specimen material specifications.

Element	Specification	Standards
Secondary beam	Grade 250 hot rolled structural steel	AS 3679 (1996)
Concrete	Grade N25, 80 mm slump, 20 mm aggregate	AS 3600 (1996)
Connection	Grade 300 Cleat plates 10 mm thick	AS 3679 (1996)
	Grade 8.8/S Structural Bolts	AS 1252 (1996)
Concrete slab	100 mm deep × 800 mm wide × 3000 mm long	BS 5950.3 (1990)
Studs	Grade 1010–1020 (410 Mpa tensile strength), ϕ 19 mm 76 mm long	AS 1443 (1994)
Reinforcement	Grade 400Y (400 MPa) Grade 250R (250 MPa)	AS 1302 (1991)

Table 1(b). Test specimen features.

Exp. No.	Specimen	Added R/F (%)
1	460UB67 + 2/250UB31 2M20 bolts	No slab Non-composite
2	460UB67 + 2/310UB40 6M20 bolts	No slab Non-composite
3	460UB67 + 2/310UB40 6M20 bolts	3Y12 (0.42%)
4	460UB67 + 2/250UB31 2M20 bolts	3R10 (0.29%)
5	460UB67 + 2/310UB40 6M20 bolts	5Y12 (0.69%)
6	460UB67 + 2/310UB40 6M20 bolts	9Y12 (1.23%)

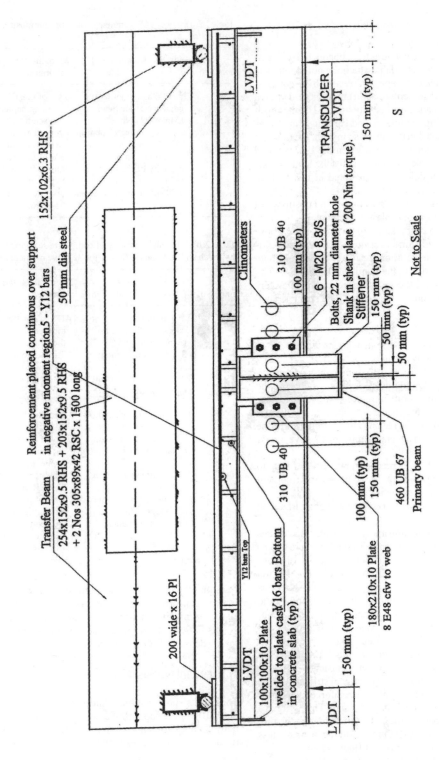

Figure 2.

2.3 Test results and observations

2.3.1 Non-composite connections

As mentioned before two tests were conducted on specimens without concrete slabs, which are non-composite steel connections, to compare the results from composite connections. Moment–rotation curve showed limited linear behaviour and non-linearity initiated by the slip at the bolts resulting the loss in load carrying capacity as rotation increases. Further increase in both, the moment capacity and rotation occurred when the bolts slipped sufficiently to enable the bolts to resist the load in bearing. Finally, bolt deformation resulted in further load stagnation and increased rotation.

It is observed that both specimens tend to slip when 7–8% of the member plastic moment capacity is reached and can carry an ultimate moment of about 33% of the plastic moment capacity. The average ultimate rotation, the rotation measured when ultimate moment is reached, is in the vicinity of 100 milRad. The ultimate rotation of the second specimen is relatively less, as expected, due to the greater joint stiffness (two rows of bolts).

2.3.2 Composite connections with additional R/F over negative regions

As summarized in Table 1(b) four composite specimens with varying additional reinforcements in the slab, over negative moment regions, were tested. For these connections, the moment–rotation graph did not have an apparent linear region. The region prior to the appearance of the fist crack appears to have linear characteristics.

Its characteristic was predominantly linear for up to 36% of the failure moment. The rotation at this moment was 3% of the failure rotation.

The moment–rotation results are summarized in Table 2(b).

For a meaningful comparison, the percentage values of moment capacities given in Table 2(b) were calculated using the plastic moment capacity of non-composite steel secondary beam. It is obvious that the ultimate moment capacity of the composite section is greater than the steel beam. The other reason one should consider is that in negative moment regions where concrete slab is on the tension zone the effectiveness of concrete slab should be ignored.

It is evident in comparing results summarized in Table 2(a) and Table 2(b) that a significant enhancement in moment capacity and rotational stiffness can be achieved by providing additional reinforcements over the negative moment region of the slab. This will have a significant impact on the deflection control at the mid span of the secondary beams, resulting a much less structural depth and an efficient floor system.

One other advantage of providing additional reinforcements as proposed is the enhancement of fire resisting capabilities of the floor system. Such reinforcements, located within a heat sink, will facilitate the slab working in catenary when the secondary steel beams undergo large plastic deformations under fire loads (de Silva & Grayson 1996).

Table 2(a). Results summary of non-composite steel connection.

Exp. No.	Moment–Rotation @ slip	Moment–Rotation (ultimate)
1	7% of plastic moment capacity of the beam (Mp = 99 kNm) 3.5% of ultimate rotation	33% of plastic moment capacity of the beam Ultimate rotation 110.2 milRad
2	8% of plastic moment capacity of the beam (Mp = 156 kNm) 2.3% of ultimate rotation	33% of plastic moment capacity of the beam Ultimate rotation 91 milRad

Table 2(b). Results summary of composite steel connection with slab and additional R/F over the negative moment region.

Exp. No.	Moment–Rotation @ first crack in the slab	Moment–Rotation (ultimate)
3	15% of plastic moment capacity of the beam (Mp = 156 kNm) 3% of ultimate rotation	41% of plastic moment capacity of the beam Ultimate rotation 49 milRad
4	12% of plastic moment capacity (Mp = 99 kNm) 1.5% of ultimate rotation	27% of plastic moment capacity of the beam Ultimate rotation 39 milRad
5	29% of plastic moment capacity (Mp = 156 kNm) 3% of ultimate rotation	72% of plastic moment capacity of the beam Ultimate rotation 57 milRad
6	N/A due to loss of data (data logger failed to record at first crack)	86% of plastic moment capacity of the beam Ultimate rotation 48 milRad

3 CONCLUSIONS

These tests show that the addition of reinforcement in the slab over the support of a composite beam substantially changes its characteristics. The major conclusions resulting from this investigation are:

1. The addition of reinforcement transforms the standard simply supported connection to a semi-rigid

moment connection through partial composite action.

2. The percentage of longitudinal reinforcement provided in the negative moment region has a profound effect on the moment capacity and also controls the rotation at the simply supported connection.

3. Resulting floor system is not only versatile in moment redistribution, resulting partial continuity, but also improve the stiffness characteristics of the floor system resulting reduced structural depths.

4. The observed failure mode for all the composite specimens was a combination of reinforcement yielding and buckling of the lower portion of the web stiffeners and cleat plates.

5. All composite specimens failed in a ductile manner.

6. When crack controlling becomes a design consideration provide 0.5% or more reinforcements through closely spaced smaller diameter bars.

REFERENCES

Standardised Structural Connections. 1985. Australian Institute of Steel Construction, Sydney, Australia.

AS 1252. 1996. *High strength steel bolts with associated nuts and washers for structural steelwork.* Standards Association of Australia, Sydney, Australia.

AS 1302. 1991. *Steel reinforcing bars for concrete*, Standards Association of Australia, Sydney.

AS 1443. 1994. *Carbon steels and carbon manganese steels – Cold Finished Bars.* Standards Association of Australia, Sydney.

AS 2327. 1996. SAA Composite Construction Code, Part 1 – Simply Supported Beams. Standards Association of Australia, Sydney.

AS 3600. 1996. *Concrete Structures.* Standards Association of Australia, Sydney.

AS 3679 – Part 1 1996. "Structural Steel – Hot Rolled Bars And Sections." Standards Association of Australia, Sydney.

AS 3679 – Part 2 1996. "Structural Steel – Welded Sections." Standards Association of Australia, Sydney.

BS 5950 – Part 3.1 1990. "Code Of Practice For Design Of Simple And Continuous Composite Beams." British Standards Institution, London.

Eurocode 4. 1985. *Common Unified Rules For Composite Steel And Concrete Structures.* EUR 9886 EN, Commission of the European Communities.

Leon, R.T. 1990. *Semi-Rigid Composite Construction.* Journal of Constructional Steel Research, Vol 15, No. 1 & 2, 99–120.

Xiao, Y., Nethercot, D.A. & Choo, B.S. 1991. *Composite Connections in Steel and Concrete.* Second Report to Building Research Establishment, Report No SR 91024, Nottingham University.

Xiao, Y., Choo, B.S. & Nethercot, D.A. 1994. *Experimental Behaviour of Composite Beam – Column Connections.* Journal of Constructional Steel Research, Volume 31, 3–30.

de Silva, K.S.P. & Grayson, W.R. 1996. *Introduction to structural aspects of fire engineering – Crown Casino experience*, Technical Report, Connell Wagner, Victoria, Australia.

Analysis of waffle composite slabs

A.M. El-Shihy, H.K. Shehab, M.K. Khalaf & S.A. Mustafa
Structural Engineering Department, Faculty of Engineering, Zagazig University, Egypt

ABSTRACT: For few decades and since the well-known composite slabs consisting of reinforced concrete poured on corrugated steel sheeting was achieved, the use of composite structures has grown widely. However and since this type of slabs acts as a one-way slab due to the ribs configurations in the steel sheeting, some problems have been created such as having two parallel heavily loaded beams in each panel while the other two perpendicular beams are almost unloaded; which is uneconomic. The steel sheeting stiffness is always satisfactorily high in one direction and relatively low in the other which requires the use propping due to the relatively high loaded beams during the construction stage.

For the above reasons and others, the idea of using two-way (or *waffle*) composite slabs was brought up. Carrying out such research required firstly choosing the most appropriate finite element modelling technique in representing the two-way or waffle composite slab and comparing its behavior to that of one-way.

Three modelling techniques named real shape, equivalent and grillage models were tested and represented by two and three-dimensional finite element techniques to detect the best of them. Since no previous experimental, numerical or empirical results of waffle composite slab analysis were yet available to evaluate the modelling results, it was necessary to study also the behavior of the well-known one-way composite slab using the pre-mentioned modelling techniques.

After the best finite element model was achieved, an extensive parametric study in which many influencing parameters such as slab aspect ratio, slab boundary conditions, steel sheet depth, corrugation cell aspect ratio and its orientation were tested, was carried out in order to investigate the overall behavior of two-way composite waffle slab under different conditions. Many encouraging results and recommendations were obtained opening the door for this new configuration of composite slabs.

1 INTRODUCTION

The two-way behavior is always preferable to that of one-way in all slab types due to its better structural behavior and economy in the cross-sectional dimensions. Many researches studied experimentally and numerically the one-way behavior of composite slabs reinforced with steel sheet corrugated in one direction; Porter & Ekberg (1977), Porter (1985), An (1993) and El-Shihy (1995). However no researchers studied slabs with steel sheeting corrugated in the two main orthogonal directions. In this paper a new suggested composite slab structural system named "waffle composite slabs" is arised. In addition to the expected and pre-mentioned better structural behavior of waffle composite slab, the way by which the corrugated steel sheet is formed prevents the horizontal slippage between the concrete and the steel sheet which is considered a very serious mode of failure in one-way composite slabs.

The main aim of this research is to analyze the structural behavior of the new suggested composite slab type under different conditions. Indeed, this analysis required a very careful choice of the used modelling technique to accurately simulate the waffle composite slabs. Therefore firstly, a modelling techniques evaluation of waffle composite slabs was achieved numerically using the finite element technique and compared to the previous available experimental results of one-way composite slab types to achieve the best modelling among three models named *real shape*, *equivalent* and *grillage* models assuming full interaction between steel sheeting and overlaying concrete.

Many influencing parameters such as the waffle rectangularity, boundary conditions, depth and corrugation cell aspect ratio and corrugation type (trapezoidal and re-entrant) which were expected to affect the structural behavior of the two-way waffle composite slabs were investigated under uniformly distributed loads.

2 MODELLING EVALUATION

Three different finite element modelling techniques were used to model composite slabs in order to detect the best of them. Since there were no previous results to compare with, it was necessary to start with modelling the traditional one-way composite slab and compare the results with those of previous models; Porter & Ekberg (1977), Porter (1985) and An (1993).

In two-way waffle composite slabs; Figure 1, the corrugations may have equal or different dimensions in two directions. Since there is no steel sheet having this shape of corrugation in the market yet, the standard corrugations dimensions of the one-way corrugated steel sheets available in market were considered. Then the dimensions of corrugations were changed gradually in one of the two directions to study their effect on the slab behaviour; as will be detailed later.

2.1 Model dimensions and loading configurations

For the sake of comparison, the dimensions and the loading configurations of previously investigated models were maintained; Figures 2 and 3. The slab had dimensions of 3.66 × 4.88 m. For one-way slab, the steel sheeting was oriented in direction of the short span. The steel sheet used had rolled embossments on the webs to provide longitudinal shear resistance at the concrete–sheet interface. Four symmetrical concentrated loads were applied to ultimate capacity. This was depending on the assumption that a slab element with concentrated loads at the quarter points behaves in a similar manner to a slab element with uniform loading; Klaiber & Porter (1981). The ultimate load on each loading point was 6 t. In waffle composite slab, the sheet dimensions were the same in the two directions of the slab. Depending on full interaction assumption between steel sheet and concrete and linear elastic uncracked behaviour of concrete, the maximum load was obtained once the stress in any steel reaches the yield strength or the compressive strain in concrete reaches the maximum compressive strain or if the maximum deflection of the slab reaches the allowable deflection value; European Code (1975).

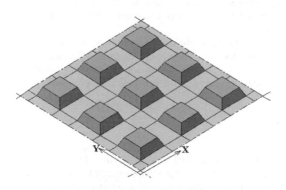

Figure 1.   Suggested shape of waffle steel sheet.

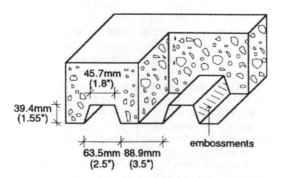

Figure 2.   Cross-sectional dimensions of composite slab.

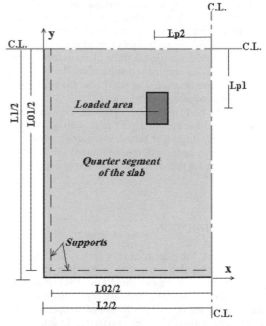

Figure 3.   Slab dimensions, loading positions and supports.

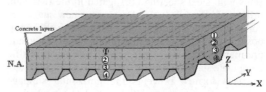

Figure 4.   Three-dimensional real shape model of waffle composite slab.

2438

2.2 *Three-dimensional real shape model*

Both one-way and waffle composite slabs were modeled using the three-dimensional finite element real technique. The concrete body was divided into four layers; as shown in Figure 4. Layers 1 and 2 were in compression zone having the concrete compressive stress.

2.3 *Two-dimensional equivalent model*

This model simulates the beam-plate bending model introduced by El-Shihy et al. (1995). For one-way composite slab, the concrete layer was modeled as a series of two adjacent plate bending elements which had the same properties and different thicknesses to simulate the depth variation of concrete layer due to the steel sheet corrugation geometry. The steel-deck was modeled as beam elements in the longitudinal direction of the ribs "strong direction".

The steel sheet was represented by two-node beam elements in the two orthogonal directions. The concrete was represented by a series of two adjacent plate

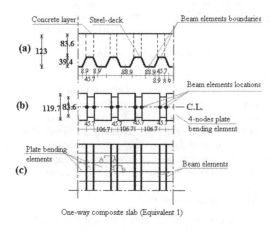

Figure 5. Two-dimensional equivalent model.

bending elements. For one-way composite slab, it had two different thicknesses of 83.6 mm and 119.7 mm, with widths of 45.7 mm and 106.7 mm respectively. In waffle composite slab, the elements representation was slightly modified to maintain the previously used procedure; Figure 5.

2.4 *Two-dimensional grillage model*

Waffle composite slab was modeled using grillage model which had grillage beams in both directions. They were chosen to coincide with the rib centre lines, to simplify data preparation. Each cycle of the waffle composite slab was idealized as an equivalent frame element having the same transformed area and moment of inertia of the real cross-section (in concrete dimensions). The remaining part of concrete over the supporting steel I-beam, and the steel I-beam itself was represented as beams having the same area and moment of inertia of the real section (in concrete dimensions); Figure 6.

2.5 *Results and discussion*

Comparison between the three models based mainly on the distribution of forces at the two orthogonal longitudinal supports, as shown in Figures 7 and 8, and Table 1. Comparison of results showed that the three-dimensional finite element real shape model gave the best and the nearest results to those obtained experimentally.

3 WAFFLE COMPOSITE SLAB SUPERIORITY

The most critical mode of failure in one-way composite slab is the slippage between the steel sheet and the

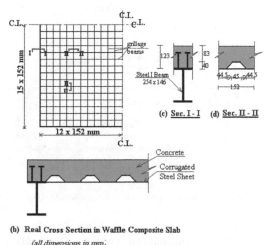

(b) **Real Cross Section in Waffle Composite Slab**
  (all dimensions in mm)

Figure 6. Two-dimensional grillage model.

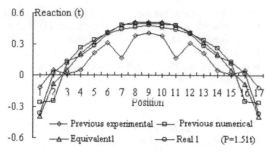

(a) Support transverse to the sheet corrugation

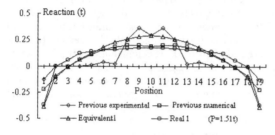

(b) Support parallel to the sheet corrugation

Figure 7. Distribution of reaction forces (one-way slab).

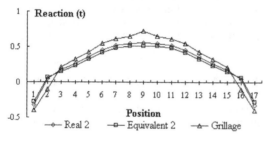

(a) Long direction

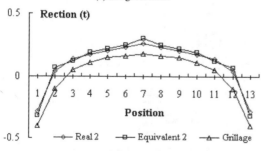

(b) Short direction

Figure 8. Distribution of reaction forces (waffle slab).

concrete. This slippage is prevented in two-way waffle slabs due to the ribs configurations. One hundred and thirty models were tested for the sake of comparing the traditional one-way composite slab and waffle

Table 1. Percentage of load transferred in each direction (load P = 2.15t (1/3Pu).

Load transferred	Main (%)	Secondary (%)
Previous experimental	71.6	28.4
Previous numerical	73.4	26.6
Real 1	73	27
Equivalent 1	69,6	30.4
Real 2	63	37
Equivalent 2	61.1	38.9
Grillage	70.1	29.9

P is one-fourth of the applied load on the slab.

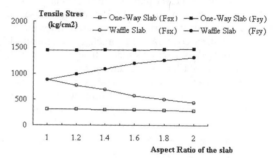

Figure 9. Compared steel tensile stress.

composite slab in both construction and composite stages. Each was examined through different slab aspect ratios (r) of 1, 1.2, 1.4, 1.6, 1.8 and 2. Slabs were examined under an increasing uniformly distributed load until failure. The considered failure limits were the concrete compressive stress or the steel sheet tensile stress or the allowable slab deflection.

3.1 Basic assumptions

1. Full interaction between steel and concrete.
2. Linear cracked elastic behavior of concrete.
3. The effect of supplementary reinforcement against shrinkage and temperature is neglected.

3.2 Construction stage

Deflection must be controlled in composite slabs construction phase. In one-way slab, loads were transferred mainly in the direction of ribs. For example in a one-way composite slab of aspect ratio of 1.4, about 79%; in average; of the total load went through the short direction and the remaining 21% went through the long direction. However, in a waffle composite slab of the same aspect ratio, about 66% of the total load went through the short direction and the remaining 34% went through the long direction. Of course, the better load distribution of the waffle composite slab enhances its structural behaviour; as shown in Figure 9.

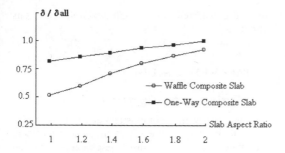

Figure 10. Maximum deflection.

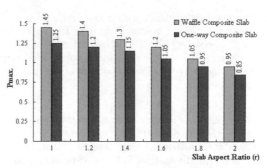

Figure 12. Slab aspect ratio influence on slab loading capacity.

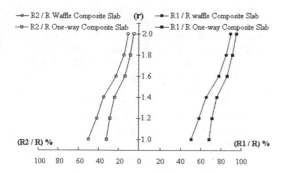

Figure 11. Percentage of load distributed on each side.

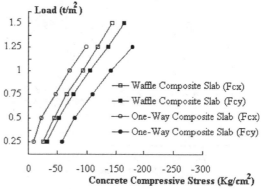

Figure 13. Compressive stress in concrete at aspect ratio of 1.2.

3.3 *Composite stage*

Seventy-two waffle and one-way models with different aspect ratios were analyzed. All slabs were assumed to be simply supported at four edges. The dead weight of the slabs was taken into account.

3.3.1 *Deflection*
Small slab aspect ratios gave better behavior of waffle composite slab. Figure 10 declares that for slab aspect ratios of 1.0, 1.2, 1.4, 1.6, 1.8 and 2.0 the maximum deflection of waffle composite slab was lower than that of one-way composite slab by 19%, 18%, 16%, 14%, 9% and 6% respectively. As the slab aspect ratio increases, the behavior of waffle composite slab becomes closer to that of the one-way slab.

3.3.2 *Load transferred in each direction*
In general, the load is distributed in a better manner in two-way slabs than one-way slabs. Figure 11 shows that the load transferred in the short direction in both types of slabs increased as aspect ratio increases. At slab aspect ratios 1.0, 1.2, 1.4, 1.6, 1.8 and 2.0, waffle slab load transferred in the short direction was reduced by 18%, 13%, 11%, 8.5%, 6.5% and 5% respectively than that recorded in one-way slab. As the slab aspect ratio increases, the behavior of both types of slabs becomes closer.

3.3.3 *Maximum loads*
Figure 12 shows that waffle composite slab could carry higher maximum load more than one-way composite slab by about 12 to 20% depending on the slab aspect ratio. This relatively higher slab loading capacity leads to reducing the total slab thickness which; of course; means economy.

3.3.4 *Compressive stress in concrete*
The maximum compressive stresses were recorded and plotted for both waffle and one-way composite slabs due to live load application; as shown in typical Figure 13 for r = 1.2 as an example. It was found that failure happened due to either yielding in steel sheet or excessive slab deflection in both types of slabs. Figure 13 shows that the concrete compressive stress in the short direction of the waffle composite slab was about 19% lower than that of its one-way analogue.

3.3.5 *Tensile stress in corrugated steel sheet*
The tensile steel stresses in waffle composite slab were found to be lower than the stresses in one-way

2441

composite slab, under the same load levels. Figure 14 shows the stresses in steel sheet in waffle slab compared to one-way slab for slab aspect ratio of 1.2 as an example. The tensile stress in the short direction in waffle slab was 17% lower than that of the one-way corrugated one.

3.3.6 Effect of generating and increasing the number of ribs in the long direction

Five three-dimensional finite element real shape slab models with dimensions 3.66 × 4.88 m were used to study the effect of variation of composite slab behavior due to changing it gradually from one-way to waffle composite slab. Table 2 shows the percentage

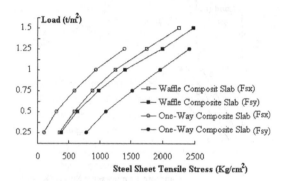

Figure 14. Tensile stress in corrugated steel sheet in waffle and one-way composite slabs.

Table 2. Effect of decreasing number of ribs in the long direction (at load (P) ≈ 25% of P_{max}).

| Slab | Load transferred [%] | | No. of ribs in the long direction |
	Short	Long	
2a	71.9	28.1	0.0 (one-way slab)
2b	68.3	31.7	4
2c	66.9	33.1	8
2d	63.5	36.5	16
2e	62.6	37.4	24 (two-way slab)

of load transferred in each slab direction due to this variation.

4 PARAMETRIC STUDY

4.1 Testing program characteristics

One hundred waffle composite slabs were tested in order to investigate the effect of some significant parameters on their behavior. These factors were slab aspect ratio, slab boundary conditions, steel sheet profiles, varying the steel sheet corrugation cells aspect ratio, and increasing the steel sheet depth with a constant total depth of the slab. The previously mentioned three-dimensional finite element real shape model was considered. The used dimensions of the slabs varied from 7.08 × 7.08 m to 7.08 × 3.48 m. A typical considered corrugated steel sheet deck section known as CF 46; PMF; was used. Except in the comparison between the different steel sheet shapes, a trapezoidal steel sheet section called PEVA 45, and a re-entrant steel sheet section were used. The details of the three different steel sheet shapes used in the study are described in Table 3 and Figure 15. The overall slab thickness was 180 mm in all models according to the ECCS for one-way composite slab. The own weight of the tested slabs was taken into account. The

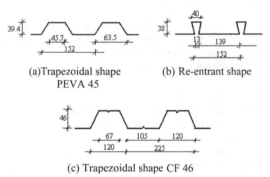

(a)Trapezoidal shape PEVA 45 (b) Re-entrant shape

(c) Trapezoidal shape CF 46

Figure 15. Cross-section details of steel sheet cycles (mm).

Table 3. Used corrugted steel-deck sections properties.

| Steel-deck section | Thickness (mm) | | Weight (kN/m²) | Moment of inertia (cm⁴/m) | Net section area (cm²/m) |
	Nominal	Design			
CF 46*	1.2	1.16	0.112	53.0	14.59
PEVA 45*	0.94	0.9	0.102	51.5	13.23
Re-entrant profile**	0.9	0.86	0.127	67.0	16.55

* Profiles for concrete for fast.
** Quick span construction limied.

live load was increased gradually till one of the failure limits is reached.

4.2 Analysis and discussion

4.2.1 Slab aspect ratio

Thirty slab models were analyzed to declare the effect of slab rectangularity. The same corrugation dimensions were typical in both slab directions (cells aspect ratio of 1.0). The maximum load is shown in Table 4. It was noticed that slab deflection decreased as slab aspect ratio increased while the percentage of load transferred in the long direction decreased, and the corresponding load value transferred in the short direction increased; Figure 16.

4.2.2 Slab boundary conditions

Slab boundary conditions were found to be a very important factor, the fixed edges are more recommended as the loads get higher. The effect of slab boundary conditions on the results is shown in Figure 17.

4.2.3 Steel sheet profile

A comparison was held between a waffle trapezoidal and a re-entrant steel sheet shape with almost the same corrugation cycle dimensions; Figures 1 and 18. Almost the same deflection, stresses and % of load transferred in each direction were recorded in the two types.

4.2.4 Steel sheet corrugation cells aspect ratio

The dimensions of the traditional steel sheet cycles of the one-way composite slab were used in the two directions of the tested slabs; Figure 19. The cells were kept constant in one direction, and varied in the other. Eighteen models divided into two were tested. *The first group* had square slabs (7 × 7 m), where the corrugation cells were oriented parallel to Y-direction.

The second group had rectangular slabs of aspect ratio r = 1.4 (7.08 × 5.055 m). The rectangular slab had two cases; *Case A*, the cell long direction was parallel to the slab long direction. *Case B*; the long

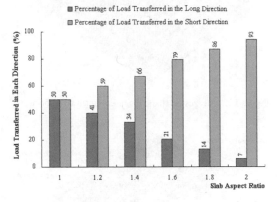

Figure 16. Effect of slab aspect ratio on the percentage of load transferred in each direction.

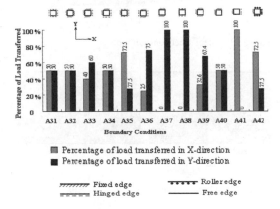

Figure 17. Percentage of load transferred in each direction.

Table 4. The geometry of the slabs with variable aspect ratio.

Slabs	Aspect ratio (r)	L_x (m)	L_y (m)	P_{max} (t/m^2)	Mode of failure
A1–A5	1.0	7.08	7.08	1.0	Deflection (serviceability)
A6–A10	1.2	7.08	5.73	1.2	Deflection (serviceability)
A11–A15	1.4	7.08	5.06	1.35	Yielding in steel sheet
A16–A20	1.6	7.08	4.38	1.45	Yielding in steel sheet
A21–A25	1.8	7.08	3.93	1.5	Yielding in steel sheet
A26–A30	2.0	7.08	3.48	1.54	Yielding in steel sheet

All slabs are simply supported at four edges.
CF 46 sheets.

Figure 18. Suggested shape for re-entrant waffle sheet.

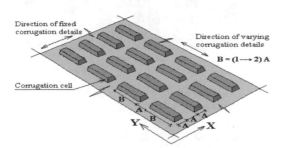

Figure 19. Steel sheet configurations.

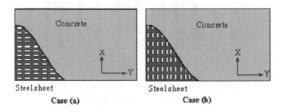

Figure 20. Orientation of the corrugation cells.

direction of the cell was perpendicular to the long direction; Figure 20a and b.

The increase of the corrugation cell aspect ratio decreased the percentage of load transferred in the long direction in both square and rectangular slab of Case B, while it was increased in rectangular slabs of Case A; Figure 21a and b. Deflections increased in both square and rectangular slabs of Case A and decreased in slabs of Case B. Both steel and concrete stresses decreased in the long direction in square and Case B slabs, and increased in the long direction in Case A slabs with the increase of the cell aspect ratio.

4.2.5 Steel sheet depth

Increasing the corrugated steel sheet depth; gradually from 46 mm to 65 mm; with the same total slab depth in square slab, decreased the slab deflection by about 15% in average, reduced the steel sheet stresses by

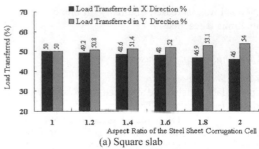

(a) Square slab

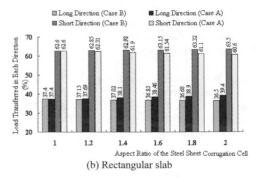

(b) Rectangular slab

Figure 21. Load transferred in each direction.

12% in average, while the concrete stresses increased by about 20% in average. There was no effect on the percentage of load transferred in each slab direction.

5 CONCLUSIONS

– The three-dimensional finite element real shape model was found to be the best modelling technique to represent both waffle and one-way composite slabs.
– In construction stage, waffle composite slab proved its superiority on its one-way analogue by reducing steel sheet stress and the percentage of load transferred in the short direction.
– In composite stage waffle composite slab sustained loads higher than its one-way analogue, and permitted lower deflection value. Steel and concrete stresses in waffle composite slab were lower and better load distribution along the two orthogonal directions was gained. In addition, the way by which waffle steel sheet is formed, prevents the slippage between the sheet and the concrete. In architectural point of view, the good appearance of the waffle shape saves the cost of constructing artificial ceiling.
– Increasing slab aspect ratio decreased slab deflection and increased steel sheet and concrete stresses in the short direction. Symmetric-fixed boundaries lead to the best behavior of waffle composite slab.

The re-entrant waffle corrugated sheet permitted lower deflection, higher steel sheet and concrete stresses than the trapezoidal shape, and it allowed the same percentage of load distribution in the two directions. Square corrugation cell of steel sheet was found the best in both square and rectangular slabs. A better behavior of waffle composite slab was obtained when the steel sheet depth was equal to one third of the total slab depth.

- In general, the behavior of waffle composite slabs is superior and it is strongly recommended to manufacture new steel sheet configurations to allow using this new type of composite slabs.

REFERENCES

Mustafa, S.A. 2002. Waffle Composite Slab, *M.Sc. Thesis* Presented to Zagazig University, Zagazig, Egypt.

An, L. 1993. Load Bearing Capacity and Behavior of Composite Slabs with Profiled Steel Sheet, *Ph.D. Thesis* Presented to Chalmers University of Technology, Goteborg, Sweden.

El-Shihy, A.M. et al. 1995. New Modeling of Composite Floors with Profiled Steel Sheeting, *Engineering Research Journal*, Faculty of Engineering Technology, Helwan University, Mataria, Cairo, Egypt, Vol. 3.

European Recommendations for Steel Construction: The Design of Composite Floors with Profiled Steel Sheet. 1975, Publication No. 14, Committee No. 11.

Klaiber, F.W. & Porter, M.L. 1981. Uniform Loading for Steel-Deck Reinforced Slabs, *ASCE, J. of the Structural Division*, Vol. 107, No. ST11, pp. 2097–2110.

Porter, M.L. & Ekberg, C.E. Jr. 1977. Behavior of Steel-Deck-Reinforced Slabs, *ASCE, J. of the Structural Division*, Vol. 103, No. ST3, pp. 663–677.

Porter, M.L. 1985. Analysis of Two-Way Acting Composite, *ASCE, J. of Structural Engineering*, Vol. 111, No. 1, pp. 1–18.

Profiles for Concrete for Fast; Long Span Flooring and Roofing, Precision Metal Forming Limited; PMF; Swindon Road, Cheltenham, Glaucestershire GL51 9LS, England.

Quick Span Construction Limited 1982. *Engineers in Reinforced Concrete*, 2Market Close, Poole, Dorset BH15 1NQ, England.

SAP 2000. A Structural Analysis Program, University of California, Berkeley, CA, Revised 1997.

NOTATION

D_s:	Total slab thickness
F_{cu}:	Ultimate concrete strength
F_{cx} and F_{cy}:	Concrete stress in long and short directions
F_{sx} and F_{sy}:	Steel sheet stress in long and short directions
F_y:	profiled steel sheet tensile strength
P_{max}:	Maximum load at which any of the considered allowable limits is reached
R_1:	Force at support in long direction
R_2:	Force at support in short direction
R:	Sum of R1 and R2
δ:	Deflection.

Structural efficiency of waffle floor systems

J. Prasad, A.K. Ahuja & S. Chander
Dept. of Civil Engg., I.I.T. Roorkee, India

ABSTRACT: The present paper elaborates the results obtained from the study carried out on waffle slab floor system with a view to achieving better load dispersion through suitable configurations of ribs. For a given floor plan, rib dimensions and the number of ribs have been incremented in an attempt to achieve better moment dispersion at the lowest dead load percentage. The waffle slab has been taken as monolithically connected to band beams. Attempts have been made to achieve structural efficiency by ascertaining the above mentioned parameters which would minimize the dead load and hence its moments for a given floor plan. Analysis has been carried out by stiffness matrix method. Feasibility of structural design of members have been ensured under the provision of IS: 456-2000.

1 INTRODUCTION

Waffle slab has had its genesis in a rather thick solid-slab floor from which the bottom layer concrete in tension is partially replaced by their ribs along orthogonal directions. The concrete replacement and rib formation is done through appropriate size waffle formwork. The ribs are reinforced with steel to resist flexural tensile stresses. At this stage, the dimensions and spacing of ribs become a matter of choice to be decided in a manner so as to achieve better load distribution (structural efficiency). The spacing of the ribs are so kept as to obviate the need for shear reinforcement as also to make the system structurally behave as a slab rather than a system of interconnected parallel beams (grid floor).

Waffle slab may be rested on a system of vertical supports (walls) in which case the floor moment and deflection get concentrated near the mid-span. The floor, on the other hand, can be framed into the vertical structural system such that there is dispersal of floor moment and deflection. There is lateral shifting of moment and deflection from the mid-span zone towards the supports. Thus, floor framing with the help of wide band-beams rigidly connected to a system of columns becomes an obvious choice.

2 PROPOSED RESEARCH WORK

Keeping the residential and office floors in mind, the waffle slab floor plans have been adopted in the range of 6×6 m to 8×8 m for square plan. Two rectangular floor plans 6×7.7 m and 6×9.4 m have also been taken. Larger spans would need much higher floor thickness which would be suitable for grid floor structural system. For rectangular plans, the aspect ratio (long-to-short span ratio) has been kept well below the limit of two so that there is meaningful transfer of the imposed load along both the spans.

Top slab of the waffle slab floors may be kept at its minimum thickness from construction point of view. It has been taken as 65 mm for all the floors studied herein. Rib width of 100 mm has been taken. The shape of ribs is taken as rectangular during analysis, although in actual practice it may be slightly tapered (Fig. 1). The rib depth has been varied in the range from 130 mm to 260 mm with a regular increment of 10/20 mm. The overall floor depth, thus varies from 195 mm to 325 mm.

The number of ribs has been taken as five at the minimum with an increment of two until all the structural requirements are adequately satisfied for the particular floor plan. It may be considered prudent to increase the rib depth after a certain number of rather closely spaced ribs is found to be inadequate in meeting a particular structural requirement. The maximum number of ribs for the largest span is, therefore, taken to be nine.

The live load values adopted are 3 kN/m^2 and 5 kN/m^2. Floor finish load has been taken as 1.5 kN/m^2 everywhere. The present study has been carried out

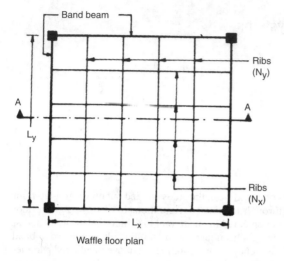

Waffle floor plan

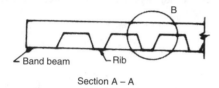

Section A – A

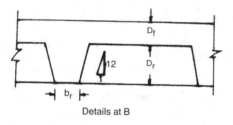

Details at B

Figure 1. Typical plan and section of waffle-slab floor.

using M20 grade concrete and Fe-415 grade steel as the materials of construction (Tables 3 and 5). However, increasing number of ribs would shift in values for higher strength materials maintaining a similar trend.

3 METHOD OF ANALYSIS

For the analysis of waffle slabs, orthotropic plate theory (Timoshenko & Kreiger 1959, Wang & Salman 1985, McCormac 1986, Nawy 1990, Abdul-Wahab & Khalil 2000), finite element method (Krishnamurthy 1987) and grid or grillage analysis (Cope & Clark 1984, Rao 1995) are generally used. In the present study, waffle slab is considered as made of grid or grillage of

beams. The loads are distributed between longitudinal beams by bending and twisting of transverse beams. The stiffness matrix is developed on the basis of writing joint equilibrium in terms of stiffness co-efficient and unknown joint displacements. Straight members of constant cross-section have been considered. The deformations considered are two orthogonal rotations in the horizontal plane and a vertical deflection at each of the node. Nodal displacements in the horizontal plane and rotations along the vertical axis are not considered keeping in view that they do not significantly contribute to the structural behaviour and hence are ignored.

The computer programme for grid analysis, written in FORTRAN 77, has been used for the present study. It results in moment, shear force and torsion for each of the elements and deflection and rotation about the two orthogonal axes at each of the nodes.

For the purpose of accepting a set of number and dimensions of ribs satisfying structural requirements, codal provisions given in IS: 456-2000 (also SP: 16-1980, SP: 24-1983) have been adopted in general. However, codal provisions of ACI: 318-1995 (also Rice & Hoffman 1985) have also been referred to with a view to highlighting certain advantages and some special features.

4 RESULTS AND DISCUSSION

4.1 *Square floor plan: 6 × 6 m*

Results of the analytical study on 6×6 m square waffle slab under a live load intensity of 3 kN/m² are presented in Tables 1–6. Whereas, Tables 1 and 2 show the results of the slab provided with 5 ribs along each of the two spans, Tables 3 and 4 show results for slab with 7 ribs. Results for slab with 9 ribs in each direction are given in Tables 5 and 6.

A close study of the deflection values (elastic deflection due to total load Δ_e and creep deflection Δ_c) in Table 1 and Figure 2 reveals that a rib depth of 130 mm exceeds the maximum permissible deflection (Δ_{max}) by 13% and hence needs to be stiffened by increasing the rib depth. Rib depth of 140 mm marginally increases the dead load percentage from 60.3 to 60.8 but brings down the deflection by 15% from 113 to 98%. This configuration of ribs is found suitable for shear capacity and bending moment resistance point of view also. It is interesting to note that a rib depth of 130 mm does not satisfy the deflection requirement even when the number of ribs is increased from 5 to 7 and finally to 9 as shown in Tables 3 and 5. However, increasing number of ribs does result in increase in dead load percentage from 60.3 to 61.1 and to 61.8. It is, therefore, inferred that increasing the rib depth is more advantageous to reducing the rib spacing.

Table 1. Deflection values for waffle slab floor of size 6 × 6 m with no. of ribs $N_x = 5$ and $N_y = 5$.

Rib depth D_r (mm)	DL as % of total load	Max. deflection (mm) Δ_e	Δ_c	Deflection as % of Δ_{max}
130	60.3	11.3	13.1	113
140	60.8	9.8	11.4	98
150	61.4	8.6	10.1	87
160	61.9	7.5	8.8	76

Table 2. Bending moment and shear force values for waffle slab floor of size 6 × 6 m with no. of ribs $N_x = 5$ and $N_y = 5$.

Rib depth D_r (mm)	DL as % of total load	Max. bending moment (kN-m) −ve	+ve	Max. shear (kN)	Value as % of balanced capacity −ve BM	Shear
130	60.3	86.6	49.3	68.6	115	68
		9.4	10.9	10.3	118	99
140	60.8	87.9	50.0	69.7	104	65
		9.6	10.9	10.4	108	95
150	61.4	89.1	50.7	70.7	95	63
		9.8	11.0	10.5	99	90
160	61.9	90.4	51.4	71.8	86	60
		10.0	11.0	10.5	91	86

Note: Every first row values are for band beams and second row values are for ribs.

Table 3. Deflection values for waffle slab floor of size 6 × 6 m with no. of ribs $N_x = 7$ and $N_y = 7$.

Rib depth D_r (mm)	DL as % of total load	Max. deflection (mm) Δ_e	Δ_c	Deflection as % of Δ_{max}
130	61.1	10.6	12.4	107
140	61.7	9.2	10.8	93
150	62.3	8.0	9.4	81
160	62.9	6.9	8.1	70

Table 4. Bending moment and shear force values for waffle slab floor of size 6 × 6 m with no. of ribs $N_x = 7$ and $N_y = 7$.

Rib depth D_r (mm)	DL as % of total load	Max. bending moment (kN-m) −ve	+ve	Max. shear (kN)	Value as % of balanced capacity −ve BM	Shear
130	61.1	84.7	45.8	69.8	113	69
		8.6	8.9	9.5	108	92
140	61.7	86.1	46.5	71.0	102	66
		8.8	10.0	9.6	99	87
150	62.3	87.4	47.2	72.2	93	64
		9.0	10.0	9.7	90	83
160	62.9	88.8	47.9	73.4	85	62
		9.2	10.1	9.8	83	80

Note: Every first row values are for band beams and second row values are for ribs.

Table 5. Deflection values for waffle slab floor of size 6 × 6 m with no. of ribs $N_x = 9$ and $N_y = 9$.

Rib depth D_r (mm)	DL as % of total load	Max. deflection (mm) Δ_e	Δ_c	Deflection as % of Δ_{max}
130	61.8	10.5	12.3	106
140	62.5	9.1	10.8	92
150	63.1	7.9	9.4	80
160	63.8	6.8	8.1	69

Table 6. Bending moment and shear force values for waffle slab floor of size 6 × 6 m with no. of ribs $N_x = 7$ and $N_y = 7$.

Rib depth D_r (mm)	DL as % of total load	Max. bending moment (kN-m) −ve	+ve	Max. shear (kN)	Value as % of balanced capacity −ve BM	Shear
130	61.8	84.5	44.5	70.5	113	70
		8.2	8.1	9.0	103	87
140	62.5	86.0	45.2	71.8	102	67
		8.4	8.2	9.1	94	83
150	63.1	87.4	45.9	73.1	93	65
		8.6	8.3	9.2	86	80
160	63.8	88.0	46.6	74.4	84	63
		8.8	8.4	9.3	80	76

Note: Every first row values are for band beams and second row values are for ribs.

Consumption and distribution of reinforcement may be effected by having a better dispersion of moment along the span from negative to positive moment. For this, attention is drawn to the moment values in Tables 2, 4 and 6 and Figures 3–8. It is seen that moment values per rib decease with the increase in the number of ribs and steel requirement per rib decreases. This would be advantageous if accommodating the bars in the thin (100 mm) rib becomes difficult. The amount of steel consumption would, however, be more for the entire floor since the number of ribs would be more.

Rib depth of 140 mm satisfies the deflection and shear requirements but bending moment value

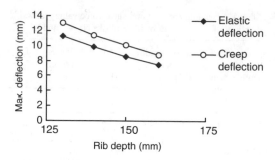

Figure 2. Variation of max. deflection with depth of rib.

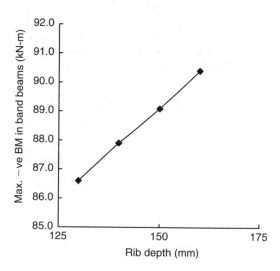

Figure 3. Variation of max. −ve BM in band beams with depth of rib.

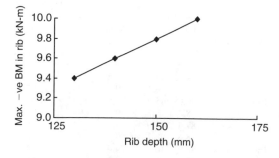

Figure 4. Variation of max. −ve BM in ribs with depth of rib.

marginally exceeds the capacity (8%) for the floor having 5 ribs. At this stage, a choice becomes available from among the following.

(i) making the ribs doubly reinforced

(ii) going in for higher rib depth (say 150 mm)
(iii) increase the rib number (say 7)

Amongst these choices, the first one is advantageous in the sense that it needs marginal increase in steel requirement from the balanced design to doubly reinforced design. It would meet all the requirements at the lowest percentage of dead load. Thus, it may be concluded that 5 ribs of 140 mm depth would be structurally most advantageous for the 3 kN/m² live load category.

Behaviour of 6 × 6 m waffle floor under 5 kN/m² live load intensity, has also been studied in the similar manner. Results of this case are not included in this paper due to paucity of space. For this loading, the ribs are required to be made deeper as well as increased in number. It is seen that 5 ribs are unsuitable. A choice of 150 mm depth with 9 ribs or 160 mm depth with 7 ribs appears to be working.

4.2 *Square floor plan: 7 × 7 m and 8 × 8 m*

Square waffle slabs with dimensions 7 × 7 m and 8 × 8 m have also been analysed to find out the structurally most suitable dimensions and number of ribs under live load intensities of 3 kN/m² and 5 kN/m². Due to paucity of space, the results of these studies are not included here. However, conclusions drawn for these slabs are reported in the section "Conclusions".

4.3 *Rectangular floor plan: 6 × 7.7 m*

Due to the reason given above, results of the study on rectangular waffle floor of 6 × 7.7 m are also not given in this paper. Conclusions drawn are, however, described in Section 5.

4.4 *Rectangular floor plan: 6 × 9.4 m*

The results of the response of rectangular waffle floor of 6 × 9.4 m under live load intensity of 3 kN/m² are presented in Tables 7 and 8. Comparison of the values in these tables with corresponding values of square floor of size 6 × 6 m indicates that the value of deflection for a particular depth is almost three times for a rectangular floor plan than that of a square floor plan. The load dispersal is also higher in 6 × 9.4 m floor along short span in comparison to 6 × 7.7 m floor. Further, the ratio of maximum bending moments in band beams along long span and short span is about 1.50 and shear force is about 1.75. From above, it becomes clear that with increase in aspect ratio the design parameters for band beams along the longer span and ribs along both directions increase rapidly making their design difficult and uneconomical.

Further it is observed that along the long span (YY-direction), rib depth of 220 mm can be managed

Table 7. Deflection values for waffle slab floor of size 6 × 9.4 m with live load = 3 kN/m², no. of ribs $N_x = 9$ and $N_y = 5$.

Rib depth D_r (mm)	DL as % of total load	Max. deflection (mm) Δ_e	Δ_c	Deflection as % of Δ_{max}
160	61.2	24.6	28.7	158
180	62.2	19.3	22.7	124
200	63.1	15.5	18.4	100
220	64.0	12.7	15.2	83

Table 9. Deflection values for waffle slab floor of size 6 × 9.4 m with live load = 5 kN/m², no. of ribs $N_x = 9$ and $N_y = 5$.

Rib depth D_r (mm)	DL as % of total load	Max. deflection (mm) Δ_e	Δ_c	Deflection as % of Δ_{max}
160	48.6	31.2	32.0	187
180	49.7	24.5	25.4	148
200	50.7	19.6	20.5	119
220	51.6	15.9	16.8	97

Table 8. Bending moment and shear force values for waffle slab floor of size 6 × 9.4 m with live load = 3 kN/m², no. of ribs $N_x = 9$ and $N_y = 5$.

Rib depth D_r (mm)	DL as % of total load	Max. bending moment (kN-m) −ve	+ve	Max. shear (kN)	Value as % of balanced capacity −ve BM	Shear
Short span (XX-direction)						
160	61.2	101.6	58.3	80.1	97	67
		15.4	16.5	17.5	140	143
180	62.2	104.8	60.0	82.8	82	63
		15.8	16.6	17.7	118	132
200	63.1	108.0	61.7	85.5	71	60
		16.4	16.7	18.0	103	123
220	64.0	111.2	63.4	88.0	62	57
		16.9	16.9	18.3	91	115
Long span (YY-direction)						
160	61.2	251.4	128.0	140.0	240	118
		23.4	22.2	13.8	212	113
180	62.2	257.5	131.0	143.4	202	109
		24.3	22.7	14.2	182	106
200	63.1	263.6	134.0	146.9	173	102
		25.3	23.3	14.6	159	100
220	64.0	269.8	137.1	150.5	150	97
		26.2	23.9	15.0	141	95

Note: Every first row values are for band beams and second row values are for ribs.

Table 10. Bending moment and shear force values for waffle slab floor of size 6 × 9.4 m with live load = 5 kN/m², no. of ribs $N_x = 9$ and $N_y = 5$.

Rib depth D_r (mm)	DL as % of total load	Max. bending moment (kN-m) −ve	+ve	Max. shear (kN)	Value as % of balanced capacity −ve BM	Shear
Short span (XX-direction)						
160	48.6	127.3	73.0	100.2	121	84
		19.4	21.2	22.2	176	182
180	49.7	130.5	74.7	103.0	102	79
		19.9	21.3	22.4	149	167
200	50.7	133.7	76.5	105.8	88	74
		20.4	21.3	22.6	128	154
220	51.6	136.9	78.2	108.5	76	70
		20.9	21.4	22.8	112	144
Long span (YY-direction)						
160	48.6	315.6	160.9	175.4	301	148
		29.6	28.1	17.5	268	143
180	49.7	321.4	163.8	178.8	252	136
		30.7	28.6	17.9	230	133
200	50.7	327.3	166.7	182.3	215	127
		31.7	29.2	18.4	200	126
220	51.6	333.4	169.6	185.8	186	119
		32.7	29.7	18.8	175	119

Note: Every first row values are for band beams and second row values are for ribs.

by making the band beam and ribs doubly reinforced. Along short span (XX-direction), rib depth of 220 mm is to be managed by making the rib section suitably doubly reinforced so that shear capacity can be increased. Hence, it may be concluded that for 3 kN/m² live load, 220 mm rib depth is structurally most efficient for 6 × 9.4 m floor plan with 9 and 5 ribs along short and long spans respectively.

Analytical results for 5 kN/m² live load (Tables 9 and 10) indicate that along long span, a rib depth of 220 mm can be managed by making the band beam and ribs suitably doubly reinforced as the shear capacity is increased

along with bending capacity. But along short span, the shear requirement is not satisfied with a rib depth of 220 mm. To avoid the shear reinforcement in ribs along short span, the rib depth is required to be increased by a large amount resulting in very stiff floor with reduced deflection, higher dead load percentage and reduced head room. Another alternative can be the use of M-40 grade concrete. Since deflections will be reduced by the use of M-40 grade concrete, 200 mm rib depth will satisfy the deflection criteria. It can also be seen that rib depth of 200 mm meets the bending capacity and shear capacity criteria along the long span. However, along short span, the rib depth of 200 mm can be managed by making the ribs slightly doubly reinforced to increase their shear capacity up to desired level.

In view of above, it is concluded that for live load intensity of 5 kN/m² by using M-40 grade concrete, 200 mm rib depth may be considered as the structurally most efficient for 6 × 9.4 m floor plan with 9 and 5 ribs along short and long spans respectively.

5 CONCLUSIONS

Following conclusions are drawn from the work presented in this paper.

– For 6 × 6 m square floor plan, 5 ribs of 140 mm depth (overall depth 205 mm) is found to be structurally most efficient for 3 kN/m² live load intensity.
– For 5 kN/m² live load intensity on a 6 × 6 m square floor plan, a choice between 9 ribs of 150 mm depth and 7 ribs of 160 mm depth becomes available. Percentage of dead load is about 50 in both the cases.
– For square floor plan of 7 × 7 m, the most efficient structural system is 9 ribs of 180 mm depth for a live load intensity of 3 kN/m².
– For a live load intensity of 5 kN/m², the most efficient structural system is 9 ribs of 200 mm depth for 7 × 7 m square floor plan.
– For square floor plan of 8 × 8 m, the most efficient structural system is 9 ribs of 240 mm depth for a live load intensity of 3 kN/m².
– For square floor plan of 7 × 7 m, the most efficient structural system is 9 ribs of 220 mm depth using M-40 grade concrete for a live load intensity of 5 kN/m².
– For 6 × 7.7 m rectangular floor plan with 7 and 5 ribs along short span and long span respectively, 180 mm rib depth is found to be structurally most efficient for a live load intensity of 3 kN/m².
– For 5 kN/m² live load intensity on 6 × 7.7 m rectangular floor plan, 220 mm rib depth is structurally most suitable with 7 and 5 number of ribs along short and long span respectively.
– For 6 × 9.4 m rectangular floor plan, 220 mm rib depth is structurally most suitable for a live load

intensity of 3 kN/m² with 9 and 5 number of ribs along short and long span respectively.
– For rectangular floor plan of 6 × 9.4 m with a live load intensity of 5 kN/m², by using M-40 grade concrete, 200 mm rib depth may be considered as the structurally most efficient with 9 and 5 number of ribs along short and long span respectively.

REFERENCES

Abdul-Wahab, H.M.S. & Khalil, M.H. 2000. Rigidity and strength of orthotropic reinforced concrete waffle slabs, *J. Str. Engg.*, Feb., 219–227.
ACI: 318-1995. *Building Code Requirements for Reinforced Concrete*, American Concrete Institute, Detroit, USA.
ACI: 318R-1995. *Commentary on Building Code Requirements for Reinforced Concrete*, ACI, Detroit, USA.
Cope, R.J. & Clark, L.A. 1984. *Analysis and Design of Concrete Slabs*, Elsevier Applied Science Publishers Ltd., Essex, UK.
IS: 456-2000. *Indian Standard Code of Practice for Plain and Reinforced Concrete*, BIS, New Delhi, India.
Jain, A.K. 1998. *Reinforced Concrete – Limit State Design*, Nem Chand and Brothers, Roorkee, India.
Krishnamurthy, C.S. 1987. *Finite Element Analysis*, Tata McGraw Hill Publishing Co. Ltd., New Delhi.
McCormac, J.C. 1986. *Design of Reinforced Concrete*, Harper and Row Publishers, New York.
Nawy, E.G. 1990. *Reinforced Concrete: A Fundamental Approach*, Prentice Hall, Engle Wood Cliffs, N.J.
Rao, G. 1995. Studies in R.C. grid floor systems, *M.E. Dissertation*, University of Roorkee, Roorkee, India.
Rice, P.F. & Hoffman, E.S. 1985. *Structural Design Guide to ACI Building Code*, Van Nostrand Reinhold Company, New York.
SP: 16-1980. *Design Aids for Reinforced Concrete to IS: 456-1978*, BIS, New Delhi, India.
SP: 24-1983. *Explanatory Handbook on Indian Standard Code of Practice for Plain and Reinforced Concrete IS: 456-1978*, BIS, New Delhi, India.
Timoshenko, S. & Kreiger, S.W. 1959. *Theory of Plates and Shells*, McGraw Hill Book Co., New York.
Wang, C.K. & Salman, C.G. 1985. *Reinforced Concrete Design*, Harper and Row Publishers, New York.

38. Special session on "Advanced conceptual tools for analysis long suspended bridges"

System-based Vision for Strategic and Creative Design, Bontempi (ed.)
© 2003 Swets & Zeitlinger, Lisse, ISBN 90 5809 599 1

A conceptual framework for the design of an intelligent monitoring system for the Messina Strait Crossing Bridge

S. Loreti
Structural Engineer, Rome, Italy

G. Senaud
Graduate Student, University of Rome "La Sapienza", Rome, Italy

ABSTRACT: This work aims at describing the research connected with the design of the health and behavior monitoring system of the Messina Strait Crossing Bridge. Due to its extreme innovative characteristics, being the longest single span cable bridge and for the contemporary presence of road and railway decks, it is important to have clear not only the structure and substructure components behavior but also the condition of the overall environment. Since the structural design is based on the definition and the assessment of the performance for the evaluation of the reliability of the global structure, the health and behavior monitoring system must be developed, in this optic, as a principal part of the overall design.

1 INTRODUCTION

A long span suspended bridge is a complex system that interacts with the surrounding environment and users with whom is destined.

The environmental actions (wind, temperature, rain, earthquake, etc.) and the users one (mobile loads), can every time modify the structural behavior. Morcover, an environmental action can influence the users comfort taking to a limitation about the service condition until a partial closure of the bridge itself.

The Messina Strait Crossing Bridge will be the longest and generally most complex structure that human mind has never thought. To arrive to the construction and the respect of its life expectation, the knowledge of the entire Engineering field must fund and share together. So, the complexity of the design needs skilful and committed workers able to share all their own knowledge with other members of the *project team* and able to gain new experiences and knowledge, which should become part of the organization's knowledge base.

The bridge structure will be subjected to deterioration over time, changing its mechanical characteristics and behavior: *performance* given by the design and *reliability* of the structural system become the keywords in order to catch up the expected life.

All this, make simply understand how important is to expect a well organized *bridge management* also before studying the different technologies to install on the bridge, in fact, the most safe and durable structures are usually the best managed ones.

2 KNOWLEDGE OF PERFORMANCES

2.1 *Knowledge of performances*

The inherent complexity in planning and designing a complex structure like the Messina Strait Crossing Bridge needs a new design statement funded on the *Performance-Based Design* (P.B.D.), which aims at defining preventively and accurately the performances requested to the structure.

Nowadays, most of the structures are designed, constructed, maintained and dismantled without taking in consideration the specific knowledge pertinent to their real behavior. The recent improvement in measurement and elaboration data technologies has created the proper condition to improve the decisional tools, traditionally based on experimental and numerical simulations, through the improvement of the information connected both to the structural performances on site and to the measurement systems which, in the last few years, have become more numerous, cheaper, stronger, more stable and precise.

In Figure 1, are shown the subsequent phases of the design procedure; the first five phases are connected

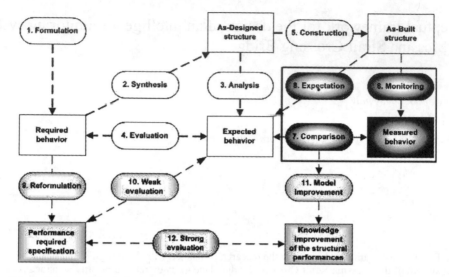

Figure 1.   Principal phases of the design process.

with the traditional design approach which takes to an "*as-built*" structure, often different from the "*as-designed*" one due to different factors such as fabrication mistakes or unexpected conditions during the construction phase, or also due to non appropriate design assumptions that could create doubts about analysis accuracy leading to a predicted behavior which does not correspond to the real one.

A way to avoid these difficulties, as already said, can be achieved using a monitoring system (phases from 6 to 8).

Once the most important task – connected with the idea of performances that the structure under examination will have to guarantee – is clearly understood, four new phases connected to a performance design procedure have been inserted in the scheme (9 to 12).

In this way it is possible to obtain an improvement of the design solution in the contest of the *Theory of Excellence*, a generalization of the *Re-Engineering*.

From Figure 1, it appears clear how important is to design an *ad hoc* monitoring system to associate with the structural design. Without measurements, it should be assumed theoretical or design code values for important parameters leading to unnecessary maintenance costs.

It has often happened that measurements on individual structures in real environments usually has revealed smaller standard deviations related to strength and loading that those adopted during design stages.

Until now, there has been a general tendency to clearly separate structural analysis from monitoring in the design phases considering the second as a consequence of the first one.

Nowadays, with the development of PBD, it must be clear how these two aspects must integrate each other

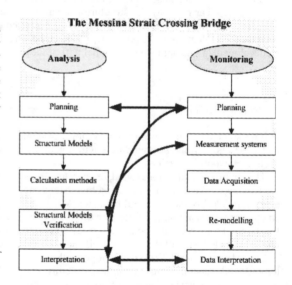

Figure 2.   The correlation between analysis and monitoring.

and as shown in Figure 2 the barrier is ready to be demolished. According to this kind of procedure, models studied and build up during the different analysis phases, aren't necessary only in the design stages, but remain useful, perhaps more useful after the construction of the structure to compare expected numerical results with the real behavior measured on the structure itself. In this way, the models can be improved and refined for a better preventive evaluation of the structural real behavior in face of extreme phenomena.

Thought in this way, the bridge will not be a simple "complex" structure, but it will be intelligent, able to learn new knowledge from its own behavior.

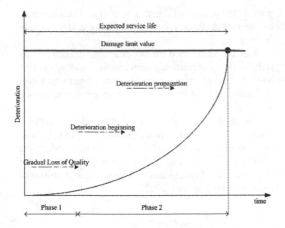

Figure 3. Time evolution of deterioration.

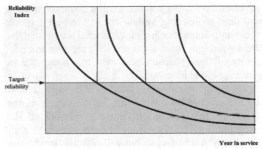

Figure 4. Reliability based inspection intervals.

2.2 *Reliability of the entire system*

As already said, the Messina Strait Crossing Bridge will be build up for an expected life of two hundred years, and during this time will be subjected to many environmental actions that will modify its behavior with vehicles continuous passages.

In this sense, very important will be the bridge reliability defined as the ability of a product to perform as intended (without failure and within specified performance limits) for a specified time, in its life cycle application environment. The objective of reliability prediction is to support decisions related to the operation and maintenance of the product to (Bentley 1999):

– Reduce output penalties including outage repair and labor costs.
– Optimize maintenance cycles and spares holdings.
– Maintain the effectiveness of equipment through optimized actions.
– Help in the design of future products, by improved safety margins and reduced failures.
– Increase profitability.

Structural systems are subjected to deterioration over time, changing their mechanical characteristics and modifying their way to function. One of the most important aims of reliability is to control deterioration in order to prevent unexpected failures during the expected life (Fig. 3).

Deterioration owns to the reliability of the system; not only the global behavior is important to control, but also the local one, strictly referred with the single element; the reliability of an element is independent of the ones of the other elements in the structure and single elements can fail in different ways. For instance for an incorrect design, for defects or inappropriate materials used in element manufacture, for incorrect installation, operation and maintenance procedures.

All this makes simply understand how much complex is a structure, and how much important is to expect all this problems.

Inspection must be expected because it represents an important part of the integrity management process as a means of monitoring the performances of structures to ensure their safety and serviceability.

However, inspections can represent a significant cost for the structure. Traditionally, inspection planning was based on general guidelines and engineering judgment based on the prescriptive codes, without considering the structure specific characteristics.

This way of doing has led in the past to an amount of inspection completely useless because not focused on most critical areas, or using inadequate techniques.

Today the tendency is to evaluate through the performances and the reliability, the real behavior of the bridge in order to maximize the structural efficiency and control the safety levels maintained. To achieve this, a model of the deterioration process needs to be developed. The model is used to develop initial reliability profiles for the areas considered for inspection optimization (Fig. 4).

It's clear that when reliability index reaches the *target reliability* an inspection is recommended. Target reliability can be derived based on a series of redundancy analyses of the specific structure in question. This kind of analyses can take to maintain the safety level inside a fixed level.

The need for bridge system redundancy is widely recognized. System redundancy has been defined as availability of warning before collapse. In a certain way the warning can be obtained with a monitoring automatic and intelligent system able to measure any structural variation beyond a fixed limit.

3 THE MESSINA MONITORING SYSTEM

3.1 *General aspects*

Due to the long expected life of the Messina Strait Crossing Bridge (two hundred years) it's necessary for the owner to require high performances also looking for costs. To achieve this last goal, it could seem that a real time monitoring system is not the best solution because in short time high investments are needed; by the way, in long time, thanks to the precision and reliability of these measurement systems, some useless repairing could be avoided.

The health monitoring system will be organized on the use of different equipments for the same aim; the choice of the monitoring devices to install on the bridge, will not be done from an economically or simply blindly point of view, but will directly derive from the specific needs understood through modelling analyses during the structural design.

Because of it's impossiblity to be sure which are the best devices to install on a structure above all because, it's simple to forgive to consider some variable, it's surely better to expect a redundant solution than a fair one, and so to install a number of devices major than needed.

A performance and reliable approach will not lead surely to a good health monitoring system without a good management of the structure. It will be necessary to operate a fusion between human and computer control.

Aspects that need to be considered include the redundancy, dominant deteriorations modes, loading history, history of operation, consequences and cost of failure, relevant methods of inspection and associated costs.

The inspection optimization methods are equally applicable to both new and existing structures. In both cases safety benefits can be achieved as the underlying philosophy of this approach is maintaining more consistent levels of safety across a structure and groups of structures (Frangopol 2002).

3.2 Bridge monitoring needs

Before starting the design of the monitoring system, it must be clear which are the main points to control and measure. So, it is important first of all to know in detail how the structure will be when built up. To achieve this

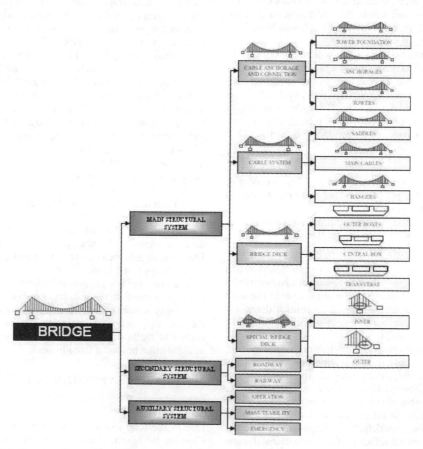

Figure 5.   Main decomposition of the bridge structure design.

task has been organized a decomposition of the structure into three levels (Fig. 5):

– Macrolevel (substructures)
– Mesolevel (components)
– Microlevel (elements)

For each one of these categories, subsequently, the variables of detail identified with the three levels named above have been defined.

Operating in this way, it's possible to define all the singular variables playing a role on the bridge, until defining the single thickness of a rib for instance. All this must be done also because the monitoring system of the bridge can be approached either from the material or from the structural point of view.

In the first case, monitoring will concentrate on the properties of the materials used in the bridge construction observing the behavior under loading and aging.

In the structural approach, the bridge is studied from the geometrical point of view. Also in this way is possible to detect the material behavior and degradation, but only if it has direct impacts on form of the bridge.

It's clear that having a redundant solution, in this case represented by the use of different kind of instruments that from a different point of view monitor the same aspects, would seem too much expensive, but sometimes happens that in the maintenance of a structure without an integrate real time monitoring system, the investments can exceed the cost of a new structure! Continuous monitoring allows to increase the safety margins without any useless intervention on the structure.

The Messina Strait Crossing Bridge, will necessarily be an intelligent and active structure able to improve its performance during its service life.

Thought as an intelligent structure, the Messina Bridge will react to the changes in its environment linking the knowledge derived by the measurements systems obtained by the computational control with the use of past events saved in its database.

3.3 Knowledge management

All the importance given to automatic and technologic aspects must not take to forget how a good monitoring system depends also by the human designer and workers in the monitoring control station. A great help can be given in organizing the monitoring system with a *Knowledge Management* approach.

Knowledge management (KM) is the process of creating value from an organization's intangible assets (Liebowitz 1999). As such, knowledge management combines various concepts from numerous disciplines, including organizational behavior, human resources managements, artificial intelligence and others. The focus is how best to share knowledge to create value-added benefits to the organization. The idea of sharing is the central point, because a monitoring system is based on a series of multiple aims. In fact, it must be based on precedent experiences of other similar bridges, on the comparative evaluation of measurements with those obtained in modelling and above all in exchanging of data.

In the monitoring system, both humans and computers must be funded forming the teamwork. The control station team members gain new experiences and knowledge, which must become part of the organization's knowledge base. These knowledge-creating teams play key roles in the process of value creation of an organization. Knowledge work is based on knowledge exchange, which requires a certain degree of trust as pre-condition. Approaching in this way, all the measurements obtained can be used to modify some of the parameters of the finite elements models for instance in order to make the model nearer to real behavior of the bridge.

In the optic of the knowledge management, a great help, not only for the monitoring system, but also referring to all the phases connected with the entire Messina Strait Crossing Bridge project, could be given by the use of the Information Technologies.

In this case could be useful to develop an internet/intranet system organized in order to involve all the participants of the entire design phase in a first phase (exchanging all the information needed) and then all the people involved in the construction and monitoring phases of the bridge. Thought in this way through internet, it will be possible to put acquaintance of the state of advance of the construction of the bridge all people interested directly and indirectly.

3.4 The use of the GPS technology

Today, using a GPS system, in parallel with other common sensors cheaper than GPS, is possible to monitor in real time the bridge starting from the early phases of the construction.

The importance of the GPS technology goes really beyond the simple functionality of the instrument, but it represents means in order to calibrate all the monitoring, control and maintenance system of the bridge and therefore its global reliability.

The GPS system is now worldwide known for its ability to monitor the displacements of the main suspended cables, decks and the bridge towers. With the displacements values obtained, the relevant stress status acting on the all bridge can be derived.

Displacements of the bridge structure, serves as an effective indicator of its structural performance condition. Using Finite Elements Models (FEM) with reference to GPS measurements can be created a robust link between reality and modelling which enables to identify critical structural components, for the purpose of long-term inspection and maintenance. Today, real-time bridge monitoring has a level precision up to centimeter and time of delay of approximately two seconds for display of bridge monitoring room due to data transmission, processing, and graphical conversion times. The Messina health monitoring system will operate 24 hours a day with automatic control for data, acquisition, processing, archiving, display and storage. Post-processing the GPS data with relevant data arriving from the other sensor built up on the structure and into grating the results, the structural evaluation needs for the bridge will be simpler to understand for all the engineers involved in the bridge continuous assessment.

Intercrossing the different values, stress/strain results of the structural response are obtained in critical structural components in order to validate the design values of environmental loading and design parameters, confirming the confidence level of safe operation. For this reason, a high level quality of data transfer must be guaranteed in order to assure stability and reliability. To achieve this aim, an optical fiber line must be prevented. This technology is insensitive to electromagnetic waves and lightning effects, provides high speed for data transmission under bad weather and severe environments such as the strong electromagnetic field generated by high voltage circuits. Moreover, advanced fiber transceivers can detect any interruption of data transmission and send an alarm to alert maintenance personnel and provide problems locations.

The Messina Strait Crossing Bridge health monitoring system will at least evaluate three main aspects:

– Serviceability
– Load-carrying
– Durability

Serviceability relates to the deformation, crack and vibration of the bridge under normal loading conditions.

Load-carrying relates to structural stability and ultimate strength of the materials. Estimating this parameter, the bridge's actual safety capacity, avoiding catastrophic damages can be evaluated.

Durability assessment focuses on damage to the bridge and its influence on physical properties of the materials.

4 CONCLUSIONS

The development of new approaches in structural engineering based on Performance-based Design and reliability, and the use of new technologies connected with monitoring, will allow to assess more exactly and with a lower degree of uncertainty the behavior of the Messina Strait Crossing Bridge.

Information technology and a good management based on knowledge will permit a more efficient evaluation of the behavior of the entire structural system increasing the safety level. This approach will make all the bridge workers, real-time involved in all activities improving their knowledge and exchanging information.

In long term, an health monitoring system thought in this way will contribute to lower costs and higher engineering quality.

The use of GPS technology and of the fiber optic line, integrated with other sensors installed on the bridge, will allow a more accurate and reliable monitoring evaluating structural performance and health conditions of the different bridge components. The results obtained can be applied in this way in planning and implementing inspection and maintenance activities.

ACKNOWLEDGMENTS

The financial support of Stretto di Messina Spa is acknowledged. Anyway, the opinions and the results presented here are the responsibility of the Authors and cannot be assumed to reflect the ones of Stretto di Messina Spa.

REFERENCES

Smith, I.F.C. 2001. *Increasing Knowledge of Structural Performance*. Structural Engineering International 3/2001.

Onoufriou, T. & Frangopol, D.M. 2002. *Reliability-based inspection optimization of complex structures: a brief retrospective*. Computer and Structures 80 (2002). Elseiver Science.

Imai, K. & Frangopol, D.M. 2002. *System reliability of suspension bridges*. Structural safety 24 (2002). Elseiver Science.

Ciampoli, M. 1998. *Time dependent reliability of structural systems subject to deterioration*. Computer and Structures 67 (1998). Elseiver Science.

Bentley, J.P. 1993. *Introduction to reliability and quality engineering*. Addison-Wesley.

Bailey, S.F. 1997. *Evaluation of existing steel and railway bridges*. Structural Engineering International 7/1997.

Nonaka, I. & Takeuchi, H. 1995. *The knowledge-creating company*.

Damm, D. & Schindler, M. 2002. *Security issues of a knowledge medium for distributed project work*. Int. journal of Project Management.

Kim, S.K. 2001. *Intelligent Bridge Maintenance System Development for Seohae Grand Bridge*. International Association of Bridge and Structural Engineering. Seoul 2001.

Mita, A. & Iwaki, H. 2000. *Performance-based Design and health monitoring system*. Third U.S.-Japan Workshop on Nonlinear System Identification and Structural Health Monitoring.

Wong, K.Y. 2002. *Monitoring Hong Kong's Bridges*. GPS World. Leica Geosystems.

Yanaka, Y. & Kitagawa, M. 2002. *Maintenance of steel bridges on Honshu-Shikoku crossing*. Journal of Constructional Research 58 (2002). Elseiver Science.

System-based Vision for Strategic and Creative Design, Bontempi (ed.)
© *2003 Swets & Zeitlinger, Lisse, ISBN 90 5809 599 1*

Conceptual framework for the aerodynamic optimization of the long span bridge deck sections

F. Giuliano
Structural Engineer, Master Mathematic Student, Rome, Italy

D. Taddei
Structural Engineer, Ph.D. Student, Polytechnic of Milan

ABSTRACT: Wind action is very important in the design of long span bridge deck sections. The challenge is the optimization of the aerodynamic behavior of deck sections. The use of Computational Fluid Dynamic allows to set the optimization process in an economical way, in terms of time and cost. In this work to represent input and output data a synthesis of principal design and state variables is done, where the latter are preferable quantitative measures of the performance of structure, changing when the design variables change. Besides, general conceptual formulation of optimization process is illustrated, with particular reference to the design of Messina Strait Bridge. Finally, some first steps of the aerodynamic optimization process for the same bridge have been done, by using a commercial finite element code.

1 INTRODUCTION

In the design of great suspended and cable-stayed bridges, the problem of interaction between the structure and the local winds is of great relevance and complexity. The wind action is very often responsible of design choices: geometry, strength and minimal rigidity of structure. The aim of engineering is to obtain the optimal structural configuration for the specific case.

The challenge of aerodynamic optimization in bridge design is matter of the last years, for three main reasons. First, the great results obtained by structural engineers with the construction of some exceptional bridges in Denmark, France, Japan, with great economic cost. Second, the development and commercialization of powered and reliable computational tools, that allow effective numerical experiments at low cost. Third, the amount of information obtained in monitoring real structures, that gives fundamental insight into the behavior of long span bridges under wind actions.

In this work, after a brief exposition of the most important design factors in great span suspension bridges, a conceptual formulation for the aerodynamic optimization process is highlighted. In this process, non-structural details assume great importance: the aerodynamic behavior, in fact, is strongly influenced by them, and the modelling difficulties makes the design process need a sensitivity analysis.

It is important to mark that some design choices, related to the necessary structural performances, are preliminary for any optimization and must be obtained by economical and risk analysis, where the role of political decisor is as important as the designers'.

Finally, the development of the first optimization process has been started for the Messina Strait Bridge deck, by using several numerical models. The calculation is conducted by a commercial finite element code.

2 OPTIMIZATION PROBLEM FOR LONG SPAN BRIDGE DECK

2.1 *Definition of state variables*

We can define "state variable" any one of the output data of a numerical (but even physical) model, representing the behaviour or an attended performance of the structure. For the aerodynamic optimization of bridge deck sections two types of state variables should be considered. The firsts, concerning structural safety, are:

- Wind speed for flutter instability;
- Amplitude and strength of vibration, induced by vortex shedding and buffeting, for fatigue life determination;
- Quality of polar curves for the section, responsible of static instability and galloping.

The second group of variables allows to control structural response:

– Lateral wind on vehicles;
– Vertical oscillations of the bridge deck (displacements and accelerations) produced by vortex shedding and buffeting, for vehicles comfort.

For each variable, a limit value should have been stated. The minimum wind of flutter, for example, is stated in dependence of local wind characteristics.

The statement of a maximum lateral wind speed is more complex, involving an economical analysis that must take into account the traffic data on the bridge. By investigating on the value, direction and frequency of local maximum winds (days per year), the hours of closure of bridge for different categories of vehicles can be esteemed. The definition of these limit values is preliminary to the optimization process, and the structural engineer can't be the only responsible of all the statements.

2.2 Definition of objectives

Just like each optimization process, the judgement of "optimality" is based on the minimization of some functions. For our purposes, the research of minimum concerns:

– Aerodynamic drag and moment;
– Aerodynamic lift;
– Deck width;
– Structural complexity.

Drag and moment resultant effects cause displacements and rotations of the bridge and consequently make the designers increase the structural strength and stiffness or, alternatively, need to be reduced: their minimization is one of the traditional aerodynamicists' objective. A positive lift is dangerous, if an inversion in tension of the secondary cables is possible. Deck width is conditioning for the dimensions (cost) of towers and its foundations. Structural complexity increases the costs and the occurrence of human errors in construction.

2.3 Definition of design variables

2.3.1 Structural details

No matter is on structural typology (suspended or cable stayed bridge, number and dimension of lanes, number and position of cables). We refer to structural details intending essentially the geometry of deck section. Obviously, only a single/multiple closed box girder deck typology is here considered, that is (today) the only analyzable with CFD.

The structural characteristics and details are then:

2.3.1.1 Number of box girders

The adoption of a multiple box girders section has been necessary to prevent flutter instability phenomena on super-long span bridge. Theorical and experimental analyses show aeroelastic instability can be avoided if the traffic lanes are subdivided in two or three box girders, separated by a wide.

Numerical simulations of aeroelastic behavior need fluid-structure interaction simulations.

This is a challenge field of research, with high computational costs and a very complex modelling, so that it is still necessary to use approximated analytical formulations. For single box girder section the Selberg's formula is very useful:

$$\frac{U_{crit}}{f_\alpha D} = 4\left[1 - \frac{f_h}{f_\alpha}\right]\left[\frac{mr}{\rho D^3}\right]^{\frac{1}{2}}$$

For multiple box girders, Larsen's semi empirical formula can be adopted:

$$U_c^{twn} = 3.72 \cdot C_{D/B} \cdot f_\alpha \sqrt{\frac{mI}{\rho B}}\left[1 - \left(\frac{f_h}{f_\alpha}\right)^2\right]$$

with the parameter $C_{D/B}$ dependent from the presence of wind barriers.

2.3.1.2 Geometrical shape of section

For long span bridges, the structural material is generally steel. For this reason, a geometrical variation of deck section, just for aerodynamic necessity, can be explored with no particular restrictions. The weight of structure, and the strength characteristics, are determined afterwards, based on the "steel design" techniques.

Only in the Messina Strait Bridge project a wing-shaped section has been proposed, and in a such pioneeristic choice the experimental and the computational processes for the optimization are strategic. Besides, a non-polygonal section introduces several new problems: some recent analyses show a strong sensibility of this kind of section to flutter instability; moreover, in the wind tunnel tests a Reynolds similitude problem is evidenced, and the flow separation, that is not caused by corner-edges of section, may occur in a not realistic point. Consequently the interpretation of wind tunnel test results is more difficult than in other circumstances.

The most important element for aerodynamic behavior seems to be leading edge of section. Some optimization study are conducted for this element, in particular for the project of Great Belt Bridge [2].

2.3.2 *Non-structural details*

Among the non-structural details we can distinguish:

i. Elements that are part of highway section (guard-rails)
ii. Elements necessary to vehicular traffic (wind barriers and walls)
iii. Elements for aerodynamic control (grids, walls, guide vanes for vortex shedding)

2.3.2.1 Guard-rails

For bridge of some importance, and for all the long span bridges, *ad hoc* design of guard-rail is recommended.

The criterion of aerodynamic design of guard-rails should be the minimum disturb of the regular fluid motion (minimum drag), respecting the strength and ductility performances necessary for the safety of vehicles and, often, architectonic choices.

Consequently, a definition of guard-rails independent from the whole section context is not possible.

Besides the disturb produced by the guard-rails cannot be neglected, and the global aerodynamic characteristics of the structural system are often substantially modified, sometimes in a positive sense: for example, one of the effects of the introduction of guard-rail is often a negative lift, that improves the performances of suspension bridges. So the integration of guard-rail design in the optimization process is strongly recommended.

2.3.2.2 Wind barriers and walls

It has already been related about the importance and difficulties in defining maximum transversal wind on vehicles.

Wind barriers must assure a reduction factor for lateral wind velocity:

$$R_w = \frac{v_{wl}}{v_{wf}}$$

where v_{wf} is the free stream wind velocity orthogonal to the bridge longitudinal axis, and v_{wl} is the transversal wind on vehicles.

In the last years, the use of wind barriers increased, consequently to the evolution of the characteristics of cars and trains, and the increase traffic velocities.

Wind barriers are, generally, elements of some complexity.

However, two principal design variables can be considered: permeability for wind barrier and gap between barrier and upper surface of the bridge deck section. Low permeability and small gaps generally increase aerodynamic drag and make worse aeroelastic behavior.

2.3.2.3 Aerodynamic devices

In aerodynamic phenomena, a small detail variation can generate a substantially different behavior of the whole system.

The insertion of aerodynamic devices has the aim of improving the aerodynamic characteristics, even with only apparently little modifications in boundary layer.

Two types of control devices can be distinguished:

– Boundary layer control devices: they are generally used in wind tunnel tests to remedy a non respected *Re* similitude (superficial roughness, trip wires).
– Streamline control devices: guide vanes, generally in proximity of corner-edges; porous devices, for flux lamination (horizontal and vertical grids, which effect is like and often complementary to this of wind barriers).

2.4 *Criterion of optimization*

Numerical and experimental simulations of the phenomenon checking the effects of the devices are always subjected to great uncertainties, due to scale and similitude problems.

Consequently this kind of strategies is not reliable to correct the aeroelastic behavior of the bridge and should be considered only as the farthest possibility of controlling and optimizing the structural performances.

The bridge bare section must have in its intrinsic shape all the warranties against the instability problems: the devices role is confined in a restricted field, concerning the adjusting of some response problems (e.g. vortex shedding oscillations), but not the prevention of destructive structural instability (e.g. flutter).

It is always preferable a low sensitivity of aeroelastic behavior of the section to small detail changes, that can occur for constructive reasons.

So during the optimization process the sensitivity analysis has a central role for the individuation of the best performing section. After imposing a hierarchy for all the aerodynamic critical aspects, the behaviour of each section has to be analyzed and evaluated as regards them: the sections sensible to small changes have generally unstable behavior and should be excluded, while the less sensible ones can be subjected to further optimization processes to improve the other non dimensioning aspects.

By fixing appropriate evaluation-criteria, it is possible to get a spectre of acceptable sectional solutions and then to order them in relation to their performances.

The scheme in Fig. 1 shows this process with regard to three different section topologies, while the fourth one, adopted for the Storebaelt, is proposed as validation model as explained further.

3 USE OF COMPUTATIONAL FLUID DYNAMIC

3.1 *Characters of used numerical code*

The Messina Strait Bridge deck section optimization process has been led by a finite element computational code.

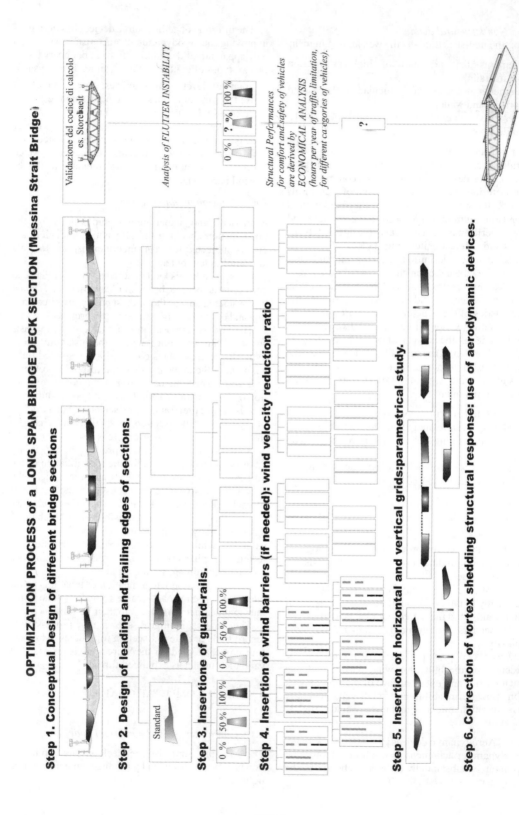

Figure 1. General schema of optimization process for the Messina Strait Bridge Section.

The deck section has been modelled with all its details, in a (250 m × 80 m) domain, divided in about 55,000 bi-dimensional elements, quadrilateral, with four nodes, and velocity, pressure, turbulent kinetic energy, dissipation of turbulent energy as degrees of freedom.

The velocity field and the distribution of pressure on the deck have been computed for only one velocity value at the border of the domain ($|v| = 46.6$ m/s), but with nine different angles of attack, between $-8°$ and $+8°$.

The fluid dynamic problem consists in solving differential partial equations, after discretizing them in finite elements, on the hypothesis of stationary domain and incompressible newtonian fluid.

From the mass conservation law comes the continuity equation for incompressible fluid:

$$\nabla \underline{v} = 0$$

The momentum equations are

$$\rho \frac{Dv_x}{Dt} = \frac{\partial(-p+\tau_{xx})}{\partial x} + \frac{\partial \tau_{yx}}{\partial y} + S_{Mx}$$

$$\rho \frac{Dv_y}{Dt} = \frac{\partial \tau_{xy}}{\partial x} + \frac{\partial(-p+\tau_{yy})}{\partial x} + S_{My}$$

The newtonian fluid conditions let the momentum equations be expressed in the Navier-Stokes equation form:

$$\underline{\underline{\sigma}} = -p\underline{I} + \underline{\underline{\tau}} = -p\delta_{ij} + \tau_{ij} = -p\delta_{ij} + \mu\left(\frac{\partial v_x}{\partial y} + \frac{\partial v_y}{\partial x}\right) + \delta_{ij}\lambda\nabla\underline{v}$$

$$\rho \frac{Dv}{Dt} = -\nabla p + \nabla\underline{\underline{\tau}} + \rho\underline{f}$$

The system of differential equations is closed, for it is composed by three equations (in 2D) and three unknown quantities: p, v_x, v_y. By imposing in the relation stress-deformation

$$\lambda = -\frac{2}{3}\mu \qquad \text{(Schlichting, 1979)}$$

$$\begin{cases} \nabla\underline{v} = 0 \\ \rho\frac{Dv_x}{Dt} = -\frac{\partial p}{\partial x} + div(\mu grad v_x) + S_{Mx} \\ \rho\frac{Dv_y}{Dt} = -\frac{\partial p}{\partial y} + div(\mu grad v_y) + S_{My} \end{cases}$$

Beyond a critical value of wind speed, inertial effects become great enough respect to viscous effects, and the flow becomes turbulent, so that the instantaneous velocity is not constant at every point in the flowfield.

If we express the velocity in terms of a mean value and a fluctuating component:

$$v_x = \overline{v_x} + v_x'$$

$$\frac{1}{\Delta t}\int_0^{\Delta t} v_x' = 0 \quad ; \quad \frac{1}{\Delta t}\int_0^{\Delta t} v_x = \overline{v_x}$$

After the substitution of the expression of v_x, v_y in the momentum equations, the time averaging operation introduces additional terms (Reynolds stress):

$$\begin{cases} \sigma_x^R = -\frac{\partial}{\partial x}(\rho v_x' v_x') - \frac{\partial}{\partial y}(\rho v_x' v_y') \\ \sigma_y^R = -\frac{\partial}{\partial x}(\rho v_y' v_x') - \frac{\partial}{\partial y}(\rho v_y' v_y') \end{cases}$$

In the eddy viscosity approach to turbulence modelling, the Reynolds stress terms are written in the form of a viscous stress one, with an unknown coefficient, the turbulent viscosity:

$$-\rho v_i v_j = \mu_t \frac{\partial v_i}{\partial x_j}$$

This approach let two terms of the moment equations be combined so that a new effective viscosity can be defined, as the sum of laminar and turbulent viscosity:

$$\mu_e = \mu + \mu_t$$

The two standard k-ε model (Launder and Spalding, 1974) equations, one for k and one for ε, are

$$\frac{\partial(\rho k)}{\partial t} + div(\rho k\underline{v}) = div\left[\frac{\mu_t}{\sigma_k}grad k\right] + 2\mu_t E_{ij} \cdot E_{ij} - \rho\varepsilon$$

$$\frac{\partial(\rho\varepsilon)}{\partial t} + div(\rho\varepsilon\underline{v}) = div\left[\frac{\mu_t}{\sigma_\varepsilon}grad\varepsilon\right] +$$

$$+ C_{1\varepsilon}\frac{\varepsilon}{k}2\mu_t E_{ij} \cdot E_{ij} - C_{2\varepsilon}\rho\frac{\varepsilon^2}{k}$$

with E_{ij} = mean components of the rate of deformation of the fluid element; and C_m, σ_k, σ_ε, $C_{1\varepsilon}$, $C_{2\varepsilon}$ constants.

The solution of these equations lets to evaluate the turbulent viscosity by the expression

$$\mu_t = C_\mu \rho \frac{k^2}{\varepsilon}$$

k = turbulent kinetic energy;
ε = turbulent kinetic energy dissipation rate.

2465

Then it is possible to calculate the effective viscosity:

$$\mu_e = \mu + C_\mu \rho \frac{k^2}{\varepsilon}$$

The k-ε model is not valid close to the walls: so a wall turbulence model one must be introduced, in which the wall shear stresses are determined by an approximate iterative solution from the value of the velocity component parallel to the wall at a defined distance (*Log Law of the wall*):

$$\frac{v_{tan}}{\sqrt{\frac{\tau}{\rho}}} = \frac{1}{k} \left(\ln \frac{E\delta}{v} \sqrt{\frac{\tau}{\rho}} \right)$$

where v_{tan} = velocity parallel to the wall;
τ = shear stress;
v = cinematic viscosity;
k, E = law of the wall constant;
δ = distance from the wall.

Close to walls the value of the turbulent kinetic energy is obtained from the k-ε model; the dissipation rate is given by the following expression:

$$\varepsilon = \frac{C_\mu^{0.75} k^{1.5}}{k\delta}$$

The momentum and turbulence equations have the form of a scalar transport equation, with transient, advection, diffusion and source terms, the general form of which is

$$\frac{\partial}{\partial t}(\rho C_\Phi \Phi) + \frac{\partial}{\partial x}(\rho v_x C_\Phi \Phi) + \frac{\partial}{\partial y}(\rho v_y C_\Phi \Phi)$$

$$= \frac{\partial}{\partial x}\left(\Gamma_\Phi \frac{\partial \Phi}{\partial x}\right) + \frac{\partial}{\partial y}\left(\Gamma_\Phi \frac{\partial \Phi}{\partial y}\right) + S_\Phi$$

The pressure equation is derived by using the continuity equation.

All the degrees of freedom are solved in sequence: the equations are coupled, so that it's necessary to solve them by using intermediate values of the other degrees of freedom.

At every *global iteration* all the properties are updated.

The algorithm used needs many solution of global iteration, obtained by a *tridiagonal method*: the problem is decomposed into several tridiagonal problems, in which every element outside the tridiagonal position is treated as a source term by using the previous values.

This approach, for unstructured meshes, consists in the Gauss-Seidel Method.

The algorithm used controls the convergence of the solution, by several convergence monitoring parameters: for each degree of freedom Φ, they are in the form:

$$Cm = \frac{\sum_{i=1}^{N} |\Phi_i^k - \Phi_i^{k-i}|}{\sum_{i=1}^{N} |\Phi_i^k|}$$

Various relaxation techniques are used to stabilize the solution.

After computing the distribution of pressure and velocity in the domain, by defining on the domain paths around the single deck and the whole one, it has been possible to obtain the lift, drag and moment partial and global coefficient, with different configuration for the bridge, to lead the advanced optimization process.

3.2 Validation of computing model

The challenge of CFD is to extend the computer aided design to the aerodynamic design of bridge deck sections, in substitution of any experimental test. For this reason, the aim of scientific research is to obtain reliable numerical codes, that allow the described independent deck section optimization.

However, for aerodynamic numerical models, the complexity of physical behavior brings to great difficulties in obtaining an accurate solution.

It is necessary to take account of an error that is often some 30–40%. If the numerical discretization or other artificial parameters (turbulence, boundary layer) are inadequate, an unphysical solution can be obtained.

If a direct validation of the specific numerical model is uneconomical, it is possible to validate the choice of fundamental parameters of the model, by modelling existent bridge deck section and comparing the numerical and the experimental values or, when available, even the real ones, derived by monitoring of existent structures (Humber Bridge, Great Belt Bridge, etc. ....).

In the spirit of the Performance Based Design, a progressive refinement of model is possible, focused on obtaining realistic results for any structural response of interest. In this way, we can define quantitatively the performances of the structure, and proceed to the optimization with the use of numerical model.

The major developments in this field will be consequence both of improving of computer technologies and CFD codes, and of a systematic monitoring of new and existent bridges, that are the only available database.

4 THE MESSINA STRAIT BRIDGE DECK SECTION

In the optimization process the use of finite element codes allows to know the flow paths, the pressure distribution on the deck surfaces and the effective disturb of non structural elements on the global behavior of a streamlined body just like Messina Strait Bridge in its first design appearance.

The authors remark that the process must be conducted through a synergic comparison between sperimental and computational results, validating each other and giving further design guides.

The streamlined deck sections show have no great attitude to separation of boundary layer and generation of Von Karman vortex wake.

The barrier on the decks and the grids anyway can disturb the flow and its physical nature, and generate vortex phenomena on the roadway.

Consequently the barrier influence on the deck pressure distribution and on the velocity field must be investigated, concerning both serviceability and durability response of the bridge.

In the Figs 2, 3, 4 and 5 it's shown the comparison between the pressure distribution and velocity field between the bare deck and the equipmented ones for wind speed of 46.6 m/s.

The comparison of the paths shows the low attitude of streamed profiles to the vortex shedding at least at high wind speed, while the barriers create vortexes but reduce the wind velocity upon the roadways,

getting the bridge suitable to the transit of vehicles in optimal comfort conditions.

In the Fig. 5 the effects of the horizontal grids on the lift resultant can be qualitatively observed.

The distributions of the horizontal velocity on the first deck is compared for the three different sections.

It is possible to observe the protection effects of the barrier for about 4 meters: the velocities are lower but vorticous motion with inversion of the direction of the particle motion occurs (Fig. 6).

This phenomenon has positive effects for the static verify of the serviceability condition, but can have negative impact in the dynamic analysis: inadequate vertical and horizontal acceleration may occur and decay the comfort levels on the bridge; so the installation of appropriate devices could be necessary.

By integrating the pressure and the shear stresses alternatively on the whole bridge deck and on the three

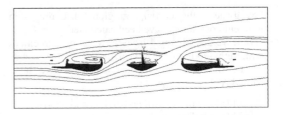

Figure 4. Streamlines for the MSB section with barriers.

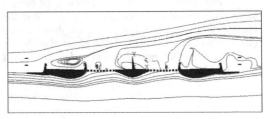

Figure 5. Streamlines for the MSB section with barriers and horizontal grids.

Figure 2. Vortex shedding for MSB section and wind speed of 10 m/s.

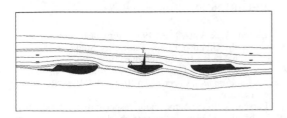

Figure 3. Streamlines for the MSB section without equipment.

Vx DISTRIBUTION ON DECK1 SURFACE

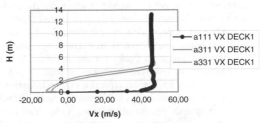

Figure 6. Distribution of Vx on the directly invested deck for three sections for MSB (v = 46.6 m/s).

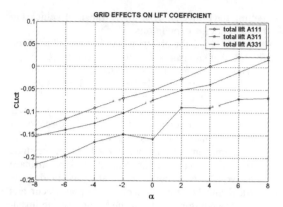

Figure 7. Global lift coefficient diagram for experimental and numerical models.

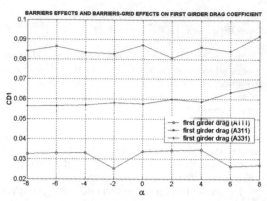

Figure 8. First girder drag coefficient for numerical models.

box girders, the partial and global lift, drag and moment coefficient have been calculated.

The values obtained have are in good agreement with the wind tunnel test results.

The three section models analyzed have been named A111, A311, A331:

– A111: bare deck, with no barriers and no grids;
– A311: only barriered deck;
– A331:equipmented deck, with barriers and grids.

The only barriered deck lift diagram is the closest to the experimental one because the grid used in the element finite model is less permeable than the real one. It is possible to remark the lift effect of grids, the subsequent hanger relieving and the reduction of global cable system stiffness (Fig. 7).

The barrier effects on drag are noticeable in the directly invested by wind girder more than in the others, for the deviation of the flow and the reduction of the horizontal component of the pressure resultant. In the above diagram (Fig. 8) it's shown the comparison of drag coefficients of the first girder for several angles of attack: the barriers can increase drag effects until three times.

The global moment coefficient diagram has been drawn after a sensitivity analysis about its dependence on the arbitrary choice of the pole: it has demonstrated the weak dependence for reasonable choices.

The comparison between moment diagrams of bare and equipmented decks for the Messina Strait Bridge shows an aerodynamic behavior with an undesired sensibility to non structural element insertion, as expected for airfoils: it's necessary to continue and provide a deeper optimization process, by scanning different design alternatives, even without streamlined profile, as presented in the scheme above and already used in other bridges.

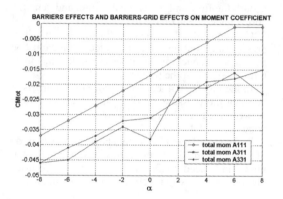

Figure 9. Global moment coefficient for experimental and numerical models.

ACKNOWLEDGEMENTS

The financial support of Stretto di Messina SpA is acknowledged. Anyway, the opinions and the results here presented are responsability only of the Authors and cannot be assumed to reflect the ones of Stretto di Messina SpA.

REFERENCES

Larsen, A. & Asitz, M.A. 1998. Aeroelastic considerations for the Gibraltar Bridge Feasibility Study. In Larsen & Esdhal (eds.) Bridge Aerodynamics, *Proceedings of the international Symposium on Advances in Bridge Aerodynamics*, Copenhagen 10–13 May 1998. Rotterdam: Balkema.

Larsen, A. 1993. Aerodynamic aspects of the final design of the 1624 m suspension bridge across the Great Belt. *Journal of Wind Engineering and Industrial Aerodynamics*, 48.

Shimodoi, H., Oryu, T. & Fumoto, K. 2002. Aerodynamic Characteristics of 2_box Girder with Windshield, *Proceedings of The Second International Symposium on Advances in Wind & Structures* (AWAS'02), August 2002, Busan, Korea, Techno.Press.

Larsen, A., Esdahl, Sren, Andersen, Jacob, E. & Vejrum, T. 2000. Storebælt Suspension Bridge – Vortex Shedding and Mitigation by Guide Vanes. *Journal of Wind Engineering and Industrial Aerodynamics*, n° 88.

Blevins, R.D. 1977. *Flow-induced vibration*. Van Nostrand Reinhold Company.

Versteeg, H.K. *An introduction to computational Fluid-Dynamics*. Longman Scientific & Technical.

Shimodoi, H. Oryu, T. & Fumoto, K. 2002. Aerodynamic Characteristics of 2_box Girder with Windshield, *Proceedings of The Second International Symposium on Advances in Wind & Structures* (AWAS'02), August 2002, Busan, Korea, Techno.Press.

System-based Vision for Strategic and Creative Design, Bontempi (ed.)
© 2003 Swets & Zeitlinger, Lisse, ISBN 90 5809 599 1

Aspects for the determination of the complex stress states in suspension bridge for the fatigue-analysis

L. Catallo
University of Rome "La Sapienza", Rome, Italy

V. Di Mella & M. Silvestri
Structural Engineer, Rome, Italy

ABSTRACT: The sensitiveness to the problem of fatigue on metal bridges has remarkably grown in the last years, in relation to the reveal of breaches due to repeated loads in several bridges built since 1950. Purpose of the present work is to show, with reference to a concrete example (the Messina Strait Bridge), some methods of analysis of steel-bridges, with particular attention to the Eurocodes. In the specific, the fatigue-analysis will be developed for the railway box-girder of the Messina Strait Bridge in the region close to the Sicilian tower. The fatigue-analysis requires the detailed knowledge, at local level, of the local stress states: it has been necessary to model the bridge deck with bidimensional finite elements able to represent the local stress state. In the next and last phase of the present work, the fatigue-analysis on the railway box girder is conducted both applying the Wöhler-method and bearing the EC3 prescriptions in mind.

1 INTRODUCTION

Nowadays, bridge decks with moderate spans are very recurring structure, whose structural behaviour is rather known, so that their design and installation can be considered a problem with few uncertaintics. The situation becomes different if span grows, with many primacies which really belong to the sector of suspended bridges: new problems take place and those already known have greater incidence. Among these, the corrosion ones can be mentioned in the case of bridges placed in aggressive environments, due to the wind, seismic, thermic and, at last, the fatigue-ones (more meaningful for steel structures). In the work presented here, precisely these last ones are object of a deeper analysis.

The sensitiveness to the problem of the fatigue on metal bridge has remarkably grown these last years, in relation to the reveal of breaches due to this phenomenon in several bridges built since 1950. This has given cue to several studies and searches, which have permitted both to improve the knowledge about the behaviour against fatigue of constructive details, and the analysis methodology, fixing the design evolution and the modern provisions.

Purpose of the present work is to show, also with reference to concrete examples (Fig. 1), some methods

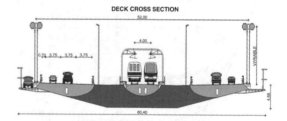

Figure 1.   Cross section of Messina Strait Bridge.

of analysis of steel-bridges, with particular attention to the Eurocodes. In the specific, the fatigue-analysis will be developed for the box-girder railway on the Messina Strait Bridge in the zone in correspondence to the Sicilian tower.

The fatigue analysis needs the detailed knowledge, at local level, of the stress state. It is necessary to model the bridge deck with bidimensional finite elements; such a modelling allowed to obtain tensional maps able to locally provide both the stress and the stress variations in relation to the different load combinations. See for example the maps of Figure 2.

In the next and last phase of the present work, the analysis has been conducted on the railway box girder. Such analysis has been carried out both with

Figure 2. Map of longitudinal stress in the zone of the bridge deck corresponding to the tower.

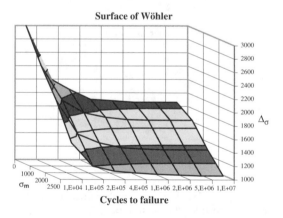

Surface of Wöhler

σ_m

Cycles to failure

Figure 3. Surface of Wöhler used for the fatigue analysis.

Wöhler method (Fig. 3) and bearing the Eurocode 3 prescriptions in mind.

2 THE MODELLING

2.1 *Introduction*

The cross sections of the highway box girder and of the railway box girder, in the Messina Strait Bridge,

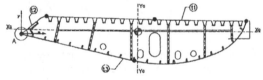

Figure 4. Cross section of the highway box girder.

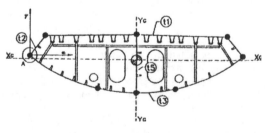

Figure 5. Cross section of the railway box girder.

have a streamline shape, that bounds the cropping up of the aeroelastic phenomenons (Figs 4 and 5).

At a global level a modelling executed with linear frames elements is sufficient, while to catch stress or deformation states at local level, it's necessary to model the bridge deck in the smallest details and so a tridimensional modelling is made necessary.

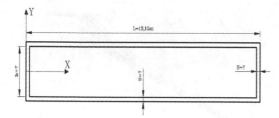

Figure 6.   Equivalent cross section of the highway box girder.

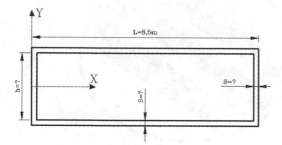

Figure 7.   Equivalent cross section of the railway box girder.

As one can see on Figures 4 and 5, the box girders (both highway and railway) have a rather complex shape, obtained by an aerodynamic optimization in the wind gallery. So it has been thought, for box girders realized this way, to adopt a simplification of their shapes, maintaining however the same global mechanical characteristics found in literature. In practice, the beam of the highway and railway box girders have been represented by box girders (rectangular hollow), with a width equal to the real box girder one, and height and thickness values calculated. For the valuation of the unknown it has simply been presumed that between the real beam and the "simplified or equivalent" one, the moment of inertia as regard the x-axis and the area remain unvaried (Figs 6 and 7). In short the following system of equations has to be solved:

$$\begin{cases} 2bs + 2hs = A \\ \dfrac{1}{12}\left[h(b+2s)^3 - (h-2s)b^3\right] = I_Y \end{cases} \quad (1)$$

where:
L: width of the box girder,
h: height of the box girder,
s: thickness of the cross-section.

Concerning the transverse box girder, instead, the section is simplified assuming a rectangular hollow shape, this time of fixed height, equal to 3.40 m, with width and thickness to be determined, imposing the same area and the same moment of inertia I_y of the

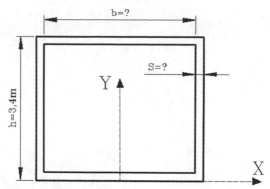

Figure 8.   Equivalent cross section of the transverse box girder.

real transverse box girder (Fig. 8). The system of equations to be solved in this case is:

$$\begin{cases} 2bs + 2hs = A \\ \dfrac{1}{12}\left[h(b+2s)^3 - (h-2s)b^3\right] = I_Y \end{cases} \quad (2)$$

where:
b: width of the transverse box girder,
h: height of the transverse box girder,
s: thickness of the cross-section.

2.2   *The model*

The model of the box girder bridge close to the towers (Fig. 6) is refined using two-dimensional finite elements working according to the shell theory. This zone, built with special elements, can be considered as a large scale D-regions, needing detailed modelling to obtain the realistic description of the behaviour of the deck, specifically regarding the fatigue analysis.

The considerations made in the previous paragraph have allowed to obtain the reproduced model in Figure 9.

3   MODEL VALIDATION

3.1   *The model evolution*

The starting point for the realization of the structural model, in which part of the bridge deck is modelled by shells elements, is represented by the "global model", which counterfeits the behaviour of the Bridge in its whole, through finite element method, bearing in mind the presence of the main cables, the hangers, the towers, and the bridge deck.

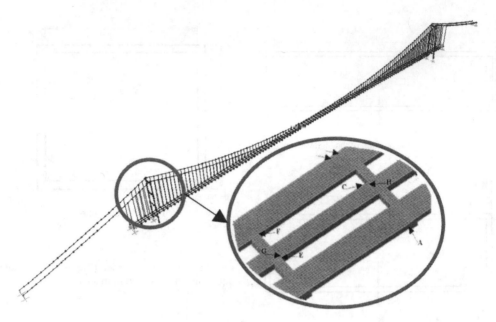

Figure 9.   Substructuring process leading the shell based modelled zone.

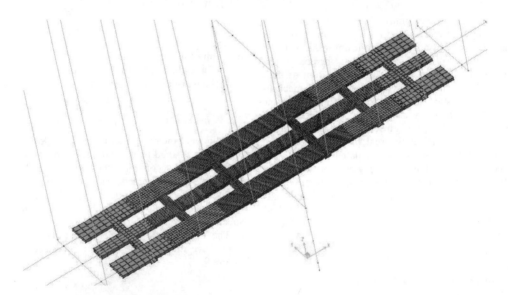

Figure 10.   Detail of the zone modelled by shells elements, in which is appreciable the refinement of the discretization.

Bearing in mind, that the purpose of modellation is to study the span behaviour close to the tower, which, otherwise from the other (30 m long) has got a 50 m of length, before starting the pure modelling-work, the possible ways to join the purpose were considered, the alternatives taken in considerations have been two:

- To model the desired zone of the bridge deck, in a quite independent way from the rest of the Bridge;

- To model the desired zone of the bridge deck, put it in the larger context of the global model of the Bridge.

The first alternative, at the beginning, seemed the easiest, but then there was the problem of constraining the model, or rather, there was the problem of realizing, at the boundary of the isolated zone, some constraints able to simulate the remaining part of the

2474

MODEL NUMBER	PERIOD [s]		
	FRAME MODEL	SHELL MODEL	Δ [%]
I	30.0646	29.9671	0.32
II	16.4066	16.4248	0.11
III	15.4397	15.3418	0.63
IV	12.2738	12.2693	0.04
V	10.9257	10.9174	0.08
VI	10.6841	10.6804	0.03
VII	10.3958	10.3048	0.88
VIII	10.0839	10.1573	0.73
IX	10.0321	10.0881	0.56
X	9.9110	9.9018	0.09

Figure 11. Comparison among the periods belonging respectively to the *frame model* and the *shell model*.

structure. It has been useful, at this point, to take in consideration the second alternative.

Afterwards, the following question has been asked: *"how many spans must be modelled, beyond the one in correspondence to the tower, so that the effect of the attack constraint among the shell elements and frame elements doesn't invalidate the obtained results?"*.

After several trials, the conclusion is that, beyond the two spans, the boundary conditions don't produce relevant effects on the span close to the tower, thus the tridimensional modelling of the bridge deck, has been developed on the seven spans close to the tower, for a total length of 220 m (Fig. 7).

From the same picture, you can see also another peculiarity of the model, that is: the mesh of the shells isn't constant, but it's realized in a such a way to provide sufficiently precise results in the interested zone, that is the central span among the seven modelled.

3.2 Model validation

In the previous paragraph the process which led to define the model has been described; some results, which have allowed to consider the *shell model* rather reliable now will be shown. From now one will indicate with the expression "shell model", the model obtained by the starting model having only frame elements, which instead will indicate with the expression "frame model".

The shell model, by the mechanical point of view, is equal to the frame model; in both models, infact, the same loads are considered: steady loads and permanent overloads.

A synthetic way to compare these two structural modellization, is by develope a model analysis. In this way, at the same time, both the stiffness and the inertia are considered.

As one can see on Figure 11, there are small differences between the frame model and the shell model: the main difference, equal to the 0.88%, is found for the VII model, and so it can be considered negligible.

4 THE FATIGUE-ANALYSIS IN THE STEEL BOX GIRDERS BRIDGES

4.1 Introduction

The connections of the orthotropic plate bridge deck of the steel bridges are among the details more exposed to the danger of fatigue-crisis.

In succession, the fatigue-analysis of the connections of the orthotropic plate bridge deck of the box girder railway of the suspension bridge on the Messina Strait are shown: in particular, the fatigue analysis will be executed in the central section of the box girder and in the section in which the box girder intersects the transverse box girder.

Because of the large number of cycles whose details are subject during the useful life of the work, fixed in 200 years, the local analysis are led for illimitated life. For ferriferous materials the useful life can be considered illimitated when the 2×10^6 cycles are overcome.

Bearing of in mind the fact that the structure in question is of an extraordinary type, the analysis will be faced following two alternatives ways:

- Through Wöhler's experimental curves;
- According to the Eurocode 3 criterions.

4.2 The analysis according to Wöhler

Considering the joints of the modelling to the finite elements, you can associate to each one, for a prefixed number of cycles (depending on the fatigue life asked for the work), a value of σ_m and a value of $\Delta\sigma$.

After elaborating the model, to each joint three coordinates remain associated, which permit the tridimensional diagram collocation of the Figure 3. The analysis, so, will be satisfied when all the points remain under Wöhler's aforesaid surface.

For the structure under study, one will to execute the fatigue analysis for illimitated life, and this condition allows to simplify clearly the analysis operations, because Wöhler's curves remain steady for a number of cycles N larger than 2×10^6 cycles.

Intersecting Wöhler's surface with a vertical plan passing through the x-axis, at any point on the right of the value 2×10^6, on the plan σ_m-$\Delta\sigma$ one find the curve of Figure 12, which will be conventionally named *"perpendicular Wöhler's curve"*.

The fatigue-analysis will be satisfied if the representative points of the fatigue condition in the joints

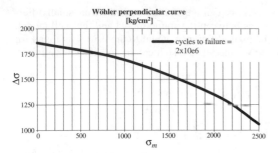

Figure 12. Perpendicular Wöhler's curve.

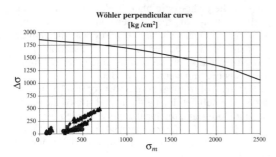

Figure 13. Graphical representation of the fatigue-analysis for the mid section of the railway box girder.

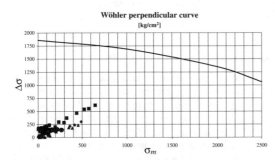

Figure 14. Graphical representation of the fatigue-analysis in the next section to the transverse box girder of the railway box girder.

of the modelling, represented in the graphic by a couple of coordinates σ_m–$\Delta\sigma$ stay under the curve.

To execute the analysis, it's necessary to find for each load combination, and for any joint, the value of σ_m and of $\Delta\sigma$.

The analysis will be executed comparing all the load combinations, with the load combination only consisting of steady loads and permanent overloads (Figs 13 and 14).

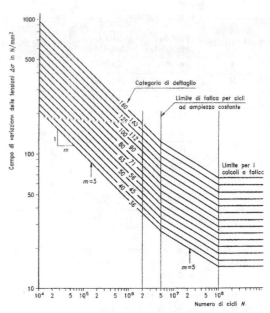

Figure 15. Curves of the fatigue resistance for variations ranges of the normal stress.

4.3 The analysis according to the EC3 criterions

On the diagram $\Delta\sigma$-N (Fig. 15) proposed by EC3, to each curve doesn't correspond a value of medium tension, as in the case of Wöhler's curves, but a certain category of structural details. In practice there is an implicit reference to the resistence of a certain structural detail and not to the effective stress state whose is subjected to a specific load condition.

The operation which associates to each detail the characteristic resistance to two millions of cycles ($\Delta\sigma_c$) is named fatigue classification; thus, its characteristic resistance is named class of a detail, expressed in N/mm², corresponding to a fatigue vision of two millions of cycles.

The criterion of valuation of the fatigue resistance is the following:

$$\gamma_{Ff}\Delta\sigma \leq \Delta\sigma_R / \gamma_{Mf} \tag{3}$$

Adopting all the prescriptions furnished by the EC3, the analysis comes to verify that in no point of the discretization (Fig. 16), varying the load combination, there is a tension variation superior to 628 daN/cm².

In the next figures (Figs 17 and 18) the obtained results will be shown for the analyses conducted in according to the EC3 criterions.

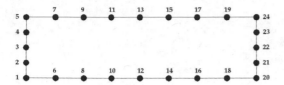

Figure 16. Geometrical disposition of the mesh points, in the railway box girder.

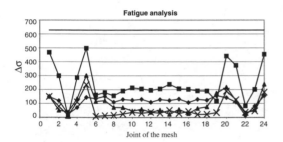

Figure 17. Graphical representation of the fatigue-analysis in according to EC3 in the mid section.

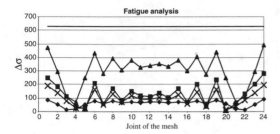

Figure 18. Graphical representation of the fatigue-analysis in according to EC3 in the section close to the transverse box girder.

5 CONCLUSIONS

In this work, some aspects of the fatigue analysis of the box girders which compose the Messina Strait Bridge decks are considered.

Specifically, some details of the substructuring techniques that allow to consider a bidimensional discretization of the girders inside the overall frame model of the Bridge are developed, and two methods of fatigue analysis, i.e. Wöhler curve and EC3 criterions, are illustrated.

A specific remark must be direct to punctualize the uncertainty related to the overall safety factor for the fatigue analysis.

The safety factor, infact, has been previously considered for the definition of the limit range in the case of EC3 analysis; in the case of Wöhler's analysis instead, no factor has been considered and that's why there is a quite large margin.

ACKNOWLEDGEMENTS

The financial support of *Messina Strait di S.p.A.* is acknowledged. Anyway, the opinions and the results here presented are responsability only of the Authors and cannot be assumed to reflect the ones of *Messina Strait di S.p.A.*

REFERENCES

Radogna, E.F. 1998. Tecnica delle costruzioni Vol. 2 – Fondamenti delle costruzioni in acciaio. MASSON. (In Italian).

EUROCODICE 3. Progettazione delle strutture di acciaio.

Bannantine, J.A, Comer J.J. & Handrock J. Fundamentals of metal fatigue analysis. Prentice Hall, Inc.

Collins, Jack, A. 1993. Failure of material in mechanical design – analysis, prediction, prevention. Wiley-Interscience.

Gimsing, Niels J. 1998.Cable supported Bridges – Concept & Design. Wiley-Interscience.

Calzona, R. & Dolara, E. Fatica e decadimento dei materiali e delle strutture sottoposte ad azioni cicliche. Ferrocemento. (In Italian).

Caramelli, S. & Croce, P. November–December 2000. Le verifiche a fatica nei ponti in acciaio. Costruzioni metalliche. (In Italian).

System-based Vision for Strategic and Creative Design, Bontempi (ed.)
© 2003 Swets & Zeitlinger, Lisse, ISBN 90 5809 599 1

A hybrid probabilistic and fuzzy model for risk assessment in a large engineering project

F. Petrilli
Structural Engineer, Rome, Italy

ABSTRACT: In this paper an operative tool for risk assessment and management in large engineering projects is presented. The method uses a *grey-box* approach to quantify uncertainties, namely a white-box probabilistic approach for aleatoric uncertainties and a black-box fuzzy system to quantify epistemic uncertainty. The case study is the Messina Strait Bridge whose economic performances are investigated according to several risk scenarios.

1 INTRODUCTION

In recent years, many large engineering projects showed poor performance records in terms of economy, environment and public support. Cost overruns and lower-than-predicted revenues frequently place project viability at risk and redefine projects that were initially promoted as effective vehicles to economic growth as possible obstacles to such growth (Flyvbjerg et al., 2003). These considerations are valid for recent projects such as the Channel tunnel and the links across Great Belt and Oresund. A study of the World Bank with a sample of 1778 projects, shows that in six out of ten projects costs are underestimated and in nine out of ten time forecasts are consistently and significantly unachieved (Figs 1–2). Moreover, traffic previsions are generally inflated (Flyvbjerg et al., 2003). As far regards regional and economical growth, it is common for proponents of major transport infrastructure projects to claim that such projects will result in substantial regional development effects. The empirical evidence shows that such claims are not well founded, the main reason being that in modern economies, transport costs constitute a marginal part of the final price of most goods and services. These poor performance records can be the natural consequence of assuming, in the feasibility studies, that projects exist in a predictable deterministic world of cause and effect where things go according to plan. In reality, the world of large projects planning and implementation is highly stochastic one where things happen only with certain probability and rarely turn out as originally intended. In other words, project viability has to be treated from a risk-based point of view. Moreover, the uncertainties involved in

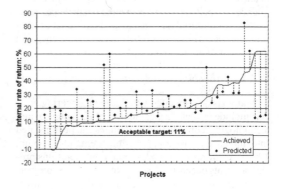

Figure 1. 43 Projects ordered in increasing achieved IRR.

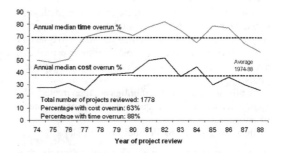

Figure 2. Cost and time overruns in 1778 projects review by the World Bank (1990).

the feasibility analyses of project are rarely estimable in terms of probability, due to the lack of input data and to the epistemic nature of most of such uncertainties. A better way to treat epistemic uncertainties is the well known fuzzy theory. In this paper a hybrid fuzzy-probabilistic methodology will be presented. The model uses a probabilistic modelling of costs and duration of the project activities and a fuzzy representation of the impact of such variables onto the project internal rate of return.

The case study is the project for the Messina Strait Bridge. The Messina Strait Bridge will link Sicily to mainland Italy. It will be a highway and railway suspension bridge with a span length of 3300 m. Construction costs are estimated to be 4.842 millions of Euro. The target completion time is 11 years and the concession period will last 30 years; the expected IRR is 15, 6%.

2 THE RISK ASSESSMENT MODEL

The concept of risk is used to assess and evaluate uncertainties associated with an event. Risk can be defined as the potential for loss as a result of a system failure. Risk can be measured as a pair of the probability of occurrence of an event, and the outcomes or consequences associated with the event's occurrence.

Risk is commonly evaluated as the product of likelihood of occurrence or probability of occurrence and the impact of an accident, i.e.:

$$Risk\ k = Probability \times Impact \qquad (1)$$

Major risks associated with large engineering projects include cost and schedule overruns. Technical and technological construction challenges can be translated into increased cost and schedule demands (Ayyub, 1999). As aforementioned, should considerations prevail or events occur that lead to excessive costs, projects may prove impractical. Schedule risk is also critical because of the time value of money and because the project must be finished within a reasonable amount of time to perhaps meet strategic goals. The methodology developed in this study can be defined as a grey-box model as it uses a white-box approach to quantify the likelihood of the event (cost or time overrun) and a black-box approach in assessing the impact on project goals.

Available data and uncertainties are then treated efficiently according to their quantity and nature (aleatoric or epistemic).

In particular, the model include four steps:

A. In the first phase, a network representing the project's activities is constructed taking into account the logical dependences among the activities. For this purpose, the project's work breakdown structure (WBS) is a basic information.
B. Secondly, time and cost analyses are carried out at different levels of uncertainties. In particular, the critical path method (CPM), the probabilistic

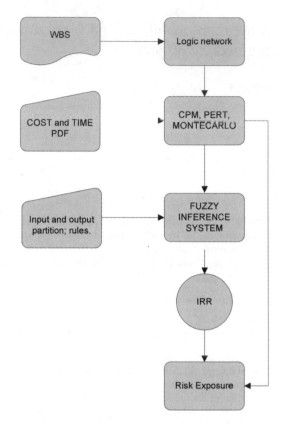

Figure 3. The proposed model for risk assessment.

evaluation and review technique (PERT) and a Monte Carlo simulation have been applied. The output of this phase is the probability of occurrence of the scheduled completion time and construction costs.
C. Thirdly, a fuzzy inference system has been implemented in order to assess the impact of cost and time overrun on the project, the latter being expressed in terms of the *internal rate of return* (IRR). A special care is needed at this stage in order to translate experts' judgments in quantified values: in this work Saaty's method is used to achieve this issue.
D. Risk exposure for cost and time overruns is eventually calculated by the product of probability and impact.

3 ASSESSING THE PROBABILITY TO MEET COST AND TIME PREVISIONS

Cost and schedule risks are then evaluated by a quantitative analysis of the probability to meet the previsions and the impact on the project's goals.

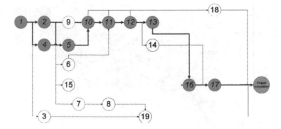

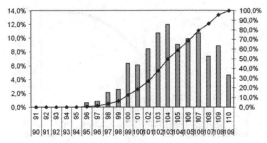

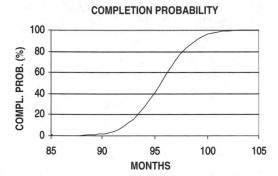

Figure 4. Logic network of the Messina Strait Bridge project with indication of the critical path.

Figure 6. Monte Carlo simulation: completion time probability density function and cumulated probability function for the Messina Strait Bridge project.

COMPLETION PROBABILITY

Figure 5. Completion probability from PERT analysis.

In order to estimate the overall project's duration and cost, three levels of analysis have been carried out. Namely,

– CPM is performed on the project to identify critical and near critical paths;
– Critical and near critical paths are modelled using PERT;
– A Monte Carlo simulation is used to validate the results.

Note that in simulating total schedule duration, only activities on the critical path are considered while in simulating total cost, all activities are considered. In this way, each activity is represented by a deterministic measure in the first model, by a Beta probability density function in the second model and by a generic distribution function in last model.

In the estimation of schedule duration and cost of activities, either historical data are available and may be used to categorize a statistical distribution, or complete data are not available and the underlying distribution has to base on the subjective information provided by an expert.

As far regards the Messina Bridge, input data are available on deterministic bases and can be used for the evaluation of the critical path. According to the logic network provided in Figure 4, representing a macro-level of analysis, the critical path includes erection of

pylons and catwalk, cable and hangars erection and, eventually, the erection of the stiffening girder. The total construction duration is evaluated in 84 months.

With results derived from the CPM, a Pert analysis was carried out. The schedule duration for each activity has been modelled by a Beta distribution. The upper and lower bounds were taken as the 90% and 125% of the modal value according to the findings of AbouRizk & Halpin (1992). For example, if a particular construction duration was assumed to be 30 months in the CPM analysis, the lower limit would be set to 27 months and the upper limit would be set to 37.5 months. Schedule duration is then approximated by a Gaussian probability distribution with mean and variance the sums of the means and variances of the activities on the critical path, respectively. In the same way, total cost is approximated by a normal distribution probability with mean and variance the sums of the means and variances of all activities, respectively. The probability associated with completing the project within certain time or cost limits can be computed using standard normal distribution formulae. The central limit theorem is used as a basis to justify computed schedule and cost probability.

In order to verify if the central limit theorem could be applied in the specific case under development, a Monte Carlo simulation was carried out. As in the previous case, real data were not available so each activity has been modelled according to some distribution recommended by other Authors. In particular, Law & Kelton (1991) recommend the Triangular distribution to estimate a random duration when sample data does not exist. The Triangular distribution is also suggested when a most likely value can be given (Banks et al. 1995). Abourizk & Halpin (1992) recommend using a Beta distribution in repetitive construction processes, while Ayyub suggests using a Gamma distribution for transportation or assembly activities.

Results analysis shows that the mean value of the total schedule duration shifts from 84 months of the CPM to 95,58 months of the PERT and to 103,95 months of the simulation (Fig. 6). The variance is 6,46%

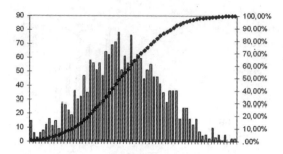

Figure 7. Monte Carlo simulation: cost probability density function and cumulated probability function for the Messina Strait Bridge project.

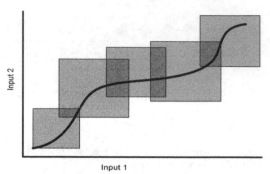

Figure 8. Local mapping of the objective function by local mapping according to the FAT.

in the PERT analysis and 11,61% in the Monte Carlo simulation; moreover, the output distribution is quite different from the Gaussian assumed in PERT analysis.

Similar conclusions apply in the analysis of cost risk as shown in Figure 7.

4 ASSESSING IMPACT

The viability of a project is commonly expressed in terms of the internal rate of return of the investment over the project's life cycle. The IRR represents the break-even return level that equates the Net Present Value of cash flow to 0, i.e. the rate of return that a firm can reasonably expect to earn on the project. IRR satisfies the following equation:

$$\sum_{t=0}^{N} \frac{Benefits - Costs}{(1+IRR)^t} = 0 \qquad (2)$$

In order to quantify the IRR of a project one should know the expected cash flow, namely both benefits and costs. For the specific case of the Messina Strait Bridge, the only available data were supposed to be construction costs and time, no data were supposed to be available for quantify traffic flows. Under these circumstances, a fuzzy inference system was developed using heuristics and Saaty's method to optimize the solution. This system allows mapping the objective function (IRR) by fuzzyfication of input variables, according to the Fuzzy Approximation Theorem (FAT), shown in Figure 8 (Kasabov, 1998).

A Mamdani system was used in order to include linguistic description of the variables (Fig. 9). The input variables include:

– Cost overrun, C;
– Time overrun, T;
– Variation in the economic growth of Sicily, PIL.

The latter variable is intended to represent a variation of traffic flows, being these two aspects strictly

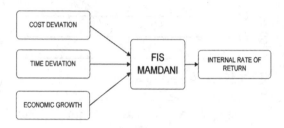

Figure 9. The proposed FIS's architecture.

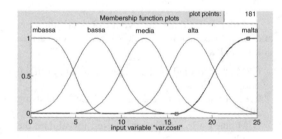

Figure 10. Membership functions for the input variable "cost overrun".

related. The output variable is the IRR expected over 30 years, which is the duration of the concession.

In the specific, the domain of input variables was partitioned in different intervals, namely:

– C = 5 partitions (*very low, low, average, high, very high*);
– T = 5 partitions (*very low, low, average, high, very high*);
– PIL = 3 partitions (*low, average, high*);
– IRR = 5 partitions (*very low, low, average, high, very high*).

The position of the MFs on the *x*-axis, their width and their shape has been determined according to expert's judgment (Ayyub, 1999; PMI, 2000) (Fig. 10).

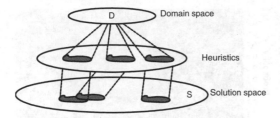

Figure 11. Mapping the solution space using heurisitics.

Table 1. Saaty's scale for pair wise comparison.

Numerical value	Linguistic definition
1	Equal importance
3	Weak importance of one over another
5	Essential or strong importance
7	Demonstrated importance
9	Absolute importance
2, 4, 6, 8	Intermediate judgments

Table 2. Weights obtained by the AHP method.

	PIL	Cost	Time	w
PIL	1	2	4	1
Cost	0.5	1	4	0.7
Time	0.25	0.25	1	0.2

A number of simulation has been carried out in order to check the robustness of the model.

Mapping the output function resulted from a system of rules based on heuristics (Fig. 11). A general form of a heuristic rule is:

IF <condition> THEN <conclusions>.

Every rule needs to be weighted: in this work, the rules weight has been determined by AHP method (Saaty, 1980).

The AHP method determines weights in the system rules by pair wise comparison between each pair of rules. Each comparison is transformed into a numerical value expressed according to a 9-values scale, reported in Table 1.

The comparison results are then composed into a positive reciprocal matrix $A = [a_{ij}]$, in which $a_{ij} = w_i/w_j$, where w_i is the weight of the ith rule. By multiplying A with the transpose of vector $w^t = (w_1, w_2, \ldots, w_n)$, one obtains nw. The problem becomes an eigenvector problem with the following linear equation:

$$Aw = nw \qquad (3)$$

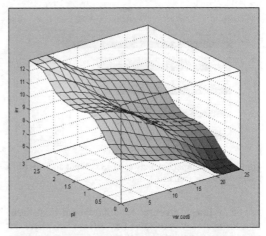

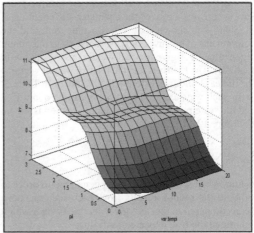

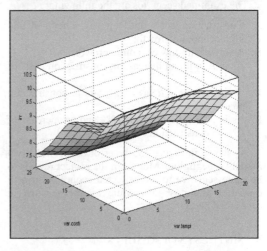

Figure 12. Output function (IRR) generated by the FIS: three-dimensional representations of the 4-D function.

2483

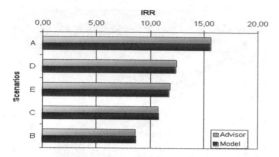

Figure 13. Comparison between FIS's output and Advisor's results for each scenario.

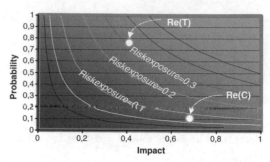

Figure 14. Risk exposure for the two examined scenarios.

Using linear algebra, to obtain nontrivial solution for vector w is to solve $(A - nI)w = 0$, where det $(A - nI) = 0$. This is a well known eigenvalue problem, where n is a root of the characteristic equation of A. Saaty has shown that A has a unit rank and the components of eigenvector corresponding to maximum eigenvalue of A give the real weights associated with each rule. The matrix A and the relative weight vector w for the case under development are shown in Table 2.

In such way, experts' judgments can be quantified and the system calibrated: the output function is shown in Figure 12.

5 SENSITIVITY ANALYSIS

A number of scenarios has been generated in order to validate the model; results from each scenario have been compared to the ones given by the Advisor which examined the true expected cash flow.

The examined scenarios are the following:

A. Construction costs and time as scheduled and high economic growth;
B. As in A, with low economic growth;
C. As in A, with an increase in construction costs of 20%;
D. As in A, with a delay of 2 years in construction phases;
E. As in A, with a delay of 10% and an increase of 15% in construction costs.

As shown in Figure 13, the error between the two models is negligible. Moreover, the project reveals to be quite sensitive to the expected economic growth (variation of almost 45%) while it is less sensitive in cases C, E and D.

6 SINTHESYS OF RESULTS

At this extent, the probability to meet the project's goals has been quantify by use of Monte Carlo simulation; on the other hand, the impact of cost and time overruns has been quantify by use of a fuzzy inference system. The expected risks associated with such events are then quantifiable. For example, if one considers the following scenarios and a linear utility function:

A. Increase in construction costs (C) of 20%;
B. Increase in construction time (T) of 20%;

the associated risk exposures can be represented in a classic probability-impact chart as shown in Figure 14, thus supporting decision makers in planning the best risk management strategy.

7 CONCLUSIONS

In this paper an operative tool to manage risks in large engineering projects has been presented. Such methodology may be useful to represent both kinds of uncertainties involved in the feasibility analysis of a project when data are few or lacking. The proposed model can be a valid instrument in the decision-making phases of a project in order to assess its viability and to support decision-makers in the best risk management strategy when data are poor or unavailable.

ACKNOWLEDGEMENTS

The financial support of *Stretto di Messina Spa* is acknowledged. Anyway, the opinions and the results here presented are responsibility only of the Authors and cannot be assumed to reflect the ones of Stretto di Messina Spa.

REFERENCES

AbouRizk, S.M., Halpin, D.W. & Sawhney, A. 1994. Modelling uncertainty in construction simulation. In: Ayyub, B.M. & Gupta, M.M. (eds) 1994. *Uncertainty Modelling and Analysis, Theory and Applications*. New York: Elsevier.

Ayyub, B.M. 1999. *Assessment of the construction feasibility of the Mobile Offshore Base*. University of Maryland.

Banks, J., Carson, J.S. & Nelson, B.L. 1995. *Discrete- Event System Simulation*. Upper Saddle River, NJ: Prentice-Hall.

Flyvbjerg, B., Bruzelius, N. & Rothengatter, W. 2003. *Megaprojects and risk: an anatomy of ambition*. Cambridge: Cambridge University Press.

Kasabov, N.K. 1998. *Foundation of Neural Networks, Fuzzy Systems and Knowledge Engineering*. The MIT Press.

Law, A. & Kelton, W.D. 1991. Simulation modelling and analysis. New York: McGraw Hill.

Project Management Institute, 2000. *A Guide to the Project Management Body of Knowledge*. Upper Darby, PA.

Saaty, T. 1980. *The analytic hierarchy process: planning, priority setting, resource allocation*. London: McGraw Hill.

World Bank, 1990. *Annual review of project performance review*. Operations Evaluation Department: World Bank.

General aspects of the structural behavior in the Messina Strait Bridge design

L. Catallo & L. Sgambi
University of Rome "La Sapienza", Rome, Italy

M. Silvestri
Structural Engineer, Rome, Italy

ABSTRACT: The behavior of long span suspended bridges is influenced by different and complex aspects that need to be treated and evaluated deeply, organic and systematic. The Performance approach in the design process, strictly linked with the new philosophy of the Performance-based Design, allows a rational approach of the design problematic, with the possibility of optimizating the different parameters and levels of detail. To do that, it must be clear which are the peculiar aspects that the design of this typology of bridge must concern, above all referring to the geometrical non linearity of the system. Moreover, big attention must be paid in the realization of the main topics of the entire design process, represented by the structural modelling of the system, which allow to estimate realistically the structural response to the action that act during the exercise phase.

1 INTRODUCTION

The subject of this study is the Messina Strait Bridge, linking mainland Italy with Sicily. The main span of the Messina bridge (*3300 m*) will approximatly equal the two world longest span bridges added together: the current longest span bridge, the Japan's Akashi Kaikyo Bridge (*1991 m* main span) and the Humber Bridge (*1410 m* main span).

The total length of the deck, *60 m* wide, is *3666 m* (including side spans).

The deck is formed by three box sections, outer ones for the roadway and the central one for the railway. The roadway deck is composed of three lanes for each carriageway (two driving lanes and one emergency lane), each *3.75 m* wide, while the railway section is composed of two tracks.

The height of the two towers has been increased by more than *6 m* (up to *382.6 m*), in the new preliminary project for the Bridge over the Strait of Messina, in order to provide a minimum vertical clearance for navigation of *65 m* – with the most unfavorable load conditions – over a width of *600 m*.

The bridge suspension system relies on two pairs of steel cables, each with a diameter of *1.24 m* and a total length, between the anchor blocks, of *5300 m*.

2 PERFORMANCE-BASED DESIGN

For advanced systems, like long span suspension bridges (i.e. Messina Bridge), it is necessary to set the global design process in the philosophy of the

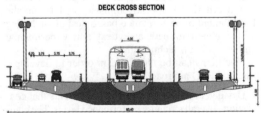

Figure 2. Deck cross section of the Messina Bridge.

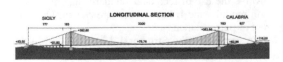

Figure 1. Longitudinal section of the Messina Strait Bridge.

Figure 3. Performance Matrix for the Messina Strait Bridge.

Performance decomposition		Supporting condition	Suspension system	Bridge deck	Aproaching span	Highway	Railway	Monitoring	Control
		Main structural system				**Secondary structural system**		**Auxiliary structural system**	
Structural Quality	Structural robustness								
	Environmental impact								
	Aesthetics								
Structural Reliability	Manmade loads								
	Environmental loads								
	Exceptional loads								
Serviceability	Highway runnability								
	Railway runnability								
	Life-Cycle Cost								
Constructive Efficiency	Constructability								

Performance-based Design (P.B.D.), rather than in the conventional specifications-based design.

This study utilizes a performance-based design methodologies to evaluate the performances of a suspension bridge that meet, as economically as possible, the uncertain future demands that both owner-users and nature will put upon them.

Performance-based design is the application of accurate simulation models to arrive at a system that satisfies all requirements. The premise is that performance levels and objectives can be quantified, that performances can be analytically predicted, and that the costs for performance improving can be evaluated, so that rational trade-offs can be made based on life-cycle considerations rather than construction costs alone.

There is a need to treat these structures and the systems that service them as complete optimized entities and not as the sum of a number of separately designed and optimized sub-systems or components. The challenge will be to identify critical areas needing to be studied in order to develop new and more advanced design criteria. A special emphasis will be on the development of performance-based design, including issues related to new analytical and experimental studies, and new technologies.

Five items can be identified in order to develop a practical PBD approach:

1. Detailed objectives for bridge components that constitute overall structure performance levels. The engineer must create an inventory of all elements (Fig. 4) – both structural and nonstructural – according to a standard categorization system of assemblies, referred as an assembly taxonomy.

2. Numerical values for these objectives and the minimum allowable probabilities of achieving them.

3. A theoretical framework for determining whether a design meets its objectives.

4. Detailed procedures to implement this theoretical framework.

5. A plain language to formalize or simplify the implementation procedures.

The first of these points, can be represented by a Performance Matrix, reporting the Performance requirements into the lines and the bridge's components into the columns.

Ideally, objective statements and quantified performance criteria for each cell or group of cells in the Performance Matrix, shown in Figure 3, should be prepared.

Obviously, the more complete and rational list of parts and attributes given in the matrix is, the better it is to be used as a guide for the preparation of objective statements and quantified criteria. The key performance parameters need to be clearly defined, and adequate methods of measuring or calculating them should be identified or specified. The latter may involve the use of technical standards and/or prediction models, and it has to take into account the context of performance.

In particular, the Structural decomposition in the Performance Matrix can be achieved by the substructuring of the structure at different levels: *macro*-level, *meso*-level and *micro*-level. The flowchart of

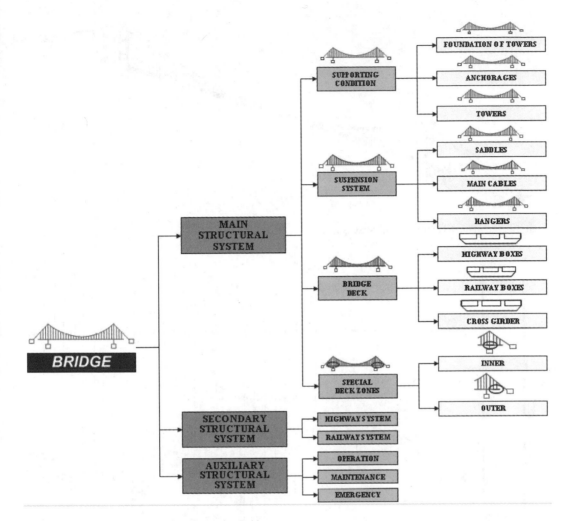

Figure 4.   Breakdown of the messina Suspension Bridge.

Figure 4 represents the structural breakdown of the Messina Suspension Bridge: the whole system is firstly divided into substructures (macro-level), then into components (meso-level) and finally into elements (micro-level). For every level, all variables are eventually recognized.

In this way, for each variable, it is possible to modify and optimize the structural behavior in order to achieve a specified performance objective, including the relationships between global and local performances.

The performance matrix must be a support during the design process and it will be able to permit to check, in every phase of the design, if the performance requirements are identified and achieved.

3   MODELLING

The most important phase of the design process is the construction of the analytical model, which should be able to predict the most realistic system behavior as far as possible, as it is necessary in the PBD. In this context, before the beginning of modelling phase, it is also necessary to plan a management model for human activity reliability. So, through the control of human factor, a potential source of mistake, and through a proactive attitude, it is possible to perform the design in according to the Theory of Excellence. Moreover, with regard to the analytical phase, computed results are rarely exact, because modelling error exists, which refers to the difference between a physical system and

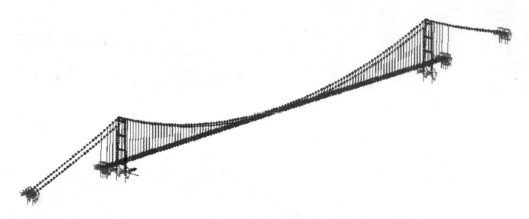

Figure 5.   Global model of the Messina Strait Bridge.

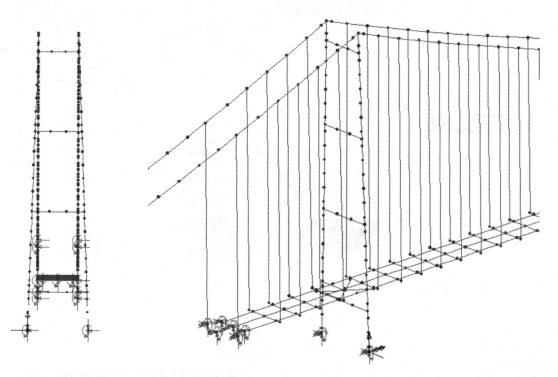

Figure 6.   Details of the global model of the Messina Strait Bridge.

its mathematical model. On the other hand, modelling error can be referred to reasonable and considered approximations made deliberately rather than by mistake, and to uncertainties, often about the actual nature of loads and boundary conditions.

Small changes in requirements entail large changes in the structure and configuration, and small errors in the programs that prescribe the behavior of the system can lead to large errors in the desired behavior. May require multiple, redundant design to reduce the risk assigned to potential additional unknown reply of the structure.

In this work, some relevant aspects are briefly reported:

- *Geometric nonlinearity*, which arises when deformations are large enough to alter the distribution or orientation of applied loads, or the orientation of internal resisting forces and moments.

2490

- *Research of the initial undeformed shape*, when the bridge is loaded with gravity and permanent loads.
- *Redundant analysis*, which permits to analyze the three-dimensional model in different finite element analyses programs, and to confront the results in terms of kinematics and static quantities.
- *Validation of the results* obtained by analyses.

4 GEOMETRIC NONLINEARITY

4.1 Introduction

For structural systems with non linear behavior a realistic description of the response under all load levels can be obtained only by taking into account the actual nonlinearities.

In structural mechanics, nonlinearities include (Cook 1997):

- *Material nonlinearity*, in which material properties are functions of the state of stress or strain. Examples include nonlinear elasticity, plasticity and creep.
- *Geometric nonlinearity*, in which deformation is large enough that equilibrium equation must be written with respect to the deformed structural geometry. Moreover, loads may change direction as they increase, as when pressure inflates a membrane.
- *Contact nonlinearity*, in which a gap between adjacent parts may open or close, the contact area between parts changes as the contact force changes, or there is sliding contact with frictional forces.

Because of the slenderness of geometrically nonlinear elastic structures, such as suspension bridges, large displacements and large rotations cannot be ignored. In this work, only the geometric nonlinearity have been considered.

For these structures, responses are nonlinear even if strains are within elastic range. Therefore, the analysis of geometrically nonlinear elastic structures has to consider the nonlinear relation between strains and displacements (Imai & Frangopol 2002).

The essential difficulty of geometrically nonlinear analysis is that equilibrium equations must be written with respect to the deformed geometry, which is not known in advance.

Nonlinearities make the problem more complicated because equations that describe the solution must incorporate conditions not fully know until the solution in know: the actual configuration, loading condition, state of stress and support condition. The solution cannot be obtained in a single step of analysis. The analysis has to be developed through several steps, updating the tentative solution after each step and repeating until a convergence test is satisfied. The usual linear analysis is only the first step in this sequence. Nonlinear analysis can treat a great variety of problems, but in a sense it is more restrictive than linear analysis because the principle of superposition does not apply; it cannot scale results in proportion to load or combine results from different load cases as in linear analysis. Accordingly, each different load case requires a separate analysis.

4.2 Modelling consideration

In the study of nonlinear geometric problems, it is necessary to define the approach with which it follows the structural response, during the load history.

It is possible to distinguish:

- *Eulerian approach*: the knowledge of the displacement field, for successive time instants t, permits to form successive states of motion, from which is not possible to observe the motion of a singular elementary particle.
- *Lagrangian approach*: it exams the motion of the elementary particles, assuming the points coordinates of them as the unknown quantities in the successive time instants. With regard to the configuration in which it is defined the point coordinates, it is usual to distinguish between *Total Lagrangian* and *Update Lagrangian* formulation. In both, the kinematics and static variables are referred respectively to the initial and undeformed configuration, and to the deformed calculated configuration.

5 RESEARCH OF THE INITIAL UNDEFORMED SHAPE

One of the main problems to consider, before the model construction, is the research of the initial undeformed configuration, that the bridge deck assumes after the gravity and permanent loads application.

Such configuration is better represented by a sixth order parabola with upwards concavity (Fig. 7).

The achievement of such configuration, when the gravity and permanent loads are applied, is a problem which is strongly afflicted by geometric nonlinearity.

The problem has been performed and solved by three different methods:

- *Impressed displacements method (SI)*.
- *Sag method (CF)*.
- *Temperature variation method (TE)*.

5.1 Impressed displacements method

This method starts with the implementation of the bridge finite element model, in which the initial bridge deck geometric configuration, with no loads, is the *initial undeformed design configuration (IUDC)*.

After loads (gravity and permanent) application, the bridge deck will be deformed, but in reality, after these loads application, it shall arrive at the IUDC.

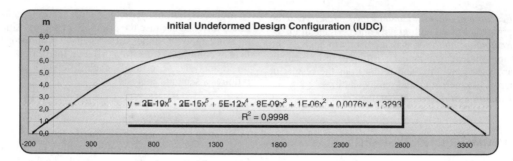

Figure 7. Sixth order parabola.

So, it has been added a further load condition which returns the deck configuration in the IUDC. In this particular case, it has impressed vertical and horizontal displacements on each anchor blocks (main cable extremities), to simulate a tug in the cable. It puts particular attention to do coincide both these configurations in the middle of the main span.

After a sufficient number of iterations, necessary to calibrate the four displacements (horizontal Δx, vertical Δy), it has been established the following values:

- Calabria's anchorage: $\Delta x = -8,30$ m;
                          $\Delta y = -1,00$ m;
- Sicilian's anchorage: $\Delta x = -8,70$ m;
                          $\Delta y = -1,00$ m.

Also not having the same long profile perfectly, this method has permitted to understand the cinematic deck behavior, after the application of most of loads conditions.

The tie of inner joint is responsible of the dragging of the tower towards the anchorages, because of impressed displacements. It studies displacement influence on the deck behavior and on the remaining model, by releasing horizontal displacement in the connection node, between main cable and the top of the tower.

It is come to conclusion, that this method meaningfully influences only static and cinematic aspects of the towers.

5.2 Sag method

In order to better understand the towers behaviour, it has been constructed a different three-dimensional bridge model, using the *sag method* to research the initial undeformed shape.

This method provides the definition of a initial geometric deck and cable configuration, which present a rise of the parabola such as, under gravity and permanent loads, the bridge deck will arrive on the IUDC.

Also in this case, it is necessary a sufficient number of iterations to calibrate the rise of parabola, because the problem is strongly geometric nonlinear and the principle of superposition does not apply.

In Figure 8, it is represented the procedure to define the rise of the parabola:

- First of all it is considered the unloaded bridge configuration (CI_0), corresponding with the IUDC.
- Then, gravity and permanent loads have been applied, so the deck configuration deforms and becomes (CD_0).
- The midspan deflection f is correlated to the sag ratio $k = H_c/L_c$, where H_c is the sag of main cable unloaded and L_c is the main span length. Then, the sag f_0 is correlated to the configuration (CI_0) and to the parameter k_0.
- the model has been analyzed again with a new deformed configuration (CD_1), obtained by summing the tipped deformed configuration CD_0 to the IUDC: $CI_1 = CI_0 + f_0$.
- At this point, an iterative process starts, that assumes $CI_{(i)}$ as initial unloaded configuration. This is equal to the sum of the previous configuration $CI_{(i-1)}$ and the sag of the previous step $f_{(i-1)}$ multiplied for $k_{(i-2)}/k_{(i-1)}$. This parameter is necessary to take into account the geometric nonlinearity of the system.
- After a sufficient number of iterations, it achieves an initial configuration with a middle span sag equal to *81 m*, that must be sum to IUDC altitude.

5.3 Temperature variation method

In the previous cases, both methods permit to understand, with particular regard, meaningfully cinematic deck behaviour. This method permits to understand also the static behaviour.

Gravity and permanent load application involves stress state establishment, which correspond a strain state. Using the axial extension of the main cable and of the hangers as references for the application of temperature variation that it causes the same shrinkage, a first step till the IUDC under loaded is obtained.

Also in this case, the process is iterative.

To each span, it is associated the pertinent main cable temperature variation, while the hanger temperature

2492

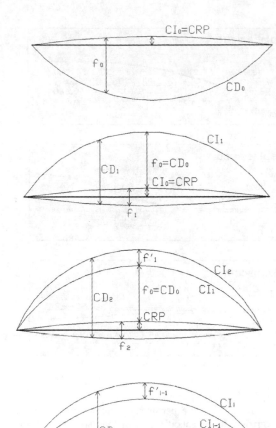

Figure 8. Determination of the IUDC with *unsag method*.

Figure 9. Loads applied.

Lusas two different models have been implemented: in the first the *"impressed displacements method"* has been used for the research of the initial undeformed configuration, while in the second the *sag method* has been used.

In Sap 2000 other two different models have been implemented: in one the *"impressed displacements method"* has been used for the research of the initial undeformed configuration, while in the second the *"temperature variation method"* has been used.

$$CI_0 = CRP$$

$$CD_0 = f_0$$

$$CI_1 = CI_0 + CD_0 = CRP + CD_0$$

$$f_1 = CD_1 - (CI_1 + CRP)$$

$$f_1' = f_1 \cdot \frac{k_0}{k_1}$$

$$CI_2 = CI_1 + f_1'$$

$$CI_2 = CI_1 + [CD_1 - (CI_1 + CRP)] \cdot \frac{k_0}{k_1}$$

$$f_2 = CD_2 - (CI_2 + CRP)$$

$$f_2' = f_2 \cdot \frac{k_1}{k_2}$$

$$CI_i = CI_{i-1} + f_{i-1}'$$

$$CI_i = CI_{i-1} + [CD_{i-1} - (CI_{i-1} + CRP)] \cdot \frac{k_{i-2}}{k_{i-1}}$$

$$f_i = CD_i - (CI_i + CRP)$$

$$f_i' = f_i \cdot \frac{k_{i-1}}{k_i}$$

In Ansys, only one model has been implemented, using *temperature variation method*.

Results (vertical displacement, horizontal displacement, longitudinal displacement) have been reported in the following diagrams, represented by different lines. As one can see in Figure 10, different calculation

variations are calibrated in the way of coinciding the deck configuration with IUDC.

The models, constructed in this way, are reliable

6 RESULTS VALIDATION

Results validation has been operated in order to compare the structural response, when environmental and anthropic loads are applied and using different research procedures for the initial undeformed configuration.

A combination of three different loads (Fig. 9) has been applied:

- Highway load
- Railway load;
- Wind load for the first level.

In particular, three different commercial calculation codes have been used: LUSAS, Sap 2000, ANSYS. In

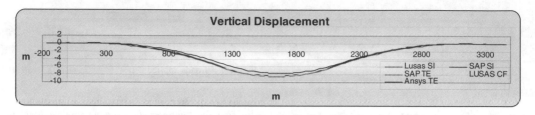

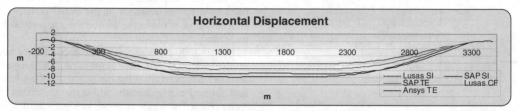

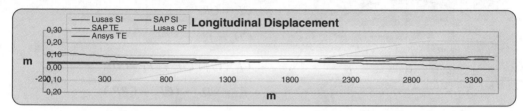

Figure 10. Response diagrams.

codes lead to different results, according to the particular solution method adopted and to the technique used to search the initial undeformed shape. In particular, as far as regards vertical displacements, results from different codes show a good accordance; on the other hand, the horizontal displacements show quite different trends, with a disagreement of some meters. In the specific case of longitudinal displacements, among all the methods used to search the initial shape, the sag method is the only one leading to unacceptable results. As afore mentioned, method's validation passes through a critic analysis of the obtained results.

7 CONCLUSIONS

This paper synthesizes the main aspects involved in the design of complex structures, such as long span suspension bridges, starting from the conceptual phase to the critic evaluation of the results. In this context, the paper shows that a traditional prescriptive approach reveals unsatisfactory, while a more general performance-based approach leads to a more accurate design.

This methodology allows to determine all performance requirements at different levels of structural substructuring. The core process is represented by the modelling phase which should predict the mechanical behavior of the structure in a realistic manner.

In this way the performance achievement can be verified. In the specific case of long span suspension bridges, the modeling process is characterized by some peculiar aspects, among which the research for the initial undeformed shape is one of the most important. With regard to this problem, the paper shows the results from three methods, each of them being implemented with three different commercial codes.

ACKNOWLEDGEMENTS

The financial support of *Stretto di Messina Spa* is acknowledged.Anyway, the opinions and the results here presented are responsibility only of the Authors and cannot be assumed to reflect the ones of Stretto di Messina SpA.

REFERENCES

Cook, R.D., 1995. *Finite element modeling for stress analysis*. Wiley & Sons. United States.
Cook, R.D., Malkus, D.S. & Plesha, M.E. 1989. *Concepts and application of finite element analysis*. Wiley & Sons. United States.
Gimsing, N.J. 1997. *Cable supported bridges*. Wiley & Sons. United States.

System-based Vision for Strategic and Creative Design, Bontempi (ed.)
© 2003 Swets & Zeitlinger, Lisse, ISBN 90 5809 599 1

Strategy and formulation levels of the structural performance analysis of advanced systems

M. Silvestri
Structural Engineer, Rome, Italy

F. Bontempi
University of Rome "La Sapienza", Rome, Italy

ABSTRACT: The modern cable suspended bridges are advanced systems of fundamental importance by social, environmental, technical and scientific view point. It is absolutely critic to predict their structural performances in advance, in such a way as defining the qualitative and quantitative element for their definition. In this work a performance approach (Performance-based Design) to this problem is represented, dividing the general treatment of this subject in two several aspects: one related to the performance approach and its main arguments, in special regard to the Messina Strait Bridge, the other related to modelling phase, the topic phase of the whole process of design. In fact, this point is central in the performance-based design, because it is necessary to simulate the real future structural behavior as better as possible. Therefore, it is fundamental the management, construction and validation of the model.

1 INTRODUCTION

The performance analysis for the evaluation of the reliability of complex structures like the Messina Strait Bridge, results from three main points (Fellows & Liu 1997):

- formulation of the structural problem and organization of the relevant data and information;
- structural modelling and numerical analysis;
- synthesis of the results and critical evaluations.

The inherent complexity in planning and designing this kind of structures leads to the necessity of a new design statement in the framework of the so-called *Performance-Based Design* (P.B.D.), which aims at defining preventively and accurately the performances requested to the structure. Absolutely critic it appears the ability to predict the performance required for such structures and the necessity to provide the qualitative and quantitative elements able to support the following decisional processes needed to specifically define such an important structure, including both limit and service behaviors (Blockley 1980, Calzona 2001).

In the specific case, it has been chosen to make use of all the advantages that a performance-based design can guarantee with respect to the more traditional prescriptive approaches. Figure 1 synthetically, but efficiently, shows the relationship between prescriptive and performance content.

Nowadays, most of the structures are designed, constructed, maintained and dismantled without taking in consideration the specific knowledge pertinent to their real behavior. The recent improvement in measurement and elaboration data technologies has created the proper condition to improve the decisional tools, traditionally based on experimental and numerical simulations, through the improvement of the information connected to both the structural *performances* on site and the measurement systems. These systems have become in the last few years more numerous, cheaper, stronger, stable and precise: some of those ones, that in the past were reliable only in laboratory applications, today are powerfully applied on site problems. The enormous advantages that can be drawn by the results of these measurements and the development of these technologies have led to the birth of activities inside the domain of the so-called *Performance-Based Structural Engineering* (Smith 2001).

The actual organization of the entire design process (generally based on *prescriptive* concepts) pays attention in defining loads and external forces applied and in the definition of the admissible stress and strain conditions. On the other hand, the P.B.D. starts the design process fixing the performance guideline required to

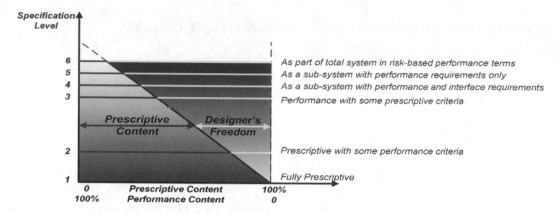

Figure 1. Prescriptive content and performance content.

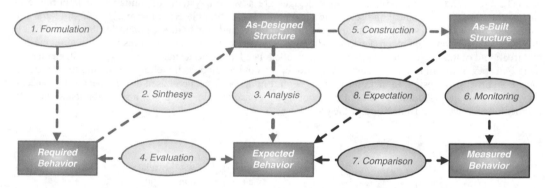

Figure 2. Extended scheme of the design process principal phases.

the structure (Hirashi *et al*, 1998). The substantial difference between these two approaches, can be explicated as follows:

"A prescriptive approach describes an acceptable solution, while a performance approach describes the performances which are required to the structure. A performance *approach is, essentially, the attitude to think and to work in terms of finality, rather than in terms of meanings. It concerns what a structure has to achieve rather than the way it is built."* (Foliente 2000).

In Figure 2 a schematic view of the subsequent phases of a traditional design are shown.

1. *Formulation* is needed by the Owner for the functions that the structure would absolve, as concerns the required behavior.
2. *Synthesis* of the performance required within the structural information.
3. *Analysis*, aimed at obtaining the structural expected behavior.

4. *Valuation*: the expected behavior must be compared with the required one.
5. *Construction*: if the evaluation phase is satisfactory, one can pass to the construction phase in which a transformation of the structural information from "as designed" to "as-built" is executed.

Difficulties associated with this kind of approach are evident. The "as-built" structure could be different from the "as designed" one due to different factors like fabrication mistakes or unexpected conditions during the construction phase, or also due to non appropriate design assumptions that could create doubts about analysis accuracy leading to a predicted behavior which does not correspond to the real one. A way to avoid these difficulties, as already said, can be achieved using a monitoring system. In this case, to the traditional designing phases must be added three other steps (Fig. 2).

6. *Structural monitoring* and data information acquisition about serviceability behavior.
7. *Comparison* between monitoring and expected behavior results.

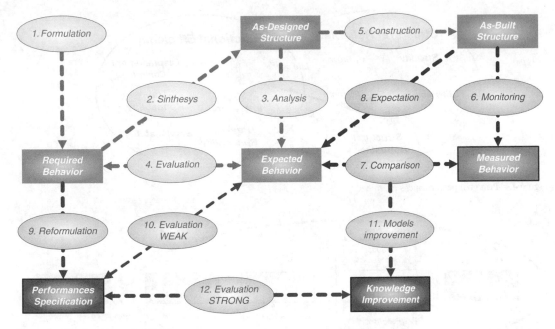

Figure 3. Further phases in P.D.B.

8. More accurate *expectation* of the future structural behavior.

Once the most important task – connected with the idea of performances that the structure under examination will have to guarantee – is clearly understood, four new phases connected to a performance design procedure have been inserted in the scheme of Figure 3.

9. *Reformulation*: development of advanced probabilistic methods for a more accurate description of the required behavior and of the required performance.
10. *Evaluation*: this methodology assumes that the analysis is exact and that all the actions are known exactly, from the probabilistic viewpoint. This evaluation is defined weak.
11. *Models improvement*: the necessity connected with the models improvement comes from experiences based on monitoring, where expected and measured behaviors on "as-built" structures are directly compared.
12. *Evaluation*: a third kind of evaluation, called strong, becomes possible when the improvement (see point 11) aims at assigning more accurate values to used parameters and to define more accurate modelling hypothesis (boundary conditions, nodes stability, section properties, etc.).

In this way, it is possible to obtain an improvement of the design solution in the context of the Theory of Excellence, a really comprehensive generalization of the *Re-engineering technique*.

The main role of modelling can be understood, both from theoretical-numerical and experimental-physical viewpoints.

2 PERFORMANCE IDENTIFICATION AND MEASURE

The performance approach to the structural design is based on the verification of the exact suitability of the structure with respect to fixed requisites connected to serviceability and safety conditions, as shown with reference to the Messina Strait Bridge in Figure 4.

In the planning of the structural design, which concerns the qualitative identification of the necessary performance, one can draw up a performance matrix in which the general outline, about the subdivision of the complex problem in exam, is described. For each structural part of the bridge, the performances have been listed, operating divisions among Structural Safety, Functional Efficiency and Constructive Efficiency.

Among all the performances pointed out in the Performance Matrix, particularly interesting for a complex structure like the Messina Strait Crossing Bridge are:

- structural strength;
- runability;
- maintenance of a free level above the sea in frequent condition;
- analysis and aerodynamic optimization of the bridge cross-section;

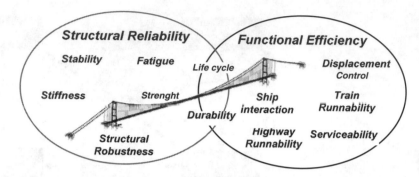

Figure 4.   Principal performances examined.

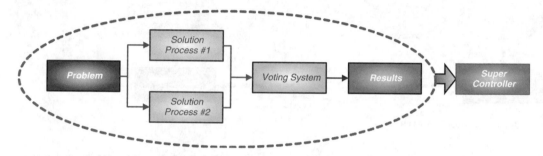

Figure 5.   Model management for human activity reliability: redundant scheme with two independent solution processes.

- limitation of deck accelerations induced by vortex shedding for frequent wind conditions;
- aeroelastic stability.

The central point of the entire process here indicated is represented by the quantification of the performances, before indicated only from a qualitative viewpoint. Quantify means to establish a boundary limit for the chosen parameters, out of which it is impossible to guarantee a fixed performance condition. This can be done only fixing the attention to the direct experience of already existent bridges, or paying attention on deeper studies made to solve specific problems.

3   MODELLING

The first point to face concerns the general consideration about all the activities connected with the construction of the model. One must speak of *modelling management*, underlining:

- the cultural context in which modelling, analysis, verifying and structural synthesis of the bridge are developed (Simon 1998; Vincenti);
- the organization of the people who carry out these activities and their personal approach;

- the importance to control the human factor: the source of human error;
- the need to improve human activity reliability, as shown in Figure 5.

For a suspended bridge, like the Messina Bridge, which presents unique and exceptional characteristics, the control of the complete design and of the solution activities start anyway with the Quality control procedure connected with industrial products and all the control aspects, as affirmed clearly in ISO9000 norms.

Quality is the characteristic of a good able to satisfy the needs implicit or explicit of a client. The introduction of a Quality System in a productive organization aims to minimizing the non conformity of a product, looking, in the productive process, for the reasons and the responsibilities of an eventual unsuccessful.

It must be underlined that the different professional figures, working in the project, can adopt different attitudes with the common` aim to obtain a better solution in agreement with the required performances requested. Specifically with:

- a passive attitude, non conformable solution of the structural problem (proposed/chosen by the client) can be avoided;

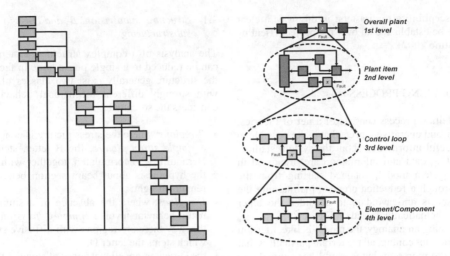

Figure 6. Block functional diagrams: decomposition of the problem/aggregation of the solution (Bentley 1999).

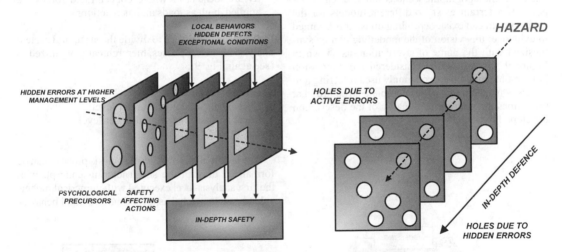

Figure 7. Causes of possible structural failure (Reason 1990; Catino 2002).

- an active attitude, lead to the same working way, but there is an attempt to improve the proposed solution, always inside the given definition of the structural problem;
- a *proactive* attitude, can lead, if necessary and needed, to a change of the definition given by the client about the structural problem, showing new aspects and different viewpoint to obtain an excellent solution, ever outside the given definition of the structural problem.

The latter attitude, well known in the Theory of Excellence, is the most interesting from the results point of view, exceeding also the quality statement,

and it is strongly based on the individual commitment of each person of the team.

At the basic level, through the operative methods connected with the quality control, it is possible to concentrate the attention to the verifying problem, driving the detailed analysis; it is possible, moreover, to extrapolate the best level of data information also considering the mistakes developed during the conceptual and design phases of such structures. These mistakes could be the starting points of the performance incapacity or for the damage of the performance capacity.

Particularly through the logical and operative resolution of Figure 6, the failure dynamics of Figure 7 can be avoided, where it results evident the design and opera susceptibility to disasters: in this way, a clear logic can be established all over the project (Bentley 1999; Catino 2002).

4 MODELLING PROCESS

The modelling process consists of a net of theories, decisions, and operative aspects that make possible to extract useful information about the system studied. The existing data and information are organized in the real system model, obtained starting from the reality through a reduction process, maintaining the useful aspects and avoiding to consider the other ones, as schematically quoted on Figure 8.

Introducing an analogy, the model is like a terrain map: it does not contain all the terrain particulars that one wants to represent, but is useful, for example, to know the distance between two points. It's evident that different topographic sections and different viewpoint of a terrain exist, so different models for the same system can exist, depending on what it is aimed to evidence: this vision of the modelling can be synthesized with the name of layerization, as shown in Figure 9. Useful can be considered the expression holographic modelling, commonly used referring it to the risks management of the complete structure, that underlines the multidimensionality of the description developed through the model.

4.1 Structural subdivision: B- and D-regions substructuring

The analysis of a complex structural system hardly can be reduced to a single phase only. In fact, inside the structure generally exist two classes of regions with strongly different mechanical behavior. One considers the so-called:

- *B-regions*: where the stress picture is congruent to a *simple strain picture*; the B letter, stems from Bernoulli, who formulated, together with Navier, the hypothesis about beam section behavior that remain straight;
- *D-regions*: where the absence of a simple cinematic behavior involve *complex stress picture*; in this way there are regions with diffusive stress, by which stems the letter D;
- the D-regions are all that zones of singularity for the structure, where geometric or material discontinuity are located, or where concentrated forces are applied, both as loads and as reactions.

It appears necessary to divide the structural system analysis in several phases, hierarchically organized as (see again Fig. 9):

- Global analysis;
- Local analysis at a first under level;
- Local analysis at a second under level;
- ....

The layout of this multilevel analysis process, can be formalized through the substructuring concept, with the first analysis level executed with numerical models able to gain the global aspect of the structural behavior

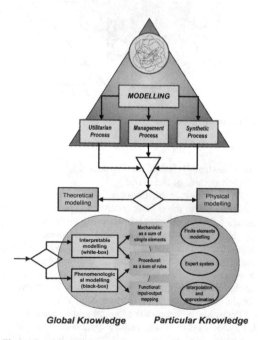

Figure 8. Modelling.

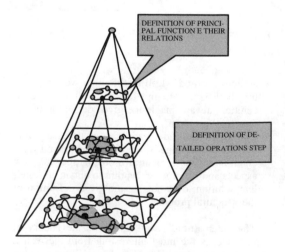

Figure 9. Subdivision of structural system analysis.

(for example, with Bernoulli beam elements) (Bontempi & Paolini 2001). It is important to note that:

- all the connections of the structural elements are configured as D-regions;
- the global behavior of the structure is the result, from a macroscopic point of view, of the integration of the local behavior of these D-regions;
- the local possible crisis, can be dangerous for the integrity (robustness) of the overall structural organism and this must be avoided with an accurate design;
- the local behavior, particularly the local strain, can lead to an unacceptable behavior (for example, regarding the second order effects).

It must be noticed how the complexity of structures such as the Messina Bridge results just from the matching and the interactions peculiar of this system at local behavior: in fact, it is possible the arise of secondary effects to jeopardize the design. The development of these mechanisms in fact, must be identified by the modelling and opportunely dominated in the design strategy.

5 MODEL VALIDATION

A specific aspect which arises from critical considerations concerns the *validation* of the solution obtained by modelling. For example, in the diagram of Figure 10, a significant unknown quantity is shown in ordinate as obtained by different numerical models developed in a process of successive refined analyses. A typical trend can then be shown in this way:

(a) a first phase of toe-in of the unknown value;
(b) a zone in which the value is stabilized;
(c) another phase in which, introducing too much refined models, there is an explosion of numerical

errors which lead to an incorrect evaluation of the solution.

Starting from this characteristic trend, empirical and heuristic considerations about the validation of the process analysis can be derived. Moreover, it must be remembered that the Finite Element Method is, for its own nature, an approximated solution method and that in a non linear field, instead of what generally happens in the linear one, the results obtainable, also if correct from a numerical point of view, can result absolutely out of signification and totally incorrect from the mechanical point of view.

Generally speaking, the numerical error is the difference between the solution obtained and the exact one, considering for the exact solution the one referred to the differential formulation that govern the assigned problem. The error sources in a finite element analysis are traceable back to the following aspects

(a) *nature* of the adopted solution and the *resolvence* of the elements used in the discretization;
(b) *errors* given by a rough *geometric representation*;
(c) *errors* given by the *numerical procedure* adopted: among these must be considered particularly the numeric integration techniques, the system solution equations.
(d) *interpretation errors*, particularly due to erroneous evaluation of the stress path.

The (a) and (b) voices concern the general aspects connected with the model creation and definition, while (c) and (d) concern numerical aspects above all.

It is worth to observe that the solution can depend from the size, the shape and the discretization orientation and the model given to non objective representations of the structural problem. With a *robust formulation*, it is defined an analysis procedure which gives large guarantees of convergence to a mechanically correct solution, if it exists or that, otherwise, is

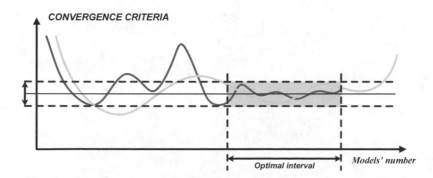

Figure 10. Convergence process of the numerical results from the modelling activity.

able to pounce the impossibility to obtain an equilibrated state conformable to the structural bearing capacity. Of course, the robustness is a fundamental condition for an effective applicability of the numerical modelling to the engineering problems, and represents a fundamental characteristic to pursue in the development of any condition techniques (Malerba 2001).

6 CONCLUSION

In this work, the main aspects of the performance-based design are exposed in concise statement, with regard to the advanced systems, like cable suspension bridge. The project of these structures involves several general aspects, and therefore, their consideration and optimization needed to be organized and synthesized with a performance approach, in opposite to the prescriptive approach.

Although in a brief form, connections with different and more ample concepts and theories are highlighted. Among these, the Theory of Excellence and the process of Re-engineering are relevant.

ACKNOWLEDGMENTS

The financial support of *Stretto di Messina Spa* is acknowledged. Anyway, the opinions and the results here presented are responsibility only of the Authors and cannot be assumed to reflect the ones of Stretto di Messina SpA.

REFERENCES

Bentley J.P., 1999. Reliability & Quality Engineering. Addison Wesley England, 1999.

Blockley D.I., 1980. "The nature of structural design and safety", Ellis Horwood Limited.

Bontempi F., Radogna E.F. & Biondini F., 2002. La modellazione realistica delle strutture esistenti in c.a./c.a.p. Giornate A.I.C.A.P. 2002.

Bontempi F. & Paolini M., 2001. Innovazione nello sviluppo e nella progettazione degli ancoraggi. Hilti Italia S.p.A.

Calzona R. & Bontempi. F, 2001. Remarks on the approval process of designs of structures provided with innovative anti-seismic systems in Italy. 7th International Seminar on Seismic Isolation, Passive Energy Dissipation and Active Control of Vibrations of Structures. Assisi, 2001.

Catino M., 2002. Da Chernobyl a Linate. Incidenti tecnologici o errori organizzativi?, Carocci Editore.

Factory Mutual Research, 2000. Performance-Based Design: From Local Needs to Global Needs. Frontiers, Vol. 14, No. 3, 1–5.

Fellows R. & Liu A., 1997. Research Methods for Construction. 1997, Blackwell Science.

Foliente G.C., 2000. "Developments in Performance-Based Building Codes and Standards." Forest Products Journal, 50 (7/8), 12–21.

Foliente G.C., Leicester R.H. & Pham L., 1998. "Development of the CIB Proactive Program on Performance Based Building Codes and Standards.", BCE Doc.98/232.

Gemeny D.F., 2000. Performance-Based Design: an Overview. Security Technology & Design.

Hiraishi H., Teshigawara M., Fukuyama H., Saito T., Gojo W., Fujitani H., Okawa I. & Okada H., 1998. "New Framework for Performance Based Design of Building Structures – Design Flow and Social System" NIST SP 931, Proceedings of the 30th joint meeting of the U.S.-Japan cooperative program in natural resources panel on wind and seismic effect.

Malerba P.G. (ed.) Analisi non lineare e limite delle strutture in cemento armato. CISM, 2000.

Silvestri M., 2002. Gli aspetti concettuali del Performance-Based Design e la loro applicazione nel progetto dei ponti sospesi. Graduate thesis, Facoltà di Ingegneria dell' Università degli Studi di Roma "La Sapienza", A.A. 2001/2002.

Simon H.A., 1998, "The Sciences of the Artificial.", The MIT Press, Cambridge.

Smith I., 2001. "Increasing Knowledge of Structural Performance." Structural Engineering International, 12 (3), 191–195.

Vincenti, W.G., 1990, "What Engineers know and how they know it." The John Hopkins University Press, Baltimore.

System-based Vision for Strategic and Creative Design, Bontempi (ed.)
© *2003 Swets & Zeitlinger, Lisse, ISBN 90 5809 599 1*

Evaluation and results' comparisons in dynamic structural response of Messina Cable-Suspended Bridge

V. Barberi & M. Ciani
Structural Engineer, Rome, Italy

L. Catallo
University of Rome "La Sapienza", Rome, Italy

ABSTRACT: Object of this article is to critically consider the dynamic results obtained by finite elements code SAP2000 on different models of the Messina Strait Bridge. A modal analysis was performed and results obtained were: natural periods, modal shapes and modal participating mass ratios. Next, a dynamic analysis was carried out: 3 acceleration's time histories were applied simultaneously on the bridge and under this load case summed to the static one, some entities were evaluated in order to determine the structural behaviour. At the end a study on results' comparison between SAP2000 and ANSYS was effected to evaluate code's reliability. An analysis of the damping mechanism was also effected to study differences between results obtained by Rayleigh damping method of ANSYS program and constant damping ratio method of SAP2000 code.

1   INTRODUCTION

This article is the second section of a study about the structural behaviour of Messina Strait Bridge; in particular this paper explains the dynamic behaviour.

At the beginning a modal analysis was performed; by this one some general characteristics were evaluated like:

- *structure's natural periods,*
- *modal shapes,*
- *modal excited mass ratios.*

Next a dynamic analysis was performed; this one was developed to simulate a real seismic attack, being earthquake simulated by 3 different acceleration time-histories, one for each direction, acting at the same time on the bridge. During this analysis some entities were evaluated in specified points, in particular:

- *displacements,*
- *external reactions,*
- *internal forces,*

to represent global structure's behaviour.

Studies about structural response's variation by damping change were developed and, eventually, comparisons of results through 2 different finite element codes, in particular SAP2000 and ANSYS were

considered. All these results are connected to the goal to evaluate model's and code's reliability.

2   MODAL ANALYSIS

Modal analysis is a way to evaluate any differences in models' behaviour regarding modal participation mass ratios, natural periods and modal deformed shapes.

Natural frequencies and mode shapes are important parameters in a structural design for dynamic load conditions; they are also required if a spectral analysis or a mode superposition harmonic or transient analysis want to be done.

In modal analysis structural, masses and live load contributions were inserted in the model to represent the exact bridge's behaviour during a seismic attack.

Structural masses were self calculated by the program by material density and geometrical characteristics of the sections; their positioning, as lumped masses, is made according to the code defaults, i.e. coincident to the nodes at the end of each frame element used in bridge's modellation.

Masses representing live load contribution were also positioned on the deck's nodes, for these masses were introduced new elements, which are mass element having the same translational characteristic in each direction, but not rotational one.

MODAL PARTICIPATING MASS RATIOS vs MODE NUMBER			
MODE	DIRECTION		
	LONGITUDINAL	TRASVERSE	VERTICAL
50	61,65	79,69	60,71
100	86,24	87,67	67,14
200	86,81	91,90	68,23
300	92,23	93,97	68,57
400	92,36	95,37	70,76
500	94,02	96,13	71,91
600	94,04	96,21	77,60
700	95,13	96,34	82,67
800	96,18	96,52	92,96

Figures 1–2. Modal participating mass ratios vs. mode number.

ENTITY	DIRECTION		
	LONGITUDINAL	TRASVERSE	VERTICAL
TOTAL EXCITED MASS	86,81%	91,9%	68,23%
SIGNIFICANT EXCITED MASS	83,08%	78,74%	63,48%
EXCITED MASS' LOSS	3,73	13,16	4,75
% LOSS	4,30%	14,32%	6,96%

Figure 3. Mass excited percentage's loss.

MODE	SAP			ANSYS		
	PERIOD	SHAPE	DIRECTION	PERIOD	SHAPE	DIRECTION
1	32,115	TRASL I	TRANSV	34,339	TRASL I	TRANSV
2	17,583	TRASL II	VERT	18,536	TRASL II	TRANSV
3	17,132	TRASL II	TRANSV	16,595	TRASL II	VERT
4	13,105	TRASL I	VERT	12,968	TRASL III	TRANSV
5	11,753	TORS + TRASL	TRANSV	12,484	TRASL I	VERT
6	11,400	TRASL	LONG	11,802	TORS I	
7	11,118	TORS I		11,423	TORS + TRASL	TRANSV
8	10,574	CABLE ONLY		10,959	TORS + TRASL	TRANSV
9	10,205	TORS II		10,861	TRASL	LONG
10	9,997	CABLE ONLY		9,943	TORS II	

Figure 4. Modal shapes' comparison SAP2000 – ANSYS.

only these significant modes: such results, referred to 200 modes analysis, are summarized in Figure 3.

To check code's reliability, specifically the emergence of numerical errors, the model of the Messina Strait Bridge was also implemented by commercial finite element code ANSYS.

For the first 10 modes periods and mode shape of the bridge models by the two commercial codes are in Figure 4.

In a modal analysis, the accuracy of structural response was evaluated considering the excited mass as the reference parameter; so modes number was increased till to excite at least 85% of the total mass in every direction. This result was obtained by considering about 800 natural modes.

Achieving this minimum excited mass value is important for results reliability since the code uses a mode superposition method to solve the dynamic analysis.

A study to evaluate model's sensitivity regarding the variation of modal participation mass percentage vs. mode number considered during the analysis was also effected, these results are presented in Figures 1–2.

It's evident that 85% of excited mass is already reached in longitudinal and transverse direction by only 100 modes, but to obtain the same value also in vertical direction about 800 modes are required.

When there are a lot of mode to consider, not all of them have the same importance in structural behaviour. So it was decided to consider only modes having a modal participation mass percentage greater than 1.5% in a direction at once, these modes are called significant. A study was effected to evaluate the loss of mass excited percentage coming from considering

3 DYNAMIC ANALYSIS

3.1 General considerations

Transient dynamic analysis, sometimes called time history analysis, is a technique used to determinate the dynamic response of a structure under the action of any general time-dependent loads.

The basic equation of motion solved by a transient dynamic analysis is:

$$[M]\{d^2u/dt^2\} + [C]\{du/dt\} + [K]\{u\} = \{F(t)\}$$

where:
$[M]$ = mass matrix
$[C]$ = damping matrix
$[K]$ = stiffness matrix
$\{d^2u/dt^2\}$ = nodal acceleration vector
$\{du/dt\}$ = nodal velocity vector
$\{u\}$ = nodal displacement vector
$\{F(t)\}$ = load vector

At any given time t, these equations can be thought as a set of "static" equilibrium equations that also take into account inertia forces, $[M]\{d^2u/dt^2\}$, and damping forces, $[C]\{du/dt\}$.

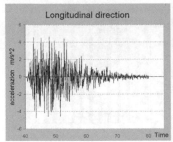
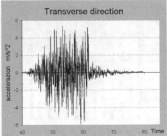
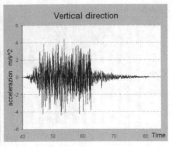

Figure 5.   Acceleration diagrams.

| DYNAMIC DISPLACEMENT | | | | STATIC CONTRIBUTION | | | TOTAL DISPLACEMENT | | |
| | | [cm] | | | | [cm] | | | [cm] | |
NODE	POSITIONING	LONG	TRANSV	VERT	LONG	TRANSV	VERT	LONG	TRANSV	VERT
Joint29	1/2 left cable	32,79	104,18	76,54	-7,31	0,48	6,61	25,48	104,65	83,15
Joint30	1/2 left cable	32,94	104,21	75,90	-7,33	-0,59	6,74	25,61	103,62	82,64
Joint167	up left tower	23,88	55,17	1,42	1,90	1,41	-23,44	25,78	56,58	-22,02
Joint168	up left tower	24,09	55,06	1,27	1,95	-1,71	-23,44	26,04	53,35	-22,17
Joint570	1/2 central cable	30,64	124,60	108,17	0,81	-4,18	9,80	31,44	120,42	117,97
Joint571	1/2 central cable	29,83	124,53	97,70	0,79	-6,18	9,80	30,63	118,35	107,50
Joint1008	up right tower	21,75	53,92	1,30	-0,07	1,36	-22,69	21,68	55,28	-21,39
Joint1009	up right tower	22,04	54,07	1,24	-0,12	-1,64	-22,69	21,92	52,43	-21,45
Joint1064	1/2 right cable	25,48	115,25	82,50	8,71	0,64	7,76	34,19	115,89	90,26
Joint1065	1/2 right cable	24,95	115,23	82,08	8,71	-0,79	7,85	33,66	114,44	89,93
Joint105	down left tower	9,36	2,32	0,31	0,01	0,01	-2,75	9,37	2,33	-2,44
Joint106	down left tower	9,22	2,35	0,47	0,01	0,05	-2,75	9,23	2,40	-2,28
Joint946	down right tower	8,94	2,30	0,42	0,05	0,01	-2,67	8,99	2,31	-2,25
Joint947	down right tower	8,96	2,34	0,28	0,05	0,04	-2,67	9,00	2,38	-2,39

Figure 6.   Table of SAP2000 dispalacements.

Two methods are available to do a transient dynamic analysis:

- *full transient analysis,*
- *mode superposition transient analysis.*

The *full method* uses the full system matrices to calculate the transient response, the *mode superposition method* sums factored mode shapes (eigenvectors) from a modal analysis to calculate the structure's response.

3.2   Obtained results

The bridge was subjected to a load combination constituted by self weight added to a superimposed load consisting of railings, road topping and other.

Used codes are different in the way to performance a dynamic analysis, infact SAP2000 uses a modal superposition analysis, instead ANSYS use a full transient analysis.

Dynamic analysis was developed to simulate bridge's behaviour under a real seismic action.

This *time history analysis* was effected using 3 different seismic histories, one for each direction, acting at the same time Figure 5.

Cheeking parameters taken in structural behaviour evaluation were:

- *displacements of signifiant points*
- *element strain forces.*

Some significant points, describing bridge's global behaviour, were chosen; in particular were evaluated middle cables points, in every of three spans, also points situated on the top of the towers and other located at the connection between towers and lower transverse tower's element were considered.

In the next table these results are summarized, these were also divided in static and dynamic contribution, so it can be evaluated seismic structural impact on the bridge, Figure 6.

To have a better idea of displacements' dynamic contribution some diagrams were portrayed reporting the percentage value of static and dynamic displacement on the total one, Figure 7.

The same evaluation was conducted considering reactions of all external bridge's restrains; this study allowed us to estimate structural weight under various load conditions: only under the static load case, time history contribution and the sum of the previous ones.

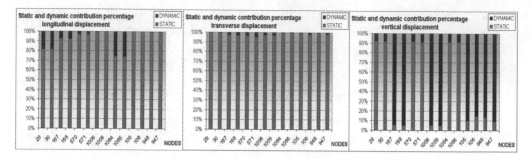

Figure 7. Static and dynamic displacements contribution percentage.

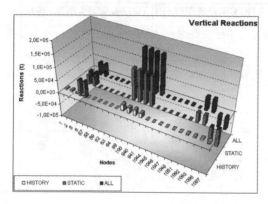

Figure 8. Static, dynamic and total comparisons of vertical reactions.

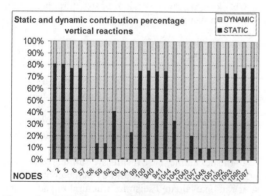

Figure 9. Static and dynamic reactions contribution percentage.

Even this time, to evaluate how much seismic action increases these reactions, Figures 8–9.

At the end of this study also axial forces, shears and moments in some significant places of the structure were evaluated.

In particular were calculated: axial forces at the connection between cable and anchor bock, longitudinal

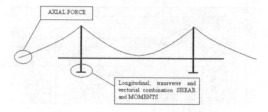

Figure 10. Considered points.

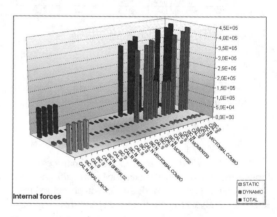

Figure 11. Static, dynamic and total comparisons of vertical internal forces.

and transverse shears and moments and their vectorial combination at the base of the towers, Figure 10.

These results are summarized in the first graphic Figure 11, and in the second one is also presented results' difference between static and dynamic load case, Figure 12.

Considered point are on each side of the bridge, so in the figure there are end points on Calabria's side and on Sicilia one, are also considered towers' and cables' points end on the northern side of the bridge and in the southern one, so total displayed points are 8.

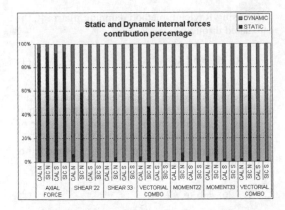

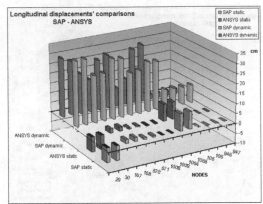

Figure 12. Static and dynamic internal forces contribution percentage.

3.3 Results' comparisons

At this point a study regarding results' comparisons obtained by different commercial finite elements codes was effected; in particular were confronted results obtained by SAP2000's modellation and those obtained by a similar modellation by ANSYS code.

Evaluated entities are:

- Global entities ⇒ DISPLACEMENTS
⇒ REACTIONS

Points in which measures were effected are the same considered in SAP2000 static modellation, so it's possible to have a direct check between different results.

It must be noted that possible differences in results' comparisons may be caused by different solution method between SAP2000 and ANSYS in non linear analysis, infact the first code uses the PΔ method, while the second solve a non linear analysis stepping or ramping total load during the time.

These differences are presented only in the static non linear analysis, that must be performed before the dynamic one. In static analysis, infact, self weight and superimposed load case effects on the bridge are calculated, starting from these results a liner dynamic analysis is performed.

In Figures 13-14-15 are presented displacements results' differences between SAP2000 e ANSYS, checked point are the same previously evaluated.

In Figures 16-17-18 are presented reactions results' differences between SAP2000 e ANSYS, even this time, checked point are the same previously evaluated.

3.4 Damping's evaluations

At the end was evaluated how different damping values or considerating methods could change structural response.

Damping is present in most systems and should be specified in a dynamic analysis.

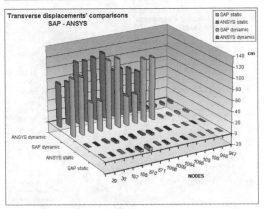

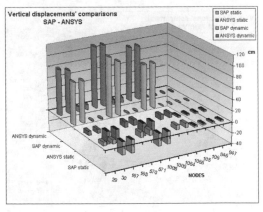

Figures 13-14-15. Comparison of displacement results obtained by SAP2000 and ANSYS.

Four forms of damping are available in the ANSYS program:

- Alpha and beta Damping (Rayleigh Damping)
- Material-Dependent Damping
- Constant Damping Ratio
- Modal Damping
- Damping Elements

2507

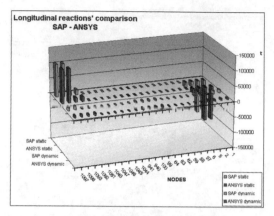

Longitudinal reactions' comparison SAP - ANSYS

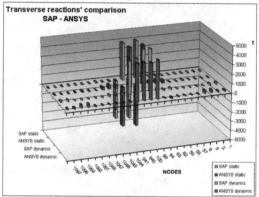

Transverse reactions' comparison SAP - ANSYS

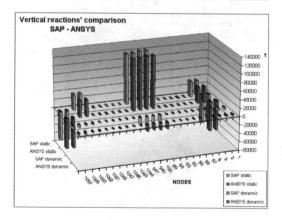

Vertical reactions' comparison SAP - ANSYS

Figure 16-17-18. Comparison of reaction results obtained by SAP2000 and ANSYS.

Instead only the last three are available in the SAP2000 program.

Comparisons were made between Rayleigh damping method (ANSYS) and constant damping ratio (SAP2000).

By using Rayleigh method damping matrix [C] is calculated through:

$$[C] = \alpha \, [M] + \beta \, [K] \qquad (1)$$

in this equation α and β value are calculated from modal damping ratios ξ_i. This is the ratio of actual damping to critical damping for a particular mode of vibration, i.

If ω_i is the natural circular frequency of mode i α and β satisfy the relation:

$$\xi_i = \alpha \, / \, 2\omega_i + \beta \, \omega_i \, / \, 2. \qquad (2)$$

Next figures present some results obtained by changing α and β value in ANSYS code to comprise Rayleigh damping method whit constant damping ratio = 2% of SAP2000 program.

Results' comparisons were effected considering a middle span node's displacement; another point monitored was the top of calabrian tower.

Also internal forces (cable's axial force) and reactions (vertical reaction at the base of calabrian tower) were considered.

Next figures portray these results, Figures 19-20-21-22.

4 CONCLUSIONS

At the end of this study on effects of dynamic load on Messina Strait Bridge some considerations may be done: in modal analysis even 800 modes are necessary to excite 85% of total mass at least, 200 modes are already enough to have a reasonably correct structural behaviour.

Regarding displacements, static contribution is the most important component and in longitudinal and in transverse direction, only in the vertical one and on the top of the tower, dynamic contribution is significant.

Also in case of reactions the static component is the most one, in the towers and in the cable at least, for the other external reaction, instead, dynamic contribution is more important that the static one.

Regarding internal forces dynamic component is more important that the static one, only in cable axial forces static component is more important.

In results' comparisons between SAP 2000 and ANSYS there is a good similarity, this is an evidence modellation and code's reliability.

Damping study has shown a good similarity too, both in displacements case and regarding internal forces.

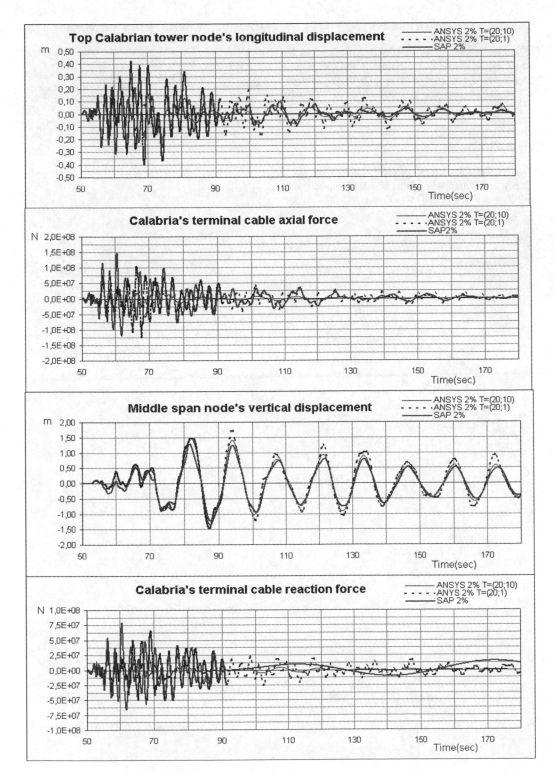

Figure 19-20-21-22.   Damping results' comparison SAP2000-ANSYS.

ACKNOWLEDGMENTS

The financial support of Stretto di Messina Spa is acknowledged. Anyway, the opinions and the results here presented are responsibility only of the authors and cannot be assumed to reflect the ones of Stretto di Messina Spa.

REFERENCES

Chopra, K. *Dynamic of Structures*, Prentice-Hall, New Jersey SAP 2000 – Analysis Reference, Computers and Structures, Berkeley California 1998.

Mcguire W. & Gallagher, R.H. *Matrix Structural Analysis*, John Wiley & Soons, New York 2000.

Ginmsing, N.J. *Cable Supported Bridge*, John Wiley & Soons, West Sussex 1998.

Chen-Duan, *Bridge Engineering HandBook*, Crc Press 2000.

Schrefler, B.A. & Cannarozzi, A.A. *Analisi per elementi finiti: modellazione strutturale e controllo dei risultati*, International Centre for Mechanical sciences.

Martinez, Y., Cabrera, F., Gentile, C. & Malerba, P.G. *Ponti e viadotti: concezione, progetto, analisi e gestione*, Pitagora editrice.

System-based Vision for Strategic and Creative Design, Bontempi (ed.)
© 2003 Swets & Zeitlinger, Lisse, ISBN 90 5809 599 1

A reference framework for the aerodynamic and aeroelastic analysis of long span bridges with Computational Fluid Dynamic

D. Taddei
Structural Engineer, Ph.D. Student, Polytechnic of Milan

F. Bontempi
University of Rome "La Sapienza"

ABSTRACT: Static and dynamic wind actions play a fundamental role in super-long span bridge engineering. After an historical retrospective on suspension bridge design, with emphasis on the influence exerted by computational methods on aesthetic and economy of such kind of bridge, a general framework of methods and application of CFD to structural engineering is done. A review of different numerical methods adopted in aerodynamic calculations is made.

1   INTRODUCTION

Static and dynamic wind actions have great importance in super-long span bridge engineering. In structural design of such bridges both the response of structure to low intensity frequent winds (horizontal displacements of deck, oscillations due to vortex shedding and buffeting) and the safety for exceptional wind conditions (flutter instability) must to be analyzed.

A analytical approach is theoretically complex (Bontempi & Malerba '94) and a closed solution is not available.

The application of Computational Fluid Dynamic (CFD) to wind-structure interaction is now possible, extending the results obtained in various field of engineering (aeronautic, applied mathematics) with the great experience on matrix structural analysis and finite element formulation own of structural engineering.

CFD is a versatile tool for qualitative visualization of phenomena. Nowadays the challenge is to obtain a good quantitative solution of aerodynamic and aeroelastic problems. The last objective is the application of CFD to not only exceptional, but even standard structural problems, if environment-structure interaction is important.

After a brief historical retrospective on suspension bridge engineering, with particular emphasis on the influence exerted by computational methods on aesthetic and economy of construction, an actual framework of CFD methods and challenges is done.

2   SUSPENSION BRIDGE ENGINEERING: AN HISTORICAL RETROSPECTIVE

2.1   *Historical tendencies in the suspension bridge design*

The attitude of structural engineer on the design of a great suspension bridge strongly reflect its technical and scientific refinement, that derive by formerly constructed bridges of its age, available computational tools, degrees of confidence reposed in calculations.

Regarding the relatively recent history of modern engineering, an evidence is that the introduction of new calculation methods, or its drastic failure, is cause of important changes in design approach and results, aesthetically and economically speaking.

Suspension bridges are, there is some 150 years, exceptional visualization of engineering at the eyes of nontechnical world. For this reason, these are subjected to collective emotions of various eras.

A structural parameter that may be assumed to synthesize economical and esthetical characteristics of a suspension bridge is, for example, *structural slenderness*, that is the principal span of bridge over its deck depth. A oscillation of this parameter during years had shown, suiting cycles run through the tree phases of introduction, refinement, excess of confidence of new calculation models.

At the '50 Steinmann (Steinmann & Watson, '57) suggested the image of *pendulum* for the history of

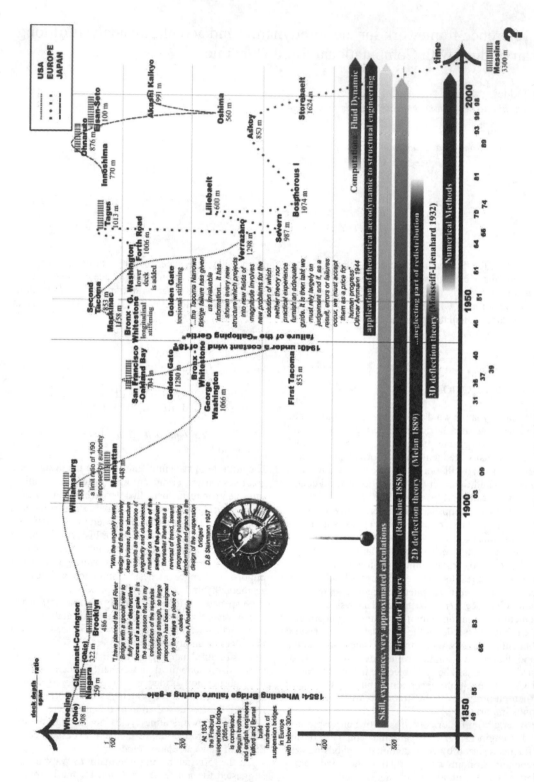

Figure 1.   Suspension bridges: a history. Slenderness variation during time. The Steinman's pendulum suggestion.

suspension bridge engineering. This is even a powerful visualization, including recent developments.

In the first cycle, that is a pioneeristic phase of engineering, calculation is a secondary element with respect of designer intuition and experience. Roebling produce an innovative design (concurrence of cable and stays to carry structural load obtaining a sufficiently rigid structure) taking this to the extreme consequences with the Brooklyn Bridge. Historical teaching of Wheeling Bridge failure is not forget.

A second cycle is this of classic engineering calculation, based on first order and linearized second order analytical methods. The *ungainly towers* and *excessively deep trusses* of the Williamsburg Bridge are then regarded as a direct product of the application of first order (Rankine) theories. Suddenly new refined theories are formulated that, in some thirty years, will aid engineers to build up beautiful and elegant bridges.

It's of interest to remind that not only theoretic advancement, but even the Wall Street Crack of '29 contributed to George Washington Bridge challenging slenderness. On the same way, twenty years back, only a standard requirement was to limit the slenderness of Manhattan Bridge to 1/90.

The increasing of confidence in calculation, the skill and experience of great designers as Moisseif and Lienahard will carry to the singular case of First Tacoma Narrows Bridge.

2.2 Aerodynamic and aeroelastic problem

After the Tacoma failure, ten years of pause in suspension bridge construction are preparatory for, on a side, a kind of "rewind" suspension bridge engineering: the Mackinac Bridge is a second Williamsburg. On the other side, new studies and a new perception of the design problem is rising. The necessity of a multidisciplinary approach to great system design is stated. This is a tendency common to various fields of civil engineering. For suspension bridge engineering, aerodynamic and aeroelastic is being a central problem.

Knowledge inherited from aeronautic engineering, however, allow only a rational explication of the phenomenon. The existent approximated theories can't to be a direct design tool: aerodynamic of a bluff body is substantially different from these of a wing or airplane; insufficient, and very expensive, are experimental tests in this innovative field.

The flutter problem for bridges are effectively highlighted by Scanlan. The definition of *flutter derivatives* can be viewed as an important tool to check the aeroelastic stability, with some (limited) wind tunnel tests. The examination of real flutter derivatives for different bridge sections get some guideline to the engineer, but not yet a design tool, neither an optimization tool.

This is the situation, as long as new advancements and new engineering schools are coming soon.

2.3 Modern tendencies of suspension bridge engineering

The last decades are characterized by the affirmation of different national schools in suspension bridge engineering. The "American School" is, obviously, the mother school, from that the others assume the necessary practical know how. In the USA, however, the impact of Tacoma failure is not overcome until the Verrazano Narrows Bridge construction. The traditional trussed deck section is always conserved.

The "European School" involve a first series of traditional trussed deck section bridges, but, already at the Sixties, English engineers proposed a new type of bridge, with a closed box girder section, that will be the typical character of this school. In reality, this character oversteps the particular field of suspension bridges, and is typical of all European bridge design.

The adoption of closed box girder deck section revolution the aerodynamic approach. Diminution of geometric section parameters clarify the problem, making easier, if not a theoretical approach, surely the experimental and, new field of research, numerical approach.

Third and more recent is the "Japanese School". We can see, even, a direct derivation from the American types. The great number of bridges constructed in the Nineties make the Japan a vanguard of suspension bridge design. However, the tendency is here for the conservation of traditional trussed deck section, that is justified, partially, by the double nature of highway and railway bridge of the most important works.

The numerical modelling of the aerodynamic of a truss is even a too complex problem. Wind tunnel test can assure from flutter instability. However, the high drag coefficient of this kind of section make it subjected to oscillation, in presence of wind with particular distribution of turbulence.

New important realizations, not mentioned in the figure, are recent Chinese bridges. The realization of Tsing Ma Bridge, in particular, proposed a modern approach to aerodynamic problem.

3 CFD AND STRUCTURAL ENGINEERING

As shown in the precedent paragraph, the new bridge engineering tendencies allow CFD to play an important role in the suspension bridge design. The methods of computational fluid dynamic are not yet regarded with confidence by the structural engineering world. This is understandable, because nonlinearity in fluid physic impose a not easy theoretical and numerical approach and a great dependence of results from calculation parameters. However, CFD can give us important insight on the studied phenomena and, if not "exact" results, surely numerical values for various interesting quantities. Above all, numerical results permit:

– Comparison with experimental values (if any);
– Comparison between proposed different designs.

A critical evaluation of wind tunnel test values is important if wind tunnel test results are affected by great uncertainty, that is when the examined phenomenon is in a not experimentable field. To test a bridge section, 1:60 scaled, subjected, for example, to a real 40 m/s lateral wind in Reynolds' similitude, if the fluid in wind tunnel is air:

$$\lambda_v = \frac{1}{\lambda_l} \rightarrow V_{test} = 60 \cdot V_{wind} = 2400 \tfrac{m}{s}$$

a velocity not obtainable in wind tunnels. This problem is common to the aeronautic tests, but in the aeronautic field a great number of tests results was compared with these obtained from real scale prototypes.

The second, most important purpose of the CFD is the possibility of analyzing, in a cost and time effective way, various different proposed designs. This is the first step of a challenging aerodynamic optimization process, as treated in other work.

4 METHODS AND CHALLENGING IN COMPUTATIONAL FLUID DYNAMIC

4.1 Numerical methods of CFD

Approximate numerical solution of the analytical fluid motion problem born with theoretical fluid dynamic. Differently from other field of science, even for flows that need, for its description, a simple linear Laplace equation, the complexity of boundary condition in the great part of real problems not allow an analytical solution. A good numerical solver for Laplace equation is, in fact, an even actual problem of the applied mathematics.

Moreover, considering the non linear turbulent Navier-Stokes equation, numerical approach is the unique way to solve, approximately, the problem.

4.1.1 Finite Difference Method (FDM)
This is the first, and more intuitive, fluid dynamic numerical method. The solution is collocated on a

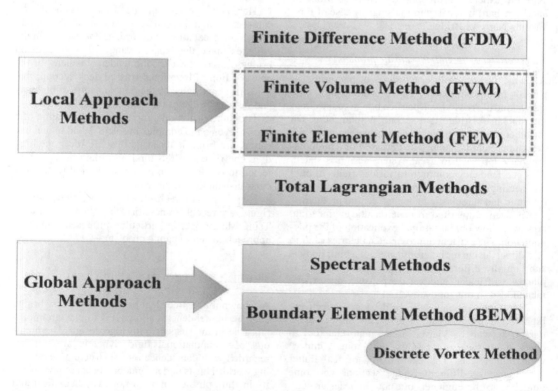

Figure 2.   Numerical methods in Computational Fluid Dynamic.

grid of points, and the finite difference equations that govern the phenomenon are developed in a Taylor's series way. The collocation point must suite a coordinate grid. Result a very rigid grid, that is the weakness of the method, not applicable in most real cases. For this reason, FDM is not used in practice. However, it is the only method that permit analytical demonstration of convergence, in some cases. The theorems conceived for FDM are often extended to other numerical methods.

4.1.2 *Finite Volumes Methods*
This is a typical method of fluid dynamic. PDEs are substituted by a integral weak formulation, that is the imposition, over small discrete subdomains, of the balance equations (mass, momentum, energy).

The theoretical merit of method is the respect of conservation laws everywhere in the fluid domain. Grid requirement are not rigid, making this method useful for real problems with complex boundary shape.

4.1.3 *Spectral Method and Boundary Elements Method (BEM)*
The general solution is expressed by superposition of global functions of simple analytical form. Spectral Method are well known in structural mechanics, that even utilize harmonic analysis in some practical problem. They are indicated in solution of wave propagation problems.

BEMs, on the contrary, express the global solution by superposition of singular solutions collocated at the boundary of the domain. Singularities, representing a discontinuity on the flow, are distributed over particular elements (segments in 2D models, surfaces in 3D) describing the boundary of the domain. A typical, useful formulation of this kind of methods are the Discrete Vortex Method (DVM), where the singular solutions are a series of simple vortex (Rankine, Lamb).

These are *grid-independent methods*. The cist of model construction and the number of equation to be solved are minimized. Some nuimerical problem is however, posed in the solution of full matrix linear system.

4.1.4 *Finite Elements Method*
FEM is the most familiar method for structural engineer. Birth for framed structures, the FEM was extended to other field of science. The adaptation of this method to non-structural problems needed of coherent mathematical reformulation. The Petrov-Galerkin weak formulation is a generalization of minimum potential energy principles typical of structural analysis. The integration of weak formulation are conducted over small discrete subdomains, over that the solution is approximated by a polynomial form. Grid flexibility is the same as in FVM, that can be viewed, now, as particular case of FEM.

Generally, FEMs adopt an Eulerian formulation, or, if a mesh deformation is needed (free surface flows, moving boundary for simulation of fluid-structure interaction) a mixed Eulerian-Lagrangian formulation.

4.1.5 *Total Lagrangian Methods*
This is a promising field of research. Discrete particles of fluid are directly modeled, and your interaction is simulated as a molecular interaction law. There are interesting applications in hydrodynamic simulations (Smoothed Particle Hydrodynamic, SPH) The different needs of simulation in the farfield and in the boundary layer add a difficult in application of these methods to bluff body aerodynamic.

4.2 *Commercial codes for CFD*
A statement of numerical aerodynamic calculation as standard part of design of medium and great bridges is nowadays possible, using a commercial CFD codes. Between the different types of numerical approach, robustness and stability necessary for commercial calculations is assured only by finite elements and finite volumes methods. FVM are specific of original aeronautical codes. FEM commercial codes, born for structural analysis, are after extended to multiphysic numerical simulations (structural, fluid dynamic, electromagnetism, acoustic). Both methods use well-stated numerical algorithms, that make easy an accurate solution of non-linear equation systems. Practical difficulties are connected with mesh generation: a time expensive process. Theoretical difficulties are principally connected with turbulence modelling, introducing a great number of arbitrary parameters that influence the solution.

A Discrete Vortex Method was used for flutter simulation of Great Belt Bridge (Larsen, '98; Larsen & Walther, '98). However, a commercialization of such kind of code are even not possible. DVM codes need of multiple smoothing techniques (Sarpkaya, '89), to obtain convergence, for which control an advanced knowledge of theoretical and numerical approach is necessary. In spite of these difficulties, DVM seem to be a powerful tool, and a challenging research field.

5 CONCLUSIONS

Historically, computational tools had a great influence in suspension bridge design. Aesthetic and economy of such type of structure are varied during years.

The introduction of CFD methods will produce fundamental advancement in super-long span bridge design. Perhaps, its applicability will be extended to medium-span bridge, substituting largely approximated rule's calculation.

Moreover, CFD is a time and cost effective tool, if compared with wind tunnel tests. It's important to promote new researches in this new field of structural engineering, that is coming to be a well defined field with the name of Computational Wind Engineering.

ACKNOWLEDGEMENTS

The financial support of Stretto di Messina SpA is acknowledged. Anyway, the opinions and the results here presented are responsibility only of the Authors and cannot be assumed to reflect the ones of Stretto di Messina SpA.

REFERENCES

Billah, K.Y., Scanlan, R.H. 1991. Resonance, Tacoma Narrows bridge failure, and undergraduate physics textbooks. *American Journal of Physics,* Vol. 59 n°2.

Bontempi, F., Malerba, P.G. 1994. Forzanti Aerodinamiche nei Modelli Aeroelastici Semplificati dei Ponti Sospesi. *Studi e Ricerche, Scuola di Specializzazione per le costruzioni in Cemento Armato F.lli Pesenti*, Politecnico di Milano, Italia, Vol. 15.

Fletcher, C.A.J. 1991 Computational Techniques for Fluid Dynamics, 2nd Edition, Springer-Verlag.

Gimsing, N.J. 1993 Wind design of Great Belt East Bridge: A Historic Retrospect. *Journal of Wind Engineering and Industrial Aerodynamics*, 48.

Gimsing, N.J. 1997 Cable-Supported Bridges, Construction and Design, 2nd edition, JohneWiley & Sons, New York.

Hadlow, R.W. 1993. Tacoma Narrows Bridge. *Historic American Engineering Record*, n°WA-99.

Katz, J., Plotkin, A. 1991. Low Speed Aerodynamics, From Wing Theory to Panel Methods. *McGraw-Hill Series in Aeronautical and Aerospace Engineering*; McGraw-Hill International Edition, Singapore.

Larsen, A. 1998. Advances in Aeroelastic Analyses of Suspension and Cable-Stayed Bridges. *Journal of Wind Engineering and Industrial Aerodynamics*.

Larsen, A. 1993 Aerodynamic aspects of the final design of the 1624 m suspension bridge across the Great Belt. *Journal of Wind Engineering and Industrial Aerodynamics*, 48.

Larsen, A., Veirum, T., Esdhal, S. 1998 Vortex Models for Aeroelastic Assessment of Multi-Element Bridge Decks. *Bridge Aerodynamics, Proceedings of the international Symposium on Advances in Bridge Aerodynamics, Copenhagen 10–13 May 1998*, Larsen & Esdhal eds, Balkema, Rotterdam.

Larsen, A., Walther, J.H. 1998. Discrete Vortex Simulation of Flow Around Five Generic Bridge Deck Sections. *Journal of Wind Engineering and Industrial Aerodynamics*.

Leonard, A. 1980. Vortex Methods for Flow Simulation. *Journal of computational physics*, Vol. 37.

Morghental, G. 2000. Comparison of Numerical Methods for Bridge-Deck Aerodynamics. *PhD thesis*, University of Cambridge.

Morghental, G. 2000. Fluid-Structure Interaction in Bluff-Body Aerodynamics and Long Span Bridge Design: Phenomena and Methods. *Technical Report CUED/ DSTRUCT/ TR.187 at the Departement of Engineering*, University of Cambridge.

Morghental, G., Mcrobie, A., Jang, S. 2001. Numerical Models for Bridge Deck Aerodynamics and Their Validation. *IABSE Conference Seoul 2001, Cable-Supported Bridge Challenging Technical Limits*, in IABSE REPORTS, Vol. 84.

Sarpkaya, T. 1989. Computational Methods with Vortices – The 1988 Freeman Scholar Lecture. *Journal of Fluids Engineering*, Vol.111/5, March 1989.

Simiu, E., Scanlan, R.H. 1996 Wind Effects on Structures: Fundamentals and Applications to Design. John Wiley and Sons, 3rd edition.

Steinman, D.B., Watson, S.R. 1957. Bridges and Their Builders, Dover Pubblications, New York.

Walter, J.H. 1998 Discrete Vortex Methods in Bridge Aerodynamics and prospect for parallel computing techniques. *Bridge Aerodynamics, Proceedings of the international Symposium on Advances in Bridge Aerodynamics, Copenhagen 10–13 May 1998*, Larsen & Esdhal eds, Balkema, Rotterdam.

Wyatt, T.A., Scruton, C. 1981. A Brief Survey of aerodynamic stability Problems of Bridges. *Bridge Aerodynamics, Proceedings of a conference held at the Institut of Civil Engineers*, London, 25–26 March 1981, Thomas Telford Limited, London.

Author index

Jain, R.K. 325
Jannadi, O. 419
Jarquio, R.V. 817, 1363, 1997
Jayyousi, K. 1599
Jeary, A. 1973
Jia, H.M. 927
Jirsa, J.O. 1459
Jo, J.-S. 617
Jommi, C. 2409
Jubeer, S.K. 729
Junag, D.S. 507
Jung, H.-J. 617
Junica, M.I. 1227
Junnila, S. 1685

Kaita, T. 885
Kale, S. 1545
Kamerling, M.W. 557, 563
Kanakubo, T. 2427
Kanhawattana, Y. 69
Kanisawa, H. 335
Kanoglu, A. 2221
Karim, U.F.A. 1259, 1263
Kastner, R. 1273
Kato, Y. 669
Kawauti, Y. 797
Kayama, N. 845
Kayashima, N. 2427
Kegl, M. 527
Kelly, W.L. 413, 1589
Keskküla, T. 25
Khairy Hassan, A. 1201
Khalaf, M.K. 2437
Kim, D.H. 1303
Kim, H.C. 1303
Kim, S.C. 677
Kim, Y.M. 899
Kimura, Y. 335
Kiroff, L. 1613
Kitagawa, T. 1291
Kitoh, H. 1369
Klotz, S. 1393
Klug, Y. 1393, 1907, 1913
Knezevic, M. 237
Ko, M.G. 2063
Ko, M.-G. 617
Kocaturk, T. 557, 563
Koehn, E. 65, 303
Kolesnikov, A.O. 1235
Koo, J. 1145
Kopecky, M. 893
Korenromp, W. 2227
Kotsikas, L. 2209
König, G. 1907
Kubo, Y. 965
Kudzys, A. 31
Kulbach, V. 951

Kumar, R. 1983
Kumaraswamy, M.M. 217
Kunugi, T. 99
Kunz, C.U. 1029
Kunz, J. 823, 1217
Kupfer, H. 823
Kuroda, I. 1285
Kusama, R. 1291
Kushida, M. 1899
Kuwamura, H. 1129
Kwan, A.S.K. 309
Kwon, Y.B. 1303

Lagae, G. 2275, 2281, 2285,
    2293
Lambert, P. 1817
Lamberts, R. 383
Lanc, D. 639
Langford, D. 85
Lawanwisut, W. 1803, 1811
Lazzari, M. 1511
Lee, I.W. 597, 611, 2063
Lee, I.-W. 617
Lee, J.H. 597, 611
Lee, K.H. 1303
Lee, S.G. 677
Letko, I. 893
Levialdi, S. 2127
Lewis, T.M. 1729
Ley, J. 309
Li, C.Q. 1803, 1811, 1943
Li, H. 2105
Li, Q. 499
Li, S.Y. 927
Liang, Z.Y. 927
Lilley, D.M. 1327
Liu, A.M.M. 425, 431
Liu, B. 1291
Liu, C.-T. 2383
Liu, T. 1279
Liu, Y.W. 1383
Loo, Y.C. 631
Loreti, S. 1003, 1581, 2455
Lu, Po-C. 2383
Luciani, G. 2127
Lueprasert, K. 69

Maegawa, K. 1423
Magaia, S.J. 1067
Magonette, G. 839, 2037
Mahmoud, K.M. 947
Malerba, P.G. 17, 981
Maloney, W.F. 397
Malécot, Y. 2393
Manca, S. 575
Mancini, G. 721, 1777
Mangat, P.S. 1817

Mangin, J.C. 481
Manoliadis, O.G. 2239
Mansour, M. 2079
Mantegazza, G. 761
Marazzi, F. 839, 2037
March, A.V. 1327
Marchiondelli, A. 533
Marino Duffy, B.M. 343
Maroldi, F. 1195
Marsalova, J. 1649
Martínez, P. 1631
Masera, M. 2257
Mastroianni, G. 2401
Masullo, A. 861
Materazzi, A.L. 2317
Matrisciano, A. 2129, 2163
Matsumoto, M. 2345
Matthys, S. 2267
Mattioli, D. 2155
Mawdesley, M. 1259, 1263
Mazzotti, C. 1375
McNamara, C. 329
Mecca, S. 2257
Meda, A. 1413
Mehta, K.C. 2369
Mendis, P. 1837
Menshari, M. 923
Meszlényi, R. 2123, 2171
Mezher, T. 229
Mezzanotte, R. 2155
Mezzetti, C. 2163
Mihara, T. 797
Mine, N. 99
Minkarah, I. 1607
Minter, A.R. 357
Mitrani, J.D. 1745
Miura, K. 603
Miura, N. 99
Miyamura, A. 1709
Mofid, M. 923
Mohamed, S. 1419
Mokha, A. 2079
Mokos, V.G. 583
Moon, Y.J. 611, 677
Moonen, S.P.G. 123, 169, 449,
    1621
Moore, D.R. 2191
Morbiducci, R. 1109
Morcos, S. 1145
Moriguchi, Y. 669
Moscogiuri, C. 2163
Moser, S. 1523
Motawa, I.A. 2185
Motoki, M. 1227
Muenger, F. 823
Mullins, G. 1467
Musso, A. 2417